05-24-85 #51875 $76⁰⁰ Den Chen

Ion Implantation Range and Energy Deposition Distributions

Volume 2
Low Incident Ion Energies

ION IMPLANTATION RANGE AND ENERGY DEPOSITION DISTRIBUTIONS

Volume 1. *High Incident Ion Energies*
David K. Brice

Volume 2. *Low Incident Ion Energies*
K. Bruce Winterbon

Ion Implantation Range and Energy Deposition Distributions

Volume 2
Low Incident Ion Energies

K. Bruce Winterbon

Atomic Energy of Canada Limited
Chalk River Nuclear Laboratories
Chalk River, Ontario, Canada

IFI/PLENUM • NEW YORK · WASHINGTON · LONDON

Library of Congress Cataloging in Publication Data (Revised)

Brice, David K
    Ion implantation range and energy deposition distributions.

    Includes bibliographical references.
    CONTENTS: v. 1. Brice, D. K. High incident ion energies. — v. 2. Winter-
bon, K. B. Low incident ion energies.
    1. Particle range (Nuclear physics) 2. Ion implantation. I. Title.
QC793.3.E5B74   541'.372   74-34119
ISBN 0-306-67402-5 (v. 2)

© 1975 IFI/Plenum Data Company
A Division of Plenum Publishing Corporation
227 West 17th Street, New York, N. Y. 10011

United Kingdom edition published by Plenum Press, London
A Division of Plenum Publishing Company, Ltd.
Davis House (4th Floor), 8 Scrubs Lane, Harlesden, London, NW10 6SE, England

Printed in the United States of America

PREFACE

The present level of understanding of ion implantation is sufficient that implantation is being used not only as a tool in various fields of research, but also as an industrial process. In these applications one uses either the implanted ions, or their energy, to modify some properties of the target substance, and is therefore concerned with the spatial distribution of the ions or of their energy. Following the pioneering work of Bohr [1], Lindhard and his collaborators have evolved a general description of the behaviour of swift ions slowing down in amorphous targets [2,3,4], a description which has been the basis of much other work in the field. Various approximate calculations have been based on this theory, but it has not always been clear whether any disagreement between experiment and theory is real or can be attributed to deficiencies in calculation. It is the purpose of this volume to present the results of the Lindhard theory, calculated in an exact manner, to serve as a guide to the users of implantation, as a tabulation of theoretical results for experimentalists to compare with, and as a statement of the theoretical results either as a standard for comparison for approximate calculations or as a starting point for a more detailed theory. Results are presented in tables and in graphs, the graphs being intended to display the qualitative features so as to illustrate the competition of the various physical processes determining the spatial distribution of the collision cascade.

The data presented here are parameters of the depth distributions of i) the implanted ions, ii) the fraction of the ion energy deposited in atom motion, and iii) the (complementary) fraction deposited in electronic excitation. These three distributions are called "range", "damage", and "ionization" distributions, respectively. (The first is often called "projected range" by others.) Calculations are given for 23 projectile-target combinations, over a range of 10^4 in energy, for energies up to $\varepsilon \sim 20$-50, where ε is Lindhard's dimensionless energy unit. These 23 cases are sufficient to cover all projectile- (monatomic) target combinations with projectile and target atom masses differing by a factor of less than 26. The "natural" quantities to calculate are spatial moments of the distributions; moments up to the fourth are given for each of the three distributions. In addition to the moments along the beam direction, the second moment in the transverse direction is given - this moment gives the sideways spreading of the distribution.

From the moments some secondary quantities are calculated. They are i) the fraction of the distribution which is outside the target surface, i.e., the fraction of backscattered ions (the reflection coefficient, R) or of backscattered or sputtered energy, denoted by γ, and ii) a quantity α which is essentially the value of the damage distribution at the target surface, and which determines the sputtering yield, the number of sputtered target atoms per incoming projectile ion. The moments and the secondary quantities are both tabulated and plotted. Furthermore iii), distributions calculated from the moments are plotted for 5 energies for each case.

This volume covers the low-energy regime, namely the range of energies in which electronic stopping is proportional to particle speed. A companion volume has been prepared by D.K. Brice, to cover a higher-energy range, from 20 KeV/amu to 1 MeV/amu. His methods, quantities tabulated, and format are different. There is some overlap between the two volumes.

I have been influenced in the preparation of this volume by discussions with many people, among whom I must mention my colleagues at Chalk River, as well as Roger Kelly, Jens Lindhard, Jim Mayer, and Peter Sigmund.

CONTENTS

1. INTRODUCTION

The theoretical description of the slowing down of swift ions in amorphous solids is based on the work of Lindhard and his collaborators [2,3,4]. They were primarily concerned with particle ranges; Sanders [5] and Sigmund and Sanders [6] went on to discuss the damage distribution, and Brice and the present author have discussed also the ionization distribution [7,8]. An extensive treatment of range and damage distributions, including plots of range and damage moments in the power-law approximation, has been given by Winterbon, Sigmund, and Sanders [9]. Sigmund has recently reviewed virtually the entire field of slowing-down problems in a series of summer-school lectures [10,11,12]. The beginner interested in going beyond this volume would be well-advised to start with the latter reviews: the Lindhard papers in particular contain many subtleties which will not be appreciated by the novice.

The names of two of the distributions, damage and ionization, are more figurative than literal. There is a fair amount of experimental evidence, primarily from backscattering measurements (e.g. [13]) that the "damage" distribution is indeed closely connected with radiation damage. The "ionization" distribution as used herein may have less direct connection to experiment. In particular, since energetic electrons have relatively enormous ranges, the neglect of transport of energy by recoiling electrons may be a serious deficiency. If on the other hand most of the electronic-stopping energy goes into bond-breaking instead of into creating high-energy electrons, the so-called ionization distribution would be the more relevant quantity for more-or-less high energy damage of insulators - say $\varepsilon \sim 1$ or larger.

As a consequence of a scaling property which allows us to reduce drastically the necessary number of cases there is some overlap between this volume and Brice's, and there are some gaps. For example, in the case in which projectile and target atoms are identical, the example used is Ge; Brice's lower limit of 20 keV/amu for Ge on Ge corresponds to the same ε value as 0.450 keV/amu for H on H and 72.2 keV/amu for U on U. As a compromise with space limitations it was decided to have a maximum energy of $\varepsilon \sim 20-50$. In the equal-mass case the maximum value of ε was 35, corresponding to 10 MeV (i.e. 137 keV/amu) Ge on Ge, 3.08 keV H on H, and 117.5 MeV U on U. For Be on Be, the lightest combination from the 12 elements selected as examples, this maximum corresponds to 78.1 keV, i.e., 8.68 keV/amu.

For each case calculations have been made for 5 values of electronic stopping.

2. OUTLINE

The projectiles and targets have been chosen from the 12 solid targets used by Northcliffe and Schilling [14] in their compilation of stopping powers and mean ranges. In order of increasing mass they are $^{9}_{4}Be$, $^{12}_{6}C$, $^{27}_{13}Al$, $^{48}_{22}Ti$, $^{59}_{28}Ni$, $^{73}_{32}Ge$, $^{91}_{40}Zr$, $^{108}_{47}Ag$, $^{152}_{63}Eu$, $^{181}_{73}Ta$, $^{197}_{79}Au$, $^{238}_{92}U$. With the use of the scaling, and allowing K, the electronic stopping constant, to vary, it is found that the range distributions depend on projectile and target masses only through their ratio. The scaling is not exact for damage and ionization distributions, but is still close enough for all practical purposes. Thus the 144 projectile-target combinations can be reduced to a smaller number; we have chosen 23, a symmetric case, Ge-Ge, and 11 pairs. They are listed in Table 0.1. The tables start with Be on U and proceed through Ge on Ge to U on Be. In the energy range covered by Brice's parallel volume, the scaling is not applicable, and he includes all 144 cases.

A subscript "1" will denote the projectile, which will often be called an "ion", without prejudice as to its charge state. A subscript "2" will denote a target atom. Atomic mass and nuclear charge are M and Z respectively, and E is energy. The dimensionless energy ε is the centre-of-mass kinetic energy in units of the Coulomb repulsion,

$$\varepsilon = E \frac{M_2}{M_1+M_2} \left/ \frac{Z_1 Z_2 e^2}{a} \right. , \tag{1}$$

where e is the electronic charge and a is a screening length. The screening length used throughout is the Bohr-Lindhard one [2],

$$a = 0.8853 \, a_0 Z^{-1/3}, \tag{2}$$

where

$$Z^{2/3} = Z_1^{2/3} + Z_2^{2/3} \tag{3}$$

and a_0 is the Bohr radius $\hbar^2/me^2 = 0.0529$ nm.

It is convenient to define a dimensionless length,

$$\rho = xN\pi a^2 \gamma, \tag{4}$$

where x is a length, N is the number density of target atoms, and

$$\gamma = 4M_1 M_2/(M_1+M_2)^2. \tag{5}$$

Electronic stopping is generally considered to be proportional to the particle velocity, up to a velocity of

$$v_1 = Z_1^{2/3} v_0, \tag{6}$$

where v_0 is the Bohr velocity, $e^2/\hbar = c/137$. (This corresponds to an energy of about 25 keV for protons.) Thus for $v < v_1$, the electronic stopping is

$$S \equiv \left(-\frac{\partial E}{\partial x}\right)_{el} = KE^{1/2}, \tag{7a}$$

or, equivalently,

$$s \equiv \left(-\frac{\partial \varepsilon}{\partial \rho}\right)_{el} = k\varepsilon^{1/2}. \tag{7b}$$

The standard Lindhard-Scharff [15] values for k and K are

$$k = \frac{0.0793 \; z_1^{2/3} \; z_2^{1/2} \; (A_1 + A_2)^{3/2}}{(z_1^{2/3} + z_2^{2/3})^{3/4} \; A_1^{3/2} \; A_2^{1/2}} \qquad (8a)$$

and

$$K = k(E/\epsilon)^{1/2} \; (\rho/x). \qquad (8b)$$

All energies in this volume are in keV, and all lengths in $\mu g \cdot cm^{-2}$. Tables 0.2 and 0.3 list values of ϵ corresponding to 1 keV, values of ρ corresponding to 1 $\mu g \cdot cm^{-2}$, and values of k and of K, for several projectile-target combinations. A more extensive table may be found in ref. [16].

To convert from $\mu g \cdot cm^{-2}$ length units to nm, multiply by 10/(density). Note that the density of a target should not be assumed to be the bulk density given in a handbook. The author is indebted to Prof. U. Hauser of the University of Cologne and his group for a discussion of this difficulty: they found discrepancies of 10-20% between values of densities (of metal foils) taken from handbooks, from diffraction measurements, and from weighing. Two recent papers [17] show that up to 17% of the lattice sites in titanium oxide may be vacant; perhaps such enormous numbers of vacancies occur in metal (and other) lattices as well.

The tabulated moments are derived from quantities

$$\langle x^n \rangle = \int dx \; x^n \; F(x, E, \cos\theta), \qquad (9)$$

where x is the depth into the target, E the ion energy, and θ the angle the incident ion makes with the surface normal. F is any of the distributions: range, damage, or ionization. The calculation of moments assumes an infinite medium, in which the surface is merely a reference plane; this will be discussed shortly.

With n=0, one obtains simply the normalization of F. For ranges this is 1; for damage and ionization it is the energy going into damage or electronic stopping, respectively, and these are commonly denoted by ν and η, so that

$$E = \nu(E) + \eta(E), \qquad (10)$$

with the boundary condition $\nu(E)/E \to 1$ as $E \to 0$. The dimensionless quantity ν/E is tabulated.

The first moment, i.e. n=1 in (9), gives the mean depth of the distributed quantity, $\langle x \rangle$. This will contain a factor $\cos\theta$; mean depths are tabulated for normal implantation, $\cos\theta = 1$.

The second moment determines the breadth of the distribution:

$$\langle \Delta x^2 \rangle = \langle x^2 \rangle - \langle x \rangle^2 \qquad (11)$$

is called the variance. This quantity, or, more commonly, its square root, is called the straggling. The quantity tabulated is the relative square straggling,

$$\langle \Delta x^2 \rangle / \langle x \rangle^2 \qquad (12)$$

i.e., the square of the ratio of the (longitudinal) width to the depth. The variance depends on θ as a sum of P_0 and P_2 Legendre coefficient terms. Besides the relative square straggling for normal implantation, the ratio of transverse to longitudinal straggling,

$$\langle y^2 \rangle / \langle \Delta x^2 \rangle \tag{13}$$

is also tabulated. This indicates the shape of the distribution as if emanating from a point source. A value greater than 1 would be obtained for oblate distributions, a value less than 1 for prolate.

By making a linear transformation, $x \to Ax + B$, arbitrary values could be given to the first two (longitudinal) moments, subject only to the inequality

$$\langle x^2 \rangle > \langle x \rangle^2, \tag{14}$$

so these are called location parameters of the distribution - they do not in any way specify its shape.

If x is transformed in this way so that $\langle x \rangle = 0$ and $\langle x^2 \rangle = 1$, then the third and fourth moments $\langle x^3 \rangle$ and $\langle x^4 \rangle$ do depend on the shape, and so are called shape parameters [18]. The third moment, known as the skewness, may take positive or negative values, and the fourth, the kurtosis, is positive. The skewness may be considered an asymmetry parameter, and the kurtosis indicates the relative importance of the tails. For a sharply cut-off distribution the kurtosis is small. The skewness contains P_1 and P_3 components, and the kurtosis P_0, P_2, and P_4 components, but only the values for normal incidence have been shown - i.e., the sum of all components with $\cos\theta = 1$.

If the ion beam is not normal to the surface, but rather has an angle of incidence θ, the mean range is $\langle x \rangle \cos\theta$, and the square straggling is $\langle \Delta x^2 \rangle \cos^2\theta + \langle y^2 \rangle \sin\theta$. The best estimates for skewness and kurtosis, in the absence of calculations, would be the tabulated values for normal incidence, with the skewness multiplied by $\cos\theta$.

Various characteristic depths other than the mean could be used. One is the modal, or most probable, depth. For a smooth, unimodal (single-peaked) distribution, the following expression, although an approximation, is usually accurate [18]:

$$\frac{(\text{mean} - \text{mode})}{\sigma} = \frac{S(K+3)}{2(5K - 6S^2 - 9)} \tag{15}$$

where for the sake of brevity S has been written for skewness and K for kurtosis. σ^2 is the variance. The median depth, beyond which half the distribution lies, may be approximated reasonably well by the relation [18]

$$(\text{mean} - \text{median}) = 1/3 \ (\text{mean} - \text{mode}). \tag{16}$$

With these simple analytic relations available there is no point in tabulating modal or median depths.

These two relations do not apply to the ionization distribution, which has a discontinuity at the surface [8]. Approximate values of the modal ionization depth are given in the figures, but are not tabulated. These values are obtained from distributions calculated from the moments by the inverse Fourier transform method which will be described in the section on moments and distributions.

Other quantities of interest relate to the intersection of the distributions with the surface. They are the number of backscattered or reflected ions, the amount of backscattered or sputtered energy, and the number of sputtered particles. The number of reflected ions is quite simply the fraction of the (semi-infinite target) range distribution lying outside the surface. This has been examined both theoretically and experimentally [19,20, 21,22], and good agreement has been found. It is necessary to apply a surface correction to the calculated values, because of the possibility of repeated scattering through the surface. The calculation is made for particles moving in an infinite medium, with the "surface"

5

being simply the plane where ion motion starts. In the calculation there is no bar to particles passing through this plane several times, whereas in experiment of course they can pass back through the surface only once. Hence the fraction of the distribution lying outside the target will be greater than that calculated, while the value of the distribution at the surface will be less. This surface correction has been evolved in refs. [20,21,22]. It is felt to be less reliable than the calculation of the uncorrected value, and so it has not been included in the tabulation. It is discussed in a later section.

The fraction of energy either reflected or sputtered from the target is in principle slightly more complicated. Once an atom has left the target it does not matter how its energy would have been shared between atom motion and electronic excitation. Hence the quantity desired is

$$\gamma_E = (\gamma_\nu \cdot \nu + \gamma_\eta \cdot \eta)/E, \tag{17}$$

where γ is the fraction of the appropriate distribution lying outside the surface. The quantities γ_ν and γ_η are tabulated as sputtering efficiency and ionization deficiency, respectively.

The sputtering yield has been shown [23] to be proportional to the value of the damage distribution at the surface. It is convenient to express this value in a dimensionless form, and so it is divided by S_n, the nuclear stopping power for the incident ion. Thus the quantity tabulated is

$$\alpha(E) = F_0(x=0)/S_n(E). \tag{18}$$

This is related to the sputtering yield S by [23]

$$S(E) = \Lambda \alpha N S_n(E) \tag{19}$$

where Λ is a function of the target surface only, depending on surface binding energy among other things. Both γ and α should be subject to surface corrections.

Finally, it is often desired to have calculated distributions, rather than just moments. With the present method of calculation, moments are the primary calculated quantities and distributions secondary. Hence the moments are more accurately calculated than the distributions and it is primarily for that reason that it is the moments that have been tabulated. However for each case range, damage, and ionization distributions are plotted for 5 energies, E_{min}, $E_{min} \times 10, \cdots, E_{min} \times 10^4 = E_{max}$. The Lindhard-Scharff value of electronic stopping appropriate for the particular projectile-target pair is used in each case (i.e., $k/k_L = 1$).

3. ACCURACY OF THE TABLES

Accuracy has several meanings in this context: how nearly the results are accurate solutions of the given equations for the given cases, how nearly they are accurate solutions for other cases with the same mass ratio or for corresponding polyatomic-target cases, and how closely the theory should be expected to correspond to reality. The first point is the easiest. It is believed that for ν/E and for mean depths the maximum errors, in the first sense, will be $\sim 1\%$. This will probably double for straggling, double again for skewness, and yet again for kurtosis. All numbers are calculated for energies of 1,2, and 5×10^n keV and interpolated (Laguerre (cubic) interpolation) for other energies. Most of the calculated values have been calculated correctly to more significant figures than are given in the tables, but where moments are varying rapidly with energy, they are more difficult to calculate accurately. Interpolated values will be less accurate than the calculated values. However, the tables of moments are believed to have greater precision than will be required.

Some idea of the difficulties may be given by the following example. Suppose the straggling, as given in the tables, is 0.01 for some particular case, and the corresponding reduced fourth moment (kurtosis) is 10. Then a 1% error in the second moment corresponds to a 100% error in the straggling, and a 1% error in the fourth moment corresponds to a factor-of-10 error in the kurtosis.

The qualitative behaviour of the transverse straggling and the higher moments (i.e., skewness and kurtosis) as functions of energy do not depend on the potential to any great degree, but rather reflect the competition of nuclear and electronic stopping on both projectile and recoil ranges, and on the relative values of these ranges. (Range distributions of course are unaffected by recoil ranges.)

The secondary quantities which are calculated from the moments, namely the surface-related quantities R, γ (γ_ν or γ_η) and α are rather less accurate.

CAUTION: At high energies in some cases the number of moments is inadequate to give a good value for these quantities, and various unphysical "bumps" appear. The tables should be used in conjunction with figures 9 and 10 to verify that the tabulated values are physically reasonable.

The second question is how faithfully the results may be translated or scaled to apply to other projectile-target combinations. For range distributions, with the appropriate choice of electronic stopping, the scaling is exact. For damage and ionization distributions, where the scaling is not exact, see the discussion under Mass Scaling. The method of scaling is outlined in that section also, and an example given. Another section is devoted to some comments on applicability to compound targets.

It remains to estimate how closely the results echo reality. Breakdown of the theory at low and high energies is discussed in the succeeding section, which also contains comments on reliability of various aspects of the calculation. No short answer is possible. For truly amorphous targets, with the electronic stopping considered as a parameter to be determined by experiment, rather than one prescribed by theory, accuracy to 10% doesn't seem unreasonable.

4. THEORY

At high energies, the velocity-proportionality of electronic stopping breaks down, and also it becomes more and more difficult with increasing energy to achieve sufficient accuracy from the computing methods used here. At the other end of the energy scale, it will be seen that many of the approximations in the theory break down as the projectile energy decreases; however as long as the particle ranges decrease sufficiently rapidly with decreasing energy the theory is valid for higher energies where the low-energy errors on the terminal parts of the distributions can be neglected. If for example the range is only a few times the interatomic spacing in the target, or the energy is comparable with the target atom binding energy, the theory is inapplicable.

The projectile slows down in the target by scattering from individual target atoms more or less elastically, and by excitation of target electrons. In the case of the damage or ionization distributions, where one is interested in some portion (ν or η) of the energy of the projectile, the motion of the recoils, both primary recoils and all generations of secondaries, must also be followed.

At very high energies, or if projectile and target atoms are light, the scattering cross section approaches the Rutherford cross section for scattering from an unscreened Coulomb potential. For heavier atoms and lower energies the screening is more important and the divergence of the cross section at forward angles becomes less acute. Of course the effective radius of the interaction increases as energy is lowered.

The quantities discussed here are all given as the solution of linear Boltzmann equations with highly anisotropic energy-dependent cross sections. Consider the distribution F of some quantity, ψ, say, which is transported by (some or all of) the moving atoms in a collision cascade. Assume the target to be infinite in extent with point scattering centres distributed randomly throughout all space with constant average density. If the scattering cross sections are assumed to be finite, then a Boltzmann equation for F can be simply derived. Consider a particle moving at $\underline{x} = 0$, with velocity $\underline{v}$, at time $t = 0$, and then again at a short time δt later. The distribution at $t = \infty$, as a function of $\underline{x}$, is

$$F(\underline{x},\underline{v}) = \left|\underline{v}\delta t\right| N\Sigma(\underline{v})\,\delta(\underline{x}) + (1-N\left|\underline{v}\delta t\right|\Sigma(\underline{v})\,\delta(\underline{x}))\{\left|\underline{v}\delta t\right| N \int d\sigma\,[F(\underline{x},\underline{v}') + \xi\bar{F}(\underline{x},\underline{v}'')]$$

$$+\ (1 - N\left|\underline{v}\delta t\right| \int d\sigma) F(\underline{x}-\underline{v}\delta t,\underline{v})\}, \tag{20}$$

where N is the number density of target atoms, Σ is the cross section for ψ to be deposited at $\underline{x}=\underline{0}$, e.g., a reaction cross section, and $d\sigma$ is the scattering cross section. The incident ion is scattered from velocity $\underline{v}$ to velocity $\underline{v}'$, and the recoil is scattered from rest to velocity $\underline{v}''$. The factor ξ is 1 if the quantity ψ is transported by recoil atoms, and 0 if it is not. If the projectile is not of the same species as the target atoms, the recoil atoms will initiate different cascades which must be calculated separately, so the recoil term has been distinguished by a bar. Extension to polyatomic targets is obvious.

It is customary to separate from the cross section $d\sigma$ the portion due to electronic excitations, which in general cause only small-angle deflections of the ions, but do retard them. This term, called electronic stopping, usually depends only weakly on the impact parameter for atom-atom collisions, so the electronic stopping is approximated by a continuous, path-independent, frictional force: in the last term of (20), v is replaced by $v-\delta v$, $\delta v = S_e/M_1 \cdot \delta t$ where S_e is the electronic stopping, eq. (7). (This continuous-slowing approximation appears to be adequate. For example, it is well-known that characteristic molecular-orbital X-rays are emitted in certain collisions with well-defined (small) impact parameters. Finneman (unpublished) has observed that the dependence of these processes on impact parameter is fitted reasonably well by Firsov's formula for electronic energy loss as a function of impact parameter [24]. Calculations using such a correlation of nuclear and electronic stopping show negligible effect on range and damage distributions (Winterbon, unpublished).

For the cross sections used here, particle ranges vanish with vanishing energy, so that if the normalization of F can be specified, and the source term Σ is simply a zero-energy delta function, it can be omitted. For example, in the range equation the source term can be replaced by the condition $\int d\underline{x}\, F(\underline{x},\underline{v}) = 1$.

Dropping the source term, including electronic stopping, and expanding to first order in δt, eq. (20) becomes

$$-\frac{\underline{v}}{v}\frac{\partial F}{\partial \underline{x}} = N \int d\sigma (F(\underline{x},\underline{v}) - F(\underline{x},\underline{v}') - \xi\overline{F}(\underline{x},\underline{v}'')) + \frac{S_e}{|v|}\frac{\partial F}{\partial |\underline{v}|} \quad . \tag{21}$$

This equation has been obtained assuming finite total cross sections. Because of the subtraction in the integrand, (21) still makes sense if dσ diverges for small angles. Equation (21) can be derived somewhat more rigorously from a Smoluchowski equation, as was done in refs. [25] and [26]. However it is even in that case assumed that only one collision occurs at one time, and each collision is of zero duration. On the other hand, at moderate or high energies, for any appreciable particle deflection, the impact parameter is much smaller than the interparticle spacing, and for small-angle deflections the impulse approximation is adequate, so scattering from different particles is additive. For energies sufficiently small that large-angle scattering can occur at large impact parameters, the residual particle ranges are so small that any error will not appreciably affect the total cascade dimensions. Hence the error appears more formal than real.

It is essential in the derivation of (21) that the target be infinite and uniform. There are two approximations here. First, the target surface has been reduced to a reference surface within the infinite target, distinguished only by being the place at which atom motion starts. Hence in the calculation particles can be scattered back and forth across this surface arbitrarily often. An approximate correction for this "surface effect" can be made, as will be seen later. The second approximation is that the target atom positions are assumed to be uncorrelated. This has two effects. In the model, target atoms may be arbitrarily close together, whereas in a real solid or liquid target there is some minimum spacing. Again this approximation is not expected to be serious, because except at very low energies the mean free path between large-angle collisions is large compared with the interatomic spacing. Finally, most solid targets will have some degree of crystalline structure, and this may cause serious errors in the calculations. There is of course the phenomenon of channeling, in single-crystal targets in particular, which can cause enormous increases in particle ranges. Even when channeling is not important there is some evidence, e.g., from multiple scattering experiments [27] that a simple transparency effect may occur: one may speculate that a crystallite of 5-10 lattice constants diameter may be significantly more transparent than an amorphous region of the same size (and density).

In the derivation of (21) it was assumed that the target atom was at rest before the collision. Sigmund [28,29] has shown that at moderately low energies with heavy projectile and target atoms the energy density within a cascade becomes sufficiently high that all atoms will be set in motion, so that the cascade region behaves like a hot gas for a short time. In such cases the linear theory cascade may be considered as defining an initial condition for a thermal spike [29].

A fundamental distinction between the range and the energy distributions is that any single range distribution, from the implantation of one projectile, is a delta function, whereas the energy distribution is spread over some volume. However the distributions calculated are in both cases averaged over an infinite number of trials, and so the calculated widths bear little or no relation to the width of any one collision cascade. (Some attempts to determine sizes of individual cascades by calculations of correlations have been made by Sigmund et al. [30,31].) The range distribution, as we have said, is a delta function at the end of the particle track, located according to a probability distribution which is just the calculated range distribution. The ionization distribution is the set of all tracks of moving atoms, weighted by the value of the electronic stopping at the particle speed. The damage distribution is the same set of tracks as the ionization distribution, with the weight depending less strongly on energy than for the latter.

The next step is to specialize eq. (21) to the plane-source case. Hence the coordinate $\underline{x}$ reduces to the depth x, and the solid angle reduces to the angle of the beam with respect to the surface normal. Eq. (21) now becomes

$$-\cos\theta \frac{\partial F}{\partial x} = S_e \frac{\partial F}{\partial E} + N \int d\sigma [F(x,E,\theta) - F(x,E',\theta') - \xi \overline{F}(x,E'',\theta'')].$$ (22)

This is still an equation in three variables, depth, angle, and energy. To dispose of the angular variable the obvious transformation is to a Legendre polynomial representation, $F(\cos\theta) = \Sigma (2\ell+1) P_\ell (\cos\theta) F_\ell$. Substituting, we obtain

$$-\ell \frac{\partial F_{\ell-1}}{\partial x} - (\ell+1) \frac{\partial F_{\ell+1}}{\partial x} = (2\ell+1) S_e \frac{\partial F_\ell}{\partial E} + (2\ell+1) N \int d\sigma [F_\ell (x,E) - P_\ell (\cos\phi') F_\ell (x,E')$$

$$-\xi P_\ell (\cos\phi'') \overline{F}_\ell (x,E'')]$$ (23)

where ϕ' and ϕ'' are the lab. scattering angle and recoil angle. To reduce this equation to a single-variable one we take moments in x:

$$F_\ell^n = \int_{-\infty}^{\infty} dx \, x^n \, F_\ell (x),$$

obtaining

$$\delta F_\ell^n (E) = S_e \frac{\partial F_\ell^n}{\partial E} + N \int d\sigma [F_\ell^n (E) - P_\ell (\cos\phi') F_\ell^n (E') - \xi P_\ell (\cos\phi'') \overline{F}_\ell^n (E'')].$$ (24)

Here δF_ℓ^n has been written as an abbreviation for

$$n(\ell \, F_{\ell-1}^{n-1} + (\ell+1) F_{\ell+1}^{n-1})/(2\ell+1).$$

We now consider the scattering cross section. Lindhard et al. [2] have shown that for several screened Coulomb cross sections the energy and angular dependence of the cross section may be combined:

$$d\sigma = \pi a^2 \frac{d\phi}{\phi^2} f(\phi)$$ (25)

where f is a function that depends on the screening, and ϕ is an energy transfer variable,

$$\phi^2 = \epsilon^2 T/(\gamma E)$$

where T is the energy transfer in the collision, and γE is the maximum energy transfer.

The value of a, the functional form of f, and the validity of eq. (25) itself, have all been subject to argument. In the present work, the Bohr-Lindhard [2] value for a, eq. (2), has been used. Some workers prefer Firsov's value [32]

$$a = 0.8853 \, a_0 (z_1^{1/2} + z_2^{1/2})^{-2/3}.$$ (26)

11

Both are obtained from Thomas-Fermi arguments and in both cases it has been assumed that projectile and target atom masses are comparable, and not especially small. A fast light projectile looks more like a point charge than like a Thomas-Fermi atom, so the screening radius should be reduced somewhat in that case. There is some evidence from multiple scattering measurements [33] that screening radii can be smaller than those given by eq. (2). In some of these measurements target crystallinity undoubtedly played some role [33], but there still remain measurements with H and He ions in particular which appear to indicate a smaller value of a than that of (2). If in keeping with the above argument, Z_1 in (2) were replaced by some Z_1^{eff}, agreement with these experiments would be obtained with $Z_1^{eff} \sim 0$ for the H and He ions.

The error introduced by (25) has been discussed by Lindhard et al. [2], and by Latta and Scanlon [34]. This question is connected with the choice of screening function f. For the choice used here, the approximation seems adequate. A particularly useful simple form for f is

$$f(\phi) = \lambda \phi^{2-2m}, \quad 0 < m < 1, \tag{27}$$

the so-called power-law approximation. This scattering law is an approximation to that from an $r^{-1/m}$ potential, and has the advantage that when $S_e = 0$, eq. (24) reduces to quadratures. Furthermore, all lengths in the cascade then scale as E^{2m}, and there is no other energy dependence.

Lindhard et al. [2] have used a Thomas-Fermi potential in calculating f numerically, giving their results in a table. In this book the analytic form is used,

$$f(\phi) = \lambda \phi^{1-2m} (1 + (2\lambda \phi^{2-2m})^q)^{-1/q}, \tag{28}$$

with the parameter values $\lambda = 1.309$, $m = 1/3$, $q = 2/3$. This function, which is an adequate fit to the tabulated values [9], was first proposed by Sigmund. Parameter values to fit other interactions have been given by Winterbon [35]. This function approaches power-law scattering with exponent m at small energy transfers, and power-law scattering with exponent 1, i.e., Coulomb scattering, at large energy transfers.

In the $\phi = 0$ limit, the corresponding $(r = \infty)$ limit of the potential defined by (28) is r^{-3}, although for the Thomas-Fermi potential the limit is r^{-4}. However the Thomas-Fermi potential does not attain this r^{-4} behaviour until very large distances, greater than typical interatomic distances in solids or liquids, so the r^{-3} is actually more appropriate [10]. One knows that the potential of isolated atoms decreases exponentially, suggesting that the large-r behaviour, if forced into the power-law form, should approach m=0. However the problem is not that simple, and some counter arguments can be put forth. First, the value of the potential for r greater than half the nearest-neighbour distance is not very important, so the choice of potential is not as critical for spatial distributions as it is for, say, detailed damage studies. If the target is solid or liquid, rather than gaseous, the large-r electron densities will not be the free-atom densities. Finally, bearing in mind the complexity of the problem, it should be remembered that reasonable agreement has been obtained between calculation and experiment for amorphous targets, especially amorphous oxides and semiconductors, with most of the cases of disagreement (other than those in which channeling has been obvious, so that the theory is manifestly inapplicable) being with metallic targets which are almost always crystalline.

As was said earlier, the electronic stopping has been chosen to be proportional to particle speed, with the constant of proportionality that due to Lindhard and Scharff [15], eq. (8). Both the velocity proportionality and that constant of proportionality are approximations. It has been known for several years that the electronic stopping depends non-monotonically on projectile Z, the so-called "Z_1 oscillations", in contradiction to (8). Although very large deviations from (8) have been observed in channeling conditions, in amorphous targets the deviations are up to perhaps 20-30%. This "oscillation" of electronic stopping has been explained by several authors (e.g. [36]) to be due to atomic shell effects. The "amplitude" and "phase" of the Z_1 oscillations vary slowly with energy; equivalently, writing $S_e = kE^p$, this corresponds to departures of p from the value $1/2$. Also, Moak et al. [37] have found electronic stopping values depending linearly on particle speed, but with a non-zero intercept, $S = kE^{1/2}-a$. Calculations of electronic stopping for slow atoms, either from a dielectric formulation [15], or by looking at the overlap of electron clouds [24] appear to lead inexorably to velocity proportionality. At least two possible reconciliations exist. The charge state of a moving ion is not a well-defined concept, but it seems likely that the number of electrons in the vicinity of the moving atom fluctuates, and is less the faster that atom moves. The stopping power will depend on the number of such electrons. Calculations of Z_1 oscillations [36] incorporating such ideas in a crude but self-consistent form were in rough agreement with experimental measurements of p. The second effect is that some fraction of the electronic stopping depends on the electron density along the particle path rather than just that near the path. At low energies the projectile is energetically excluded from high density - and hence high stopping power - regions which it can enter more readily at higher energies. This would be consistent with the Moak et al. [37] observations.

In the power-law approximation, with electronic stopping neglected, the transport equation (24) becomes

$$\delta F_\ell^n = \frac{NC}{E^{2m}} \int_0^\gamma \frac{dt}{t^{1m}} \ [F_\ell^n(E) - P_\ell(\cos\phi')F_\ell^n(E-Et) - \xi P_\ell(\cos\phi'')\overline{F}_\ell^n(Et)], \tag{29}$$

with $C = \frac{\pi}{2}\lambda a^2 (\frac{M_1}{M_2})^m (2Z_1Z_2e^2/a)^{2m}$, and $t = T/E$. With this change of variables,

$$\cos\phi' = (1-t)^{1/2} + \frac{1}{2}(1-M_2/M_1)t(1-t)^{-1/2}, \tag{30a}$$

and

$$\cos\phi'' = (t/\gamma)^{1/2}. \tag{30b}$$

Writing

$$F_\ell^n = (E^{2m}/NC)^n \ E^\xi A_\ell^n, \tag{31}$$

then

$$\delta A_\ell^n = \int_0^\gamma \frac{dt}{t^{1m}} \ [A_\ell^n - P_\ell(\cos\phi')(1-t)^{2mn+\xi}A_\ell^n - \xi P_\ell(\cos\phi'')t^{2mn+\xi}\overline{A}_\ell^n (C/\overline{C})^n], \tag{32}$$

which may be rewritten as

$$A_\ell^n = [\delta A_\ell^n + \xi\overline{A}_\ell^n (C/\overline{C})^n \int_0^\gamma \frac{dt}{t^{1+m}} t^{2mn+\xi}P_\ell(\cos\phi'')] \ / \int_0^\gamma \frac{dt}{t^{1+m}} \ [1-P_\ell(\cos\phi')(1-t)^{2mn+\xi}] \tag{33}$$

to make explicit that the problem has been reduced to quadratures. The integrals appearing in (32) or (33) may be obtained from incomplete beta functions by recursion [9]. In writing (31) we have recalled that for range distributions $\xi = 0$ and the normalization is $F_o^o = 1$, while for the damage distribution the normalization is $F_o^o = E$ (since there is no electronic stopping). Equation (32) is independent of E, proving the energy independence mentioned above. The power-law equation is of great value because it combines simplicity and energy independence with reasonable accuracy. Power law moments up to the fourth for the range and damage distributions have been given in [9] for projectile-target mass ratios between 0.1 and 10.

If one simplifies still further, to path-length equations, where the "depth" coordinate is measured along the particle path, and all the angular variables drop out, one can make analytic calculations of some properties of the distributions. For example, it has been shown [38] that the asymptotic behaviour of the path length distributions at large distances is

$$-\log F = Ax^{1/(1-m)} + \text{smaller terms},$$ (34)

and it has been argued that the depth distributions will behave similarly.

Now consider the effect of including (velocity-proportional) electronic stopping in (29). First, the stopping is continuous, not stochastic, so there is a well-defined maximum path length, and hence maximum range, namely $2E^{1/2}/K$, found by neglecting elastic collisions. At low energies, it is expected that the nuclear stopping will be more important than the electronic, so the substitution (31) is used again, giving

$$\delta A_\ell^n(E) = \frac{KE^\zeta}{NC}(E\frac{\partial A_\ell^n(E)}{\partial E} + (\xi+2mn)A_\ell^n(E)) + \int_0^\gamma \frac{dt}{t^{1m}} [A_\ell^n(E) - P_\ell(\cos\phi')A_\ell^n(E-Et)(1-t)^{\xi+2mn}$$

$$- \xi(C/\overline{C})^n P_\ell(\cos\phi'') \quad \overline{A}_\ell^n(Et)t^{\xi+2mn}].$$ (35)

In order that the electronic stopping term be small at small energies it is required that $\zeta = 2m-1/2 > 0$, i.e., $m > 1/4$, or that there be a low-energy cutoff.

It is easy to see that a low energy cutoff is needed in the definition of deposited energy [26]. Consider the rate of energy loss to electronic stopping,

$$- \frac{\partial E}{\partial t} = - \frac{\partial E}{\partial x} \cdot \frac{\partial x}{\partial t} = K E^{1/2} V \propto E.$$ (36)

Energy is conserved in collisions, so the rate of energy loss is unaffected by collisions, provided both collision partners remain above any threshold. If there is no threshold, the energy available for damage decreases exponentially with time to zero, regardless of the nuclear scattering law. Thus without a threshold, $\nu(E) = 0$. It may be seen from (35) that for $m > 1/4$, as the threshold decreases to 0, the zero-energy limit of $\nu(E)/E$ is 1, whereas for $m < 1/4$ it is 0.

Now the power cross section is replaced by the algebraic form (28), in (24) or its successor (35), to get

$$\delta A_\ell^n(E) = \frac{KE^\zeta}{NC}(E\frac{\partial A^n}{\partial E}(E) + (\xi+2mn)A_\ell^n(E)) + \int_0^\gamma \frac{dt}{t^{1+m}} \psi(E^2t) [A_\ell^n(E) - P_\ell(\cos\phi')A_\ell^n(E-Et)(1-t)^{\xi+2mn}$$

$$- \xi(C/\overline{C})^n P_\ell(\cos\phi'')\overline{A}_\ell^n(Et)\cdot t^{\xi+2mn}],$$ (37)

with $\psi(E^2t) = (1 + (2\lambda(\epsilon/E)^{2-2m} (E^2t/\gamma)^{1-m})^q)^{-1/q}$. With the equation written in this form, so that its relation to the power-law equation (32) is apparent, an obvious way to solve the equation is by series methods [35]. Two powers of E appear in (37), E^ζ in the electronic stopping term, and E^ν in the new factor in the elastic cross section, where ν has been written for $2q(1-m)$. These two powers of E are considered as independent variables, and the moment A_ℓ^n is written as a double power series. The radius of convergence from the second variable is certainly not greater than $\epsilon = 0.486$, from (28), and it is not obvious what determines the radius of convergence for the first variable. At any rate it is necessary to sum the series beyond the known radius of convergence. Continued fraction methods have been used, summing first the E^ζ series for each power of E^ν, and then the resulting E^ν series. It was found that this gave better results than summing in the opposite order. Series of 30×30 terms were used; because of the finite computer word length, longer series did not appear to give increased convergence.

[It is possible to attempt a large-E expansion of the moments. Each integral is divergent, except perhaps the first few, but the integrals may be regarded as analytic continuations of the corresponding integrals defined by the small-E expansion. However because the large-E limit of the cross section is the $m=1$ power, the "power" series contain logarithmic terms as well, and very rapidly get very messy. One could change the cross section so the large-E power limit is some $m < 1$, but this has not yet been carried out. (Such high and low energy expansions have been used in a very similar problem [39].)]

In fact the series for the moment coefficients are not themselves summed, but instead they are combined as formal power series to obtain the desired quantities, namely standardized central moments, and it is these latter series that are summed.

Moments have also been constructed by a numerical integration technique, using splines [39]. A spline is a piecewise continuous polynomial:

$$S_n(y) = \sum_{i=0}^{N-1} \alpha_i y^i + \sum_{j=0}^{n-1} \gamma_j (y-y_j)_+^N , \qquad (38)$$

where $y_+ = \max(y,0)$, is an N-th order spline with $n+1$ nodes y_j. In the present calculations cubic ($N=3$) splines are used. Such splines are continuous and have continuous first and second derivatives. A cubic spline with $n+1$ nodes has $n+3$ parameters, so if it is to be determined by its values at $n+1$ points two supplementary conditions must be used. We require $S''(y_0) = S''(y_n) = 0$. It will be assumed from now on that all splines mentioned will have these end conditions. In that case a spline is uniquely determined by its values at the nodes. It is convenient to construct a set of cardinal splines, $C_n(y)$, which are defined to be splines having the values $C_n(y_m) = \delta_{mn}$. Then any spline may be written as a linear combination of the C_n, with the coefficients being the values of the spline at its nodes.

Now return to the moment equation (37). For small E the solution is a power series in E^ζ and E^ν, with $\nu/\zeta \gg 1$: with the parameter values used here, $\zeta = 1/6$ and $\nu = 8/9$. Small fractional powers are approximated poorly by sums of integer powers, so it is desirable to make a change of variable, $y = E^\zeta$. With this change, and writing $A_\ell^n(E) = B(y)$, (the ℓ and n suffices can be dropped at this point without confusion,) (37) becomes

$$\delta B(y) = \frac{Ky}{NC} \left(\zeta y \frac{\partial B(y)}{\partial y} + (\xi+2mn) B(y) \right) + \int \frac{dt}{t^{1+m}} \psi(y^{2/s}t) [B(y) - P_\ell(\cos\phi') B(y(1-t)^\zeta)(1-t)^{\xi+2mn}$$

$$- \xi P_\ell(\cos\phi'') (C/\bar{C})^n B(yt^\zeta) t^{\xi+2mn} . \qquad (39)$$

Because of the high accuracy required in the solution, it is desirable to subtract the "exactly known" constant $X = B(0)$, writing $B(y) = X+D(y)$. Then adding and subtracting some terms in the integrand, one obtains the recursion relation (33) for X, and the integro-differential equation,

$$\delta D - X\left[\frac{Ky}{NC}(\xi+2mn) + \int \frac{dt}{t^{1+m}}(\psi-1)\left[1 - P_\ell(\cos\phi')(1-t)^{\xi+2mn} - \xi(C/\overline{C})^n P_\ell(\cos\phi'')t^{\xi+2mn}\right](\overline{X}/X)\right.$$

$$= D\left\{(\xi+2mn)\frac{Ky}{NC} + \int \frac{dt}{t^{1+m}}\psi\left[1 - P_\ell(\cos\phi')(1-t)^{\xi+2mn}\right]\right\} + \zeta y\frac{\partial D}{\partial y}\left\{\frac{Ky}{NC} + \int \frac{dt}{t^{1+m}}\psi P_\ell(\cos\phi')(1-t)^{\xi+2mn}\cdot t\right\}$$

$$- \int \frac{dt}{t^{1+m}}\psi P_\ell(\cos\phi')(1-t)^{\xi+2mn}\left[D(y(1-t)^\zeta) - D(y) - \zeta yt\frac{\partial D(y)}{\partial y}\right]$$

$$- \xi(C/\overline{C})^n \int \frac{dt}{t^{1+m}}\psi P_\ell(\cos\phi'')t^{\xi+2mn}\overline{D}(yt^\zeta) \tag{40}$$

for D. A set of nodes is chosen, and D is written as a sum of cardinal splines in (40). This reduces (40) to a set of linear algebraic equations. In these calculations 20 nodes were used, linearly spaced in y, with $y_0=0$, and the last one, y_{19}, corresponding to the largest energy required.

For each case the computational procedure was as follows. Calculate the series for the central moments, using the Lindhard-Scharff value k_L(8) of electronic stopping. From this series construct the series for the difference between this quantity and the corresponding quantity calculated with any other value of k. This will be a series in $k-k_L$. Choose a maximum energy, 1, 2, or 5×10^m keV, corresponding to $\epsilon \sim 30$. Sum the first series for $k = k_L$, and the difference series for the other 4 values of k, for 13 values of projectile energy, 1, 2, and 5×10^n keV, up to the maximum energy. Calculate the moments using the spline program, with y_{19} corresponding to the same maximum energy, for 3 k values, $k = k_L$, and the maximum and minimum values. Find the moments for the other two k values by quadratic interpolation in k. Combine the two sets of results, using the series where convergence is good, and going over to the spline at high energies. Plot all moments to see that the patching between the two computations has been done properly, and repeat the last two steps as required.

5. MOMENTS AND DISTRIBUTIONS

The first two moments are location parameters: they specify where the distribution is, and how broad it is. The relative transverse straggling indicates how prolate or oblate the the distribution is, considered as a point-source distribution. The higher moments, skewness and kurtosis in this instance, determine the shape. A symmetric distribution has skewness=0, although the converse is not necessarily true. The kurtosis is a measure of the importance of the tails of the distribution, although again not by itself an infallible measure. To gain some idea of the meaning of kurtosis, consider the distribution

$$f = N \exp(-\lambda |x|^p);$$

The skewness is zero, and the kurtosis is $\Gamma(1/p)\Gamma(5/p)\Gamma^{-2}(3/p)$. For p=1 this is 6, for p=2 (gaussian), 3, and for $p \to \infty$ (rectangular), 1.8. For $p \to 0$, the kurtosis increases without limit. Thus a large value of kurtosis indicates that the distribution has substantial tails. The minimum value of kurtosis possible for a unimodal distribution (one with only a single peak) is 1.6875 [40]. For eq. (14) to be valid the kurtosis must be greater than $1.8 + 1.2 \times (\text{skewness})^2$.

There are many ways of constructing distributions from the moments. It is reassuring to know that the infinite sequence of moments would in fact determine the distribution uniquely: it has been shown that in the absence of electronic stopping, power-law distributions have the asymptotic behaviour (34); with electronic stopping, there is a maximum range. Hence in either case the distributions fall off faster than an exponential, and this guarantees the uniqueness [41]. The given finite sequence of course does not uniquely determine a distribution, and other information is useful. We know that the distributions are smooth, except perhaps at the surface, and are unimodal. The ionization distribution has a discontinuity at the surface equal to the value of electronic stopping for the incident ion. The question of the order of discontinuity of the other distributions is currently being examined.

The simplest method of obtaining an approximate distribution is by an orthogonal polynomial expansion, based on e.g., a Gaussian weight function. (If a Gaussian is used the expansion is called a Gram-Charlier expansion.) In this method a "suitable" non-negative weight function is chosen, and a set of polynomials is constructed orthogonal with respect to this weight function. Because the polynomials are orthogonal, the coefficient of the n-th order polynomial is independent of the values of moments higher than the n-th. The distributions obtained by this method usually have negative excursions.

It is usual to choose a weight function which has no free parameters, other than the location parameters, so for example the Gram-Charlier Gaussian will have the correct mean range and straggling, but will have skewness=0 and kurtosis=3.

From here one can go in two directions. One may try to choose a more appropriate density as a weight function for an orthogonal polynomial series - this was examined at some length in [9]. Alternatively one may try to guess a functional form which has the correct first few moments, and just stop there. This is the basis for approximating a range distribution by a Gaussian. With four moments available, one's horizons are wider. On looking through a table of Mellin transforms, e.g., [42], two likely candidates emerge,

$$f(x) \sim x^p \exp(-\alpha x^h), \quad x > 0, \quad h > 0, p \geq 1 \tag{41a}$$

and

$$f(x) \sim x^p \exp(-\alpha x^h - \beta x^{-h}), \quad x > 0, \quad h > 0 \tag{42a}$$

which have normalized moments

$$\Gamma(\frac{n+p+1}{h}) / (\alpha^{n/h} \, \Gamma(\frac{n+1}{h})) \tag{41b}$$

and

$$(\beta/\alpha)^{n/2h} \, K_{\frac{n+p+1}{h}} (2\sqrt{\alpha\beta}) / K_{\frac{p+1}{h}} (2\sqrt{\alpha\beta}) \tag{42b}$$

respectively. Here K is the modified Bessel function of the third kind. However either choice requires a non-linear search to determine the parameter values, so they are not overpoweringly attractive.

The classical choices for distributions determined by four moments are the Pearson distributions [18]. They are defined by the differential equation,

$$\frac{\partial f}{\partial x} = \phi(x) f(x), \tag{43}$$

with $\phi = \dfrac{ax+b}{x^2+cx+d}$. Writing $\phi = P/Q$, (43) implies

$$\int dx \, x^n [P(x) f(x) - Q(x) \frac{\partial f}{\partial x}] \equiv 0. \tag{44}$$

From this identity one gets a set of equations bilinear in the parameters of P and Q, and the moments of f. Equation (44) and the resulting bilinear relations would be obtained for arbitrary polynomials P and Q. The Pearson distributions are of different types according to the positions of the roots of the denominator Q (in the complex plane) relative to the root of P. These distributions have been used by Hofker et al. [43] to deduce moments of measured range distributions (for B in Si). However they obtained their best fits with those Pearson distributions which decrease asymptotically as powers of x, and at high energies these exponents became sufficiently small that they seriously overestimated the higher moments. When they extrapolated from the limit of their measurement to infinite (positive and negative) depths with an exponential, rather than the power, they obtained much better agreement with the calculated values [44].

The differential equation (43) has been suggested as the basis for a sequence of approximate distributions, with successively more complicated ϕ's, starting from $\phi = (ax^2+bx+c)/(x+d)$, and increasing the order of numerator and denominator polynomial by one at each step [45]. In the zeroth order, $\phi = ax + b$, and f is the Gram-Charlier Gaussian. In n-th order it turns out that n extra conditions are needed to determine the parameters.

Consider the first order. Then the density is

$$f(x) = N(x+b)^\gamma \exp(\frac{\alpha x^2}{2} + \beta x). \tag{45}$$

If we require $\gamma > 1$, which seems reasonable, then the kurtosis and the absolute value of the skewness must lie within the lune shown in Fig. 1. If a fifth moment, or one of the parameters of (45) is specified, a point in this lune is chosen. If the given skewness and kurtosis do not take values within this region, the distribution constructed does have skewness and kurtosis within this region, but has all its moments wrong.

Another method of constructing distributions is through inverse Fourier transforms. This method was suggested by Chisholm and Common [46], and applied to distributions related to the present ones by Littmark [47]. This is the method used to construct the distributions that appear with the tables.

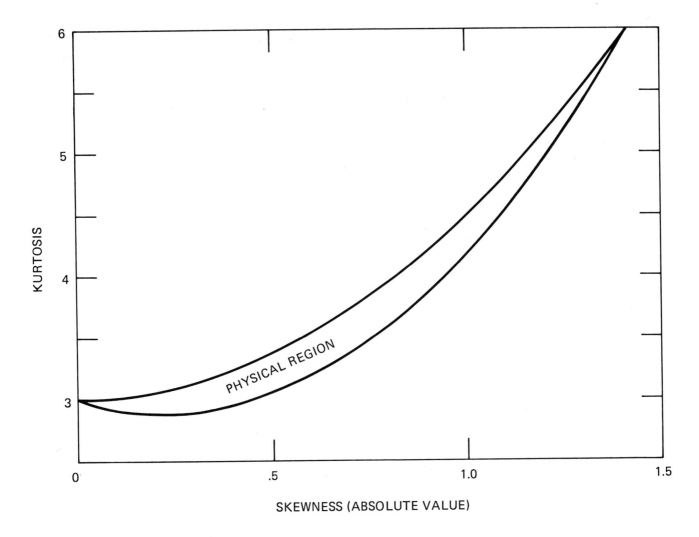

Figure 1.   Physical Region for Density (45).

The Fourier transform of a distribution f is

$$g(k) = \int dx\ e^{ikx}\ f(x),$$

(46)

which may be formally expanded to

$$g(k) = \sum_{n=0}^{\infty} \frac{(ik)^n}{n!}\ \langle x \rangle^n.$$

(47)

The inverse transformation is to be performed on this series, so it is necessary to have some method of extrapolating the sum to infinity. The method used is to write Pade approximants to g, which is equivalent to writing g as a quotient of polynomials. This quotient is then written as a sum of partial fractions, and each partial fraction, on inverse transformation, gives an exponential. Thus the inverse Fourier transform is a sum of exponentials. This method is not constrained to give positive definite functions.

Finally there are constructions based on the classical solutions of the moment problem [41]. It can be shown that certain sets of sums of delta functions are extremal densities, in the sense that their indefinite integrals give upper and lower bounds on the indefinite integrals of any (non-negative) densities having the same first few moments.

There are algorithms for constructing such extremal densities [48]. Thus one is immediately able to construct a density consisting of a sum of other densities having known moments, by convoluting the known density with the sum-of-delta-functions extremal density. For example the density can be approximated by a sum of Gaussians, all of the same width, located to have the given moments. These algorithms can be modified to construct extremal unimodal densities [40], and these latter densities, which are rectangles, can be smoothed with given densities. The surface quantities, R, γ, and α, are obtained from distributions which are unimodal extremal distributions, smoothed by replacing the step functions by error functions. Thus the distribution is of the form

$$\sum_i A_i \, \text{erf}((x-x_2)/a).$$

$$(48)$$

The breadth a of the error functions is the largest value consistent with the given moments.

6. QUALITATIVE BEHAVIOUR OF MOMENTS

The first observation is that the power-law results of [9] are borne out. The moments start at the m=1/3 value, move toward the m=1/2 value, and at some energy above $\varepsilon=1$ become strongly affected by electronic stopping. The energy distribution moments, damage and ionization, are similar in their behaviour in the low-energy regime where nuclear stopping dominates, then diverge in the high-energy, or electronic-stopping, region. When the projectile is appreciably heavier than the target atoms, the recoils can have longer ranges than the original projectile. This effect occurs as the effective m value increases from 1/3 to 1/2. At higher energies, the electronic stopping dominates, and it has a greater effect on the light recoils than on the heavy projectile, thus reducing the recoil ranges. Also at these higher energies the probability of creating long-range recoils is reduced, thus reducing their effects on the moments.

Fractional Deposited Energy, $\nu(E)/E$ (Fig. 2)

This fraction is $1 - O(E^{1/6})$, i.e., even at low energies a noticeable fraction of the energy goes into electronic stopping. It is not obvious how much ν/E would change if a threshold were introduced. Measurements of $\varepsilon(E) = E - \nu(E)$ for Ge in Ge from 254 eV to 2 KeV show no signs of a threshold [49].

Mean and Modal Ranges (Figs. 3,4)

Modal depths were calculated from the Pearson formula, eq. (15), for range and damage distributions, and from the inverse Fourier transform calculations for the ionization distribution.

The modal depth is smaller than the mean depth at low energies, and for range and damage, larger at high energies. The low-energy ratio of the two is largest near unit mass ratios, and tends to vanish for both very large and very small mass ratios.

While the energy is in the nuclear stopping range, the mean ionization depth is smaller than, but follows, the mean damage depth in its energy dependence, as was commented on above; at larger energies it starts to decrease, for the electronic stopping is occurring more at the beginning of the particle track, and the multiplication of moving particles at depth is not enough to counteract this. Conversely the damage is approximately proportional to nuclear stopping, so is less at the surface, and the mean damage depth approaches the mean range. With very light projectiles at low energies the maximum in the ionization distribution is at the surface, but otherwise it is at a finite depth. The modal depth of the damage distribution increases rapidly near $\varepsilon=1$, relative to the mean range, or, for that matter, the mean damage depth. This corresponds to the damage distribution changing from the relatively symmetric form of the low-energy regime to the characteristic high-energy form, a track of light damage intensity from the surface in to a peak of damage intensity near the end of the particle tracks. The apparent sharpness of this low-energy to high-energy transition may be partly due to the small number of moments used.

Even in the mean and modal depths, one can already see for heavy projectiles the effect of energy transport by recoils. It can be seen from the cross-section constant C (eq. (29)) that the maximum-energy recoils have larger ranges than the projectile, and the ratio of recoil range to projectile range increases with increasing m, and hence with increasing energy. Because these recoil ranges are cut off by the electronic stopping, the resulting "bump" in the damage and ionization moments is sensitive to the value of electronic stopping.

Straggling (along the beam direction) (Fig. 5)

Relative longitudinal straggling of ranges is large for light projectiles, small for heavy projectiles. As the energy increases, first the cross section becomes ever more strongly peaked toward small angle deflections, making the path approach a straight line. The path length straggling decreases at the same time, as the probability of large-energy-loss collisions decreases. As the energy continues to increase, the continuous nature of the electronic stopping decreases straggling still further. (It should be remembered that we are talking about relative straggling, a ratio. The absolute straggling, a length, will increase continuously with increasing energy, approaching a constant limit (with continuous electronic stopping).) The damage distribution is spread out along the collision cascade, so the relative damage straggling varies more slowly with both mass ratio and energy than the range straggling. The decrease in straggling as the damage distribution evolves into its high-energy form is clearly seen. The ionization distribution remains spread out over the entire path, so the "straggling" (breadth) increases at large energies. The value of the ionization distribution at a given depth is the product of the number of moving particles with their speeds, divided by their direction cosines. The effect of recoil ranges on the damage and ionization distributions is becoming more apparent.

Transverse Straggling (Fig. 6)

At low energies, range distributions are prolate (relative transverse straggling less than 1), approaching sphericity as the mass ratio increases. At high energies, the small angle multiple scattering spreads the beam transversely more than it affects energy loss, so the distribution becomes oblate. Very heavy ions need to transfer a large fraction of the maximum energy transfer to be scattered through an appreciable angle, so the decrease in large angle scattering with increasing energy makes those range distributions more prolate at first. Damage and ionization distributions, being weighted by particle energies, are more prolate, the ionization more so than the damage, and become increasingly prolate at high energy. (At extremely high energy, one recalls, the damage track, which is actually an ionization distribution, is simply the straight-line path of the projectile (e.g. fission tracks in mica).) Note that the energy distributions are broadened transversely, as well as longitudinally, in the $m=1/2$ regime by the recoil particles.

Skewness (Fig. 7)

Positive skewness corresponds to the mean depth being greater than the most probable depth (q.v. eq. (15)). Range skewness is positive in the nuclear-stopping regime, negative in the electronic-stopping. In the first case there is a tail of particles which have undergone less stopping than average, a tail extending to comparatively large depths. In the second, the electronic stopping cuts off this tail, but now large angle collisions near the surface are unlikely, so there is a tail consisting of the few particles which have undergone these violent collisions and so have substantially smaller ranges than average. The distribution includes the backscattered particles, so once the skewness becomes negative it can decrease more rapidly for the easily-backscattered light projectiles than for heavy projectiles. Damage distributions have the same transition from low-energy to high-energy forms, and the damage skewness also goes negative. Ionization skewness stays positive, as the peak stays near the surface. Recoils tend to be forward directed, so they increase damage and ionization skewness in the $m=1/2$ region.

Kurtosis (Fig. 8)

Large kurtosis implies long-tailed distributions. Range kurtosis generally decreases as the large-depth tail is cut off, and then increases as the small-depth tail grows. Damage kurtosis also increases at high energies, but more slowly than range kurtosis, perhaps because the straggling is larger. Heavy-projectile damage and ionization kurtosis are large in the m=1/2 region, because the high energy recoils form a long tail extending beyond the range distribution. Damage and ionization kurtosis are very small near $\varepsilon \sim 5$, indicating that the distribution approaches a rectangular distribution.

7. SURFACE QUANTITIES, R, γ, AND α

(Figs. 9, 10)

These are secondary quantities, less reliable than the moments. They are calculated from the moments assuming some approximate form for the distribution, and the number of moments is too small to give really reliable values. (See under Moments and Distributions for the construction of the distributions.) It should also be remembered that a surface correction should be included, before they are applied. From comparison with power-law calculations using 20 moments [20], it is suspected that R values for $M_2 \leq M_1$ are overestimated. The irregularities in γ and α, especially when the skewness is near zero, are also artifacts. *These curves should be examined before accepting any value from the tables.*

The surface correction for R has been worked out in some detail, and verified by comparison with experiment [20,21,22]. Analogous surface corrections for γ and α have been obtained (Sigmund, private communication) but have not yet been published.

The surface correction is constructed by comparing reflection values for infinite (R) and semi-infinite (R_s) targets. Define

$$R(E,\Omega;E_0,\Omega_0)dE_0 d\Omega_0$$

to be the number of ions with energy (E_0,dE_0) backscattered into the solid angle ($\Omega_0,d\Omega_0$), for one atom incident with energy E and angle Ω and let

$$R(E,\Omega) = \int dE_0 d\Omega_0\, R(E,\Omega;E_0,\Omega_0); \tag{49}$$

R_s is defined similarly. Then R and R_s are related by [20]

$$R(E,\Omega) = \int dE d\Omega_0 R_s(E,\Omega;E_0,\Omega_0)(1-R(E_0,\Omega_0)). \tag{50}$$

To solve (50), it is assumed that the E_0 and Ω_0 dependence of R_s are separable:

$$R_s(E,\Omega;E_0,\Omega_0) = R_s(E,\Omega)\rho(E_0)\sigma(\Omega_0), \tag{51}$$

with

$$\int dE_0\, \rho(E_0) \int d\Omega_0\, \sigma(\Omega_0) = 1,$$

so that

$$R_s(E,\Omega) = \frac{R(E,\Omega)}{1-\delta}, \tag{52}$$

where

$$\delta = \int dE_0 d\Omega_0 \rho(E_0)\sigma(\Omega_0)R(E_0,\Omega_0) = \frac{1}{2}\int d\theta \cos\theta \sigma(\cos\theta) \int dE_0\, \rho(E_0)R(E_0,\cos\theta).$$

The infinite-target value $R(E,\cos\theta)$ is 1/2 for $\theta = \pi/2$, and is given in the tables for $\theta = 0$. An interpolation formula that has been used is $R(\theta) = 1/2\,(2r)^{\cos\theta}$, where r is the value of R at $\theta = 0$, i.e., the tabulated value. The E dependence of R will come from the tables. The function σ is approximately $\cos\theta$. The spectrum $\rho(E_0)$ can be approximated [21, 22] by a power E^ν, chosen to give a reasonable value to the mean energy of the sputtered particles, say 0.1E to 0.3E for light ions [21]. The corrected value is not especially sensitive to this value.

Figures 2 to 10.   Moments and Surface Quantities given in Tables.

Solid line, $k = k_L$.   Shorter dashes for small values of k/k_L, longer dashes for large values.   For values of k/k_L for each case see Table 0.1.   Letters R, D, and I at the sides of figures 3-9 denote $k = k_L$ range, damage, and ionization curves respectively.   The energy corresponding to $\epsilon = 1$ is indicated in each frame.

Figure 2. ν/E. The fraction of energy going into damage.

Figure 3.   Mean (and Modal) Range, divided by $E^{2/3}$. The 'R' denotes the mean range.   The ordinates are in units of $\mu g \cdot cm^{-2} \cdot keV^{-2/3}$.

Figure 4.   Relative Mean (and modal) Depth for Damage and Ionization, $<x>/<x>_R$.   The 'D' and 'I' denote mean depths.   The modal ionization depth is less than the modal damage depth.   In each case the depth has been divided by the mean range for the value of electronic stopping.

Figure 5.   Straggling, $<\Delta x^2>/<x>^2$.

Figure 6.   Transverse Straggling, $<y^2>/<\Delta x^2>$.

Figure 7.   Skewness, $<\Delta x^3>/<\Delta x^2>^{3/2}$.

Figure 8.   Kurtosis, $<\Delta x^4>/<\Delta x^2>^2$.

Figure 9.   R,γ. The fraction of the distribution extending outside the surface.   No surface correction.

Figure 10. α. The sputtering yield coefficient.   The value of the damage distribution at the surface, divided by the value of nuclear stopping for the incident ion.   No surface correction.

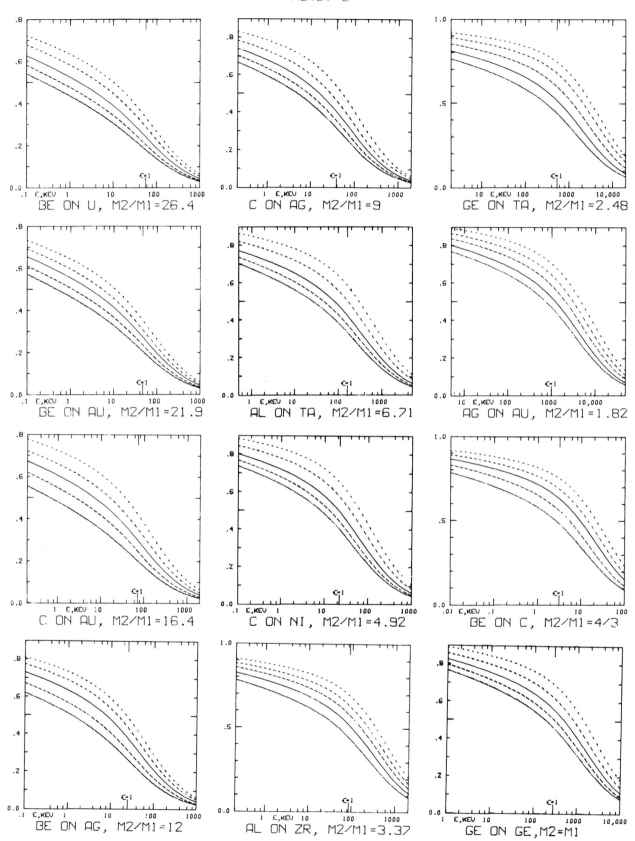

NU(E)/E

BE ON U, M2/M1=26.4

C ON AG, M2/M1=9

GE ON TA, M2/M1=2.48

BE ON AU, M2/M1=21.9

AL ON TA, M2/M1=6.71

AG ON AU, M2/M1=1.82

C ON AU, M2/M1=16.4

C ON NI, M2/M1=4.92

BE ON C, M2/M1=4/3

BE ON AG, M2/M1=12

AL ON ZR, M2/M1=3.37

GE ON GE, M2=M1

Figure 2

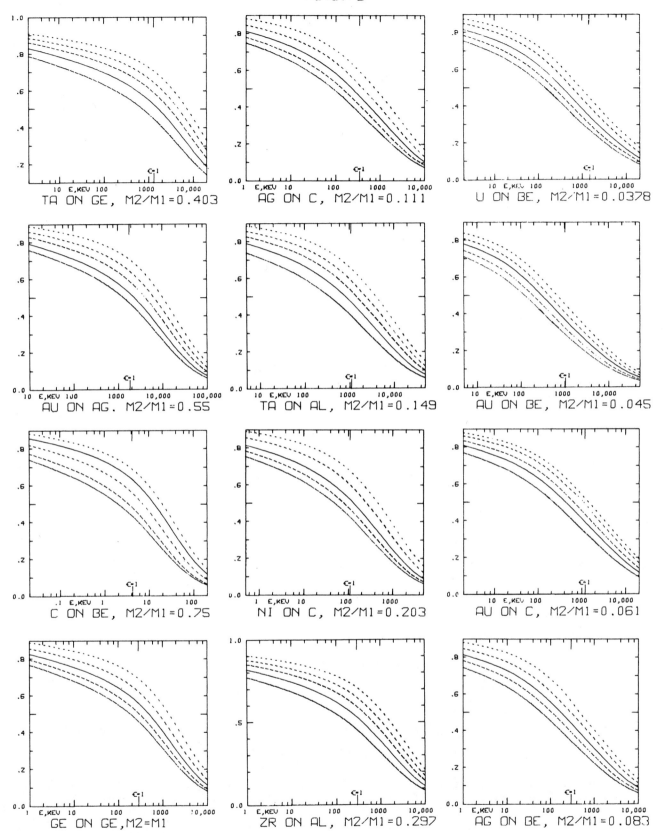

Figure 2 (continued)

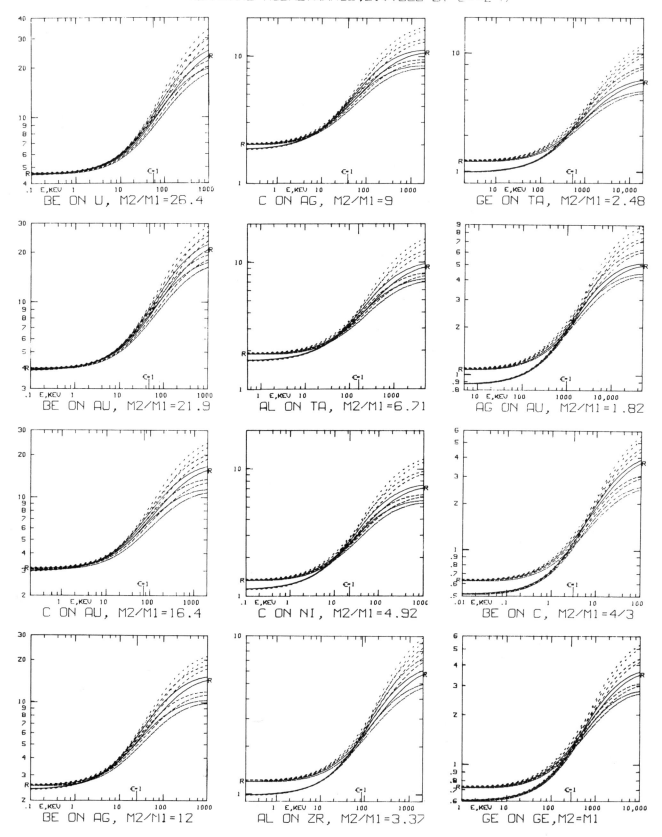

Figure 3

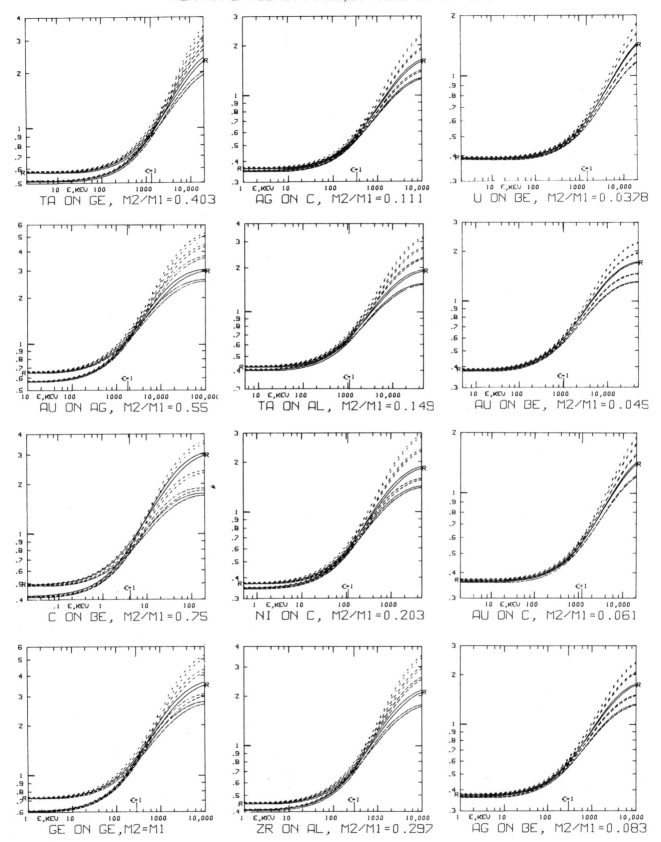

Figure 3 (continued)

31

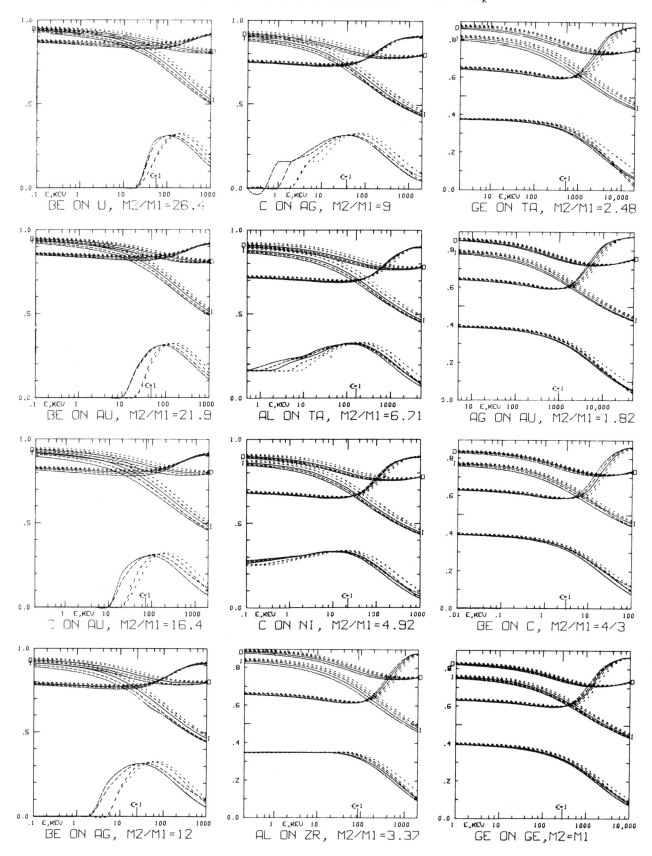

Figure 4

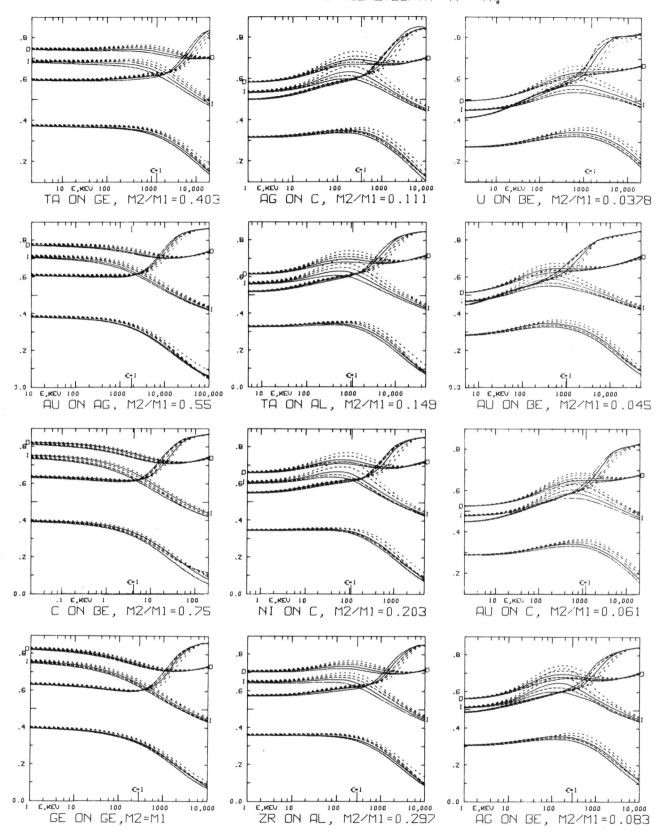

Figure 4 (continued)

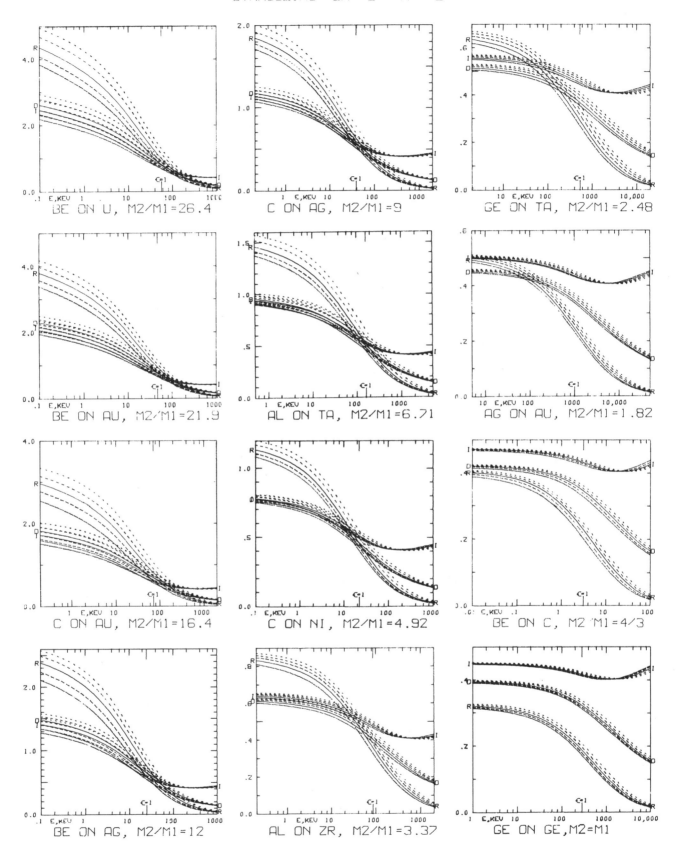

Figure 5

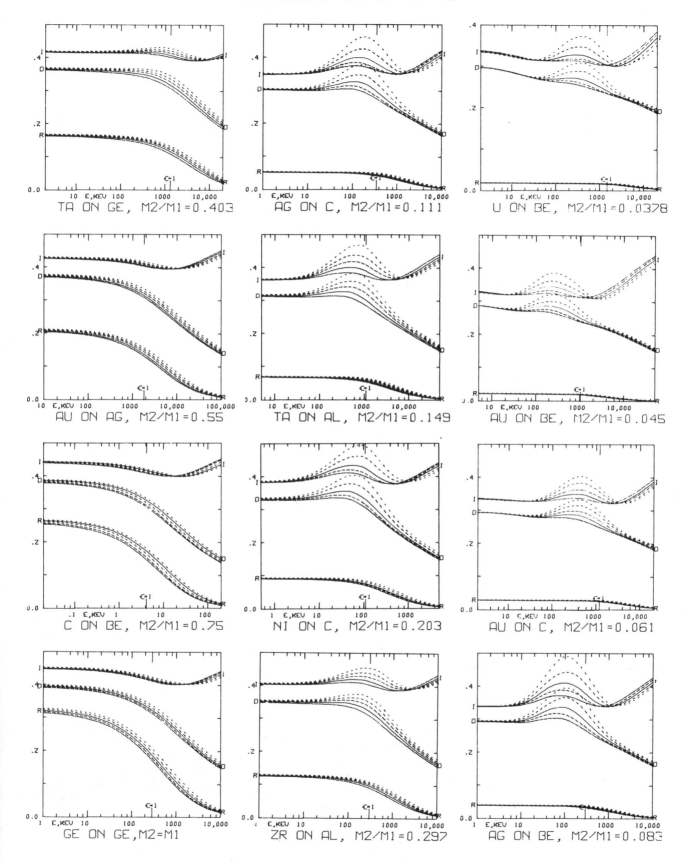

Figure 5 (continued)

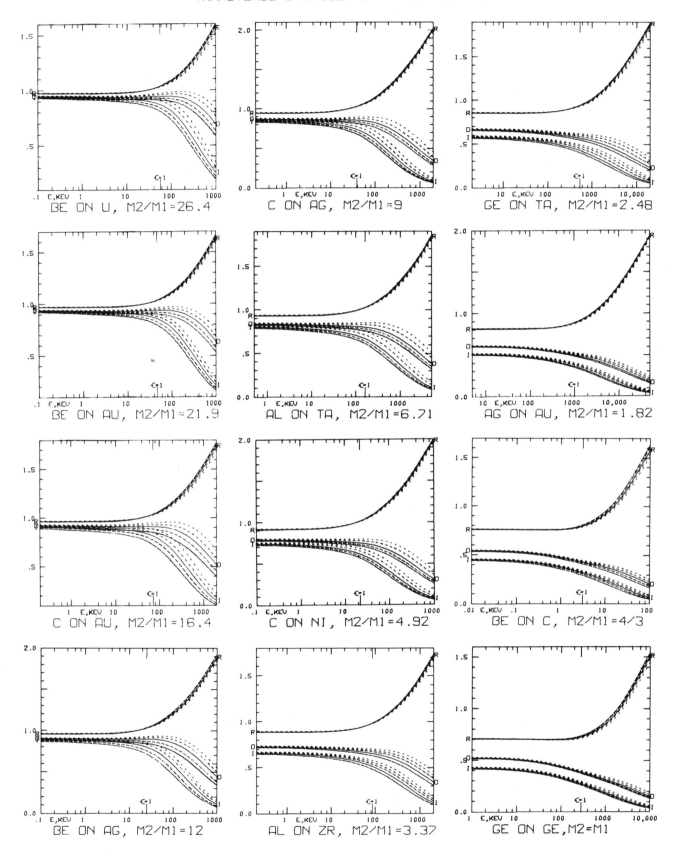

Figure 6

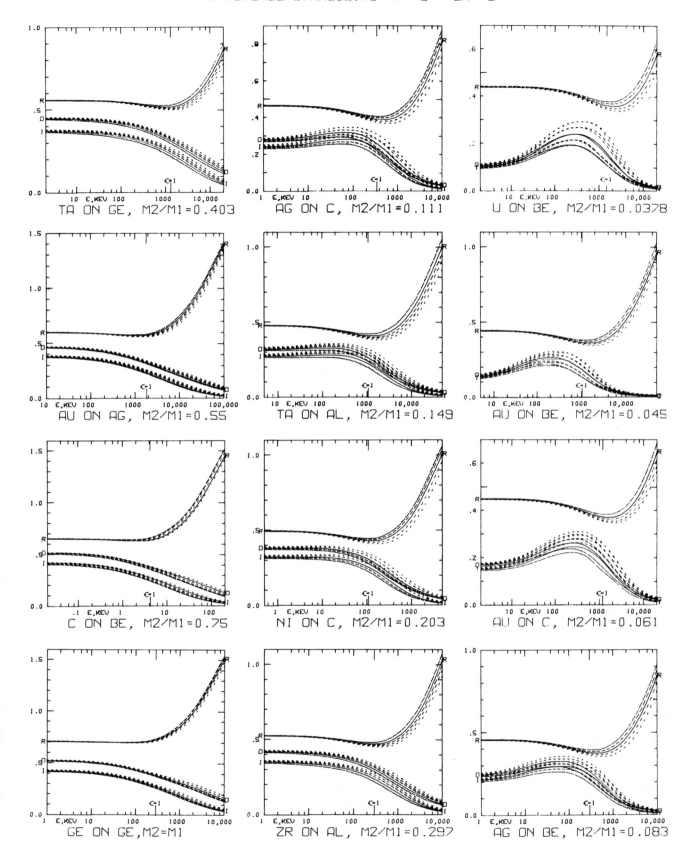

Figure 6 (continued)

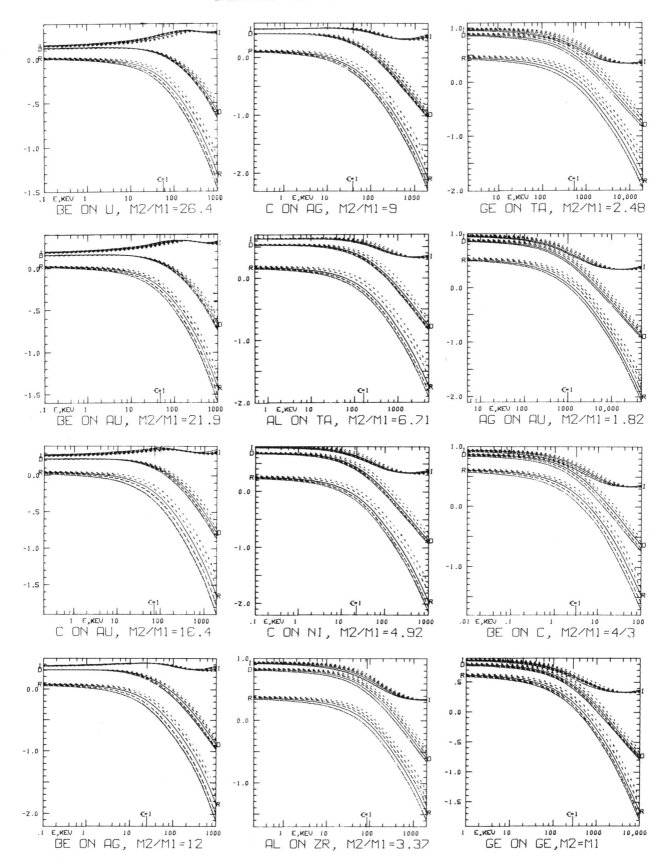

Figure 7

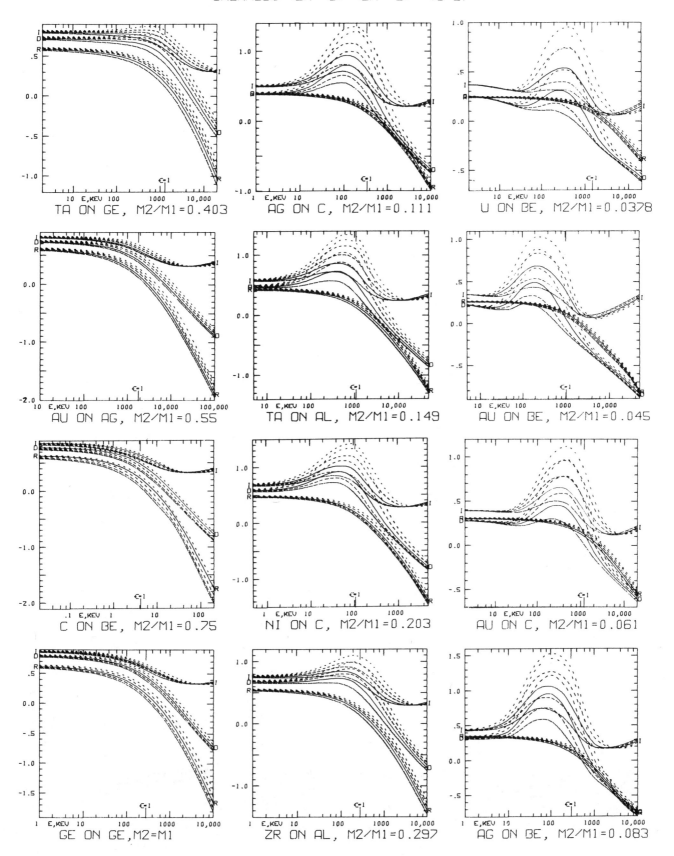

Figure 7 (continued)

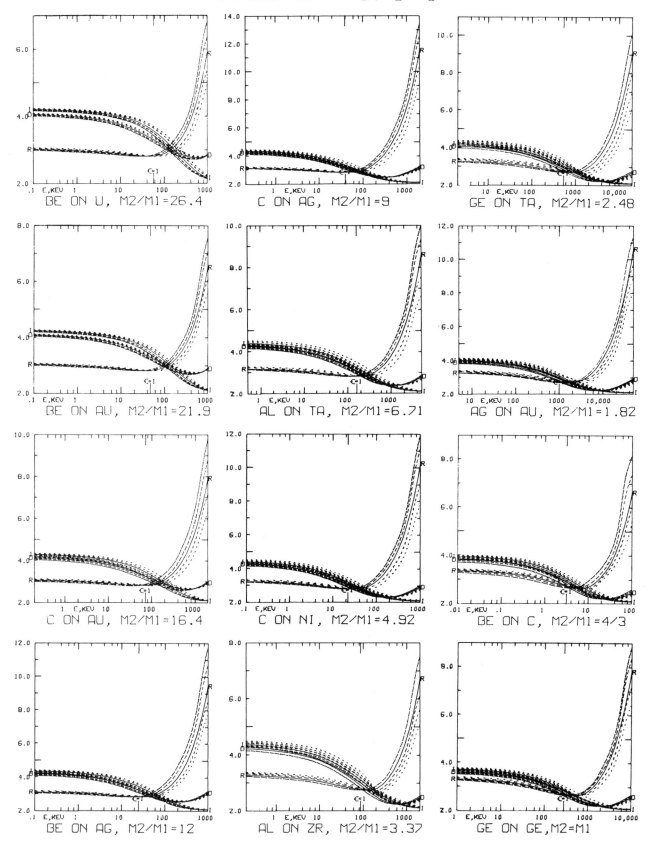

BE ON U, M2/M1=26.4

C ON AG, M2/M1=9

GE ON TA, M2/M1=2.48

BE ON AU, M2/M1=21.9

AL ON TA, M2/M1=6.71

AG ON AU, M2/M1=1.82

C ON AU, M2/M1=16.4

C ON NI, M2/M1=4.92

BE ON C, M2/M1=4/3

BE ON AG, M2/M1=12

AL ON ZR, M2/M1=3.37

GE ON GE, M2=M1

Figure 8

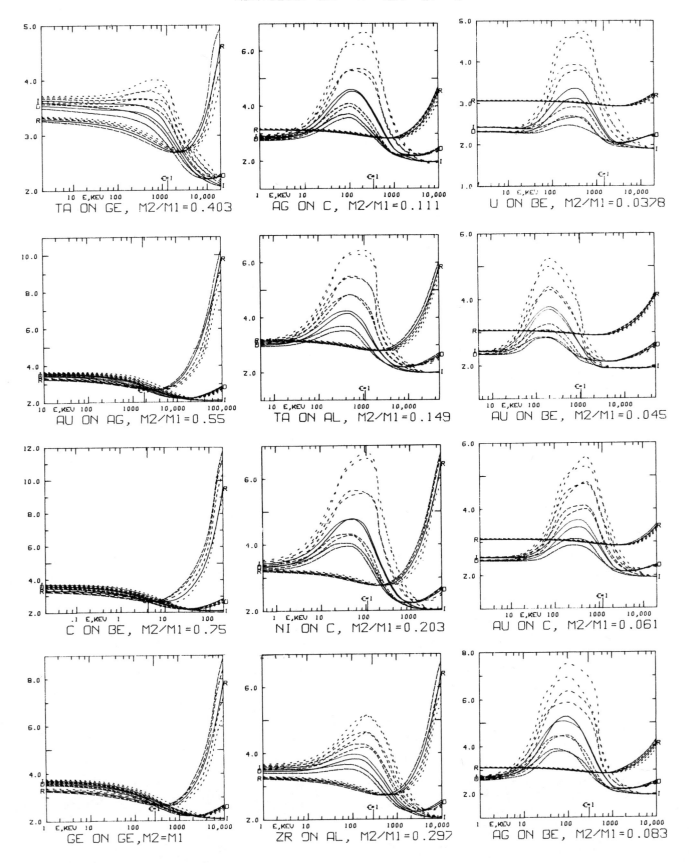

Figure 8 (continued)

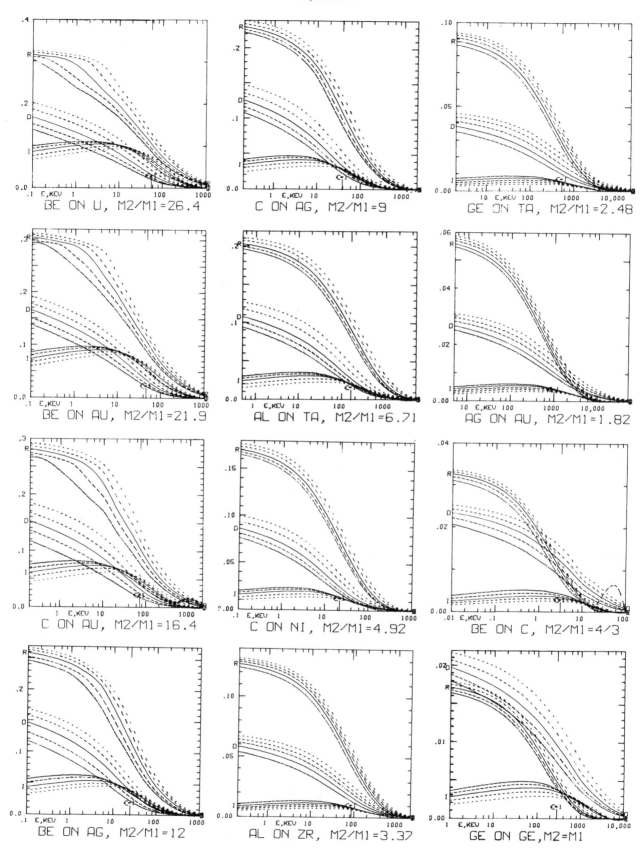

Figure 9

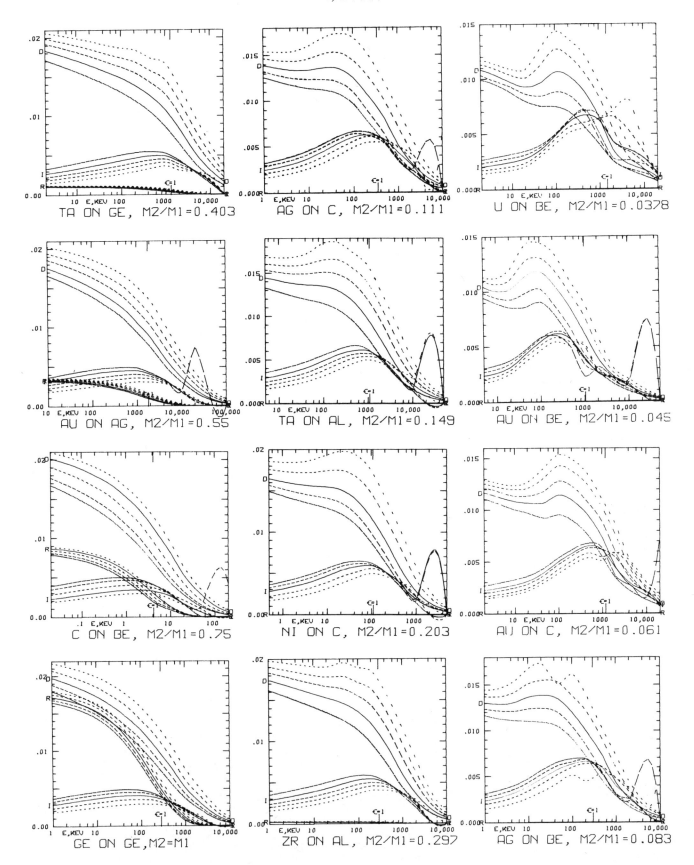

Figure 9 (continued)

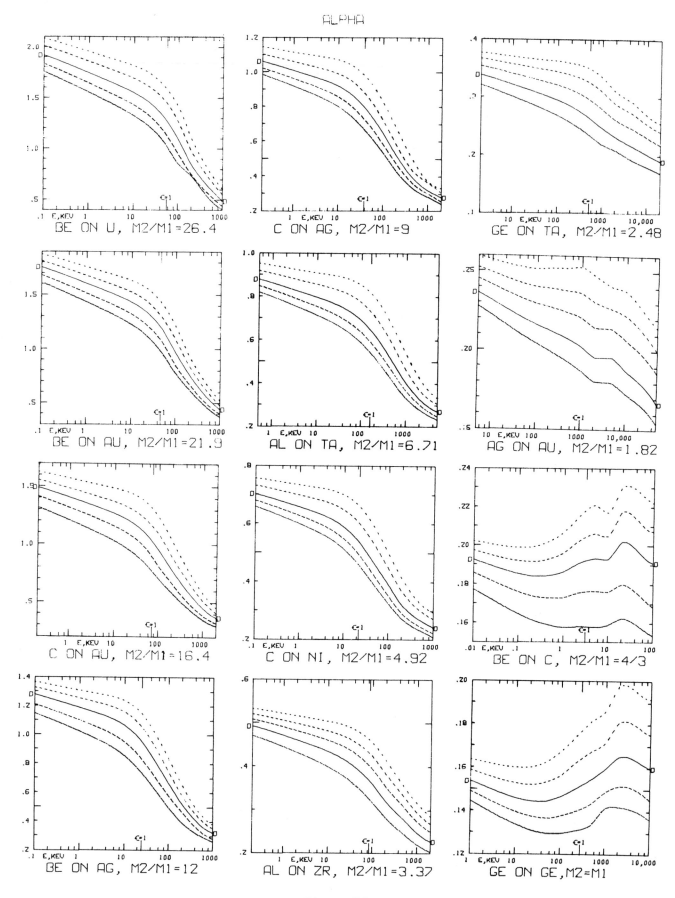

Figure 10

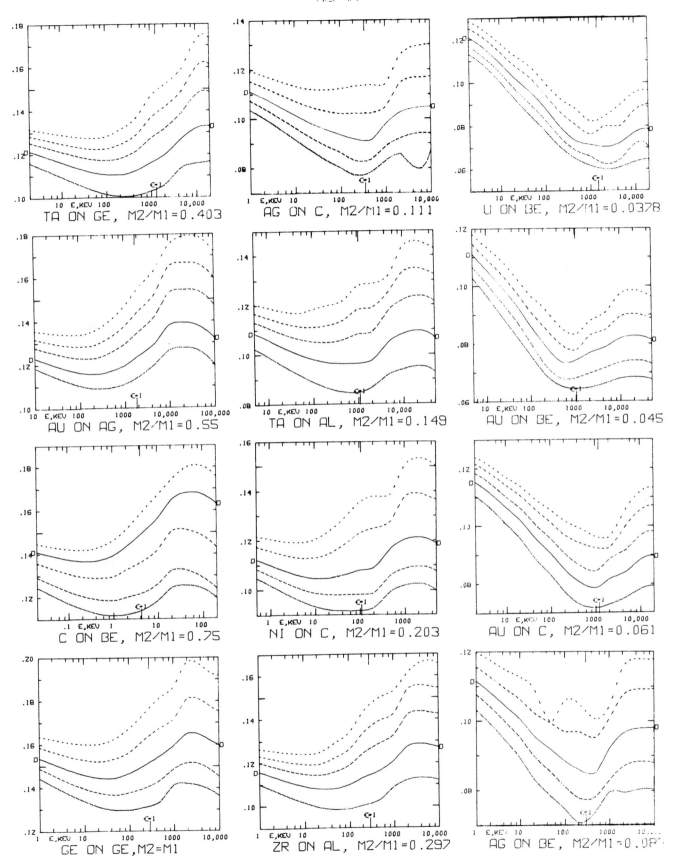

Figure 10 (continued)

8. MASS SCALING
(Fig. 11)

It is implicit in the selection of cases for this volume that the ratio of projectile and target atom masses is the main parameter, and that the values of the masses are otherwise irrelevant, except as they affect the electronic stopping. However the mass values affect also the ratio of electronic and nuclear stopping for the recoils, and so they do affect the values of the damage and ionization moments. A calculation has been done to see how large this effect may be. The Ag-Au pair (solid curves) is compared to a B-Ne pair (dashed curves), with the k value for the latter adjusted to agree with the former. The Ge-Ge results are replotted beside them to show how large the effects of varying electronic stopping are. Range moments are unchanged, as expected, and damage and ionization moments are seen to scale quite well. In the latter distributions one is following simultaneously the slowing of the primary particles and that of all the recoils. For light ions in heavy targets, the motion of the recoils does not have much effect on the distributions, and the electronic stopping value for the projectile is the important parameter. When the projectile is heavier than the target atoms, there is an energy range, approximately $0.1 < \epsilon < 1$, in which many of the recoil atoms have ranges greater than the residual range of the projectile. Within this energy range the mean damage and ionization depths will still depend mainly on the stopping power for the projectile, but straggling and higher moments will be more dependent on the stopping power for the recoils. These cases, and the energy range, may be identified from the plotted moments as those in which the kurtosis has an intermediate-energy maximum.

Scaling of the tables for different projectile-target combinations is done as follows. Suppose the range distribution for 100 keV Au in Au is desired. The appropriate table is that for Ge in Ge. From Tables 0.2 and 0.3 we find

	Ge in Ge	Au in Au
ϵ_0	.0035381	.0004295
ρ_0	.0282193	.0057247
k_L	.1573322	.1749433

From the k values, we see

$$k/k_L = .1749/.1573 = 1.11.$$

For Au in Au, 100 keV corresponds to $\epsilon = .04295$.

For Ge in Ge, $\epsilon = .04295$ corresponds to E = 12.18 keV.

The appropriate portion of the Ge on Ge table is

k/k_L	1.0	1.2
E = 12 keV	3.968	3.927
E = 14 keV	4.426	4.378

After 2-dimensional linear interpolation, we get a mean range of 3.986 $\mu g \cdot cm^{-2}$, which corresponds to 0.112 ρ units. For Au in Au, 0.112 ρ units corresponds to 19.65 $\mu g \cdot cm^{-2}$.

Other moments may be found by similar interpolation but since they are dimensionless, no change of units, corresponding to the ρ change, is necessary.

In general one will not find a table which has exactly the same target-projectile mass-ratio as the desired one, and so will have to do a further interpolation between two sets of tables.

Figure 11. Dependence on Masses, with Mass Ratio Fixed.

$M_2/M_1 = 0.55$:   Au → Ag   solid line

    ^{20}Ne → ^{11}B,   dashed line.

$M_1/M_2 = 0.55$:   Ag → Au,   solid line,

    ^{11}B → ^{20}Ne,   dashed line.

Curves for Ge on Ge repeated for comparison. Value of k for B ↔ Ne changed to agree with Ag ↔ Au.

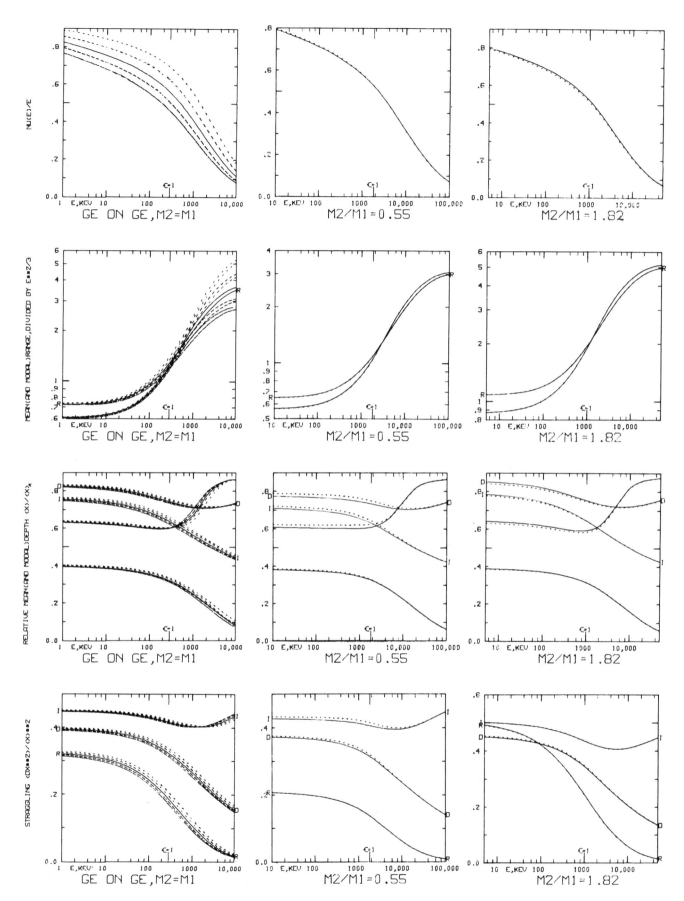

Figure 11

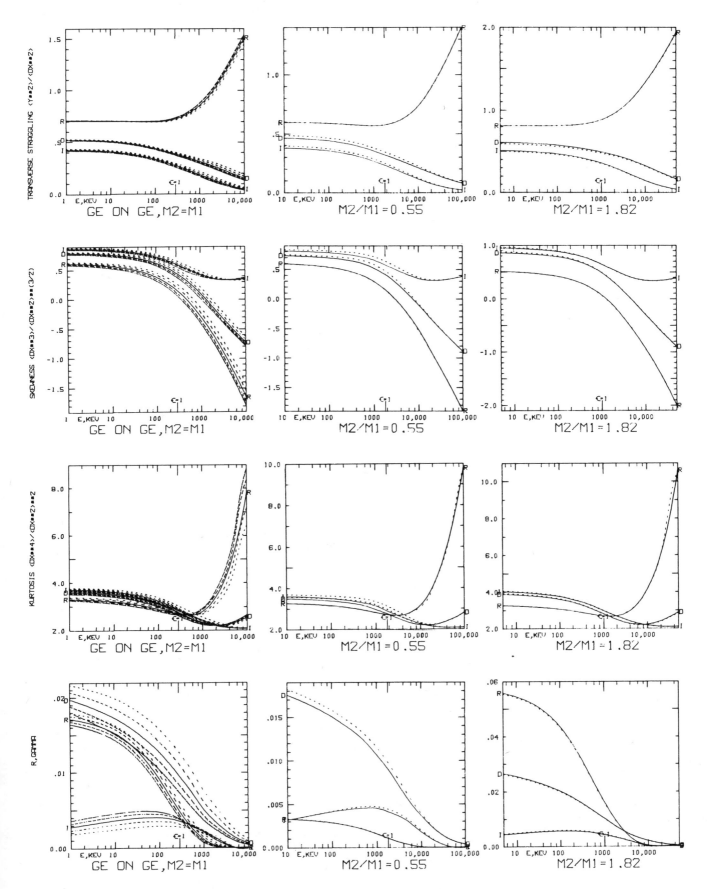

Figure 11 (continued)

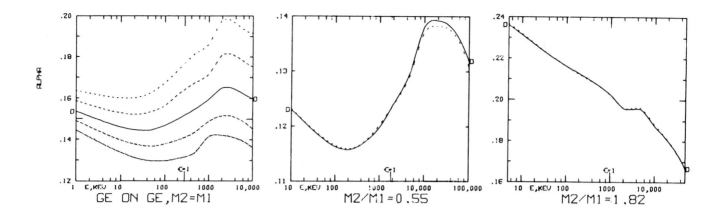

Figure 11 (continued)

9. POLYATOMIC TARGETS

(Fig. 12)

The standard recipe in estimating results for polyatomic target cases from monatomic-target calculations is to assume a fictitious monatomic target with Z and A equal to the averaged values for the real target material. This can not be a uniformly valid approximation, because for example the backscattering of, say Ag from tantalum oxide would be expected to be greater than the backscattering from the corresponding monatomic target, ^{63}Co. Here calculations have been done for projectile-target combinations corresponding to Al-Zr and Ge-Ge (solid curves). The polyatomic targets were ZnS (short dashes) and AlSb (long dashes), with target constituent mass ratios 2.06:1 and 3.92:1 respectively. Results for the 2:1 target agree surprisingly well with the corresponding monatomic-target results, while results for the 4:1 target differ. In all cases one can see the effects of the long-range recoils on the damage and ionization distributions.

Figure 12. Polyatomic Targets

$M_2/M_1 = 0.297$

Zr → Aℓ   solid line
^{165}Ho → ZnS   short dashes
^{251}Cf → AℓSb   long dashes

$M_2 = M_1$

Ge → Ge   solid line
^{48}Ti → ZnS   short dashes
^{75}As → AℓSb   long dashes

$M_2/M_1 = 3.37$

Aℓ → Zr   solid line
^{15}N → ZnS   short dashes
^{22}Na → AℓSb   long dashes

Values of k for polyatomic targets changed to agree with corresponding monatomic value. In each case, e.g. Ho → ZnS, all 9 electronic stopping constants, e.g. Ho → Zn, S: Zn → Zn, S: S → Zn, S, are changed by the same factor.

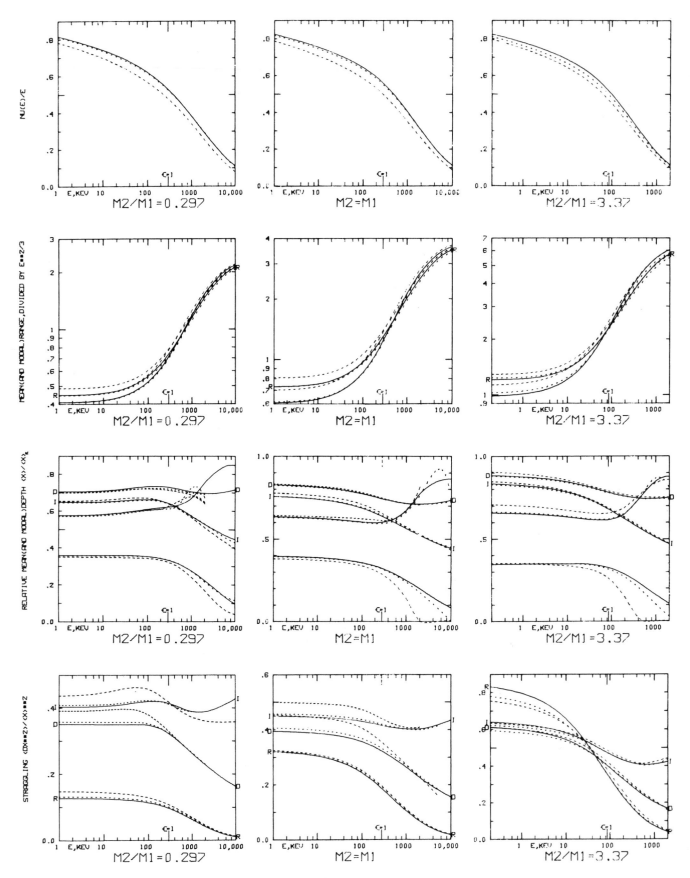

Figure 12

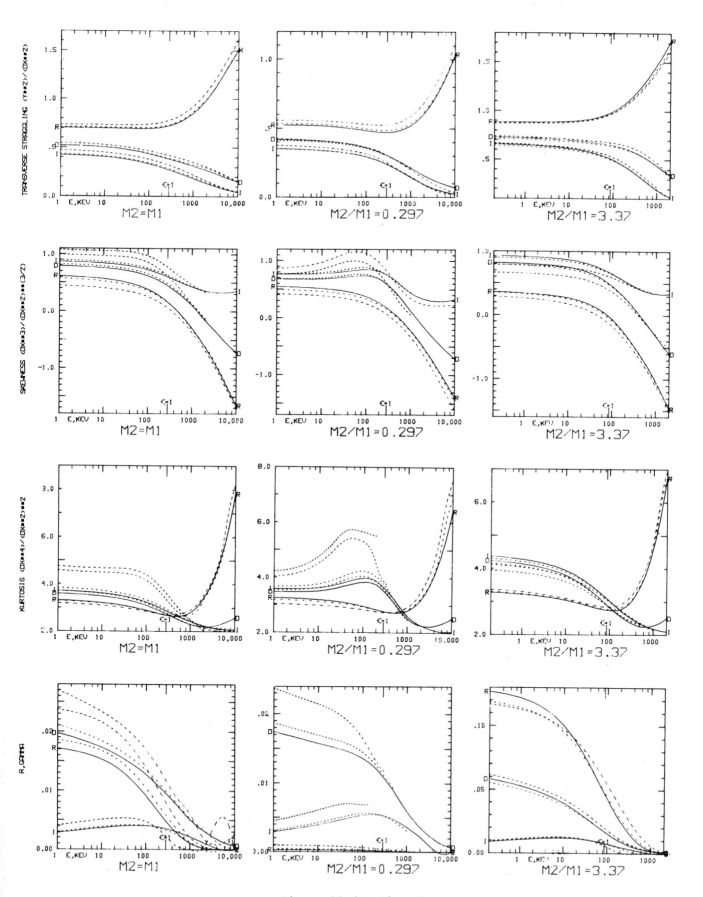

Figure 12 (continued)

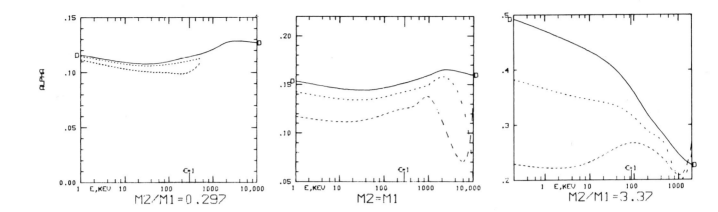

Figure 12 (continued)

10. CALCULATED DISTRIBUTIONS

Each set of tables is accompanied by a figure showing the range, damage, and ionization distributions at each of 5 energies, for $k = k_L$. Each figure consists of 5 independent subfigures. These distributions are calculated from the inverse Fourier transform, as described near eq. (46). This method has been found to *underestimate* modal depths.

The three distributions in each subfigure are normalized to the same peak height. The abscissa is depth, centred on the mean range, and the width of the figure is twice the mean range (2r) or 6 times the range standard deviation (6σ), whichever is the larger. The three mean depths are marked on each subfigure. In each case $\langle x \rangle_I < \langle x \rangle_D < \langle x \rangle_R$. The vertical line in a subfigure denotes the position of the surface, if $6\sigma > 2r$. Otherwise the surface is at the left-hand side of the figure.

The ionization distribution has a step discontinuity at the surface equal to the value of the electronic stopping. The damage distribution is allowed to have an unspecified slope discontinuity at the surface, and the range distribution is allowed to have a discontinuous fourth derivative.

11. REFERENCES

1. N. Bohr, Mat.-Fys. Medd. Dan. Vid. Selsk. 18(1948)8.

2. J. Lindhard, V. Nielsen, and M. Scharff, Mat.-Fys. Medd. Dan. Vis. Selsk. 36(1968)10.

3. J. Lindhard, M. Scharff, and H.E. Schiøtt, Mat.-Fys. Medd. Dan. Vid. Selsk. 33(1963)14.

4. J. Lindhard, V. Nielsen, M. Scharff, and P.V. Thomsen, Mat.-Fys. Medd. Dan. Vid. Selsk. 33(1963)10.

5. J.B. Sanders, Thesis, University of Leiden, The Netherlands, 1968.

6. P. Sigmund and J.B. Sanders, in Proc. Int. Conf. on Application of Ion Beams to Science and Technology, Ed. P. Glotin, Eds. Ophrys, p.215, 1967.

7. D.K. Brice, in Proc. III Int. Conf. on Ion Implantation,..., Yorktown Heights, N.Y., (Ed. B. Crowder) Plenum, N.Y., 1973.

8. K.B. Winterbon, in Atomic Collisions in Solids, Ed. S. Datz et al., Plenum, N.Y., p.35, 1975.

9. K.B. Winterbon, P. Sigmund, and J.B. Sanders, Mat.-Fys. Medd. Dan. Vid. Selsk. 37 (1970)14.

10. P. Sigmund, Rev. Roum. Phys. 17(1972)823,969, and 1079.

11. P. Sigmund, in Phys. Ionized Gases (M. Kurepa, Ed.) Inst. Phys. Belgrade, 13(1972)7.

12. P. Sigmund, Corsican Summer School Lecture Notes, 1973.

13. E. Bøgh, P. Høgild, and I. Stensgaard, Rad. Eff. 7(1971)115.

14. L.C. Northcliffe and R.F. Schilling, Nuclear Data 7(1970)233.

15. J. Lindhard and M. Scharff, Phys. Rev. 124(1961)128.

16. K.B. Winterbon, Atomic Energy of Canada Limited Report AECL-3194, 1968.

17. A.S. Samoilov, A.N. Mosevich, L.L. Makarov, and B.F. Ormont, Dokl. Akad. Nauk. SSSR 213(1973)1076; (English Transl. Sov. Phys. Dokl. 18(1974)766.) A.S. Samoilov, A.N. Mosevich, V.I. Smirnova, and B.F. Ormont, Dokl. Akad. Nauk. SSSR 213(1973)1307; (English Transl. Sov. Phys. Dokl. 18(1974)773.)

18. M.G. Kendall and A. Stuart, The Advanced Theory of Statistics, Vol. 1, Third Edition, Griffin, London, 1969.

19. J. Bøttiger and J.A. Davies, Rad. Eff. 11(1971)61.

20. J. Bøttiger, J.A. Davies, P. Sigmund, and K.B. Winterbon, Rad. Eff. 11(1971)69.

21. J. Bøttiger, H. Wolder Jørgensen, and K.B. Winterbon, Rad. Eff. 11(1971)133.

22. J. Bøttiger and K.B. Winterbon, Rad. Eff. 20(1973)65.

23. P. Sigmund, Phys. Rev. 184(1969)383.

24. O.B. Firsov, J. Exptl. Theoret. Phys. (USSR) 36(1959)1517. (English Transl. Sov. Phys. JETP 36(9)(1959)1076.)

25. P. Mazur and J.B. Sanders, Physica 44(1969)444.

26. J.B. Sanders and K.B. Winterbon, Rad. Eff. 22(1974)109.

27. H.H. Andersen, J. Bøttiger, H. Knudsen, P. Møller Petersen, and T. Wohlenberg, Phys. Rev. A10(1974)1568.

28. P. Sigmund, in Fourth Int. Conf. on Atomic Collisions in Solids, Gausdal, Norway, 1971.

29. P. Sigmund, Appl. Phys. Lett. 25(1974)169.

30. P. Sigmund, G.P. Scheidler, and G. Roth, BNL 50083 (C-52)(1968)374.

31. J.E. Westmoreland and P. Sigmund, Rad. Eff. 6(1970)187.

32. O.B. Firsov, J. Exptl. Theoret. Phys. (USSR) 33(1957)696; (English Transl. Sov. Phys. JETP 6(33)(1958)534.)

33. S. Schwabe, at a meeting on Multiple Scattering, Copenhagen, Dec. 1974.

34. B.M. Latta and P.J. Scanlon, Phys. Rev. A10(1974)1638.

35. K.B. Winterbon, Rad. Eff. 13(1972)215.

36. K.B. Winterbon, Can. J. Phys. 46(1968)2429.

37. C.D. Moak, B.R. Appleton, J.A. Biggerstaff, S. Datz, and T.S. Noggle, in Atomic Collisions in Solids, Ed. S. Datz et al., Plenum, N.Y., p.57, 1975.

38. K.B. Winterbon, Rad. Eff. 15(1972)73.

39. K.B. Winterbon, Atomic Energy of Canada Limited Report AECL-4829, 1974, and Nucl. Phys., in press.

40. C.L. Mallows, J. Roy. Stat. Soc., Ser. B, 18(1956)139.

41. J.A. Shohat and J.D. Tamarkin, The Problem of Moments, American Mathematical Society, Providence, 1943.

42. A. Erdelyi, W. Magnus, F. Oberhettinger, and F.G. Tricomi, Tables of Integral Transforms, Vol. 1, McGraw-Hill, New York, 1954.

43. W.K. Hofker, D.P. Oosthoek, N.J. Koeman, H.A.M. de Grefte, Rad. Eff., in press.

44. W.K. Hofker, submitted to Rad. Eff.

45. K.B. Winterbon, Phys. Lett. 32A(1970)265.

46. J.S.R. Chisholm and A.K. Common, in The Pade Approximant in Theoretical Physics, (Ed. Baker and Gammel), p.183, 1970.

47. U. Littmark, Thesis, University of Copenhagen, 1974.

48. K.B. Winterbon, Atomic Energy of Canada Limited Report AECL-4832, 1974.

49. K.W. Jones and H.W. Kraner, Phys. Rev. C4(1971)125, and Phys. Rev. A11(1975)1347.

TABLE 0.1

Ion-target combinations used in the calculations, with ion-target mass ratios and values of electronic stopping.

Ion	Target	M_2/M_1	k/k_L				
^{9}Be	^{238}U	26.44	0.65	0.8	1.0	1.2	1.4
^{9}Be	^{197}Au	21.89	0.7	0.85	1.0	1.2	1.4
^{12}C	^{197}Au	16.42	0.6	0.8	1.0	1.25	1.6
^{9}Be	^{108}Ag	12.00	0.65	0.8	1.0	1.2	1.45
^{12}C	^{108}Ag	9.00	0.6	0.8	1.0	1.2	1.4
^{27}Al	^{181}Ta	6.70	0.55	0.75	1.0	1.2	1.4
^{12}C	^{59}Ni	4.92	0.55	0.75	1.0	1.2	1.4
^{27}Al	^{91}Zr	3.37	0.5	0.65	0.8	1.0	1.3
^{73}Ge	^{181}Ta	2.48	0.4	0.55	0.75	1.0	1.3
^{108}Ag	^{197}Au	1.82	0.5	0.65	0.8	1.0	1.2
^{9}Be	^{12}C	1.33	0.6	0.8	1.0	1.3	1.7
^{73}Ge	^{73}Ge	1.00	0.6	0.8	1.0	1.2	1.4
^{12}C	^{9}Be	0.750	0.8	1.0	1.3	1.7	2.0
^{197}Au	^{108}Ag	0.548	0.5	0.65	0.8	1.0	1.2
^{181}Ta	^{73}Ge	0.403	0.5	0.65	0.8	1.0	1.3
^{91}Zr	^{27}Al	0.297	0.5	0.65	0.8	1.0	1.3
^{59}Ni	^{12}C	0.203	0.55	0.75	1.0	1.2	1.4
^{181}Ta	^{27}Al	0.149	0.5	0.65	0.8	1.0	1.3
^{108}Ag	^{12}C	0.111	0.6	0.8	1.0	1.2	1.4
^{108}Ag	^{9}Be	0.0833	0.65	0.8	1.0	1.2	1.45
^{197}Au	^{12}C	0.0609	0.6	0.7	0.85	1.0	1.25
^{197}Au	^{9}Be	0.0457	0.7	0.85	1.0	1.2	1.4
^{238}U	^{9}Be	0.0378	0.65	0.8	1.0	1.2	1.4

TABLE 0.2

EPSILON CORRESPONDING TO 1 KEV

TARGETS

PROJECTILES			BE 4 9	C 6 12	AL 13 27	TI 22 48	NI 28 59	GE 32 73
H	1	1	3.9016918	2.4131657	.9444491	.4369091	.3573804	.3013115
D	1	2	3.5469926	2.2407963	.9118019	.4771710	.3515217	.2972940
T	1	3	3.2514099	2.0314103	.8814859	.4578147	.3458520	.2933822
HE	2	4	1.3892039	.9195755	.4085427	.2221561	.1654921	.1410941
BE	4	9	.4528771	.3210371	.1653976	.0365718	.0735546	.0637469
C	6	12	.2407778	.1758332	.0971715	.0593399	.0454743	.0397819
AL	13	27	.0551325	.0431873	.0289458	.0199026	.0159659	.0144500
TI	22	48	.0181260	.0147603	.0111953	.0084814	.0070481	.0065841
NI	28	59	.0112202	.0092490	.0073064	.0057341	.0048315	.0045708
GE	32	73	.0078592	.0065395	.0053447	.0043293	.0036942	.0035381
ZR	40	91	.0048536	.0040780	.0034493	.0028876	.0024982	.0024245
AG	47	108	.0033784	.0028553	.0024724	.0021189	.0018518	.0018151
EU	63	152	.0016846	.0014397	.0012904	.0011576	.0010303	.0010286
TA	73	181	.0011805	.0010134	.0009280	.0008439	.0007574	.0007620
AU	79	197	.0009832	.0008457	.0007804	.0007156	.0006448	.0006519
U	92	238	.0006732	.0005813	.0005445	.0005077	.0004611	.0004700

			ZR 40 91	AG 47 108	EU 63 152	TA 73 181	AU 79 197	U 92 238
H	1	1	.2257828	.1831475	.1250444	.1131382	.0929991	.0761597
D	1	2	.2233551	.1814825	.1242324	.1125746	.0925317	.0758424
T	1	3	.2209790	.1798477	.1234309	.1120172	.0920691	.0755277
HE	2	4	.1068819	.0873111	.0602771	.0493447	.0451278	.0371016
BE	4	9	.0490757	.0405171	.0284511	.0237412	.0215203	.0173014
C	6	12	.0309249	.0256975	.0182359	.0152853	.0138639	.0115299
AL	13	27	.0116255	.0098695	.0072983	.0062210	.0056940	.0047997
TI	22	48	.0054743	.0047675	.0036657	.0031821	.0029371	.0025175
NI	28	59	.0038531	.0033893	.0026545	.0023236	.0021531	.0018602
GE	32	73	.0030223	.0026853	.0021417	.0018909	.0017592	.0015323
ZR	40	91	.0021021	.0018890	.0015391	.0013729	.0012835	.0011292
AG	47	108	.0015916	.0014423	.0011958	.0010755	.0010094	.0008925
EU	63	152	.0009214	.0008495	.0007283	.0006663	.0006305	.0005692
TA	73	181	.0006302	.0006417	.0005530	.0005165	.0004908	.0004473
AU	79	197	.0005929	.0005534	.0004863	.0004510	.0004295	.0003932
U	92	238	.0004318	.0004064	.0003635	.0003402	.0003255	.0003010

RHO CORRESPONDING TO 1 MICROGRAM/CM**2

TARGETS

PROJECTILES			BE 4 9	C 6 12	AL 13 27	TI 22 48	NI 28 59	GE 32 73
H	1	1	.4719204	.2284765	.0324522	.0178160	.0045144	.0027379
D	1	2	.7800337	.3940060	.0605054	.0150130	.0087352	.0053307
T	1	3	.9831675	.5148345	.0848064	.0216451	.0126836	.0077670
HE	2	4	.9572267	.5308383	.0971587	.0269332	.0154807	.0096055
BE	4	9	.9155585	.5822920	.1433205	.0443663	.0275373	.0176467
C	6	12	.7763894	.5240264	.1484621	.0490438	.0315758	.0206176
AL	13	27	.4299614	.3339114	.1390943	.0595896	.0411128	.0287346
TI	22	48	.2366201	.1985753	.1059371	.0550930	.0407919	.0303713
NI	28	59	.1305226	.1552473	.0898391	.0501400	.0381601	.0291429
GE	32	73	.1431343	.1254235	.0776097	.0461896	.0360582	.0282193
ZR	40	91	.1063307	.0950004	.0630292	.0400233	.0321183	.0258094
AG	47	108	.0843133	.0763107	.0530577	.0353134	.0289175	.0237023
EU	63	152	.0530741	.0490595	.0360808	.0260511	.0226359	.0192471
TA	73	181	.0416698	.0388637	.0302203	.0226446	.0195585	.0169177
AU	79	197	.0368409	.0345017	.0272410	.0207580	.0180702	.0157553
U	92	238	.0282978	.0267105	.0217290	.0171204	.0151412	.0134195

			ZR 40 91	AG 47 108	EU 63 152	TA 73 181	AU 79 197	U 92 238
H	1	1	.0015458	.0009970	.0004216	.0002716	.0002183	.0001380
D	1	2	.0030254	.0019573	.0008322	.0005372	.0004322	.0002698
T	1	3	.0044421	.0028841	.0012324	.0007970	.0006418	.0004013
HE	2	4	.0055423	.0036252	.0015673	.0010189	.0008224	.0005165
BE	4	9	.0105162	.0070261	.0031425	.0020720	.0016831	.0010701
C	6	12	.0125275	.0084790	.0038731	.0025760	.0021016	.0013467
AL	13	27	.0186998	.0132644	.0065526	.0045080	.0037337	.0024651
TI	22	48	.0211112	.0156943	.0084151	.0060052	.0050578	.0034528
NI	28	59	.0208240	.0157975	.0087863	.0063754	.0054119	.0037535
GE	32	73	.0207043	.0160210	.0092453	.0066232	.0056383	.0041161
ZR	40	91	.0195084	.0154413	.0092933	.0070059	.0060531	.0043538
AG	47	108	.0183259	.0147620	.0091900	.0070446	.0061348	.0044862
EU	63	152	.0155322	.0129433	.0086278	.0068376	.0060534	.0045841
TA	73	181	.0139348	.0116062	.0081422	.0065676	.0058648	.0045250
AU	79	197	.0131039	.0111904	.0078455	.0063833	.0057247	.0044581
U	92	238	.0113869	.0098862	.0071777	.0059500	.0053860	.0042809

TABLE 0.3

ELECTRONIC STOPPING CONSTANT k, DIMENSIONLESS

TARGETS

PROJECTILES			BE 4 9	C 5 12	AL 13 27	TI 22 48	NI 28 59	GE 32 73
H	1	1	.6505633	.8798850	1.9960723	3.5383778	4.4415759	5.5036154
D	1	2	.2653589	.3476625	.7438600	1.3177178	1.6097557	1.9653972
T	1	3	.1645809	.2098775	.4260291	.7332928	.8976750	1.1023995
HE	2	4	.1704240	.2165698	.4325310	.7417834	.9093194	1.1150813
BE	4	9	.1120206	.1343710	.2321322	.3730885	.4567422	.5440760
C	6	12	.1078240	.1271225	.2078065	.3247374	.3899719	.4671293
AL	13	27	.0941733	.1050611	.1419034	.1964229	.2281240	.2634854
TI	22	48	.0929409	.1008133	.1206131	.1511379	.1699689	.1893441
NI	28	59	.0950973	.1025330	.1180305	.1439510	.1601001	.1753420
GE	32	73	.0948624	.1014991	.1132397	.1325230	.1453999	.1573322
ZR	40	91	.0972324	.1036256	.1123951	.1275104	.1382507	.1473857
AG	47	108	.0991736	.1053114	.1120366	.1242872	.1335612	.1407828
EU	63	152	.1029059	.1087553	.1120498	.1194973	.1263252	.1303540
TA	73	181	.1050639	.1108493	.1127512	.1182350	.1240013	.1267908
AU	79	197	.1063588	.1121552	.1134948	.1181662	.1236251	.1257735
U	92	238	.1088147	.1146014	.1147309	.1176783	.1222969	.1232734

			ZR 40 91	AG 47 108	EU 63 152	TA 73 181	AU 79 197	U 92 238
H	1	1	6.8978258	8.2149467	11.6262506	13.3303711	15.1290008	18.3222360
D	1	2	2.4786198	2.9444823	4.1506051	4.9480251	5.3094827	6.5185670
T	1	3	1.3710092	1.6246731	2.2814850	2.7154700	2.9558630	3.5704694
HE	2	4	1.3882921	1.6463547	2.3140005	2.7371173	3.0025080	3.6297181
BE	4	9	.6702101	.7392913	1.0977271	1.3022045	1.4160192	1.7063474
C	6	12	.5728079	.6729535	.9309251	1.1324739	1.1981087	1.4420470
AL	13	27	.3141031	.3617379	.4849005	.5509429	.6130058	.7298822
TI	22	48	.2188129	.2464127	.3175467	.3550558	.3919032	.4598966
NI	28	59	.2009277	.2242653	.2843050	.3244083	.3472531	.4045087
GE	32	73	.1770139	.1953555	.2424402	.2739985	.2919585	.3370418
ZR	40	91	.1635178	.1784783	.2160036	.2425510	.2572656	.2941172
AG	47	108	.1544813	.1571343	.1994780	.2212661	.2337977	.2650022
EU	63	152	.1400053	.1488272	.1712712	.1365201	.1954098	.2173343
TA	73	181	.1348061	.1420793	.1605204	.1731497	.1805784	.1907631
AU	79	197	.1331226	.1397610	.1565687	.1581108	.1749433	.1913249
U	92	238	.1292031	.1344965	.1478437	.1371177	.1626927	.1761770

ELECTRONIC STOPPING CONSTANT K, KEV**.5*CM**2/MICROGRAM

TARGETS

PROJECTILES			BE 4 9	C 6 12	AL 13 27	TI 22 48	NI 28 59	GE 32 73
H	1	1	.1554289	.1294120	.0660548	.0401940	.0335408	.0274507
D	1	2	.1099048	.0915081	.0471320	.0284214	.0237109	.0194106
T	1	3	.0897369	.0747150	.0384831	.0232060	.0193648	.0158487
HE	2	4	.1384085	.1198855	.0657477	.0409709	.0346212	.0285149
BE	4	9	.1524031	.1380920	.0818048	.0532372	.0457565	.0380271
C	6	12	.1706030	.1588640	.0989324	.0563476	.0577438	.0482671
AL	13	27	.1724461	.1588087	.1160130	.0329070	.0742232	.0629822
TI	22	48	.1633458	.1647784	.1207602	.0904174	.0825800	.0708700
NI	28	59	.1620689	.1655168	.1246899	.0953164	.0879077	.0758415
GE	32	73	.1531611	.1574232	.1203309	.0930311	.0862597	.0740411
ZR	40	91	.1484322	.1541588	.1206132	.0949715	.0888397	.0772547
AG	47	108	.1439009	.1503960	.1195494	.0953473	.0897512	.0783236
EU	63	152	.1330681	.1406181	.1147981	.0935048	.0890050	.0782307
TA	73	181	.1274212	.1353285	.1118550	.0921050	.0881609	.0776730
AU	79	197	.1249660	.1330599	.1106705	.0915929	.0879726	.0770135
U	92	238	.1186811	.1269570	.1068366	.0894112	.0862300	.0763053

			ZR 40 91	AG 47 108	EU 63 152	TA 73 181	AU 79 197	U 92 238
H	1	1	.0224395	.0191374	.0138598	.0117369	.0108287	.0090506
D	1	2	.0158671	.0135322	.0090004	.0082992	.0076571	.0063856
T	1	3	.0129555	.0110490	.0080020	.0067763	.0062520	.0052138
HE	2	4	.0235354	.0201987	.0147709	.0125697	.0116233	.0097326
BE	4	9	.0318154	.0275507	.0204515	.0175111	.0162452	.0136855
C	6	12	.0408056	.0355733	.0267000	.0229762	.0213710	.0180864
AL	13	27	.0544758	.0482494	.0371931	.0324036	.0303312	.0259701
TI	22	48	.0624339	.0560107	.0441412	.0388625	.0365754	.0316347
NI	28	59	.0674055	.0606505	.0484855	.0429141	.0405007	.0352120
GE	32	73	.0666675	.0603973	.0484207	.0429929	.0406399	.0354395
ZR	40	91	.0695784	.0634100	.0513879	.0453612	.0434710	.0381605
AG	47	108	.0709614	.0649530	.0530517	.0475297	.0451450	.0397283
EU	63	152	.0716377	.0660870	.0547544	.0494072	.0471073	.0417575
TA	73	181	.0715003	.0662161	.0552550	.0500395	.0478026	.0425316
AU	79	197	.0716429	.0664844	.0556904	.0505316	.0483229	.0430807
U	92	238	.0708035	.0659615	.0556556	.0505880	.0485697	.0434885

K/KL= E,KEV	FRACTIONAL DEPOSITED ENERGY					MEAN RANGE, MICROGRAM/SQ.CM.				
	0.65	0.80	1.0	1.2	1.4	0.65	0.80	1.0	1.2	1.4
.10	.7241	.6796	.6274	.5820	.5422	1.003	.9986	.9929	.9873	.9818
.12	.7178	.6726	.6199	.5741	.5340	1.139	1.134	1.127	1.121	1.114
.14	.7123	.6666	.6134	.5674	.5271	1.263	1.257	1.250	1.242	1.235
.17	.7053	.6590	.6052	.5588	.5184	1.435	1.428	1.419	1.410	1.402
.20	.6993	.6525	.5983	.5516	.5111	1.596	1.588	1.578	1.568	1.558
.23	.6941	.6468	.5922	.5454	.5047	1.750	1.742	1.730	1.719	1.708
.26	.6895	.6418	.5869	.5398	.4991	1.899	1.889	1.877	1.864	1.852
.30	.6840	.6359	.5806	.5333	.4925	2.091	2.080	2.066	2.052	2.038
.35	.6780	.6294	.5738	.5263	.4853	2.322	2.309	2.293	2.277	2.261
.40	.6727	.6238	.5678	.5201	.4791	2.544	2.530	2.512	2.494	2.476
.45	.6680	.6187	.5624	.5147	.4736	2.760	2.744	2.723	2.703	2.684
.50	.6637	.6141	.5576	.5097	.4686	2.968	2.951	2.929	2.907	2.885
.60	.6562	.6061	.5492	.5011	.4600	3.356	3.336	3.310	3.284	3.259
.70	.6497	.5992	.5420	.4938	.4527	3.730	3.707	3.677	3.648	3.619
.80	.6440	.5931	.5357	.4874	.4463	4.093	4.067	4.033	4.000	3.967
.90	.6388	.5877	.5300	.4817	.4406	4.446	4.417	4.379	4.342	4.305
1.0	.6341	.5827	.5249	.4765	.4354	4.790	4.758	4.716	4.674	4.634
1.2	.6259	.5741	.5159	.4674	.4264	5.428	5.390	5.341	5.293	5.246
1.4	.6187	.5665	.5082	.4597	.4187	6.049	6.005	5.949	5.893	5.839
1.7	.6094	.5568	.4982	.4497	.4089	6.958	6.905	6.836	6.769	6.703
2.0	.6014	.5484	.4897	.4412	.4005	7.844	7.781	7.699	7.619	7.542
2.3	.5943	.5410	.4821	.4337	.3932	8.642	8.571	8.479	8.389	8.302
2.6	.5878	.5344	.4754	.4270	.3866	9.432	9.352	9.249	9.149	9.052
3.0	.5801	.5264	.4674	.4191	.3789	10.48	10.39	10.27	10.15	10.04
3.5	.5715	.5176	.4585	.4104	.3704	11.79	11.68	11.54	11.40	11.27
4.0	.5638	.5097	.4507	.4027	.3629	13.09	12.96	12.80	12.64	12.48
4.5	.5568	.5026	.4435	.3957	.3562	14.39	14.24	14.05	13.86	13.68
5.0	.5503	.4960	.4370	.3894	.3501	15.68	15.51	15.29	15.08	14.87
6.0	.5388	.4843	.4255	.3782	.3393	18.02	17.82	17.55	17.30	17.06
7.0	.5285	.4740	.4154	.3684	.3299	20.42	20.17	19.86	19.55	19.26
8.0	.5192	.4647	.4063	.3597	.3216	22.86	22.56	22.18	21.82	21.48
9.0	.5108	.4563	.3981	.3518	.3141	25.33	24.98	24.53	24.11	23.71
10	.5029	.4485	.3906	.3446	.3073	27.81	27.41	26.89	26.40	25.94
12	.4887	.4343	.3770	.3318	.2952	32.35	31.86	31.23	30.63	30.07
14	.4760	.4219	.3651	.3206	.2846	37.10	36.49	35.72	34.99	34.30
17	.4592	.4056	.3497	.3061	.2710	44.57	43.74	42.70	41.73	40.81
20	.4444	.3914	.3363	.2935	.2594	52.36	51.27	49.91	48.65	47.48
23	.4308	.3788	.3243	.2821	.2491	59.26	58.00	56.43	54.96	53.59
26	.4185	.3675	.3135	.2719	.2399	66.49	65.02	63.19	61.46	59.85
30	.4037	.3541	.3007	.2598	.2291	76.66	74.83	72.54	70.41	68.44
35	.3873	.3392	.2868	.2467	.2175	90.09	87.68	84.72	81.99	79.48
40	.3728	.3261	.2745	.2352	.2073	104.2	101.1	97.30	93.88	90.77
45	.3597	.3142	.2636	.2251	.1983	118.7	114.8	110.2	106.0	102.2
50	.3479	.3035	.2538	.2161	.1903	133.6	128.8	123.8	118.2	113.8
60	.3270	.2840	.2366	.2006	.1765	161.4	155.2	147.9	141.6	135.8
70	.3092	.2674	.2221	.1878	.1650	190.9	182.8	173.6	165.5	158.3
80	.2937	.2531	.2097	.1768	.1551	221.7	211.5	199.9	190.0	181.1
90	.2800	.2405	.1988	.1672	.1466	253.5	241.0	226.7	214.6	204.0
100	.2678	.2293	.1892	.1589	.1391	286.0	270.9	253.9	239.5	227.0
120	.2467	.2106	.1730	.1447	.1266	349.0	329.3	306.7	287.7	271.7
140	.2291	.1952	.1597	.1332	.1164	414.3	389.3	360.3	336.2	316.5
170	.2074	.1765	.1436	.1194	.1041	515.1	480.9	441.4	408.9	383.1
200	.1899	.1614	.1307	.1084	.0944	617.9	573.6	522.6	481.1	449.0
230	.1757	.1491	.1204	.0996	.0867	718.3	663.7	601.4	551.0	512.7
260	.1637	.1387	.1117	.0922	.0802	819.3	753.9	679.7	619.9	575.3
300	.1503	.1271	.1021	.0841	.0731	954.3	873.7	782.9	710.4	657.2
350	.1366	.1153	.0924	.0759	.0660	1123.	1022.	909.9	820.9	756.9
400	.1253	.1056	.0844	.0693	.0602	1290.	1169.	1034.	928.5	853.8
450	.1158	.0975	.0778	.0638	.0554	1455.	1313.	1156.	1033.	947.9
500	.1078	.0906	.0723	.0592	.0513	1618.	1455.	1275.	1136.	1039.
600	.0948	.0795	.0633	.0518	.0449	1937.	1731.	1505.	1332.	1215.
700	.0849	.0711	.0565	.0461	.0399	2245.	1997.	1725.	1519.	1382.
800	.0770	.0644	.0511	.0417	.0361	2543.	2253.	1936.	1698.	1541.
900	.0706	.0590	.0468	.0381	.0330	2832.	2499.	2139.	1869.	1693.
1000	.0654	.0547	.0433	.0352	.0304	3110.	2738.	2334.	2034.	1838.

K/KL =	MEAN DAMAGE DEPTH,MICROGRAM/SQ.CM.					MEAN IONZN.DEPTH,MICROGRAM/SQ.CM.				
E,KEV	0.65	0.80	1.0	1.2	1.4	0.65	0.80	1.0	1.2	1.4
.10	.9636	.9545	.9427	.9313	.9203	.9550	.9450	.9320	.9194	.9072
.12	1.094	1.083	1.069	1.056	1.043	1.084	1.072	1.056	1.042	1.027
.14	1.212	1.199	1.184	1.169	1.154	1.200	1.187	1.169	1.153	1.137
.17	1.375	1.361	1.343	1.325	1.308	1.362	1.347	1.326	1.307	1.288
.20	1.529	1.513	1.492	1.472	1.453	1.514	1.497	1.474	1.451	1.430
.23	1.676	1.658	1.634	1.612	1.590	1.659	1.639	1.614	1.589	1.565
.26	1.817	1.797	1.772	1.747	1.723	1.799	1.777	1.749	1.722	1.695
.30	1.999	1.977	1.948	1.920	1.894	1.979	1.954	1.923	1.892	1.862
.35	2.218	2.193	2.160	2.128	2.098	2.196	2.167	2.131	2.096	2.063
.40	2.429	2.400	2.364	2.328	2.294	2.404	2.372	2.331	2.292	2.255
.45	2.633	2.601	2.560	2.521	2.484	2.605	2.570	2.525	2.482	2.440
.50	2.830	2.796	2.751	2.708	2.667	2.800	2.761	2.712	2.665	2.619
.60	3.197	3.157	3.105	3.055	3.008	3.162	3.117	3.060	3.005	2.952
.70	3.550	3.504	3.445	3.389	3.335	3.510	3.459	3.394	3.331	3.271
.80	3.892	3.840	3.774	3.711	3.650	3.847	3.789	3.716	3.646	3.579
.90	4.224	4.167	4.093	4.023	3.956	4.174	4.110	4.029	3.951	3.877
1.0	4.547	4.484	4.403	4.326	4.253	4.492	4.422	4.332	4.247	4.165
1.2	5.147	5.074	4.979	4.890	4.804	5.083	5.001	4.896	4.797	4.702
1.4	5.730	5.645	5.538	5.435	5.337	5.655	5.561	5.442	5.328	5.220
1.7	6.580	6.479	6.350	6.228	6.111	6.490	6.377	6.234	6.098	5.970
2.0	7.406	7.288	7.137	6.995	6.859	7.299	7.167	7.000	6.842	6.692
2.3	8.153	8.019	7.851	7.691	7.539	8.033	7.885	7.698	7.520	7.352
2.6	8.889	8.741	8.553	8.375	8.206	8.757	8.592	8.383	8.185	7.998
3.0	9.865	9.695	9.480	9.277	9.084	9.712	9.522	9.283	9.058	8.845
3.5	11.08	10.88	10.63	10.39	10.17	10.89	10.67	10.39	10.13	9.884
4.0	12.28	12.05	11.76	11.49	11.23	12.07	11.81	11.49	11.19	10.90
4.5	13.47	13.21	12.89	12.58	12.29	13.23	12.93	12.57	12.23	11.91
5.0	14.66	14.36	14.00	13.65	13.33	14.37	14.04	13.63	13.25	12.89
6.0	16.82	16.47	16.03	15.62	15.24	16.48	16.08	15.60	15.14	14.72
7.0	19.02	18.60	18.09	17.61	17.16	18.60	18.14	17.56	17.02	16.53
8.0	21.24	20.76	20.16	19.60	19.08	20.74	20.20	19.52	18.90	18.33
9.0	23.49	22.93	22.23	21.59	20.99	22.89	22.26	21.49	20.77	20.11
10	25.74	25.10	24.31	23.58	22.91	25.04	24.32	23.44	22.63	21.89
12	29.87	29.10	28.15	27.27	26.46	29.02	28.15	27.09	26.10	25.20
14	34.15	33.22	32.08	31.03	30.07	33.10	32.06	30.78	29.60	28.52
17	40.84	39.63	38.16	36.81	35.58	39.41	38.05	36.40	34.90	33.54
20	47.77	46.24	44.38	42.70	41.18	45.86	44.15	42.09	40.24	38.58
23	53.95	52.15	50.02	48.09	46.30	51.79	49.77	47.37	45.30	43.41
26	60.39	58.29	55.82	53.60	51.53	57.86	55.51	52.73	50.38	48.25
30	69.38	66.80	63.82	61.14	58.64	66.20	63.33	59.96	57.19	54.70
35	81.17	77.91	74.16	70.82	67.73	76.94	73.34	69.14	65.75	62.74
40	93.44	89.42	84.79	80.70	76.97	87.94	83.52	78.39	74.29	70.71
45	106.1	101.2	95.62	90.71	86.31	99.10	93.81	87.68	82.80	78.60
50	118.9	113.2	106.6	100.8	95.70	110.4	104.1	96.96	91.25	86.39
60	143.1	135.9	127.4	119.9	113.6	131.8	124.0	114.8	107.4	101.1
70	168.5	159.6	148.8	139.5	131.7	153.8	144.0	132.7	123.4	115.6
80	194.9	183.9	170.7	159.4	150.1	176.1	164.2	150.5	139.2	130.0
90	222.0	208.7	192.9	179.5	168.6	198.6	184.5	168.2	154.8	144.1
100	249.5	233.8	215.3	199.6	187.0	221.1	204.7	185.7	170.3	158.0
120	302.8	282.4	258.6	238.6	222.8	265.2	244.1	219.9	200.3	185.1
140	357.6	331.9	302.4	277.8	258.4	309.2	283.0	253.4	229.5	211.3
170	441.8	407.4	368.4	336.3	311.6	374.4	340.4	302.0	271.6	249.1
200	527.4	483.5	434.3	394.3	364.0	438.5	396.3	349.0	311.9	285.0
230	610.6	557.5	498.1	450.5	414.6	501.0	450.7	394.2	350.3	319.2
260	694.0	631.5	561.3	505.8	464.6	562.1	503.7	437.6	387.2	351.9
300	805.4	729.6	644.7	578.3	529.5	641.3	571.8	493.9	434.3	393.6
350	944.1	851.1	747.2	666.9	608.8	736.8	653.6	560.6	490.2	442.8
400	1082.	970.9	847.7	753.2	686.0	828.8	732.0	624.2	543.1	489.5
450	1217.	1089.	945.9	837.3	761.1	917.4	807.1	684.8	593.4	533.7
500	1351.	1204.	1042.	919.2	834.2	1003.	879.3	742.9	641.5	576.0
600	1612.	1429.	1228.	1077.	974.8	1166.	1016.	852.3	731.9	655.2
700	1865.	1645.	1406.	1227.	1108.	1319.	1144.	954.2	815.7	728.5
800	2108.	1854.	1576.	1370.	1236.	1463.	1264.	1050.	894.0	796.9
900	2344.	2054.	1740.	1508.	1358.	1600.	1378.	1140.	967.7	861.3
1000	2572.	2248.	1898.	1640.	1475.	1731.	1486.	1225.	1037.	922.0

	RELATIVE RANGE STRAGGLING					RELATIVE DAMAGE STRAGGLING				
K/KL=	0.65	0.80	1.0	1.2	1.4	0.65	0.80	1.0	1.2	1.4
E,KEV										
.10	4.950	4.678	4.359	4.083	3.841	2.941	2.792	2.618	2.467	2.335
.12	4.907	4.631	4.310	4.032	3.788	2.918	2.767	2.591	2.439	2.307
.14	4.869	4.590	4.267	3.987	3.743	2.898	2.745	2.568	2.415	2.282
.17	4.820	4.538	4.211	3.930	3.686	2.872	2.717	2.538	2.385	2.252
.20	4.778	4.493	4.164	3.882	3.637	2.849	2.693	2.513	2.359	2.226
.23	4.740	4.453	4.123	3.839	3.594	2.830	2.672	2.491	2.337	2.203
.26	4.707	4.418	4.085	3.801	3.556	2.812	2.653	2.471	2.316	2.182
.30	4.666	4.375	4.041	3.756	3.510	2.790	2.631	2.448	2.292	2.158
.35	4.621	4.327	3.992	3.707	3.461	2.767	2.606	2.422	2.266	2.132
.40	4.580	4.285	3.948	3.662	3.417	2.746	2.584	2.399	2.243	2.109
.45	4.543	4.246	3.909	3.623	3.377	2.726	2.564	2.378	2.222	2.088
.50	4.509	4.211	3.873	3.586	3.341	2.709	2.545	2.360	2.203	2.069
.60	4.447	4.148	3.808	3.522	3.278	2.677	2.512	2.326	2.169	2.036
.70	4.393	4.092	3.752	3.466	3.222	2.649	2.484	2.297	2.140	2.007
.80	4.344	4.042	3.701	3.416	3.173	2.624	2.458	2.271	2.114	1.981
.90	4.299	3.996	3.655	3.371	3.129	2.601	2.434	2.247	2.090	1.958
1.0	4.257	3.953	3.613	3.329	3.088	2.580	2.412	2.225	2.069	1.937
1.2	4.181	3.877	3.537	3.254	3.015	2.541	2.373	2.186	2.030	1.899
1.4	4.113	3.809	3.470	3.188	2.951	2.507	2.339	2.152	1.997	1.866
1.7	4.022	3.718	3.381	3.102	2.868	2.462	2.293	2.106	1.952	1.823
2.0	3.940	3.637	3.302	3.026	2.794	2.421	2.252	2.066	1.913	1.785
2.3	3.866	3.564	3.231	2.957	2.729	2.384	2.215	2.030	1.878	1.751
2.6	3.798	3.497	3.166	2.895	2.669	2.350	2.182	1.997	1.846	1.720
3.0	3.714	3.415	3.087	2.820	2.597	2.308	2.140	1.957	1.807	1.683
3.5	3.620	3.323	2.999	2.736	2.517	2.261	2.094	1.912	1.764	1.642
4.0	3.534	3.240	2.920	2.660	2.445	2.218	2.052	1.871	1.725	1.605
4.5	3.455	3.163	2.847	2.591	2.380	2.178	2.013	1.834	1.690	1.571
5.0	3.382	3.093	2.780	2.528	2.321	2.142	1.977	1.800	1.657	1.540
6.0	3.247	2.963	2.658	2.413	2.212	2.074	1.912	1.737	1.598	1.483
7.0	3.128	2.849	2.551	2.313	2.118	2.014	1.854	1.682	1.545	1.434
8.0	3.021	2.747	2.456	2.224	2.035	1.960	1.801	1.633	1.499	1.390
9.0	2.924	2.655	2.370	2.144	1.960	1.910	1.754	1.588	1.457	1.350
10	2.835	2.571	2.292	2.071	1.892	1.865	1.710	1.547	1.418	1.314
12	2.675	2.420	2.152	1.941	1.771	1.782	1.631	1.472	1.348	1.249
14	2.536	2.290	2.032	1.830	1.668	1.709	1.561	1.408	1.288	1.192
17	2.359	2.124	1.879	1.690	1.538	1.614	1.472	1.325	1.211	1.120
20	2.209	1.984	1.751	1.572	1.429	1.533	1.396	1.254	1.145	1.059
23	2.080	1.863	1.642	1.472	1.337	1.462	1.329	1.193	1.089	1.006
26	1.967	1.759	1.547	1.385	1.257	1.399	1.270	1.139	1.039	.9606
30	1.836	1.638	1.437	1.285	1.165	1.325	1.201	1.076	.9807	.9073
35	1.698	1.510	1.322	1.180	1.069	1.246	1.127	1.008	.9190	.8506
40	1.580	1.402	1.225	1.092	.9882	1.177	1.064	.9509	.8663	.8022
45	1.478	1.309	1.142	1.017	.9193	1.118	1.009	.9010	.8207	.7603
50	1.390	1.228	1.069	.9513	.8596	1.065	.9601	.8572	.7807	.7235
60	1.244	1.096	.9522	.8457	.7634	.9753	.8795	.7840	.7143	.6597
70	1.127	.9905	.8588	.7618	.6871	.9021	.8138	.7246	.6605	.6086
80	1.031	.9039	.7823	.6932	.6247	.8408	.7589	.6751	.6158	.5670
90	.9496	.8313	.7184	.6359	.5728	.7885	.7122	.6332	.5781	.5327
100	.8803	.7695	.6641	.5874	.5289	.7434	.6718	.5972	.5457	.5042
120	.7708	.6721	.5789	.5115	.4601	.6710	.6061	.5396	.4940	.4668
140	.6858	.5968	.5133	.4531	.4074	.6138	.5541	.4945	.4535	.4392
170	.5834	.5110	.4387	.3869	.3476	.5469	.4933	.4421	.4066	.4082
200	.5150	.4466	.3829	.3374	.3031	.4954	.4467	.4021	.3709	.3840
230	.4585	.3974	.3401	.2992	.2690	.4549	.4111	.3712	.3432	.3567
260	.4131	.3581	.3059	.2687	.2418	.4219	.3823	.3461	.3207	.3328
300	.3648	.3164	.2697	.2364	.2131	.3862	.3515	.3192	.2966	.3054
350	.3182	.2763	.2349	.2054	.1856	.3511	.3213	.2929	.2731	.2771
400	.2820	.2453	.2081	.1815	.1644	.3234	.2977	.2723	.2547	.2540
450	.2532	.2206	.1867	.1626	.1475	.3010	.2786	.2557	.2398	.2350
500	.2296	.2004	.1693	.1472	.1338	.2826	.2628	.2420	.2276	.2195
600	.1935	.1693	.1428	.1238	.1129	.2541	.2384	.2207	.2087	.1964
700	.1672	.1456	.1235	.1069	.0977	.2333	.2203	.2050	.1947	.1812
800	.1472	.1291	.1088	.0941	.0861	.2177	.2064	.1930	.1841	.1718
900	.1316	.1152	.0972	.0842	.0769	.2056	.1953	.1835	.1756	.1667
1000	.1191	.1039	.0878	.0763	.0695	.1961	.1863	.1758	.1689	.1650

	RELATIVE IONIZATION STRAGGLING					RELATIVE TRANSVERSE RANGE STRAGGLING				
K/KL=	0.65	0.80	1.0	1.2	1.4	0.65	0.80	1.0	1.2	1.4
E,KEV										
.10	2.755	2.615	2.453	2.312	2.190	.9789	.9778	.9763	.9748	.9734
.12	2.733	2.592	2.428	2.286	2.163	.9788	.9776	.9761	.9746	.9732
.14	2.714	2.571	2.406	2.264	2.141	.9787	.9775	.9759	.9744	.9729
.17	2.690	2.545	2.378	2.236	2.112	.9785	.9773	.9757	.9742	.9726
.20	2.668	2.522	2.355	2.211	2.087	.9784	.9772	.9755	.9739	.9724
.23	2.649	2.503	2.334	2.190	2.066	.9783	.9770	.9754	.9737	.9722
.26	2.633	2.485	2.315	2.171	2.047	.9782	.9769	.9752	.9736	.9720
.30	2.612	2.463	2.293	2.148	2.024	.9781	.9768	.9751	.9734	.9717
.35	2.590	2.440	2.269	2.124	1.999	.9780	.9767	.9749	.9732	.9715
.40	2.569	2.418	2.247	2.102	1.978	.9780	.9766	.9748	.9730	.9713
.45	2.551	2.399	2.227	2.082	1.958	.9779	.9765	.9747	.9729	.9712
.50	2.534	2.382	2.209	2.064	1.940	.9779	.9764	.9746	.9728	.9710
.60	2.504	2.351	2.178	2.032	1.909	.9778	.9763	.9744	.9726	.9708
.70	2.477	2.323	2.150	2.005	1.882	.9777	.9762	.9743	.9724	.9706
.80	2.453	2.298	2.125	1.980	1.858	.9777	.9762	.9742	.9723	.9704
.90	2.431	2.276	2.102	1.958	1.836	.9776	.9761	.9741	.9722	.9703
1.0	2.410	2.255	2.081	1.937	1.816	.9776	.9761	.9741	.9721	.9702
1.2	2.373	2.217	2.044	1.901	1.780	.9776	.9760	.9740	.9720	.9701
1.4	2.339	2.184	2.011	1.868	1.748	.9776	.9760	.9740	.9719	.9700
1.7	2.295	2.139	1.967	1.826	1.707	.9777	.9761	.9740	.9719	.9699
2.0	2.255	2.100	1.929	1.788	1.672	.9778	.9762	.9740	.9720	.9699
2.3	2.219	2.064	1.894	1.755	1.640	.9780	.9763	.9741	.9720	.9700
2.6	2.186	2.032	1.863	1.725	1.611	.9781	.9764	.9742	.9721	.9701
3.0	2.146	1.992	1.825	1.689	1.576	.9783	.9766	.9744	.9723	.9702
3.5	2.100	1.948	1.782	1.648	1.537	.9786	.9769	.9747	.9726	.9705
4.0	2.059	1.907	1.743	1.611	1.502	.9790	.9773	.9751	.9729	.9709
4.5	2.020	1.870	1.708	1.578	1.471	.9793	.9776	.9754	.9733	.9713
5.0	1.985	1.836	1.675	1.547	1.442	.9797	.9780	.9758	.9738	.9717
6.0	1.919	1.773	1.616	1.491	1.389	.9804	.9788	.9766	.9746	.9726
7.0	1.861	1.717	1.563	1.441	1.342	.9812	.9796	.9775	.9755	.9736
8.0	1.808	1.666	1.516	1.397	1.301	.9821	.9805	.9785	.9765	.9746
9.0	1.761	1.621	1.474	1.358	1.264	.9830	.9814	.9794	.9775	.9757
10	1.717	1.580	1.435	1.322	1.231	.9839	.9824	.9805	.9786	.9769
12	1.638	1.505	1.366	1.258	1.171	.9856	.9842	.9825	.9808	.9792
14	1.570	1.441	1.307	1.203	1.120	.9874	.9862	.9846	.9830	.9816
17	1.482	1.359	1.232	1.134	1.056	.9902	.9892	.9879	.9866	.9854
20	1.408	1.289	1.168	1.075	1.002	.9932	.9924	.9913	.9903	.9893
23	1.342	1.228	1.113	1.025	.9561	.9960	.9952	.9945	.9939	.9930
26	1.285	1.175	1.065	.9812	.9159	.9990	.9981	.9978	.9975	.9968
30	1.218	1.113	1.009	.9308	.8698	1.003	1.002	1.002	1.002	1.002
35	1.147	1.048	.9504	.8776	.8214	1.008	1.007	1.008	1.009	1.009
40	1.086	.9926	.9008	.8329	.7808	1.013	1.012	1.014	1.015	1.015
45	1.033	.9448	.8583	.7947	.7462	1.018	1.018	1.020	1.021	1.022
50	.9874	.9033	.8214	.7616	.7163	1.023	1.023	1.025	1.028	1.028
60	.9114	.8351	.7615	.7081	.6686	1.033	1.034	1.037	1.040	1.042
70	.8502	.7806	.7138	.6658	.6310	1.042	1.045	1.049	1.052	1.055
80	.7998	.7359	.6750	.6316	.6008	1.051	1.055	1.060	1.064	1.067
90	.7575	.6988	.6429	.6032	.5759	1.060	1.065	1.071	1.075	1.079
100	.7216	.6673	.6158	.5795	.5551	1.069	1.075	1.081	1.086	1.091
120	.6655	.6188	.5743	.5433	.5235	1.086	1.093	1.101	1.107	1.113
140	.6225	.5820	.5430	.5162	.5002	1.103	1.110	1.120	1.127	1.134
170	.5742	.5411	.5087	.4868	.4750	1.126	1.135	1.147	1.155	1.163
200	.5386	.5113	.4841	.4661	.4574	1.148	1.158	1.172	1.182	1.190
230	.5124	.4894	.4665	.4517	.4453	1.170	1.181	1.196	1.207	1.216
260	.4920	.4726	.4533	.4411	.4365	1.199	1.203	1.219	1.231	1.240
300	.4711	.4556	.4403	.4310	.4282	1.216	1.230	1.248	1.261	1.271
350	.4522	.4404	.4291	.4226	.4215	1.246	1.262	1.281	1.295	1.305
400	.4395	.4296	.4214	.4173	.4174	1.274	1.291	1.311	1.327	1.337
450	.4284	.4219	.4162	.4139	.4150	1.301	1.318	1.340	1.356	1.367
500	.4209	.4163	.4127	.4118	.4137	1.325	1.344	1.366	1.383	1.394
600	.4111	.4094	.4089	.4102	.4131	1.370	1.390	1.414	1.431	1.443
700	.4058	.4061	.4078	.4105	.4142	1.410	1.432	1.456	1.475	1.487
800	.4032	.4049	.4081	.4119	.4160	1.447	1.470	1.495	1.514	1.527
900	.4023	.4052	.4094	.4137	.4182	1.481	1.505	1.530	1.549	1.562
1000	.4027	.4063	.4111	.4159	.4205	1.513	1.537	1.563	1.582	1.595

K/KL= E,KEV	RELATIVE TRANSV. DAMAGE STRAGGLING					RELATIVE TRANSV. IONZN. STRAGGLING				
	0.65	0.80	1.0	1.2	1.4	0.65	0.80	1.0	1.2	1.4
.10	.9564	.9533	.9493	.9452	.9411	.9512	.9476	.9428	.9379	.9330
.12	.9560	.9529	.9487	.9445	.9402	.9507	.9470	.9420	.9370	.9319
.14	.9557	.9525	.9482	.9438	.9395	.9503	.9465	.9414	.9362	.9310
.17	.9553	.9519	.9475	.9430	.9385	.9498	.9459	.9406	.9353	.9299
.20	.9549	.9515	.9469	.9423	.9377	.9494	.9453	.9399	.9344	.9288
.23	.9546	.9511	.9464	.9417	.9370	.9490	.9449	.9393	.9336	.9279
.26	.9544	.9508	.9460	.9412	.9364	.9487	.9444	.9387	.9330	.9271
.30	.9541	.9504	.9455	.9406	.9356	.9483	.9439	.9381	.9322	.9262
.35	.9537	.9500	.9449	.9399	.9348	.9479	.9434	.9374	.9313	.9251
.40	.9534	.9496	.9444	.9393	.9340	.9475	.9429	.9367	.9305	.9242
.45	.9532	.9493	.9440	.9387	.9334	.9471	.9425	.9362	.9298	.9233
.50	.9530	.9490	.9436	.9382	.9328	.9468	.9421	.9356	.9291	.9226
.60	.9526	.9485	.9429	.9374	.9318	.9463	.9414	.9347	.9280	.9212
.70	.9523	.9480	.9423	.9366	.9309	.9459	.9408	.9339	.9270	.9200
.80	.9520	.9477	.9418	.9360	.9301	.9455	.9403	.9332	.9261	.9190
.90	.9518	.9474	.9414	.9354	.9294	.9451	.9398	.9326	.9253	.9180
1.0	.9516	.9471	.9410	.9349	.9288	.9448	.9394	.9320	.9246	.9171
1.2	.9513	.9466	.9403	.9341	.9277	.9443	.9386	.9310	.9233	.9156
1.4	.9510	.9462	.9398	.9333	.9268	.9438	.9380	.9302	.9222	.9142
1.7	.9507	.9458	.9391	.9324	.9257	.9433	.9372	.9290	.9208	.9124
2.0	.9505	.9454	.9386	.9317	.9248	.9428	.9365	.9281	.9195	.9109
2.3	.9504	.9452	.9381	.9311	.9240	.9424	.9360	.9272	.9184	.9096
2.6	.9503	.9450	.9378	.9305	.9233	.9421	.9355	.9265	.9175	.9083
3.0	.9502	.9447	.9374	.9299	.9225	.9417	.9349	.9256	.9163	.9069
3.5	.9502	.9446	.9370	.9293	.9216	.9413	.9342	.9246	.9149	.9052
4.0	.9503	.9445	.9367	.9288	.9209	.9410	.9337	.9238	.9137	.9036
4.5	.9504	.9444	.9364	.9284	.9203	.9407	.9331	.9229	.9126	.9022
5.0	.9505	.9444	.9363	.9281	.9198	.9404	.9327	.9222	.9116	.9009
6.0	.9508	.9445	.9360	.9275	.9189	.9400	.9319	.9209	.9097	.8985
7.0	.9511	.9447	.9360	.9271	.9183	.9397	.9312	.9197	.9081	.8963
8.0	.9516	.9449	.9359	.9269	.9177	.9394	.9306	.9186	.9065	.8942
9.0	.9520	.9452	.9360	.9267	.9173	.9392	.9300	.9175	.9049	.8922
10	.9525	.9455	.9361	.9265	.9169	.9389	.9294	.9165	.9034	.8903
12	.9535	.9463	.9364	.9265	.9164	.9386	.9284	.9148	.9009	.8869
14	.9545	.9471	.9368	.9265	.9160	.9382	.9273	.9130	.8983	.8835
17	.9561	.9482	.9374	.9266	.9155	.9376	.9258	.9103	.8943	.8783
20	.9576	.9494	.9381	.9268	.9151	.9369	.9244	.9074	.8903	.8731
23	.9593	.9506	.9390	.9271	.9150	.9368	.9242	.9049	.8866	.8681
26	.9609	.9517	.9398	.9275	.9148	.9365	.9238	.9023	.8827	.8630
30	.9628	.9530	.9407	.9278	.9146	.9358	.9228	.8985	.8774	.8561
35	.9651	.9546	.9417	.9281	.9141	.9343	.9209	.8934	.8704	.8472
40	.9672	.9560	.9426	.9281	.9134	.9324	.9182	.8879	.8631	.8383
45	.9692	.9573	.9432	.9280	.9126	.9301	.9149	.8821	.8557	.8293
50	.9709	.9585	.9437	.9278	.9116	.9274	.9112	.8762	.8481	.8203
60	.9746	.9613	.9449	.9275	.9099	.9217	.9020	.8643	.8331	.8025
70	.9776	.9634	.9454	.9266	.9076	.9151	.8919	.8519	.8178	.7849
80	.9800	.9649	.9454	.9252	.9048	.9076	.8811	.8390	.8025	.7676
90	.9819	.9660	.9449	.9234	.9018	.8995	.8700	.8259	.7872	.7506
100	.9835	.9667	.9441	.9213	.8984	.8910	.8586	.8127	.7719	.7338
120	.9862	.9675	.9419	.9167	.8912	.8725	.8346	.7844	.7397	.6992
140	.9876	.9671	.9387	.9111	.8834	.8535	.8111	.7570	.7093	.6669
170	.9880	.9647	.9324	.9017	.8709	.8249	.7770	.7183	.6671	.6229
200	.9866	.9607	.9249	.8914	.8580	.7968	.7446	.6824	.6288	.5837
230	.9843	.9553	.9160	.8796	.8438	.7664	.7115	.6472	.5925	.5477
260	.9811	.9491	.9066	.8675	.8297	.7373	.6805	.6146	.5596	.5154
300	.9755	.9400	.8936	.8515	.8114	.7009	.6423	.5753	.5202	.4771
350	.9672	.9278	.8773	.8320	.7896	.6591	.5994	.5317	.4772	.4357
400	.9578	.9150	.8611	.8131	.7689	.6213	.5612	.4936	.4399	.4003
450	.9477	.9019	.8451	.7948	.7494	.5869	.5270	.4599	.4075	.3696
500	.9370	.8887	.8294	.7773	.7309	.5555	.4961	.4301	.3790	.3428
600	.9147	.8622	.7990	.7443	.6968	.5004	.4430	.3796	.3315	.2985
700	.8916	.8362	.7701	.7137	.6658	.4536	.3989	.3387	.2938	.2638
800	.8682	.8106	.7425	.6851	.6375	.4133	.3617	.3053	.2635	.2362
900	.8449	.7857	.7162	.6584	.6115	.3783	.3301	.2776	.2389	.2140
1000	.8217	.7615	.6910	.6332	.5873	.3475	.3029	.2545	.2189	.1962

K/KL= E,KEV	RANGE SKEWNESS					DAMAGE SKEWNESS				
	0.65	0.80	1.0	1.2	1.4	0.65	0.80	1.0	1.2	1.4
.10	.0268	.0230	.0178	.0125	.0070	.1235	.1253	.1275	.1297	.1318
.12	.0263	.0223	.0169	.0114	.0057	.1237	.1255	.1279	.1301	.1322
.14	.0258	.0217	.0162	.0105	.0046	.1239	.1258	.1282	.1304	.1326
.17	.0252	.0210	.0152	.0092	.0032	.1242	.1261	.1286	.1309	.1331
.20	.0247	.0203	.0143	.0082	.0019	.1245	.1264	.1289	.1312	.1334
.23	.0242	.0197	.0136	.0072	.0008	.1247	.1267	.1292	.1315	.1338
.26	.0238	.0192	.0128	.0063	-.0003	.1249	.1269	.1294	.1318	.1341
.30	.0232	.0185	.0120	.0053	-.0016	.1251	.1271	.1297	.1321	.1344
.35	.0226	.0177	.0110	.0041	-.0030	.1254	.1274	.1300	.1324	.1348
.40	.0221	.0170	.0101	.0030	-.0043	.1256	.1276	.1303	.1327	.1351
.45	.0216	.0164	.0093	.0020	-.0055	.1258	.1278	.1305	.1330	.1353
.50	.0211	.0158	.0085	.0010	-.0066	.1259	.1280	.1307	.1332	.1356
.60	.0203	.0147	.0071	-.0007	-.0086	.1262	.1284	.1310	.1336	.1359
.70	.0195	.0138	.0059	-.0022	-.0105	.1265	.1286	.1313	.1338	.1362
.80	.0188	.0129	.0047	-.0036	-.0121	.1267	.1289	.1315	.1341	.1365
.90	.0182	.0120	.0036	-.0050	-.0137	.1269	.1290	.1317	.1343	.1367
1.0	.0175	.0112	.0026	-.0062	-.0152	.1271	.1292	.1319	.1344	.1368
1.2	.0164	.0098	.0007	-.0085	-.0179	.1274	.1295	.1322	.1347	.1371
1.4	.0153	.0084	-.0010	-.0106	-.0203	.1276	.1297	.1323	.1348	.1372
1.7	.0138	.0065	-.0034	-.0135	-.0238	.1278	.1299	.1325	.1349	.1373
2.0	.0124	.0047	-.0057	-.0163	-.0270	.1279	.1300	.1325	.1349	.1372
2.3	.0111	.0031	-.0077	-.0187	-.0299	.1281	.1300	.1325	.1349	.1371
2.6	.0099	.0016	-.0097	-.0211	-.0326	.1281	.1301	.1325	.1348	.1369
3.0	.0083	-.0004	-.0122	-.0241	-.0361	.1281	.1300	.1323	.1345	.1366
3.5	.0064	-.0028	-.0152	-.0278	-.0404	.1281	.1298	.1320	.1341	.1360
4.0	.0045	-.0052	-.0182	-.0313	-.0444	.1279	.1295	.1316	.1335	.1354
4.5	.0026	-.0075	-.0211	-.0347	-.0484	.1276	.1292	.1311	.1329	.1346
5.0	.0007	-.0098	-.0239	-.0380	-.0522	.1274	.1288	.1305	.1322	.1339
6.0	-.0027	-.0140	-.0291	-.0442	-.0592	.1268	.1280	.1295	.1310	.1324
7.0	-.0061	-.0181	-.0342	-.0502	-.0661	.1261	.1271	.1283	.1295	.1307
8.0	-.0094	-.0222	-.0392	-.0561	-.0727	.1253	.1260	.1269	.1279	.1288
9.0	-.0128	-.0263	-.0442	-.0618	-.0792	.1243	.1248	.1255	.1262	.1269
10	-.0162	-.0304	-.0490	-.0675	-.0856	.1233	.1235	.1239	.1244	.1249
12	-.0226	-.0381	-.0583	-.0782	-.0976	.1211	.1208	.1207	.1207	.1208
14	-.0291	-.0457	-.0674	-.0887	-.1093	.1186	.1179	.1173	.1168	.1166
17	-.0387	-.0569	-.0808	-.1038	-.1261	.1147	.1134	.1121	.1109	.1102
20	-.0481	-.0680	-.0937	-.1185	-.1423	.1107	.1088	.1067	.1050	.1038
23	-.0571	-.0785	-.1061	-.1324	-.1579	.1067	.1042	.1015	.0994	.0975
26	-.0660	-.0889	-.1181	-.1459	-.1730	.1026	.0996	.0963	.0938	.0914
30	-.0778	-.1025	-.1337	-.1633	-.1924	.0971	.0935	.0893	.0865	.0832
35	-.0922	-.1191	-.1527	-.1843	-.2155	.0902	.0859	.0808	.0773	.0733
40	-.1063	-.1352	-.1711	-.2045	-.2376	.0833	.0783	.0723	.0683	.0635
45	-.1202	-.1509	-.1889	-.2240	-.2587	.0765	.0709	.0640	.0595	.0540
50	-.1338	-.1663	-.2062	-.2428	-.2789	.0697	.0636	.0559	.0508	.0447
60	-.1603	-.1963	-.2406	-.2796	-.3178	.0564	.0491	.0400	.0335	.0267
70	-.1857	-.2246	-.2726	-.3139	-.3538	.0435	.0353	.0249	.0170	.0096
80	-.2101	-.2515	-.3026	-.3459	-.3873	.0312	.0220	.0104	.0013	-.0068
90	-.2335	-.2770	-.3308	-.3760	-.4187	.0194	.0093	-.0035	-.0138	-.0224
100	-.2561	-.3013	-.3573	-.4044	-.4482	.0080	-.0029	-.0168	-.0282	-.0375
120	-.3002	-.3472	-.4057	-.4565	-.5024	-.0131	-.0257	-.0415	-.0548	-.0656
140	-.3409	-.3892	-.4497	-.5038	-.5515	-.0328	-.0471	-.0647	-.0797	-.0920
170	-.3965	-.4466	-.5095	-.5679	-.6178	-.0606	-.0772	-.0972	-.1144	-.1287
200	-.4462	-.4983	-.5634	-.6253	-.6772	-.0864	-.1052	-.1275	-.1467	-.1626
230	-.4894	-.5450	-.6132	-.6778	-.7314	-.1108	-.1317	-.1565	-.1777	-.1953
260	-.5290	-.5881	-.6592	-.7260	-.7812	-.1339	-.1568	-.1838	-.2069	-.2260
300	-.5774	-.6409	-.7157	-.7849	-.8420	-.1629	-.1883	-.2180	-.2432	-.2640
350	-.6325	-.7009	-.7798	-.8517	-.9109	-.1968	-.2248	-.2575	-.2852	-.3076
400	-.6829	-.7555	-.8382	-.9123	-.9734	-.2285	-.2589	-.2941	-.3238	-.3476
450	-.7296	-.8058	-.8918	-.9680	-1.031	-.2583	-.2907	-.3283	-.3596	-.3845
500	-.7732	-.8525	-.9415	-1.020	-1.084	-.2865	-.3208	-.3603	-.3930	-.4188
600	-.8531	-.9371	-1.031	-1.113	-1.180	-.3389	-.3762	-.4189	-.4539	-.4808
700	-.9253	-1.013	-1.111	-1.195	-1.264	-.3868	-.4266	-.4719	-.5084	-.5359
800	-.9916	-1.081	-1.183	-1.270	-1.341	-.4311	-.4729	-.5202	-.5578	-.5856
900	-1.053	-1.144	-1.249	-1.338	-1.411	-.4725	-.5159	-.5648	-.6031	-.6308
1000	-1.111	-1.202	-1.309	-1.400	-1.475	-.5113	-.5562	-.6062	-.6450	-.6723

	IONIZATION SKEWNESS					RANGE KURTOSIS				
K/KL=	0.65	0.80	1.0	1.2	1.4	0.65	0.80	1.0	1.2	1.4
E,KEV										
.10	.1464	.1498	.1543	.1588	.1631	3.065	3.048	3.027	3.007	2.989
.12	.1469	.1505	.1551	.1597	.1641	3.063	3.045	3.023	3.004	2.985
.14	.1474	.1510	.1558	.1604	.1650	3.061	3.043	3.021	3.001	2.982
.17	.1480	.1517	.1566	.1614	.1661	3.058	3.039	3.017	2.997	2.978
.20	.1485	.1524	.1574	.1623	.1671	3.055	3.037	3.014	2.993	2.974
.23	.1490	.1529	.1581	.1631	.1680	3.053	3.034	3.011	2.990	2.971
.26	.1495	.1535	.1587	.1638	.1688	3.051	3.032	3.009	2.987	2.968
.30	.1500	.1541	.1594	.1646	.1697	3.049	3.029	3.006	2.984	2.965
.35	.1506	.1548	.1602	.1656	.1708	3.046	3.026	3.002	2.981	2.961
.40	.1512	.1554	.1610	.1664	.1718	3.044	3.024	3.000	2.978	2.958
.45	.1517	.1560	.1617	.1672	.1726	3.042	3.022	2.997	2.975	2.955
.50	.1522	.1566	.1623	.1679	.1734	3.040	3.019	2.994	2.972	2.952
.60	.1530	.1576	.1635	.1692	.1749	3.037	3.015	2.990	2.968	2.947
.70	.1538	.1584	.1645	.1704	.1762	3.033	3.012	2.986	2.964	2.943
.80	.1545	.1593	.1654	.1714	.1773	3.031	3.009	2.983	2.960	2.939
.90	.1552	.1600	.1663	.1724	.1784	3.028	3.006	2.980	2.957	2.936
1.0	.1558	.1607	.1671	.1733	.1793	3.026	3.004	2.977	2.954	2.933
1.2	.1569	.1619	.1684	.1748	.1811	3.022	2.999	2.972	2.948	2.927
1.4	.1579	.1630	.1697	.1762	.1826	3.018	2.995	2.968	2.943	2.922
1.7	.1593	.1645	.1714	.1781	.1847	3.013	2.989	2.961	2.937	2.916
2.0	.1606	.1659	.1730	.1799	.1866	3.008	2.984	2.956	2.932	2.910
2.3	.1618	.1673	.1745	.1815	.1884	3.004	2.980	2.951	2.927	2.905
2.6	.1630	.1686	.1759	.1831	.1901	3.000	2.975	2.947	2.922	2.900
3.0	.1644	.1702	.1776	.1850	.1921	2.995	2.970	2.941	2.916	2.895
3.5	.1661	.1720	.1796	.1872	.1945	2.989	2.964	2.935	2.910	2.888
4.0	.1676	.1736	.1815	.1891	.1966	2.984	2.959	2.929	2.904	2.883
4.5	.1690	.1751	.1831	.1910	.1986	2.979	2.954	2.924	2.899	2.878
5.0	.1703	.1766	.1847	.1927	.2005	2.975	2.949	2.919	2.894	2.873
6.0	.1724	.1788	.1872	.1955	.2035	2.966	2.940	2.910	2.885	2.864
7.0	.1743	.1809	.1896	.1981	.2064	2.958	2.932	2.902	2.878	2.857
8.0	.1762	.1830	.1919	.2006	.2091	2.951	2.924	2.895	2.871	2.851
9.0	.1780	.1849	.1940	.2030	.2117	2.944	2.917	2.888	2.864	2.845
10	.1798	.1868	.1962	.2053	.2143	2.938	2.911	2.882	2.859	2.840
12	.1835	.1909	.2006	.2102	.2195	2.925	2.899	2.870	2.848	2.831
14	.1870	.1946	.2048	.2148	.2245	2.914	2.888	2.861	2.840	2.823
17	.1919	.1999	.2106	.2211	.2312	2.900	2.874	2.849	2.829	2.816
20	.1963	.2047	.2160	.2269	.2373	2.887	2.863	2.839	2.822	2.811
23	.2002	.2091	.2208	.2322	.2430	2.875	2.852	2.829	2.815	2.806
26	.2038	.2131	.2254	.2371	.2481	2.864	2.842	2.822	2.809	2.803
30	.2083	.2180	.2309	.2431	.2543	2.851	2.832	2.814	2.805	2.802
35	.2133	.2237	.2372	.2498	.2612	2.839	2.822	2.808	2.803	2.805
40	.2179	.2288	.2428	.2558	.2673	2.829	2.816	2.806	2.805	2.811
45	.2222	.2336	.2480	.2613	.2728	2.822	2.812	2.806	2.809	2.820
50	.2262	.2380	.2528	.2663	.2777	2.816	2.810	2.809	2.816	2.831
60	.2332	.2461	.2619	.2757	.2871	2.807	2.808	2.816	2.832	2.856
70	.2396	.2533	.2696	.2835	.2946	2.803	2.811	2.829	2.854	2.886
80	.2455	.2597	.2763	.2901	.3009	2.804	2.820	2.846	2.881	2.921
90	.2511	.2655	.2821	.2956	.3059	2.810	2.832	2.868	2.910	2.958
100	.2564	.2708	.2872	.3003	.3101	2.818	2.847	2.892	2.942	2.997
120	.2684	.2816	.2962	.3075	.3156	2.841	2.883	2.943	3.008	3.077
140	.2786	.2904	.3032	.3127	.3192	2.871	2.925	3.000	3.080	3.162
170	.2908	.3004	.3105	.3177	.3222	2.925	2.997	3.094	3.194	3.296
200	.2997	.3074	.3151	.3204	.3233	2.987	3.076	3.195	3.313	3.433
230	.3030	.3088	.3147	.3188	.3212	3.049	3.154	3.293	3.430	3.566
250	.3047	.3090	.3134	.3166	.3187	3.115	3.236	3.395	3.549	3.701
300	.3054	.3079	.3108	.3132	.3153	3.207	3.349	3.533	3.710	3.881
350	.3048	.3057	.3072	.3090	.3113	3.327	3.493	3.708	3.912	4.107
400	.3032	.3031	.3037	.3054	.3081	3.450	3.641	3.884	4.114	4.332
450	.3012	.3004	.3006	.3023	.3055	3.576	3.789	4.060	4.315	4.554
500	.2989	.2978	.2979	.2999	.3037	3.702	3.937	4.235	4.514	4.774
600	.2942	.2933	.2939	.2969	.3021	3.954	4.231	4.580	4.904	5.204
700	.2897	.2897	.2917	.2960	.3026	4.204	4.521	4.917	5.283	5.621
800	.2855	.2871	.2909	.2968	.3047	4.449	4.803	5.245	5.651	6.024
900	.2817	.2853	.2913	.2990	.3083	4.690	5.079	5.563	6.008	6.414
1000	.2783	.2842	.2927	.3023	.3129	4.925	5.348	5.873	6.354	6.791

K/KL= E,KEV	DAMAGE KURTOSIS					IONIZATION KURTOSIS				
	0.65	0.80	1.0	1.2	1.4	0.65	0.80	1.0	1.2	1.4
.10	4.094	4.077	4.055	4.036	4.017	4.240	4.224	4.206	4.188	4.172
.12	4.091	4.073	4.052	4.031	4.013	4.237	4.222	4.202	4.185	4.168
.14	4.088	4.070	4.048	4.028	4.009	4.235	4.219	4.200	4.182	4.165
.17	4.085	4.066	4.044	4.023	4.004	4.232	4.216	4.196	4.178	4.160
.20	4.082	4.063	4.040	4.019	3.999	4.230	4.213	4.193	4.174	4.157
.23	4.079	4.060	4.036	4.015	3.995	4.228	4.211	4.190	4.171	4.153
.26	4.076	4.057	4.033	4.011	3.991	4.225	4.208	4.187	4.168	4.150
.30	4.073	4.053	4.029	4.007	3.987	4.223	4.206	4.184	4.165	4.146
.35	4.069	4.049	4.025	4.002	3.981	4.220	4.202	4.181	4.161	4.142
.40	4.065	4.045	4.020	3.998	3.977	4.218	4.200	4.178	4.157	4.138
.45	4.062	4.042	4.017	3.994	3.972	4.215	4.197	4.175	4.154	4.135
.50	4.059	4.038	4.013	3.990	3.968	4.213	4.194	4.172	4.151	4.132
.60	4.053	4.032	4.006	3.983	3.961	4.209	4.190	4.167	4.146	4.126
.70	4.048	4.026	4.000	3.976	3.954	4.206	4.187	4.163	4.141	4.121
.80	4.043	4.021	3.995	3.970	3.947	4.202	4.183	4.159	4.137	4.116
.90	4.039	4.016	3.989	3.964	3.941	4.199	4.179	4.155	4.133	4.112
1.0	4.034	4.012	3.984	3.959	3.936	4.196	4.176	4.152	4.129	4.108
1.2	4.026	4.003	3.975	3.949	3.926	4.190	4.169	4.144	4.121	4.099
1.4	4.018	3.995	3.966	3.940	3.916	4.184	4.163	4.137	4.114	4.091
1.7	4.008	3.984	3.954	3.928	3.903	4.176	4.154	4.128	4.104	4.081
2.0	3.998	3.973	3.943	3.916	3.891	4.169	4.147	4.120	4.095	4.071
2.3	3.988	3.963	3.933	3.905	3.879	4.164	4.141	4.113	4.088	4.064
2.6	3.979	3.953	3.923	3.894	3.868	4.159	4.136	4.107	4.081	4.057
3.0	3.968	3.941	3.910	3.881	3.855	4.152	4.129	4.100	4.073	4.048
3.5	3.954	3.927	3.895	3.866	3.839	4.145	4.121	4.091	4.064	4.038
4.0	3.941	3.914	3.881	3.851	3.824	4.137	4.112	4.082	4.054	4.027
4.5	3.929	3.901	3.868	3.837	3.809	4.130	4.104	4.073	4.045	4.017
5.0	3.917	3.889	3.855	3.824	3.796	4.122	4.096	4.065	4.035	4.007
6.0	3.895	3.866	3.831	3.799	3.771	4.105	4.078	4.045	4.015	3.986
7.0	3.875	3.844	3.809	3.777	3.747	4.089	4.061	4.027	3.996	3.966
8.0	3.855	3.824	3.788	3.755	3.725	4.074	4.046	4.010	3.978	3.947
9.0	3.837	3.806	3.769	3.735	3.705	4.061	4.031	3.995	3.961	3.929
10	3.820	3.788	3.750	3.716	3.685	4.048	4.018	3.981	3.946	3.913
12	3.786	3.754	3.715	3.680	3.648	4.029	3.997	3.958	3.921	3.886
14	3.756	3.722	3.683	3.647	3.613	4.012	3.978	3.937	3.898	3.861
17	3.715	3.680	3.639	3.603	3.568	3.987	3.951	3.907	3.866	3.826
20	3.678	3.642	3.601	3.564	3.527	3.963	3.925	3.879	3.835	3.792
23	3.644	3.608	3.565	3.527	3.490	3.937	3.899	3.849	3.802	3.756
26	3.614	3.577	3.533	3.494	3.456	3.913	3.873	3.820	3.770	3.722
30	3.576	3.539	3.493	3.454	3.416	3.881	3.839	3.783	3.730	3.679
35	3.534	3.496	3.450	3.409	3.371	3.843	3.799	3.739	3.682	3.627
40	3.497	3.458	3.410	3.369	3.330	3.807	3.761	3.696	3.636	3.578
45	3.463	3.423	3.375	3.333	3.294	3.773	3.725	3.656	3.592	3.532
50	3.431	3.392	3.343	3.300	3.261	3.740	3.690	3.618	3.551	3.488
60	3.374	3.334	3.285	3.241	3.201	3.679	3.622	3.544	3.472	3.404
70	3.325	3.285	3.235	3.190	3.150	3.623	3.560	3.476	3.400	3.329
80	3.283	3.242	3.192	3.146	3.105	3.571	3.503	3.413	3.333	3.260
90	3.246	3.205	3.154	3.108	3.066	3.522	3.449	3.356	3.272	3.197
100	3.212	3.171	3.120	3.074	3.032	3.477	3.398	3.301	3.216	3.139
120	3.157	3.115	3.062	3.015	2.973	3.391	3.304	3.200	3.110	3.033
140	3.111	3.068	3.015	2.967	2.925	3.313	3.220	3.110	3.017	2.941
170	3.056	3.012	2.958	2.911	2.868	3.210	3.110	2.995	2.900	2.825
200	3.012	2.968	2.914	2.867	2.826	3.119	3.015	2.897	2.802	2.730
230	2.974	2.931	2.878	2.833	2.794	3.032	2.929	2.813	2.721	2.654
260	2.942	2.900	2.848	2.805	2.770	2.954	2.853	2.740	2.652	2.590
300	2.908	2.867	2.818	2.778	2.746	2.863	2.765	2.658	2.574	2.518
350	2.875	2.837	2.791	2.755	2.727	2.764	2.673	2.572	2.495	2.446
400	2.850	2.815	2.774	2.741	2.717	2.681	2.595	2.501	2.431	2.387
450	2.833	2.800	2.763	2.734	2.713	2.610	2.529	2.442	2.378	2.340
500	2.820	2.791	2.758	2.732	2.714	2.549	2.474	2.393	2.334	2.301
600	2.807	2.785	2.760	2.741	2.725	2.451	2.385	2.315	2.266	2.241
700	2.805	2.789	2.773	2.760	2.746	2.377	2.319	2.259	2.219	2.200
800	2.810	2.802	2.793	2.787	2.773	2.321	2.271	2.219	2.185	2.170
900	2.821	2.820	2.819	2.819	2.804	2.279	2.234	2.190	2.161	2.150
1000	2.837	2.843	2.849	2.854	2.838	2.248	2.208	2.170	2.145	2.137

TABLE 1- 9

BE ON U, M2/M1=26.4

K/KL=	REFLECTION COEFFICIENT					SPUTTERING EFFICIENCY				
	0.65	0.80	1.0	1.2	1.4	0.65	0.80	1.0	1.2	1.4
E,KEV										
.10	.3267	.3221	.3163	.3107	.3055	.2021	.1861	.1677	.1519	.1384
.12	.3260	.3213	.3149	.3111	.3012	.1997	.1835	.1650	.1492	.1356
.14	.3254	.3205	.3140	.3105	.2975	.1977	.1814	.1627	.1469	.1333
.17	.3245	.3196	.3131	.3087	.2926	.1951	.1786	.1598	.1439	.1303
.20	.3238	.3188	.3124	.3064	.2883	.1929	.1763	.1574	.1414	.1278
.23	.3232	.3180	.3118	.3039	.2844	.1910	.1742	.1552	.1392	.1257
.26	.3226	.3174	.3113	.3013	.2810	.1893	.1724	.1533	.1373	.1238
.30	.3219	.3166	.3106	.2979	.2769	.1872	.1703	.1511	.1351	.1215
.35	.3211	.3156	.3096	.2939	.2723	.1850	.1679	.1487	.1327	.1191
.40	.3204	.3148	.3085	.2901	.2682	.1831	.1659	.1466	.1305	.1170
.45	.3197	.3140	.3074	.2866	.2646	.1813	.1640	.1447	.1286	.1152
.50	.3191	.3133	.3062	.2833	.2612	.1797	.1623	.1430	.1269	.1135
.60	.3179	.3121	.3035	.2774	.2548	.1769	.1594	.1400	.1239	.1106
.70	.3169	.3110	.3006	.2723	.2494	.1744	.1569	.1374	.1214	.1081
.80	.3159	.3100	.2978	.2678	.2448	.1723	.1546	.1351	.1192	.1059
.90	.3150	.3089	.2950	.2638	.2409	.1703	.1526	.1331	.1172	.1040
1.0	.3142	.3079	.2921	.2602	.2375	.1685	.1508	.1313	.1154	.1022
1.2	.3129	.3065	.2861	.2541	.2328	.1654	.1475	.1280	.1122	.0992
1.4	.3117	.3048	.2804	.2490	.2291	.1626	.1447	.1252	.1095	.0966
1.7	.3097	.3019	.2727	.2426	.2245	.1590	.1411	.1216	.1060	.0932
2.0	.3076	.2984	.2658	.2373	.2205	.1558	.1379	.1185	.1030	.0903
2.3	.3056	.2930	.2598	.2330	.2165	.1530	.1351	.1157	.1003	.0878
2.6	.3036	.2875	.2543	.2293	.2127	.1505	.1325	.1133	.0980	.0856
3.0	.3006	.2805	.2479	.2250	.2082	.1474	.1294	.1103	.0951	.0829
3.5	.2968	.2724	.2409	.2201	.2032	.1439	.1260	.1070	.0920	.0800
4.0	.2927	.2650	.2349	.2158	.1988	.1408	.1230	.1041	.0893	.0774
4.5	.2885	.2584	.2295	.2119	.1948	.1380	.1202	.1014	.0868	.0751
5.0	.2842	.2524	.2248	.2082	.1912	.1354	.1176	.0990	.0845	.0730
6.0	.2735	.2434	.2173	.2008	.1853	.1306	.1130	.0947	.0805	.0692
7.0	.2638	.2358	.2109	.1943	.1801	.1264	.1090	.0909	.0770	.0660
8.0	.2550	.2291	.2053	.1885	.1754	.1227	.1054	.0875	.0739	.0632
9.0	.2471	.2233	.2002	.1833	.1712	.1192	.1021	.0845	.0711	.0606
10	.2400	.2180	.1956	.1787	.1672	.1160	.0990	.0817	.0685	.0583
12	.2294	.2091	.1867	.1710	.1597	.1102	.0935	.0767	.0640	.0542
14	.2206	.2014	.1789	.1643	.1530	.1051	.0887	.0723	.0601	.0507
17	.2094	.1914	.1689	.1558	.1443	.0984	.0825	.0668	.0551	.0462
20	.1998	.1828	.1603	.1484	.1368	.0926	.0772	.0620	.0509	.0425
23	.1906	.1742	.1531	.1414	.1302	.0873	.0725	.0579	.0473	.0393
26	.1823	.1666	.1467	.1351	.1243	.0827	.0683	.0542	.0440	.0366
30	.1725	.1574	.1392	.1275	.1175	.0772	.0635	.0500	.0403	.0334
35	.1618	.1475	.1311	.1194	.1100	.0712	.0583	.0454	.0364	.0300
40	.1527	.1389	.1241	.1122	.1035	.0660	.0538	.0416	.0330	.0272
45	.1447	.1315	.1178	.1060	.0978	.0615	.0498	.0383	.0302	.0248
50	.1376	.1249	.1122	.1004	.0927	.0575	.0464	.0353	.0277	.0227
60	.1262	.1143	.1018	.0911	.0836	.0506	.0405	.0304	.0235	.0192
70	.1168	.1057	.0933	.0834	.0761	.0450	.0357	.0265	.0202	.0165
80	.1089	.0985	.0860	.0770	.0699	.0403	.0317	.0233	.0176	.0143
90	.1021	.0923	.0799	.0716	.0646	.0364	.0283	.0207	.0154	.0126
100	.0962	.0869	.0746	.0669	.0601	.0329	.0255	.0184	.0136	.0111
120	.0863	.0780	.0667	.0595	.0535	.0276	.0212	.0152	.0111	.0093
140	.0784	.0709	.0605	.0538	.0484	.0234	.0179	.0127	.0093	.0080
170	.0690	.0624	.0534	.0470	.0425	.0187	.0142	.0101	.0074	.0066
200	.0616	.0558	.0479	.0419	.0381	.0152	.0115	.0082	.0061	.0057
230	.0561	.0504	.0433	.0378	.0343	.0128	.0096	.0068	.0051	.0048
260	.0516	.0460	.0395	.0344	.0312	.0109	.0081	.0057	.0043	.0041
300	.0466	.0412	.0353	.0308	.0278	.0090	.0067	.0047	.0035	.0033
350	.0417	.0364	.0311	.0271	.0244	.0072	.0054	.0037	.0027	.0025
400	.0378	.0325	.0277	.0242	.0216	.0059	.0044	.0030	.0022	.0020
450	.0345	.0293	.0250	.0218	.0194	.0050	.0037	.0025	.0018	.0016
500	.0317	.0267	.0226	.0198	.0175	.0042	.0031	.0021	.0015	.0012
600	.0272	.0225	.0189	.0165	.0145	.0032	.0024	.0016	.0011	.0008
700	.0236	.0193	.0161	.0140	.0122	.0025	.0018	.0012	.0009	.0005
800	.0206	.0169	.0140	.0120	.0105	.0020	.0015	.0010	.0007	.0004
900	.0181	.0150	.0122	.0104	.0091	.0016	.0012	.0008	.0006	.0003
1000	.0159	.0134	.0108	.0090	.0080	.0013	.0010	.0007	.0005	.0004

TABLE 1-10

BE ON U, M2/M1=26.4

K/KL=	IONIZATION DEFICIENCY					SPUTTERING YIELD ALPHA				
E,KEV	0.65	0.80	1.0	1.2	1.4	0.65	0.80	1.0	1.2	1.4
.10	.0671	.0761	.0858	.0933	.0993	2.087	2.011	1.919	1.835	1.758
.12	.0684	.0774	.0870	.0945	.1004	2.077	2.000	1.906	1.821	1.743
.14	.0695	.0785	.0881	.0955	.1013	2.068	1.990	1.895	1.808	1.729
.17	.0709	.0799	.0895	.0968	.1024	2.057	1.977	1.880	1.792	1.712
.20	.0721	.0811	.0906	.0978	.1033	2.048	1.967	1.868	1.779	1.698
.23	.0731	.0821	.0915	.0986	.1040	2.040	1.957	1.858	1.768	1.686
.26	.0740	.0830	.0923	.0994	.1046	2.033	1.949	1.848	1.757	1.675
.30	.0750	.0840	.0933	.1002	.1053	2.024	1.939	1.837	1.745	1.662
.35	.0761	.0851	.0943	.1011	.1060	2.015	1.929	1.825	1.732	1.648
.40	.0771	.0860	.0951	.1018	.1066	2.007	1.920	1.815	1.721	1.636
.45	.0779	.0868	.0958	.1024	.1071	2.000	1.912	1.806	1.711	1.626
.50	.0787	.0875	.0965	.1029	.1075	1.994	1.905	1.798	1.702	1.616
.60	.0800	.0888	.0975	.1038	.1082	1.983	1.892	1.783	1.686	1.599
.70	.0811	.0898	.0984	.1045	.1087	1.974	1.881	1.771	1.673	1.585
.80	.0820	.0907	.0991	.1051	.1091	1.966	1.872	1.761	1.662	1.573
.90	.0828	.0914	.0998	.1055	.1095	1.958	1.864	1.751	1.651	1.562
1.0	.0835	.0920	.1003	.1059	.1097	1.952	1.857	1.743	1.642	1.552
1.2	.0848	.0931	.1011	.1065	.1101	1.941	1.844	1.728	1.626	1.535
1.4	.0858	.0940	.1018	.1070	.1103	1.932	1.833	1.716	1.613	1.521
1.7	.0870	.0951	.1026	.1074	.1104	1.920	1.819	1.700	1.596	1.503
2.0	.0880	.0959	.1031	.1077	.1104	1.911	1.808	1.687	1.581	1.488
2.3	.0889	.0966	.1035	.1079	.1104	1.902	1.798	1.676	1.569	1.474
2.6	.0896	.0971	.1039	.1079	.1102	1.895	1.790	1.666	1.558	1.463
3.0	.0904	.0977	.1041	.1079	.1099	1.887	1.780	1.655	1.545	1.449
3.5	.0912	.0982	.1043	.1078	.1095	1.878	1.769	1.642	1.531	1.434
4.0	.0918	.0986	.1044	.1075	.1090	1.870	1.759	1.631	1.519	1.421
4.5	.0923	.0989	.1043	.1072	.1084	1.863	1.751	1.621	1.508	1.410
5.0	.0927	.0991	.1042	.1068	.1078	1.857	1.743	1.612	1.498	1.399
6.0	.0933	.0992	.1038	.1060	.1065	1.846	1.730	1.596	1.481	1.381
7.0	.0937	.0992	.1033	.1050	.1051	1.837	1.719	1.583	1.466	1.365
8.0	.0939	.0990	.1027	.1040	.1038	1.829	1.709	1.571	1.453	1.350
9.0	.0940	.0988	.1020	.1029	.1024	1.821	1.699	1.560	1.441	1.338
10	.0939	.0984	.1013	.1019	.1011	1.814	1.691	1.550	1.429	1.326
12	.0938	.0976	.0997	.0997	.0984	1.801	1.675	1.531	1.409	1.304
14	.0934	.0967	.0981	.0975	.0959	1.790	1.660	1.515	1.391	1.285
17	.0926	.0950	.0956	.0944	.0922	1.773	1.641	1.492	1.367	1.259
20	.0915	.0933	.0931	.0913	.0887	1.757	1.622	1.471	1.344	1.235
23	.0903	.0914	.0905	.0883	.0853	1.741	1.606	1.451	1.321	1.214
26	.0890	.0894	.0881	.0854	.0822	1.726	1.591	1.432	1.300	1.195
30	.0873	.0869	.0849	.0818	.0782	1.705	1.571	1.408	1.274	1.170
35	.0850	.0838	.0811	.0776	.0736	1.681	1.547	1.380	1.243	1.141
40	.0828	.0808	.0776	.0737	.0695	1.657	1.523	1.354	1.214	1.114
45	.0805	.0780	.0744	.0702	.0656	1.635	1.500	1.328	1.187	1.088
50	.0783	.0753	.0713	.0669	.0621	1.612	1.478	1.304	1.161	1.064
60	.0740	.0701	.0654	.0606	.0551	1.572	1.432	1.256	1.110	1.008
70	.0699	.0655	.0603	.0552	.0492	1.533	1.389	1.212	1.065	.9608
80	.0660	.0612	.0557	.0505	.0441	1.495	1.348	1.172	1.024	.9202
90	.0624	.0574	.0517	.0464	.0398	1.458	1.309	1.135	.9878	.8857
100	.0590	.0539	.0480	.0428	.0362	1.422	1.272	1.101	.9553	.8566
120	.0521	.0475	.0418	.0368	.0311	1.344	1.198	1.035	.8996	.8221
140	.0462	.0421	.0367	.0321	.0273	1.273	1.133	.9779	.8532	.7961
170	.0388	.0354	.0305	.0265	.0229	1.181	1.050	.9068	.7964	.7654
200	.0328	.0300	.0256	.0222	.0196	1.104	.9816	.8491	.7508	.7396
230	.0288	.0255	.0218	.0193	.0168	1.049	.9342	.8085	.7163	.7064
260	.0257	.0217	.0187	.0170	.0144	1.003	.8950	.7752	.6875	.6760
300	.0223	.0178	.0153	.0145	.0119	.9518	.8519	.7386	.6556	.6397
350	.0191	.0142	.0120	.0122	.0093	.8997	.8083	.7017	.6231	.6005
400	.0165	.0116	.0095	.0104	.0074	.8568	.7726	.6716	.5965	.5672
450	.0145	.0099	.0075	.0089	.0058	.8207	.7423	.6461	.5739	.5389
500	.0128	.0089	.0059	.0076	.0046	.7896	.7162	.6242	.5545	.5148
600	.0102	.0084	.0035	.0056	.0027	.7385	.6725	.5874	.5224	.4767
700	.0082	.0093	.0020	.0039	.0015	.6976	.6366	.5573	.4964	.4403
800	.0066	.0112	.0010	.0025	.0007	.6638	.6060	.5315	.4746	.4298
900	.0053	.0138	.0003	.0012	.0003	.6351	.5792	.5087	.4559	.4164
1000	.0042	.0170	0.0000	0.0000	0.0000	.6102	.5551	.4883	.4395	.4079

BE ON U, M2/M1=26.4

1000 KEV

100 KEV

10 KEV

1 KEV

.10 KEV

TABLE 2- 1

BE ON AU, M2/M1=21.9

E,KEV	FRACTIONAL DEPOSITED ENERGY					MEAN RANGE,MICROGRAM/SQ.CM.				
K/KL=	0.7	0.85	1.0	1.2	1.4	0.7	0.85	1.0	1.2	1.4
.10	.7344	.6935	.6565	.6123	.5731	.8663	.8625	.8587	.8537	.8488
.12	.7281	.6866	.6492	.6045	.5650	.9845	.9799	.9754	.9695	.9637
.14	.7227	.6807	.6429	.5979	.5581	1.092	1.086	1.081	1.075	1.068
.17	.7158	.6732	.6349	.5895	.5494	1.240	1.234	1.228	1.220	1.213
.20	.7099	.6668	.6281	.5823	.5421	1.379	1.373	1.366	1.357	1.348
.23	.7047	.6612	.6222	.5761	.5357	1.513	1.505	1.498	1.488	1.478
.26	.7001	.6562	.6169	.5706	.5300	1.642	1.633	1.625	1.614	1.603
.30	.6947	.6503	.6107	.5641	.5234	1.808	1.798	1.789	1.776	1.764
.35	.6887	.6439	.6040	.5571	.5162	2.008	1.997	1.986	1.972	1.958
.40	.6835	.6382	.5980	.5509	.5099	2.201	2.189	2.177	2.161	2.145
.45	.6788	.6332	.5928	.5454	.5043	2.389	2.375	2.361	2.343	2.325
.50	.6745	.6286	.5880	.5405	.4992	2.570	2.555	2.540	2.520	2.501
.60	.6670	.6206	.5796	.5318	.4905	2.907	2.890	2.872	2.849	2.826
.70	.6605	.6137	.5724	.5244	.4829	3.233	3.213	3.192	3.166	3.140
.80	.6548	.6076	.5660	.5179	.4764	3.550	3.527	3.504	3.474	3.444
.90	.6496	.6021	.5604	.5120	.4705	3.858	3.832	3.807	3.773	3.740
1.0	.6449	.5971	.5552	.5068	.4652	4.159	4.130	4.102	4.065	4.029
1.2	.6365	.5883	.5461	.4975	.4559	4.717	4.683	4.650	4.606	4.564
1.4	.6293	.5807	.5382	.4895	.4479	5.262	5.223	5.184	5.134	5.085
1.7	.6198	.5707	.5281	.4792	.4376	6.062	6.014	5.967	5.906	5.846
2.0	.6116	.5621	.5193	.4703	.4288	6.845	6.788	6.732	6.659	6.588
2.3	.6043	.5545	.5115	.4625	.4211	7.548	7.484	7.421	7.339	7.259
2.6	.5977	.5477	.5046	.4556	.4142	8.247	8.175	8.104	8.012	7.923
3.0	.5898	.5395	.4963	.4473	.4060	9.179	9.095	9.013	8.906	8.802
3.5	.5809	.5303	.4870	.4381	.3970	10.35	10.25	10.15	10.02	9.897
4.0	.5729	.5221	.4788	.4299	.3890	11.51	11.40	11.28	11.13	10.99
4.5	.5656	.5147	.4713	.4225	.3817	12.68	12.54	12.41	12.23	12.07
5.0	.5589	.5078	.4644	.4157	.3752	13.84	13.68	13.53	13.33	13.14
6.0	.5467	.4955	.4521	.4037	.3635	15.96	15.76	15.58	15.34	15.11
7.0	.5360	.4846	.4413	.3932	.3534	18.13	17.89	17.67	17.38	17.10
8.0	.5262	.4748	.4317	.3838	.3443	20.35	20.07	19.80	19.45	19.12
9.0	.5173	.4659	.4229	.3753	.3362	22.60	22.27	21.95	21.54	21.15
10	.5090	.4577	.4148	.3675	.3287	24.88	24.49	24.12	23.64	23.19
12	.4939	.4427	.4001	.3534	.3154	29.05	28.58	28.12	27.54	26.98
14	.4804	.4294	.3873	.3412	.3038	33.43	32.84	32.27	31.55	30.88
17	.4626	.4120	.3704	.3253	.2888	40.34	39.52	38.76	37.79	36.89
20	.4469	.3968	.3559	.3116	.2760	47.56	46.48	45.47	44.22	43.06
23	.4325	.3830	.3428	.2996	.2648	54.00	52.75	51.57	50.11	48.74
26	.4195	.3705	.3311	.2889	.2548	60.75	59.28	57.90	56.17	54.57
30	.4039	.3556	.3171	.2763	.2431	70.23	68.39	66.66	64.52	62.55
35	.3866	.3393	.3019	.2624	.2303	82.73	80.31	78.06	75.31	72.79
40	.3713	.3250	.2885	.2503	.2192	95.79	92.69	89.84	86.38	83.25
45	.3576	.3122	.2766	.2395	.2095	109.3	105.4	101.9	97.64	93.85
50	.3452	.3008	.2660	.2299	.2007	123.0	118.3	114.1	109.0	104.5
60	.3233	.2810	.2474	.2125	.1855	149.0	142.8	137.3	130.8	124.9
70	.3047	.2643	.2317	.1980	.1728	176.4	168.4	161.4	153.1	145.7
80	.2886	.2499	.2183	.1855	.1620	204.8	194.7	185.9	175.7	166.7
90	.2744	.2373	.2066	.1748	.1527	233.9	221.6	210.9	198.5	187.7
100	.2518	.2261	.1963	.1654	.1445	263.5	248.8	236.0	221.4	208.8
120	.2402	.2069	.1789	.1501	.1311	321.3	302.0	285.1	265.9	249.8
140	.2223	.1910	.1648	.1378	.1202	380.6	356.0	334.6	310.4	290.6
170	.2004	.1715	.1476	.1231	.1072	471.0	437.9	408.9	376.7	351.1
200	.1828	.1561	.1340	.1116	.0969	562.2	519.8	482.8	442.1	410.5
230	.1686	.1437	.1231	.1023	.0888	651.2	599.4	554.5	505.2	467.7
260	.1567	.1333	.1141	.0946	.0820	739.9	678.5	625.3	567.2	523.7
300	.1434	.1218	.1040	.0860	.0746	857.6	782.8	718.3	648.2	596.6
350	.1299	.1101	.0939	.0775	.0672	1003.	911.0	832.0	746.6	684.9
400	.1189	.1006	.0856	.0706	.0611	1146.	1037.	942.8	842.0	770.3
450	.1097	.0927	.0788	.0648	.0562	1287.	1160.	1051.	934.5	853.0
500	.1019	.0860	.0731	.0600	.0520	1426.	1280.	1156.	1024.	933.1
600	.0893	.0753	.0639	.0524	.0454	1694.	1512.	1359.	1197.	1085.
700	.0797	.0672	.0569	.0465	.0403	1952.	1734.	1552.	1360.	1231.
800	.0722	.0608	.0514	.0420	.0363	2200.	1948.	1736.	1516.	1369.
900	.0661	.0556	.0470	.0383	.0331	2439.	2152.	1912.	1664.	1500.
1000	.0611	.0514	.0434	.0353	.0305	2669.	2349.	2082.	1807.	1626.

TABLE 2- 2

BE ON AU, M2/M1=21.9

E,KEV	MEAN DAMAGE DEPTH,MICROGRAM/SQ.CM. K/KL= 0.7	0.85	1.0	1.2	1.4	MEAN IONZN.DEPTH,MICROGRAM/SQ.CM. 0.7	0.85	1.0	1.2	1.4
.10	.8278	.8198	.8121	.8021	.7924	.8191	.8103	.8017	.7907	.7800
.12	.9397	.9303	.9212	.9094	.8980	.9295	.9191	.9090	.8960	.8834
.14	1.041	1.031	1.020	1.007	.9939	1.030	1.018	1.006	.9917	.9775
.17	1.182	1.169	1.157	1.142	1.127	1.169	1.155	1.142	1.124	1.108
.20	1.314	1.300	1.286	1.268	1.251	1.299	1.283	1.268	1.249	1.230
.23	1.440	1.424	1.409	1.389	1.370	1.423	1.406	1.389	1.367	1.346
.26	1.562	1.544	1.527	1.506	1.485	1.544	1.524	1.506	1.482	1.458
.30	1.718	1.699	1.680	1.655	1.632	1.698	1.676	1.655	1.628	1.602
.35	1.907	1.885	1.863	1.835	1.809	1.884	1.859	1.835	1.805	1.775
.40	2.089	2.064	2.040	2.009	1.979	2.063	2.036	2.009	1.974	1.941
.45	2.265	2.237	2.210	2.176	2.143	2.236	2.206	2.176	2.138	2.101
.50	2.435	2.405	2.375	2.338	2.301	2.404	2.370	2.338	2.296	2.256
.60	2.752	2.717	2.682	2.638	2.596	2.716	2.677	2.639	2.590	2.544
.70	3.057	3.017	2.978	2.928	2.880	3.016	2.971	2.928	2.873	2.820
.80	3.353	3.308	3.264	3.208	3.154	3.307	3.257	3.208	3.146	3.086
.90	3.641	3.591	3.542	3.480	3.420	3.590	3.534	3.480	3.411	3.345
1.0	3.922	3.866	3.813	3.744	3.679	3.865	3.803	3.744	3.668	3.596
1.2	4.442	4.377	4.315	4.235	4.159	4.376	4.304	4.235	4.146	4.062
1.4	4.949	4.874	4.803	4.711	4.624	4.873	4.790	4.711	4.609	4.513
1.7	5.691	5.601	5.515	5.406	5.302	5.599	5.499	5.404	5.283	5.168
2.0	6.415	6.309	6.208	6.081	5.959	6.305	6.188	6.076	5.934	5.800
2.3	7.066	6.948	6.835	6.691	6.555	6.944	6.813	6.687	6.528	6.377
2.6	7.712	7.580	7.454	7.294	7.142	7.576	7.429	7.288	7.111	6.943
3.0	8.570	8.418	8.274	8.090	7.917	8.412	8.243	8.082	7.879	7.687
3.5	9.640	9.462	9.292	9.078	8.876	9.451	9.253	9.064	8.826	8.603
4.0	10.71	10.50	10.30	10.06	9.826	10.48	10.25	10.04	9.762	9.505
4.5	11.77	11.53	11.31	11.03	10.77	11.51	11.25	11.00	10.69	10.39
5.0	12.82	12.56	12.31	11.99	11.69	12.52	12.23	11.95	11.60	11.27
6.0	14.75	14.43	14.13	13.76	13.40	14.39	14.03	13.70	13.28	12.89
7.0	16.72	16.34	15.99	15.54	15.13	16.28	15.86	15.46	14.97	14.51
8.0	18.72	18.27	17.86	17.34	16.86	18.19	17.69	17.23	16.65	16.12
9.0	20.74	20.23	19.74	19.15	18.59	20.11	19.53	19.00	18.33	17.72
10	22.78	22.19	21.64	20.95	20.32	22.03	21.37	20.76	20.01	19.31
12	26.52	25.80	25.14	24.31	23.55	25.61	24.81	24.06	23.14	22.31
14	30.41	29.55	28.74	27.75	26.83	29.28	28.31	27.41	26.30	25.31
17	36.51	35.37	34.32	33.04	31.87	34.95	33.68	32.52	31.10	29.83
20	42.83	41.38	40.06	38.45	37.00	40.75	39.15	37.69	35.93	34.34
23	48.50	46.82	45.29	43.43	41.73	46.11	44.24	42.53	40.54	38.61
26	54.41	52.44	50.67	48.52	46.55	51.59	49.43	47.43	45.17	42.88
30	62.65	60.24	58.08	55.48	53.11	59.09	56.47	54.05	51.37	48.58
35	73.43	70.38	67.67	64.42	61.50	68.71	65.44	62.41	59.13	55.69
40	84.63	80.86	77.51	73.55	70.02	78.53	74.53	70.83	66.87	62.77
45	96.13	91.57	87.53	82.78	78.62	88.46	83.67	79.27	74.55	69.80
50	107.8	102.4	97.66	92.08	87.26	98.46	92.83	87.68	82.17	76.75
60	130.0	123.1	117.0	109.9	103.8	117.7	110.5	103.9	96.73	90.17
70	153.2	144.5	136.8	127.9	120.5	137.2	128.2	120.1	111.1	103.3
80	177.1	166.4	157.0	146.2	137.3	156.8	145.9	136.1	125.3	116.3
90	201.5	188.7	177.4	164.6	154.2	176.4	163.5	152.0	139.2	128.9
100	226.2	211.1	198.0	183.0	171.0	196.0	181.0	167.7	152.9	141.3
120	274.1	254.7	237.8	218.6	203.6	234.5	215.1	198.2	179.6	165.3
140	323.0	298.8	277.8	254.2	235.9	272.4	248.5	227.9	205.3	188.4
170	397.4	365.4	337.7	307.1	283.8	328.0	297.3	270.8	242.3	221.3
200	472.2	432.0	397.3	359.2	330.8	382.1	344.3	312.0	277.5	252.4
230	545.1	496.6	454.8	409.5	376.0	434.3	389.7	351.5	310.9	281.7
260	617.7	560.6	511.7	459.0	420.3	485.0	433.6	389.5	343.0	309.8
300	713.9	645.1	586.3	523.7	478.0	550.3	489.8	437.9	383.7	345.3
350	832.7	748.9	677.7	602.3	548.1	628.6	556.8	495.5	431.8	387.3
400	949.6	850.7	766.8	678.6	616.0	703.4	620.6	550.0	477.2	427.0
450	1064.	950.2	853.7	752.8	681.9	775.2	681.5	602.0	520.3	464.6
500	1177.	1048.	938.4	824.8	745.9	844.1	739.9	651.6	561.3	500.4
600	1395.	1236.	1102.	963.1	868.5	974.6	849.8	744.7	638.1	567.5
700	1605.	1416.	1258.	1094.	984.8	1097.	952.0	831.1	709.1	629.9
800	1806.	1589.	1406.	1219.	1095.	1211.	1048.	911.9	775.2	687.9
900	2000.	1755.	1549.	1339.	1201.	1320.	1138.	987.8	837.3	742.5
1000	2187.	1915.	1686.	1454.	1302.	1423.	1224.	1060.	895.8	794.1

	RELATIVE RANGE STRAGGLING					RELATIVE DAMAGE STRAGGLING				
K/KL=	0.7	0.85	1.0	1.2	1.4	0.7	0.85	1.0	1.2	1.4
E,KEV										
.10	4.188	3.985	3.802	3.584	3.390	2.505	2.394	2.295	2.177	2.072
.12	4.152	3.947	3.762	3.542	3.347	2.486	2.374	2.274	2.154	2.049
.14	4.121	3.914	3.727	3.505	3.310	2.469	2.357	2.255	2.135	2.029
.17	4.081	3.871	3.682	3.459	3.262	2.448	2.334	2.232	2.111	2.004
.20	4.046	3.834	3.643	3.419	3.221	2.430	2.315	2.211	2.090	1.983
.23	4.015	3.801	3.609	3.383	3.185	2.414	2.297	2.193	2.071	1.964
.26	3.987	3.771	3.578	3.352	3.153	2.399	2.282	2.177	2.054	1.947
.30	3.953	3.735	3.542	3.314	3.115	2.382	2.263	2.158	2.035	1.927
.35	3.915	3.695	3.500	3.272	3.072	2.362	2.243	2.137	2.013	1.905
.40	3.880	3.659	3.464	3.234	3.035	2.344	2.224	2.118	1.993	1.885
.45	3.849	3.627	3.430	3.200	3.001	2.328	2.207	2.101	1.976	1.868
.50	3.819	3.597	3.399	3.169	2.970	2.313	2.192	2.085	1.960	1.852
.60	3.767	3.543	3.345	3.114	2.915	2.287	2.164	2.057	1.931	1.823
.70	3.720	3.495	3.296	3.065	2.866	2.263	2.140	2.032	1.906	1.798
.80	3.678	3.451	3.252	3.022	2.823	2.242	2.118	2.009	1.884	1.776
.90	3.638	3.411	3.212	2.982	2.784	2.222	2.098	1.989	1.863	1.756
1.0	3.602	3.374	3.175	2.945	2.748	2.204	2.079	1.970	1.845	1.737
1.2	3.535	3.307	3.108	2.879	2.683	2.171	2.045	1.936	1.811	1.704
1.4	3.475	3.247	3.048	2.820	2.625	2.141	2.015	1.906	1.781	1.675
1.7	3.395	3.167	2.969	2.742	2.550	2.101	1.975	1.866	1.742	1.637
2.0	3.323	3.095	2.898	2.674	2.483	2.066	1.940	1.831	1.707	1.603
2.3	3.257	3.030	2.834	2.612	2.423	2.033	1.907	1.799	1.676	1.572
2.6	3.196	2.970	2.775	2.555	2.368	2.003	1.877	1.769	1.647	1.544
3.0	3.121	2.897	2.704	2.486	2.302	1.966	1.841	1.733	1.612	1.510
3.5	3.037	2.814	2.624	2.409	2.229	1.924	1.799	1.693	1.573	1.473
4.0	2.960	2.739	2.551	2.340	2.163	1.886	1.762	1.656	1.538	1.439
4.5	2.890	2.671	2.485	2.277	2.102	1.851	1.727	1.622	1.505	1.408
5.0	2.824	2.608	2.424	2.218	2.047	1.818	1.695	1.591	1.475	1.379
6.0	2.704	2.492	2.313	2.113	1.947	1.758	1.637	1.535	1.421	1.327
7.0	2.599	2.390	2.215	2.021	1.860	1.705	1.585	1.485	1.373	1.282
8.0	2.504	2.300	2.128	1.939	1.783	1.657	1.539	1.440	1.331	1.241
9.0	2.418	2.218	2.050	1.866	1.714	1.613	1.496	1.399	1.292	1.204
10	2.340	2.143	1.979	1.799	1.652	1.573	1.457	1.362	1.257	1.171
12	2.200	2.010	1.853	1.681	1.541	1.500	1.387	1.294	1.193	1.111
14	2.080	1.896	1.745	1.581	1.447	1.436	1.326	1.236	1.138	1.059
17	1.926	1.751	1.608	1.454	1.328	1.353	1.248	1.161	1.068	.9933
20	1.798	1.630	1.494	1.348	1.230	1.283	1.181	1.098	1.009	.9378
23	1.687	1.527	1.398	1.258	1.150	1.222	1.123	1.043	.9578	.8900
26	1.592	1.438	1.314	1.181	1.081	1.167	1.071	.9947	.9132	.8483
30	1.481	1.335	1.219	1.092	1.002	1.104	1.012	.9388	.8616	.8000
35	1.365	1.228	1.118	.9991	.9190	1.036	.9485	.8793	.8067	.7489
40	1.267	1.137	1.034	.9215	.8498	.9783	.8943	.8285	.7599	.7056
45	1.182	1.059	.9618	.8555	.7905	.9277	.8474	.7847	.7195	.6683
50	1.109	.9917	.8995	.7986	.7391	.8832	.8062	.7462	.6842	.6357
60	.9894	.8821	.7984	.7086	.6528	.8086	.7381	.6825	.6253	.5822
70	.8938	.7948	.7182	.6376	.5840	.7477	.6829	.6309	.5777	.5392
80	.8152	.7234	.6526	.5798	.5279	.6969	.6369	.5881	.5383	.5036
90	.7494	.6639	.5981	.5316	.4813	.6537	.5980	.5520	.5051	.4737
100	.6934	.6134	.5519	.4908	.4421	.6166	.5645	.5210	.4768	.4482
120	.6051	.5347	.4798	.4257	.3829	.5570	.5104	.4717	.4324	.4076
140	.5370	.4743	.4247	.3756	.3381	.5100	.4679	.4331	.3979	.3760
170	.4592	.4058	.3622	.3187	.2880	.4554	.4185	.3886	.3583	.3396
200	.4009	.3546	.3157	.2763	.2512	.4137	.3807	.3547	.3283	.3121
230	.3562	.3150	.2800	.2445	.2225	.3812	.3518	.3285	.3049	.2907
260	.3205	.2832	.2515	.2193	.1997	.3550	.3283	.3073	.2859	.2734
300	.2827	.2497	.2215	.1929	.1757	.3267	.3032	.2846	.2656	.2549
350	.2464	.2173	.1927	.1678	.1528	.2991	.2788	.2624	.2458	.2368
400	.2183	.1923	.1705	.1485	.1351	.2774	.2596	.2451	.2304	.2227
450	.1959	.1724	.1529	.1333	.1212	.2599	.2441	.2311	.2179	.2112
500	.1777	.1562	.1386	.1209	.1099	.2454	.2313	.2195	.2076	.2017
600	.1498	.1315	.1168	.1021	.0926	.2231	.2115	.2016	.1918	.1869
700	.1294	.1136	.1009	.0884	.0801	.2067	.1968	.1883	.1801	.1759
800	.1140	.1000	.0889	.0779	.0705	.1941	.1855	.1781	.1711	.1672
900	.1018	.0893	.0793	.0695	.0630	.1842	.1764	.1699	.1640	.1603
1000	.0919	.0807	.0716	.0625	.0568	.1763	.1691	.1633	.1583	.1545

K/KL= E,KEV	RELATIVE IONIZATION STRAGGLING					RELATIVE TRANSVERSE RANGE STRAGGLING				
	0.7	0.85	1.0	1.2	1.4	0.7	0.85	1.0	1.2	1.4
.10	2.351	2.248	2.156	2.045	1.948	.9753	.9742	.9730	.9716	.9701
.12	2.334	2.229	2.136	2.024	1.927	.9752	.9740	.9729	.9714	.9699
.14	2.318	2.213	2.118	2.006	1.908	.9750	.9739	.9727	.9712	.9697
.17	2.298	2.191	2.096	1.983	1.884	.9749	.9737	.9725	.9709	.9694
.20	2.281	2.173	2.077	1.963	1.864	.9748	.9735	.9723	.9707	.9691
.23	2.265	2.157	2.060	1.946	1.846	.9747	.9734	.9722	.9705	.9689
.26	2.251	2.142	2.044	1.930	1.831	.9746	.9733	.9720	.9704	.9687
.30	2.234	2.124	2.026	1.912	1.812	.9745	.9732	.9719	.9702	.9685
.35	2.216	2.104	2.006	1.891	1.791	.9745	.9731	.9718	.9700	.9683
.40	2.199	2.087	1.988	1.873	1.773	.9744	.9730	.9717	.9699	.9682
.45	2.183	2.071	1.971	1.856	1.756	.9743	.9729	.9716	.9698	.9680
.50	2.169	2.056	1.956	1.841	1.741	.9743	.9729	.9715	.9697	.9679
.60	2.143	2.030	1.930	1.814	1.714	.9742	.9728	.9714	.9695	.9677
.70	2.121	2.006	1.906	1.790	1.690	.9742	.9727	.9713	.9694	.9675
.80	2.100	1.985	1.885	1.769	1.669	.9742	.9727	.9712	.9693	.9674
.90	2.081	1.966	1.865	1.749	1.650	.9742	.9727	.9712	.9692	.9673
1.0	2.063	1.948	1.847	1.732	1.633	.9742	.9727	.9712	.9692	.9673
1.2	2.031	1.915	1.815	1.700	1.602	.9743	.9727	.9712	.9692	.9672
1.4	2.002	1.886	1.786	1.671	1.574	.9743	.9728	.9712	.9692	.9672
1.7	1.963	1.848	1.748	1.634	1.537	.9745	.9729	.9713	.9693	.9673
2.0	1.929	1.813	1.714	1.601	1.505	.9747	.9731	.9715	.9694	.9674
2.3	1.897	1.782	1.683	1.571	1.477	.9749	.9733	.9717	.9696	.9676
2.6	1.868	1.753	1.655	1.543	1.450	.9752	.9735	.9719	.9698	.9678
3.0	1.832	1.718	1.620	1.510	1.419	.9755	.9739	.9722	.9702	.9681
3.5	1.792	1.678	1.582	1.473	1.383	.9760	.9743	.9727	.9706	.9686
4.0	1.755	1.642	1.547	1.440	1.351	.9765	.9749	.9733	.9712	.9692
4.5	1.721	1.609	1.515	1.409	1.322	.9771	.9754	.9738	.9718	.9698
5.0	1.689	1.579	1.485	1.381	1.296	.9776	.9760	.9744	.9724	.9704
6.0	1.632	1.523	1.432	1.330	1.247	.9787	.9771	.9756	.9736	.9717
7.0	1.580	1.474	1.384	1.286	1.205	.9798	.9783	.9768	.9749	.9731
8.0	1.535	1.430	1.342	1.246	1.168	.9810	.9796	.9781	.9763	.9745
9.0	1.493	1.390	1.304	1.210	1.134	.9823	.9808	.9795	.9777	.9760
10	1.455	1.353	1.269	1.178	1.104	.9835	.9822	.9809	.9792	.9776
12	1.386	1.288	1.207	1.120	1.050	.9859	.9847	.9836	.9821	.9806
14	1.327	1.232	1.154	1.071	1.003	.9884	.9874	.9864	.9850	.9838
17	1.251	1.160	1.087	1.008	.9452	.9923	.9915	.9907	.9897	.9887
20	1.187	1.100	1.030	.9557	.8968	.9962	.9957	.9951	.9944	.9936
23	1.131	1.048	.9815	.9114	.8561	.9999	.9997	.9994	.9990	.9985
26	1.083	1.003	.9395	.8730	.8210	1.004	1.004	1.004	1.004	1.003
30	1.027	.9509	.8913	.8292	.7809	1.009	1.009	1.009	1.010	1.010
35	.9675	.8962	.8407	.7833	.7392	1.015	1.016	1.017	1.018	1.018
40	.9173	.8501	.7982	.7448	.7043	1.021	1.023	1.024	1.025	1.026
45	.8743	.8108	.7619	.7122	.6748	1.028	1.029	1.031	1.033	1.035
50	.8368	.7767	.7306	.6840	.6495	1.034	1.036	1.038	1.041	1.042
60	.7754	.7217	.6802	.6384	.6088	1.046	1.049	1.052	1.056	1.058
70	.7264	.6782	.6404	.6027	.5770	1.058	1.062	1.066	1.070	1.073
80	.6864	.6429	.6083	.5739	.5517	1.070	1.074	1.079	1.084	1.088
90	.6531	.6137	.5819	.5505	.5310	1.081	1.086	1.092	1.097	1.102
100	.6250	.5892	.5598	.5310	.5139	1.092	1.098	1.104	1.110	1.115
120	.5818	.5514	.5265	.5023	.4888	1.113	1.120	1.127	1.135	1.141
140	.5491	.5230	.5018	.4815	.4706	1.132	1.141	1.149	1.157	1.164
170	.5130	.4919	.4751	.4594	.4517	1.160	1.170	1.180	1.189	1.197
200	.4869	.4698	.4564	.4444	.4389	1.186	1.197	1.208	1.219	1.227
230	.4682	.4542	.4434	.4341	.4304	1.211	1.223	1.235	1.247	1.256
260	.4539	.4425	.4339	.4267	.4245	1.235	1.248	1.261	1.274	1.283
300	.4398	.4312	.4249	.4200	.4192	1.265	1.279	1.293	1.307	1.317
350	.4274	.4215	.4175	.4148	.4155	1.299	1.315	1.329	1.344	1.355
400	.4189	.4152	.4129	.4119	.4136	1.331	1.347	1.363	1.378	1.390
450	.4130	.4110	.4101	.4105	.4129	1.360	1.377	1.394	1.410	1.422
500	.4089	.4083	.4085	.4099	.4130	1.387	1.405	1.422	1.439	1.451
600	.4043	.4057	.4076	.4106	.4145	1.437	1.456	1.474	1.492	1.504
700	.4027	.4055	.4084	.4125	.4168	1.481	1.502	1.520	1.538	1.550
800	.4027	.4065	.4102	.4149	.4196	1.521	1.543	1.561	1.579	1.592
900	.4038	.4082	.4124	.4176	.4224	1.558	1.580	1.598	1.617	1.629
1000	.4055	.4104	.4148	.4204	.4251	1.591	1.614	1.632	1.651	1.663

TABLE 2- 5

BE ON AU, M2/M1=21.9

K/KL= E,KEV	RELATIVE TRANSV. DAMAGE STRAGGLING					RELATIVE TRANSV. IONZN. STRAGGLING				
	0.7	0.85	1.0	1.2	1.4	0.7	0.85	1.0	1.2	1.4
.10	.9478	.9445	.9413	.9370	.9327	.9411	.9373	.9334	.9282	.9230
.12	.9473	.9440	.9407	.9363	.9318	.9406	.9366	.9326	.9273	.9219
.14	.9470	.9436	.9402	.9356	.9311	.9401	.9361	.9320	.9265	.9210
.17	.9465	.9430	.9395	.9348	.9301	.9396	.9354	.9311	.9254	.9197
.20	.9462	.9425	.9389	.9341	.9292	.9391	.9347	.9304	.9245	.9187
.23	.9458	.9421	.9384	.9334	.9285	.9386	.9342	.9297	.9237	.9177
.26	.9455	.9417	.9380	.9329	.9278	.9383	.9337	.9292	.9230	.9169
.30	.9452	.9413	.9374	.9322	.9270	.9378	.9332	.9285	.9222	.9158
.35	.9448	.9408	.9369	.9315	.9262	.9373	.9325	.9277	.9212	.9147
.40	.9445	.9404	.9364	.9309	.9254	.9369	.9320	.9270	.9204	.9137
.45	.9442	.9401	.9359	.9303	.9248	.9365	.9315	.9264	.9197	.9128
.50	.9440	.9398	.9355	.9298	.9242	.9361	.9310	.9259	.9190	.9120
.60	.9436	.9392	.9348	.9290	.9231	.9355	.9302	.9249	.9178	.9106
.70	.9432	.9387	.9342	.9282	.9222	.9350	.9296	.9241	.9167	.9093
.80	.9429	.9383	.9337	.9276	.9214	.9346	.9290	.9233	.9158	.9082
.90	.9427	.9380	.9333	.9270	.9207	.9342	.9284	.9227	.9149	.9072
1.0	.9425	.9377	.9329	.9265	.9201	.9338	.9280	.9221	.9142	.9062
1.2	.9422	.9372	.9323	.9257	.9190	.9332	.9271	.9210	.9128	.9046
1.4	.9419	.9368	.9317	.9249	.9181	.9327	.9264	.9201	.9117	.9032
1.7	.9416	.9364	.9311	.9241	.9170	.9320	.9255	.9189	.9101	.9013
2.0	.9415	.9360	.9306	.9233	.9161	.9314	.9247	.9179	.9088	.8996
2.3	.9413	.9358	.9302	.9227	.9153	.9310	.9240	.9170	.9076	.8982
2.6	.9413	.9356	.9299	.9222	.9146	.9306	.9234	.9162	.9066	.8968
3.0	.9412	.9354	.9295	.9217	.9138	.9301	.9227	.9153	.9053	.8952
3.5	.9413	.9352	.9292	.9211	.9130	.9296	.9219	.9142	.9038	.8933
4.0	.9414	.9352	.9290	.9206	.9123	.9292	.9212	.9132	.9024	.8916
4.5	.9415	.9352	.9288	.9203	.9118	.9288	.9206	.9123	.9012	.8900
5.0	.9417	.9352	.9287	.9200	.9113	.9285	.9200	.9114	.9000	.8885
6.0	.9421	.9354	.9286	.9196	.9105	.9279	.9190	.9100	.8979	.8859
7.0	.9426	.9356	.9286	.9193	.9099	.9274	.9181	.9086	.8960	.8833
8.0	.9432	.9360	.9287	.9191	.9094	.9269	.9172	.9073	.8941	.8809
9.0	.9437	.9363	.9289	.9189	.9089	.9265	.9163	.9060	.8923	.8785
10	.9443	.9367	.9291	.9188	.9086	.9260	.9154	.9048	.8905	.8762
12	.9454	.9376	.9296	.9189	.9081	.9253	.9140	.9026	.8873	.8719
14	.9466	.9385	.9301	.9189	.9077	.9246	.9125	.9003	.8841	.8677
17	.9483	.9398	.9309	.9191	.9072	.9232	.9100	.8967	.8792	.8612
20	.9500	.9410	.9317	.9192	.9066	.9215	.9073	.8929	.8740	.8546
23	.9518	.9423	.9326	.9195	.9064	.9203	.9047	.8895	.8690	.8483
26	.9536	.9436	.9334	.9197	.9060	.9188	.9019	.8858	.8639	.8418
30	.9558	.9450	.9344	.9199	.9054	.9165	.8980	.8806	.8568	.8331
35	.9582	.9466	.9352	.9198	.9044	.9130	.8926	.8736	.8477	.8221
40	.9604	.9479	.9358	.9194	.9031	.9090	.8869	.8663	.8383	.8111
45	.9622	.9489	.9362	.9189	.9016	.9046	.8810	.8587	.8288	.8001
50	.9639	.9498	.9363	.9181	.9000	.8999	.8748	.8509	.8193	.7891
60	.9672	.9517	.9366	.9165	.8965	.8904	.8634	.8351	.8001	.7671
70	.9696	.9528	.9361	.9143	.8925	.8800	.8512	.8189	.7811	.7457
80	.9713	.9532	.9351	.9115	.8882	.8689	.8383	.8027	.7624	.7251
90	.9723	.9529	.9335	.9083	.8835	.8573	.8250	.7866	.7441	.7052
100	.9728	.9522	.9314	.9048	.8786	.8453	.8112	.7705	.7262	.6860
120	.9727	.9496	.9266	.8972	.8685	.8193	.7802	.7369	.6893	.6473
140	.9713	.9459	.9209	.8889	.8581	.7936	.7501	.7051	.6551	.6121
170	.9673	.9387	.9110	.8757	.8421	.7567	.7079	.6613	.6088	.5650
200	.9617	.9302	.9001	.8620	.8262	.7220	.6691	.6219	.5678	.5240
230	.9549	.9203	.8877	.8468	.8091	.6869	.6326	.5849	.5308	.4878
260	.9473	.9100	.8751	.8318	.7924	.6543	.5992	.5514	.4977	.4559
300	.9365	.8960	.8584	.8123	.7712	.6145	.5591	.5116	.4588	.4187
350	.9223	.8784	.8380	.7891	.7463	.5701	.5152	.4683	.4170	.3791
400	.9078	.8610	.8183	.7673	.7232	.5309	.4769	.4309	.3814	.3456
450	.8931	.8439	.7994	.7466	.7017	.4961	.4433	.3985	.3508	.3170
500	.8784	.8273	.7812	.7271	.6816	.4650	.4136	.3700	.3242	.2923
600	.8494	.7954	.7471	.6911	.6452	.4120	.3638	.3229	.2806	.2521
700	.8210	.7651	.7155	.6587	.6128	.3686	.3237	.2856	.2468	.2211
800	.7935	.7365	.6861	.6291	.5838	.3327	.2911	.2559	.2202	.1970
900	.7668	.7093	.6586	.6019	.5575	.3027	.2642	.2319	.1991	.1780
1000	.7410	.6833	.6328	.5768	.5334	.2774	.2419	.2124	.1824	.1631

K/KL=	RANGE SKEWNESS					DAMAGE SKEWNESS				
E,KEV	0.7	0.85	1.0	1.2	1.4	0.7	0.85	1.0	1.2	1.4
.10	.0359	.0318	.0275	.0217	.0159	.1564	.1581	.1597	.1617	.1636
.12	.0353	.0310	.0266	.0206	.0145	.1567	.1584	.1600	.1620	.1640
.14	.0348	.0303	.0257	.0196	.0133	.1569	.1586	.1602	.1623	.1643
.17	.0340	.0294	.0247	.0182	.0117	.1572	.1589	.1606	.1627	.1647
.20	.0334	.0286	.0237	.0171	.0103	.1574	.1592	.1609	.1631	.1651
.23	.0328	.0279	.0229	.0160	.0091	.1577	.1595	.1612	.1633	.1654
.26	.0323	.0272	.0221	.0151	.0080	.1579	.1597	.1614	.1636	.1657
.30	.0317	.0265	.0211	.0139	.0066	.1581	.1599	.1617	.1639	.1659
.35	.0310	.0256	.0201	.0126	.0050	.1583	.1602	.1619	.1642	.1663
.40	.0303	.0247	.0191	.0114	.0036	.1586	.1604	.1622	.1644	.1665
.45	.0297	.0240	.0182	.0103	.0023	.1587	.1606	.1624	.1646	.1667
.50	.0291	.0233	.0173	.0092	.0011	.1589	.1608	.1626	.1648	.1669
.60	.0281	.0220	.0158	.0074	-.0011	.1592	.1611	.1629	.1651	.1672
.70	.0272	.0208	.0144	.0057	-.0032	.1595	.1613	.1631	.1653	.1675
.80	.0263	.0197	.0131	.0041	-.0050	.1597	.1615	.1633	.1655	.1676
.90	.0255	.0187	.0119	.0026	-.0068	.1598	.1617	.1634	.1657	.1677
1.0	.0247	.0177	.0107	.0012	-.0084	.1600	.1618	.1636	.1658	.1678
1.2	.0233	.0160	.0086	-.0014	-.0114	.1602	.1620	.1638	.1659	.1679
1.4	.0219	.0143	.0066	-.0037	-.0142	.1604	.1622	.1639	.1660	.1679
1.7	.0201	.0120	.0039	-.0071	-.0181	.1606	.1623	.1639	.1659	.1678
2.0	.0183	.0098	.0013	-.0102	-.0217	.1607	.1623	.1638	.1657	.1675
2.3	.0167	.0078	-.0011	-.0130	-.0250	.1607	.1622	.1637	.1655	.1672
2.6	.0151	.0059	-.0034	-.0157	-.0282	.1607	.1621	.1635	.1652	.1669
3.0	.0131	.0034	-.0063	-.0193	-.0322	.1605	.1619	.1631	.1647	.1662
3.5	.0106	.0004	-.0099	-.0235	-.0371	.1602	.1614	.1626	.1640	.1654
4.0	.0081	-.0026	-.0134	-.0276	-.0418	.1599	.1609	.1619	.1632	.1644
4.5	.0057	-.0056	-.0168	-.0316	-.0464	.1594	.1603	.1611	.1622	.1633
5.0	.0033	-.0084	-.0201	-.0355	-.0508	.1589	.1596	.1603	.1613	.1622
6.0	-.0012	-.0138	-.0263	-.0428	-.0591	.1579	.1584	.1588	.1594	.1600
7.0	-.0056	-.0191	-.0324	-.0500	-.0672	.1568	.1569	.1571	.1574	.1577
8.0	-.0101	-.0243	-.0384	-.0569	-.0750	.1554	.1553	.1552	.1552	.1552
9.0	-.0145	-.0295	-.0443	-.0637	-.0826	.1540	.1535	.1532	.1529	.1527
10	-.0189	-.0347	-.0502	-.0704	-.0901	.1524	.1517	.1511	.1505	.1500
12	-.0273	-.0445	-.0613	-.0831	-.1043	.1491	.1478	.1468	.1455	.1445
14	-.0356	-.0542	-.0723	-.0954	-.1180	.1456	.1438	.1423	.1404	.1388
17	-.0480	-.0685	-.0882	-.1133	-.1377	.1401	.1376	.1354	.1328	.1304
20	-.0601	-.0824	-.1037	-.1306	-.1566	.1345	.1313	.1285	.1252	.1222
23	-.0718	-.0956	-.1186	-.1474	-.1745	.1290	.1252	.1219	.1183	.1143
26	-.0832	-.1086	-.1330	-.1636	-.1918	.1236	.1191	.1153	.1114	.1067
30	-.0981	-.1254	-.1517	-.1844	-.2138	.1163	.1111	.1066	.1023	.0967
35	-.1164	-.1458	-.1742	-.2092	-.2400	.1074	.1013	.0960	.0912	.0846
40	-.1343	-.1655	-.1958	-.2329	-.2650	.0986	.0917	.0857	.0804	.0728
45	-.1517	-.1846	-.2165	-.2555	-.2888	.0899	.0824	.0756	.0697	.0614
50	-.1687	-.2030	-.2365	-.2772	-.3116	.0814	.0732	.0658	.0593	.0503
60	-.2031	-.2398	-.2755	-.3193	-.3558	.0648	.0553	.0469	.0385	.0288
70	-.2352	-.2740	-.3116	-.3580	-.3965	.0490	.0384	.0289	.0188	.0085
80	-.2651	-.3058	-.3452	-.3938	-.4340	.0339	.0223	.0118	.0002	-.0108
90	-.2931	-.3356	-.3766	-.4272	-.4688	.0195	.0069	-.0045	-.0176	-.0292
100	-.3193	-.3635	-.4060	-.4584	-.5013	.0058	-.0079	-.0201	-.0346	-.0467
120	-.3658	-.4138	-.4593	-.5147	-.5592	-.0196	-.0352	-.0492	-.0659	-.0798
140	-.4080	-.4593	-.5075	-.5655	-.6113	-.0433	-.0606	-.0763	-.0949	-.1104
170	-.4652	-.5210	-.5725	-.6337	-.6813	-.0764	-.0961	-.1140	-.1351	-.1528
200	-.5170	-.5767	-.6308	-.6948	-.7441	-.1071	-.1290	-.1489	-.1721	-.1916
230	-.5656	-.6289	-.6845	-.7508	-.8025	-.1364	-.1606	-.1824	-.2078	-.2289
260	-.6106	-.6771	-.7339	-.8022	-.8562	-.1640	-.1903	-.2137	-.2411	-.2635
300	-.6659	-.7361	-.7943	-.8650	-.9219	-.1986	-.2272	-.2527	-.2822	-.3061
350	-.7290	-.8030	-.8627	-.9360	-.9962	-.2386	-.2698	-.2974	-.3292	-.3545
400	-.7864	-.8636	-.9248	-1.000	-1.063	-.2756	-.3090	-.3383	-.3720	-.3983
450	-.8393	-.9190	-.9818	-1.059	-1.125	-.3102	-.3454	-.3762	-.4113	-.4384
500	-.8883	-.9701	-1.035	-1.114	-1.182	-.3426	-.3793	-.4114	-.4477	-.4753
600	-.9770	-1.062	-1.130	-1.212	-1.283	-.4022	-.4412	-.4752	-.5132	-.5414
700	-1.056	-1.143	-1.214	-1.299	-1.373	-.4560	-.4966	-.5319	-.5709	-.5992
800	-1.127	-1.215	-1.290	-1.378	-1.454	-.5052	-.5468	-.5831	-.6227	-.6505
900	-1.192	-1.281	-1.359	-1.449	-1.527	-.5507	-.5929	-.6298	-.6695	-.6968
1000	-1.251	-1.341	-1.423	-1.515	-1.594	-.5930	-.6355	-.6728	-.7123	-.7387

TABLE 2- 7

BE ON AU, M2/M1=21.9

K/KL=	IONIZATION SKEWNESS					RANGE KURTOSIS				
E,KEV	0.7	0.85	1.0	1.2	1.4	0.7	0.85	1.0	1.2	1.4
.10	.1853	.1888	.1922	.1966	.2009	3.080	3.062	3.045	3.024	3.005
.12	.1859	.1895	.1930	.1975	.2019	3.077	3.059	3.041	3.020	3.000
.14	.1865	.1901	.1936	.1983	.2028	3.075	3.056	3.038	3.017	2.997
.17	.1872	.1909	.1945	.1993	.2040	3.071	3.052	3.034	3.012	2.992
.20	.1878	.1916	.1953	.2002	.2050	3.069	3.049	3.031	3.009	2.988
.23	.1883	.1922	.1960	.2010	.2059	3.066	3.046	3.028	3.005	2.985
.26	.1888	.1928	.1967	.2017	.2067	3.064	3.044	3.025	3.002	2.982
.30	.1894	.1935	.1974	.2026	.2077	3.061	3.041	3.022	2.999	2.978
.35	.1901	.1943	.1983	.2036	.2088	3.058	3.037	3.018	2.995	2.974
.40	.1908	.1950	.1991	.2045	.2098	3.056	3.034	3.015	2.991	2.970
.45	.1914	.1957	.1998	.2053	.2107	3.053	3.032	3.012	2.988	2.967
.50	.1919	.1963	.2005	.2061	.2115	3.051	3.029	3.009	2.985	2.964
.60	.1929	.1974	.2017	.2074	.2130	3.047	3.025	3.005	2.980	2.958
.70	.1938	.1984	.2028	.2087	.2143	3.043	3.021	3.001	2.976	2.954
.80	.1947	.1993	.2038	.2098	.2155	3.040	3.017	2.997	2.972	2.950
.90	.1954	.2002	.2048	.2108	.2166	3.037	3.014	2.993	2.968	2.946
1.0	.1962	.2010	.2056	.2117	.2177	3.034	3.011	2.990	2.965	2.942
1.2	.1975	.2024	.2072	.2135	.2196	3.029	3.006	2.984	2.959	2.936
1.4	.1987	.2037	.2086	.2150	.2212	3.025	3.001	2.979	2.953	2.931
1.7	.2004	.2055	.2106	.2171	.2235	3.019	2.994	2.972	2.946	2.923
2.0	.2019	.2072	.2123	.2190	.2255	3.013	2.988	2.966	2.940	2.917
2.3	.2033	.2087	.2139	.2207	.2274	3.008	2.983	2.961	2.934	2.911
2.6	.2046	.2101	.2154	.2223	.2291	3.004	2.978	2.956	2.929	2.906
3.0	.2063	.2118	.2172	.2243	.2312	2.998	2.972	2.949	2.923	2.899
3.5	.2081	.2138	.2193	.2266	.2336	2.991	2.965	2.942	2.915	2.892
4.0	.2099	.2156	.2213	.2287	.2359	2.985	2.959	2.936	2.909	2.886
4.5	.2115	.2174	.2231	.2306	.2380	2.979	2.953	2.929	2.903	2.880
5.0	.2131	.2190	.2248	.2325	.2399	2.973	2.947	2.924	2.897	2.875
6.0	.2159	.2220	.2280	.2359	.2437	2.963	2.936	2.913	2.887	2.865
7.0	.2185	.2248	.2309	.2390	.2470	2.953	2.927	2.904	2.878	2.857
8.0	.2209	.2273	.2336	.2419	.2501	2.945	2.918	2.896	2.870	2.850
9.0	.2231	.2296	.2361	.2446	.2529	2.937	2.910	2.888	2.863	2.844
10	.2251	.2318	.2384	.2471	.2554	2.929	2.903	2.881	2.857	2.838
12	.2287	.2357	.2426	.2517	.2597	2.916	2.889	2.868	2.846	2.829
14	.2320	.2392	.2464	.2558	.2635	2.903	2.877	2.857	2.837	2.821
17	.2365	.2441	.2516	.2615	.2687	2.888	2.863	2.844	2.827	2.814
20	.2407	.2486	.2564	.2667	.2736	2.875	2.851	2.835	2.820	2.811
23	.2450	.2535	.2615	.2721	.2794	2.862	2.841	2.825	2.814	2.808
26	.2491	.2580	.2663	.2771	.2848	2.851	2.832	2.819	2.810	2.807
30	.2541	.2636	.2721	.2830	.2913	2.839	2.823	2.813	2.808	2.809
35	.2598	.2698	.2785	.2896	.2984	2.828	2.816	2.810	2.810	2.817
40	.2649	.2752	.2842	.2952	.3045	2.820	2.813	2.811	2.816	2.828
45	.2694	.2801	.2892	.3002	.3097	2.815	2.813	2.814	2.824	2.842
50	.2735	.2844	.2936	.3045	.3141	2.813	2.815	2.820	2.835	2.859
60	.2803	.2917	.3011	.3119	.3209	2.809	2.820	2.835	2.860	2.894
70	.2861	.2977	.3071	.3176	.3258	2.813	2.832	2.856	2.891	2.935
80	.2912	.3027	.3119	.3220	.3295	2.822	2.850	2.882	2.927	2.980
90	.2959	.3069	.3158	.3253	.3320	2.835	2.872	2.911	2.967	3.028
100	.3001	.3103	.3189	.3278	.3338	2.852	2.897	2.944	3.008	3.078
120	.3097	.3164	.3233	.3301	.3345	2.893	2.952	3.012	3.094	3.180
140	.3173	.3207	.3259	.3309	.3339	2.942	3.014	3.088	3.185	3.286
170	.3255	.3248	.3277	.3304	.3318	3.024	3.116	3.208	3.328	3.450
200	.3305	.3266	.3266	.3278	.3292	3.113	3.223	3.333	3.476	3.617
230	.3285	.3237	.3241	.3246	.3251	3.202	3.329	3.455	3.619	3.778
250	.3252	.3203	.3201	.3204	.3213	3.294	3.437	3.580	3.763	3.940
300	.3201	.3153	.3149	.3152	.3169	3.419	3.583	3.747	3.956	4.155
350	.3135	.3095	.3090	.3097	.3124	3.580	3.769	3.958	4.198	4.422
400	.3072	.3043	.3040	.3053	.3091	3.742	3.956	4.169	4.438	4.688
450	.3017	.2998	.3000	.3021	.3069	3.904	4.143	4.379	4.675	4.949
500	.2969	.2962	.2969	.2998	.3056	4.066	4.328	4.586	4.910	5.207
600	.2896	.2911	.2932	.2978	.3053	4.385	4.694	4.994	5.368	5.710
700	.2849	.2885	.2921	.2986	.3074	4.697	5.050	5.390	5.812	6.196
800	.2824	.2879	.2931	.3014	.3114	5.001	5.397	5.774	6.241	6.665
900	.2818	.2889	.2959	.3058	.3167	5.296	5.734	6.146	6.656	7.118
1000	.2827	.2913	.2999	.3114	.3231	5.583	6.061	6.507	7.057	7.556

K/KL= E,KEV	DAMAGE KURTOSIS					IONIZATION KURTOSIS				
	0.7	0.85	1.0	1.2	1.4	0.7	0.85	1.0	1.2	1.4
.10	4.130	4.110	4.092	4.069	4.048	4.279	4.261	4.245	4.223	4.204
.12	4.126	4.106	4.087	4.064	4.042	4.276	4.258	4.241	4.219	4.199
.14	4.123	4.102	4.083	4.060	4.037	4.274	4.255	4.237	4.215	4.195
.17	4.118	4.097	4.078	4.054	4.031	4.270	4.251	4.233	4.210	4.189
.20	4.114	4.093	4.073	4.049	4.026	4.267	4.247	4.229	4.206	4.185
.23	4.111	4.089	4.069	4.044	4.021	4.264	4.244	4.225	4.202	4.181
.26	4.108	4.086	4.065	4.040	4.016	4.261	4.241	4.222	4.199	4.177
.30	4.104	4.081	4.061	4.035	4.011	4.258	4.238	4.218	4.195	4.172
.35	4.099	4.076	4.055	4.029	4.005	4.254	4.234	4.214	4.190	4.167
.40	4.095	4.072	4.050	4.024	3.999	4.251	4.230	4.210	4.185	4.162
.45	4.091	4.067	4.046	4.019	3.994	4.248	4.227	4.206	4.181	4.158
.50	4.087	4.063	4.041	4.014	3.989	4.245	4.223	4.203	4.178	4.154
.60	4.080	4.056	4.033	4.006	3.980	4.240	4.218	4.197	4.171	4.147
.70	4.074	4.049	4.026	3.998	3.972	4.235	4.212	4.191	4.165	4.140
.80	4.068	4.043	4.019	3.991	3.964	4.231	4.208	4.186	4.159	4.134
.90	4.062	4.036	4.013	3.984	3.957	4.226	4.203	4.181	4.154	4.128
1.0	4.056	4.031	4.007	3.977	3.950	4.223	4.199	4.177	4.149	4.123
1.2	4.046	4.020	3.996	3.966	3.938	4.215	4.191	4.168	4.140	4.113
1.4	4.037	4.010	3.985	3.955	3.927	4.209	4.184	4.160	4.131	4.104
1.7	4.024	3.996	3.971	3.940	3.911	4.199	4.173	4.150	4.120	4.092
2.0	4.012	3.984	3.958	3.926	3.896	4.190	4.164	4.140	4.109	4.081
2.3	4.000	3.971	3.945	3.912	3.882	4.182	4.155	4.130	4.099	4.070
2.6	3.989	3.960	3.933	3.900	3.869	4.175	4.147	4.122	4.090	4.060
3.0	3.975	3.945	3.917	3.884	3.853	4.165	4.137	4.111	4.078	4.048
3.5	3.958	3.928	3.900	3.865	3.834	4.154	4.125	4.098	4.064	4.033
4.0	3.942	3.911	3.883	3.848	3.816	4.143	4.113	4.085	4.051	4.019
4.5	3.928	3.896	3.867	3.831	3.799	4.132	4.102	4.074	4.038	4.006
5.0	3.913	3.881	3.851	3.816	3.783	4.122	4.091	4.062	4.026	3.993
6.0	3.887	3.854	3.823	3.786	3.753	4.103	4.071	4.041	4.004	3.969
7.0	3.862	3.828	3.797	3.759	3.725	4.086	4.052	4.021	3.983	3.946
8.0	3.839	3.804	3.773	3.734	3.699	4.069	4.035	4.003	3.963	3.925
9.0	3.817	3.782	3.750	3.711	3.675	4.053	4.018	3.985	3.944	3.905
10	3.797	3.761	3.728	3.689	3.653	4.038	4.002	3.968	3.925	3.886
12	3.758	3.720	3.687	3.647	3.610	4.010	3.972	3.936	3.892	3.850
14	3.722	3.684	3.650	3.608	3.571	3.984	3.944	3.907	3.860	3.817
17	3.674	3.635	3.600	3.557	3.519	3.948	3.905	3.865	3.815	3.770
20	3.631	3.591	3.555	3.512	3.473	3.913	3.868	3.826	3.773	3.726
23	3.592	3.551	3.515	3.471	3.431	3.881	3.834	3.789	3.733	3.683
26	3.557	3.515	3.478	3.434	3.393	3.851	3.801	3.754	3.695	3.642
30	3.514	3.472	3.434	3.389	3.348	3.812	3.759	3.709	3.647	3.590
35	3.467	3.424	3.386	3.340	3.298	3.767	3.710	3.657	3.590	3.530
40	3.424	3.381	3.342	3.296	3.253	3.723	3.663	3.607	3.538	3.474
45	3.387	3.343	3.304	3.257	3.214	3.683	3.620	3.561	3.489	3.422
50	3.352	3.308	3.268	3.221	3.178	3.644	3.578	3.517	3.442	3.373
60	3.291	3.246	3.206	3.157	3.113	3.567	3.496	3.431	3.352	3.280
70	3.239	3.193	3.152	3.103	3.059	3.498	3.423	3.354	3.271	3.198
80	3.194	3.148	3.106	3.056	3.013	3.434	3.356	3.284	3.198	3.124
90	3.155	3.108	3.066	3.016	2.973	3.375	3.294	3.220	3.132	3.058
100	3.120	3.073	3.031	2.981	2.938	3.321	3.238	3.162	3.072	2.998
120	3.060	3.013	2.972	2.922	2.880	3.222	3.134	3.056	2.965	2.892
140	3.012	2.965	2.924	2.875	2.833	3.135	3.044	2.964	2.873	2.803
170	2.955	2.909	2.868	2.821	2.781	3.023	2.930	2.849	2.760	2.693
200	2.910	2.866	2.827	2.782	2.744	2.928	2.834	2.754	2.667	2.605
230	2.873	2.831	2.795	2.753	2.718	2.844	2.753	2.675	2.593	2.536
260	2.844	2.804	2.770	2.732	2.700	2.770	2.683	2.609	2.532	2.479
300	2.813	2.778	2.747	2.713	2.685	2.685	2.603	2.535	2.464	2.417
350	2.787	2.756	2.729	2.700	2.677	2.596	2.521	2.459	2.396	2.355
400	2.769	2.743	2.720	2.696	2.677	2.523	2.455	2.398	2.341	2.307
450	2.759	2.737	2.718	2.699	2.684	2.462	2.399	2.348	2.297	2.268
500	2.754	2.737	2.721	2.707	2.695	2.410	2.353	2.307	2.262	2.236
600	2.756	2.747	2.739	2.732	2.726	2.329	2.281	2.244	2.207	2.189
700	2.770	2.768	2.766	2.766	2.766	2.269	2.230	2.199	2.170	2.157
800	2.791	2.795	2.800	2.806	2.811	2.226	2.193	2.168	2.145	2.136
900	2.818	2.829	2.839	2.850	2.858	2.195	2.167	2.146	2.127	2.122
1000	2.848	2.865	2.880	2.896	2.908	2.173	2.149	2.131	2.117	2.114

TABLE 2- 9

BE ON AU, M2/M1=21.9

E,KEV	REFLECTION COEFFICIENT					SPUTTERING EFFICIENCY				
K/KL=	0.7	0.85	1.0	1.2	1.4	0.7	0.85	1.0	1.2	1.4
.10	.3127	.3084	.3043	.2991	.2941	.1927	.1787	.1662	.1515	.1386
.12	.3120	.3076	.3034	.2973	.2941	.1904	.1763	.1636	.1489	.1360
.14	.3113	.3069	.3025	.2963	.2934	.1885	.1742	.1615	.1466	.1337
.17	.3105	.3059	.3015	.2954	.2916	.1861	.1716	.1588	.1438	.1309
.20	.3098	.3050	.3005	.2948	.2894	.1840	.1694	.1564	.1414	.1285
.23	.3091	.3043	.2997	.2943	.2871	.1822	.1674	.1544	.1393	.1264
.26	.3085	.3036	.2989	.2939	.2847	.1805	.1657	.1526	.1375	.1246
.30	.3077	.3027	.2980	.2931	.2816	.1786	.1636	.1505	.1353	.1224
.35	.3069	.3018	.2969	.2921	.2780	.1764	.1614	.1482	.1330	.1200
.40	.3061	.3009	.2960	.2909	.2745	.1745	.1594	.1461	.1309	.1180
.45	.3054	.3001	.2951	.2894	.2713	.1728	.1576	.1443	.1291	.1161
.50	.3047	.2994	.2943	.2879	.2683	.1713	.1560	.1426	.1274	.1145
.60	.3035	.2980	.2932	.2833	.2631	.1686	.1531	.1397	.1245	.1116
.70	.3024	.2968	.2921	.2790	.2585	.1662	.1507	.1372	.1220	.1091
.80	.3013	.2956	.2908	.2750	.2544	.1641	.1485	.1350	.1198	.1070
.90	.3004	.2946	.2894	.2713	.2507	.1621	.1465	.1330	.1178	.1050
1.0	.2995	.2936	.2880	.2679	.2473	.1604	.1447	.1312	.1160	.1033
1.2	.2978	.2922	.2845	.2627	.2417	.1573	.1415	.1280	.1128	.1002
1.4	.2963	.2908	.2810	.2581	.2369	.1545	.1387	.1252	.1101	.0975
1.7	.2941	.2883	.2757	.2520	.2307	.1509	.1351	.1216	.1065	.0941
2.0	.2921	.2855	.2705	.2466	.2253	.1478	.1319	.1184	.1035	.0912
2.3	.2908	.2816	.2651	.2411	.2204	.1449	.1290	.1156	.1008	.0886
2.6	.2895	.2776	.2600	.2361	.2160	.1424	.1265	.1131	.0983	.0863
3.0	.2875	.2723	.2536	.2299	.2107	.1393	.1234	.1101	.0954	.0835
3.5	.2847	.2660	.2464	.2232	.2049	.1358	.1199	.1067	.0922	.0804
4.0	.2816	.2601	.2399	.2172	.1997	.1326	.1168	.1037	.0893	.0777
4.5	.2782	.2545	.2340	.2119	.1951	.1297	.1140	.1009	.0867	.0753
5.0	.2745	.2492	.2287	.2072	.1909	.1270	.1114	.0984	.0844	.0731
6.0	.2647	.2397	.2201	.1998	.1834	.1222	.1066	.0939	.0801	.0692
7.0	.2553	.2312	.2126	.1934	.1769	.1179	.1025	.0899	.0764	.0658
8.0	.2467	.2236	.2060	.1878	.1712	.1140	.0988	.0864	.0732	.0627
9.0	.2389	.2169	.2000	.1827	.1662	.1105	.0954	.0832	.0702	.0601
10	.2317	.2107	.1945	.1779	.1616	.1072	.0923	.0803	.0676	.0576
12	.2200	.2002	.1841	.1689	.1537	.1013	.0867	.0750	.0628	.0533
14	.2100	.1913	.1752	.1610	.1468	.0960	.0818	.0705	.0587	.0496
17	.1973	.1799	.1637	.1508	.1381	.0892	.0755	.0646	.0535	.0449
20	.1865	.1703	.1541	.1421	.1306	.0834	.0701	.0597	.0491	.0410
23	.1770	.1621	.1464	.1345	.1237	.0782	.0653	.0554	.0453	.0377
26	.1686	.1548	.1396	.1278	.1176	.0736	.0611	.0517	.0421	.0348
30	.1587	.1463	.1318	.1200	.1104	.0681	.0563	.0473	.0383	.0316
35	.1482	.1371	.1235	.1117	.1027	.0623	.0511	.0427	.0344	.0282
40	.1392	.1291	.1163	.1046	.0961	.0573	.0466	.0388	.0311	.0253
45	.1314	.1221	.1100	.0984	.0903	.0529	.0428	.0354	.0282	.0229
50	.1245	.1158	.1044	.0929	.0852	.0491	.0395	.0325	.0258	.0208
60	.1129	.1044	.0942	.0837	.0769	.0427	.0340	.0276	.0218	.0176
70	.1035	.0951	.0858	.0762	.0702	.0374	.0297	.0238	.0187	.0151
80	.0957	.0872	.0787	.0700	.0646	.0332	.0261	.0207	.0161	.0130
90	.0891	.0806	.0727	.0647	.0598	.0296	.0232	.0181	.0141	.0114
100	.0835	.0749	.0676	.0603	.0557	.0265	.0207	.0160	.0124	.0101
120	.0748	.0665	.0598	.0535	.0492	.0219	.0170	.0130	.0100	.0081
140	.0681	.0601	.0538	.0482	.0441	.0184	.0142	.0109	.0083	.0067
170	.0602	.0527	.0469	.0422	.0381	.0146	.0112	.0086	.0064	.0052
200	.0541	.0471	.0416	.0375	.0335	.0118	.0090	.0070	.0052	.0042
230	.0489	.0424	.0374	.0336	.0299	.0098	.0075	.0058	.0043	.0034
260	.0445	.0386	.0339	.0304	.0269	.0083	.0063	.0049	.0035	.0028
300	.0396	.0344	.0301	.0269	.0238	.0067	.0051	.0039	.0028	.0023
350	.0347	.0301	.0264	.0233	.0206	.0053	.0040	.0031	.0022	.0018
400	.0307	.0268	.0234	.0205	.0181	.0044	.0033	.0025	.0018	.0014
450	.0274	.0240	.0209	.0182	.0161	.0036	.0027	.0021	.0015	.0012
500	.0247	.0217	.0189	.0162	.0144	.0031	.0023	.0017	.0012	.0010
600	.0204	.0180	.0156	.0132	.0117	.0023	.0017	.0013	.0009	.0008
700	.0172	.0152	.0131	.0109	.0097	.0018	.0013	.0010	.0007	.0006
800	.0148	.0130	.0112	.0091	.0081	.0014	.0011	.0008	.0006	.0005
900	.0130	.0112	.0096	.0077	.0068	.0011	.0009	.0006	.0005	.0004
1000	.0116	.0098	.0082	.0066	.0058	.0009	.0007	.0005	.0004	.0003

BE ON AU, M2/M1=21.9   TABLE   2-10

	IONIZATION DEFICIENCY					SPUTTERING YIELD ALPHA				
K/KL=	0.7	0.85	1.0	1.2	1.4	0.7	0.85	1.0	1.2	1.4
E,KEV										
.10	.0599	.0675	.0740	.0810	.0867	1.877	1.814	1.755	1.683	1.615
.12	.0611	.0687	.0751	.0822	.0877	1.868	1.804	1.744	1.670	1.602
.14	.0621	.0697	.0761	.0831	.0886	1.861	1.795	1.735	1.659	1.590
.17	.0634	.0710	.0774	.0843	.0897	1.851	1.784	1.722	1.646	1.575
.20	.0644	.0721	.0784	.0852	.0905	1.843	1.775	1.712	1.634	1.563
.23	.0653	.0730	.0793	.0861	.0913	1.836	1.767	1.703	1.624	1.552
.26	.0661	.0738	.0801	.0867	.0919	1.829	1.759	1.695	1.615	1.542
.30	.0671	.0747	.0810	.0875	.0926	1.822	1.751	1.685	1.605	1.531
.35	.0681	.0757	.0819	.0884	.0933	1.814	1.742	1.675	1.593	1.519
.40	.0689	.0765	.0827	.0891	.0939	1.807	1.734	1.666	1.583	1.508
.45	.0697	.0773	.0834	.0897	.0944	1.800	1.726	1.658	1.575	1.499
.50	.0704	.0779	.0840	.0902	.0948	1.795	1.720	1.651	1.567	1.490
.60	.0715	.0790	.0850	.0910	.0955	1.785	1.709	1.639	1.553	1.476
.70	.0725	.0799	.0858	.0917	.0960	1.777	1.700	1.628	1.542	1.463
.80	.0733	.0807	.0865	.0923	.0964	1.770	1.691	1.619	1.532	1.452
.90	.0741	.0814	.0870	.0927	.0967	1.764	1.684	1.611	1.523	1.443
1.0	.0747	.0819	.0875	.0931	.0970	1.758	1.678	1.604	1.515	1.434
1.2	.0758	.0829	.0884	.0937	.0973	1.749	1.666	1.591	1.501	1.419
1.4	.0767	.0837	.0890	.0941	.0976	1.741	1.657	1.581	1.489	1.407
1.7	.0778	.0846	.0897	.0946	.0977	1.730	1.645	1.567	1.474	1.390
2.0	.0786	.0853	.0902	.0948	.0978	1.722	1.635	1.556	1.461	1.377
2.3	.0793	.0858	.0906	.0950	.0977	1.715	1.626	1.546	1.450	1.365
2.6	.0799	.0863	.0909	.0951	.0976	1.708	1.619	1.538	1.441	1.355
3.0	.0806	.0867	.0912	.0950	.0973	1.701	1.610	1.528	1.429	1.342
3.5	.0812	.0871	.0913	.0949	.0968	1.693	1.600	1.517	1.417	1.329
4.0	.0817	.0874	.0913	.0946	.0963	1.686	1.592	1.507	1.406	1.317
4.5	.0820	.0875	.0913	.0943	.0957	1.680	1.584	1.498	1.396	1.306
5.0	.0823	.0876	.0911	.0939	.0951	1.674	1.577	1.490	1.387	1.297
6.0	.0827	.0876	.0907	.0930	.0938	1.664	1.565	1.476	1.372	1.280
7.0	.0828	.0874	.0902	.0920	.0925	1.655	1.554	1.464	1.358	1.265
8.0	.0829	.0871	.0895	.0910	.0912	1.647	1.544	1.453	1.345	1.251
9.0	.0828	.0866	.0888	.0900	.0898	1.640	1.535	1.442	1.334	1.239
10	.0826	.0862	.0881	.0889	.0885	1.633	1.527	1.433	1.323	1.227
12	.0821	.0851	.0864	.0867	.0858	1.620	1.511	1.415	1.303	1.207
14	.0815	.0839	.0848	.0845	.0832	1.608	1.497	1.399	1.285	1.188
17	.0803	.0820	.0823	.0814	.0796	1.591	1.476	1.376	1.261	1.162
20	.0790	.0800	.0798	.0784	.0762	1.574	1.457	1.355	1.238	1.138
23	.0775	.0780	.0773	.0754	.0729	1.557	1.437	1.335	1.217	1.116
26	.0759	.0759	.0748	.0726	.0698	1.540	1.419	1.315	1.198	1.095
30	.0739	.0733	.0718	.0692	.0661	1.518	1.395	1.291	1.173	1.069
35	.0714	.0701	.0681	.0651	.0618	1.492	1.366	1.262	1.144	1.039
40	.0690	.0672	.0648	.0614	.0579	1.467	1.339	1.234	1.116	1.011
45	.0666	.0644	.0617	.0579	.0545	1.442	1.314	1.207	1.089	.9855
50	.0644	.0617	.0588	.0547	.0513	1.419	1.289	1.182	1.064	.9614
60	.0599	.0567	.0533	.0482	.0455	1.372	1.242	1.128	1.008	.9126
70	.0558	.0522	.0486	.0427	.0406	1.328	1.198	1.080	.9597	.8705
80	.0520	.0482	.0444	.0381	.0365	1.287	1.159	1.037	.9172	.8340
90	.0487	.0447	.0408	.0341	.0329	1.249	1.122	.9996	.8802	.8022
100	.0456	.0415	.0376	.0308	.0299	1.214	1.089	.9659	.8478	.7745
120	.0400	.0361	.0324	.0264	.0253	1.144	1.026	.9115	.8002	.7331
140	.0354	.0316	.0282	.0231	.0218	1.082	.9718	.8666	.7624	.7000
170	.0296	.0262	.0233	.0195	.0178	1.005	.9041	.8119	.7173	.6605
200	.0250	.0219	.0194	.0169	.0148	.9413	.8489	.7678	.6814	.6289
230	.0217	.0187	.0166	.0147	.0126	.8953	.8084	.7320	.6497	.6011
250	.0190	.0160	.0143	.0129	.0108	.8569	.7747	.7018	.6227	.5774
300	.0161	.0131	.0119	.0110	.0090	.8145	.7373	.6678	.5923	.5505
350	.0133	.0102	.0095	.0091	.0073	.7712	.6991	.6329	.5610	.5227
400	.0110	.0080	.0077	.0076	.0060	.7356	.6677	.6040	.5352	.4996
450	.0092	.0063	.0062	.0064	.0050	.7055	.6411	.5797	.5136	.4800
500	.0077	.0049	.0051	.0054	.0041	.6794	.6182	.5587	.4953	.4633
600	.0053	.0029	.0033	.0038	.0028	.6362	.5800	.5244	.4660	.4360
700	.0035	.0016	.0021	.0025	.0019	.6012	.5490	.4974	.4437	.4147
800	.0021	.0008	.0012	.0015	.0011	.5718	.5228	.4754	.4265	.3978
900	.0009	.0003	.0005	.0007	.0005	.5463	.5002	.4570	.4131	.3841
1000	0.0000	0.0000	0.0000	0.0000	0.0000	.5239	.4803	.4415	.4024	.3728

BE ON AU, M2/M1=21.9

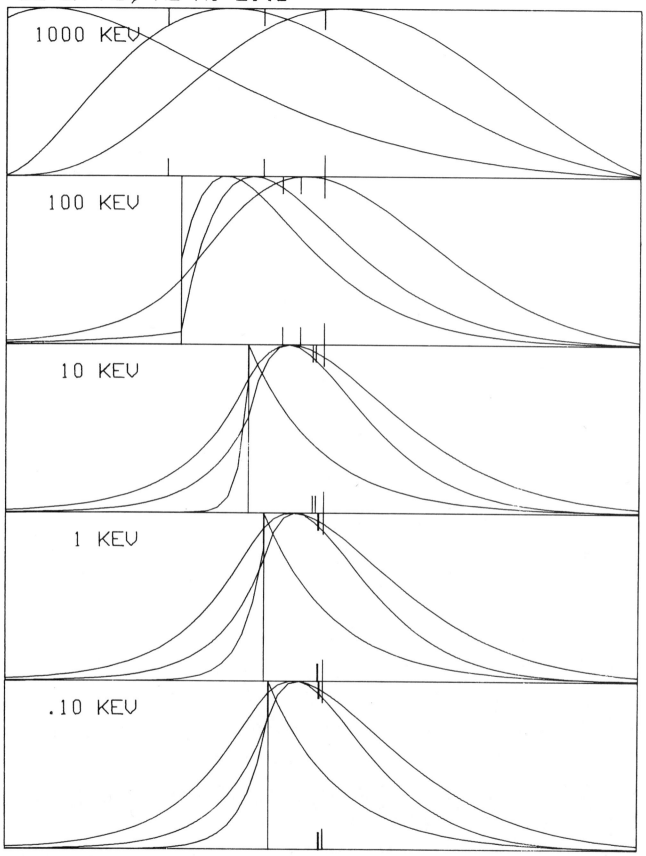

TABLE 3- 1

C ON AU, M2/M1=16.4

K/KL= E,KEV	FRACTIONAL DEPOSITED ENERGY					MEAN RANGE,MICROGRAM/SQ.CM.				
	0.6	0.8	1.0	1.25	1.6	0.6	0.8	1.0	1.25	1.6
.20	.7786	.7232	.6743	.6206	.5568	1.081	1.074	1.067	1.058	1.046
.23	.7744	.7182	.6688	.6147	.5504	1.197	1.189	1.180	1.170	1.156
.26	.7706	.7138	.6639	.6094	.5448	1.303	1.294	1.284	1.273	1.257
.30	.7661	.7085	.6581	.6031	.5382	1.434	1.423	1.413	1.400	1.382
.35	.7611	.7028	.6518	.5963	.5311	1.587	1.574	1.562	1.548	1.528
.40	.7568	.6977	.6462	.5904	.5248	1.731	1.718	1.705	1.688	1.666
.45	.7529	.6932	.6413	.5851	.5193	1.871	1.856	1.841	1.823	1.799
.50	.7493	.6891	.6368	.5803	.5143	2.006	1.990	1.974	1.954	1.928
.60	.7430	.6819	.6289	.5720	.5056	2.266	2.248	2.229	2.207	2.176
.70	.7376	.6757	.6222	.5648	.4981	2.517	2.496	2.475	2.449	2.414
.80	.7328	.6702	.6162	.5585	.4916	2.760	2.735	2.712	2.683	2.643
.90	.7285	.6653	.6109	.5529	.4859	2.995	2.968	2.942	2.909	2.866
1.0	.7245	.6608	.6061	.5478	.4806	3.224	3.194	3.165	3.129	3.081
1.2	.7176	.6529	.5977	.5390	.4716	3.650	3.615	3.581	3.539	3.483
1.4	.7116	.6461	.5904	.5313	.4638	4.062	4.022	3.983	3.935	3.870
1.7	.7037	.6373	.5810	.5215	.4539	4.663	4.614	4.567	4.509	4.432
2.0	.6969	.6297	.5729	.5132	.4454	5.245	5.188	5.132	5.064	4.974
2.3	.6909	.6230	.5658	.5058	.4380	5.772	5.708	5.645	5.569	5.467
2.6	.6854	.6170	.5594	.4993	.4315	6.291	6.219	6.149	6.064	5.951
3.0	.6789	.6097	.5518	.4915	.4237	6.976	6.893	6.813	6.715	6.585
3.5	.6716	.6017	.5434	.4828	.4151	7.825	7.727	7.632	7.518	7.365
4.0	.6650	.5944	.5358	.4752	.4076	8.665	8.551	8.441	8.308	8.132
4.5	.6589	.5879	.5290	.4683	.4008	9.496	9.365	9.239	9.088	8.887
5.0	.6534	.5818	.5228	.4620	.3947	10.32	10.17	10.03	9.856	9.629
6.0	.6434	.5711	.5116	.4508	.3838	11.82	11.64	11.47	11.27	11.00
7.0	.6345	.5615	.5019	.4411	.3743	13.34	13.13	12.93	12.68	12.36
8.0	.6264	.5529	.4931	.4324	.3660	14.87	14.62	14.38	14.10	13.73
9.0	.6190	.5451	.4851	.4245	.3584	16.42	16.12	15.85	15.52	15.08
10	.6121	.5378	.4778	.4173	.3516	17.96	17.62	17.30	16.93	16.43
12	.5996	.5247	.4646	.4043	.3393	20.78	20.38	19.99	19.53	18.93
14	.5884	.5131	.4530	.3930	.3287	23.71	23.21	22.74	22.18	21.46
17	.5733	.4976	.4376	.3782	.3150	28.26	27.60	26.97	26.25	25.32
20	.5599	.4839	.4242	.3654	.3032	32.98	32.12	31.31	30.39	29.21
23	.5475	.4713	.4121	.3541	.2927	37.12	36.13	35.21	34.14	32.78
26	.5362	.4599	.4011	.3438	.2833	41.44	40.30	39.23	37.99	36.41
30	.5224	.4461	.3878	.3316	.2721	47.50	46.09	44.78	43.26	41.34
35	.5067	.4307	.3731	.3180	.2599	55.50	53.67	51.98	50.06	47.65
40	.4925	.4170	.3601	.3058	.2492	63.86	61.53	59.41	57.02	54.06
45	.4796	.4046	.3483	.2949	.2397	72.50	69.60	67.00	64.09	60.54
50	.4676	.3934	.3376	.2849	.2312	81.34	77.81	74.69	71.23	67.05
60	.4456	.3737	.3185	.2663	.2162	97.78	93.20	89.23	84.90	79.49
70	.4263	.3567	.3021	.2504	.2035	115.3	109.4	104.3	98.88	92.12
80	.4091	.3417	.2877	.2366	.1925	133.6	126.2	119.8	113.1	104.9
90	.3938	.3283	.2750	.2245	.1829	152.5	143.4	135.6	127.4	117.6
100	.3798	.3162	.2636	.2139	.1745	171.9	160.9	151.6	141.7	130.4
120	.3553	.2945	.2440	.1966	.1601	209.1	194.7	182.4	169.4	155.0
140	.3344	.2760	.2276	.1823	.1483	248.0	229.5	213.9	197.2	179.5
170	.3081	.2529	.2073	.1651	.1340	308.5	283.2	261.7	239.0	216.1
200	.2863	.2336	.1908	.1514	.1224	370.9	337.9	309.8	280.5	252.3
230	.2678	.2178	.1772	.1400	.1131	431.5	391.1	356.6	321.0	287.6
260	.2519	.2042	.1656	.1304	.1052	493.0	444.7	403.2	361.1	322.3
300	.2337	.1888	.1526	.1196	.0964	575.9	516.3	465.0	413.8	367.7
350	.2146	.1728	.1392	.1086	.0875	680.3	605.5	541.5	478.2	423.0
400	.1987	.1595	.1281	.0996	.0802	784.9	694.1	616.7	541.2	476.7
450	.1851	.1483	.1187	.0921	.0741	889.1	781.7	690.6	602.5	528.9
500	.1734	.1386	.1108	.0856	.0689	992.5	868.0	763.1	662.4	579.6
600	.1547	.1232	.0981	.0755	.0607	1194.	1035.	902.6	776.9	676.1
700	.1398	.1111	.0882	.0677	.0543	1391.	1197.	1037.	886.4	768.2
800	.1277	.1013	.0802	.0614	.0493	1584.	1355.	1167.	991.5	856.3
900	.1177	.0932	.0737	.0562	.0451	1772.	1508.	1293.	1093.	940.9
1000	.1092	.0863	.0681	.0519	.0416	1957.	1658.	1415.	1190.	1022.
1200	.0956	.0754	.0594	.0451	.0361	2312.	1945.	1648.	1375.	1177.
1400	.0852	.0671	.0527	.0399	.0320	2652.	2218.	1869.	1549.	1322.
1700	.0736	.0578	.0453	.0342	.0274	3135.	2604.	2179.	1792.	1524.
2000	.0651	.0510	.0399	.0301	.0242	3589.	2966.	2468.	2018.	1712.

MEAN DAMAGE DEPTH,MICROGRAM/SQ.CM. MEAN IONZN.DEPTH,MICROGRAM/SQ.CM.

K/KL= E,KEV	0.6	0.8	1.0	1.25	1.6	0.6	0.8	1.0	1.25	1.6
.20	1.028	1.013	.9983	.9810	.9580	1.014	.9974	.9815	.9624	.9371
.23	1.137	1.120	1.103	1.083	1.057	1.121	1.102	1.084	1.062	1.033
.26	1.237	1.217	1.199	1.177	1.148	1.219	1.198	1.178	1.153	1.121
.30	1.360	1.338	1.317	1.293	1.260	1.341	1.317	1.294	1.267	1.231
.35	1.504	1.479	1.456	1.428	1.391	1.482	1.455	1.429	1.399	1.358
.40	1.640	1.613	1.587	1.556	1.516	1.616	1.587	1.558	1.524	1.479
.45	1.771	1.741	1.713	1.679	1.635	1.745	1.713	1.681	1.644	1.595
.50	1.898	1.866	1.835	1.798	1.750	1.870	1.835	1.801	1.760	1.707
.60	2.143	2.105	2.070	2.027	1.971	2.111	2.070	2.030	1.983	1.922
.70	2.378	2.335	2.294	2.246	2.183	2.342	2.295	2.250	2.197	2.127
.80	2.605	2.557	2.511	2.457	2.387	2.565	2.512	2.461	2.402	2.324
.90	2.825	2.772	2.721	2.661	2.583	2.781	2.722	2.666	2.600	2.514
1.0	3.038	2.980	2.925	2.859	2.774	2.990	2.926	2.865	2.792	2.698
1.2	3.436	3.368	3.304	3.228	3.129	3.380	3.305	3.234	3.150	3.041
1.4	3.820	3.742	3.669	3.582	3.470	3.757	3.671	3.590	3.494	3.370
1.7	4.377	4.285	4.198	4.095	3.961	4.302	4.200	4.103	3.990	3.843
2.0	4.916	4.809	4.707	4.588	4.434	4.829	4.710	4.598	4.466	4.296
2.3	5.406	5.285	5.172	5.038	4.866	5.309	5.175	5.049	4.902	4.711
2.6	5.887	5.753	5.626	5.478	5.287	5.779	5.630	5.490	5.326	5.115
3.0	6.519	6.366	6.223	6.054	5.837	6.396	6.226	6.067	5.881	5.641
3.5	7.300	7.122	6.955	6.761	6.510	7.156	6.958	6.773	6.558	6.282
4.0	8.070	7.866	7.676	7.454	7.169	7.904	7.677	7.466	7.221	6.907
4.5	8.830	8.599	8.384	8.135	7.815	8.641	8.384	8.146	7.870	7.518
5.0	9.580	9.322	9.082	8.804	8.449	9.366	9.080	8.814	8.506	8.116
6.0	10.95	10.65	10.36	10.04	9.617	10.70	10.36	10.05	9.683	9.223
7.0	12.33	11.97	11.64	11.26	10.77	12.04	11.64	11.27	10.85	10.31
8.0	13.72	13.30	12.92	12.48	11.92	13.37	12.91	12.48	12.00	11.38
9.0	15.11	14.63	14.20	13.69	13.06	14.70	14.17	13.69	13.13	12.44
10	16.50	15.96	15.47	14.90	14.19	16.03	15.43	14.88	14.26	13.48
12	19.05	18.40	17.81	17.14	16.29	18.48	17.76	17.11	16.36	15.43
14	21.67	20.90	20.19	19.39	18.40	20.98	20.12	19.34	18.46	17.37
17	25.72	24.73	23.84	22.83	21.58	24.80	23.70	22.71	21.60	20.24
20	29.89	28.65	27.54	26.30	24.78	28.70	27.32	26.10	24.74	23.09
23	33.58	32.16	30.89	29.46	27.71	32.22	30.64	29.23	27.66	25.75
26	37.40	35.77	34.31	32.67	30.68	35.83	34.01	32.38	30.58	28.41
30	42.72	40.76	39.00	37.05	34.68	40.78	38.58	36.64	34.50	31.93
35	49.67	47.22	45.04	42.65	39.77	47.15	44.42	42.03	39.42	36.32
40	56.89	53.87	51.23	48.34	44.90	53.69	50.35	47.47	44.35	40.68
45	64.31	60.67	57.51	54.10	50.07	60.32	56.33	52.92	49.26	45.00
50	71.85	67.55	63.85	59.88	55.23	67.02	62.33	58.36	54.15	49.27
60	85.90	80.41	75.80	70.87	65.04	79.71	73.72	68.78	63.62	57.47
70	100.7	93.84	88.13	82.08	74.97	92.77	85.31	79.25	72.98	65.54
80	116.2	107.7	100.7	93.43	84.96	106.1	97.02	89.70	82.21	73.47
90	132.0	121.9	113.5	104.9	94.96	119.6	108.8	100.1	91.30	81.27
100	148.2	136.3	126.4	116.3	104.9	133.1	120.6	110.4	100.2	88.91
120	179.4	164.1	151.4	138.5	124.1	159.6	143.6	130.5	117.4	103.7
140	211.8	192.6	176.7	160.7	143.3	186.2	166.5	150.3	134.0	117.9
170	261.9	236.3	215.1	194.0	171.9	226.1	200.4	179.2	158.1	138.4
200	313.2	280.6	253.5	226.9	200.0	265.7	233.7	207.3	181.2	158.0
230	363.0	323.4	290.7	258.6	227.3	304.4	266.0	234.5	203.4	176.8
260	413.2	366.4	327.7	289.9	254.0	342.5	297.6	260.8	224.8	194.8
300	480.8	423.7	376.7	331.0	289.1	392.3	338.6	294.9	252.3	217.7
350	565.7	495.2	437.3	381.4	332.0	453.1	388.2	335.6	284.9	244.9
400	650.5	566.1	497.0	430.6	373.8	512.0	436.1	374.7	315.8	270.6
450	734.8	636.2	555.6	478.7	414.5	569.3	482.2	412.1	345.4	295.1
500	818.3	705.4	613.2	525.6	454.1	624.9	526.9	448.1	373.6	318.5
600	979.9	838.9	723.7	615.3	529.6	731.7	611.9	516.2	426.7	362.1
700	1138.	969.0	830.6	701.5	602.0	832.9	692.1	579.9	476.1	402.6
800	1294.	1096.	934.2	784.4	671.5	929.2	768.1	639.9	522.4	440.6
900	1446.	1219.	1035.	864.5	738.5	1021.	840.4	696.8	566.0	476.3
1000	1594.	1339.	1132.	941.9	803.2	1109.	909.5	750.9	607.3	510.2
1200	1882.	1571.	1319.	1090.	926.5	1275.	1039.	852.1	684.2	573.2
1400	2157.	1791.	1496.	1229.	1043.	1429.	1160.	945.4	754.8	631.0
1700	2548.	2104.	1747.	1425.	1205.	1643.	1326.	1074.	851.2	710.1
2000	2917.	2397.	1981.	1607.	1357.	1839.	1478.	1190.	938.7	782.0

C ON AU, M2/M1=16.4

TABLE 3- 3

K/KL=	RELATIVE RANGE STRAGGLING					RELATIVE DAMAGE STRAGGLING				
E,KEV	0.6	0.8	1.0	1.25	1.6	0.6	0.8	1.0	1.25	1.6
.20	3.346	3.147	2.971	2.778	2.548	2.019	1.912	1.818	1.715	1.593
.23	3.328	3.126	2.948	2.754	2.523	2.010	1.901	1.806	1.702	1.580
.26	3.311	3.107	2.927	2.732	2.500	2.001	1.891	1.795	1.691	1.568
.30	3.290	3.084	2.903	2.706	2.474	1.990	1.879	1.783	1.678	1.554
.35	3.268	3.059	2.876	2.678	2.444	1.979	1.867	1.769	1.663	1.539
.40	3.247	3.036	2.852	2.652	2.418	1.969	1.855	1.756	1.650	1.526
.45	3.228	3.015	2.830	2.630	2.395	1.959	1.844	1.745	1.638	1.514
.50	3.211	2.996	2.810	2.609	2.373	1.950	1.835	1.735	1.628	1.503
.60	3.180	2.962	2.774	2.571	2.335	1.935	1.818	1.717	1.609	1.483
.70	3.152	2.932	2.742	2.539	2.302	1.921	1.802	1.701	1.592	1.467
.80	3.126	2.904	2.714	2.509	2.273	1.908	1.788	1.686	1.577	1.452
.90	3.102	2.879	2.687	2.483	2.246	1.896	1.776	1.673	1.564	1.438
1.0	3.080	2.856	2.663	2.458	2.222	1.885	1.764	1.661	1.551	1.426
1.2	3.040	2.814	2.620	2.415	2.178	1.865	1.743	1.639	1.529	1.404
1.4	3.004	2.776	2.582	2.376	2.140	1.848	1.724	1.620	1.510	1.384
1.7	2.955	2.725	2.530	2.325	2.090	1.824	1.699	1.594	1.484	1.359
2.0	2.911	2.680	2.484	2.279	2.046	1.802	1.677	1.571	1.461	1.336
2.3	2.870	2.638	2.442	2.238	2.006	1.782	1.656	1.551	1.441	1.316
2.6	2.832	2.599	2.404	2.200	1.969	1.764	1.638	1.532	1.422	1.298
3.0	2.785	2.551	2.356	2.153	1.925	1.742	1.615	1.509	1.399	1.276
3.5	2.731	2.497	2.303	2.101	1.875	1.716	1.589	1.483	1.373	1.252
4.0	2.681	2.448	2.254	2.054	1.831	1.693	1.565	1.459	1.350	1.229
4.5	2.635	2.402	2.210	2.011	1.790	1.671	1.542	1.437	1.329	1.209
5.0	2.593	2.360	2.169	1.972	1.753	1.650	1.522	1.417	1.309	1.190
6.0	2.514	2.283	2.094	1.900	1.686	1.612	1.484	1.379	1.273	1.156
7.0	2.444	2.215	2.027	1.837	1.627	1.577	1.449	1.346	1.240	1.125
8.0	2.380	2.152	1.968	1.780	1.574	1.546	1.418	1.315	1.211	1.098
9.0	2.321	2.095	1.913	1.728	1.526	1.517	1.390	1.287	1.184	1.073
10	2.267	2.043	1.862	1.680	1.483	1.490	1.363	1.262	1.160	1.050
12	2.166	1.946	1.770	1.593	1.403	1.440	1.314	1.215	1.115	1.009
14	2.076	1.860	1.688	1.517	1.333	1.396	1.271	1.173	1.076	.9723
17	1.960	1.749	1.583	1.419	1.244	1.337	1.215	1.119	1.025	.9253
20	1.860	1.654	1.494	1.336	1.168	1.286	1.166	1.072	.9811	.8849
23	1.772	1.571	1.416	1.264	1.104	1.240	1.122	1.031	.9419	.8491
26	1.694	1.498	1.347	1.201	1.047	1.199	1.082	.9932	.9070	.8172
30	1.602	1.412	1.267	1.127	.9812	1.149	1.036	.9489	.8658	.7799
35	1.503	1.320	1.182	1.049	.9112	1.095	.9844	.9009	.8213	.7395
40	1.417	1.241	1.108	.9818	.8515	1.048	.9399	.8591	.7827	.7046
45	1.342	1.172	1.044	.9236	.7999	1.006	.9005	.8223	.7488	.6740
50	1.276	1.111	.9880	.8725	.7546	.9681	.8653	.7895	.7186	.6470
60	1.162	1.007	.8933	.7869	.6789	.9026	.8049	.7336	.6675	.6011
70	1.069	.9228	.8161	.7174	.6176	.8477	.7545	.6871	.6252	.5633
80	.9906	.8520	.7517	.6595	.5667	.8008	.7117	.6478	.5895	.5315
90	.9235	.7917	.6970	.6105	.5238	.7600	.6746	.6139	.5588	.5042
100	.8653	.7396	.6499	.5685	.4871	.7242	.6422	.5843	.5321	.4806
120	.7710	.6558	.5745	.5015	.4291	.6646	.5885	.5357	.4885	.4420
140	.6961	.5896	.5152	.4489	.3839	.6161	.5452	.4966	.4536	.4112
170	.6080	.5124	.4464	.3882	.3318	.5579	.4935	.4501	.4122	.3749
200	.5399	.4530	.3938	.3419	.2922	.5119	.4530	.4139	.3799	.3467
230	.4866	.4069	.3530	.3062	.2613	.4752	.4215	.3854	.3542	.3248
260	.4431	.3694	.3200	.2774	.2364	.4448	.3955	.3620	.3330	.3069
300	.3959	.3290	.2846	.2465	.2097	.4113	.3672	.3365	.3099	.2874
350	.3494	.2894	.2500	.2164	.1836	.3777	.3389	.3110	.2868	.2680
400	.3125	.2583	.2228	.1927	.1633	.3507	.3163	.2907	.2685	.2526
450	.2825	.2332	.2009	.1736	.1469	.3284	.2977	.2741	.2535	.2399
500	.2577	.2125	.1829	.1579	.1335	.3098	.2821	.2602	.2410	.2293
600	.2195	.1812	.1553	.1332	.1131	.2810	.2575	.2386	.2221	.2127
700	.1910	.1581	.1349	.1150	.0981	.2592	.2388	.2223	.2079	.2000
800	.1690	.1404	.1193	.1010	.0867	.2421	.2240	.2095	.1969	.1899
900	.1515	.1263	.1070	.0900	.0777	.2283	.2121	.1992	.1880	.1817
1000	.1373	.1149	.0970	.0811	.0705	.2169	.2022	.1906	.1806	.1749
1200	.1156	.0973	.0817	.0677	.0594	.1992	.1868	.1772	.1690	.1641
1400	.0998	.0843	.0706	.0582	.0515	.1860	.1753	.1671	.1602	.1559
1700	.0828	.0700	.0586	.0484	.0428	.1716	.1627	.1559	.1501	.1465
2000	.0708	.0593	.0500	.0419	.0365	.1612	.1536	.1475	.1423	.1393

K/KL = E,KEV	RELATIVE IONIZATION STRAGGLING					RELATIVE TRANSVERSE RANGE STRAGGLING				
	0.6	0.8	1.0	1.25	1.6	0.6	0.8	1.0	1.25	1.6
.20	1.906	1.806	1.718	1.622	1.510	.9695	.9678	.9661	.9641	.9613
.23	1.897	1.795	1.707	1.610	1.497	.9694	.9676	.9660	.9639	.9611
.26	1.888	1.786	1.697	1.600	1.486	.9693	.9676	.9658	.9637	.9609
.30	1.878	1.775	1.685	1.587	1.473	.9692	.9674	.9657	.9635	.9607
.35	1.867	1.762	1.672	1.574	1.459	.9691	.9673	.9655	.9633	.9604
.40	1.857	1.751	1.660	1.561	1.447	.9691	.9672	.9654	.9632	.9602
.45	1.848	1.741	1.649	1.550	1.435	.9690	.9671	.9653	.9630	.9600
.50	1.840	1.732	1.640	1.540	1.425	.9690	.9670	.9652	.9629	.9598
.60	1.825	1.716	1.622	1.522	1.407	.9689	.9669	.9650	.9627	.9595
.70	1.811	1.701	1.607	1.507	1.391	.9689	.9668	.9649	.9625	.9593
.80	1.799	1.688	1.593	1.493	1.377	.9688	.9668	.9648	.9624	.9591
.90	1.787	1.676	1.581	1.480	1.365	.9688	.9667	.9647	.9623	.9589
1.0	1.777	1.665	1.569	1.468	1.353	.9688	.9667	.9647	.9622	.9588
1.2	1.758	1.644	1.549	1.447	1.332	.9689	.9667	.9646	.9621	.9586
1.4	1.740	1.626	1.530	1.429	1.314	.9689	.9667	.9646	.9620	.9585
1.7	1.717	1.602	1.506	1.405	1.291	.9690	.9668	.9646	.9620	.9584
2.0	1.696	1.581	1.484	1.383	1.270	.9692	.9669	.9647	.9620	.9584
2.3	1.677	1.561	1.464	1.364	1.251	.9693	.9670	.9647	.9620	.9584
2.6	1.659	1.543	1.446	1.346	1.234	.9694	.9671	.9648	.9621	.9585
3.0	1.637	1.520	1.424	1.324	1.213	.9697	.9673	.9650	.9623	.9586
3.5	1.612	1.495	1.399	1.300	1.190	.9700	.9676	.9653	.9626	.9589
4.0	1.589	1.472	1.376	1.278	1.169	.9704	.9680	.9657	.9629	.9592
4.5	1.567	1.451	1.355	1.258	1.150	.9707	.9683	.9660	.9633	.9596
5.0	1.547	1.431	1.336	1.239	1.133	.9711	.9687	.9664	.9637	.9600
6.0	1.510	1.394	1.300	1.205	1.101	.9719	.9695	.9672	.9644	.9608
7.0	1.477	1.362	1.269	1.175	1.073	.9727	.9703	.9680	.9653	.9617
8.0	1.447	1.332	1.240	1.148	1.048	.9735	.9711	.9689	.9662	.9627
9.0	1.419	1.305	1.214	1.123	1.026	.9744	.9721	.9698	.9672	.9637
10	1.393	1.280	1.190	1.101	1.005	.9753	.9730	.9708	.9682	.9648
12	1.345	1.234	1.146	1.060	.9672	.9770	.9748	.9727	.9703	.9670
14	1.302	1.193	1.108	1.024	.9345	.9788	.9767	.9748	.9724	.9693
17	1.247	1.140	1.058	.9770	.8924	.9816	.9797	.9779	.9758	.9729
20	1.199	1.095	1.015	.9374	.8567	.9844	.9827	.9811	.9792	.9767
23	1.156	1.055	.9773	.9030	.8255	.9871	.9856	.9842	.9825	.9803
26	1.118	1.019	.9440	.8727	.7982	.9898	.9886	.9874	.9859	.9840
30	1.073	.9770	.9051	.8374	.7665	.9935	.9925	.9916	.9904	.9889
35	1.025	.9319	.8633	.7996	.7327	.9980	.9974	.9969	.9961	.9950
40	.9828	.8930	.8275	.7672	.7041	1.002	1.002	1.002	1.002	1.001
45	.9457	.8590	.7963	.7390	.6795	1.007	1.007	1.007	1.007	1.007
50	.9128	.8290	.7689	.7141	.6580	1.011	1.012	1.013	1.013	1.013
60	.8562	.7784	.7227	.6714	.6228	1.020	1.021	1.022	1.024	1.025
70	.8096	.7371	.6852	.6369	.5946	1.028	1.030	1.032	1.034	1.036
80	.7703	.7026	.6541	.6083	.5715	1.036	1.039	1.042	1.044	1.047
90	.7367	.6734	.6278	.5844	.5523	1.044	1.048	1.051	1.054	1.058
100	.7076	.6483	.6053	.5642	.5361	1.052	1.057	1.060	1.064	1.069
120	.6605	.6082	.5698	.5337	.5109	1.068	1.075	1.079	1.084	1.090
140	.6234	.5770	.5425	.5106	.4919	1.083	1.092	1.097	1.104	1.110
170	.5803	.5412	.5115	.4852	.4710	1.105	1.116	1.123	1.130	1.138
200	.5476	.5144	.4886	.4670	.4560	1.125	1.138	1.146	1.156	1.164
230	.5230	.4944	.4721	.4539	.4457	1.145	1.159	1.169	1.179	1.189
260	.5034	.4786	.4593	.4440	.4381	1.164	1.179	1.190	1.202	1.212
300	.4829	.4623	.4464	.4342	.4308	1.187	1.204	1.216	1.229	1.240
350	.4636	.4471	.4348	.4258	.4247	1.215	1.233	1.247	1.261	1.273
400	.4491	.4361	.4266	.4202	.4208	1.241	1.259	1.275	1.291	1.303
450	.4380	.4278	.4207	.4163	.4184	1.265	1.284	1.302	1.318	1.331
500	.4294	.4216	.4165	.4138	.4170	1.288	1.308	1.326	1.343	1.357
600	.4181	.4139	.4117	.4115	.4162	1.330	1.352	1.372	1.390	1.404
700	.4111	.4097	.4095	.4112	.4168	1.369	1.392	1.414	1.433	1.446
800	.4067	.4074	.4089	.4119	.4183	1.404	1.429	1.451	1.471	1.484
900	.4041	.4065	.4092	.4132	.4201	1.436	1.462	1.485	1.505	1.519
1000	.4027	.4064	.4100	.4149	.4221	1.466	1.493	1.517	1.537	1.551
1200	.4020	.4076	.4126	.4187	.4262	1.521	1.549	1.573	1.594	1.607
1400	.4030	.4098	.4158	.4225	.4302	1.569	1.599	1.623	1.644	1.657
1700	.4060	.4139	.4206	.4280	.4357	1.633	1.663	1.688	1.709	1.721
2000	.4097	.4180	.4252	.4327	.4404	1.689	1.719	1.744	1.765	1.776

K/KL=	RELATIVE TRANSV. DAMAGE STRAGGLING					RELATIVE TRANSV. IONZN. STRAGGLING				
E,KEV	0.6	0.8	1.0	1.25	1.6	0.6	0.8	1.0	1.25	1.6
.20	.9339	.9287	.9236	.9171	.9081	.9245	.9182	.9118	.9039	.8928
.23	.9336	.9283	.9230	.9164	.9071	.9241	.9176	.9111	.9030	.8916
.26	.9333	.9279	.9225	.9157	.9063	.9237	.9171	.9105	.9022	.8905
.30	.9330	.9274	.9219	.9150	.9053	.9233	.9165	.9097	.9012	.8892
.35	.9326	.9269	.9213	.9141	.9042	.9228	.9158	.9088	.9001	.8878
.40	.9323	.9265	.9207	.9134	.9033	.9224	.9152	.9081	.8991	.8866
.45	.9321	.9261	.9202	.9128	.9024	.9220	.9147	.9074	.8983	.8854
.50	.9318	.9258	.9197	.9122	.9017	.9217	.9142	.9068	.8975	.8844
.60	.9314	.9252	.9189	.9111	.9003	.9211	.9134	.9057	.8961	.8826
.70	.9311	.9246	.9182	.9102	.8991	.9205	.9127	.9048	.8949	.8810
.80	.9308	.9242	.9176	.9095	.8981	.9201	.9120	.9039	.8938	.8796
.90	.9305	.9238	.9171	.9088	.8972	.9197	.9114	.9032	.8928	.8783
1.0	.9303	.9235	.9166	.9082	.8963	.9193	.9109	.9025	.8919	.8771
1.2	.9299	.9229	.9158	.9071	.8949	.9187	.9100	.9013	.8903	.8750
1.4	.9296	.9224	.9151	.9061	.8936	.9182	.9092	.9002	.8889	.8732
1.7	.9293	.9218	.9143	.9050	.8920	.9175	.9082	.8988	.8871	.8708
2.0	.9290	.9213	.9136	.9040	.8907	.9170	.9073	.8976	.8855	.8687
2.3	.9288	.9209	.9130	.9032	.8895	.9165	.9066	.8966	.8842	.8668
2.6	.9287	.9206	.9125	.9024	.8885	.9161	.9059	.8957	.8829	.8651
3.0	.9286	.9203	.9119	.9016	.8873	.9156	.9050	.8945	.8814	.8630
3.5	.9285	.9199	.9114	.9007	.8860	.9150	.9041	.8932	.8796	.8607
4.0	.9285	.9197	.9109	.9000	.8848	.9145	.9033	.8921	.8781	.8585
4.5	.9285	.9195	.9105	.8993	.8838	.9141	.9026	.8910	.8766	.8566
5.0	.9285	.9193	.9101	.8987	.8829	.9138	.9019	.8900	.8752	.8547
6.0	.9287	.9191	.9096	.8978	.8814	.9132	.9008	.8883	.8729	.8514
7.0	.9289	.9191	.9092	.8970	.8801	.9127	.8997	.8868	.8707	.8484
8.0	.9292	.9191	.9089	.8963	.8789	.9122	.8988	.8853	.8686	.8455
9.0	.9296	.9191	.9087	.8958	.8779	.9118	.8979	.8840	.8667	.8428
10	.9299	.9192	.9085	.8953	.8770	.9114	.8970	.8826	.8648	.8402
12	.9307	.9195	.9084	.8946	.8755	.9107	.8955	.8803	.8615	.8354
14	.9315	.9199	.9083	.8940	.8741	.9101	.8941	.8781	.8582	.8308
17	.9327	.9205	.9083	.8932	.8724	.9092	.8919	.8747	.8535	.8243
20	.9340	.9212	.9084	.8926	.8708	.9081	.8897	.8714	.8488	.8181
23	.9353	.9220	.9087	.8922	.8695	.9074	.8879	.8685	.8447	.8130
26	.9365	.9227	.9089	.8918	.8683	.9066	.8860	.8655	.8405	.8080
30	.9381	.9236	.9092	.8913	.8667	.9053	.8832	.8614	.8349	.8013
35	.9399	.9247	.9094	.8906	.8647	.9033	.8796	.8561	.8277	.7927
40	.9415	.9255	.9095	.8898	.8627	.9011	.8756	.8506	.8204	.7840
45	.9431	.9263	.9096	.8888	.8606	.8987	.8715	.8449	.8130	.7751
50	.9444	.9270	.9095	.8879	.8585	.8961	.8672	.8391	.8056	.7661
60	.9472	.9283	.9095	.8862	.8549	.8911	.8588	.8276	.7909	.7465
70	.9495	.9292	.9090	.8841	.8510	.8853	.8499	.8159	.7762	.7276
80	.9513	.9297	.9082	.8817	.8467	.8791	.8406	.8040	.7617	.7094
90	.9528	.9299	.9070	.8789	.8421	.8724	.8311	.7921	.7474	.6921
100	.9539	.9297	.9056	.8759	.8373	.8655	.8214	.7803	.7333	.6754
120	.9559	.9293	.9024	.8694	.8262	.8508	.8010	.7559	.7049	.6436
140	.9569	.9279	.8984	.8623	.8152	.8356	.7810	.7324	.6780	.6145
170	.9569	.9244	.8916	.8514	.7991	.8126	.7518	.6991	.6409	.5754
200	.9555	.9198	.8838	.8402	.7838	.7896	.7240	.6681	.6071	.5408
230	.9531	.9139	.8750	.8287	.7696	.7644	.6956	.6371	.5745	.5087
260	.9499	.9074	.8660	.8174	.7562	.7402	.6688	.6084	.5448	.4799
300	.9447	.8982	.8538	.8025	.7392	.7094	.6356	.5733	.5090	.4458
350	.9372	.8862	.8385	.7846	.7193	.6737	.5979	.5342	.4698	.4089
400	.9289	.8740	.8235	.7673	.7007	.6409	.5640	.4996	.4356	.3772
450	.9201	.8617	.8087	.7507	.6832	.6106	.5334	.4688	.4055	.3497
500	.9109	.8494	.7944	.7347	.6667	.5827	.5056	.4412	.3790	.3256
600	.8910	.8241	.7653	.7027	.6341	.5310	.4563	.3944	.3353	.2868
700	.8712	.8001	.7384	.6736	.6052	.4866	.4147	.3556	.2999	.2557
800	.8518	.7775	.7136	.6473	.5793	.4482	.3794	.3231	.2706	.2302
900	.8330	.7562	.6907	.6233	.5560	.4147	.3489	.2955	.2459	.2089
1000	.8148	.7361	.6695	.6013	.5349	.3854	.3225	.2717	.2249	.1909
1200	.7802	.6993	.6312	.5626	.4981	.3365	.2793	.2333	.1915	.1623
1400	.7478	.6660	.5976	.5294	.4670	.2979	.2457	.2040	.1663	.1409
1700	.7029	.6218	.5541	.4875	.4285	.2538	.2082	.1718	.1393	.1178
2000	.6617	.5828	.5169	.4528	.3972	.2217	.1817	.1496	.1211	.1022

TABLE 3- 6

C ON AU, M2/M1=16.4

K/KL=	RANGE SKEWNESS					DAMAGE SKEWNESS				
E,KEV	0.6	0.8	1.0	1.25	1.6	0.6	0.8	1.0	1.25	1.6
.20	.0587	.0515	.0442	.0350	.0220	.2195	.2213	.2230	.2249	.2274
.23	.0581	.0508	.0433	.0338	.0204	.2197	.2215	.2232	.2251	.2276
.26	.0577	.0501	.0424	.0328	.0191	.2198	.2216	.2233	.2253	.2278
.30	.0571	.0493	.0414	.0315	.0174	.2200	.2218	.2235	.2255	.2280
.35	.0564	.0484	.0403	.0300	.0155	.2202	.2220	.2238	.2258	.2282
.40	.0559	.0476	.0393	.0288	.0139	.2203	.2222	.2240	.2259	.2284
.45	.0553	.0469	.0384	.0276	.0123	.2205	.2224	.2241	.2261	.2286
.50	.0548	.0462	.0375	.0265	.0109	.2206	.2225	.2243	.2263	.2287
.60	.0539	.0450	.0359	.0245	.0084	.2209	.2228	.2245	.2265	.2290
.70	.0531	.0439	.0345	.0227	.0061	.2211	.2230	.2247	.2267	.2291
.80	.0524	.0429	.0332	.0211	.0040	.2213	.2231	.2249	.2268	.2292
.90	.0517	.0419	.0320	.0196	.0020	.2214	.2233	.2250	.2269	.2293
1.0	.0511	.0410	.0309	.0182	.0002	.2216	.2234	.2251	.2270	.2293
1.2	.0500	.0395	.0289	.0156	-.0030	.2218	.2236	.2253	.2271	.2294
1.4	.0489	.0381	.0271	.0133	-.0059	.2220	.2238	.2254	.2271	.2293
1.7	.0475	.0361	.0246	.0102	-.0100	.2223	.2239	.2254	.2271	.2292
2.0	.0461	.0342	.0222	.0072	-.0137	.2224	.2240	.2254	.2270	.2289
2.3	.0448	.0324	.0200	.0044	-.0172	.2226	.2241	.2254	.2269	.2287
2.6	.0435	.0307	.0179	.0018	-.0205	.2227	.2241	.2253	.2267	.2284
3.0	.0419	.0285	.0152	-.0015	-.0246	.2228	.2240	.2252	.2264	.2279
3.5	.0400	.0260	.0120	-.0054	-.0295	.2228	.2239	.2248	.2259	.2272
4.0	.0381	.0235	.0090	-.0091	-.0340	.2227	.2236	.2244	.2253	.2263
4.5	.0364	.0212	.0061	-.0126	-.0384	.2226	.2233	.2239	.2247	.2255
5.0	.0347	.0190	.0033	-.0160	-.0425	.2224	.2229	.2234	.2239	.2245
6.0	.0321	.0153	-.0013	-.0217	-.0496	.2220	.2221	.2223	.2225	.2227
7.0	.0294	.0117	-.0058	-.0273	-.0566	.2214	.2212	.2211	.2209	.2207
8.0	.0266	.0089	-.0104	-.0328	-.0633	.2206	.2201	.2197	.2192	.2186
9.0	.0238	.0042	-.0149	-.0383	-.0699	.2198	.2190	.2183	.2175	.2165
10	.0209	.0004	-.0195	-.0438	-.0765	.2199	.2178	.2168	.2157	.2143
12	.0147	-.0077	-.0291	-.0549	-.0897	.2173	.2155	.2139	.2122	.2101
14	.0085	-.0156	-.0384	-.0658	-.1024	.2154	.2129	.2108	.2086	.2058
17	-.0008	-.0273	-.0519	-.0814	-.1205	.2122	.2089	.2060	.2029	.1992
20	-.0097	-.0384	-.0648	-.0962	-.1376	.2088	.2046	.2010	.1972	.1926
23	-.0177	-.0483	-.0764	-.1095	-.1528	.2050	.2000	.1958	.1912	.1859
26	-.0255	-.0578	-.0874	-.1222	-.1674	.2012	.1954	.1906	.1853	.1792
30	-.0358	-.0702	-.1017	-.1386	-.1861	.1960	.1892	.1836	.1775	.1705
35	-.0485	-.0853	-.1190	-.1581	-.2082	.1894	.1815	.1749	.1679	.1597
40	-.0609	-.1000	-.1356	-.1769	-.2294	.1828	.1739	.1665	.1585	.1493
45	-.0732	-.1143	-.1518	-.1950	-.2498	.1762	.1664	.1581	.1494	.1392
50	-.0853	-.1283	-.1675	-.2124	-.2694	.1698	.1590	.1500	.1405	.1293
60	-.1093	-.1559	-.1983	-.2463	-.3081	.1574	.1449	.1345	.1238	.1107
70	-.1323	-.1822	-.2274	-.2782	-.3440	.1454	.1313	.1197	.1078	.0929
80	-.1544	-.2073	-.2550	-.3083	-.3775	.1337	.1182	.1054	.0924	.0758
90	-.1757	-.2313	-.2812	-.3369	-.4088	.1224	.1056	.0916	.0776	.0595
100	-.1962	-.2543	-.3062	-.3641	-.4381	.1115	.0934	.0783	.0632	.0438
120	-.2357	-.2986	-.3540	-.4161	-.4918	.0904	.0701	.0530	.0354	.0138
140	-.2724	-.3394	-.3978	-.4637	-.5404	.0704	.0482	.0292	.0093	-.0141
170	-.3228	-.3951	-.4574	-.5280	-.6059	.0425	.0176	-.0041	-.0272	-.0529
200	-.3686	-.4453	-.5110	-.5856	-.6644	.0164	-.0109	-.0350	-.0608	-.0886
230	-.4102	-.4902	-.5592	-.6366	-.7177	-.0078	-.0378	-.0641	-.0924	-.1224
260	-.4487	-.5315	-.6034	-.6832	-.7666	-.0307	-.0630	-.0915	-.1219	-.1538
300	-.4959	-.5821	-.6574	-.7400	-.8263	-.0593	-.0946	-.1256	-.1585	-.1926
350	-.5496	-.6392	-.7185	-.8040	-.8936	-.0927	-.1312	-.1650	-.2004	-.2369
400	-.5986	-.6913	-.7738	-.8619	-.9546	-.1237	-.1651	-.2013	-.2389	-.2772
450	-.6437	-.7391	-.8246	-.9150	-1.010	-.1529	-.1968	-.2351	-.2745	-.3142
500	-.6855	-.7836	-.8716	-.9640	-1.062	-.1804	-.2266	-.2667	-.3077	-.3484
600	-.7609	-.8639	-.9559	-1.052	-1.154	-.2323	-.2827	-.3258	-.3694	-.4112
700	-.8282	-.9355	-1.031	-1.131	-1.236	-.2793	-.3329	-.3785	-.4239	-.4662
800	-.8893	-1.000	-1.099	-1.202	-1.309	-.3222	-.3784	-.4259	-.4727	-.5150
900	-.9454	-1.060	-1.161	-1.267	-1.377	-.3618	-.4201	-.4689	-.5167	-.5588
1000	-.9973	-1.115	-1.218	-1.327	-1.439	-.3985	-.4584	-.5083	-.5567	-.5984
1200	-1.091	-1.215	-1.322	-1.435	-1.551	-.4650	-.5269	-.5782	-.6272	-.6678
1400	-1.175	-1.303	-1.415	-1.532	-1.651	-.5239	-.5869	-.6387	-.6877	-.7268
1700	-1.286	-1.421	-1.537	-1.659	-1.783	-.6017	-.6649	-.7165	-.7646	-.8013
2000	-1.384	-1.524	-1.645	-1.772	-1.900	-.6698	-.7321	-.7826	-.8291	-.8632

K/KL=	IONIZATION SKEWNESS					RANGE KURTOSIS				
E,KEV	0.6	0.8	1.0	1.25	1.6	0.6	0.8	1.0	1.25	1.6
.20	.2594	.2641	.2687	.2743	.2817	3.120	3.090	3.063	3.033	2.995
.23	.2599	.2647	.2693	.2750	.2826	3.118	3.088	3.060	3.029	2.991
.26	.2603	.2652	.2699	.2757	.2834	3.116	3.085	3.057	3.026	2.988
.30	.2608	.2658	.2706	.2765	.2844	3.113	3.082	3.054	3.022	2.984
.35	.2614	.2665	.2714	.2774	.2854	3.110	3.079	3.050	3.018	2.979
.40	.2619	.2671	.2721	.2782	.2864	3.108	3.076	3.047	3.014	2.975
.45	.2624	.2676	.2728	.2789	.2872	3.106	3.073	3.044	3.011	2.971
.50	.2628	.2682	.2733	.2796	.2880	3.103	3.070	3.041	3.008	2.968
.60	.2636	.2691	.2744	.2808	.2894	3.099	3.066	3.036	3.002	2.962
.70	.2644	.2700	.2754	.2819	.2907	3.096	3.062	3.031	2.997	2.957
.80	.2651	.2708	.2763	.2829	.2918	3.093	3.058	3.027	2.993	2.952
.90	.2657	.2715	.2771	.2838	.2928	3.090	3.055	3.024	2.989	2.948
1.0	.2663	.2722	.2778	.2847	.2938	3.088	3.052	3.021	2.986	2.945
1.2	.2675	.2734	.2792	.2862	.2955	3.084	3.047	3.015	2.980	2.939
1.4	.2685	.2746	.2805	.2876	.2970	3.080	3.043	3.011	2.975	2.934
1.7	.2699	.2761	.2821	.2894	.2991	3.075	3.037	3.004	2.969	2.927
2.0	.2712	.2775	.2837	.2910	.3008	3.070	3.032	2.998	2.962	2.920
2.3	.2724	.2788	.2850	.2924	.3024	3.065	3.026	2.992	2.956	2.914
2.6	.2734	.2799	.2862	.2937	.3038	3.059	3.020	2.986	2.950	2.907
3.0	.2747	.2813	.2877	.2954	.3056	3.053	3.013	2.979	2.942	2.900
3.5	.2763	.2829	.2894	.2972	.3076	3.046	3.005	2.971	2.934	2.892
4.0	.2777	.2845	.2910	.2989	.3094	3.040	2.999	2.964	2.927	2.885
4.5	.2791	.2859	.2925	.3005	.3111	3.034	2.993	2.958	2.921	2.879
5.0	.2804	.2873	.2940	.3020	.3128	3.029	2.987	2.952	2.915	2.874
6.0	.2831	.2901	.2969	.3051	.3160	3.023	2.981	2.945	2.909	2.868
7.0	.2856	.2926	.2994	.3078	.3188	3.018	2.975	2.939	2.903	2.863
8.0	.2877	.2948	.3018	.3102	.3213	3.012	2.969	2.933	2.897	2.859
9.0	.2897	.2968	.3039	.3123	.3236	3.006	2.963	2.927	2.892	2.854
10	.2915	.2987	.3057	.3143	.3257	3.000	2.956	2.921	2.886	2.850
12	.2943	.3016	.3088	.3175	.3290	2.983	2.940	2.905	2.871	2.838
14	.2967	.3041	.3114	.3203	.3320	2.968	2.924	2.890	2.858	2.827
17	.3000	.3075	.3150	.3240	.3358	2.947	2.904	2.872	2.842	2.814
20	.3030	.3106	.3182	.3273	.3392	2.930	2.888	2.857	2.829	2.804
23	.3061	.3139	.3216	.3306	.3426	2.920	2.878	2.847	2.821	2.796
26	.3090	.3170	.3248	.3336	.3456	2.911	2.870	2.840	2.815	2.790
30	.3125	.3208	.3287	.3373	.3494	2.902	2.861	2.833	2.810	2.785
35	.3165	.3249	.3330	.3416	.3536	2.893	2.854	2.827	2.808	2.785
40	.3200	.3286	.3367	.3455	.3573	2.884	2.848	2.824	2.809	2.789
45	.3231	.3318	.3400	.3490	.3606	2.876	2.843	2.822	2.811	2.796
50	.3259	.3347	.3429	.3523	.3637	2.869	2.840	2.822	2.816	2.806
60	.3311	.3400	.3481	.3604	.3710	2.847	2.827	2.818	2.822	2.832
70	.3353	.3441	.3521	.3666	.3764	2.829	2.820	2.821	2.834	2.863
80	.3387	.3474	.3552	.3712	.3802	2.818	2.818	2.828	2.852	2.899
90	.3414	.3499	.3575	.3744	.3826	2.810	2.821	2.839	2.873	2.936
100	.3437	.3518	.3593	.3764	.3838	2.807	2.827	2.854	2.897	2.976
120	.3470	.3538	.3611	.3743	.3800	2.811	2.849	2.891	2.952	3.053
140	.3490	.3546	.3616	.3707	.3749	2.824	2.878	2.936	3.013	3.135
170	.3502	.3544	.3608	.3644	.3666	2.856	2.933	3.011	3.113	3.262
200	.3499	.3531	.3589	.3582	.3587	2.898	2.996	3.094	3.218	3.392
230	.3467	.3501	.3547	.3538	.3520	2.942	3.060	3.176	3.321	3.519
260	.3431	.3469	.3503	.3498	.3461	2.991	3.128	3.262	3.426	3.647
300	.3381	.3425	.3446	.3451	.3395	3.061	3.223	3.379	3.567	3.817
350	.3320	.3372	.3378	.3400	.3328	3.155	3.346	3.528	3.746	4.031
400	.3264	.3322	.3318	.3356	.3276	3.253	3.472	3.679	3.925	4.242
450	.3214	.3275	.3266	.3318	.3238	3.353	3.598	3.830	4.102	4.451
500	.3169	.3233	.3220	.3284	.3210	3.454	3.725	3.981	4.279	4.657
600	.3102	.3150	.3144	.3210	.3191	3.650	3.969	4.268	4.614	5.048
700	.3049	.3085	.3090	.3157	.3191	3.849	4.213	4.553	4.945	5.431
800	.3007	.3036	.3053	.3123	.3203	4.048	4.456	4.836	5.271	5.808
900	.2975	.2999	.3030	.3103	.3223	4.248	4.698	5.116	5.593	6.177
1000	.2949	.2974	.3018	.3096	.3249	4.446	4.937	5.392	5.909	6.540
1200	.2915	.2951	.3020	.3112	.3312	4.836	5.406	5.931	6.526	7.244
1400	.2897	.2956	.3049	.3158	.3383	5.217	5.861	6.453	7.121	7.921
1700	.2891	.2999	.3125	.3265	.3497	5.771	6.519	7.205	7.975	8.892
2000	.2903	.3072	.3228	.3404	.3614	6.301	7.148	7.922	8.787	9.812

C ON AU, M2/M1=16.4 — TABLE 3-8

K/KL = E,KEV	DAMAGE KURTOSIS					IONIZATION KURTOSIS				
	0.6	0.8	1.0	1.25	1.6	0.6	0.8	1.0	1.25	1.6
.20	4.201	4.165	4.133	4.095	4.048	4.354	4.320	4.289	4.252	4.206
.23	4.197	4.161	4.128	4.090	4.042	4.351	4.316	4.284	4.248	4.200
.26	4.194	4.157	4.124	4.085	4.036	4.348	4.313	4.281	4.243	4.195
.30	4.190	4.153	4.119	4.080	4.030	4.345	4.309	4.276	4.238	4.190
.35	4.186	4.148	4.113	4.073	4.023	4.341	4.305	4.271	4.232	4.183
.40	4.182	4.143	4.108	4.068	4.017	4.338	4.301	4.267	4.227	4.177
.45	4.178	4.139	4.103	4.063	4.011	4.335	4.297	4.262	4.223	4.172
.50	4.175	4.135	4.099	4.058	4.006	4.332	4.294	4.259	4.218	4.167
.60	4.168	4.128	4.091	4.049	3.996	4.327	4.287	4.252	4.211	4.158
.70	4.163	4.121	4.084	4.041	3.987	4.322	4.282	4.245	4.204	4.150
.80	4.157	4.115	4.077	4.034	3.979	4.317	4.277	4.240	4.197	4.143
.90	4.152	4.110	4.071	4.027	3.972	4.313	4.272	4.235	4.191	4.137
1.0	4.148	4.104	4.065	4.021	3.965	4.310	4.268	4.230	4.186	4.130
1.2	4.139	4.095	4.055	4.009	3.952	4.303	4.260	4.221	4.177	4.120
1.4	4.131	4.086	4.045	3.999	3.941	4.297	4.253	4.213	4.168	4.110
1.7	4.119	4.073	4.031	3.984	3.925	4.288	4.243	4.203	4.156	4.097
2.0	4.109	4.062	4.019	3.971	3.911	4.280	4.234	4.193	4.145	4.085
2.3	4.099	4.051	4.008	3.959	3.898	4.272	4.225	4.183	4.134	4.073
2.6	4.090	4.041	3.997	3.948	3.886	4.265	4.217	4.173	4.124	4.062
3.0	4.079	4.029	3.984	3.934	3.871	4.255	4.206	4.162	4.111	4.048
3.5	4.065	4.014	3.968	3.917	3.854	4.244	4.193	4.148	4.097	4.032
4.0	4.052	4.000	3.953	3.902	3.837	4.233	4.182	4.136	4.083	4.017
4.5	4.040	3.986	3.939	3.887	3.821	4.223	4.171	4.124	4.071	4.003
5.0	4.028	3.974	3.926	3.873	3.807	4.214	4.161	4.113	4.059	3.990
6.0	4.005	3.949	3.900	3.846	3.779	4.200	4.145	4.095	4.039	3.969
7.0	3.983	3.926	3.877	3.821	3.753	4.186	4.129	4.078	4.021	3.949
8.0	3.963	3.905	3.854	3.798	3.729	4.172	4.114	4.062	4.003	3.929
9.0	3.944	3.885	3.833	3.776	3.706	4.159	4.099	4.046	3.986	3.911
10	3.926	3.866	3.814	3.756	3.685	4.145	4.084	4.030	3.969	3.892
12	3.892	3.830	3.777	3.718	3.646	4.117	4.054	3.997	3.934	3.854
14	3.861	3.798	3.743	3.683	3.610	4.090	4.024	3.966	3.901	3.819
17	3.819	3.753	3.697	3.636	3.562	4.053	3.984	3.923	3.855	3.770
20	3.780	3.713	3.656	3.594	3.518	4.019	3.947	3.884	3.813	3.725
23	3.743	3.675	3.617	3.553	3.476	3.990	3.916	3.850	3.777	3.686
26	3.709	3.639	3.580	3.516	3.437	3.963	3.886	3.819	3.743	3.649
30	3.667	3.596	3.536	3.472	3.390	3.930	3.850	3.780	3.701	3.604
35	3.619	3.547	3.487	3.421	3.337	3.891	3.808	3.734	3.652	3.552
40	3.577	3.503	3.442	3.376	3.290	3.855	3.768	3.691	3.606	3.504
45	3.538	3.464	3.402	3.335	3.248	3.820	3.730	3.651	3.563	3.459
50	3.502	3.427	3.365	3.298	3.210	3.787	3.694	3.612	3.523	3.416
60	3.439	3.363	3.299	3.231	3.145	3.724	3.624	3.536	3.442	3.330
70	3.384	3.307	3.242	3.173	3.089	3.665	3.559	3.467	3.369	3.254
80	3.335	3.257	3.193	3.123	3.041	3.610	3.500	3.404	3.304	3.186
90	3.292	3.214	3.149	3.079	2.998	3.559	3.445	3.346	3.243	3.125
100	3.253	3.175	3.110	3.040	2.961	3.510	3.393	3.292	3.188	3.069
120	3.183	3.106	3.040	2.970	2.892	3.415	3.296	3.193	3.087	2.969
140	3.124	3.048	2.983	2.913	2.836	3.330	3.210	3.107	3.000	2.885
170	3.051	2.977	2.913	2.845	2.770	3.219	3.099	2.996	2.890	2.780
200	2.993	2.920	2.858	2.792	2.719	3.123	3.005	2.903	2.798	2.695
230	2.943	2.873	2.812	2.749	2.681	3.041	2.924	2.824	2.722	2.626
260	2.902	2.834	2.775	2.715	2.652	2.970	2.854	2.755	2.657	2.567
300	2.856	2.792	2.736	2.680	2.623	2.887	2.773	2.677	2.584	2.502
350	2.811	2.751	2.700	2.649	2.600	2.799	2.688	2.596	2.509	2.437
400	2.777	2.721	2.674	2.628	2.586	2.725	2.617	2.530	2.447	2.384
450	2.750	2.699	2.656	2.614	2.579	2.661	2.557	2.474	2.396	2.341
500	2.729	2.682	2.644	2.607	2.577	2.605	2.506	2.426	2.354	2.305
600	2.701	2.665	2.635	2.608	2.588	2.515	2.426	2.354	2.292	2.253
700	2.687	2.660	2.639	2.620	2.608	2.444	2.363	2.300	2.246	2.215
800	2.681	2.663	2.650	2.638	2.633	2.386	2.314	2.258	2.211	2.187
900	2.683	2.673	2.666	2.661	2.661	2.339	2.274	2.225	2.184	2.166
1000	2.689	2.686	2.685	2.686	2.691	2.299	2.242	2.198	2.163	2.149
1200	2.712	2.722	2.732	2.742	2.755	2.238	2.192	2.158	2.132	2.127
1400	2.744	2.765	2.783	2.802	2.821	2.194	2.158	2.132	2.113	2.114
1700	2.803	2.838	2.866	2.894	2.918	2.149	2.124	2.107	2.098	2.105
2000	2.869	2.914	2.951	2.986	3.013	2.122	2.106	2.095	2.091	2.103

	REFLECTION COEFFICIENT					SPUTTERING EFFICIENCY				
K/KL =	0.6	0.8	1.0	1.25	1.6	0.6	0.8	1.0	1.25	1.6
E,KEV										
.20	.2925	.2867	.2813	.2748	.2664	.1856	.1679	.1527	.1363	.1175
.23	.2920	.2861	.2805	.2738	.2670	.1842	.1663	.1509	.1345	.1156
.26	.2915	.2855	.2798	.2730	.2669	.1829	.1648	.1493	.1328	.1139
.30	.2909	.2848	.2790	.2721	.2663	.1814	.1631	.1475	.1309	.1120
.35	.2903	.2840	.2781	.2711	.2649	.1797	.1612	.1455	.1288	.1099
.40	.2897	.2833	.2773	.2702	.2631	.1782	.1596	.1437	.1270	.1080
.45	.2891	.2827	.2766	.2694	.2613	.1769	.1581	.1422	.1254	.1064
.50	.2886	.2821	.2759	.2686	.2593	.1757	.1568	.1407	.1239	.1049
.60	.2877	.2810	.2746	.2673	.2555	.1735	.1544	.1382	.1213	.1024
.70	.2869	.2800	.2735	.2661	.2518	.1716	.1523	.1361	.1191	.1002
.80	.2861	.2791	.2725	.2649	.2483	.1700	.1505	.1341	.1172	.0983
.90	.2854	.2782	.2715	.2638	.2452	.1684	.1488	.1324	.1154	.0966
1.0	.2847	.2774	.2706	.2628	.2422	.1671	.1473	.1309	.1139	.0950
1.2	.2834	.2760	.2691	.2611	.2372	.1646	.1447	.1281	.1111	.0923
1.4	.2823	.2746	.2676	.2594	.2328	.1624	.1423	.1257	.1087	.0900
1.7	.2807	.2728	.2656	.2567	.2271	.1596	.1393	.1226	.1056	.0871
2.0	.2792	.2711	.2637	.2539	.2221	.1571	.1367	.1199	.1030	.0846
2.3	.2778	.2697	.2620	.2499	.2176	.1549	.1343	.1176	.1006	.0824
2.6	.2766	.2683	.2603	.2460	.2136	.1528	.1322	.1154	.0986	.0804
3.0	.2749	.2666	.2580	.2411	.2089	.1504	.1296	.1129	.0961	.0781
3.5	.2730	.2645	.2552	.2354	.2038	.1476	.1267	.1100	.0933	.0755
4.0	.2712	.2624	.2524	.2303	.1993	.1450	.1242	.1074	.0908	.0732
4.5	.2695	.2604	.2496	.2257	.1954	.1427	.1218	.1051	.0886	.0712
5.0	.2678	.2585	.2468	.2215	.1920	.1406	.1196	.1030	.0866	.0694
6.0	.2651	.2552	.2412	.2157	.1872	.1367	.1157	.0991	.0829	.0661
7.0	.2624	.2518	.2357	.2105	.1829	.1331	.1122	.0958	.0798	.0632
8.0	.2596	.2482	.2303	.2057	.1788	.1300	.1090	.0927	.0769	.0607
9.0	.2568	.2444	.2252	.2011	.1750	.1270	.1061	.0900	.0744	.0584
10	.2539	.2404	.2202	.1967	.1712	.1243	.1035	.0875	.0721	.0564
12	.2475	.2305	.2097	.1868	.1625	.1193	.0987	.0829	.0679	.0528
14	.2414	.2211	.2002	.1780	.1546	.1149	.0944	.0789	.0643	.0497
17	.2326	.2084	.1878	.1663	.1442	.1090	.0888	.0737	.0596	.0457
20	.2246	.1973	.1772	.1565	.1354	.1039	.0840	.0693	.0557	.0424
23	.2174	.1894	.1692	.1493	.1289	.0991	.0796	.0653	.0522	.0394
26	.2106	.1825	.1623	.1431	.1234	.0949	.0757	.0618	.0491	.0369
30	.2023	.1744	.1544	.1360	.1172	.0898	.0710	.0576	.0455	.0339
35	.1929	.1657	.1459	.1284	.1105	.0841	.0659	.0531	.0417	.0308
40	.1843	.1581	.1386	.1219	.1048	.0792	.0615	.0492	.0384	.0282
45	.1763	.1513	.1321	.1162	.0997	.0747	.0576	.0458	.0355	.0259
50	.1690	.1451	.1263	.1111	.0952	.0708	.0542	.0428	.0329	.0239
60	.1544	.1328	.1157	.1015	.0865	.0637	.0483	.0377	.0286	.0207
70	.1419	.1223	.1068	.0935	.0793	.0578	.0434	.0335	.0251	.0181
80	.1313	.1134	.0993	.0866	.0732	.0527	.0393	.0300	.0223	.0160
90	.1221	.1058	.0927	.0807	.0680	.0484	.0358	.0271	.0199	.0143
100	.1142	.0991	.0870	.0756	.0635	.0446	.0328	.0246	.0178	.0128
120	.1028	.0889	.0775	.0673	.0567	.0383	.0279	.0207	.0148	.0106
140	.0939	.0809	.0699	.0606	.0513	.0334	.0240	.0177	.0126	.0089
170	.0836	.0714	.0610	.0528	.0451	.0276	.0196	.0143	.0101	.0071
200	.0756	.0641	.0540	.0468	.0403	.0231	.0162	.0118	.0083	.0058
230	.0690	.0581	.0488	.0422	.0363	.0198	.0138	.0100	.0069	.0048
250	.0634	.0532	.0446	.0384	.0330	.0173	.0119	.0085	.0059	.0041
300	.0573	.0478	.0400	.0343	.0294	.0145	.0100	.0071	.0048	.0034
350	.0511	.0424	.0354	.0303	.0257	.0119	.0082	.0057	.0039	.0027
400	.0461	.0380	.0318	.0270	.0227	.0100	.0068	.0047	.0031	.0022
450	.0419	.0344	.0288	.0243	.0203	.0084	.0058	.0040	.0026	.0019
500	.0384	.0314	.0263	.0221	.0182	.0072	.0049	.0034	.0022	.0016
600	.0327	.0265	.0222	.0183	.0150	.0055	.0037	.0025	.0016	.0012
700	.0283	.0228	.0189	.0154	.0126	.0043	.0029	.0020	.0013	.0009
800	.0248	.0199	.0164	.0130	.0107	.0035	.0024	.0016	.0010	.0007
900	.0220	.0175	.0143	.0112	.0092	.0029	.0019	.0013	.0008	.0006
1000	.0196	.0155	.0125	.0096	.0080	.0024	.0016	.0011	.0007	.0005
1200	.0158	.0124	.0098	.0073	.0060	.0018	.0012	.0008	.0006	.0004
1400	.0130	.0101	.0078	.0056	.0046	.0014	.0010	.0007	.0005	.0003
1700	.0098	.0075	.0056	.0038	.0032	.0010	.0007	.0005	.0003	.0002
2000	.0075	.0057	.0041	.0027	.0022	.0008	.0005	.0004	.0002	.0002

TABLE 3-10

C ON AU, M2/M1=16.4

	IONIZATION DEFICIENCY					SPUTTERING YIELD ALPHA				
K/KL=	0.6	0.8	1.0	1.25	1.6	0.6	0.8	1.0	1.25	1.6
E,KEV										
.20	.0444	.0537	.0611	.0685	.0759	1.629	1.557	1.490	1.414	1.319
.23	.0451	.0544	.0619	.0692	.0765	1.624	1.551	1.483	1.406	1.309
.26	.0458	.0551	.0626	.0698	.0770	1.620	1.545	1.477	1.399	1.301
.30	.0465	.0559	.0633	.0705	.0776	1.614	1.539	1.469	1.390	1.291
.35	.0473	.0567	.0642	.0713	.0782	1.609	1.532	1.461	1.380	1.280
.40	.0480	.0575	.0649	.0719	.0787	1.604	1.525	1.454	1.372	1.271
.45	.0487	.0581	.0655	.0725	.0792	1.599	1.520	1.447	1.365	1.263
.50	.0492	.0587	.0661	.0730	.0795	1.595	1.515	1.441	1.358	1.255
.60	.0502	.0597	.0670	.0738	.0802	1.589	1.506	1.431	1.347	1.243
.70	.0510	.0605	.0678	.0745	.0806	1.583	1.499	1.423	1.337	1.231
.80	.0517	.0612	.0684	.0750	.0810	1.578	1.493	1.415	1.329	1.222
.90	.0523	.0618	.0690	.0755	.0813	1.573	1.487	1.409	1.321	1.213
1.0	.0529	.0624	.0695	.0759	.0816	1.569	1.482	1.403	1.314	1.206
1.2	.0539	.0633	.0703	.0765	.0819	1.562	1.473	1.393	1.302	1.192
1.4	.0547	.0641	.0709	.0770	.0822	1.557	1.466	1.384	1.292	1.181
1.7	.0557	.0650	.0717	.0776	.0824	1.550	1.456	1.373	1.280	1.167
2.0	.0565	.0657	.0723	.0779	.0825	1.544	1.448	1.363	1.269	1.155
2.3	.0572	.0663	.0728	.0782	.0825	1.538	1.442	1.355	1.260	1.144
2.6	.0578	.0668	.0732	.0784	.0825	1.534	1.436	1.348	1.252	1.135
3.0	.0585	.0674	.0736	.0786	.0824	1.529	1.429	1.340	1.242	1.124
3.5	.0592	.0679	.0739	.0787	.0821	1.523	1.421	1.331	1.231	1.112
4.0	.0598	.0684	.0742	.0787	.0818	1.519	1.415	1.323	1.222	1.102
4.5	.0603	.0687	.0743	.0787	.0815	1.514	1.409	1.316	1.214	1.093
5.0	.0607	.0690	.0744	.0785	.0811	1.511	1.403	1.309	1.207	1.084
6.0	.0614	.0694	.0745	.0782	.0803	1.504	1.394	1.298	1.194	1.070
7.0	.0619	.0696	.0745	.0778	.0794	1.499	1.387	1.289	1.182	1.057
8.0	.0623	.0697	.0743	.0773	.0785	1.494	1.380	1.280	1.172	1.046
9.0	.0626	.0698	.0740	.0768	.0776	1.490	1.373	1.272	1.163	1.036
10	.0628	.0697	.0738	.0762	.0767	1.485	1.367	1.265	1.155	1.026
12	.0631	.0695	.0730	.0749	.0749	1.479	1.357	1.252	1.140	1.010
14	.0632	.0692	.0722	.0737	.0731	1.472	1.348	1.241	1.127	.9950
17	.0632	.0684	.0709	.0717	.0705	1.463	1.334	1.225	1.109	.9753
20	.0629	.0676	.0696	.0698	.0681	1.454	1.322	1.210	1.092	.9577
23	.0626	.0667	.0681	.0679	.0657	1.445	1.310	1.197	1.078	.9413
26	.0622	.0657	.0667	.0661	.0634	1.437	1.299	1.184	1.064	.9262
30	.0615	.0644	.0648	.0637	.0606	1.426	1.284	1.167	1.046	.9076
35	.0606	.0627	.0626	.0610	.0574	1.412	1.266	1.148	1.026	.8863
40	.0596	.0610	.0604	.0584	.0544	1.399	1.250	1.129	1.006	.8668
45	.0586	.0594	.0583	.0560	.0517	1.385	1.234	1.112	.9871	.8488
50	.0575	.0578	.0564	.0537	.0493	1.371	1.219	1.095	.9688	.8320
60	.0553	.0545	.0525	.0493	.0446	1.343	1.190	1.060	.9293	.7993
70	.0531	.0515	.0491	.0453	.0407	1.315	1.163	1.029	.8940	.7707
80	.0510	.0487	.0460	.0419	.0372	1.289	1.137	1.001	.8627	.7456
90	.0489	.0461	.0432	.0389	.0343	1.264	1.113	.9744	.8348	.7233
100	.0470	.0437	.0406	.0362	.0317	1.239	1.090	.9502	.8100	.7034
120	.0432	.0394	.0361	.0318	.0275	1.190	1.042	.9059	.7705	.6707
140	.0398	.0357	.0322	.0283	.0242	1.145	.9983	.8675	.7378	.6434
170	.0354	.0309	.0275	.0240	.0203	1.085	.9419	.8185	.6974	.6096
200	.0317	.0271	.0238	.0207	.0173	1.033	.8942	.7778	.6643	.5820
230	.0287	.0235	.0205	.0178	.0151	.9906	.8571	.7450	.6354	.5586
260	.0261	.0206	.0178	.0154	.0133	.9531	.8253	.7168	.6105	.5385
300	.0231	.0174	.0149	.0129	.0114	.9096	.7889	.6846	.5819	.5155
350	.0201	.0144	.0122	.0106	.0095	.8632	.7506	.6509	.5520	.4914
400	.0175	.0123	.0104	.0090	.0080	.8237	.7183	.6225	.5270	.4711
450	.0154	.0109	.0092	.0080	.0068	.7895	.6904	.5981	.5056	.4537
500	.0136	.0101	.0085	.0074	.0058	.7595	.6659	.5768	.4872	.4386
600	.0109	.0122	.0105	.0091	.0039	.7093	.6238	.5403	.4572	.4129
700	.0087	.0143	.0123	.0107	.0024	.6687	.5897	.5111	.4334	.3923
800	.0071	.0159	.0138	.0120	.0014	.6351	.5614	.4870	.4142	.3756
900	.0057	.0169	.0147	.0128	.0006	.6067	.5376	.4669	.3984	.3617
1000	.0046	.0173	.0152	.0132	0.0000	.5825	.5171	.4499	.3850	.3501
1200	.0030	.0167	.0147	.0127	0.0000	.5433	.4841	.4230	.3641	.3318
1400	.0018	.0143	.0126	.0110	0.0000	.5131	.4587	.4028	.3486	.3184
1700	.0007	.0083	.0073	.0063	0.0000	.4794	.4302	.3811	.3320	.3044
2000	0.0000	0.0000	0.0000	0.0000	0.0000	.4552	.4097	.3664	.3209	.2954

C ON AU, M2/M1=16.4

2000 KEV

200 KEV

20 KEV

2 KEV

.20 KEV

K/KL=	FRACTIONAL DEPOSITED ENERGY					MEAN RANGE, MICROGRAM/SQ.CM.				
E,KEV	0.65	0.80	1.0	1.3	1.6	0.65	0.80	1.0	1.3	1.6
.10	.8099	.7745	.7312	.6732	.6223	.5580	.5553	.5518	.5466	.5415
.12	.8049	.7688	.7246	.6658	.6143	.6355	.6323	.6280	.6218	.6157
.14	.8005	.7638	.7190	.6594	.6075	.7052	.7015	.6967	.6896	.6826
.17	.7949	.7574	.7118	.6513	.5987	.8016	.7973	.7916	.7833	.7752
.20	.7900	.7519	.7056	.6443	.5912	.8921	.8872	.8808	.8713	.8622
.23	.7858	.7471	.7001	.6383	.5847	.9790	.9735	.9663	.9557	.9454
.26	.7819	.7427	.6953	.6328	.5790	1.063	1.057	1.049	1.037	1.026
.30	.7774	.7376	.6895	.6264	.5722	1.172	1.165	1.156	1.143	1.130
.35	.7723	.7319	.6832	.6194	.5647	1.305	1.297	1.286	1.271	1.256
.40	.7678	.7269	.6776	.6132	.5581	1.433	1.424	1.412	1.395	1.378
.45	.7638	.7223	.6725	.6076	.5523	1.558	1.548	1.535	1.515	1.496
.50	.7601	.7182	.6679	.6026	.5470	1.680	1.668	1.654	1.632	1.612
.60	.7536	.7109	.6599	.5938	.5377	1.905	1.891	1.874	1.849	1.825
.70	.7478	.7045	.6528	.5861	.5297	2.124	2.109	2.089	2.060	2.032
.80	.7427	.6988	.6466	.5793	.5226	2.340	2.322	2.299	2.266	2.234
.90	.7380	.6936	.6409	.5732	.5162	2.551	2.531	2.506	2.468	2.432
1.0	.7337	.6889	.6357	.5676	.5105	2.759	2.737	2.708	2.666	2.626
1.2	.7261	.6804	.6265	.5576	.5002	3.142	3.116	3.082	3.033	2.985
1.4	.7192	.6729	.6183	.5490	.4913	3.522	3.492	3.452	3.394	3.339
1.7	.7102	.6630	.6076	.5376	.4797	4.090	4.052	4.002	3.930	3.862
2.0	.7022	.6542	.5983	.5278	.4697	4.655	4.608	4.548	4.461	4.378
2.3	.6950	.6464	.5899	.5190	.4608	5.156	5.103	5.034	4.936	4.843
2.6	.6884	.6393	.5823	.5110	.4528	5.661	5.602	5.524	5.413	5.308
3.0	.6803	.6306	.5730	.5014	.4432	6.348	6.277	6.186	6.055	5.931
3.5	.6711	.6207	.5626	.4907	.4326	7.225	7.138	7.026	6.867	6.717
4.0	.6627	.6117	.5532	.4811	.4231	8.118	8.012	7.878	7.687	7.508
4.5	.6549	.6034	.5446	.4723	.4144	9.022	8.896	8.736	8.510	8.300
5.0	.6476	.5957	.5366	.4643	.4065	9.932	9.785	9.597	9.335	9.091
6.0	.6343	.5816	.5220	.4497	.3923	11.58	11.40	11.17	10.85	10.55
7.0	.6222	.5690	.5091	.4368	.3799	13.32	13.10	12.81	12.41	12.05
8.0	.6111	.5575	.4974	.4253	.3688	15.13	14.85	14.50	14.02	13.57
9.0	.6009	.5469	.4866	.4148	.3588	16.98	16.65	16.23	15.65	15.12
10	.5912	.5370	.4767	.4052	.3496	18.88	18.48	17.98	17.29	16.68
12	.5734	.5187	.4585	.3877	.3331	22.39	21.88	21.25	20.40	19.62
14	.5573	.5025	.4424	.3724	.3188	26.11	25.46	24.68	23.61	22.64
17	.5359	.4809	.4214	.3525	.3004	32.06	31.15	30.06	28.60	27.29
20	.5168	.4620	.4030	.3355	.2848	38.34	37.12	35.65	33.71	32.04
23	.4993	.4448	.3866	.3204	.2711	44.08	42.63	40.87	38.54	36.55
26	.4834	.4293	.3718	.3070	.2590	50.10	48.38	46.26	43.47	41.12
30	.4642	.4107	.3543	.2912	.2449	58.53	56.36	53.68	50.17	47.30
35	.4430	.3903	.3352	.2742	.2298	69.61	66.75	63.24	58.72	55.12
40	.4241	.3723	.3186	.2595	.2167	81.15	77.48	73.03	67.37	62.98
45	.4073	.3563	.3039	.2466	.2053	93.00	88.45	82.97	76.08	70.85
50	.3920	.3419	.2907	.2351	.1953	105.1	99.56	92.97	84.79	78.69
60	.3649	.3168	.2681	.2155	.1781	128.3	120.9	112.3	101.7	93.82
70	.3420	.2956	.2492	.1994	.1641	152.3	142.9	131.9	118.6	108.8
80	.3222	.2775	.2331	.1859	.1523	176.8	165.2	151.6	135.4	123.7
90	.3048	.2617	.2192	.1743	.1423	201.8	187.7	171.3	152.0	138.4
100	.2895	.2479	.2071	.1643	.1337	226.9	210.2	191.0	168.4	152.9
120	.2637	.2249	.1870	.1484	.1199	276.3	254.4	229.4	200.3	180.8
140	.2425	.2063	.1708	.1358	.1090	325.9	298.5	267.3	231.4	208.0
170	.2168	.1839	.1515	.1207	.0962	399.9	363.9	323.0	276.5	247.4
200	.1963	.1662	.1363	.1089	.0864	473.1	428.1	377.1	319.9	285.1
230	.1801	.1523	.1244	.0988	.0785	543.9	490.1	429.2	361.5	321.1
260	.1666	.1407	.1145	.0903	.0720	613.6	550.8	479.9	401.6	355.8
300	.1516	.1280	.1037	.0809	.0649	704.5	629.8	545.4	453.1	400.3
350	.1365	.1152	.0929	.0715	.0579	815.1	725.4	624.1	514.5	453.4
400	.1243	.1049	.0843	.0639	.0522	922.4	817.7	699.8	573.1	504.0
450	.1142	.0964	.0772	.0576	.0476	1026.	906.0	772.7	629.2	552.8
500	.1057	.0892	.0712	.0525	.0438	1128.	993.4	843.0	683.0	598.8
600	.0921	.0778	.0618	.0446	.0378	1321.	1159.	976.7	784.7	686.5
700	.0818	.0691	.0547	.0390	.0334	1505.	1314.	1102.	879.7	768.4
800	.0737	.0622	.0492	.0349	.0299	1680.	1462.	1221.	969.0	845.3
900	.0673	.0567	.0448	.0320	.0272	1847.	1603.	1334.	1053.	918.1
1000	.0620	.0521	.0412	.0300	.0250	2007.	1738.	1441.	1134.	987.3

K/KL=	MEAN DAMAGE DEPTH, MICROGRAM/SQ.CM.					MEAN IONZN.DEPTH, MICROGRAM/SQ.CM.				
E,KEV	0.65	0.80	1.0	1.3	1.6	0.65	0.80	1.0	1.3	1.6
.10	.5256	.5202	.5133	.5033	.4939	.5166	.5107	.5030	.4921	.4816
.12	.5979	.5915	.5832	.5714	.5603	.5874	.5803	.5712	.5582	.5459
.14	.6630	.6557	.6463	.6329	.6203	.6512	.6432	.6328	.6181	.6041
.17	.7529	.7445	.7336	.7180	.7033	.7395	.7301	.7181	.7009	.6847
.20	.8374	.8278	.8154	.7978	.7811	.8223	.8117	.7980	.7785	.7601
.23	.9184	.9076	.8938	.8740	.8554	.9016	.8897	.8745	.8527	.8322
.26	.9968	.9848	.9695	.9477	.9272	.9784	.9652	.9484	.9243	.9016
.30	1.098	1.085	1.067	1.043	1.020	1.078	1.063	1.044	1.017	.9911
.35	1.221	1.205	1.186	1.158	1.131	1.198	1.181	1.159	1.128	1.099
.40	1.340	1.322	1.300	1.269	1.239	1.314	1.295	1.270	1.235	1.203
.45	1.455	1.436	1.411	1.376	1.343	1.426	1.405	1.378	1.339	1.303
.50	1.567	1.546	1.519	1.480	1.444	1.536	1.512	1.482	1.440	1.400
.60	1.775	1.750	1.719	1.674	1.632	1.739	1.711	1.676	1.627	1.581
.70	1.977	1.949	1.912	1.861	1.813	1.936	1.905	1.864	1.808	1.755
.80	2.175	2.143	2.102	2.044	1.990	2.129	2.093	2.047	1.983	1.924
.90	2.369	2.333	2.287	2.222	2.162	2.317	2.277	2.226	2.155	2.088
1.0	2.559	2.519	2.468	2.396	2.330	2.501	2.457	2.401	2.322	2.249
1.2	2.910	2.862	2.803	2.719	2.642	2.842	2.790	2.724	2.632	2.546
1.4	3.256	3.201	3.132	3.036	2.946	3.178	3.118	3.041	2.935	2.836
1.7	3.771	3.704	3.620	3.503	3.395	3.676	3.602	3.509	3.379	3.261
2.0	4.281	4.201	4.101	3.962	3.835	4.167	4.079	3.968	3.815	3.675
2.3	4.735	4.645	4.532	4.376	4.233	4.608	4.509	4.384	4.211	4.053
2.6	5.192	5.091	4.964	4.789	4.628	5.050	4.938	4.798	4.604	4.427
3.0	5.808	5.691	5.544	5.340	5.155	5.643	5.513	5.350	5.125	4.921
3.5	6.591	6.450	6.275	6.034	5.815	6.392	6.236	6.041	5.775	5.535
4.0	7.384	7.219	7.012	6.730	6.475	7.145	6.961	6.733	6.422	6.143
4.5	8.183	7.990	7.751	7.425	7.133	7.902	7.687	7.423	7.065	6.745
5.0	8.986	8.764	8.490	8.119	7.788	8.658	8.412	8.110	7.703	7.341
6.0	10.45	10.18	9.849	9.400	9.000	10.05	9.748	9.381	8.888	8.451
7.0	11.97	11.65	11.25	10.71	10.23	11.48	11.11	10.67	10.08	9.562
8.0	13.55	13.15	12.68	12.04	11.48	12.94	12.51	11.98	11.28	10.67
9.0	15.16	14.69	14.13	13.39	12.73	14.42	13.91	13.30	12.49	11.78
10	16.79	16.25	15.60	14.74	13.99	15.92	15.33	14.62	13.69	12.88
12	19.82	19.15	18.35	17.29	16.36	18.74	18.01	17.12	15.99	15.01
14	23.01	22.13	21.19	19.90	18.78	21.66	20.75	19.67	18.29	17.12
17	28.06	26.94	25.63	23.93	22.48	26.18	24.97	23.54	21.75	20.25
20	33.34	31.89	30.20	28.04	26.23	30.82	29.26	27.45	25.18	23.33
23	38.17	36.44	34.46	31.93	29.77	35.21	33.32	31.20	28.43	26.18
26	43.21	41.15	38.83	35.88	33.35	39.70	37.44	34.96	31.66	28.99
30	50.23	47.67	44.83	41.23	38.18	45.79	43.00	39.98	35.93	32.72
35	59.39	56.12	52.52	48.01	44.26	53.53	50.03	46.27	41.22	37.32
40	68.88	64.83	60.37	54.84	50.38	61.37	57.11	52.53	46.44	41.87
45	78.59	73.71	68.30	61.69	56.49	69.25	64.20	58.74	51.59	46.36
50	88.45	82.69	76.29	68.52	62.58	77.12	71.27	64.88	56.66	50.79
60	107.4	100.1	91.69	81.63	74.31	92.35	85.08	76.78	66.55	59.56
70	127.0	117.8	107.3	94.73	85.95	107.6	98.72	88.43	76.10	68.00
80	146.9	135.7	122.9	107.7	97.48	122.7	112.1	99.80	85.33	76.12
90	167.0	153.8	138.5	120.6	108.9	137.6	125.3	110.9	94.26	83.93
100	187.2	171.8	154.0	133.4	120.1	152.4	138.3	121.7	102.9	91.46
120	226.6	206.8	184.2	158.1	141.7	181.5	163.2	142.6	119.4	105.6
140	266.2	241.7	213.9	182.2	162.8	209.6	187.2	162.6	134.9	119.0
170	325.2	293.5	257.7	217.3	193.4	250.2	221.7	190.9	156.7	137.7
200	383.6	344.5	300.4	251.3	222.8	288.9	254.4	217.6	177.1	155.0
230	440.2	393.7	341.4	283.5	250.9	325.8	285.8	242.8	196.1	171.1
260	495.8	442.0	381.3	314.8	278.0	361.1	315.8	266.8	214.0	186.3
300	568.5	504.9	433.2	355.1	313.0	406.1	353.8	297.0	236.6	205.4
350	657.2	581.3	495.8	403.5	354.9	459.3	398.7	332.5	262.8	227.7
400	743.3	655.2	556.1	449.9	395.0	509.5	441.0	365.9	287.4	248.6
450	827.0	726.9	614.4	494.5	433.6	557.2	481.0	397.4	310.4	268.3
500	908.3	796.3	670.8	537.5	470.7	602.7	519.0	427.4	332.2	286.9
600	1065.	929.3	778.4	619.2	541.3	688.0	590.0	482.9	372.2	321.6
700	1213.	1055.	879.9	696.1	607.5	766.8	655.4	534.2	409.7	353.4
800	1354.	1175.	976.2	768.7	670.1	840.2	716.2	581.7	443.9	383.0
900	1490.	1289.	1068.	837.6	729.4	909.3	773.1	626.2	475.9	410.6
1000	1619.	1398.	1156.	903.3	786.0	974.4	826.7	668.1	505.9	436.6

K/KL=	RELATIVE RANGE STRAGGLING					RELATIVE DAMAGE STRAGGLING				
E,KEV	0.65	0.80	1.0	1.3	1.6	0.65	0.80	1.0	1.3	1.6
.10	2.575	2.488	2.381	2.238	2.112	1.581	1.535	1.479	1.404	1.338
.12	2.558	2.469	2.360	2.215	2.088	1.572	1.525	1.468	1.392	1.326
.14	2.543	2.453	2.343	2.196	2.068	1.565	1.517	1.459	1.383	1.316
.17	2.522	2.431	2.319	2.170	2.041	1.555	1.507	1.448	1.370	1.303
.20	2.504	2.411	2.298	2.148	2.017	1.546	1.497	1.437	1.359	1.291
.23	2.488	2.394	2.279	2.128	1.996	1.538	1.488	1.428	1.349	1.281
.26	2.473	2.377	2.262	2.109	1.977	1.531	1.481	1.420	1.340	1.271
.30	2.454	2.358	2.241	2.087	1.954	1.522	1.471	1.409	1.329	1.260
.35	2.433	2.335	2.217	2.062	1.929	1.512	1.460	1.398	1.317	1.247
.40	2.413	2.314	2.195	2.039	1.905	1.502	1.450	1.387	1.305	1.236
.45	2.394	2.294	2.175	2.018	1.884	1.493	1.441	1.377	1.295	1.225
.50	2.377	2.276	2.156	1.999	1.864	1.485	1.432	1.368	1.286	1.215
.60	2.345	2.243	2.121	1.963	1.829	1.470	1.416	1.352	1.268	1.198
.70	2.315	2.212	2.090	1.931	1.797	1.456	1.401	1.337	1.253	1.182
.80	2.288	2.184	2.061	1.902	1.768	1.443	1.388	1.323	1.239	1.168
.90	2.262	2.158	2.035	1.875	1.741	1.431	1.376	1.310	1.226	1.155
1.0	2.238	2.134	2.010	1.850	1.716	1.420	1.364	1.298	1.214	1.143
1.2	2.193	2.088	1.964	1.804	1.671	1.399	1.343	1.276	1.192	1.121
1.4	2.152	2.047	1.922	1.763	1.630	1.380	1.323	1.256	1.172	1.102
1.7	2.097	1.990	1.866	1.708	1.577	1.354	1.297	1.230	1.145	1.075
2.0	2.046	1.940	1.816	1.659	1.529	1.331	1.273	1.205	1.121	1.052
2.3	2.000	1.893	1.769	1.613	1.485	1.309	1.250	1.183	1.099	1.030
2.6	1.956	1.850	1.727	1.572	1.445	1.288	1.230	1.162	1.079	1.010
3.0	1.903	1.797	1.675	1.522	1.397	1.263	1.204	1.137	1.054	.9863
3.5	1.844	1.738	1.617	1.466	1.344	1.234	1.176	1.109	1.026	.9592
4.0	1.789	1.684	1.564	1.416	1.296	1.208	1.149	1.083	1.001	.9348
4.5	1.740	1.636	1.516	1.370	1.252	1.184	1.125	1.059	.9776	.9126
5.0	1.694	1.591	1.473	1.328	1.213	1.162	1.103	1.037	.9563	.8922
6.0	1.611	1.509	1.393	1.253	1.141	1.120	1.062	.9966	.9176	.8552
7.0	1.539	1.438	1.324	1.188	1.080	1.084	1.026	.9613	.8838	.8230
8.0	1.474	1.375	1.264	1.131	1.027	1.051	.9938	.9298	.8538	.7944
9.0	1.417	1.319	1.210	1.080	.9791	1.022	.9646	.9014	.8267	.7688
10	1.365	1.268	1.161	1.035	.9367	.9948	.9380	.8756	.8022	.7456
12	1.274	1.180	1.077	.9566	.8636	.9463	.8904	.8295	.7587	.7046
14	1.197	1.105	1.006	.8908	.8025	.9044	.8494	.7900	.7216	.6696
17	1.100	1.012	.9117	.8092	.7269	.8507	.7971	.7399	.6746	.6257
20	1.020	.9352	.8450	.7425	.6655	.8054	.7532	.6979	.6356	.5893
23	.9526	.8708	.7846	.6873	.6148	.7664	.7156	.6623	.6027	.5588
26	.8945	.8155	.7329	.6403	.5718	.7323	.6828	.6313	.5742	.5325
30	.8282	.7527	.6743	.5873	.5234	.6928	.6451	.5958	.5416	.5025
35	.7590	.6873	.6136	.5326	.4736	.6508	.6051	.5583	.5074	.4711
40	.7011	.6328	.5633	.4875	.4326	.6151	.5712	.5267	.4788	.4448
45	.6518	.5867	.5208	.4495	.3982	.5842	.5420	.4996	.4543	.4224
50	.6092	.5470	.4844	.4171	.3689	.5572	.5166	.4761	.4330	.4030
60	.5409	.4836	.4265	.3659	.3229	.5123	.4747	.4377	.3987	.3715
70	.4867	.4337	.3813	.3261	.2873	.4760	.4410	.4069	.3713	.3465
80	.4424	.3932	.3447	.2942	.2587	.4459	.4131	.3816	.3489	.3260
90	.4054	.3597	.3146	.2679	.2354	.4204	.3897	.3604	.3302	.3088
100	.3740	.3315	.2893	.2459	.2158	.3987	.3697	.3424	.3143	.2942
120	.3236	.2872	.2499	.2118	.1856	.3641	.3381	.3138	.2891	.2710
140	.2847	.2534	.2200	.1861	.1628	.3371	.3134	.2917	.2695	.2531
170	.2406	.2153	.1865	.1575	.1376	.3059	.2851	.2663	.2472	.2325
200	.2077	.1871	.1619	.1365	.1191	.2821	.2636	.2471	.2303	.2171
230	.1839	.1655	.1430	.1203	.1049	.2635	.2468	.2321	.2172	.2054
260	.1652	.1483	.1280	.1076	.0938	.2485	.2333	.2200	.2067	.1961
300	.1458	.1303	.1123	.0943	.0821	.2323	.2188	.2070	.1954	.1861
350	.1274	.1130	.0973	.0817	.0710	.2164	.2046	.1943	.1843	.1764
400	.1134	.0999	.0859	.0720	.0626	.2038	.1934	.1843	.1755	.1687
450	.1024	.0894	.0769	.0644	.0560	.1937	.1843	.1762	.1684	.1624
500	.0934	.0810	.0696	.0582	.0507	.1853	.1768	.1694	.1624	.1572
600	.0796	.0682	.0585	.0489	.0427	.1721	.1651	.1587	.1528	.1487
700	.0693	.0589	.0505	.0421	.0369	.1623	.1563	.1506	.1453	.1420
800	.0610	.0519	.0444	.0369	.0325	.1546	.1493	.1442	.1393	.1364
900	.0541	.0464	.0396	.0327	.0291	.1485	.1437	.1390	.1342	.1316
1000	.0481	.0419	.0357	.0292	.0262	.1434	.1391	.1345	.1297	.1274

K/KL= E,KEV	RELATIVE IONIZATION STRAGGLING					RELATIVE TRANSVERSE RANGE STRAGGLING				
	0.65	0.80	1.0	1.3	1.6	0.65	0.80	1.0	1.3	1.6
.10	1.503	1.460	1.408	1.338	1.278	.9610	.9599	.9584	.9562	.9541
.12	1.495	1.451	1.398	1.328	1.266	.9610	.9598	.9583	.9560	.9538
.14	1.488	1.443	1.389	1.318	1.257	.9609	.9597	.9582	.9559	.9536
.17	1.478	1.433	1.378	1.306	1.244	.9609	.9596	.9580	.9557	.9534
.20	1.470	1.424	1.368	1.296	1.233	.9609	.9596	.9579	.9555	.9532
.23	1.462	1.415	1.360	1.286	1.223	.9608	.9595	.9579	.9554	.9530
.26	1.455	1.408	1.351	1.278	1.215	.9608	.9595	.9578	.9553	.9529
.30	1.446	1.399	1.342	1.267	1.204	.9609	.9595	.9578	.9552	.9528
.35	1.436	1.388	1.330	1.256	1.192	.9609	.9595	.9578	.9552	.9527
.40	1.427	1.378	1.320	1.245	1.181	.9610	.9596	.9578	.9551	.9526
.45	1.418	1.369	1.311	1.235	1.172	.9610	.9596	.9578	.9551	.9526
.50	1.410	1.361	1.302	1.226	1.163	.9611	.9597	.9578	.9552	.9526
.60	1.396	1.346	1.286	1.210	1.146	.9613	.9598	.9579	.9552	.9526
.70	1.382	1.332	1.272	1.196	1.132	.9615	.9600	.9581	.9553	.9527
.80	1.370	1.319	1.259	1.182	1.119	.9617	.9602	.9583	.9555	.9528
.90	1.358	1.307	1.247	1.170	1.106	.9620	.9604	.9585	.9557	.9530
1.0	1.347	1.296	1.235	1.159	1.095	.9622	.9607	.9587	.9559	.9532
1.2	1.326	1.275	1.214	1.138	1.074	.9627	.9611	.9592	.9563	.9535
1.4	1.308	1.256	1.195	1.119	1.056	.9632	.9617	.9597	.9568	.9540
1.7	1.282	1.230	1.169	1.093	1.031	.9640	.9625	.9605	.9576	.9548
2.0	1.259	1.207	1.146	1.071	1.009	.9649	.9634	.9614	.9585	.9557
2.3	1.238	1.185	1.125	1.050	.9893	.9657	.9642	.9622	.9593	.9566
2.6	1.218	1.166	1.106	1.031	.9712	.9666	.9651	.9631	.9602	.9575
3.0	1.194	1.142	1.082	1.008	.9494	.9678	.9662	.9643	.9615	.9588
3.5	1.166	1.114	1.055	.9828	.9249	.9693	.9678	.9659	.9631	.9605
4.0	1.142	1.090	1.031	.9598	.9031	.9708	.9693	.9675	.9648	.9623
4.5	1.119	1.067	1.009	.9389	.8832	.9724	.9709	.9691	.9666	.9642
5.0	1.098	1.046	.9889	.9196	.8651	.9739	.9726	.9708	.9683	.9660
6.0	1.059	1.008	.9517	.8845	.8320	.9769	.9757	.9741	.9718	.9697
7.0	1.024	.9745	.9193	.8540	.8034	.9800	.9788	.9774	.9753	.9734
8.0	.9939	.9448	.8908	.8273	.7785	.9830	.9820	.9807	.9788	.9771
9.0	.9666	.9182	.8653	.8035	.7564	.9860	.9851	.9840	.9824	.9809
10	.9418	.8942	.8423	.7822	.7366	.9890	.9883	.9873	.9859	.9846
12	.8983	.8521	.8024	.7454	.7026	.9947	.9942	.9936	.9927	.9918
14	.8613	.8165	.7687	.7145	.6743	1.000	1.000	.9999	.9995	.9990
17	.8148	.7719	.7268	.6764	.6395	1.009	1.009	1.009	1.009	1.010
20	.7762	.7351	.6924	.6454	.6113	1.017	1.017	1.018	1.019	1.020
23	.7434	.7042	.6638	.6199	.5881	1.024	1.026	1.027	1.029	1.030
26	.7152	.6776	.6393	.5983	.5685	1.032	1.034	1.036	1.038	1.040
30	.6831	.6475	.6118	.5741	.5467	1.042	1.044	1.047	1.050	1.053
35	.6496	.6164	.5836	.5494	.5247	1.054	1.057	1.061	1.065	1.069
40	.6218	.5908	.5604	.5293	.5070	1.066	1.070	1.074	1.079	1.084
45	.5983	.5693	.5411	.5127	.4925	1.077	1.082	1.087	1.093	1.098
50	.5781	.5510	.5248	.4988	.4806	1.088	1.094	1.100	1.107	1.112
60	.5461	.5227	.4998	.4774	.4634	1.111	1.117	1.125	1.133	1.139
70	.5213	.5010	.4809	.4615	.4510	1.132	1.139	1.148	1.158	1.165
80	.5016	.4840	.4662	.4493	.4418	1.152	1.160	1.170	1.181	1.189
90	.4856	.4703	.4546	.4399	.4348	1.171	1.180	1.191	1.203	1.211
100	.4724	.4592	.4453	.4325	.4294	1.188	1.199	1.210	1.224	1.232
120	.4532	.4430	.4323	.4230	.4222	1.221	1.232	1.246	1.261	1.270
140	.4395	.4316	.4235	.4170	.4179	1.250	1.263	1.278	1.295	1.305
170	.4254	.4202	.4153	.4121	.4145	1.291	1.306	1.322	1.341	1.352
200	.4162	.4131	.4105	.4099	.4133	1.328	1.344	1.362	1.382	1.394
230	.4103	.4088	.4080	.4092	.4134	1.364	1.381	1.400	1.421	1.433
260	.4065	.4063	.4069	.4095	.4143	1.397	1.415	1.435	1.456	1.468
300	.4034	.4046	.4066	.4107	.4161	1.438	1.457	1.477	1.500	1.511
350	.4016	.4041	.4075	.4129	.4187	1.484	1.504	1.525	1.548	1.560
400	.4012	.4047	.4091	.4155	.4216	1.526	1.546	1.568	1.591	1.603
450	.4017	.4059	.4111	.4182	.4245	1.564	1.585	1.607	1.631	1.641
500	.4027	.4075	.4133	.4209	.4273	1.599	1.620	1.643	1.666	1.677
600	.4056	.4112	.4178	.4261	.4325	1.661	1.683	1.706	1.730	1.739
700	.4091	.4151	.4221	.4307	.4371	1.716	1.738	1.761	1.784	1.793
800	.4126	.4189	.4261	.4349	.4410	1.764	1.786	1.809	1.832	1.840
900	.4161	.4224	.4298	.4385	.4445	1.807	1.829	1.852	1.875	1.882
1000	.4194	.4257	.4330	.4417	.4474	1.847	1.868	1.891	1.913	1.919

K/KL= E,KEV	RELATIVE TRANSV. DAMAGE STRAGGLING					RELATIVE TRANSV. IONZN. STRAGGLING				
	0.65	0.80	1.0	1.3	1.6	0.65	0.80	1.0	1.3	1.6
.10	.9125	.9087	.9036	.8960	.8885	.8986	.8939	.8875	.8781	.8687
.12	.9121	.9082	.9029	.8951	.8874	.8980	.8931	.8866	.8768	.8671
.14	.9118	.9077	.9023	.8943	.8864	.8975	.8925	.8858	.8758	.8658
.17	.9113	.9071	.9016	.8933	.8851	.8969	.8917	.8847	.8744	.8641
.20	.9110	.9067	.9009	.8924	.8840	.8963	.8910	.8838	.8732	.8626
.23	.9107	.9062	.9004	.8917	.8830	.8958	.8903	.8830	.8721	.8612
.26	.9104	.9059	.8999	.8910	.8822	.8954	.8898	.8823	.8711	.8600
.30	.9101	.9055	.8993	.8902	.8812	.8949	.8891	.8815	.8700	.8586
.35	.9098	.9050	.8987	.8894	.8801	.8944	.8884	.8805	.8687	.8570
.40	.9095	.9046	.8982	.8886	.8792	.8939	.8878	.8797	.8676	.8556
.45	.9093	.9043	.8977	.8880	.8783	.8935	.8872	.8789	.8666	.8543
.50	.9091	.9040	.8973	.8874	.8775	.8931	.8867	.8782	.8656	.8531
.60	.9088	.9036	.8966	.8863	.8762	.8924	.8858	.8770	.8640	.8511
.70	.9086	.9032	.8961	.8855	.8751	.8918	.8850	.8760	.8625	.8492
.80	.9084	.9029	.8956	.8847	.8741	.8913	.8843	.8750	.8612	.8475
.90	.9083	.9026	.8952	.8841	.8732	.8909	.8837	.8742	.8600	.8460
1.0	.9082	.9024	.8948	.8835	.8724	.8905	.8831	.8734	.8589	.8446
1.2	.9081	.9021	.8943	.8826	.8711	.8898	.8821	.8720	.8569	.8421
1.4	.9081	.9019	.8938	.8818	.8699	.8892	.8812	.8707	.8551	.8398
1.7	.9082	.9018	.8933	.8808	.8685	.8884	.8801	.8691	.8527	.8367
2.0	.9084	.9017	.8930	.8800	.8673	.8877	.8790	.8676	.8506	.8339
2.3	.9086	.9017	.8927	.8794	.8663	.8871	.8782	.8663	.8487	.8314
2.6	.9088	.9018	.8925	.8788	.8654	.8867	.8774	.8651	.8469	.8291
3.0	.9092	.9020	.8924	.8782	.8644	.8860	.8764	.8636	.8447	.8261
3.5	.9098	.9022	.8923	.8776	.8632	.8853	.8752	.8618	.8420	.8227
4.0	.9103	.9025	.8922	.8771	.8622	.8846	.8740	.8601	.8394	.8193
4.5	.9109	.9029	.8922	.8766	.8613	.8839	.8729	.8584	.8370	.8161
5.0	.9115	.9032	.8923	.8762	.8605	.8833	.8718	.8567	.8346	.8130
6.0	.9127	.9040	.8925	.8756	.8592	.8822	.8700	.8539	.8302	.8073
7.0	.9140	.9048	.8928	.8751	.8580	.8811	.8681	.8510	.8260	.8018
8.0	.9151	.9056	.8931	.8747	.8568	.8798	.8661	.8480	.8217	.7963
9.0	.9162	.9063	.8933	.8742	.8557	.8785	.8640	.8450	.8174	.7909
10	.9173	.9070	.8935	.8738	.8546	.8771	.8619	.8420	.8131	.7856
12	.9195	.9085	.8942	.8731	.8528	.8746	.8579	.8362	.8049	.7753
14	.9214	.9098	.8946	.8724	.8509	.8717	.8536	.8303	.7967	.7651
17	.9238	.9113	.8948	.8709	.8478	.8668	.8468	.8210	.7843	.7501
20	.9258	.9124	.8947	.8692	.8446	.8614	.8395	.8115	.7719	.7354
23	.9277	.9135	.8947	.8675	.8415	.8559	.8320	.8017	.7594	.7204
26	.9293	.9143	.8943	.8656	.8382	.8501	.8243	.7919	.7470	.7059
30	.9308	.9149	.8934	.8627	.8336	.8420	.8139	.7788	.7308	.6873
35	.9321	.9150	.8916	.8587	.8276	.8314	.8006	.7626	.7113	.6652
40	.9327	.9144	.8893	.8543	.8214	.8205	.7874	.7467	.6925	.6445
45	.9328	.9133	.8867	.8496	.8151	.8094	.7741	.7311	.6743	.6249
50	.9326	.9118	.8837	.8447	.8087	.7981	.7609	.7158	.6569	.6063
60	.9316	.9081	.8771	.8343	.7955	.7748	.7339	.6848	.6219	.5701
70	.9294	.9034	.8698	.8237	.7825	.7520	.7082	.6559	.5900	.5379
80	.9264	.8981	.8621	.8130	.7698	.7300	.6838	.6290	.5609	.5089
90	.9226	.8922	.8541	.8024	.7574	.7087	.6607	.6040	.5343	.4827
100	.9184	.8859	.8459	.7918	.7454	.6882	.6388	.5806	.5099	.4589
120	.9083	.8721	.8285	.7702	.7214	.6471	.5960	.5366	.4657	.4169
140	.8973	.8580	.8113	.7496	.6991	.6096	.5578	.4980	.4277	.3814
170	.8802	.8370	.7866	.7208	.6686	.5599	.5080	.4486	.3801	.3373
200	.8628	.8166	.7632	.6945	.6413	.5168	.4655	.4072	.3411	.3015
230	.8440	.7956	.7399	.6692	.6158	.4795	.4297	.3734	.3102	.2734
260	.8258	.7757	.7180	.6459	.5926	.4467	.3986	.3444	.2841	.2499
300	.8025	.7507	.6911	.6177	.5649	.4088	.3630	.3115	.2551	.2238
350	.7752	.7219	.6606	.5863	.5344	.3686	.3257	.2777	.2256	.1975
400	.7499	.6956	.6332	.5586	.5077	.3349	.2946	.2498	.2017	.1763
450	.7263	.6715	.6084	.5339	.4842	.3061	.2684	.2265	.1820	.1589
500	.7043	.6493	.5859	.5118	.4632	.2815	.2461	.2068	.1654	.1443
600	.6645	.6096	.5464	.4737	.4274	.2415	.2101	.1755	.1394	.1215
700	.6293	.5751	.5129	.4420	.3980	.2110	.1830	.1521	.1202	.1047
800	.5979	.5447	.4840	.4153	.3734	.1874	.1621	.1343	.1057	.0921
900	.5695	.5177	.4587	.3924	.3525	.1690	.1460	.1207	.0947	.0825
1000	.5436	.4934	.4365	.3727	.3347	.1547	.1335	.1103	.0863	.0751

K/KL=	RANGE SKEWNESS					DAMAGE SKEWNESS				
E,KEV	0.65	0.80	1.0	1.3	1.6	0.65	0.80	1.0	1.3	1.6
.10	.0937	.0880	.0803	.0688	.0573	.3139	.3142	.3146	.3151	.3154
.12	.0929	.0869	.0790	.0671	.0551	.3140	.3143	.3147	.3151	.3154
.14	.0922	.0860	.0778	.0656	.0533	.3141	.3144	.3147	.3151	.3153
.17	.0912	.0848	.0763	.0636	.0509	.3143	.3145	.3148	.3151	.3153
.20	.0904	.0838	.0750	.0619	.0487	.3144	.3146	.3149	.3151	.3153
.23	.0896	.0829	.0738	.0603	.0468	.3145	.3147	.3149	.3151	.3152
.26	.0889	.0820	.0727	.0589	.0451	.3146	.3148	.3150	.3151	.3151
.30	.0881	.0809	.0714	.0571	.0429	.3147	.3148	.3150	.3151	.3150
.35	.0871	.0797	.0698	.0551	.0404	.3148	.3149	.3150	.3150	.3148
.40	.0862	.0786	.0684	.0532	.0381	.3149	.3149	.3149	.3148	.3146
.45	.0854	.0775	.0671	.0515	.0360	.3150	.3150	.3149	.3147	.3144
.50	.0846	.0765	.0658	.0498	.0340	.3150	.3150	.3148	.3145	.3142
.60	.0832	.0747	.0635	.0468	.0304	.3151	.3150	.3147	.3143	.3137
.70	.0818	.0730	.0614	.0441	.0270	.3152	.3149	.3145	.3139	.3132
.80	.0805	.0714	.0594	.0415	.0239	.3152	.3148	.3143	.3135	.3127
.90	.0793	.0699	.0575	.0391	.0209	.3152	.3147	.3141	.3131	.3121
1.0	.0781	.0684	.0556	.0367	.0181	.3151	.3146	.3138	.3127	.3115
1.2	.0760	.0658	.0523	.0324	.0129	.3151	.3143	.3134	.3119	.3104
1.4	.0739	.0632	.0491	.0283	.0080	.3150	.3140	.3128	.3110	.3093
1.7	.0709	.0595	.0445	.0225	.0011	.3146	.3133	.3118	.3095	.3074
2.0	.0679	.0558	.0400	.0169	-.0055	.3140	.3125	.3106	.3079	.3054
2.3	.0652	.0525	.0359	.0118	-.0116	.3134	.3116	.3094	.3062	.3033
2.6	.0625	.0493	.0320	.0068	-.0175	.3126	.3106	.3080	.3045	.3012
3.0	.0590	.0449	.0267	.0003	-.0251	.3115	.3091	.3061	.3021	.2984
3.5	.0545	.0396	.0203	-.0076	-.0342	.3099	.3071	.3036	.2989	.2947
4.0	.0500	.0343	.0140	-.0152	-.0431	.3082	.3049	.3010	.2957	.2910
4.5	.0456	.0290	.0077	-.0227	-.0517	.3063	.3027	.2984	.2925	.2872
5.0	.0412	.0239	.0016	-.0300	-.0600	.3044	.3004	.2956	.2892	.2835
6.0	.0325	.0137	-.0103	-.0443	-.0762	.3009	.2962	.2905	.2831	.2765
7.0	.0240	.0038	-.0218	-.0579	-.0915	.2970	.2917	.2853	.2769	.2694
8.0	.0156	-.0058	-.0329	-.0709	-.1061	.2930	.2870	.2799	.2706	.2624
9.0	.0075	-.0152	-.0436	-.0835	-.1202	.2888	.2822	.2744	.2642	.2553
10	-.0005	-.0243	-.0541	-.0956	-.1336	.2844	.2773	.2689	.2579	.2483
12	-.0152	-.0411	-.0734	-.1180	-.1586	.2753	.2671	.2575	.2450	.2343
14	-.0295	-.0574	-.0919	-.1392	-.1821	.2662	.2570	.2463	.2324	.2208
17	-.0506	-.0811	-.1186	-.1695	-.2154	.2527	.2423	.2301	.2142	.2012
20	-.0712	-.1041	-.1442	-.1983	-.2467	.2398	.2282	.2146	.1970	.1828
23	-.0924	-.1275	-.1700	-.2269	-.2775	.2275	.2149	.2000	.1808	.1654
26	-.1130	-.1501	-.1948	-.2542	-.3068	.2157	.2021	.1861	.1654	.1489
30	-.1395	-.1789	-.2262	-.2887	-.3435	.2005	.1858	.1683	.1459	.1279
35	-.1709	-.2129	-.2631	-.3289	-.3863	.1825	.1664	.1473	.1227	.1033
40	-.2006	-.2450	-.2976	-.3663	-.4259	.1653	.1480	.1274	.1009	.0800
45	-.2287	-.2752	-.3300	-.4012	-.4628	.1490	.1305	.1084	.0802	.0580
50	-.2554	-.3037	-.3606	-.4341	-.4974	.1333	.1137	.0903	.0605	.0370
60	-.3058	-.3575	-.4178	-.4952	-.5619	.1036	.0820	.0560	.0230	-.0027
70	-.3517	-.4062	-.4694	-.5500	-.6194	.0761	.0525	.0243	-.0114	-.0391
80	-.3938	-.4505	-.5164	-.5998	-.6713	.0503	.0251	-.0052	-.0434	-.0727
90	-.4326	-.4916	-.5595	-.6454	-.7186	.0260	-.0008	-.0329	-.0733	-.1040
100	-.4687	-.5295	-.5994	-.6874	-.7619	.0031	-.0252	-.0590	-.1013	-.1334
120	-.5332	-.5969	-.6697	-.7611	-.8363	-.0396	-.0707	-.1077	-.1536	-.1877
140	-.5907	-.6569	-.7321	-.8264	-.9019	-.0785	-.1121	-.1519	-.2007	-.2363
170	-.6672	-.7365	-.8149	-.9132	-.9891	-.1315	-.1682	-.2113	-.2635	-.3008
200	-.7348	-.8068	-.8882	-.9900	-1.067	-.1794	-.2185	-.2641	-.3189	-.3572
230	-.7954	-.8701	-.9548	-1.061	-1.140	-.2246	-.2655	-.3131	-.3694	-.4080
260	-.8507	-.9278	-1.016	-1.125	-1.207	-.2661	-.3085	-.3576	-.4151	-.4536
300	-.9178	-.9979	-1.090	-1.204	-1.289	-.3168	-.3606	-.4112	-.4698	-.5079
350	-.9935	-1.077	-1.173	-1.293	-1.382	-.3736	-.4187	-.4706	-.5299	-.5674
400	-1.062	-1.148	-1.248	-1.373	-1.465	-.4244	-.4705	-.5231	-.5827	-.6195
450	-1.125	-1.214	-1.317	-1.446	-1.541	-.4704	-.5171	-.5702	-.6297	-.6657
500	-1.183	-1.274	-1.381	-1.513	-1.611	-.5124	-.5593	-.6126	-.6720	-.7070
600	-1.287	-1.383	-1.495	-1.633	-1.736	-.5865	-.6335	-.6866	-.7451	-.7784
700	-1.381	-1.480	-1.596	-1.740	-1.845	-.6504	-.6970	-.7494	-.8065	-.8383
800	-1.465	-1.567	-1.687	-1.835	-1.943	-.7063	-.7522	-.8035	-.8592	-.8894
900	-1.542	-1.647	-1.770	-1.921	-2.031	-.7560	-.8009	-.8510	-.9050	-.9338
1000	-1.613	-1.720	-1.846	-2.000	-2.112	-.8006	-.8444	-.8930	-.9453	-.9728

TABLE 4- 7

BE ON AG, M2/M1=12

K/KL = E,KEV	IONIZATION SKEWNESS					RANGE KURTOSIS				
	0.65	0.80	1.0	1.3	1.6	0.65	0.80	1.0	1.3	1.6
.10	.3683	.3707	.3738	.3782	.3825	3.168	3.145	3.116	3.078	3.044
.12	.3689	.3713	.3745	.3790	.3834	3.164	3.141	3.112	3.073	3.038
.14	.3694	.3719	.3751	.3797	.3842	3.161	3.137	3.108	3.068	3.033
.17	.3701	.3726	.3759	.3806	.3852	3.157	3.132	3.102	3.062	3.026
.20	.3707	.3733	.3766	.3814	.3861	3.153	3.128	3.098	3.057	3.021
.23	.3713	.3739	.3773	.3822	.3869	3.150	3.125	3.094	3.052	3.016
.26	.3718	.3745	.3779	.3828	.3876	3.147	3.121	3.090	3.048	3.011
.30	.3725	.3752	.3787	.3837	.3885	3.143	3.117	3.085	3.043	3.006
.35	.3733	.3760	.3795	.3846	.3895	3.139	3.113	3.080	3.037	3.000
.40	.3741	.3768	.3803	.3855	.3904	3.136	3.109	3.076	3.032	2.995
.45	.3748	.3775	.3811	.3863	.3912	3.132	3.105	3.072	3.028	2.990
.50	.3755	.3782	.3818	.3870	.3920	3.129	3.101	3.068	3.024	2.985
.60	.3768	.3796	.3832	.3885	.3935	3.124	3.095	3.061	3.016	2.978
.70	.3780	.3808	.3844	.3897	.3948	3.118	3.089	3.055	3.009	2.971
.80	.3791	.3819	.3856	.3909	.3960	3.114	3.084	3.049	3.003	2.964
.90	.3801	.3829	.3866	.3919	.3970	3.109	3.079	3.044	2.998	2.959
1.0	.3811	.3839	.3875	.3929	.3980	3.105	3.075	3.039	2.992	2.953
1.2	.3827	.3855	.3891	.3944	.3995	3.097	3.066	3.030	2.983	2.944
1.4	.3841	.3869	.3905	.3958	.4009	3.090	3.059	3.022	2.975	2.935
1.7	.3861	.3888	.3924	.3977	.4028	3.080	3.048	3.011	2.963	2.924
2.0	.3880	.3906	.3942	.3994	.4044	3.071	3.039	3.001	2.953	2.915
2.3	.3901	.3927	.3962	.4014	.4063	3.063	3.030	2.992	2.944	2.906
2.6	.3921	.3947	.3981	.4032	.4081	3.055	3.022	2.983	2.936	2.898
3.0	.3945	.3970	.4003	.4053	.4102	3.045	3.011	2.973	2.925	2.888
3.5	.3971	.3995	.4027	.4076	.4124	3.034	2.999	2.961	2.914	2.878
4.0	.3994	.4016	.4048	.4095	.4142	3.023	2.989	2.950	2.904	2.869
4.5	.4013	.4035	.4065	.4112	.4157	3.013	2.979	2.940	2.895	2.861
5.0	.4029	.4050	.4080	.4125	.4170	3.004	2.969	2.931	2.887	2.855
6.0	.4049	.4067	.4096	.4140	.4183	2.984	2.950	2.914	2.872	2.842
7.0	.4065	.4081	.4109	.4152	.4192	2.967	2.934	2.899	2.860	2.832
8.0	.4079	.4094	.4120	.4161	.4200	2.953	2.920	2.886	2.850	2.825
9.0	.4092	.4106	.4130	.4169	.4207	2.941	2.909	2.876	2.842	2.820
10	.4104	.4118	.4140	.4176	.4213	2.931	2.899	2.867	2.835	2.816
12	.4136	.4154	.4169	.4197	.4234	2.918	2.886	2.854	2.824	2.811
14	.4164	.4185	.4192	.4212	.4251	2.909	2.876	2.844	2.817	2.811
17	.4195	.4220	.4218	.4227	.4265	2.898	2.866	2.836	2.814	2.817
20	.4217	.4244	.4234	.4233	.4269	2.889	2.860	2.833	2.817	2.829
23	.4228	.4250	.4236	.4227	.4255	2.870	2.847	2.827	2.821	2.843
26	.4234	.4249	.4232	.4217	.4235	2.853	2.837	2.825	2.829	2.860
30	.4234	.4241	.4220	.4198	.4204	2.836	2.829	2.827	2.845	2.887
35	.4226	.4223	.4199	.4169	.4161	2.822	2.826	2.837	2.872	2.927
40	.4212	.4199	.4172	.4135	.4115	2.815	2.830	2.854	2.905	2.973
45	.4193	.4173	.4142	.4099	.4069	2.814	2.839	2.875	2.941	3.022
50	.4171	.4143	.4110	.4062	.4024	2.819	2.853	2.900	2.981	3.073
60	.4114	.4077	.4032	.3972	.3924	2.844	2.894	2.961	3.067	3.180
70	.4055	.4010	.3957	.3888	.3834	2.880	2.944	3.029	3.159	3.292
80	.3997	.3945	.3884	.3809	.3753	2.922	3.000	3.102	3.254	3.407
90	.3940	.3883	.3816	.3736	.3681	2.969	3.060	3.178	3.352	3.522
100	.3885	.3823	.3752	.3670	.3616	3.019	3.123	3.257	3.450	3.638
120	.3777	.3705	.3625	.3543	.3504	3.121	3.247	3.409	3.640	3.859
140	.3680	.3601	.3517	.3438	.3416	3.229	3.377	3.565	3.831	4.080
170	.3552	.3469	.3384	.3316	.3318	3.399	3.576	3.802	4.119	4.410
200	.3444	.3361	.3281	.3227	.3252	3.572	3.778	4.040	4.404	4.735
230	.3340	.3272	.3209	.3179	.3225	3.738	3.972	4.268	4.676	5.043
260	.3250	.3199	.3155	.3149	.3213	3.905	4.166	4.494	4.946	5.347
300	.3152	.3122	.3105	.3129	.3215	4.129	4.424	4.795	5.301	5.746
350	.3058	.3054	.3066	.3126	.3233	4.409	4.745	5.166	5.737	6.235
400	.2989	.3009	.3049	.3138	.3263	4.687	5.063	5.532	6.166	6.714
450	.2941	.2982	.3045	.3161	.3300	4.962	5.377	5.892	6.586	7.182
500	.2909	.2969	.3053	.3191	.3341	5.234	5.685	6.246	6.998	7.640
600	.2885	.2976	.3093	.3265	.3431	5.766	6.288	6.934	7.796	8.526
700	.2900	.3013	.3155	.3351	.3524	6.281	6.870	7.597	8.562	9.375
800	.2942	.3071	.3230	.3441	.3618	6.780	7.432	8.235	9.299	10.19
900	.3004	.3145	.3315	.3535	.3710	7.263	7.975	8.851	10.01	10.97
1000	.3082	.3229	.3406	.3629	.3801	7.731	8.501	9.446	10.69	11.73

K/KL=	DAMAGE KURTOSIS					IONIZATION KURTOSIS				
E,KEV	0.65	0.80	1.0	1.3	1.6	0.65	0.80	1.0	1.3	1.6
.10	4.273	4.243	4.206	4.155	4.107	4.428	4.399	4.361	4.310	4.263
.12	4.267	4.237	4.199	4.146	4.098	4.423	4.393	4.355	4.302	4.254
.14	4.262	4.231	4.192	4.139	4.090	4.418	4.387	4.349	4.295	4.254
.17	4.255	4.223	4.183	4.129	4.079	4.412	4.380	4.341	4.286	4.236
.20	4.249	4.216	4.176	4.120	4.069	4.406	4.374	4.333	4.278	4.227
.23	4.243	4.210	4.169	4.112	4.061	4.401	4.368	4.327	4.270	4.219
.26	4.238	4.204	4.162	4.105	4.053	4.396	4.363	4.321	4.263	4.211
.30	4.231	4.197	4.154	4.096	4.043	4.391	4.356	4.314	4.255	4.202
.35	4.224	4.189	4.145	4.086	4.032	4.384	4.349	4.305	4.246	4.192
.40	4.216	4.181	4.136	4.076	4.022	4.378	4.342	4.298	4.237	4.182
.45	4.210	4.173	4.128	4.067	4.012	4.372	4.336	4.291	4.229	4.173
.50	4.203	4.166	4.121	4.059	4.003	4.367	4.330	4.284	4.222	4.165
.60	4.191	4.153	4.106	4.043	3.986	4.358	4.320	4.273	4.209	4.152
.70	4.180	4.141	4.093	4.029	3.971	4.349	4.311	4.263	4.198	4.139
.80	4.169	4.130	4.081	4.016	3.957	4.341	4.302	4.253	4.187	4.127
.90	4.159	4.119	4.070	4.003	3.944	4.333	4.293	4.243	4.176	4.115
1.0	4.149	4.109	4.059	3.992	3.932	4.325	4.284	4.234	4.166	4.104
1.2	4.132	4.090	4.039	3.970	3.909	4.308	4.266	4.214	4.144	4.082
1.4	4.115	4.073	4.020	3.950	3.888	4.292	4.249	4.196	4.125	4.061
1.7	4.092	4.048	3.994	3.923	3.859	4.271	4.226	4.171	4.098	4.032
2.0	4.070	4.025	3.970	3.897	3.833	4.251	4.205	4.149	4.073	4.006
2.3	4.049	4.003	3.947	3.872	3.807	4.236	4.188	4.131	4.054	3.986
2.6	4.029	3.981	3.924	3.849	3.783	4.222	4.173	4.115	4.036	3.967
3.0	4.003	3.955	3.896	3.820	3.753	4.204	4.154	4.094	4.014	3.943
3.5	3.973	3.923	3.864	3.786	3.718	4.183	4.131	4.069	3.987	3.915
4.0	3.945	3.894	3.834	3.755	3.686	4.162	4.109	4.046	3.962	3.888
4.5	3.919	3.867	3.806	3.725	3.656	4.142	4.088	4.023	3.937	3.862
5.0	3.895	3.842	3.779	3.698	3.628	4.122	4.066	4.000	3.913	3.836
6.0	3.849	3.795	3.731	3.648	3.577	4.077	4.019	3.951	3.861	3.782
7.0	3.808	3.752	3.686	3.603	3.531	4.036	3.976	3.906	3.813	3.733
8.0	3.769	3.712	3.646	3.561	3.489	3.998	3.937	3.864	3.769	3.687
9.0	3.734	3.676	3.609	3.523	3.450	3.964	3.900	3.826	3.729	3.645
10	3.700	3.641	3.573	3.487	3.414	3.932	3.867	3.790	3.692	3.606
12	3.637	3.577	3.508	3.420	3.347	3.879	3.811	3.731	3.629	3.539
14	3.580	3.519	3.449	3.361	3.288	3.832	3.761	3.679	3.573	3.479
17	3.506	3.444	3.373	3.284	3.211	3.767	3.693	3.608	3.498	3.400
20	3.441	3.378	3.307	3.218	3.145	3.709	3.632	3.543	3.431	3.330
23	3.385	3.321	3.249	3.160	3.087	3.648	3.570	3.478	3.363	3.262
26	3.335	3.271	3.198	3.109	3.036	3.592	3.513	3.418	3.301	3.200
30	3.276	3.211	3.139	3.049	2.978	3.523	3.443	3.345	3.225	3.126
35	3.212	3.147	3.075	2.985	2.915	3.445	3.364	3.263	3.141	3.045
40	3.157	3.092	3.020	2.931	2.863	3.374	3.293	3.190	3.068	2.974
45	3.108	3.044	2.972	2.885	2.818	3.310	3.229	3.124	3.002	2.911
50	3.065	3.001	2.930	2.844	2.779	3.251	3.170	3.065	2.943	2.855
60	2.987	2.926	2.858	2.776	2.716	3.147	3.065	2.962	2.843	2.760
70	2.924	2.866	2.800	2.722	2.667	3.057	2.975	2.876	2.760	2.682
80	2.871	2.816	2.754	2.679	2.628	2.980	2.899	2.802	2.690	2.617
90	2.827	2.775	2.715	2.645	2.599	2.912	2.832	2.738	2.630	2.561
100	2.790	2.741	2.684	2.618	2.575	2.853	2.773	2.682	2.578	2.514
120	2.731	2.687	2.636	2.579	2.544	2.755	2.678	2.592	2.495	2.438
140	2.688	2.649	2.604	2.555	2.527	2.676	2.602	2.520	2.430	2.380
170	2.644	2.611	2.575	2.536	2.517	2.580	2.511	2.435	2.354	2.313
200	2.616	2.590	2.561	2.532	2.520	2.505	2.441	2.370	2.298	2.264
230	2.601	2.582	2.561	2.541	2.536	2.444	2.385	2.321	2.256	2.228
260	2.594	2.581	2.568	2.557	2.556	2.394	2.340	2.282	2.224	2.201
300	2.594	2.589	2.584	2.583	2.588	2.339	2.291	2.240	2.190	2.173
350	2.604	2.607	2.612	2.621	2.632	2.284	2.243	2.200	2.160	2.148
400	2.621	2.631	2.644	2.662	2.678	2.242	2.207	2.170	2.138	2.130
450	2.643	2.659	2.679	2.705	2.725	2.208	2.178	2.148	2.121	2.118
500	2.668	2.690	2.716	2.748	2.772	2.180	2.155	2.130	2.109	2.109
600	2.725	2.756	2.791	2.834	2.863	2.139	2.122	2.106	2.094	2.099
700	2.786	2.823	2.867	2.917	2.949	2.112	2.101	2.091	2.087	2.096
800	2.849	2.892	2.941	2.997	3.032	2.095	2.088	2.083	2.084	2.096
900	2.913	2.960	3.013	3.073	3.110	2.083	2.080	2.079	2.084	2.098
1000	2.977	3.027	3.083	3.147	3.184	2.077	2.077	2.078	2.087	2.103

TABLE 4- 9

BE ON AG, M2/M1=12

	REFLECTION COEFFICIENT					SPUTTERING EFFICIENCY				
K/KL=	0.65	0.80	1.0	1.3	1.6	0.65	0.80	1.0	1.3	1.6
E,KEV										
.10	.2666	.2632	.2587	.2523	.2463	.1690	.1589	.1467	.1308	.1172
.12	.2659	.2624	.2578	.2512	.2451	.1674	.1571	.1447	.1287	.1151
.14	.2653	.2617	.2570	.2503	.2441	.1660	.1556	.1431	.1269	.1132
.17	.2645	.2608	.2559	.2491	.2427	.1643	.1536	.1409	.1246	.1108
.20	.2638	.2600	.2550	.2480	.2415	.1627	.1519	.1391	.1226	.1087
.23	.2631	.2592	.2542	.2470	.2404	.1614	.1504	.1375	.1209	.1070
.26	.2625	.2585	.2534	.2461	.2394	.1601	.1491	.1360	.1193	.1054
.30	.2617	.2576	.2524	.2450	.2382	.1586	.1475	.1343	.1175	.1035
.35	.2608	.2566	.2513	.2437	.2368	.1570	.1457	.1323	.1154	.1015
.40	.2600	.2557	.2502	.2426	.2355	.1555	.1441	.1306	.1136	.0996
.45	.2592	.2548	.2492	.2414	.2343	.1541	.1426	.1291	.1120	.0980
.50	.2584	.2540	.2483	.2404	.2331	.1529	.1412	.1276	.1105	.0965
.60	.2570	.2524	.2466	.2385	.2313	.1506	.1388	.1251	.1079	.0939
.70	.2557	.2510	.2450	.2368	.2295	.1486	.1367	.1229	.1056	.0917
.80	.2544	.2496	.2436	.2351	.2278	.1468	.1348	.1209	.1036	.0897
.90	.2533	.2484	.2422	.2336	.2260	.1451	.1330	.1191	.1018	.0878
1.0	.2521	.2471	.2408	.2322	.2242	.1436	.1314	.1174	.1001	.0862
1.2	.2500	.2448	.2384	.2296	.2203	.1407	.1285	.1143	.0971	.0833
1.4	.2480	.2427	.2361	.2272	.2166	.1382	.1258	.1117	.0944	.0807
1.7	.2452	.2397	.2329	.2237	.2113	.1348	.1223	.1081	.0909	.0774
2.0	.2426	.2369	.2299	.2203	.2064	.1318	.1192	.1050	.0879	.0745
2.3	.2401	.2344	.2274	.2166	.2020	.1290	.1164	.1021	.0851	.0719
2.6	.2377	.2320	.2250	.2131	.1979	.1264	.1138	.0996	.0826	.0695
3.0	.2348	.2290	.2219	.2085	.1928	.1233	.1106	.0964	.0796	.0667
3.5	.2313	.2253	.2181	.2031	.1869	.1197	.1070	.0929	.0763	.0636
4.0	.2280	.2218	.2143	.1980	.1815	.1165	.1038	.0897	.0733	.0609
4.5	.2249	.2184	.2106	.1932	.1764	.1135	.1008	.0868	.0707	.0585
5.0	.2219	.2151	.2068	.1886	.1717	.1107	.0980	.0842	.0682	.0562
6.0	.2166	.2086	.1985	.1798	.1622	.1056	.0930	.0794	.0638	.0522
7.0	.2116	.2025	.1907	.1719	.1539	.1011	.0886	.0751	.0600	.0488
8.0	.2069	.1968	.1836	.1648	.1466	.0970	.0846	.0714	.0566	.0459
9.0	.2023	.1914	.1770	.1583	.1402	.0933	.0810	.0680	.0536	.0432
10	.1979	.1864	.1710	.1525	.1345	.0898	.0777	.0650	.0509	.0409
12	.1892	.1769	.1605	.1422	.1254	.0836	.0718	.0595	.0462	.0368
14	.1812	.1684	.1514	.1334	.1179	.0782	.0667	.0549	.0422	.0334
17	.1703	.1571	.1398	.1223	.1086	.0713	.0603	.0491	.0373	.0292
20	.1605	.1472	.1299	.1131	.1009	.0655	.0549	.0443	.0333	.0259
23	.1512	.1378	.1214	.1052	.0938	.0604	.0502	.0403	.0300	.0232
26	.1428	.1295	.1140	.0984	.0876	.0559	.0463	.0368	.0273	.0210
30	.1328	.1196	.1055	.0907	.0805	.0508	.0417	.0329	.0242	.0185
35	.1219	.1091	.0965	.0825	.0730	.0454	.0370	.0289	.0210	.0160
40	.1126	.1001	.0889	.0757	.0668	.0409	.0330	.0256	.0185	.0140
45	.1046	.0925	.0823	.0699	.0614	.0371	.0297	.0229	.0164	.0123
50	.0976	.0859	.0766	.0649	.0568	.0338	.0269	.0206	.0146	.0109
60	.0868	.0760	.0674	.0569	.0498	.0286	.0225	.0171	.0120	.0089
70	.0781	.0683	.0601	.0505	.0442	.0245	.0191	.0144	.0100	.0074
80	.0711	.0620	.0541	.0454	.0397	.0212	.0164	.0123	.0085	.0062
90	.0651	.0567	.0491	.0411	.0360	.0186	.0143	.0106	.0073	.0053
100	.0600	.0523	.0450	.0375	.0328	.0164	.0125	.0093	.0063	.0045
120	.0519	.0453	.0386	.0320	.0278	.0131	.0099	.0073	.0050	.0035
140	.0456	.0398	.0338	.0279	.0240	.0108	.0081	.0059	.0040	.0028
170	.0384	.0336	.0284	.0232	.0197	.0082	.0061	.0044	.0030	.0021
200	.0329	.0289	.0243	.0197	.0164	.0065	.0048	.0035	.0024	.0016
230	.0287	.0250	.0209	.0168	.0138	.0053	.0039	.0028	.0019	.0013
260	.0252	.0218	.0181	.0144	.0117	.0043	.0032	.0023	.0015	.0010
300	.0215	.0184	.0151	.0119	.0096	.0035	.0025	.0018	.0012	.0008
350	.0180	.0151	.0122	.0094	.0075	.0027	.0020	.0014	.0009	.0006
400	.0152	.0125	.0100	.0075	.0059	.0022	.0016	.0011	.0008	.0005
450	.0130	.0105	.0082	.0061	.0047	.0018	.0013	.0010	.0006	.0005
500	.0112	.0088	.0068	.0049	.0037	.0015	.0012	.0008	.0005	.0004
600	.0085	.0064	.0047	.0032	.0024	.0012	.0009	.0007	.0004	.0003
700	.0065	.0047	.0034	.0022	.0016	.0009	.0007	.0005	.0003	.0003
800	.0051	.0036	.0025	.0015	.0011	.0008	.0006	.0004	.0003	.0002
900	.0040	.0028	.0019	.0011	.0008	.0006	.0005	.0004	.0002	.0002
1000	.0031	.0024	.0016	.0009	.0007	.0005	.0004	.0003	.0002	.0001

TABLE 4-10

BE ON AG, M2/M1=12

E,KEV	IONIZATION DEFICIENCY K/KL= 0.65	0.80	1.0	1.3	1.6	SPUTTERING YIELD ALPHA 0.65	0.80	1.0	1.3	1.6
.10	.0326	.0378	.0438	.0511	.0567	1.366	1.327	1.278	1.210	1.148
.12	.0333	.0386	.0446	.0519	.0574	1.361	1.321	1.271	1.202	1.139
.14	.0339	.0392	.0453	.0525	.0581	1.357	1.317	1.266	1.195	1.131
.17	.0347	.0401	.0462	.0534	.0588	1.352	1.310	1.258	1.186	1.120
.20	.0354	.0408	.0469	.0541	.0594	1.348	1.305	1.252	1.178	1.112
.23	.0360	.0414	.0475	.0547	.0600	1.344	1.300	1.246	1.172	1.104
.26	.0365	.0419	.0480	.0552	.0604	1.340	1.296	1.241	1.166	1.097
.30	.0371	.0426	.0486	.0557	.0609	1.337	1.292	1.236	1.159	1.090
.35	.0377	.0432	.0493	.0563	.0614	1.332	1.287	1.229	1.151	1.081
.40	.0383	.0439	.0498	.0568	.0618	1.329	1.282	1.224	1.145	1.074
.45	.0387	.0443	.0503	.0572	.0621	1.326	1.278	1.219	1.139	1.067
.50	.0392	.0447	.0507	.0576	.0623	1.323	1.275	1.215	1.134	1.061
.60	.0399	.0454	.0514	.0581	.0628	1.318	1.269	1.208	1.125	1.051
.70	.0405	.0460	.0520	.0586	.0631	1.314	1.264	1.202	1.117	1.042
.80	.0410	.0466	.0525	.0589	.0633	1.311	1.260	1.196	1.111	1.035
.90	.0415	.0470	.0528	.0592	.0634	1.308	1.256	1.191	1.105	1.028
1.0	.0419	.0474	.0532	.0594	.0635	1.305	1.252	1.187	1.099	1.022
1.2	.0426	.0480	.0537	.0597	.0636	1.301	1.247	1.180	1.090	1.011
1.4	.0431	.0485	.0541	.0599	.0635	1.297	1.242	1.173	1.082	1.002
1.7	.0438	.0490	.0545	.0600	.0634	1.293	1.235	1.165	1.072	.9901
2.0	.0443	.0495	.0547	.0600	.0631	1.289	1.230	1.158	1.063	.9800
2.3	.0447	.0498	.0549	.0599	.0627	1.286	1.225	1.152	1.055	.9713
2.6	.0450	.0500	.0549	.0597	.0623	1.283	1.221	1.147	1.049	.9634
3.0	.0454	.0502	.0550	.0594	.0618	1.279	1.217	1.141	1.040	.9540
3.5	.0457	.0504	.0549	.0590	.0610	1.276	1.211	1.133	1.031	.9434
4.0	.0459	.0504	.0547	.0585	.0602	1.272	1.206	1.127	1.023	.9338
4.5	.0460	.0504	.0544	.0579	.0594	1.269	1.201	1.120	1.015	.9250
5.0	.0461	.0503	.0541	.0573	.0586	1.265	1.197	1.114	1.008	.9168
6.0	.0461	.0500	.0535	.0561	.0569	1.259	1.188	1.104	.9944	.9020
7.0	.0461	.0497	.0527	.0548	.0552	1.254	1.180	1.094	.9822	.8885
8.0	.0459	.0492	.0519	.0536	.0537	1.248	1.173	1.084	.9708	.8760
9.0	.0456	.0487	.0511	.0524	.0521	1.242	1.165	1.075	.9599	.8644
10	.0454	.0482	.0503	.0512	.0507	1.236	1.157	1.066	.9496	.8534
12	.0447	.0470	.0485	.0488	.0479	1.224	1.143	1.048	.9295	.8324
14	.0439	.0458	.0469	.0466	.0453	1.211	1.128	1.031	.9110	.8133
17	.0427	.0441	.0445	.0436	.0418	1.193	1.107	1.008	.8856	.7877
20	.0415	.0423	.0423	.0409	.0388	1.175	1.086	.9855	.8626	.7650
23	.0401	.0406	.0401	.0384	.0361	1.155	1.065	.9631	.8406	.7443
26	.0388	.0389	.0381	.0362	.0337	1.136	1.045	.9421	.8204	.7256
30	.0371	.0368	.0356	.0334	.0309	1.112	1.019	.9162	.7959	.7031
35	.0351	.0344	.0328	.0305	.0279	1.083	.9895	.8867	.7687	.6781
40	.0333	.0322	.0304	.0279	.0253	1.057	.9623	.8601	.7445	.6559
45	.0316	.0302	.0282	.0257	.0231	1.032	.9371	.8359	.7229	.6360
50	.0300	.0285	.0263	.0237	.0212	1.008	.9137	.8139	.7034	.6180
60	.0270	.0253	.0230	.0204	.0181	.9649	.8707	.7748	.6697	.5856
70	.0244	.0226	.0203	.0177	.0156	.9266	.8332	.7413	.6413	.5582
80	.0222	.0203	.0180	.0155	.0136	.8925	.8002	.7121	.6169	.5346
90	.0203	.0183	.0161	.0137	.0119	.8619	.7709	.6864	.5957	.5141
100	.0186	.0166	.0145	.0122	.0105	.8343	.7448	.6635	.5771	.4960
120	.0158	.0140	.0121	.0101	.0085	.7865	.7003	.6247	.5468	.4661
140	.0136	.0119	.0102	.0085	.0070	.7461	.6634	.5926	.5220	.4419
170	.0110	.0094	.0081	.0067	.0054	.6960	.6183	.5534	.4916	.4132
200	.0090	.0076	.0065	.0054	.0043	.6551	.5821	.5221	.4669	.3909
230	.0072	.0060	.0051	.0043	.0033	.6213	.5532	.4968	.4448	.3743
250	.0057	.0047	.0040	.0034	.0026	.5927	.5291	.4756	.4257	.3609
300	.0041	.0034	.0029	.0024	.0018	.5607	.5024	.4522	.4042	.3464
350	.0026	.0021	.0018	.0015	.0011	.5281	.4756	.4286	.3820	.3321
400	.0015	.0012	.0010	.0008	.0006	.5016	.4540	.4097	.3639	.3207
450	.0006	.0005	.0004	.0003	.0003	.4798	.4363	.3941	.3489	.3114
500	0.0000	0.0000	0.0000	0.0000	0.0000	.4615	.4216	.3812	.3363	.3035
600	0.0000	0.0000	0.0000	0.0000	0.0000	.4331	.3987	.3612	.3169	.2910
700	0.0000	0.0000	0.0000	0.0000	0.0000	.4124	.3819	.3466	.3030	.2812
800	0.0000	0.0000	0.0000	0.0000	0.0000	.3971	.3695	.3358	.2930	.2732
900	0.0000	0.0000	0.0000	0.0000	0.0000	.3859	.3601	.3277	.2860	.2666
1000	0.0000	0.0000	0.0000	0.0000	0.0000	.3777	.3530	.3217	.2812	.2608

BE ON AG, M2/M1=12

1000 KEV

100 KEV

10 KEV

1 KEV

.10 KEV

E,KEV	FRACTIONAL DEPOSITED ENERGY					MEAN RANGE,MICROGRAM/SQ.CM.				
K/KL=	0.6	0.80	1.0	1.2	1.4	0.6	0.80	1.0	1.2	1.4
.20	.8296	.7830	.7405	.7015	.6656	.7035	.6983	.6932	.6882	.6833
.23	.8260	.7787	.7355	.6960	.6598	.7812	.7751	.7691	.7632	.7575
.26	.8228	.7748	.7311	.6912	.6546	.8513	.8445	.8377	.8312	.8247
.30	.8190	.7702	.7259	.6855	.6485	.9374	.9297	.9221	.9147	.9074
.35	.8148	.7652	.7201	.6792	.6417	1.038	1.029	1.020	1.012	1.004
.40	.8111	.7607	.7151	.6736	.6358	1.133	1.123	1.114	1.104	1.095
.45	.8078	.7567	.7105	.6687	.6305	1.224	1.214	1.203	1.193	1.183
.50	.8047	.7530	.7064	.6642	.6258	1.314	1.302	1.291	1.280	1.269
.60	.7993	.7465	.6990	.6562	.6173	1.487	1.473	1.460	1.447	1.434
.70	.7945	.7408	.6927	.6493	.6100	1.654	1.639	1.623	1.609	1.594
.80	.7903	.7358	.6871	.6432	.6036	1.817	1.800	1.782	1.766	1.749
.90	.7864	.7312	.6820	.6377	.5978	1.977	1.957	1.937	1.919	1.900
1.0	.7829	.7271	.6773	.6327	.5926	2.132	2.110	2.089	2.068	2.047
1.2	.7767	.7198	.6692	.6240	.5834	2.420	2.394	2.369	2.345	2.321
1.4	.7713	.7133	.6621	.6164	.5754	2.703	2.672	2.643	2.615	2.587
1.7	.7641	.7049	.6527	.6064	.5650	3.119	3.082	3.046	3.011	2.977
2.0	.7577	.6975	.6446	.5978	.5560	3.527	3.483	3.440	3.399	3.359
2.3	.7521	.6909	.6374	.5901	.5481	3.893	3.843	3.795	3.748	3.703
2.6	.7469	.6850	.6308	.5832	.5409	4.258	4.202	4.148	4.095	4.044
3.0	.7405	.6777	.6229	.5749	.5323	4.746	4.681	4.618	4.557	4.498
3.5	.7334	.6695	.6140	.5655	.5228	5.360	5.282	5.206	5.133	5.063
4.0	.7268	.6620	.6060	.5571	.5142	5.978	5.884	5.795	5.709	5.627
4.5	.7208	.6552	.5986	.5495	.5064	6.596	6.487	6.383	6.283	6.187
5.0	.7151	.6488	.5918	.5424	.4992	7.213	7.087	6.968	6.853	6.744
6.0	.7049	.6373	.5795	.5297	.4863	8.337	8.184	8.039	7.900	7.768
7.0	.6956	.6269	.5686	.5185	.4750	9.496	9.311	9.135	8.967	8.808
8.0	.6870	.6175	.5587	.5083	.4648	10.69	10.46	10.25	10.05	9.860
9.0	.6791	.6087	.5495	.4990	.4555	11.90	11.63	11.38	11.14	10.92
10	.6716	.6006	.5410	.4904	.4470	13.13	12.81	12.52	12.24	11.98
12	.6578	.5856	.5256	.4749	.4315	15.38	14.99	14.63	14.29	13.97
14	.6453	.5722	.5118	.4611	.4179	17.75	17.26	16.82	16.40	16.00
17	.6284	.5542	.4935	.4429	.4001	21.50	20.84	20.23	19.67	19.14
20	.6131	.5381	.4773	.4269	.3846	25.45	24.57	23.76	23.03	22.36
23	.5987	.5233	.4625	.4125	.3706	28.97	27.94	27.00	26.13	25.34
26	.5855	.5097	.4490	.3994	.3581	32.67	31.46	30.35	29.32	28.39
30	.5693	.4932	.4328	.3838	.3432	37.87	36.35	34.97	33.69	32.55
35	.5509	.4747	.4148	.3665	.3268	44.76	42.76	40.96	39.33	37.88
40	.5343	.4581	.3988	.3513	.3124	51.98	49.40	47.13	45.10	43.30
45	.5190	.4431	.3844	.3377	.2996	59.44	56.21	53.41	50.95	48.77
50	.5049	.4293	.3713	.3254	.2881	67.07	63.14	59.77	56.84	54.26
60	.4790	.4045	.3480	.3037	.2679	81.54	76.32	71.99	68.28	64.85
70	.4563	.3830	.3281	.2852	.2510	96.77	90.02	84.50	79.86	75.54
80	.4363	.3643	.3107	.2693	.2364	112.6	104.1	97.22	91.51	86.25
90	.4183	.3476	.2955	.2554	.2237	128.7	118.4	110.0	103.2	96.94
100	.4020	.3327	.2819	.2431	.2125	145.2	132.9	122.9	114.7	107.6
120	.3735	.3069	.2588	.2223	.1936	177.3	161.1	147.9	137.2	128.2
140	.3493	.2854	.2397	.2052	.1782	210.1	189.7	172.9	159.4	148.5
170	.3190	.2588	.2162	.1845	.1596	260.1	232.8	210.2	192.2	178.4
200	.2941	.2372	.1973	.1678	.1448	310.5	275.7	247.0	224.3	207.6
230	.2735	.2193	.1820	.1544	.1330	359.9	317.4	282.7	255.3	235.6
260	.2559	.2042	.1692	.1432	.1232	409.0	358.6	317.7	285.6	262.9
300	.2359	.1872	.1548	.1308	.1123	474.1	412.7	363.4	325.0	298.2
350	.2151	.1699	.1401	.1181	.1013	554.3	479.1	418.9	372.5	340.8
400	.1978	.1557	.1281	.1078	.0924	633.2	543.8	472.8	418.4	381.8
450	.1831	.1440	.1181	.0992	.0850	710.5	606.9	525.0	462.7	421.3
500	.1705	.1340	.1097	.0920	.0788	786.2	668.5	575.7	505.5	459.5
600	.1497	.1195	.0964	.0808	.0692	931.8	786.4	672.1	586.7	532.0
700	.1334	.1082	.0861	.0722	.0618	1072.	899.2	763.8	663.5	600.4
800	.1204	.0992	.0779	.0653	.0559	1207.	1007.	851.3	736.6	665.3
900	.1096	.0918	.0712	.0597	.0511	1337.	1111.	935.1	806.3	727.3
1000	.1008	.0854	.0656	.0549	.0471	1463.	1212.	1016.	873.2	786.7
1200	.0870	.0749	.0568	.0474	.0408	1704.	1402.	1168.	999.5	898.5
1400	.0772	.0665	.0501	.0417	.0359	1930.	1581.	1310.	1117.	1003.
1700	.0671	.0562	.0428	.0352	.0304	2248.	1831.	1509.	1281.	1147.
2000	.0610	.0477	.0376	.0303	.0262	2544.	2063.	1692.	1431.	1280.

	MEAN DAMAGE DEPTH,MICROGRAM/SQ.CM.					MEAN IONZN.DEPTH,MICROGRAM/SQ.CM.				
K/KL=	0.6	0.80	1.0	1.2	1.4	0.6	0.80	1.0	1.2	1.4
E,KEV										
.20	.6561	.6461	.6365	.6273	.6184	.6407	.6297	.6191	.6090	.5993
.23	.7274	.7159	.7048	.6942	.6840	.7100	.6972	.6850	.6734	.6622
.26	.7921	.7791	.7668	.7550	.7436	.7728	.7586	.7450	.7320	.7195
.30	.8716	.8570	.8431	.8299	.8171	.8502	.8342	.8189	.8043	.7903
.35	.9641	.9477	.9321	.9171	.9028	.9404	.9223	.9051	.8886	.8729
.40	1.052	1.034	1.016	.9998	.9839	1.026	1.006	.9868	.9685	.9511
.45	1.136	1.116	1.097	1.079	1.062	1.108	1.086	1.065	1.045	1.026
.50	1.219	1.197	1.176	1.156	1.138	1.188	1.164	1.141	1.120	1.099
.60	1.378	1.352	1.328	1.305	1.284	1.343	1.315	1.288	1.263	1.239
.70	1.531	1.502	1.475	1.449	1.424	1.492	1.460	1.430	1.401	1.374
.80	1.680	1.648	1.617	1.588	1.560	1.637	1.601	1.567	1.534	1.504
.90	1.826	1.790	1.755	1.723	1.691	1.777	1.738	1.700	1.664	1.630
1.0	1.967	1.928	1.890	1.854	1.820	1.915	1.871	1.829	1.790	1.752
1.2	2.230	2.184	2.139	2.098	2.058	2.169	2.118	2.069	2.023	1.980
1.4	2.487	2.433	2.382	2.334	2.289	2.417	2.358	2.303	2.250	2.200
1.7	2.863	2.798	2.737	2.680	2.626	2.781	2.709	2.642	2.579	2.519
2.0	3.231	3.155	3.083	3.016	2.952	3.135	3.051	2.972	2.898	2.829
2.3	3.562	3.476	3.396	3.321	3.249	3.455	3.361	3.272	3.189	3.111
2.6	3.891	3.795	3.706	3.621	3.542	3.772	3.667	3.568	3.475	3.388
3.0	4.329	4.219	4.116	4.019	3.928	4.193	4.071	3.958	3.852	3.752
3.5	4.877	4.747	4.626	4.513	4.406	4.717	4.574	4.441	4.316	4.200
4.0	5.424	5.274	5.134	5.003	4.880	5.240	5.073	4.919	4.776	4.642
4.5	5.971	5.798	5.638	5.489	5.350	5.759	5.568	5.392	5.229	5.077
5.0	6.515	6.319	6.138	5.970	5.814	6.275	6.059	5.860	5.677	5.506
6.0	7.507	7.272	7.055	6.855	6.669	7.222	6.963	6.725	6.506	6.303
7.0	8.523	8.243	7.986	7.749	7.530	8.184	7.876	7.594	7.335	7.095
8.0	9.560	9.229	8.928	8.651	8.395	9.158	8.795	8.465	8.162	7.884
9.0	10.61	10.23	9.877	9.557	9.263	10.14	9.718	9.336	8.988	8.669
10	11.67	11.23	10.83	10.46	10.13	11.13	10.64	10.20	9.809	9.448
12	13.62	13.09	12.60	12.16	11.75	12.96	12.37	11.83	11.35	10.92
14	15.66	15.01	14.42	13.88	13.40	14.84	14.12	13.48	12.91	12.39
17	18.86	18.00	17.23	16.54	15.92	17.76	16.82	15.99	15.25	14.59
20	22.19	21.08	20.11	19.25	18.48	20.75	19.55	18.51	17.60	16.78
23	25.18	23.89	22.76	21.75	20.85	23.51	22.10	20.89	19.82	18.83
26	28.29	26.80	25.48	24.31	23.26	26.34	24.69	23.29	22.05	20.87
30	32.64	30.81	29.20	27.79	26.53	30.22	28.21	26.51	25.03	23.59
35	38.35	36.01	34.00	32.24	30.69	35.21	32.68	30.57	28.75	26.98
40	44.27	41.37	38.91	36.77	34.91	40.30	37.21	34.64	32.44	30.35
45	50.37	46.84	43.88	41.35	39.14	45.46	41.76	38.70	36.11	33.70
50	56.57	52.38	48.90	45.94	43.39	50.65	46.32	42.74	39.73	37.01
60	68.32	62.86	58.48	54.79	51.51	60.66	55.14	50.55	46.72	43.46
70	80.59	73.71	68.26	63.72	59.69	70.81	63.97	58.28	53.58	49.77
80	93.26	84.82	78.17	72.70	67.89	81.02	72.77	65.91	60.30	55.92
90	106.2	96.10	88.15	81.66	76.08	91.25	81.51	73.42	66.88	61.91
100	119.3	107.5	98.15	90.59	84.21	101.4	90.15	80.82	73.32	67.75
120	144.9	129.7	117.6	107.8	99.97	121.4	106.9	95.15	85.75	78.91
140	170.9	152.2	137.0	124.9	115.5	141.0	123.4	109.0	97.69	89.58
170	210.5	186.0	165.9	150.1	138.5	170.0	147.3	129.0	114.8	104.8
200	250.4	219.7	194.5	174.9	160.9	198.1	170.3	148.0	130.9	119.1
230	289.2	252.2	222.0	198.7	182.3	225.4	192.5	166.2	146.2	132.7
260	327.9	284.4	249.1	222.0	203.3	252.0	214.0	183.7	160.8	145.6
300	379.1	326.7	284.5	252.4	230.5	286.2	241.5	206.0	179.4	162.0
350	442.4	378.7	327.6	289.3	263.5	327.2	274.3	232.3	201.2	181.3
400	504.6	429.5	369.6	325.0	295.3	366.5	305.5	257.3	221.9	199.5
450	565.7	479.1	410.5	359.6	326.2	404.2	335.3	281.0	241.4	216.7
500	625.5	527.6	450.2	393.2	356.1	440.5	363.8	303.6	260.0	233.1
600	740.3	620.2	525.9	456.9	412.7	509.1	417.3	345.8	294.7	263.7
700	851.1	709.2	598.3	517.6	466.6	573.3	467.1	385.0	326.8	292.0
800	958.3	795.0	667.8	575.6	518.1	633.7	513.9	421.7	356.7	318.4
900	1062.	877.8	734.6	631.3	567.5	690.9	558.2	456.2	384.8	343.1
1000	1163.	957.9	799.0	684.9	615.1	745.3	600.2	489.0	411.4	366.4
1200	1356.	1111.	921.7	786.6	705.3	846.9	678.5	549.8	460.6	409.7
1400	1538.	1255.	1037.	881.9	789.9	940.4	750.6	605.6	505.7	449.2
1700	1795.	1459.	1199.	1015.	908.1	1069.	849.6	682.0	567.1	503.0
2000	2035.	1648.	1349.	1138.	1018.	1186.	939.6	751.3	622.6	551.7

TABLE 5- 3

C ON AG, M2/M1=9

K/KL=	RELATIVE RANGE STRAGGLING					RELATIVE DAMAGE STRAGGLING				
E,KEV	0.6	0.80	1.0	1.2	1.4	0.6	0.80	1.0	1.2	1.4
.20	1.999	1.918	1.843	1.774	1.711	1.254	1.211	1.173	1.138	1.106
.23	1.989	1.907	1.831	1.762	1.698	1.249	1.206	1.167	1.132	1.099
.26	1.981	1.897	1.821	1.750	1.686	1.245	1.202	1.162	1.126	1.093
.30	1.970	1.885	1.808	1.737	1.672	1.240	1.196	1.156	1.120	1.087
.35	1.958	1.871	1.793	1.721	1.656	1.234	1.189	1.149	1.112	1.079
.40	1.946	1.859	1.780	1.707	1.641	1.229	1.184	1.143	1.106	1.072
.45	1.936	1.848	1.767	1.694	1.628	1.224	1.178	1.137	1.099	1.065
.50	1.926	1.837	1.756	1.683	1.616	1.219	1.173	1.131	1.094	1.060
.60	1.908	1.817	1.735	1.661	1.593	1.211	1.164	1.122	1.083	1.049
.70	1.891	1.799	1.716	1.641	1.573	1.203	1.155	1.113	1.074	1.039
.80	1.875	1.782	1.699	1.623	1.555	1.196	1.148	1.105	1.066	1.031
.90	1.861	1.767	1.682	1.607	1.538	1.190	1.141	1.097	1.058	1.023
1.0	1.847	1.752	1.667	1.591	1.522	1.183	1.134	1.090	1.051	1.015
1.2	1.821	1.725	1.639	1.563	1.493	1.172	1.121	1.077	1.037	1.002
1.4	1.797	1.700	1.614	1.537	1.467	1.161	1.110	1.065	1.025	.9896
1.7	1.764	1.666	1.580	1.502	1.433	1.146	1.094	1.049	1.009	.9732
2.0	1.734	1.636	1.548	1.471	1.401	1.132	1.080	1.035	.9943	.9584
2.3	1.706	1.607	1.519	1.442	1.373	1.120	1.067	1.021	.9808	.9449
2.6	1.680	1.580	1.493	1.415	1.346	1.108	1.055	1.009	.9684	.9325
3.0	1.648	1.548	1.460	1.383	1.314	1.093	1.040	.9936	.9530	.9171
3.5	1.611	1.510	1.422	1.345	1.277	1.077	1.023	.9761	.9355	.8997
4.0	1.577	1.476	1.388	1.312	1.244	1.061	1.007	.9600	.9194	.8838
4.5	1.545	1.444	1.357	1.281	1.213	1.046	.9918	.9450	.9045	.8690
5.0	1.516	1.414	1.327	1.252	1.185	1.033	.9778	.9311	.8907	.8553
6.0	1.461	1.360	1.274	1.199	1.133	1.007	.9519	.9051	.8650	.8300
7.0	1.412	1.311	1.226	1.152	1.088	.9839	.9286	.8819	.8420	.8074
8.0	1.368	1.268	1.183	1.110	1.047	.9629	.9074	.8609	.8213	.7871
9.0	1.328	1.228	1.144	1.073	1.011	.9435	.8880	.8417	.8024	.7686
10	1.291	1.192	1.109	1.038	.9771	.9256	.8701	.8240	.7850	.7516
12	1.224	1.126	1.045	.9763	.9173	.8929	.8375	.7918	.7535	.7208
14	1.166	1.070	.9901	.9231	.8660	.8641	.8088	.7636	.7260	.6940
17	1.092	.9972	.9199	.8555	.8009	.8263	.7714	.7270	.6904	.6595
20	1.029	.9361	.8610	.7988	.7464	.7937	.7392	.6957	.6599	.6300
23	.9752	.8838	.8107	.7507	.7004	.7646	.7107	.6680	.6331	.6041
26	.9277	.8381	.7670	.7089	.6604	.7387	.6854	.6434	.6095	.5813
30	.8727	.7852	.7165	.6608	.6146	.7081	.6556	.6147	.5818	.5547
35	.8139	.7290	.6630	.6100	.5663	.6749	.6234	.5837	.5521	.5262
40	.7637	.6813	.6178	.5672	.5257	.6460	.5955	.5570	.5265	.5018
45	.7202	.6400	.5788	.5304	.4908	.6205	.5710	.5337	.5043	.4805
50	.6819	.6039	.5448	.4983	.4605	.5978	.5494	.5130	.4847	.4618
60	.6178	.5439	.4887	.4457	.4110	.5590	.5126	.4783	.4518	.4307
70	.5656	.4953	.4436	.4035	.3713	.5268	.4823	.4499	.4250	.4052
80	.5220	.4551	.4063	.3687	.3388	.4994	.4567	.4259	.4025	.3840
90	.4850	.4211	.3749	.3396	.3116	.4757	.4347	.4054	.3833	.3659
100	.4531	.3919	.3481	.3148	.2885	.4550	.4156	.3877	.3667	.3502
120	.4016	.3453	.3055	.2755	.2520	.4208	.3844	.3587	.3397	.3249
140	.3610	.3089	.2724	.2451	.2239	.3931	.3594	.3356	.3183	.3049
170	.3136	.2668	.2344	.2104	.1918	.3600	.3297	.3084	.2931	.2813
200	.2772	.2347	.2056	.1842	.1677	.3341	.3066	.2873	.2736	.2631
230	.2489	.2100	.1835	.1641	.1492	.3136	.2885	.2707	.2582	.2488
250	.2259	.1899	.1657	.1480	.1344	.2967	.2736	.2571	.2455	.2370
300	.2011	.1685	.1467	.1308	.1187	.2781	.2573	.2422	.2317	.2242
350	.1758	.1477	.1283	.1143	.1036	.2596	.2411	.2275	.2179	.2115
400	.1576	.1314	.1140	.1014	.0919	.2446	.2281	.2156	.2069	.2013
450	.1422	.1183	.1025	.0912	.0826	.2323	.2173	.2059	.1978	.1929
500	.1295	.1076	.0932	.0828	.0750	.2220	.2083	.1978	.1902	.1858
600	.1100	.0911	.0788	.0699	.0633	.2059	.1941	.1849	.1783	.1745
700	.0956	.0790	.0682	.0605	.0547	.1936	.1831	.1750	.1692	.1658
800	.0845	.0698	.0602	.0534	.0483	.1838	.1744	.1671	.1619	.1589
900	.0758	.0625	.0539	.0478	.0432	.1758	.1673	.1607	.1559	.1531
1000	.0686	.0566	.0488	.0432	.0391	.1692	.1614	.1552	.1509	.1483
1200	.0578	.0477	.0411	.0364	.0329	.1587	.1519	.1466	.1428	.1405
1400	.0499	.0413	.0355	.0314	.0284	.1507	.1446	.1399	.1365	.1343
1700	.0413	.0343	.0294	.0259	.0235	.1417	.1363	.1321	.1291	.1271
2000	.0351	.0293	.0250	.0219	.0200	.1349	.1299	.1260	.1232	.1214

K/KL= E,KEV	RELATIVE IONIZATION STRAGGLING					RELATIVE TRANSVERSE RANGE STRAGGLING				
	0.6	0.80	1.0	1.2	1.4	0.6	0.80	1.0	1.2	1.4
.20	1.207	1.168	1.132	1.100	1.071	.9508	.9491	.9475	.9459	.9443
.23	1.203	1.163	1.127	1.094	1.065	.9508	.9491	.9474	.9458	.9442
.26	1.199	1.159	1.122	1.089	1.059	.9508	.9490	.9473	.9457	.9441
.30	1.194	1.153	1.116	1.083	1.053	.9507	.9490	.9473	.9456	.9440
.35	1.188	1.147	1.110	1.076	1.046	.9507	.9490	.9472	.9455	.9438
.40	1.183	1.141	1.104	1.070	1.039	.9508	.9489	.9472	.9454	.9437
.45	1.178	1.136	1.098	1.064	1.033	.9508	.9489	.9471	.9454	.9437
.50	1.174	1.131	1.093	1.059	1.028	.9508	.9489	.9471	.9453	.9436
.60	1.166	1.122	1.084	1.049	1.018	.9509	.9490	.9471	.9453	.9435
.70	1.158	1.114	1.075	1.040	1.009	.9511	.9491	.9472	.9453	.9435
.80	1.151	1.107	1.067	1.032	1.001	.9512	.9492	.9473	.9454	.9436
.90	1.145	1.100	1.060	1.025	.9931	.9514	.9493	.9474	.9455	.9436
1.0	1.139	1.093	1.054	1.018	.9861	.9515	.9495	.9475	.9456	.9437
1.2	1.127	1.082	1.041	1.005	.9735	.9518	.9498	.9478	.9458	.9439
1.4	1.117	1.071	1.030	.9941	.9622	.9522	.9501	.9481	.9461	.9442
1.7	1.103	1.056	1.015	.9789	.9469	.9528	.9506	.9486	.9466	.9447
2.0	1.089	1.042	1.001	.9651	.9332	.9534	.9512	.9492	.9472	.9452
2.3	1.077	1.030	.9884	.9523	.9205	.9540	.9518	.9497	.9477	.9458
2.6	1.066	1.018	.9765	.9405	.9088	.9546	.9524	.9503	.9483	.9463
3.0	1.051	1.003	.9620	.9261	.8946	.9554	.9532	.9511	.9491	.9472
3.5	1.035	.9866	.9453	.9096	.8784	.9565	.9543	.9522	.9502	.9483
4.0	1.020	.9713	.9302	.8946	.8637	.9576	.9554	.9533	.9514	.9494
4.5	1.005	.9572	.9162	.8809	.8502	.9587	.9566	.9545	.9525	.9507
5.0	.9923	.9440	.9032	.8681	.8377	.9598	.9577	.9557	.9537	.9519
6.0	.9680	.9198	.8793	.8448	.8150	.9620	.9599	.9579	.9560	.9542
7.0	.9463	.8983	.8582	.8242	.7949	.9642	.9621	.9602	.9584	.9567
8.0	.9266	.8788	.8392	.8057	.7770	.9664	.9644	.9626	.9609	.9592
9.0	.9087	.8611	.8219	.7889	.7608	.9686	.9668	.9650	.9634	.9618
10	.8921	.8448	.8060	.7736	.7460	.9708	.9691	.9675	.9659	.9644
12	.8617	.8151	.7772	.7458	.7193	.9753	.9737	.9723	.9709	.9695
14	.8352	.7893	.7523	.7218	.6963	.9797	.9783	.9771	.9759	.9747
17	.8010	.7561	.7204	.6913	.6671	.9862	.9852	.9842	.9833	.9824
20	.7719	.7280	.6935	.6657	.6428	.9925	.9919	.9913	.9906	.9900
23	.7466	.7038	.6705	.6439	.6222	.9986	.9983	.9980	.9977	.9973
26	.7244	.6827	.6505	.6250	.6045	1.004	1.005	1.005	1.005	1.004
30	.6986	.6582	.6275	.6034	.5842	1.012	1.013	1.013	1.014	1.014
35	.6711	.6323	.6033	.5808	.5630	1.021	1.023	1.024	1.025	1.025
40	.6476	.6104	.5829	.5618	.5453	1.030	1.032	1.034	1.035	1.036
45	.6273	.5916	.5655	.5457	.5304	1.039	1.042	1.044	1.045	1.047
50	.6096	.5752	.5505	.5319	.5175	1.048	1.051	1.053	1.055	1.057
60	.5800	.5483	.5261	.5097	.4969	1.063	1.068	1.071	1.074	1.077
70	.5562	.5270	.5069	.4924	.4808	1.078	1.084	1.089	1.093	1.096
80	.5366	.5096	.4914	.4786	.4681	1.093	1.099	1.105	1.110	1.113
90	.5201	.4952	.4787	.4674	.4579	1.107	1.114	1.121	1.127	1.131
100	.5062	.4832	.4682	.4581	.4496	1.120	1.129	1.137	1.143	1.147
120	.4843	.4650	.4524	.4443	.4381	1.148	1.158	1.167	1.174	1.179
140	.4676	.4515	.4410	.4345	.4301	1.173	1.185	1.195	1.203	1.208
170	.4491	.4370	.4290	.4244	.4223	1.209	1.223	1.234	1.243	1.249
200	.4358	.4269	.4209	.4179	.4175	1.242	1.257	1.269	1.279	1.285
230	.4267	.4201	.4159	.4141	.4148	1.272	1.288	1.302	1.312	1.319
260	.4199	.4153	.4125	.4118	.4133	1.299	1.317	1.332	1.343	1.350
300	.4134	.4109	.4097	.4102	.4125	1.334	1.353	1.368	1.380	1.388
350	.4080	.4076	.4080	.4097	.4126	1.373	1.394	1.410	1.422	1.430
400	.4046	.4058	.4076	.4101	.4136	1.409	1.431	1.448	1.461	1.469
450	.4025	.4050	.4079	.4111	.4149	1.442	1.465	1.483	1.496	1.504
500	.4014	.4049	.4087	.4125	.4165	1.473	1.496	1.515	1.528	1.536
600	.4012	.4062	.4111	.4156	.4200	1.529	1.554	1.573	1.587	1.595
700	.4023	.4083	.4140	.4190	.4235	1.580	1.605	1.625	1.638	1.646
800	.4040	.4108	.4170	.4224	.4269	1.625	1.651	1.671	1.684	1.691
900	.4061	.4135	.4200	.4256	.4302	1.666	1.692	1.712	1.725	1.732
1000	.4083	.4161	.4229	.4286	.4332	1.703	1.730	1.750	1.763	1.769
1200	.4129	.4211	.4283	.4341	.4386	1.770	1.796	1.816	1.828	1.834
1400	.4173	.4258	.4330	.4388	.4432	1.828	1.854	1.873	1.885	1.890
1700	.4232	.4318	.4390	.4447	.4488	1.903	1.928	1.947	1.957	1.961
2000	.4284	.4368	.4438	.4493	.4533	1.967	1.991	2.009	2.019	2.021

C ON AG, M2/M1=9   TABLE 5- 5

	RELATIVE TRANSV. DAMAGE STRAGGLING					RELATIVE TRANSV. ION7N. STRAGGLING				
K/KL =	0.6	0.80	1.0	1.2	1.4	0.6	0.80	1.0	1.2	1.4
E,KEV										
.20	.8860	.8800	.8741	.8682	.8623	.8647	.8571	.8495	.8420	.8345
.23	.8857	.8795	.8734	.8674	.8614	.8642	.8564	.8486	.8409	.8333
.26	.8854	.8791	.8729	.8667	.8606	.8637	.8557	.8478	.8399	.8321
.30	.8850	.8786	.8722	.8659	.8597	.8632	.8550	.8468	.8388	.8308
.35	.8847	.8780	.8715	.8650	.8587	.8626	.8542	.8458	.8375	.8293
.40	.8843	.8776	.8709	.8643	.8577	.8621	.8534	.8449	.8364	.8280
.45	.8840	.8771	.8703	.8636	.8569	.8616	.8528	.8440	.8354	.8268
.50	.8838	.8768	.8698	.8630	.8562	.8611	.8522	.8433	.8345	.8257
.60	.8834	.8761	.8689	.8619	.8549	.8604	.8511	.8419	.8328	.8238
.70	.8830	.8756	.8682	.8609	.8538	.8597	.8501	.8407	.8313	.8221
.80	.8827	.8751	.8676	.8601	.8528	.8591	.8493	.8396	.8300	.8206
.90	.8825	.8747	.8670	.8594	.8520	.8586	.8485	.8386	.8288	.8192
1.0	.8823	.8744	.8665	.8588	.8512	.8581	.8478	.8377	.8277	.8179
1.2	.8820	.8738	.8656	.8577	.8498	.8572	.8466	.8361	.8258	.8156
1.4	.8817	.8733	.8649	.8567	.8486	.8565	.8455	.8347	.8240	.8135
1.7	.8815	.8727	.8640	.8555	.8472	.8555	.8441	.8328	.8217	.8108
2.0	.8813	.8722	.8633	.8545	.8459	.8547	.8428	.8311	.8196	.8083
2.3	.8813	.8719	.8627	.8537	.8448	.8539	.8417	.8296	.8178	.8061
2.6	.8812	.8716	.8622	.8529	.8439	.8532	.8406	.8282	.8160	.8041
3.0	.8812	.8713	.8616	.8521	.8428	.8524	.8393	.8265	.8139	.8016
3.5	.8813	.8711	.8610	.8512	.8416	.8514	.8379	.8245	.8114	.7986
4.0	.8815	.8709	.8605	.8504	.8405	.8506	.8365	.8226	.8091	.7959
4.5	.8816	.8708	.8601	.8497	.8395	.8497	.8352	.8209	.8069	.7933
5.0	.8819	.8707	.8598	.8491	.8387	.8489	.8339	.8192	.8048	.7908
6.0	.8824	.8707	.8593	.8481	.8372	.8476	.8317	.8162	.8011	.7863
7.0	.8829	.8707	.8588	.8472	.8359	.8463	.8296	.8133	.7975	.7821
8.0	.8835	.8708	.8585	.8465	.8348	.8450	.8275	.8105	.7940	.7780
9.0	.8840	.8709	.8582	.8458	.8337	.8438	.8255	.8078	.7906	.7740
10	.8846	.8711	.8579	.8451	.8327	.8425	.8235	.8051	.7873	.7701
12	.8858	.8715	.8576	.8441	.8310	.8403	.8199	.8002	.7812	.7629
14	.8869	.8718	.8572	.8430	.8293	.8381	.8163	.7954	.7752	.7559
17	.8884	.8723	.8566	.8415	.8269	.8344	.8107	.7880	.7663	.7456
20	.8898	.8725	.8560	.8400	.8245	.8306	.8050	.7807	.7576	.7356
23	.8912	.8730	.8555	.8386	.8224	.8270	.7995	.7736	.7490	.7257
26	.8924	.8733	.8549	.8372	.8203	.8233	.7940	.7665	.7406	.7161
30	.8938	.8734	.8539	.8352	.8173	.8180	.7864	.7570	.7295	.7036
35	.8950	.8732	.8524	.8325	.8135	.8111	.7769	.7453	.7160	.6885
40	.8959	.8727	.8505	.8295	.8095	.8039	.7673	.7336	.7027	.6740
45	.8964	.8718	.8485	.8264	.8055	.7966	.7577	.7222	.6898	.6600
50	.8966	.8707	.8463	.8232	.8013	.7891	.7481	.7109	.6771	.6464
60	.8967	.8684	.8417	.8167	.7932	.7738	.7287	.6881	.6515	.6195
70	.8960	.8653	.8366	.8098	.7849	.7585	.7099	.6665	.6276	.5947
80	.8946	.8617	.8311	.8028	.7766	.7434	.6918	.6461	.6053	.5717
90	.8927	.8577	.8254	.7956	.7682	.7285	.6744	.6268	.5845	.5504
100	.8904	.8534	.8195	.7884	.7600	.7140	.6577	.6084	.5650	.5306
120	.8850	.8440	.8070	.7733	.7430	.6846	.6246	.5728	.5283	.4936
140	.8787	.8342	.7944	.7585	.7267	.6570	.5943	.5408	.4958	.4611
170	.8682	.8191	.7759	.7373	.7036	.6191	.5537	.4987	.4536	.4194
200	.8569	.8041	.7580	.7172	.6823	.5849	.5180	.4623	.4177	.3842
230	.8442	.7882	.7397	.6973	.6615	.5524	.4856	.4304	.3868	.3547
260	.8314	.7728	.7223	.6786	.6423	.5229	.4567	.4024	.3599	.3291
300	.8147	.7532	.7005	.6555	.6186	.4877	.4227	.3698	.3289	.3000
350	.7945	.7302	.6753	.6291	.5920	.4491	.3861	.3352	.2964	.2696
400	.7752	.7087	.6522	.6053	.5682	.4157	.3549	.3060	.2692	.2443
450	.7567	.6886	.6310	.5836	.5467	.3865	.3279	.2811	.2461	.2230
500	.7391	.6698	.6114	.5638	.5271	.3608	.3044	.2596	.2264	.2048
600	.7050	.6344	.5756	.5283	.4924	.3189	.2670	.2261	.1961	.1769
700	.6743	.6032	.5446	.4978	.4630	.2849	.2373	.1998	.1726	.1553
800	.6467	.5756	.5174	.4714	.4376	.2569	.2130	.1786	.1538	.1381
900	.6216	.5509	.4934	.4482	.4155	.2335	.1928	.1611	.1384	.1241
1000	.5988	.5287	.4720	.4277	.3960	.2136	.1759	.1465	.1256	.1125
1200	.5588	.4903	.4355	.3931	.3632	.1818	.1490	.1236	.1056	.0945
1400	.5246	.4584	.4055	.3649	.3367	.1579	.1290	.1066	.0909	.0813
1700	.4819	.4193	.3693	.3313	.3054	.1321	.1076	.0886	.0752	.0673
2000	.4467	.3880	.3409	.3053	.2813	.1147	.0931	.0764	.0646	.0579

	RANGE SKEWNESS					DAMAGE SKEWNESS				
K/KL=	0.6	0.80	1.0	1.2	1.4	0.6	0.80	1.0	1.2	1.4
E,KEV										
.20	.1369	.1269	.1171	.1073	.0975	.4214	.4198	.4182	.4166	.4150
.23	.1361	.1259	.1157	.1057	.0957	.4213	.4196	.4180	.4163	.4147
.26	.1354	.1249	.1145	.1042	.0940	.4212	.4195	.4178	.4161	.4144
.30	.1346	.1238	.1131	.1025	.0921	.4212	.4194	.4176	.4158	.4140
.35	.1336	.1225	.1115	.1006	.0899	.4211	.4192	.4174	.4155	.4136
.40	.1328	.1214	.1101	.0989	.0879	.4211	.4191	.4171	.4152	.4133
.45	.1320	.1203	.1088	.0974	.0861	.4211	.4190	.4170	.4149	.4129
.50	.1313	.1194	.1076	.0959	.0844	.4210	.4189	.4168	.4147	.4126
.60	.1300	.1176	.1053	.0932	.0813	.4210	.4187	.4164	.4141	.4119
.70	.1288	.1159	.1032	.0907	.0784	.4209	.4184	.4160	.4136	.4113
.80	.1276	.1144	.1013	.0885	.0758	.4208	.4182	.4156	.4131	.4107
.90	.1266	.1130	.0995	.0863	.0733	.4207	.4179	.4153	.4126	.4101
1.0	.1256	.1116	.0978	.0843	.0710	.4206	.4177	.4149	.4121	.4095
1.2	.1237	.1091	.0948	.0807	.0668	.4204	.4172	.4142	.4112	.4083
1.4	.1220	.1068	.0919	.0773	.0629	.4202	.4167	.4135	.4103	.4072
1.7	.1196	.1035	.0878	.0725	.0575	.4198	.4160	.4123	.4089	.4055
2.0	.1172	.1004	.0840	.0680	.0524	.4193	.4151	.4112	.4074	.4038
2.3	.1151	.0976	.0806	.0639	.0477	.4189	.4144	.4102	.4061	.4023
2.6	.1131	.0949	.0772	.0600	.0433	.4185	.4137	.4091	.4048	.4007
3.0	.1104	.0914	.0729	.0550	.0376	.4178	.4126	.4077	.4030	.3986
3.5	.1071	.0870	.0677	.0489	.0307	.4168	.4111	.4057	.4007	.3959
4.0	.1038	.0828	.0626	.0430	.0241	.4157	.4094	.4036	.3982	.3931
4.5	.1005	.0786	.0576	.0373	.0178	.4145	.4077	.4015	.3957	.3903
5.0	.0973	.0745	.0527	.0318	.0116	.4131	.4059	.3993	.3932	.3874
6.0	.0914	.0670	.0438	.0215	.0002	.4103	.4022	.3948	.3880	.3817
7.0	.0855	.0595	.0350	.0115	-.0109	.4073	.3983	.3903	.3829	.3760
8.0	.0794	.0521	.0262	-.0017	-.0217	.4041	.3944	.3856	.3777	.3703
9.0	.0733	.0446	.0176	-.0080	-.0323	.4009	.3904	.3810	.3725	.3647
10	.0672	.0372	.0090	-.0175	-.0427	.3976	.3864	.3764	.3674	.3591
12	.0543	.0219	-.0083	-.0367	-.0635	.3913	.3787	.3676	.3577	.3485
14	.0417	.0071	-.0250	-.0550	-.0832	.3848	.3710	.3589	.3481	.3381
17	.0236	-.0141	-.0488	-.0809	-.1110	.3759	.3595	.3460	.3340	.3230
20	.0064	-.0341	-.0710	-.1051	-.1368	.3650	.3480	.3333	.3202	.3084
23	-.0089	-.0519	-.0909	-.1266	-.1600	.3543	.3363	.3204	.3064	.2945
26	-.0235	-.0689	-.1098	-.1470	-.1818	.3438	.3248	.3079	.2930	.2811
30	-.0423	-.0905	-.1338	-.1728	-.2093	.3301	.3099	.2918	.2758	.2638
35	-.0648	-.1163	-.1621	-.2032	-.2415	.3138	.2922	.2726	.2554	.2431
40	-.0865	-.1409	-.1889	-.2320	-.2718	.2982	.2753	.2544	.2360	.2234
45	-.1074	-.1645	-.2146	-.2594	-.3004	.2834	.2592	.2370	.2176	.2045
50	-.1278	-.1872	-.2392	-.2857	-.3278	.2693	.2438	.2205	.2001	.1864
60	-.1681	-.2320	-.2872	-.3375	-.3804	.2436	.2152	.1896	.1676	.1514
70	-.2057	-.2735	-.3314	-.3850	-.4286	.2197	.1887	.1611	.1375	.1190
80	-.2411	-.3121	-.3724	-.4287	-.4730	.1972	.1638	.1344	.1095	.0890
90	-.2744	-.3481	-.4105	-.4691	-.5142	.1759	.1404	.1093	.0831	.0609
100	-.3059	-.3818	-.4462	-.5067	-.5527	.1556	.1182	.0855	.0582	.0345
120	-.3651	-.4439	-.5115	-.5739	-.6228	.1172	.0763	.0409	.0113	-.0141
140	-.4187	-.4997	-.5702	-.6339	-.6854	.0819	.0381	.0003	-.0312	-.0578
170	-.4904	-.5741	-.6483	-.7135	-.7686	.0336	-.0138	-.0546	-.0885	-.1163
200	-.5537	-.6399	-.7172	-.7837	-.8418	-.0101	-.0605	-.1036	-.1394	-.1680
230	-.6089	-.6980	-.7784	-.8468	-.9067	-.0507	-.1038	-.1489	-.1861	-.2153
260	-.6590	-.7509	-.8340	-.9042	-.9655	-.0882	-.1435	-.1903	-.2285	-.2582
300	-.7195	-.8149	-.9012	-.9737	-1.037	-.1341	-.1918	-.2403	-.2797	-.3097
350	-.7872	-.8868	-.9766	-1.052	-1.116	-.1859	-.2460	-.2962	-.3364	-.3667
400	-.8482	-.9515	-1.044	-1.122	-1.188	-.2326	-.2945	-.3459	-.3868	-.4171
450	-.9039	-1.011	-1.106	-1.186	-1.253	-.2753	-.3385	-.3908	-.4320	-.4622
500	-.9552	-1.065	-1.163	-1.245	-1.313	-.3145	-.3787	-.4315	-.4729	-.5029
600	-1.047	-1.163	-1.265	-1.350	-1.420	-.3858	-.4507	-.5037	-.5447	-.5739
700	-1.128	-1.250	-1.355	-1.443	-1.515	-.4477	-.5127	-.5655	-.6060	-.6343
800	-1.202	-1.328	-1.436	-1.527	-1.600	-.5023	-.5671	-.6194	-.6592	-.6867
900	-1.270	-1.400	-1.511	-1.604	-1.678	-.5510	-.6153	-.6670	-.7061	-.7328
1000	-1.332	-1.466	-1.579	-1.674	-1.750	-.5948	-.6585	-.7096	-.7480	-.7739
1200	-1.445	-1.585	-1.703	-1.802	-1.880	-.6710	-.7332	-.7828	-.8198	-.8444
1400	-1.546	-1.691	-1.813	-1.914	-1.995	-.7352	-.7958	-.8440	-.8797	-.9031
1700	-1.680	-1.831	-1.958	-2.063	-2.146	-.8154	-.8735	-.9195	-.9534	-.9755
2000	-1.798	-1.954	-2.086	-2.195	-2.280	-.8815	-.9370	-.9810	-1.013	-1.034

TABLE 5- 7

C ON AG, M2/M1=9

K/KL= E,KEV	IONIZATION SKEWNESS					RANGE KURTOSIS				
	0.6	0.80	1.0	1.2	1.4	0.6	0.80	1.0	1.2	1.4
.20	.4941	.4955	.4968	.4981	.4994	3.208	3.172	3.139	3.109	3.081
.23	.4944	.4958	.4972	.4985	.4998	3.205	3.168	3.135	3.104	3.076
.26	.4948	.4962	.4975	.4988	.5001	3.202	3.165	3.131	3.100	3.071
.30	.4952	.4966	.4979	.4992	.5005	3.199	3.161	3.126	3.095	3.066
.35	.4956	.4970	.4984	.4997	.5010	3.195	3.156	3.121	3.089	3.060
.40	.4961	.4974	.4988	.5001	.5014	3.192	3.152	3.117	3.084	3.055
.45	.4965	.4978	.4992	.5005	.5018	3.188	3.148	3.112	3.080	3.050
.50	.4968	.4982	.4995	.5009	.5021	3.185	3.145	3.109	3.076	3.045
.60	.4976	.4989	.5002	.5015	.5028	3.180	3.139	3.102	3.068	3.038
.70	.4983	.4996	.5009	.5021	.5033	3.175	3.133	3.096	3.061	3.031
.80	.4989	.5002	.5014	.5027	.5039	3.171	3.128	3.090	3.056	3.025
.90	.4996	.5008	.5020	.5032	.5043	3.167	3.123	3.085	3.050	3.019
1.0	.5002	.5013	.5025	.5037	.5048	3.163	3.119	3.080	3.045	3.014
1.2	.5014	.5025	.5035	.5046	.5057	3.156	3.111	3.071	3.036	3.004
1.4	.5025	.5035	.5045	.5055	.5065	3.150	3.104	3.064	3.028	2.996
1.7	.5040	.5048	.5057	.5066	.5075	3.141	3.094	3.053	3.017	2.985
2.0	.5053	.5060	.5067	.5075	.5083	3.133	3.085	3.044	3.007	2.975
2.3	.5064	.5069	.5075	.5082	.5089	3.125	3.077	3.035	2.999	2.966
2.6	.5074	.5077	.5082	.5087	.5093	3.118	3.069	3.027	2.990	2.958
3.0	.5085	.5087	.5090	.5094	.5099	3.110	3.060	3.017	2.981	2.949
3.5	.5099	.5097	.5098	.5100	.5104	3.100	3.049	3.006	2.969	2.938
4.0	.5110	.5107	.5105	.5106	.5108	3.090	3.039	2.996	2.959	2.928
4.5	.5121	.5115	.5112	.5111	.5111	3.081	3.030	2.987	2.950	2.919
5.0	.5131	.5123	.5118	.5115	.5114	3.073	3.021	2.978	2.942	2.911
6.0	.5156	.5143	.5134	.5129	.5125	3.057	3.005	2.962	2.926	2.896
7.0	.5175	.5158	.5146	.5138	.5132	3.043	2.990	2.948	2.913	2.884
8.0	.5190	.5169	.5154	.5143	.5134	3.030	2.977	2.935	2.901	2.873
9.0	.5202	.5177	.5158	.5145	.5134	3.017	2.964	2.923	2.890	2.863
10	.5209	.5181	.5160	.5144	.5131	3.005	2.953	2.912	2.880	2.855
12	.5212	.5177	.5151	.5131	.5115	2.979	2.928	2.889	2.859	2.836
14	.5209	.5169	.5138	.5114	.5094	2.956	2.906	2.870	2.842	2.822
17	.5199	.5151	.5113	.5084	.5059	2.928	2.881	2.847	2.824	2.808
20	.5186	.5130	.5087	.5051	.5021	2.905	2.861	2.831	2.812	2.800
23	.5181	.5115	.5066	.5023	.4984	2.888	2.848	2.822	2.807	2.802
26	.5173	.5099	.5045	.4995	.4947	2.875	2.838	2.817	2.805	2.807
30	.5160	.5076	.5015	.4955	.4898	2.861	2.830	2.814	2.807	2.818
35	.5139	.5043	.4975	.4906	.4838	2.848	2.825	2.815	2.816	2.836
40	.5113	.5007	.4932	.4855	.4779	2.840	2.824	2.821	2.828	2.858
45	.5084	.4969	.4888	.4804	.4722	2.834	2.826	2.830	2.844	2.882
50	.5051	.4929	.4842	.4754	.4667	2.830	2.830	2.841	2.861	2.906
60	.4970	.4837	.4739	.4643	.4557	2.818	2.833	2.859	2.893	2.947
70	.4887	.4747	.4639	.4540	.4455	2.815	2.846	2.886	2.933	2.995
80	.4805	.4661	.4544	.4443	.4361	2.819	2.866	2.920	2.979	3.049
90	.4726	.4577	.4455	.4353	.4274	2.830	2.891	2.958	3.029	3.107
100	.4649	.4497	.4370	.4268	.4193	2.846	2.922	3.001	3.083	3.169
120	.4496	.4337	.4206	.4106	.4034	2.893	2.994	3.096	3.199	3.302
140	.4356	.4193	.4062	.3965	.3898	2.949	3.075	3.200	3.321	3.441
170	.4170	.4008	.3880	.3789	.3732	3.046	3.207	3.362	3.511	3.654
200	.4011	.3852	.3731	.3648	.3601	3.152	3.345	3.529	3.703	3.868
230	.3866	.3719	.3610	.3536	.3509	3.256	3.479	3.688	3.886	4.071
260	.3740	.3606	.3511	.3446	.3437	3.363	3.614	3.849	4.068	4.273
300	.3596	.3482	.3405	.3352	.3366	3.510	3.796	4.063	4.310	4.541
350	.3450	.3360	.3306	.3267	.3306	3.696	4.024	4.329	4.611	4.870
400	.3333	.3266	.3233	.3209	.3269	3.883	4.252	4.593	4.907	5.194
450	.3239	.3195	.3182	.3171	.3249	4.070	4.478	4.853	5.198	5.512
500	.3165	.3142	.3147	.3150	.3242	4.256	4.701	5.110	5.484	5.824
600	.3074	.3094	.3132	.3168	.3270	4.608	5.123	5.595	6.024	6.411
700	.3020	.3076	.3143	.3206	.3314	4.957	5.539	6.070	6.552	6.984
800	.2991	.3077	.3169	.3255	.3367	5.303	5.950	6.538	7.069	7.543
900	.2980	.3092	.3203	.3308	.3422	5.644	6.353	6.997	7.576	8.091
1000	.2982	.3116	.3243	.3364	.3479	5.981	6.750	7.447	8.072	8.627
1200	.3012	.3179	.3332	.3474	.3591	6.639	7.524	8.322	9.036	9.665
1400	.3065	.3255	.3425	.3579	.3697	7.276	8.270	9.165	9.962	10.66
1700	.3170	.3380	.3564	.3724	.3844	8.193	9.341	10.37	11.29	12.08
2000	.3292	.3510	.3697	.3852	.3974	9.067	10.36	11.52	12.54	13.43

	DAMAGE KURTOSIS					IONIZATION KURTOSIS				
K/KL=	0.6	0.80	1.0	1.2	1.4	0.6	0.80	1.0	1.2	1.4
E,KEV										
.20	4.321	4.271	4.224	4.181	4.140	4.474	4.423	4.376	4.332	4.290
.23	4.315	4.264	4.217	4.173	4.131	4.469	4.417	4.369	4.324	4.282
.26	4.310	4.258	4.210	4.166	4.124	4.464	4.412	4.363	4.317	4.274
.30	4.304	4.251	4.202	4.157	4.114	4.459	4.405	4.355	4.309	4.265
.35	4.297	4.243	4.193	4.147	4.104	4.452	4.397	4.347	4.299	4.255
.40	4.291	4.236	4.185	4.138	4.095	4.446	4.390	4.339	4.291	4.246
.45	4.285	4.229	4.178	4.130	4.086	4.441	4.384	4.332	4.283	4.237
.50	4.279	4.223	4.171	4.123	4.078	4.435	4.378	4.325	4.276	4.230
.60	4.269	4.211	4.158	4.109	4.063	4.426	4.367	4.312	4.262	4.215
.70	4.259	4.200	4.146	4.096	4.049	4.417	4.357	4.301	4.250	4.202
.80	4.250	4.190	4.135	4.084	4.037	4.409	4.347	4.291	4.239	4.190
.90	4.241	4.180	4.124	4.073	4.025	4.401	4.339	4.281	4.229	4.179
1.0	4.233	4.171	4.114	4.062	4.014	4.394	4.330	4.272	4.219	4.169
1.2	4.218	4.154	4.096	4.043	3.994	4.381	4.316	4.257	4.202	4.151
1.4	4.203	4.138	4.079	4.025	3.975	4.369	4.303	4.242	4.186	4.135
1.7	4.183	4.116	4.056	4.000	3.949	4.353	4.284	4.222	4.164	4.112
2.0	4.165	4.096	4.034	3.978	3.926	4.337	4.266	4.202	4.144	4.090
2.3	4.147	4.077	4.014	3.957	3.904	4.320	4.248	4.183	4.124	4.069
2.6	4.131	4.060	3.996	3.937	3.884	4.305	4.231	4.165	4.104	4.049
3.0	4.111	4.038	3.972	3.913	3.859	4.285	4.209	4.141	4.080	4.024
3.5	4.086	4.011	3.945	3.885	3.830	4.261	4.184	4.115	4.052	3.994
4.0	4.063	3.987	3.919	3.858	3.802	4.240	4.160	4.090	4.026	3.968
4.5	4.041	3.963	3.894	3.833	3.777	4.220	4.139	4.067	4.002	3.943
5.0	4.020	3.941	3.871	3.808	3.752	4.201	4.118	4.045	3.979	3.919
6.0	3.979	3.897	3.826	3.762	3.705	4.170	4.084	4.009	3.941	3.880
7.0	3.941	3.857	3.784	3.720	3.661	4.140	4.052	3.975	3.906	3.844
8.0	3.905	3.820	3.746	3.680	3.622	4.112	4.022	3.943	3.873	3.810
9.0	3.872	3.785	3.710	3.644	3.585	4.085	3.992	3.912	3.841	3.777
10	3.841	3.753	3.677	3.610	3.551	4.058	3.964	3.882	3.810	3.745
12	3.784	3.694	3.616	3.549	3.489	4.002	3.905	3.821	3.747	3.681
14	3.733	3.640	3.562	3.494	3.434	3.951	3.850	3.765	3.690	3.622
17	3.664	3.569	3.489	3.421	3.361	3.881	3.777	3.689	3.612	3.542
20	3.602	3.506	3.426	3.357	3.297	3.819	3.711	3.621	3.542	3.472
23	3.544	3.447	3.366	3.298	3.237	3.770	3.657	3.562	3.481	3.411
26	3.492	3.393	3.313	3.244	3.184	3.726	3.608	3.509	3.426	3.356
30	3.429	3.329	3.249	3.180	3.121	3.673	3.549	3.445	3.360	3.290
35	3.359	3.259	3.179	3.111	3.052	3.611	3.482	3.374	3.287	3.218
40	3.297	3.197	3.117	3.051	2.993	3.555	3.421	3.310	3.222	3.154
45	3.242	3.143	3.064	2.998	2.941	3.501	3.365	3.253	3.163	3.096
50	3.193	3.094	3.016	2.951	2.895	3.451	3.313	3.200	3.110	3.044
60	3.109	3.012	2.934	2.870	2.818	3.344	3.211	3.101	3.014	2.949
70	3.038	2.944	2.867	2.805	2.755	3.249	3.122	3.016	2.932	2.870
80	2.977	2.886	2.811	2.751	2.704	3.166	3.044	2.942	2.861	2.801
90	2.925	2.837	2.763	2.705	2.661	3.093	2.975	2.878	2.800	2.742
100	2.879	2.794	2.723	2.666	2.625	3.028	2.915	2.821	2.746	2.690
120	2.801	2.722	2.656	2.605	2.568	2.923	2.816	2.727	2.657	2.605
140	2.739	2.666	2.605	2.559	2.527	2.838	2.735	2.651	2.585	2.537
170	2.668	2.603	2.550	2.511	2.484	2.735	2.638	2.560	2.500	2.457
200	2.615	2.558	2.513	2.480	2.459	2.653	2.562	2.488	2.433	2.395
230	2.578	2.529	2.491	2.464	2.448	2.586	2.500	2.432	2.381	2.348
260	2.550	2.509	2.477	2.456	2.444	2.530	2.450	2.386	2.340	2.310
300	2.524	2.492	2.469	2.454	2.447	2.468	2.394	2.336	2.295	2.269
350	2.504	2.484	2.470	2.462	2.460	2.406	2.339	2.288	2.251	2.230
400	2.495	2.484	2.478	2.476	2.479	2.356	2.295	2.249	2.217	2.200
450	2.494	2.491	2.492	2.496	2.503	2.315	2.260	2.219	2.191	2.176
500	2.498	2.503	2.510	2.518	2.528	2.280	2.231	2.194	2.170	2.158
600	2.522	2.539	2.556	2.572	2.587	2.229	2.188	2.159	2.140	2.133
700	2.554	2.582	2.606	2.628	2.646	2.191	2.158	2.134	2.121	2.117
800	2.590	2.626	2.658	2.684	2.706	2.162	2.135	2.117	2.107	2.107
900	2.628	2.672	2.709	2.740	2.763	2.140	2.118	2.105	2.099	2.100
1000	2.668	2.718	2.760	2.793	2.819	2.123	2.106	2.096	2.093	2.097
1200	2.748	2.808	2.857	2.896	2.925	2.098	2.089	2.086	2.087	2.094
1400	2.826	2.893	2.949	2.992	3.023	2.083	2.080	2.082	2.087	2.096
1700	2.939	3.014	3.076	3.124	3.158	2.071	2.076	2.083	2.092	2.103
2000	3.046	3.126	3.192	3.243	3.279	2.067	2.077	2.088	2.100	2.113

C ON AG, M2/M1=9   TABLE 5- 9

| E,KEV | REFLECTION COEFFICIENT | | | | | SPUTTERING EFFICIENCY | | | | |
K/KL=	0.6	0.80	1.0	1.2	1.4	0.6	0.80	1.0	1.2	1.4
.20	.2393	.2348	.2306	.2265	.2225	.1478	.1361	.1257	.1164	.1080
.23	.2388	.2342	.2299	.2257	.2217	.1468	.1349	.1244	.1150	.1066
.26	.2383	.2336	.2292	.2250	.2209	.1458	.1339	.1233	.1138	.1053
.30	.2377	.2329	.2284	.2241	.2199	.1447	.1326	.1219	.1123	.1038
.35	.2370	.2322	.2275	.2231	.2189	.1435	.1312	.1204	.1107	.1021
.40	.2364	.2314	.2267	.2222	.2179	.1424	.1300	.1190	.1093	.1007
.45	.2358	.2308	.2260	.2214	.2170	.1414	.1289	.1178	.1080	.0994
.50	.2353	.2301	.2253	.2206	.2162	.1405	.1278	.1167	.1069	.0982
.60	.2342	.2289	.2239	.2192	.2146	.1389	.1260	.1147	.1048	.0961
.70	.2332	.2278	.2227	.2179	.2132	.1374	.1243	.1130	.1030	.0942
.80	.2323	.2268	.2216	.2166	.2119	.1361	.1229	.1114	.1014	.0926
.90	.2314	.2258	.2205	.2155	.2107	.1349	.1215	.1100	.0999	.0911
1.0	.2306	.2249	.2195	.2144	.2096	.1337	.1203	.1087	.0986	.0898
1.2	.2291	.2232	.2176	.2124	.2075	.1317	.1181	.1063	.0962	.0874
1.4	.2276	.2216	.2159	.2105	.2055	.1298	.1161	.1043	.0941	.0853
1.7	.2255	.2193	.2134	.2080	.2028	.1274	.1134	.1016	.0914	.0826
2.0	.2236	.2172	.2112	.2056	.2003	.1252	.1111	.0992	.0890	.0802
2.3	.2218	.2152	.2091	.2034	.1981	.1232	.1090	.0970	.0868	.0781
2.6	.2201	.2133	.2071	.2013	.1959	.1213	.1070	.0950	.0848	.0761
3.0	.2179	.2110	.2046	.1986	.1932	.1190	.1047	.0926	.0825	.0738
3.5	.2153	.2082	.2016	.1956	.1901	.1164	.1020	.0899	.0798	.0712
4.0	.2128	.2055	.1988	.1927	.1871	.1141	.0995	.0875	.0774	.0689
4.5	.2105	.2030	.1962	.1899	.1842	.1119	.0973	.0853	.0753	.0668
5.0	.2083	.2006	.1937	.1873	.1815	.1098	.0952	.0832	.0733	.0649
6.0	.2040	.1961	.1891	.1827	.1764	.1060	.0913	.0794	.0696	.0614
7.0	.2001	.1920	.1848	.1783	.1716	.1025	.0879	.0761	.0664	.0584
8.0	.1965	.1881	.1807	.1741	.1670	.0994	.0848	.0731	.0636	.0557
9.0	.1930	.1845	.1769	.1701	.1627	.0966	.0820	.0704	.0610	.0533
10	.1898	.1811	.1733	.1663	.1586	.0939	.0794	.0679	.0587	.0511
12	.1838	.1748	.1663	.1584	.1503	.0890	.0747	.0634	.0545	.0473
14	.1784	.1691	.1600	.1512	.1428	.0847	.0705	.0596	.0509	.0439
17	.1710	.1614	.1514	.1415	.1329	.0791	.0652	.0546	.0463	.0397
20	.1644	.1544	.1438	.1332	.1244	.0742	.0606	.0504	.0425	.0363
23	.1582	.1477	.1366	.1260	.1171	.0699	.0566	.0467	.0392	.0333
26	.1526	.1415	.1301	.1196	.1107	.0660	.0530	.0435	.0363	.0307
30	.1457	.1340	.1224	.1121	.1032	.0614	.0489	.0398	.0330	.0278
35	.1379	.1257	.1140	.1040	.0953	.0564	.0444	.0359	.0296	.0247
40	.1310	.1183	.1066	.0970	.0885	.0522	.0406	.0326	.0267	.0222
45	.1248	.1117	.1001	.0909	.0826	.0484	.0374	.0298	.0242	.0201
50	.1191	.1058	.0944	.0855	.0775	.0451	.0345	.0273	.0221	.0182
60	.1086	.0954	.0848	.0763	.0691	.0395	.0299	.0234	.0188	.0154
70	.0998	.0868	.0768	.0688	.0624	.0350	.0261	.0203	.0162	.0132
80	.0921	.0795	.0702	.0625	.0568	.0312	.0231	.0178	.0141	.0115
90	.0855	.0732	.0645	.0572	.0521	.0281	.0205	.0157	.0124	.0101
100	.0797	.0677	.0597	.0526	.0480	.0254	.0184	.0140	.0110	.0089
120	.0703	.0592	.0518	.0456	.0415	.0212	.0151	.0114	.0089	.0072
140	.0627	.0525	.0456	.0401	.0364	.0180	.0127	.0095	.0074	.0059
170	.0537	.0447	.0384	.0337	.0304	.0144	.0100	.0074	.0057	.0046
200	.0467	.0386	.0329	.0289	.0258	.0117	.0081	.0060	.0046	.0036
230	.0412	.0338	.0285	.0250	.0222	.0098	.0067	.0049	.0038	.0030
260	.0367	.0298	.0249	.0218	.0192	.0084	.0057	.0041	.0032	.0025
300	.0318	.0255	.0211	.0183	.0160	.0069	.0047	.0034	.0026	.0020
350	.0270	.0212	.0174	.0149	.0129	.0055	.0037	.0027	.0020	.0016
400	.0231	.0179	.0145	.0123	.0105	.0045	.0031	.0022	.0017	.0013
450	.0200	.0153	.0122	.0102	.0086	.0038	.0026	.0019	.0014	.0011
500	.0174	.0131	.0104	.0085	.0071	.0032	.0022	.0016	.0012	.0010
600	.0135	.0099	.0077	.0062	.0051	.0024	.0017	.0012	.0009	.0007
700	.0107	.0077	.0059	.0047	.0038	.0019	.0013	.0009	.0007	.0006
800	.0085	.0060	.0045	.0035	.0028	.0015	.0011	.0008	.0006	.0005
900	.0069	.0048	.0035	.0027	.0021	.0013	.0009	.0006	.0005	.0004
1000	.0056	.0038	.0028	.0021	.0016	.0011	.0008	.0006	.0004	.0003
1200	.0037	.0025	.0018	.0013	.0010	.0008	.0007	.0004	.0003	.0003
1400	.0026	.0017	.0011	.0008	.0006	.0007	.0005	.0004	.0003	.0002
1700	.0016	.0010	.0006	.0004	.0003	.0005	.0004	.0003	.0002	.0002
2000	.0011	.0007	.0004	.0003	.0002	.0004	.0003	.0002	.0001	.0001

TABLE 5-10

C ON AG, M2/M1=9

E,KEV	IONIZATION DEFICIENCY K/KL= 0.6	0.80	1.0	1.2	1.4	SPUTTERING YIELD ALPHA 0.6	0.80	1.0	1.2	1.4
.20	.0242	.0299	.0347	.0388	.0422	1.145	1.101	1.060	1.021	.9847
.23	.0247	.0304	.0352	.0392	.0426	1.142	1.098	1.056	1.016	.9791
.26	.0250	.0308	.0356	.0397	.0431	1.140	1.095	1.052	1.012	.9741
.30	.0255	.0313	.0361	.0402	.0435	1.137	1.091	1.047	1.007	.9683
.35	.0259	.0318	.0367	.0407	.0440	1.134	1.087	1.042	1.001	.9619
.40	.0264	.0323	.0371	.0411	.0445	1.131	1.083	1.038	.9959	.9563
.45	.0267	.0326	.0375	.0415	.0448	1.129	1.080	1.034	.9914	.9514
.50	.0271	.0330	.0379	.0419	.0451	1.127	1.077	1.031	.9874	.9469
.60	.0276	.0336	.0385	.0424	.0456	1.123	1.072	1.025	.9804	.9391
.70	.0281	.0341	.0390	.0429	.0461	1.120	1.068	1.019	.9744	.9325
.80	.0285	.0345	.0394	.0433	.0464	1.117	1.064	1.015	.9692	.9267
.90	.0289	.0349	.0397	.0436	.0466	1.115	1.061	1.011	.9646	.9215
1.0	.0292	.0352	.0400	.0438	.0469	1.113	1.058	1.007	.9605	.9169
1.2	.0297	.0358	.0405	.0443	.0472	1.109	1.053	1.001	.9534	.9089
1.4	.0302	.0362	.0409	.0446	.0474	1.106	1.049	.9961	.9473	.9021
1.7	.0307	.0367	.0413	.0449	.0477	1.103	1.044	.9895	.9396	.8935
2.0	.0311	.0371	.0416	.0451	.0478	1.100	1.039	.9839	.9330	.8861
2.3	.0315	.0374	.0419	.0453	.0478	1.098	1.036	.9791	.9273	.8797
2.6	.0318	.0377	.0421	.0454	.0478	1.096	1.032	.9749	.9222	.8740
3.0	.0321	.0379	.0422	.0454	.0477	1.093	1.028	.9698	.9162	.8671
3.5	.0325	.0381	.0423	.0453	.0475	1.090	1.024	.9641	.9095	.8596
4.0	.0327	.0383	.0423	.0452	.0473	1.088	1.020	.9591	.9035	.8529
4.5	.0329	.0384	.0423	.0451	.0470	1.086	1.017	.9545	.8980	.8467
5.0	.0331	.0384	.0422	.0449	.0467	1.084	1.013	.9502	.8930	.8411
6.0	.0333	.0385	.0420	.0445	.0460	1.081	1.008	.9428	.8842	.8313
7.0	.0334	.0384	.0417	.0439	.0453	1.078	1.003	.9361	.8763	.8224
8.0	.0335	.0382	.0414	.0434	.0446	1.075	.9979	.9299	.8690	.8143
9.0	.0335	.0381	.0410	.0428	.0439	1.072	.9933	.9239	.8621	.8068
10	.0335	.0378	.0406	.0423	.0432	1.069	.9887	.9182	.8556	.7997
12	.0333	.0373	.0397	.0410	.0417	1.063	.9801	.9075	.8434	.7864
14	.0331	.0367	.0388	.0399	.0403	1.057	.9718	.8975	.8321	.7743
17	.0327	.0358	.0375	.0382	.0383	1.048	.9597	.8832	.8164	.7578
20	.0322	.0349	.0361	.0366	.0364	1.039	.9479	.8696	.8019	.7426
23	.0316	.0339	.0348	.0350	.0347	1.029	.9359	.8561	.7876	.7280
26	.0310	.0329	.0336	.0335	.0331	1.020	.9242	.8432	.7743	.7144
30	.0302	.0317	.0320	.0317	.0311	1.007	.9093	.8271	.7577	.6978
35	.0292	.0302	.0302	.0296	.0289	.9916	.8916	.8083	.7387	.6789
40	.0283	.0289	.0285	.0278	.0270	.9765	.8748	.7909	.7214	.6620
45	.0273	.0276	.0270	.0261	.0253	.9619	.8589	.7747	.7054	.6465
50	.0264	.0264	.0256	.0246	.0237	.9476	.8438	.7596	.6906	.6323
60	.0246	.0241	.0231	.0220	.0209	.9190	.8146	.7314	.6634	.6073
70	.0230	.0221	.0209	.0198	.0185	.8927	.7882	.7063	.6396	.5855
80	.0215	.0203	.0191	.0179	.0165	.8685	.7642	.6839	.6186	.5662
90	.0202	.0188	.0174	.0162	.0148	.8462	.7423	.6636	.5997	.5490
100	.0190	.0174	.0160	.0148	.0134	.8255	.7221	.6451	.5827	.5333
120	.0168	.0148	.0138	.0126	.0113	.7882	.6854	.6118	.5528	.5049
140	.0150	.0128	.0120	.0109	.0096	.7555	.6538	.5835	.5277	.4810
170	.0127	.0106	.0098	.0088	.0078	.7134	.6138	.5480	.4964	.4512
200	.0109	.0093	.0081	.0072	.0063	.6775	.5806	.5187	.4709	.4272
230	.0096	.0099	.0065	.0058	.0051	.6461	.5527	.4948	.4498	.4084
250	.0085	.0108	.0052	.0046	.0040	.6186	.5288	.4745	.4319	.3928
300	.0073	.0119	.0038	.0033	.0029	.5868	.5021	.4517	.4119	.3756
350	.0061	.0131	.0024	.0021	.0018	.5532	.4749	.4285	.3916	.3582
400	.0051	.0139	.0014	.0012	.0010	.5249	.4530	.4095	.3751	.3443
450	.0043	.0142	.0006	.0005	.0004	.5009	.4351	.3938	.3615	.3329
500	.0037	.0142	0.0000	0.0000	0.0000	.4801	.4204	.3806	.3501	.3234
600	.0024	.0106	0.0000	0.0000	0.0000	.4465	.4026	.3614	.3345	.3100
700	.0015	.0072	0.0000	0.0000	0.0000	.4205	.3896	.3469	.3227	.2998
800	.0009	.0043	0.0000	0.0000	0.0000	.4001	.3793	.3354	.3132	.2916
900	.0004	.0019	0.0000	0.0000	0.0000	.3839	.3706	.3261	.3053	.2848
1000	0.0000	0.0000	0.0000	0.0000	0.0000	.3709	.3629	.3183	.2984	.2788
1200	0.0000	0.0000	0.0000	0.0000	0.0000	.3522	.3494	.3059	.2869	.2688
1400	0.0000	0.0000	0.0000	0.0000	0.0000	.3404	.3371	.2962	.2771	.2602
1700	0.0000	0.0000	0.0000	0.0000	0.0000	.3315	.3195	.2850	.2644	.2488
2000	0.0000	0.0000	0.0000	0.0000	0.0000	.3295	.3022	.2762	.2530	.2386

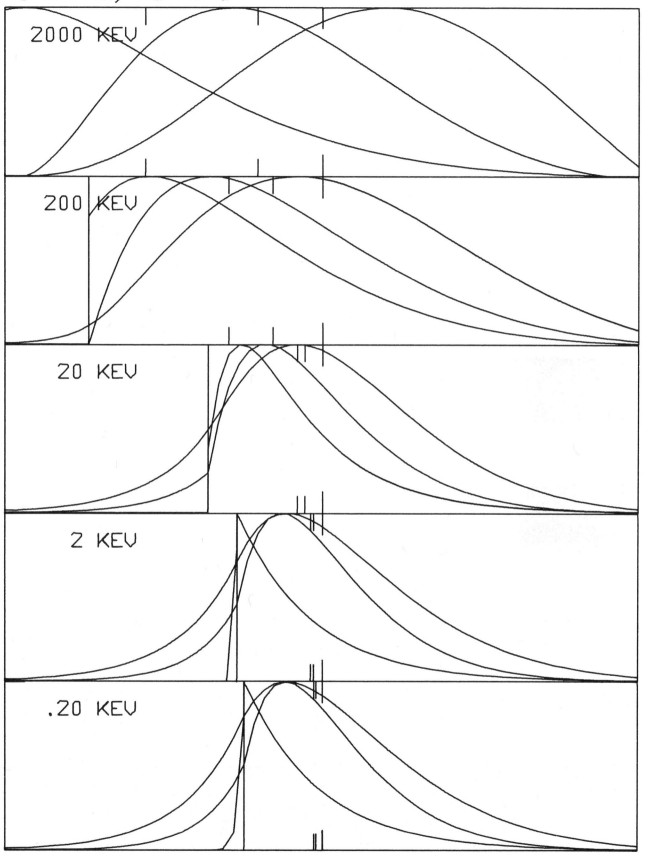

C ON AG, M2/M1=9

2000 KEV

200 KEV

20 KEV

2 KEV

.20 KEV

E,KEV	FRACTIONAL DEPOSITED ENERGY K/KL= 0.55	0.75	1.0	1.2	1.4	MEAN RANGE, MICROGRAM/SQ.CM. 0.55	0.75	1.0	1.2	1.4
.50	.8629	.8202	.7712	.7351	.7016	1.220	1.211	1.201	1.192	1.184
.60	.8591	.8154	.7653	.7286	.6945	1.386	1.376	1.363	1.353	1.343
.70	.8558	.8112	.7603	.7230	.6884	1.538	1.526	1.511	1.499	1.488
.80	.8528	.8075	.7558	.7180	.6830	1.680	1.666	1.650	1.637	1.624
.90	.8502	.8042	.7518	.7136	.6782	1.816	1.801	1.783	1.769	1.755
1.0	.8477	.8011	.7482	.7096	.6739	1.947	1.931	1.911	1.896	1.881
1.2	.8435	.7958	.7418	.7025	.6662	2.199	2.180	2.158	2.140	2.122
1.4	.8397	.7912	.7362	.6964	.6597	2.441	2.420	2.394	2.373	2.353
1.7	.8349	.7852	.7291	.6885	.6512	2.790	2.764	2.733	2.709	2.685
2.0	.8307	.7800	.7229	.6817	.6440	3.124	3.094	3.058	3.030	3.003
2.3	.8270	.7754	.7175	.6758	.6377	3.430	3.397	3.356	3.325	3.294
2.6	.8237	.7713	.7127	.6705	.6320	3.728	3.691	3.646	3.611	3.577
3.0	.8197	.7664	.7069	.6642	.6253	4.116	4.074	4.023	3.983	3.944
3.5	.8153	.7610	.7006	.6573	.6179	4.588	4.539	4.480	4.434	4.389
4.0	.8113	.7562	.6949	.6512	.6114	5.048	4.993	4.925	4.873	4.822
4.5	.8077	.7518	.6898	.6456	.6056	5.498	5.436	5.360	5.301	5.244
5.0	.8045	.7478	.6852	.6406	.6003	5.938	5.868	5.784	5.719	5.655
6.0	.7987	.7408	.6770	.6317	.5909	6.749	6.666	6.567	6.490	6.416
7.0	.7935	.7345	.6698	.6240	.5828	7.547	7.451	7.335	7.246	7.160
8.0	.7889	.7290	.6634	.6171	.5756	8.335	8.225	8.092	7.989	7.890
9.0	.7847	.7239	.6575	.6109	.5690	9.115	8.988	8.837	8.721	8.609
10	.7807	.7192	.6521	.6051	.5630	9.884	9.742	9.572	9.441	9.316
12	.7736	.7107	.6425	.5948	.5523	11.30	11.13	10.93	10.77	10.62
14	.7672	.7031	.6339	.5858	.5430	12.72	12.52	12.28	12.10	11.92
17	.7587	.6930	.6226	.5738	.5306	14.87	14.61	14.31	14.08	13.86
20	.7510	.6840	.6126	.5634	.5199	17.03	16.71	16.34	16.06	15.79
23	.7440	.6759	.6036	.5539	.5102	18.93	18.58	18.16	17.84	17.53
26	.7376	.6685	.5954	.5454	.5015	20.87	20.47	19.99	19.63	19.28
30	.7296	.6593	.5854	.5350	.4909	23.54	23.06	22.49	22.06	21.65
35	.7205	.6489	.5740	.5233	.4791	26.99	26.39	25.68	25.15	24.65
40	.7120	.6393	.5637	.5128	.4685	30.54	29.80	28.94	28.29	27.69
45	.7041	.6305	.5542	.5031	.4588	34.16	33.27	32.23	31.46	30.74
50	.6967	.6222	.5454	.4942	.4499	37.83	36.77	35.55	34.65	33.81
60	.6830	.6069	.5293	.4779	.4338	44.51	43.18	41.67	40.55	39.51
70	.6704	.5931	.5150	.4635	.4196	51.59	49.92	48.03	46.65	45.37
80	.6588	.5805	.5019	.4505	.4069	59.00	56.92	54.59	52.90	51.34
90	.6480	.5688	.4900	.4387	.3953	66.66	64.12	61.29	59.26	57.39
100	.6377	.5580	.4790	.4279	.3848	74.51	71.46	68.09	65.68	63.50
120	.6186	.5378	.4588	.4081	.3658	89.15	85.24	80.97	77.91	75.13
140	.6013	.5198	.4409	.3909	.3492	104.7	99.72	94.31	90.47	87.01
170	.5780	.4959	.4177	.3685	.3279	129.6	122.5	115.0	109.8	105.2
200	.5572	.4750	.3975	.3494	.3099	155.7	146.2	136.3	129.5	123.6
230	.5380	.4560	.3796	.3325	.2940	180.2	168.6	156.7	148.6	141.3
250	.5204	.4389	.3636	.3175	.2801	205.7	191.7	177.4	167.9	159.1
300	.4993	.4185	.3447	.3000	.2639	241.0	223.4	205.4	193.7	182.9
350	.4758	.3961	.3243	.2811	.2466	287.0	264.1	241.0	226.1	212.7
400	.4550	.3765	.3065	.2649	.2317	334.5	305.7	276.8	258.4	242.3
450	.4363	.3591	.2910	.2507	.2188	382.8	347.8	312.6	290.5	271.6
500	.4194	.3435	.2772	.2382	.2074	431.7	390.0	348.3	322.3	300.6
600	.3897	.3165	.2537	.2172	.1885	526.6	472.3	417.4	383.5	356.6
700	.3646	.2940	.2343	.1999	.1731	623.2	555.0	486.0	443.7	411.4
800	.3428	.2747	.2180	.1854	.1602	720.8	637.7	553.7	502.7	464.9
900	.3238	.2581	.2040	.1731	.1493	818.6	719.8	620.4	560.6	517.3
1000	.3070	.2435	.1918	.1625	.1399	916.2	801.3	686.0	617.3	568.5
1200	.2786	.2195	.1720	.1452	.1248	1108.	959.7	813.0	726.6	666.9
1400	.2552	.2002	.1561	.1316	.1128	1297.	1115.	935.9	831.8	761.3
1700	.2270	.1772	.1374	.1156	.0989	1574.	1341.	1113.	982.5	896.3
2000	.2047	.1592	.1229	.1031	.0881	1844.	1558.	1282.	1125.	1024.
2300	.1864	.1448	.1113	.0932	.0796	2106.	1769.	1444.	1262.	1146.
2600	.1714	.1329	.1017	.0850	.0726	2361.	1972.	1600.	1392.	1262.
3000	.1549	.1201	.0914	.0762	.0651	2688.	2232.	1798.	1557.	1409.
3500	.1387	.1077	.0814	.0674	.0578	3080.	2543.	2033.	1752.	1583.
4000	.1260	.0981	.0735	.0605	.0520	3454.	2838.	2255.	1936.	1747.
4500	.1160	.0905	.0673	.0550	.0474	3813.	3121.	2466.	2110.	1902.
5000	.1080	.0845	.0623	.0504	.0437	4157.	3391.	2668.	2276.	2049.

E,KEV	MEAN DAMAGE DEPTH,MICROGRAM/SQ.CM.					MEAN IONZN.DEPTH,MICROGRAM/SQ.CM.				
K/KL=	0.55	0.75	1.0	1.2	1.4	0.55	0.75	1.0	1.2	1.4
.50	1.129	1.112	1.092	1.077	1.062	1.094	1.076	1.054	1.037	1.021
.60	1.281	1.261	1.238	1.220	1.202	1.241	1.219	1.194	1.174	1.155
.70	1.420	1.397	1.371	1.350	1.331	1.375	1.351	1.321	1.299	1.278
.80	1.550	1.525	1.496	1.473	1.451	1.501	1.474	1.441	1.417	1.393
.90	1.675	1.647	1.615	1.590	1.566	1.622	1.592	1.556	1.529	1.503
1.0	1.795	1.765	1.730	1.703	1.677	1.738	1.705	1.667	1.637	1.609
1.2	2.026	1.991	1.950	1.919	1.889	1.961	1.923	1.878	1.844	1.811
1.4	2.247	2.208	2.161	2.126	2.092	2.174	2.131	2.080	2.042	2.005
1.7	2.564	2.518	2.463	2.421	2.381	2.481	2.430	2.370	2.324	2.281
2.0	2.868	2.815	2.751	2.703	2.657	2.774	2.715	2.646	2.593	2.543
2.3	3.147	3.087	3.016	2.963	2.911	3.043	2.977	2.900	2.841	2.785
2.6	3.418	3.351	3.273	3.214	3.157	3.304	3.231	3.145	3.080	3.019
3.0	3.769	3.694	3.606	3.539	3.475	3.642	3.560	3.463	3.390	3.320
3.5	4.196	4.110	4.009	3.932	3.859	4.053	3.958	3.848	3.764	3.684
4.0	4.611	4.514	4.399	4.313	4.231	4.452	4.345	4.220	4.125	4.036
4.5	5.015	4.907	4.779	4.683	4.592	4.840	4.721	4.581	4.476	4.377
5.0	5.410	5.290	5.149	5.044	4.943	5.219	5.087	4.933	4.817	4.708
6.0	6.139	5.998	5.834	5.710	5.593	5.919	5.764	5.584	5.450	5.322
7.0	6.853	6.691	6.502	6.360	6.226	6.603	6.425	6.218	6.063	5.917
8.0	7.556	7.371	7.156	6.995	6.844	7.275	7.073	6.837	6.662	6.497
9.0	8.248	8.040	7.798	7.618	7.449	7.936	7.708	7.444	7.247	7.062
10	8.930	8.698	8.429	8.230	8.042	8.585	8.331	8.038	7.820	7.616
12	10.19	9.912	9.596	9.361	9.141	9.784	9.485	9.139	8.883	8.643
14	11.44	11.12	10.75	10.48	10.22	10.98	10.63	10.22	9.927	9.649
17	13.32	12.92	12.47	12.14	11.83	12.76	12.32	11.83	11.47	11.13
20	15.20	14.72	14.18	13.78	13.41	14.53	14.01	13.42	12.98	12.58
23	16.87	16.33	15.72	15.27	14.85	16.12	15.53	14.86	14.37	13.92
26	18.56	17.95	17.26	16.75	16.28	17.72	17.06	16.30	15.75	15.24
30	20.87	20.15	19.34	18.75	18.20	19.89	19.11	18.23	17.58	16.99
35	23.82	22.96	21.98	21.27	20.62	22.65	21.70	20.64	19.88	19.17
40	26.84	25.81	24.65	23.82	23.05	25.45	24.32	23.07	22.17	21.34
45	29.91	28.69	27.34	26.37	25.49	28.27	26.94	25.48	24.44	23.50
50	33.00	31.59	30.03	28.92	27.91	31.10	29.57	27.89	26.71	25.63
60	38.66	36.93	35.02	33.67	32.44	36.32	34.44	32.39	30.95	29.64
70	44.59	42.47	40.16	38.52	37.04	41.73	39.43	36.95	35.21	33.64
80	50.76	48.19	45.41	43.46	41.71	47.26	44.50	41.54	39.48	37.64
90	57.09	54.02	50.74	48.45	46.41	52.90	49.62	46.14	43.74	41.61
100	63.54	59.94	56.12	53.48	51.13	58.60	54.77	50.75	47.99	45.56
120	75.60	71.08	66.30	63.01	60.09	69.43	64.61	59.58	56.15	53.14
140	88.30	82.67	76.77	72.74	69.19	80.59	74.62	68.44	64.28	60.65
170	108.4	100.7	92.90	87.63	83.04	97.80	89.83	81.76	76.40	71.77
200	129.3	119.4	109.4	102.7	96.99	115.4	105.2	95.01	88.37	82.69
230	148.9	136.8	125.0	117.2	110.3	132.2	119.7	107.8	100.1	93.22
260	169.1	154.6	140.9	131.9	123.6	149.3	134.4	120.5	111.5	103.6
300	197.1	179.1	162.3	151.4	141.5	172.3	153.9	137.3	126.5	117.0
350	233.4	210.5	189.5	175.9	163.8	201.3	178.4	157.9	144.8	133.4
400	270.6	242.5	216.7	200.4	186.1	230.3	202.8	178.0	162.6	149.3
450	308.4	275.0	244.0	224.6	208.1	259.3	227.0	197.8	179.8	164.8
500	346.6	307.5	271.2	248.6	230.0	288.1	251.0	217.2	196.5	179.8
600	420.4	371.1	323.8	294.6	272.1	344.2	297.9	254.5	228.3	208.4
700	495.5	435.0	376.0	339.8	313.4	399.3	343.6	290.3	258.6	235.6
800	571.2	498.7	427.6	384.3	353.9	453.3	388.1	324.8	287.5	261.6
900	647.1	562.1	478.5	428.1	393.6	506.3	431.3	358.0	315.3	286.5
1000	722.8	625.0	528.6	471.0	432.5	558.1	473.2	390.1	342.0	310.4
1200	870.5	746.4	625.2	553.8	507.1	658.2	553.5	451.1	392.6	355.5
1400	1017.	865.5	719.1	633.9	579.0	754.1	629.7	508.6	440.1	397.6
1700	1232.	1040.	855.2	749.4	682.4	890.8	737.5	589.2	506.3	456.3
2000	1442.	1209.	986.0	859.8	781.2	1020.	838.6	664.2	567.7	510.6
2300	1646.	1372.	1112.	965.6	875.7	1142.	933.9	734.4	625.0	561.2
2600	1845.	1531.	1233.	1067.	966.5	1259.	1024.	800.7	678.9	608.7
3000	2102.	1736.	1389.	1197.	1082.	1406.	1138.	883.7	746.3	668.1
3500	2410.	1980.	1574.	1351.	1219.	1579.	1271.	980.2	824.5	736.9
4000	2705.	2214.	1750.	1497.	1349.	1741.	1395.	1070.	897.2	800.7
4500	2988.	2438.	1918.	1636.	1473.	1894.	1512.	1154.	965.2	860.4
5000	3261.	2653.	2079.	1768.	1591.	2040.	1623.	1234.	1029.	916.6

K/KL= E,KEV	RELATIVE RANGE STRAGGLING					RELATIVE DAMAGE STRAGGLING				
	0.55	0.75	1.0	1.2	1.4	0.55	0.75	1.0	1.2	1.4
.50	1.572	1.520	1.460	1.416	1.375	1.011	.9844	.9543	.9322	.9116
.60	1.565	1.512	1.451	1.406	1.364	1.007	.9806	.9499	.9274	.9065
.70	1.559	1.505	1.443	1.397	1.355	1.005	.9773	.9461	.9232	.9021
.80	1.553	1.499	1.436	1.389	1.346	1.002	.9743	.9426	.9195	.8981
.90	1.548	1.493	1.429	1.382	1.339	.9997	.9715	.9395	.9161	.8945
1.0	1.543	1.487	1.423	1.375	1.332	.9975	.9690	.9366	.9130	.8912
1.2	1.535	1.477	1.411	1.363	1.319	.9935	.9643	.9314	.9074	.8853
1.4	1.526	1.468	1.401	1.353	1.308	.9899	.9602	.9267	.9024	.8800
1.7	1.515	1.455	1.388	1.338	1.292	.9849	.9545	.9204	.8957	.8730
2.0	1.505	1.444	1.375	1.325	1.279	.9804	.9494	.9148	.8898	.8668
2.3	1.496	1.434	1.364	1.314	1.267	.9762	.9448	.9097	.8844	.8613
2.6	1.487	1.424	1.354	1.303	1.256	.9724	.9406	.9051	.8795	.8562
3.0	1.476	1.412	1.341	1.289	1.242	.9676	.9353	.8993	.8735	.8500
3.5	1.463	1.398	1.326	1.274	1.227	.9620	.9291	.8927	.8666	.8429
4.0	1.451	1.386	1.313	1.260	1.212	.9567	.9234	.8866	.8603	.8364
4.5	1.439	1.373	1.300	1.247	1.199	.9518	.9181	.8809	.8544	.8304
5.0	1.428	1.362	1.288	1.235	1.187	.9471	.9131	.8756	.8489	.8247
6.0	1.408	1.341	1.266	1.212	1.164	.9384	.9037	.8657	.8388	.8144
7.0	1.389	1.321	1.246	1.192	1.143	.9303	.8952	.8567	.8296	.8051
8.0	1.372	1.303	1.227	1.173	1.124	.9227	.8872	.8484	.8211	.7965
9.0	1.355	1.286	1.210	1.156	1.107	.9156	.8797	.8406	.8132	.7885
10	1.340	1.270	1.193	1.139	1.090	.9089	.8726	.8333	.8057	.7810
12	1.311	1.240	1.163	1.109	1.060	.8962	.8594	.8197	.7920	.7672
14	1.284	1.213	1.135	1.081	1.033	.8846	.8474	.8073	.7795	.7547
17	1.247	1.176	1.098	1.044	.9961	.8687	.8309	.7905	.7626	.7378
20	1.215	1.143	1.065	1.012	.9636	.8541	.8159	.7754	.7474	.7227
23	1.184	1.112	1.035	.9812	.9338	.8405	.8019	.7612	.7333	.7087
26	1.156	1.084	1.007	.9536	.9067	.8278	.7890	.7481	.7203	.6957
30	1.122	1.049	.9727	.9202	.8739	.8123	.7731	.7321	.7043	.6800
35	1.083	1.011	.9348	.8830	.8374	.7945	.7550	.7141	.6864	.6623
40	1.049	.9763	.9009	.8498	.8050	.7783	.7386	.6977	.6702	.6463
45	1.017	.9451	.8703	.8198	.7758	.7635	.7236	.6827	.6554	.6317
50	.9885	.9166	.8424	.7925	.7492	.7497	.7097	.6689	.6417	.6183
60	.9369	.8654	.7925	.7439	.7019	.7246	.6843	.6438	.6170	.5941
70	.8922	.8212	.7496	.7022	.6615	.7024	.6621	.6218	.5955	.5730
80	.8529	.7825	.7121	.6659	.6263	.6826	.6422	.6023	.5764	.5544
90	.8180	.7482	.6790	.6338	.5954	.6647	.6244	.5848	.5593	.5377
100	.7867	.7175	.6494	.6052	.5678	.6483	.6081	.5689	.5438	.5226
120	.7324	.6645	.5986	.5563	.5208	.6192	.5792	.5409	.5165	.4962
140	.6866	.6200	.5562	.5156	.4817	.5940	.5544	.5169	.4933	.4737
170	.6297	.5649	.5039	.4657	.4339	.5620	.5230	.4867	.4641	.4455
200	.5828	.5199	.4615	.4252	.3954	.5350	.4967	.4615	.4399	.4223
230	.5437	.4826	.4267	.3923	.3640	.5118	.4744	.4404	.4197	.4029
260	.5102	.4507	.3971	.3644	.3376	.4916	.4550	.4221	.4022	.3862
300	.4721	.4148	.3639	.3331	.3080	.4681	.4326	.4011	.3823	.3672
350	.4324	.3777	.3298	.3011	.2779	.4432	.4090	.3791	.3614	.3473
400	.3993	.3469	.3017	.2748	.2532	.4221	.3891	.3606	.3439	.3307
450	.3711	.3211	.2781	.2528	.2326	.4038	.3720	.3448	.3290	.3166
500	.3469	.2989	.2580	.2340	.2151	.3878	.3571	.3311	.3161	.3043
600	.3074	.2640	.2261	.2042	.1877	.3612	.3325	.3088	.2952	.2846
700	.2762	.2367	.2014	.1811	.1666	.3397	.3128	.2909	.2786	.2689
800	.2508	.2147	.1815	.1626	.1497	.3218	.2965	.2763	.2649	.2560
900	.2297	.1965	.1652	.1475	.1360	.3067	.2828	.2639	.2534	.2452
1000	.2119	.1812	.1516	.1349	.1245	.2937	.2712	.2534	.2435	.2360
1200	.1840	.1569	.1304	.1157	.1067	.2729	.2528	.2367	.2277	.2212
1400	.1627	.1383	.1145	.1013	.0934	.2565	.2385	.2236	.2154	.2096
1700	.1387	.1174	.0968	.0854	.0787	.2373	.2220	.2085	.2010	.1962
2000	.1208	.1018	.0838	.0739	.0680	.2226	.2093	.1969	.1900	.1858
2300	.1069	.0899	.0739	.0652	.0599	.2109	.1992	.1876	.1812	.1775
2600	.0959	.0804	.0661	.0583	.0535	.2013	.1909	.1800	.1740	.1707
3000	.0842	.0704	.0579	.0511	.0468	.1909	.1817	.1717	.1662	.1633
3500	.0731	.0609	.0502	.0443	.0405	.1806	.1726	.1635	.1584	.1558
4000	.0645	.0538	.0442	.0391	.0358	.1725	.1651	.1568	.1522	.1498
4500	.0578	.0482	.0396	.0349	.0320	.1659	.1589	.1513	.1471	.1448
5000	.0523	.0438	.0358	.0314	.0289	.1605	.1535	.1467	.1429	.1406

AL ON TA M2/M1=6.71

TABLE 6- 4

E,KEV	RELATIVE IONIZATION STRAGGLING K/KL= 0.55	0.75	1.0	1.2	1.4	RELATIVE TRANSVERSE RANGE STRAGGLING 0.55	0.75	1.0	1.2	1.4
.50	.9909	.9667	.9392	.9191	.9004	.9375	.9360	.9340	.9325	.9311
.60	.9878	.9631	.9351	.9147	.8957	.9375	.9359	.9339	.9324	.9309
.70	.9852	.9600	.9315	.9108	.8917	.9375	.9358	.9338	.9322	.9307
.80	.9827	.9572	.9283	.9074	.8880	.9374	.9357	.9337	.9321	.9305
.90	.9805	.9546	.9254	.9042	.8847	.9374	.9357	.9336	.9320	.9304
1.0	.9784	.9522	.9227	.9013	.8817	.9374	.9357	.9335	.9319	.9303
1.2	.9745	.9478	.9178	.8961	.8762	.9374	.9356	.9335	.9318	.9301
1.4	.9710	.9438	.9134	.8914	.8713	.9374	.9356	.9334	.9317	.9300
1.7	.9662	.9385	.9075	.8852	.8649	.9375	.9356	.9334	.9316	.9299
2.0	.9618	.9337	.9022	.8797	.8592	.9376	.9357	.9334	.9316	.9298
2.3	.9579	.9293	.8975	.8747	.8541	.9377	.9357	.9334	.9316	.9298
2.6	.9541	.9252	.8931	.8702	.8493	.9378	.9358	.9334	.9316	.9298
3.0	.9494	.9201	.8877	.8645	.8436	.9380	.9359	.9335	.9316	.9298
3.5	.9440	.9142	.8814	.8581	.8370	.9382	.9361	.9337	.9318	.9299
4.0	.9389	.9088	.8757	.8522	.8310	.9384	.9363	.9338	.9319	.9301
4.5	.9341	.9037	.8703	.8468	.8255	.9387	.9366	.9340	.9321	.9302
5.0	.9296	.8989	.8653	.8416	.8203	.9389	.9368	.9343	.9323	.9304
6.0	.9212	.8901	.8561	.8323	.8109	.9394	.9373	.9347	.9327	.9308
7.0	.9134	.8819	.8477	.8238	.8023	.9400	.9378	.9352	.9332	.9312
8.0	.9062	.8743	.8400	.8159	.7945	.9405	.9383	.9357	.9337	.9318
9.0	.8993	.8672	.8327	.8086	.7872	.9411	.9389	.9363	.9342	.9323
10	.8928	.8605	.8258	.8018	.7804	.9417	.9395	.9368	.9348	.9329
12	.8805	.8479	.8130	.7890	.7677	.9428	.9406	.9379	.9359	.9340
14	.8692	.8363	.8014	.7774	.7562	.9440	.9417	.9391	.9371	.9352
17	.8538	.8207	.7858	.7619	.7409	.9458	.9436	.9409	.9389	.9370
20	.8399	.8066	.7718	.7480	.7273	.9476	.9454	.9428	.9409	.9390
23	.8270	.7936	.7589	.7353	.7148	.9493	.9472	.9446	.9427	.9409
26	.8151	.7817	.7470	.7237	.7035	.9511	.9490	.9465	.9446	.9429
30	.8005	.7671	.7327	.7097	.6898	.9534	.9514	.9490	.9472	.9455
35	.7841	.7508	.7167	.6940	.6745	.9564	.9545	.9522	.9505	.9489
40	.7693	.7360	.7023	.6800	.6609	.9594	.9575	.9554	.9538	.9523
45	.7557	.7226	.6892	.6673	.6486	.9623	.9606	.9586	.9571	.9557
50	.7433	.7103	.6773	.6557	.6374	.9653	.9637	.9618	.9604	.9591
60	.7205	.6878	.6556	.6348	.6173	.9709	.9696	.9680	.9669	.9657
70	.7007	.6684	.6370	.6169	.6001	.9765	.9755	.9742	.9732	.9723
80	.6832	.6514	.6207	.6013	.5852	.9820	.9812	.9803	.9796	.9789
90	.6677	.6363	.6064	.5876	.5721	.9874	.9869	.9863	.9858	.9853
100	.6537	.6228	.5936	.5754	.5605	.9928	.9925	.9922	.9919	.9916
120	.6294	.5995	.5718	.5547	.5410	1.003	1.003	1.004	1.004	1.004
140	.6090	.5801	.5537	.5378	.5250	1.013	1.014	1.015	1.015	1.016
170	.5837	.5562	.5317	.5172	.5058	1.027	1.029	1.031	1.032	1.033
200	.5630	.5369	.5141	.5009	.4907	1.041	1.043	1.046	1.048	1.049
230	.5458	.5210	.4999	.4879	.4787	1.053	1.057	1.061	1.063	1.065
260	.5311	.5076	.4881	.4772	.4689	1.065	1.070	1.074	1.077	1.079
300	.5146	.4928	.4752	.4655	.4582	1.081	1.087	1.092	1.095	1.098
350	.4978	.4779	.4623	.4540	.4479	1.099	1.106	1.113	1.117	1.120
400	.4840	.4658	.4521	.4450	.4399	1.117	1.125	1.133	1.138	1.141
450	.4725	.4560	.4439	.4379	.4336	1.134	1.143	1.152	1.157	1.161
500	.4629	.4479	.4372	.4321	.4286	1.150	1.161	1.170	1.176	1.180
600	.4481	.4362	.4278	.4242	.4222	1.183	1.195	1.206	1.212	1.217
700	.4370	.4277	.4213	.4189	.4181	1.213	1.226	1.238	1.246	1.251
800	.4286	.4215	.4167	.4153	.4155	1.240	1.255	1.268	1.277	1.282
900	.4220	.4169	.4134	.4129	.4139	1.266	1.281	1.296	1.305	1.311
1000	.4169	.4134	.4112	.4114	.4130	1.290	1.306	1.322	1.331	1.337
1200	.4102	.4090	.4089	.4103	.4127	1.333	1.351	1.369	1.379	1.385
1400	.4062	.4067	.4082	.4105	.4135	1.371	1.391	1.410	1.421	1.428
1700	.4029	.4054	.4088	.4121	.4156	1.423	1.445	1.466	1.477	1.484
2000	.4016	.4055	.4104	.4143	.4182	1.469	1.493	1.515	1.527	1.534
2300	.4015	.4064	.4125	.4169	.4210	1.512	1.536	1.559	1.571	1.578
2600	.4020	.4079	.4147	.4195	.4237	1.551	1.576	1.600	1.612	1.618
3000	.4035	.4101	.4178	.4230	.4273	1.599	1.625	1.648	1.660	1.666
3500	.4058	.4132	.4216	.4270	.4314	1.654	1.680	1.703	1.715	1.720
4000	.4085	.4164	.4252	.4308	.4351	1.704	1.730	1.753	1.763	1.768
4500	.4113	.4196	.4285	.4342	.4385	1.751	1.776	1.798	1.808	1.811
5000	.4141	.4227	.4316	.4372	.4415	1.794	1.818	1.839	1.848	1.851

```
AL ON TA M2/M1=6.71                                          TABLE 6- 5
      RELATIVE TRANSV. DAMAGE STRAGGLING RELATIVE TRANSV. IONZN. STRAGGLING
 K/KL=   0.55    0.75    1.0     1.2     1.4     0.55    0.75    1.0     1.2     1.4
 E,KEV
```

E,KEV	| RELATIVE TRANSV. DAMAGE STRAGGLING					RELATIVE TRANSV. IONZN. STRAGGLING				
	0.55	0.75	1.0	1.2	1.4	0.55	0.75	1.0	1.2	1.4
.50	.8531	.8471	.8396	.8338	.8280	.8218	.8140	.8043	.7967	.7892
.60	.8527	.8464	.8388	.8327	.8268	.8212	.8131	.8031	.7953	.7875
.70	.8523	.8459	.8380	.8318	.8257	.8206	.8123	.8021	.7940	.7861
.80	.8519	.8454	.8373	.8310	.8248	.8201	.8116	.8011	.7929	.7848
.90	.8516	.8449	.8367	.8303	.8240	.8196	.8109	.8003	.7919	.7837
1.0	.8513	.8445	.8362	.8297	.8232	.8192	.8103	.7995	.7910	.7826
1.2	.8508	.8438	.8352	.8285	.8219	.8184	.8093	.7981	.7894	.7807
1.4	.8504	.8432	.8344	.8275	.8208	.8177	.8084	.7969	.7879	.7791
1.7	.8499	.8425	.8334	.8263	.8193	.8168	.8072	.7953	.7860	.7769
2.0	.8494	.8418	.8325	.8252	.8180	.8161	.8061	.7940	.7844	.7750
2.3	.8491	.8412	.8317	.8242	.8169	.8154	.8052	.7927	.7830	.7733
2.6	.8487	.8407	.8310	.8234	.8159	.8148	.8044	.7916	.7816	.7718
3.0	.8483	.8402	.8302	.8224	.8147	.8141	.8034	.7903	.7800	.7700
3.5	.8479	.8395	.8293	.8212	.8134	.8132	.8022	.7888	.7783	.7679
4.0	.8476	.8390	.8285	.8203	.8122	.8125	.8012	.7874	.7766	.7661
4.5	.8473	.8385	.8278	.8194	.8112	.8118	.8003	.7862	.7751	.7644
5.0	.8471	.8381	.8271	.8186	.8103	.8112	.7994	.7850	.7738	.7628
6.0	.8466	.8373	.8260	.8172	.8086	.8101	.7978	.7829	.7713	.7599
7.0	.8463	.8367	.8251	.8160	.8071	.8090	.7964	.7810	.7690	.7573
8.0	.8460	.8362	.8242	.8149	.8059	.8081	.7951	.7793	.7670	.7550
9.0	.8458	.8357	.8235	.8140	.8047	.8073	.7939	.7777	.7651	.7528
10	.8456	.8353	.8228	.8131	.8037	.8065	.7928	.7762	.7633	.7507
12	.8453	.8346	.8217	.8116	.8018	.8050	.7908	.7735	.7601	.7470
14	.8452	.8341	.8207	.8103	.8002	.8037	.7889	.7710	.7571	.7436
17	.8450	.8334	.8194	.8086	.7981	.8019	.7863	.7675	.7529	.7388
20	.8450	.8329	.8184	.8071	.7962	.8002	.7839	.7643	.7491	.7345
23	.8450	.8325	.8175	.8059	.7946	.7988	.7818	.7614	.7457	.7305
26	.8451	.8322	.8167	.8047	.7932	.7974	.7798	.7586	.7424	.7267
30	.8453	.8319	.8158	.8034	.7914	.7956	.7771	.7551	.7382	.7219
35	.8456	.8315	.8147	.8018	.7893	.7934	.7740	.7508	.7331	.7162
40	.8459	.8312	.8137	.8003	.7874	.7913	.7709	.7467	.7283	.7107
45	.8461	.8309	.8128	.7989	.7856	.7892	.7679	.7427	.7236	.7054
50	.8464	.8307	.8119	.7976	.7839	.7871	.7650	.7389	.7191	.7002
60	.8471	.8304	.8105	.7953	.7809	.7834	.7596	.7317	.7106	.6907
70	.8478	.8300	.8090	.7931	.7780	.7797	.7543	.7247	.7024	.6815
80	.8483	.8296	.8076	.7909	.7751	.7759	.7491	.7178	.6945	.6726
90	.8488	.8292	.8061	.7888	.7724	.7721	.7438	.7110	.6867	.6640
100	.8491	.8287	.8047	.7867	.7696	.7683	.7386	.7044	.6791	.6556
120	.8501	.8279	.8021	.7828	.7646	.7608	.7283	.6913	.6643	.6393
140	.8506	.8268	.7993	.7788	.7595	.7533	.7182	.6787	.6501	.6238
170	.8508	.8248	.7949	.7728	.7521	.7421	.7033	.6606	.6300	.6022
200	.8503	.8223	.7902	.7667	.7447	.7309	.6889	.6433	.6111	.5821
230	.8495	.8195	.7854	.7606	.7372	.7202	.6744	.6260	.5924	.5623
260	.8482	.8165	.7804	.7545	.7300	.7096	.6603	.6096	.5748	.5439
300	.8461	.8122	.7737	.7463	.7205	.6957	.6425	.5889	.5530	.5212
350	.8427	.8063	.7651	.7362	.7089	.6787	.6214	.5651	.5280	.4954
400	.8389	.8002	.7566	.7263	.6978	.6622	.6017	.5431	.5051	.4722
450	.8348	.7939	.7481	.7165	.6870	.6460	.5831	.5227	.4841	.4511
500	.8303	.7875	.7396	.7069	.6766	.6302	.5655	.5038	.4648	.4318
600	.8211	.7741	.7224	.6873	.6560	.5972	.5315	.4683	.4288	.3969
700	.8114	.7610	.7060	.6690	.6369	.5669	.5013	.4373	.3978	.3671
800	.8014	.7481	.6904	.6518	.6192	.5393	.4742	.4100	.3707	.3413
900	.7914	.7355	.6756	.6357	.6028	.5142	.4498	.3858	.3469	.3187
1000	.7813	.7233	.6615	.6206	.5874	.4911	.4276	.3642	.3259	.2989
1200	.7602	.6990	.6344	.5923	.5589	.4502	.3888	.3274	.2909	.2660
1400	.7400	.6764	.6099	.5670	.5337	.4153	.3559	.2970	.2624	.2393
1700	.7116	.6457	.5772	.5339	.5009	.3714	.3153	.2600	.2281	.2074
2000	.6856	.6182	.5487	.5053	.4728	.3353	.2823	.2305	.2012	.1825
2300	.6617	.5934	.5235	.4804	.4485	.3052	.2552	.2066	.1795	.1625
2600	.6396	.5710	.5011	.4583	.4271	.2798	.2324	.1869	.1618	.1462
3000	.6127	.5440	.4746	.4325	.4022	.2514	.2074	.1655	.1427	.1288
3500	.5823	.5143	.4459	.4049	.3759	.2230	.1827	.1447	.1243	.1120
4000	.5549	.4879	.4210	.3813	.3535	.2003	.1635	.1288	.1103	.0993
4500	.5301	.4645	.3992	.3608	.3342	.1821	.1484	.1166	.0996	.0896
5000	.5073	.4433	.3799	.3428	.3175	.1674	.1364	.1072	.0914	.0823

TABLE 6- 6

AL ON TA M2/M1=6.71

K/KL=	RANGE SKEWNESS					DAMAGE SKEWNESS				
E,KEV	0.55	0.75	1.0	1.2	1.4	0.55	0.75	1.0	1.2	1.4
.50	.1997	.1886	.1748	.1640	.1533	.5505	.5463	.5412	.5372	.5334
.60	.1987	.1872	.1730	.1618	.1508	.5501	.5458	.5406	.5365	.5325
.70	.1979	.1860	.1714	.1599	.1486	.5499	.5454	.5400	.5358	.5317
.80	.1971	.1849	.1700	.1583	.1467	.5497	.5451	.5395	.5352	.5310
.90	.1964	.1839	.1687	.1567	.1450	.5494	.5448	.5391	.5347	.5304
1.0	.1957	.1830	.1675	.1553	.1433	.5493	.5445	.5387	.5342	.5298
1.2	.1945	.1814	.1654	.1528	.1404	.5489	.5439	.5379	.5333	.5288
1.4	.1935	.1800	.1635	.1505	.1378	.5486	.5435	.5373	.5325	.5278
1.7	.1921	.1781	.1609	.1475	.1344	.5482	.5429	.5364	.5314	.5265
2.0	.1908	.1763	.1587	.1448	.1313	.5479	.5423	.5356	.5304	.5254
2.3	.1897	.1748	.1566	.1425	.1285	.5476	.5418	.5348	.5295	.5243
2.6	.1887	.1734	.1548	.1402	.1260	.5473	.5413	.5342	.5287	.5234
3.0	.1874	.1716	.1525	.1375	.1229	.5470	.5407	.5333	.5276	.5222
3.5	.1859	.1696	.1498	.1344	.1193	.5466	.5400	.5323	.5264	.5207
4.0	.1845	.1677	.1473	.1314	.1159	.5461	.5394	.5313	.5252	.5193
4.5	.1832	.1659	.1450	.1287	.1128	.5457	.5387	.5304	.5241	.5180
5.0	.1819	.1642	.1427	.1261	.1098	.5453	.5381	.5295	.5230	.5167
6.0	.1796	.1611	.1387	.1214	.1045	.5446	.5369	.5278	.5210	.5144
7.0	.1775	.1582	.1349	.1169	.0995	.5439	.5357	.5262	.5190	.5121
8.0	.1754	.1554	.1313	.1128	.0947	.5431	.5346	.5246	.5171	.5099
9.0	.1734	.1527	.1279	.1088	.0902	.5423	.5334	.5230	.5152	.5078
10	.1714	.1501	.1245	.1049	.0859	.5415	.5322	.5214	.5134	.5057
12	.1678	.1453	.1184	.0978	.0779	.5402	.5302	.5186	.5100	.5019
14	.1642	.1406	.1125	.0910	.0704	.5387	.5280	.5158	.5067	.4981
17	.1590	.1338	.1040	.0814	.0596	.5363	.5247	.5114	.5017	.4925
20	.1538	.1272	.0959	.0721	.0493	.5337	.5211	.5070	.4966	.4869
23	.1490	.1210	.0882	.0634	.0397	.5309	.5175	.5025	.4915	.4813
26	.1442	.1150	.0807	.0550	.0305	.5280	.5139	.4980	.4865	.4758
30	.1379	.1070	.0711	.0441	.0185	.5240	.5089	.4920	.4798	.4685
35	.1300	.0972	.0593	.0310	.0042	.5189	.5026	.4845	.4715	.4596
40	.1221	.0876	.0478	.0183	-.0095	.5137	.4963	.4772	.4635	.4508
45	.1143	.0781	.0367	.0060	-.0228	.5085	.4901	.4699	.4555	.4423
50	.1067	.0689	.0258	-.0059	-.0356	.5033	.4839	.4628	.4478	.4340
60	.0918	.0510	.0050	-.0287	-.0602	.4934	.4722	.4493	.4331	.4183
70	.0772	.0338	-.0149	-.0504	-.0833	.4835	.4606	.4361	.4189	.4032
80	.0629	.0171	-.0340	-.0711	-.1054	.4736	.4492	.4233	.4051	.3886
90	.0490	.0009	-.0524	-.0909	-.1264	.4638	.4380	.4107	.3917	.3746
100	.0355	-.0148	-.0701	-.1099	-.1465	.4541	.4270	.3985	.3787	.3610
120	.0091	-.0450	-.1040	-.1461	-.1846	.4344	.4052	.3743	.3531	.3344
140	-.0160	-.0735	-.1358	-.1798	-.2200	.4156	.3843	.3514	.3289	.3096
170	-.0515	-.1134	-.1798	-.2265	-.2688	.3889	.3549	.3194	.2953	.2753
200	-.0846	-.1504	-.2204	-.2692	-.3133	.3641	.3274	.2898	.2642	.2439
230	-.1165	-.1854	-.2584	-.3090	-.3548	.3409	.3016	.2624	.2355	.2158
260	-.1466	-.2181	-.2938	-.3460	-.3933	.3191	.2774	.2368	.2087	.1897
300	-.1840	-.2589	-.3377	-.3917	-.4408	.2920	.2472	.2049	.1755	.1573
350	-.2272	-.3058	-.3881	-.4441	-.4949	.2608	.2123	.1683	.1373	.1200
400	-.2668	-.3490	-.4342	-.4921	-.5442	.2320	.1800	.1346	.1022	.0854
450	-.3036	-.3890	-.4769	-.5365	-.5897	.2054	.1500	.1032	.0697	.0530
500	-.3378	-.4265	-.5167	-.5778	-.6318	.1806	.1219	.0739	.0393	.0226
600	-.3989	-.4957	-.5895	-.6538	-.7080	.1370	.0695	.0193	-.0171	-.0362
700	-.4538	-.5574	-.6543	-.7214	-.7757	.0973	.0227	-.0294	-.0672	-.0887
800	-.5041	-.6130	-.7128	-.7822	-.8368	.0604	-.0195	-.0734	-.1124	-.1360
900	-.5505	-.6637	-.7661	-.8377	-.8927	.0258	-.0580	-.1136	-.1535	-.1788
1000	-.5938	-.7101	-.8152	-.8886	-.9442	-.0070	-.0934	-.1505	-.1911	-.2180
1200	-.6742	-.7910	-.9018	-.9782	-1.036	-.0718	-.1562	-.2170	-.2585	-.2867
1400	-.7460	-.8623	-.9783	-1.057	-1.118	-.1302	-.2111	-.2750	-.3170	-.3460
1700	-.8412	-.9563	-1.079	-1.161	-1.225	-.2072	-.2824	-.3499	-.3921	-.4218
2000	-.9246	-1.039	-1.167	-1.252	-1.319	-.2739	-.3439	-.4138	-.4560	-.4857
2300	-.9989	-1.114	-1.247	-1.333	-1.403	-.3323	-.3980	-.4694	-.5113	-.5408
2600	-1.066	-1.182	-1.319	-1.407	-1.478	-.3840	-.4463	-.5185	-.5600	-.5891
3000	-1.146	-1.265	-1.406	-1.496	-1.570	-.4442	-.5038	-.5760	-.6168	-.6452
3500	-1.235	-1.359	-1.504	-1.596	-1.672	-.5086	-.5667	-.6379	-.6777	-.7051
4000	-1.314	-1.444	-1.592	-1.686	-1.763	-.5635	-.6222	-.6913	-.7300	-.7563
4500	-1.385	-1.522	-1.673	-1.769	-1.846	-.6108	-.6717	-.7381	-.7756	-.8008
5000	-1.449	-1.595	-1.747	-1.845	-1.922	-.6521	-.7165	-.7796	-.8160	-.8399

K/KL=	IONIZATION SKEWNESS					RANGE KURTOSIS				
E,KEV	0.55	0.75	1.0	1.2	1.4	0.55	0.75	1.0	1.2	1.4
.50	.6403	.6387	.6368	.6354	.6340	3.267	3.228	3.184	3.151	3.121
.60	.6403	.6387	.6367	.6352	.6338	3.263	3.223	3.178	3.145	3.114
.70	.6404	.6387	.6366	.6351	.6336	3.259	3.219	3.173	3.139	3.108
.80	.6404	.6387	.6366	.6350	.6334	3.256	3.215	3.169	3.134	3.103
.90	.6405	.6387	.6365	.6349	.6333	3.254	3.212	3.165	3.130	3.098
1.0	.6406	.6387	.6365	.6348	.6332	3.251	3.209	3.161	3.126	3.094
1.2	.6407	.6387	.6364	.6346	.6329	3.246	3.203	3.154	3.119	3.087
1.4	.6408	.6388	.6363	.6345	.6328	3.242	3.198	3.149	3.113	3.080
1.7	.6411	.6389	.6363	.6344	.6325	3.237	3.192	3.141	3.105	3.071
2.0	.6413	.6389	.6362	.6342	.6323	3.232	3.186	3.134	3.097	3.064
2.3	.6415	.6390	.6362	.6341	.6321	3.228	3.181	3.128	3.091	3.057
2.6	.6417	.6391	.6362	.6340	.6319	3.224	3.176	3.123	3.085	3.051
3.0	.6419	.6393	.6362	.6339	.6317	3.219	3.170	3.117	3.078	3.044
3.5	.6423	.6394	.6361	.6337	.6314	3.213	3.164	3.109	3.070	3.036
4.0	.6426	.6396	.6361	.6336	.6312	3.208	3.158	3.102	3.063	3.028
4.5	.6429	.6397	.6361	.6334	.6309	3.203	3.152	3.096	3.057	3.021
5.0	.6432	.6399	.6361	.6333	.6307	3.198	3.147	3.090	3.051	3.015
6.0	.6439	.6403	.6362	.6332	.6304	3.190	3.137	3.080	3.040	3.004
7.0	.6445	.6406	.6363	.6331	.6301	3.182	3.129	3.070	3.030	2.994
8.0	.6450	.6409	.6362	.6328	.6297	3.175	3.121	3.062	3.021	2.985
9.0	.6455	.6410	.6361	.6325	.6292	3.168	3.113	3.054	3.013	2.977
10	.6458	.6411	.6359	.6322	.6287	3.162	3.106	3.046	3.005	2.970
12	.6462	.6410	.6353	.6312	.6275	3.150	3.093	3.032	2.991	2.956
14	.6464	.6407	.6346	.6302	.6262	3.139	3.081	3.020	2.979	2.943
17	.6465	.6402	.6334	.6285	.6242	3.124	3.065	3.003	2.962	2.927
20	.6464	.6395	.6321	.6269	.6221	3.110	3.050	2.988	2.948	2.914
23	.6468	.6393	.6313	.6257	.6207	3.096	3.036	2.974	2.934	2.901
26	.6471	.6390	.6305	.6245	.6192	3.084	3.023	2.961	2.922	2.889
30	.6471	.6384	.6292	.6228	.6170	3.068	3.007	2.946	2.908	2.876
35	.6468	.6373	.6273	.6204	.6142	3.050	2.989	2.929	2.892	2.861
40	.6460	.6358	.6251	.6177	.6112	3.034	2.973	2.914	2.879	2.850
45	.6450	.6340	.6227	.6149	.6080	3.019	2.958	2.901	2.867	2.840
50	.6436	.6321	.6201	.6119	.6047	3.005	2.945	2.890	2.857	2.832
60	.6396	.6269	.6138	.6049	.5971	2.978	2.920	2.868	2.838	2.817
70	.6354	.6216	.6075	.5980	.5897	2.955	2.899	2.851	2.824	2.807
80	.6311	.6163	.6013	.5913	.5825	2.935	2.881	2.837	2.814	2.801
90	.6268	.6112	.5954	.5848	.5757	2.917	2.867	2.827	2.808	2.798
100	.6226	.6062	.5897	.5787	.5692	2.902	2.854	2.819	2.804	2.797
120	.6150	.5971	.5793	.5675	.5573	2.874	2.833	2.807	2.799	2.800
140	.6076	.5885	.5695	.5571	.5464	2.853	2.819	2.803	2.802	2.809
170	.5968	.5761	.5557	.5425	.5313	2.831	2.808	2.805	2.815	2.833
200	.5863	.5643	.5429	.5291	.5175	2.816	2.805	2.817	2.837	2.865
230	.5753	.5517	.5293	.5151	.5031	2.804	2.806	2.831	2.861	2.899
260	.5648	.5398	.5167	.5020	.4898	2.796	2.812	2.850	2.890	2.937
300	.5515	.5250	.5011	.4862	.4738	2.794	2.826	2.883	2.934	2.994
350	.5359	.5082	.4837	.4686	.4562	2.801	2.854	2.931	2.997	3.072
400	.5214	.4929	.4682	.4532	.4410	2.817	2.889	2.986	3.066	3.154
450	.5080	.4791	.4544	.4396	.4278	2.839	2.929	3.046	3.139	3.240
500	.4954	.4665	.4420	.4275	.4163	2.866	2.974	3.109	3.215	3.327
600	.4706	.4434	.4201	.4070	.3976	2.933	3.072	3.241	3.371	3.503
700	.4492	.4240	.4023	.3905	.3830	3.009	3.178	3.379	3.531	3.681
800	.4308	.4077	.3875	.3771	.3715	3.091	3.287	3.519	3.693	3.859
900	.4149	.3938	.3753	.3663	.3622	3.177	3.400	3.660	3.854	4.037
1000	.4011	.3820	.3651	.3574	.3548	3.265	3.513	3.800	4.014	4.213
1200	.3797	.3638	.3504	.3452	.3450	3.438	3.730	4.069	4.318	4.548
1400	.3631	.3502	.3400	.3371	.3389	3.615	3.950	4.335	4.618	4.877
1700	.3446	.3356	.3299	.3299	.3341	3.883	4.279	4.732	5.062	5.362
2000	.3315	.3260	.3241	.3266	.3326	4.153	4.606	5.122	5.497	5.835
2300	.3220	.3197	.3212	.3258	.3332	4.419	4.928	5.506	5.922	6.297
2600	.3153	.3159	.3203	.3266	.3351	4.683	5.245	5.881	6.338	6.747
3000	.3095	.3134	.3213	.3295	.3391	5.028	5.658	6.369	6.876	7.329
3500	.3058	.3135	.3250	.3350	.3456	5.447	6.160	6.960	7.527	8.029
4000	.3049	.3160	.3304	.3418	.3530	5.854	6.644	7.529	8.152	8.702
4500	.3059	.3201	.3370	.3495	.3610	6.247	7.113	8.078	8.755	9.349
5000	.3085	.3254	.3444	.3576	.3693	6.628	7.566	8.609	9.338	9.973

E,KEV	DAMAGE KURTOSIS K/KL= 0.55	0.75	1.0	1.2	1.4	IONIZATION KURTOSIS 0.55	0.75	1.0	1.2	1.4
.50	4.377	4.322	4.258	4.210	4.165	4.525	4.469	4.403	4.354	4.307
.60	4.371	4.314	4.248	4.200	4.154	4.519	4.461	4.393	4.343	4.296
.70	4.365	4.307	4.240	4.191	4.144	4.513	4.454	4.385	4.334	4.286
.80	4.360	4.301	4.233	4.182	4.135	4.508	4.448	4.378	4.326	4.277
.90	4.355	4.295	4.226	4.175	4.127	4.504	4.442	4.371	4.319	4.269
1.0	4.351	4.290	4.220	4.168	4.120	4.499	4.437	4.365	4.312	4.262
1.2	4.342	4.280	4.209	4.156	4.107	4.491	4.427	4.354	4.300	4.249
1.4	4.335	4.271	4.198	4.145	4.095	4.484	4.419	4.344	4.289	4.237
1.7	4.325	4.259	4.185	4.130	4.079	4.474	4.407	4.331	4.274	4.221
2.0	4.315	4.249	4.173	4.117	4.065	4.466	4.397	4.319	4.261	4.208
2.3	4.307	4.239	4.162	4.105	4.052	4.457	4.387	4.308	4.249	4.195
2.6	4.299	4.230	4.151	4.094	4.040	4.450	4.379	4.298	4.238	4.183
3.0	4.289	4.219	4.139	4.080	4.026	4.440	4.368	4.285	4.225	4.169
3.5	4.278	4.205	4.124	4.064	4.009	4.429	4.355	4.271	4.210	4.153
4.0	4.267	4.193	4.110	4.050	3.994	4.419	4.343	4.258	4.196	4.138
4.5	4.256	4.181	4.097	4.036	3.980	4.410	4.332	4.246	4.183	4.125
5.0	4.246	4.170	4.085	4.023	3.966	4.400	4.322	4.234	4.171	4.112
6.0	4.227	4.149	4.062	3.999	3.941	4.385	4.304	4.214	4.149	4.089
7.0	4.210	4.130	4.041	3.977	3.918	4.369	4.287	4.195	4.129	4.068
8.0	4.194	4.112	4.022	3.957	3.897	4.355	4.271	4.177	4.110	4.049
9.0	4.178	4.095	4.003	3.937	3.877	4.341	4.255	4.160	4.092	4.030
10	4.163	4.079	3.986	3.919	3.858	4.327	4.240	4.144	4.075	4.012
12	4.135	4.049	3.953	3.886	3.824	4.300	4.210	4.111	4.041	3.977
14	4.110	4.021	3.924	3.855	3.792	4.274	4.182	4.081	4.010	3.945
17	4.073	3.982	3.882	3.812	3.749	4.238	4.143	4.039	3.966	3.900
20	4.040	3.946	3.844	3.773	3.709	4.205	4.107	4.001	3.927	3.860
23	4.006	3.910	3.807	3.735	3.670	4.177	4.077	3.969	3.893	3.826
26	3.975	3.877	3.773	3.700	3.634	4.150	4.048	3.938	3.862	3.794
30	3.936	3.836	3.729	3.656	3.590	4.117	4.012	3.901	3.823	3.754
35	3.890	3.788	3.680	3.605	3.539	4.078	3.971	3.857	3.778	3.708
40	3.848	3.744	3.634	3.559	3.493	4.042	3.932	3.816	3.737	3.666
45	3.809	3.703	3.592	3.517	3.450	4.007	3.895	3.777	3.697	3.626
50	3.772	3.664	3.553	3.477	3.410	3.974	3.859	3.741	3.660	3.589
60	3.705	3.594	3.482	3.406	3.339	3.906	3.788	3.667	3.586	3.514
70	3.644	3.532	3.418	3.342	3.276	3.845	3.724	3.602	3.519	3.447
80	3.589	3.475	3.361	3.285	3.219	3.789	3.666	3.542	3.459	3.387
90	3.538	3.424	3.309	3.234	3.168	3.738	3.613	3.488	3.405	3.333
100	3.491	3.376	3.262	3.186	3.121	3.691	3.565	3.439	3.356	3.284
120	3.404	3.288	3.174	3.100	3.037	3.609	3.480	3.353	3.270	3.199
140	3.328	3.212	3.099	3.026	2.964	3.537	3.406	3.279	3.196	3.127
170	3.230	3.114	3.004	2.932	2.873	3.443	3.311	3.185	3.102	3.035
200	3.147	3.033	2.924	2.855	2.799	3.361	3.229	3.104	3.022	2.958
230	3.076	2.963	2.858	2.789	2.737	3.285	3.155	3.031	2.951	2.889
260	3.014	2.903	2.801	2.734	2.684	3.216	3.088	2.967	2.889	2.830
300	2.942	2.835	2.737	2.671	2.626	3.135	3.011	2.892	2.817	2.760
350	2.867	2.764	2.672	2.608	2.567	3.047	2.927	2.812	2.740	2.688
400	2.804	2.706	2.618	2.558	2.521	2.971	2.856	2.745	2.676	2.626
450	2.751	2.657	2.575	2.517	2.484	2.905	2.794	2.687	2.621	2.574
500	2.705	2.616	2.539	2.484	2.454	2.847	2.740	2.636	2.574	2.529
600	2.630	2.551	2.483	2.437	2.413	2.752	2.653	2.556	2.499	2.459
700	2.572	2.504	2.444	2.406	2.387	2.675	2.583	2.492	2.439	2.404
800	2.528	2.469	2.417	2.385	2.370	2.612	2.525	2.440	2.392	2.359
900	2.494	2.443	2.398	2.373	2.362	2.558	2.476	2.397	2.352	2.323
1000	2.467	2.424	2.386	2.366	2.359	2.513	2.435	2.361	2.319	2.292
1200	2.433	2.403	2.378	2.367	2.366	2.441	2.371	2.304	2.267	2.246
1400	2.414	2.396	2.382	2.378	2.382	2.386	2.321	2.261	2.229	2.211
1700	2.404	2.400	2.401	2.406	2.416	2.323	2.266	2.213	2.187	2.174
2000	2.408	2.417	2.430	2.441	2.456	2.276	2.225	2.179	2.158	2.149
2300	2.421	2.440	2.463	2.480	2.499	2.239	2.194	2.154	2.137	2.131
2600	2.440	2.467	2.499	2.521	2.543	2.209	2.169	2.136	2.122	2.119
3000	2.471	2.508	2.550	2.577	2.602	2.178	2.145	2.118	2.108	2.107
3500	2.516	2.563	2.614	2.646	2.674	2.149	2.122	2.103	2.097	2.100
4000	2.565	2.620	2.678	2.714	2.744	2.126	2.107	2.094	2.092	2.097
4500	2.616	2.677	2.741	2.781	2.812	2.108	2.096	2.089	2.090	2.097
5000	2.667	2.734	2.802	2.845	2.877	2.094	2.088	2.087	2.091	2.099

TABLE 6- 9

AL ON TA M2/M1=6.71

E,KEV	REFLECTION COEFFICIENT					SPUTTERING EFFICIENCY				
K/KL=	0.55	0.75	1.0	1.2	1.4	0.55	0.75	1.0	1.2	1.4
.50	.2110	.2072	.2028	.1993	.1961	.1270	.1180	.1079	.1006	.0940
.60	.2105	.2066	.2020	.1985	.1952	.1261	.1169	.1066	.0993	.0926
.70	.2100	.2061	.2014	.1978	.1944	.1253	.1159	.1055	.0981	.0913
.80	.2096	.2056	.2008	.1972	.1937	.1246	.1151	.1045	.0970	.0903
.90	.2092	.2052	.2003	.1966	.1930	.1240	.1143	.1036	.0961	.0893
1.0	.2089	.2047	.1998	.1961	.1925	.1234	.1136	.1029	.0952	.0884
1.2	.2082	.2040	.1989	.1951	.1914	.1223	.1123	.1014	.0937	.0868
1.4	.2076	.2033	.1981	.1942	.1904	.1213	.1112	.1002	.0924	.0855
1.7	.2068	.2023	.1970	.1929	.1891	.1201	.1098	.0986	.0907	.0837
2.0	.2060	.2014	.1959	.1918	.1879	.1190	.1085	.0972	.0893	.0822
2.3	.2053	.2006	.1950	.1908	.1868	.1180	.1074	.0959	.0880	.0809
2.6	.2046	.1998	.1941	.1899	.1858	.1171	.1064	.0948	.0868	.0797
3.0	.2037	.1988	.1931	.1887	.1846	.1160	.1052	.0935	.0854	.0782
3.5	.2027	.1977	.1918	.1873	.1831	.1148	.1038	.0920	.0838	.0767
4.0	.2017	.1966	.1906	.1861	.1818	.1136	.1025	.0906	.0824	.0752
4.5	.2008	.1956	.1895	.1849	.1805	.1126	.1013	.0894	.0812	.0740
5.0	.1999	.1946	.1884	.1837	.1793	.1116	.1003	.0882	.0800	.0728
6.0	.1982	.1928	.1864	.1816	.1771	.1098	.0983	.0862	.0779	.0707
7.0	.1967	.1911	.1845	.1797	.1751	.1082	.0966	.0844	.0761	.0689
8.0	.1952	.1894	.1828	.1778	.1732	.1068	.0950	.0828	.0745	.0673
9.0	.1938	.1879	.1811	.1761	.1714	.1054	.0936	.0813	.0730	.0658
10	.1924	.1864	.1795	.1744	.1696	.1041	.0922	.0799	.0716	.0645
12	.1898	.1836	.1765	.1713	.1664	.1017	.0897	.0774	.0691	.0620
14	.1874	.1810	.1738	.1684	.1635	.0996	.0875	.0751	.0669	.0599
17	.1839	.1774	.1699	.1644	.1594	.0967	.0845	.0721	.0640	.0570
20	.1808	.1740	.1663	.1608	.1556	.0941	.0819	.0695	.0614	.0546
23	.1778	.1708	.1630	.1573	.1521	.0917	.0794	.0671	.0591	.0524
26	.1749	.1678	.1598	.1541	.1488	.0895	.0772	.0649	.0570	.0504
30	.1714	.1641	.1559	.1501	.1447	.0868	.0745	.0623	.0545	.0480
35	.1673	.1597	.1514	.1455	.1400	.0837	.0714	.0594	.0517	.0454
40	.1635	.1557	.1472	.1412	.1358	.0809	.0687	.0568	.0492	.0430
45	.1599	.1520	.1434	.1373	.1318	.0784	.0662	.0544	.0470	.0410
50	.1566	.1485	.1398	.1336	.1281	.0760	.0639	.0523	.0450	.0391
60	.1504	.1420	.1331	.1269	.1213	.0717	.0597	.0484	.0414	.0358
70	.1448	.1363	.1271	.1209	.1153	.0679	.0561	.0451	.0384	.0330
80	.1398	.1310	.1217	.1155	.1099	.0645	.0529	.0422	.0358	.0306
90	.1352	.1262	.1169	.1106	.1050	.0615	.0501	.0397	.0335	.0286
100	.1309	.1219	.1124	.1061	.1006	.0587	.0475	.0374	.0314	.0267
120	.1233	.1139	.1044	.0981	.0926	.0539	.0430	.0335	.0279	.0236
140	.1166	.1071	.0975	.0913	.0857	.0497	.0393	.0302	.0251	.0211
170	.1079	.0982	.0887	.0826	.0771	.0445	.0346	.0263	.0216	.0180
200	.1005	.0908	.0813	.0753	.0699	.0401	.0308	.0231	.0189	.0156
230	.0942	.0843	.0750	.0691	.0639	.0365	.0278	.0206	.0167	.0138
260	.0885	.0787	.0696	.0638	.0588	.0334	.0252	.0185	.0149	.0122
300	.0820	.0722	.0633	.0578	.0530	.0299	.0223	.0162	.0130	.0106
350	.0750	.0653	.0567	.0515	.0470	.0263	.0194	.0139	.0111	.0090
400	.0690	.0595	.0512	.0462	.0420	.0233	.0170	.0121	.0096	.0077
450	.0638	.0545	.0465	.0417	.0378	.0208	.0150	.0106	.0084	.0067
500	.0592	.0502	.0424	.0378	.0342	.0187	.0134	.0094	.0074	.0059
600	.0514	.0433	.0359	.0316	.0285	.0155	.0110	.0076	.0060	.0047
700	.0451	.0378	.0308	.0267	.0241	.0130	.0092	.0063	.0049	.0039
800	.0399	.0333	.0266	.0229	.0205	.0111	.0078	.0053	.0042	.0033
900	.0355	.0296	.0232	.0197	.0177	.0096	.0067	.0046	.0036	.0028
1000	.0318	.0264	.0204	.0171	.0153	.0084	.0059	.0040	.0031	.0025
1200	.0261	.0214	.0161	.0132	.0118	.0066	.0046	.0031	.0024	.0019
1400	.0218	.0177	.0130	.0105	.0093	.0054	.0038	.0025	.0020	.0016
1700	.0170	.0135	.0096	.0075	.0066	.0042	.0029	.0019	.0015	.0012
2000	.0135	.0105	.0072	.0055	.0048	.0034	.0023	.0016	.0012	.0010
2300	.0108	.0082	.0055	.0041	.0035	.0028	.0019	.0013	.0010	.0008
2600	.0088	.0065	.0042	.0031	.0026	.0024	.0016	.0011	.0008	.0007
3000	.0068	.0048	.0030	.0021	.0018	.0020	.0014	.0009	.0007	.0006
3500	.0050	.0034	.0020	.0014	.0011	.0016	.0011	.0008	.0006	.0005
4000	.0037	.0025	.0014	.0009	.0007	.0013	.0009	.0006	.0005	.0004
4500	.0029	.0019	.0011	.0007	.0005	.0011	.0008	.0005	.0004	.0003
5000	.0023	.0016	.0009	.0006	.0005	.0009	.0006	.0004	.0003	.0003

AL ON TA M2/M1=6.71   TABLE 6-10

K/KL=	IONIZATION DEFICIENCY					SPUTTERING YIELD ALPHA				
E,KEV	0.55	0.75	1.0	1.2	1.4	0.55	0.75	1.0	1.2	1.4
.50	.0157	.0200	.0246	.0277	.0303	.9513	.9176	.8779	.8479	.8194
.60	.0161	.0205	.0251	.0282	.0308	.9487	.9140	.8734	.8427	.8136
.70	.0164	.0208	.0255	.0286	.0313	.9464	.9110	.8695	.8382	.8087
.80	.0167	.0212	.0258	.0290	.0316	.9444	.9083	.8661	.8343	.8043
.90	.0170	.0215	.0262	.0293	.0319	.9426	.9059	.8631	.8308	.8004
1.0	.0172	.0217	.0264	.0296	.0322	.9410	.9038	.8603	.8277	.7969
1.2	.0176	.0222	.0269	.0301	.0327	.9382	.9000	.8555	.8222	.7907
1.4	.0179	.0226	.0273	.0305	.0331	.9358	.8968	.8514	.8174	.7855
1.7	.0184	.0231	.0278	.0309	.0335	.9328	.8928	.8462	.8114	.7788
2.0	.0187	.0235	.0282	.0313	.0339	.9303	.8893	.8417	.8063	.7731
2.3	.0190	.0238	.0286	.0317	.0342	.9282	.8864	.8379	.8019	.7681
2.6	.0193	.0241	.0289	.0320	.0345	.9263	.8837	.8344	.7979	.7637
3.0	.0196	.0244	.0292	.0323	.0348	.9241	.8807	.8304	.7933	.7586
3.5	.0200	.0248	.0296	.0326	.0350	.9218	.8773	.8261	.7883	.7530
4.0	.0202	.0251	.0299	.0328	.0352	.9198	.8745	.8223	.7839	.7481
4.5	.0205	.0253	.0301	.0331	.0354	.9181	.8719	.8190	.7800	.7437
5.0	.0207	.0256	.0303	.0332	.0355	.9166	.8697	.8160	.7765	.7398
6.0	.0211	.0259	.0306	.0335	.0357	.9141	.8659	.8109	.7705	.7331
7.0	.0214	.0262	.0308	.0337	.0359	.9120	.8627	.8065	.7653	.7273
8.0	.0216	.0264	.0310	.0338	.0359	.9102	.8599	.8026	.7608	.7222
9.0	.0218	.0266	.0312	.0338	.0359	.9086	.8574	.7992	.7568	.7177
10	.0220	.0268	.0312	.0339	.0359	.9072	.8551	.7961	.7531	.7136
12	.0223	.0270	.0314	.0339	.0358	.9049	.8513	.7907	.7467	.7064
14	.0225	.0272	.0314	.0338	.0356	.9029	.8479	.7859	.7411	.7001
17	.0227	.0273	.0314	.0337	.0353	.9003	.8434	.7797	.7338	.6919
20	.0229	.0274	.0313	.0334	.0350	.8980	.8394	.7741	.7273	.6848
23	.0230	.0274	.0311	.0332	.0346	.8961	.8361	.7694	.7217	.6785
26	.0231	.0273	.0309	.0328	.0341	.8944	.8330	.7650	.7165	.6728
30	.0231	.0272	.0306	.0324	.0335	.8921	.8291	.7595	.7102	.6657
35	.0231	.0270	.0302	.0318	.0328	.8893	.8244	.7532	.7029	.6577
40	.0231	.0268	.0298	.0312	.0321	.8864	.8199	.7472	.6961	.6504
45	.0230	.0266	.0294	.0306	.0314	.8835	.8156	.7415	.6897	.6435
50	.0229	.0263	.0289	.0301	.0307	.8807	.8113	.7361	.6837	.6370
60	.0226	.0258	.0280	.0289	.0293	.8750	.8031	.7258	.6723	.6249
70	.0224	.0252	.0271	.0278	.0280	.8693	.7952	.7161	.6618	.6139
80	.0220	.0246	.0262	.0267	.0268	.8634	.7875	.7070	.6521	.6037
90	.0217	.0240	.0254	.0258	.0257	.8575	.7799	.6983	.6428	.5942
100	.0214	.0235	.0246	.0248	.0247	.8516	.7725	.6899	.6341	.5853
120	.0207	.0224	.0231	.0231	.0228	.8394	.7578	.6736	.6174	.5686
140	.0200	.0213	.0217	.0215	.0211	.8275	.7438	.6586	.6022	.5536
170	.0190	.0199	.0199	.0195	.0189	.8102	.7243	.6382	.5819	.5338
200	.0181	.0186	.0183	.0178	.0171	.7937	.7064	.6199	.5641	.5165
230	.0172	.0174	.0169	.0163	.0156	.7770	.6895	.6029	.5481	.5011
250	.0163	.0163	.0157	.0150	.0143	.7611	.6737	.5875	.5337	.4873
300	.0153	.0150	.0142	.0134	.0127	.7413	.6544	.5688	.5165	.4710
350	.0141	.0136	.0127	.0119	.0112	.7186	.6326	.5482	.4977	.4531
400	.0130	.0124	.0114	.0106	.0099	.6978	.6129	.5299	.4811	.4375
450	.0121	.0113	.0102	.0095	.0088	.6788	.5950	.5136	.4664	.4237
500	.0113	.0104	.0093	.0085	.0079	.6612	.5786	.4988	.4532	.4113
600	.0098	.0089	.0079	.0072	.0065	.6293	.5481	.4726	.4297	.3894
700	.0086	.0077	.0067	.0061	.0054	.6017	.5222	.4507	.4102	.3712
800	.0077	.0067	.0058	.0053	.0046	.5776	.4998	.4321	.3936	.3561
900	.0068	.0058	.0050	.0046	.0039	.5564	.4805	.4161	.3795	.3432
1000	.0061	.0050	.0043	.0039	.0033	.5375	.4636	.4023	.3673	.3323
1200	.0050	.0035	.0030	.0027	.0023	.5054	.4365	.3805	.3482	.3161
1400	.0042	.0023	.0019	.0018	.0015	.4791	.4149	.3633	.3332	.3037
1700	.0033	.0009	.0008	.0007	.0006	.4476	.3899	.3433	.3158	.2895
2000	.0026	0.0000	0.0000	0.0000	0.0000	.4231	.3709	.3281	.3024	.2788
2300	.0020	0.0000	0.0000	0.0000	0.0000	.4035	.3561	.3161	.2918	.2703
2600	.0016	0.0000	0.0000	0.0000	0.0000	.3878	.3445	.3065	.2831	.2634
3000	.0011	0.0000	0.0000	0.0000	0.0000	.3714	.3326	.2964	.2738	.2558
3500	.0007	0.0000	0.0000	0.0000	0.0000	.3561	.3219	.2868	.2647	.2483
4000	.0004	0.0000	0.0000	0.0000	0.0000	.3453	.3145	.2796	.2575	.2421
4500	.0002	0.0000	0.0000	0.0000	0.0000	.3377	.3095	.2742	.2518	.2369
5000	0.0000	0.0000	0.0000	0.0000	0.0000	.3326	.3064	.2701	.2471	.2325

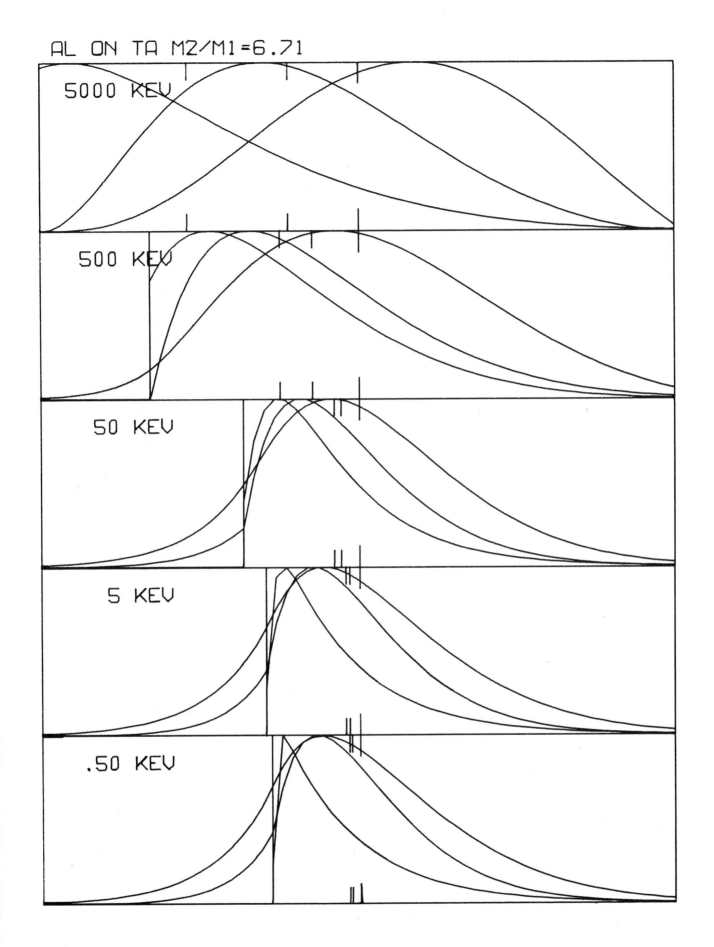

AL ON TA M2/M1=6.71

5000 KEV

500 KEV

50 KEV

5 KEV

.50 KEV

137

C ON NI, M2/M1=4.92   TABLE 7- 1

| E,KEV | FRACTIONAL DEPOSITED ENERGY | | | | | MEAN RANGE,MICROGRAM/SQ.CM. | | | | |
	0.55	0.75	1.0	1.2	1.4	0.55	0.75	1.0	1.2	1.4
.10	.8823	.8445	.8005	.7677	.7367	.2919	.2898	.2873	.2854	.2835
.12	.8789	.8402	.7952	.7616	.7301	.3325	.3301	.3271	.3247	.3224
.14	.8759	.8364	.7905	.7564	.7244	.3690	.3662	.3628	.3602	.3575
.17	.8721	.8315	.7846	.7497	.7171	.4195	.4162	.4123	.4092	.4061
.20	.8688	.8273	.7794	.7440	.7108	.4669	.4632	.4587	.4552	.4517
.23	.8659	.8236	.7749	.7389	.7052	.5124	.5083	.5032	.4992	.4954
.26	.8633	.8203	.7709	.7344	.7003	.5566	.5520	.5464	.5420	.5377
.30	.8601	.8163	.7661	.7290	.6945	.6138	.6086	.6023	.5973	.5924
.35	.8567	.8119	.7607	.7231	.6880	.6833	.6773	.6699	.6642	.6586
.40	.8535	.8080	.7560	.7178	.6823	.7507	.7439	.7356	.7291	.7227
.45	.8507	.8045	.7518	.7131	.6772	.8164	.8087	.7994	.7921	.7850
.50	.8482	.8013	.7479	.7088	.6726	.8804	.8719	.8615	.8535	.8456
.60	.8436	.7957	.7411	.7012	.6645	.9986	.9886	.9765	.9671	.9579
.70	.8396	.7908	.7351	.6946	.6574	1.114	1.103	1.089	1.078	1.067
.80	.8360	.7864	.7298	.6887	.6511	1.228	1.214	1.198	1.186	1.174
.90	.8327	.7824	.7250	.6833	.6455	1.339	1.324	1.306	1.292	1.278
1.0	.8297	.7787	.7205	.6784	.6403	1.449	1.432	1.412	1.396	1.381
1.2	.8242	.7720	.7126	.6697	.6310	1.651	1.631	1.607	1.588	1.570
1.4	.8194	.7661	.7055	.6620	.6228	1.851	1.828	1.800	1.779	1.757
1.7	.8129	.7582	.6962	.6518	.6120	2.152	2.123	2.088	2.061	2.035
2.0	.8071	.7512	.6880	.6429	.6026	2.451	2.416	2.373	2.341	2.309
2.3	.8019	.7449	.6807	.6350	.5943	2.716	2.676	2.628	2.591	2.556
2.6	.7972	.7391	.6740	.6277	.5867	2.984	2.939	2.885	2.843	2.803
3.0	.7913	.7321	.6658	.6189	.5775	3.348	3.295	3.231	3.182	3.135
3.5	.7845	.7241	.6565	.6090	.5672	3.815	3.749	3.671	3.611	3.554
4.0	.7783	.7167	.6481	.6000	.5579	4.291	4.211	4.116	4.045	3.976
4.5	.7724	.7098	.6403	.5917	.5494	4.773	4.678	4.565	4.481	4.400
5.0	.7670	.7034	.6331	.5840	.5415	5.259	5.147	5.016	4.918	4.824
6.0	.7569	.6917	.6199	.5701	.5272	6.141	6.003	5.843	5.722	5.607
7.0	.7477	.6811	.6080	.5576	.5145	7.069	6.899	6.700	6.551	6.411
8.0	.7391	.6713	.5971	.5463	.5030	8.037	7.826	7.583	7.402	7.232
9.0	.7311	.6621	.5870	.5358	.4924	9.034	8.778	8.484	8.267	8.065
10	.7234	.6535	.5776	.5261	.4827	10.05	9.747	9.398	9.143	8.905
12	.7091	.6375	.5603	.5082	.4649	11.93	11.54	11.11	10.79	10.49
14	.6959	.6230	.5447	.4923	.4492	13.93	13.44	12.89	12.49	12.12
17	.6778	.6033	.5238	.4713	.4285	17.14	16.45	15.69	15.14	14.64
20	.6613	.5856	.5054	.4528	.4105	20.54	19.61	18.59	17.87	17.22
23	.6458	.5691	.4885	.4360	.3943	23.64	22.53	21.31	20.44	19.66
26	.6313	.5540	.4731	.4208	.3797	26.90	25.58	24.11	23.07	22.15
30	.6135	.5356	.4546	.4027	.3624	31.49	29.80	27.95	26.65	25.51
35	.5932	.5150	.4340	.3828	.3435	37.54	35.31	32.88	31.21	29.77
40	.5747	.4964	.4158	.3652	.3270	43.85	41.00	37.92	35.83	34.07
45	.5578	.4795	.3995	.3496	.3123	50.35	46.81	43.01	40.47	38.38
50	.5421	.4641	.3846	.3355	.2991	56.98	52.70	48.13	45.12	42.67
60	.5132	.4362	.3584	.3109	.2764	69.71	64.03	58.03	54.11	50.96
70	.4879	.4122	.3361	.2901	.2573	82.94	75.65	68.02	63.09	59.22
80	.4655	.3911	.3169	.2723	.2411	96.52	87.45	78.03	72.03	67.39
90	.4454	.3725	.3000	.2569	.2270	110.3	99.35	88.02	80.88	75.46
100	.4272	.3558	.2850	.2432	.2146	124.3	111.3	97.95	89.64	83.42
120	.3954	.3271	.2598	.2205	.1941	151.7	134.7	117.3	106.6	98.84
140	.3685	.3031	.2391	.2020	.1775	179.4	157.9	136.3	123.2	113.8
170	.3349	.2736	.2139	.1797	.1576	220.8	192.4	164.2	147.3	135.5
200	.3074	.2497	.1938	.1621	.1420	261.8	226.3	191.1	170.4	156.3
230	.2848	.2304	.1778	.1482	.1296	301.8	259.0	217.0	192.5	176.1
260	.2657	.2141	.1645	.1367	.1194	341.0	291.0	242.0	213.7	195.1
300	.2441	.1959	.1497	.1240	.1082	392.4	332.6	274.4	241.0	219.5
350	.2218	.1773	.1348	.1112	.0970	454.8	382.8	313.1	273.6	248.6
400	.2034	.1621	.1226	.1009	.0880	515.5	431.3	350.2	304.7	276.3
450	.1880	.1493	.1126	.0925	.0805	574.3	478.1	385.9	334.4	302.8
500	.1748	.1386	.1042	.0854	.0743	631.4	523.5	420.2	363.0	328.2
600	.1535	.1212	.0907	.0741	.0645	740.9	610.0	485.4	416.9	376.2
700	.1371	.1080	.0805	.0656	.0571	844.6	691.5	546.4	467.3	421.0
800	.1242	.0976	.0725	.0590	.0513	943.3	768.9	604.0	514.7	463.1
900	.1138	.0893	.0662	.0537	.0467	1038.	842.5	658.6	559.5	502.9
1000	.1054	.0825	.0610	.0495	.0430	1128.	912.8	710.6	602.0	540.7

E,KEV	MEAN DAMAGE DEPTH,MICROGRAM/SQ.CM.					MEAN IONZN.DEPTH,MICROGRAM/SQ.CM.				
K/KL=	0.55	0.75	1.0	1.2	1.4	0.55	0.75	1.0	1.2	1.4
.10	.2659	.2623	.2580	.2546	.2514	.2554	.2515	.2468	.2432	.2397
.12	.3025	.2982	.2931	.2891	.2853	.2905	.2858	.2802	.2759	.2718
.14	.3355	.3306	.3248	.3203	.3160	.3221	.3168	.3104	.3055	.3008
.17	.3811	.3754	.3686	.3634	.3584	.3658	.3596	.3522	.3465	.3411
.20	.4239	.4175	.4098	.4038	.3981	.4068	.3998	.3914	.3850	.3788
.23	.4650	.4578	.4491	.4425	.4361	.4462	.4383	.4289	.4217	.4148
.26	.5047	.4967	.4872	.4799	.4728	.4842	.4755	.4651	.4572	.4495
.30	.5562	.5471	.5364	.5281	.5202	.5334	.5236	.5119	.5029	.4943
.35	.6184	.6080	.5957	.5863	.5773	.5929	.5817	.5682	.5580	.5482
.40	.6787	.6670	.6531	.6426	.6324	.6505	.6378	.6227	.6112	.6002
.45	.7373	.7242	.7088	.6971	.6858	.7064	.6922	.6755	.6627	.6505
.50	.7943	.7799	.7629	.7500	.7376	.7608	.7452	.7267	.7127	.6992
.60	.8997	.8830	.8631	.8481	.8337	.8615	.8432	.8217	.8053	.7897
.70	1.002	.9832	.9604	.9431	.9267	.9593	.9384	.9137	.8949	.8771
.80	1.103	1.081	1.055	1.036	1.017	1.055	1.031	1.003	.9820	.9619
.90	1.201	1.177	1.148	1.126	1.105	1.148	1.122	1.091	1.067	1.044
1.0	1.298	1.271	1.239	1.215	1.192	1.240	1.211	1.176	1.150	1.125
1.2	1.476	1.444	1.407	1.378	1.352	1.409	1.375	1.334	1.304	1.275
1.4	1.652	1.615	1.572	1.539	1.508	1.576	1.536	1.489	1.454	1.421
1.7	1.914	1.868	1.816	1.776	1.739	1.824	1.775	1.718	1.675	1.634
2.0	2.173	2.119	2.056	2.009	1.965	2.068	2.009	1.941	1.891	1.843
2.3	2.403	2.343	2.272	2.219	2.169	2.287	2.221	2.144	2.087	2.033
2.6	2.635	2.567	2.487	2.428	2.373	2.506	2.432	2.346	2.282	2.222
3.0	2.949	2.869	2.777	2.708	2.644	2.801	2.714	2.614	2.540	2.471
3.5	3.347	3.252	3.141	3.059	2.983	3.173	3.069	2.950	2.863	2.781
4.0	3.750	3.638	3.508	3.412	3.323	3.548	3.426	3.286	3.184	3.088
4.5	4.157	4.026	3.875	3.764	3.663	3.925	3.782	3.621	3.503	3.393
5.0	4.566	4.415	4.242	4.116	4.000	4.302	4.138	3.953	3.819	3.695
6.0	5.310	5.127	4.917	4.765	4.625	4.993	4.795	4.571	4.408	4.258
7.0	6.086	5.864	5.611	5.427	5.261	5.706	5.465	5.196	5.001	4.823
8.0	6.888	6.622	6.319	6.101	5.904	6.436	6.148	5.827	5.598	5.388
9.0	7.710	7.394	7.037	6.782	6.553	7.179	6.837	6.462	6.195	5.952
10	8.545	8.177	7.762	7.467	7.205	7.929	7.532	7.098	6.791	6.513
12	10.09	9.631	9.118	8.754	8.431	9.333	8.836	8.302	7.926	7.582
14	11.72	11.15	10.52	10.07	9.683	10.79	10.18	9.523	9.066	8.650
17	14.30	13.54	12.70	12.11	11.60	13.06	12.24	11.38	10.78	10.25
20	17.01	16.02	14.93	14.19	13.55	15.40	14.34	13.25	12.50	11.84
23	19.48	18.31	17.02	16.14	15.39	17.60	16.35	15.03	14.14	13.37
26	22.06	20.68	19.16	18.13	17.25	19.85	18.38	16.83	15.78	14.89
30	25.67	23.95	22.09	20.83	19.77	22.93	21.13	19.22	17.95	16.89
35	30.41	28.20	25.83	24.25	22.95	26.87	24.60	22.21	20.63	19.35
40	35.33	32.57	29.65	27.71	26.15	30.88	28.10	25.18	23.28	21.77
45	40.38	37.03	33.49	31.19	29.35	34.92	31.59	28.12	25.89	24.15
50	45.52	41.53	37.36	34.67	32.55	38.98	35.08	31.03	28.46	26.48
60	55.39	50.21	44.82	41.38	38.71	46.83	41.82	36.66	33.42	30.97
70	65.60	59.07	52.33	48.08	44.84	54.71	48.50	42.16	38.22	35.30
80	76.06	68.06	59.87	54.75	50.92	62.59	55.11	47.52	42.87	39.48
90	86.66	77.11	67.38	61.37	56.94	70.41	61.62	52.76	47.38	43.52
100	97.35	86.17	74.85	67.92	62.88	78.17	68.03	57.87	51.76	47.44
120	118.2	103.8	89.32	80.57	74.33	93.35	80.49	67.72	60.14	54.92
140	139.2	121.4	103.6	92.94	85.50	108.2	92.54	77.12	68.09	61.99
170	170.7	147.5	124.5	111.0	101.8	129.7	109.9	90.49	79.30	71.96
200	202.0	173.2	145.0	128.5	117.5	150.4	126.4	103.1	89.79	81.26
230	232.4	198.1	164.5	145.2	132.4	170.2	142.1	114.9	99.61	89.96
260	262.3	222.4	183.6	161.3	146.9	189.3	157.2	126.2	108.9	98.19
300	301.6	254.2	208.3	182.2	165.6	213.7	176.3	140.5	120.6	108.5
350	349.6	292.8	238.1	207.3	188.0	242.8	199.0	157.3	134.3	120.7
400	396.3	330.3	266.8	231.4	209.5	270.3	220.4	173.0	147.2	132.0
450	441.9	366.6	294.5	254.5	230.2	296.6	240.7	187.9	159.2	142.7
500	486.5	401.9	321.4	276.9	250.2	321.7	260.1	202.0	170.7	152.8
600	571.6	469.5	372.5	319.4	288.1	368.9	296.4	228.4	192.0	171.6
700	652.9	533.7	420.8	359.4	323.7	412.7	330.0	252.7	211.6	188.9
800	730.5	594.8	466.6	397.2	357.4	453.6	361.4	275.2	229.7	204.9
900	804.8	653.1	510.1	433.2	389.4	492.2	390.8	296.3	246.6	219.9
1000	876.1	708.9	551.8	467.4	419.9	528.7	418.6	316.2	262.6	234.0

TABLE 7- 3

C ON NI, M2/M1=4.92

K/KL= E,KEV	RELATIVE RANGE STRAGGLING					RELATIVE DAMAGE STRAGGLING				
	0.55	0.75	1.0	1.2	1.4	0.55	0.75	1.0	1.2	1.4
.10	1.199	1.169	1.134	1.107	1.082	.8079	.7930	.7757	.7628	.7506
.12	1.194	1.163	1.127	1.100	1.074	.8058	.7906	.7728	.7596	.7472
.14	1.190	1.158	1.121	1.094	1.068	.8039	.7884	.7703	.7569	.7443
.17	1.184	1.151	1.114	1.086	1.059	.8013	.7854	.7670	.7532	.7404
.20	1.178	1.145	1.107	1.078	1.051	.7990	.7828	.7640	.7500	.7369
.23	1.173	1.139	1.101	1.072	1.044	.7969	.7803	.7612	.7471	.7338
.26	1.168	1.134	1.095	1.065	1.038	.7949	.7781	.7587	.7443	.7309
.30	1.162	1.127	1.087	1.058	1.030	.7924	.7752	.7555	.7410	.7274
.35	1.155	1.120	1.079	1.049	1.020	.7895	.7720	.7519	.7372	.7234
.40	1.148	1.112	1.071	1.040	1.012	.7867	.7689	.7486	.7336	.7197
.45	1.141	1.105	1.064	1.033	1.004	.7841	.7661	.7454	.7303	.7162
.50	1.135	1.099	1.057	1.025	.9961	.7816	.7633	.7425	.7272	.7130
.60	1.124	1.086	1.044	1.012	.9823	.7770	.7583	.7370	.7214	.7070
.70	1.113	1.075	1.032	.9996	.9697	.7727	.7537	.7320	.7162	.7016
.80	1.103	1.064	1.020	.9881	.9580	.7686	.7493	.7273	.7113	.6966
.90	1.093	1.054	1.010	.9774	.9471	.7647	.7452	.7229	.7067	.6919
1.0	1.084	1.045	1.000	.9672	.9367	.7611	.7413	.7188	.7024	.6875
1.2	1.067	1.027	.9815	.9483	.9176	.7542	.7340	.7110	.6944	.6794
1.4	1.051	1.010	.9645	.9310	.9002	.7478	.7273	.7039	.6871	.6719
1.7	1.029	.9877	.9412	.9075	.8765	.7389	.7179	.6941	.6770	.6617
2.0	1.009	.9670	.9201	.8862	.8551	.7307	.7093	.6851	.6679	.6524
2.3	.9897	.9475	.9004	.8664	.8352	.7229	.7012	.6767	.6592	.6437
2.6	.9721	.9295	.8821	.8480	.8169	.7155	.6935	.6688	.6512	.6356
3.0	.9503	.9073	.8596	.8256	.7945	.7064	.6840	.6590	.6413	.6256
3.5	.9254	.8820	.8342	.8001	.7691	.6958	.6731	.6477	.6299	.6142
4.0	.9027	.8591	.8111	.7770	.7462	.6860	.6630	.6374	.6194	.6037
4.5	.8819	.8380	.7899	.7560	.7253	.6769	.6536	.6278	.6098	.5941
5.0	.8626	.8184	.7704	.7365	.7060	.6684	.6449	.6189	.6008	.5850
6.0	.8272	.7827	.7346	.7010	.6709	.6527	.6287	.6024	.5841	.5685
7.0	.7961	.7514	.7034	.6700	.6403	.6386	.6143	.5876	.5693	.5537
8.0	.7684	.7234	.6756	.6425	.6132	.6258	.6012	.5743	.5560	.5405
9.0	.7434	.6983	.6506	.6179	.5890	.6141	.5892	.5622	.5438	.5284
10	.7207	.6755	.6280	.5956	.5670	.6032	.5781	.5510	.5326	.5173
12	.6807	.6353	.5883	.5565	.5286	.5833	.5579	.5306	.5123	.4973
14	.6463	.6009	.5544	.5232	.4960	.5657	.5401	.5128	.4947	.4799
17	.6027	.5574	.5116	.4813	.4552	.5428	.5171	.4898	.4719	.4575
20	.5662	.5211	.4761	.4466	.4214	.5231	.4973	.4702	.4525	.4386
23	.5350	.4905	.4461	.4173	.3931	.5059	.4801	.4531	.4358	.4222
26	.5079	.4640	.4202	.3920	.3688	.4906	.4649	.4381	.4211	.4078
30	.4765	.4336	.3906	.3632	.3411	.4726	.4470	.4206	.4039	.3912
35	.4433	.4015	.3594	.3330	.3123	.4531	.4278	.4018	.3856	.3734
40	.4151	.3743	.3333	.3077	.2882	.4362	.4112	.3856	.3698	.3582
45	.3907	.3510	.3109	.2861	.2677	.4213	.3966	.3716	.3562	.3450
50	.3693	.3307	.2915	.2675	.2500	.4080	.3837	.3591	.3442	.3334
60	.3335	.2968	.2599	.2375	.2216	.3851	.3617	.3382	.3241	.3140
70	.3045	.2696	.2347	.2138	.1991	.3661	.3435	.3210	.3076	.2983
80	.2804	.2471	.2141	.1944	.1809	.3499	.3282	.3066	.2939	.2851
90	.2600	.2282	.1968	.1783	.1657	.3359	.3150	.2943	.2822	.2739
100	.2425	.2121	.1822	.1647	.1529	.3237	.3035	.2836	.2720	.2642
120	.2146	.1866	.1592	.1433	.1328	.3036	.2847	.2663	.2556	.2486
140	.1927	.1667	.1414	.1269	.1175	.2873	.2697	.2525	.2426	.2361
170	.1673	.1439	.1212	.1084	.1002	.2678	.2517	.2361	.2272	.2214
200	.1478	.1266	.1061	.0946	.0874	.2524	.2376	.2233	.2151	.2100
230	.1325	.1131	.0944	.0839	.0775	.2401	.2264	.2131	.2055	.2008
260	.1201	.1021	.0850	.0754	.0696	.2299	.2170	.2046	.1976	.1932
300	.1066	.0905	.0750	.0665	.0613	.2186	.2067	.1953	.1888	.1848
350	.0935	.0791	.0654	.0578	.0533	.2072	.1964	.1859	.1800	.1763
400	.0831	.0702	.0579	.0512	.0472	.1980	.1879	.1782	.1728	.1693
450	.0748	.0631	.0520	.0459	.0423	.1903	.1809	.1719	.1668	.1635
500	.0679	.0573	.0471	.0416	.0384	.1838	.1750	.1665	.1616	.1586
600	.0573	.0483	.0398	.0351	.0324	.1733	.1653	.1577	.1533	.1505
700	.0496	.0418	.0344	.0304	.0280	.1652	.1579	.1508	.1467	.1441
800	.0439	.0369	.0303	.0268	.0247	.1587	.1519	.1452	.1414	.1389
900	.0394	.0331	.0271	.0239	.0221	.1535	.1469	.1405	.1369	.1345
1000	.0359	.0301	.0246	.0216	.0199	.1491	.1427	.1366	.1330	.1307

C ON NI, M2/M1=4.92

TABLE 7-4

E,KEV	RELATIVE IONIZATION STRAGGLING					RELATIVE TRANSVERSE RANGE STRAGGLING				
K/KL=	0.55	0.75	1.0	1.2	1.4	0.55	0.75	1.0	1.2	1.4
.10	.8104	.7969	.7813	.7697	.7589	.9196	.9183	.9166	.9153	.9140
.12	.8084	.7946	.7787	.7669	.7558	.9197	.9183	.9165	.9152	.9139
.14	.8066	.7925	.7763	.7643	.7531	.9197	.9182	.9165	.9151	.9138
.17	.8042	.7898	.7732	.7610	.7496	.9197	.9182	.9165	.9151	.9137
.20	.8020	.7873	.7704	.7580	.7464	.9198	.9183	.9164	.9150	.9136
.23	.7999	.7850	.7679	.7553	.7436	.9198	.9183	.9164	.9150	.9135
.26	.7980	.7829	.7655	.7528	.7410	.9199	.9183	.9164	.9150	.9135
.30	.7956	.7802	.7626	.7497	.7378	.9200	.9184	.9165	.9150	.9135
.35	.7928	.7771	.7593	.7462	.7341	.9202	.9186	.9166	.9151	.9136
.40	.7902	.7743	.7562	.7429	.7307	.9204	.9187	.9167	.9152	.9137
.45	.7877	.7716	.7532	.7399	.7276	.9206	.9189	.9169	.9153	.9138
.50	.7853	.7690	.7505	.7370	.7246	.9208	.9191	.9170	.9154	.9139
.60	.7809	.7642	.7454	.7318	.7192	.9211	.9194	.9173	.9157	.9142
.70	.7768	.7598	.7408	.7270	.7143	.9215	.9198	.9177	.9161	.9145
.80	.7729	.7557	.7364	.7225	.7098	.9220	.9202	.9181	.9164	.9148
.90	.7692	.7518	.7323	.7183	.7055	.9224	.9206	.9185	.9168	.9152
1.0	.7657	.7481	.7285	.7144	.7015	.9228	.9210	.9189	.9172	.9156
1.2	.7591	.7412	.7213	.7070	.6941	.9237	.9219	.9197	.9180	.9164
1.4	.7529	.7348	.7147	.7003	.6873	.9246	.9227	.9206	.9189	.9173
1.7	.7445	.7260	.7056	.6912	.6781	.9259	.9241	.9219	.9203	.9187
2.0	.7367	.7179	.6974	.6829	.6698	.9273	.9255	.9233	.9217	.9201
2.3	.7293	.7104	.6897	.6751	.6621	.9286	.9268	.9247	.9231	.9215
2.6	.7224	.7033	.6825	.6680	.6549	.9300	.9282	.9261	.9245	.9229
3.0	.7139	.6945	.6737	.6591	.6461	.9318	.9300	.9279	.9264	.9249
3.5	.7041	.6846	.6637	.6491	.6362	.9341	.9324	.9304	.9288	.9274
4.0	.6951	.6755	.6545	.6400	.6272	.9364	.9347	.9328	.9313	.9299
4.5	.6868	.6671	.6461	.6317	.6189	.9387	.9371	.9353	.9339	.9325
5.0	.6791	.6592	.6383	.6239	.6113	.9410	.9395	.9377	.9364	.9351
6.0	.6648	.6448	.6239	.6097	.5974	.9455	.9441	.9425	.9413	.9401
7.0	.6521	.6320	.6112	.5973	.5852	.9499	.9487	.9473	.9462	.9451
8.0	.6407	.6206	.6000	.5862	.5744	.9544	.9533	.9520	.9511	.9502
9.0	.6304	.6103	.5899	.5763	.5647	.9587	.9578	.9567	.9559	.9551
10	.6210	.6009	.5807	.5673	.5560	.9630	.9623	.9614	.9607	.9601
12	.6041	.5842	.5643	.5515	.5406	.9714	.9710	.9705	.9701	.9697
14	.5896	.5698	.5504	.5380	.5276	.9795	.9795	.9794	.9793	.9791
17	.5711	.5516	.5330	.5212	.5115	.9913	.9918	.9923	.9926	.9928
20	.5556	.5366	.5186	.5075	.4984	1.002	1.004	1.005	1.005	1.006
23	.5423	.5239	.5065	.4959	.4876	1.013	1.014	1.016	1.017	1.018
26	.5307	.5129	.4962	.4861	.4784	1.022	1.025	1.027	1.029	1.031
30	.5174	.5005	.4846	.4751	.4682	1.035	1.038	1.042	1.044	1.046
35	.5034	.4875	.4726	.4639	.4579	1.050	1.055	1.059	1.062	1.065
40	.4917	.4768	.4628	.4548	.4496	1.065	1.070	1.076	1.079	1.082
45	.4817	.4678	.4546	.4473	.4428	1.080	1.086	1.092	1.096	1.099
50	.4730	.4600	.4478	.4411	.4372	1.094	1.100	1.107	1.112	1.116
60	.4589	.4477	.4372	.4318	.4290	1.122	1.130	1.139	1.144	1.148
70	.4478	.4382	.4294	.4251	.4232	1.149	1.158	1.167	1.173	1.177
80	.4390	.4308	.4235	.4201	.4189	1.173	1.184	1.194	1.200	1.205
90	.4319	.4249	.4189	.4163	.4158	1.196	1.207	1.219	1.225	1.230
100	.4261	.4202	.4154	.4135	.4135	1.217	1.230	1.242	1.249	1.254
120	.4180	.4139	.4110	.4103	.4111	1.254	1.269	1.283	1.291	1.297
140	.4125	.4099	.4084	.4086	.4101	1.288	1.304	1.320	1.329	1.335
170	.4071	.4063	.4066	.4079	.4102	1.333	1.351	1.369	1.380	1.387
200	.4039	.4045	.4063	.4084	.4112	1.374	1.394	1.414	1.425	1.432
230	.4021	.4039	.4068	.4096	.4128	1.414	1.436	1.456	1.468	1.475
260	.4012	.4040	.4079	.4112	.4146	1.451	1.474	1.495	1.507	1.514
300	.4008	.4048	.4097	.4135	.4172	1.496	1.520	1.542	1.554	1.561
350	.4013	.4064	.4122	.4165	.4205	1.547	1.572	1.595	1.607	1.613
400	.4024	.4083	.4149	.4196	.4237	1.593	1.618	1.642	1.654	1.660
450	.4038	.4104	.4175	.4225	.4267	1.635	1.661	1.684	1.696	1.701
500	.4054	.4125	.4201	.4252	.4295	1.673	1.699	1.723	1.734	1.739
600	.4089	.4167	.4249	.4303	.4345	1.740	1.767	1.791	1.802	1.806
700	.4126	.4207	.4293	.4347	.4389	1.798	1.825	1.849	1.860	1.863
800	.4161	.4245	.4331	.4386	.4427	1.850	1.876	1.900	1.910	1.913
900	.4196	.4280	.4366	.4420	.4460	1.895	1.922	1.945	1.955	1.957
1000	.4229	.4312	.4397	.4450	.4489	1.936	1.962	1.985	1.994	1.996

TABLE 7- 5

C ON NI, M2/M1=4.92

E,KEV	RELATIVE TRANSV. DAMAGE STRAGGLING					RELATIVE TRANSV. IONZN. STRAGGLING				
K/KL=	0.55	0.75	1.0	1.2	1.4	0.55	0.75	1.0	1.2	1.4
.10	.8060	.8004	.7935	.7881	.7828	.7616	.7542	.7452	.7381	.7311
.12	.8055	.7998	.7927	.7871	.7816	.7609	.7533	.7440	.7367	.7295
.14	.8051	.7992	.7919	.7862	.7806	.7603	.7525	.7429	.7355	.7281
.17	.8046	.7985	.7910	.7851	.7793	.7595	.7514	.7416	.7338	.7263
.20	.8042	.7979	.7902	.7841	.7782	.7588	.7505	.7404	.7324	.7247
.23	.8038	.7973	.7894	.7833	.7772	.7581	.7497	.7393	.7312	.7232
.26	.8034	.7968	.7888	.7825	.7763	.7575	.7489	.7383	.7300	.7219
.30	.8030	.7963	.7880	.7816	.7753	.7568	.7480	.7371	.7287	.7204
.35	.8026	.7956	.7872	.7806	.7741	.7560	.7469	.7358	.7271	.7186
.40	.8022	.7951	.7864	.7797	.7731	.7553	.7460	.7346	.7257	.7170
.45	.8018	.7946	.7857	.7788	.7721	.7546	.7451	.7335	.7244	.7155
.50	.8015	.7941	.7851	.7781	.7712	.7540	.7443	.7324	.7232	.7141
.60	.8009	.7933	.7840	.7768	.7697	.7529	.7428	.7306	.7210	.7117
.70	.8004	.7926	.7830	.7756	.7683	.7518	.7415	.7289	.7191	.7095
.80	.8000	.7919	.7821	.7745	.7671	.7509	.7403	.7273	.7173	.7075
.90	.7996	.7914	.7814	.7736	.7660	.7500	.7391	.7259	.7156	.7056
1.0	.7992	.7909	.7806	.7727	.7650	.7492	.7381	.7246	.7141	.7038
1.2	.7987	.7900	.7794	.7712	.7632	.7477	.7361	.7221	.7112	.7007
1.4	.7982	.7892	.7783	.7698	.7616	.7463	.7344	.7199	.7087	.6978
1.7	.7976	.7882	.7768	.7681	.7595	.7444	.7319	.7168	.7051	.6938
2.0	.7971	.7873	.7756	.7665	.7577	.7427	.7297	.7140	.7019	.6901
2.3	.7967	.7866	.7745	.7651	.7561	.7411	.7276	.7114	.6989	.6868
2.6	.7963	.7860	.7735	.7639	.7546	.7396	.7257	.7090	.6961	.6837
3.0	.7959	.7852	.7723	.7624	.7528	.7378	.7233	.7060	.6926	.6798
3.5	.7955	.7844	.7710	.7607	.7507	.7356	.7204	.7023	.6885	.6751
4.0	.7952	.7837	.7698	.7591	.7488	.7335	.7177	.6989	.6845	.6707
4.5	.7950	.7830	.7686	.7576	.7471	.7315	.7151	.6956	.6808	.6665
5.0	.7948	.7824	.7676	.7563	.7454	.7295	.7126	.6925	.6772	.6625
6.0	.7946	.7815	.7658	.7539	.7425	.7260	.7080	.6867	.6705	.6551
7.0	.7945	.7807	.7642	.7517	.7398	.7227	.7036	.6812	.6642	.6480
8.0	.7944	.7799	.7627	.7497	.7373	.7194	.6994	.6758	.6581	.6413
9.0	.7943	.7792	.7613	.7477	.7349	.7163	.6952	.6707	.6522	.6348
10	.7943	.7785	.7599	.7458	.7326	.7132	.6912	.6657	.6466	.6286
12	.7945	.7775	.7575	.7426	.7284	.7074	.6836	.6562	.6358	.6167
14	.7945	.7765	.7553	.7394	.7245	.7018	.6763	.6471	.6255	.6054
17	.7945	.7748	.7518	.7347	.7188	.6935	.6656	.6340	.6109	.5896
20	.7941	.7729	.7483	.7301	.7132	.6853	.6553	.6216	.5971	.5747
23	.7939	.7712	.7450	.7257	.7079	.6772	.6451	.6092	.5835	.5601
26	.7935	.7694	.7416	.7212	.7026	.6692	.6352	.5975	.5706	.5464
30	.7926	.7666	.7370	.7154	.6957	.6588	.6226	.5825	.5543	.5293
35	.7910	.7630	.7311	.7081	.6873	.6463	.6076	.5650	.5354	.5095
40	.7890	.7591	.7252	.7009	.6791	.6341	.5933	.5486	.5178	.4913
45	.7867	.7550	.7193	.6938	.6712	.6224	.5797	.5332	.5014	.4745
50	.7841	.7508	.7133	.6869	.6635	.6110	.5667	.5185	.4861	.4588
60	.7786	.7420	.7013	.6729	.6482	.5889	.5414	.4907	.4570	.4295
70	.7725	.7330	.6896	.6596	.6338	.5683	.5184	.4657	.4313	.4038
80	.7660	.7241	.6783	.6469	.6202	.5489	.4972	.4432	.4083	.3810
90	.7593	.7153	.6674	.6348	.6074	.5308	.4778	.4227	.3877	.3607
100	.7525	.7065	.6568	.6233	.5952	.5137	.4597	.4041	.3690	.3424
120	.7381	.6887	.6360	.6009	.5721	.4807	.4262	.3706	.3361	.3107
140	.7239	.6719	.6167	.5804	.5512	.4514	.3969	.3419	.3084	.2841
170	.7033	.6483	.5904	.5530	.5234	.4132	.3596	.3061	.2740	.2515
200	.6839	.6266	.5669	.5287	.4990	.3806	.3284	.2767	.2461	.2252
230	.6645	.6058	.5451	.5066	.4772	.3531	.3027	.2530	.2240	.2045
260	.6463	.5867	.5253	.4868	.4576	.3292	.2806	.2330	.2054	.1872
300	.6238	.5633	.5015	.4631	.4346	.3017	.2555	.2105	.1848	.1681
350	.5983	.5373	.4754	.4374	.4096	.2729	.2296	.1876	.1639	.1488
400	.5751	.5141	.4526	.4151	.3880	.2487	.2081	.1689	.1469	.1332
450	.5541	.4934	.4323	.3955	.3692	.2282	.1900	.1534	.1329	.1204
500	.5349	.4746	.4143	.3781	.3525	.2106	.1746	.1402	.1212	.1097
600	.5011	.4421	.3834	.3486	.3244	.1820	.1499	.1194	.1027	.0929
700	.4719	.4146	.3579	.3245	.3016	.1600	.1311	.1038	.0890	.0804
800	.4466	.3912	.3365	.3044	.2827	.1427	.1166	.0919	.0786	.0709
900	.4242	.3708	.3182	.2875	.2669	.1291	.1052	.0827	.0705	.0636
1000	.4043	.3530	.3025	.2731	.2534	.1182	.0962	.0754	.0643	.0579

TABLE 7- 6

C ON NI, M2/M1=4.92

K/KL= E,KEV	RANGE SKEWNESS					DAMAGE SKEWNESS				
	0.55	0.75	1.0	1.2	1.4	0.55	0.75	1.0	1.2	1.4
.10	.2802	.2681	.2534	.2419	.2305	.6987	.6917	.6832	.6767	.6704
.12	.2790	.2666	.2514	.2395	.2278	.6980	.6908	.6821	.6754	.6689
.14	.2780	.2652	.2496	.2374	.2254	.6975	.6901	.6811	.6742	.6676
.17	.2767	.2635	.2473	.2347	.2223	.6969	.6891	.6799	.6727	.6658
.20	.2756	.2620	.2453	.2323	.2196	.6963	.6883	.6788	.6714	.6643
.23	.2746	.2606	.2435	.2302	.2171	.6958	.6876	.6778	.6702	.6630
.26	.2736	.2593	.2418	.2282	.2149	.6953	.6869	.6769	.6692	.6617
.30	.2725	.2577	.2398	.2258	.2121	.6947	.6860	.6757	.6678	.6602
.35	.2711	.2559	.2374	.2230	.2089	.6940	.6850	.6744	.6662	.6584
.40	.2699	.2542	.2352	.2205	.2060	.6933	.6841	.6732	.6648	.6567
.45	.2687	.2527	.2332	.2181	.2033	.6927	.6832	.6720	.6634	.6551
.50	.2676	.2512	.2313	.2159	.2008	.6921	.6824	.6709	.6621	.6537
.60	.2657	.2485	.2279	.2118	.1962	.6910	.6809	.6689	.6597	.6510
.70	.2638	.2461	.2247	.2081	.1920	.6899	.6794	.6669	.6575	.6484
.80	.2621	.2437	.2216	.2046	.1880	.6889	.6780	.6651	.6554	.6460
.90	.2604	.2415	.2188	.2013	.1843	.6880	.6767	.6634	.6533	.6438
1.0	.2587	.2394	.2160	.1981	.1807	.6870	.6754	.6617	.6514	.6416
1.2	.2558	.2354	.2110	.1923	.1741	.6856	.6733	.6588	.6479	.6376
1.4	.2529	.2317	.2063	.1868	.1680	.6841	.6711	.6559	.6445	.6338
1.7	.2487	.2262	.1994	.1790	.1592	.6817	.6678	.6516	.6395	.6282
2.0	.2447	.2210	.1929	.1715	.1510	.6791	.6644	.6473	.6346	.6228
2.3	.2409	.2162	.1869	.1646	.1433	.6762	.6609	.6430	.6298	.6175
2.6	.2372	.2114	.1810	.1580	.1359	.6733	.6573	.6387	.6250	.6123
3.0	.2323	.2052	.1734	.1494	.1264	.6693	.6525	.6331	.6187	.6055
3.5	.2262	.1976	.1642	.1390	.1150	.6643	.6466	.6261	.6110	.5972
4.0	.2201	.1901	.1551	.1289	.1040	.6594	.6407	.6192	.6035	.5892
4.5	.2141	.1827	.1463	.1191	.0934	.6545	.6350	.6125	.5962	.5813
5.0	.2081	.1755	.1378	.1096	.0830	.6498	.6293	.6060	.5890	.5737
6.0	.1967	.1616	.1213	.0915	.0634	.6415	.6192	.5939	.5757	.5593
7.0	.1854	.1481	.1055	.0741	.0447	.6330	.6091	.5821	.5628	.5455
8.0	.1743	.1349	.0902	.0574	.0268	.6245	.5991	.5705	.5501	.5320
9.0	.1633	.1220	.0754	.0413	.0096	.6158	.5890	.5590	.5378	.5190
10	.1526	.1095	.0610	.0257	-.0070	.6070	.5791	.5478	.5257	.5062
12	.1316	.0850	.0333	-.0041	-.0387	.5883	.5583	.5250	.5016	.4812
14	.1113	.0617	.0071	-.0322	-.0684	.5701	.5384	.5032	.4788	.4575
17	.0823	.0287	-.0298	-.0715	-.1098	.5442	.5101	.4727	.4467	.4244
20	.0549	-.0022	-.0641	-.1080	-.1479	.5200	.4838	.4443	.4171	.3939
23	.0293	-.0311	-.0965	-.1425	-.1839	.4979	.4597	.4181	.3896	.3655
26	.0048	-.0585	-.1270	-.1749	-.2175	.4772	.4371	.3935	.3639	.3389
30	-.0264	-.0929	-.1651	-.2152	-.2592	.4512	.4089	.3630	.3320	.3060
35	-.0632	-.1333	-.2094	-.2618	-.3073	.4211	.3762	.3278	.2953	.2682
40	-.0981	-.1711	-.2504	-.3048	-.3515	.3929	.3458	.2953	.2614	.2335
45	-.1313	-.2067	-.2888	-.3448	-.3925	.3665	.3175	.2650	.2300	.2014
50	-.1631	-.2405	-.3248	-.3823	-.4308	.3416	.2908	.2366	.2007	.1714
60	-.2276	-.3068	-.3931	-.4519	-.5016	.2943	.2407	.1837	.1462	.1158
70	-.2854	-.3662	-.4542	-.5141	-.5647	.2515	.1956	.1364	.0976	.0664
80	-.3371	-.4195	-.5092	-.5702	-.6217	.2125	.1546	.0936	.0537	.0219
90	-.3838	-.4679	-.5593	-.6214	-.6737	.1767	.1171	.0544	.0138	-.0185
100	-.4259	-.5118	-.6051	-.6684	-.7215	.1436	.0825	.0184	-.0229	-.0555
120	-.4932	-.5849	-.6839	-.7504	-.8057	.0840	.0201	-.0464	-.0888	-.1217
140	-.5514	-.6487	-.7530	-.8226	-.8798	.0313	-.0348	-.1031	-.1462	-.1792
170	-.6279	-.7326	-.8440	-.9177	-.9775	-.0377	-.1063	-.1765	-.2203	-.2532
200	-.6957	-.8066	-.9241	-1.001	-1.063	-.0978	-.1680	-.2395	-.2836	-.3162
230	-.7616	-.8763	-.9975	-1.077	-1.140	-.1521	-.2230	-.2948	-.3388	-.3709
260	-.8224	-.9401	-1.064	-1.145	-1.210	-.2008	-.2720	-.3438	-.3875	-.4191
300	-.8966	-1.018	-1.146	-1.229	-1.295	-.2587	-.3299	-.4014	-.4447	-.4755
350	-.9803	-1.105	-1.237	-1.322	-1.390	-.3219	-.3928	-.4637	-.5063	-.5363
400	-1.056	-1.184	-1.319	-1.406	-1.475	-.3771	-.4475	-.5177	-.5596	-.5888
450	-1.125	-1.256	-1.393	-1.482	-1.553	-.4260	-.4958	-.5652	-.6064	-.6349
500	-1.188	-1.322	-1.462	-1.553	-1.624	-.4697	-.5388	-.6074	-.6479	-.6758
600	-1.301	-1.440	-1.586	-1.679	-1.753	-.5450	-.6127	-.6797	-.7190	-.7457
700	-1.400	-1.544	-1.695	-1.791	-1.866	-.6079	-.6743	-.7397	-.7780	-.8038
800	-1.488	-1.637	-1.792	-1.891	-1.968	-.6616	-.7267	-.7907	-.8281	-.8532
900	-1.567	-1.721	-1.880	-1.982	-2.061	-.7081	-.7719	-.8347	-.8713	-.8958
1000	-1.639	-1.798	-1.962	-2.066	-2.146	-.7489	-.8116	-.8733	-.9092	-.9332

TABLE 7- 7

C ON NI, M2/M1=4.92

K/KL =	IONIZATION SKEWNESS					RANGE KURTOSIS				
E,KEV	0.55	0.75	1.0	1.2	1.4	0.55	0.75	1.0	1.2	1.4
.10	.7991	.7940	.7879	.7833	.7788	3.316	3.277	3.232	3.199	3.168
.12	.7987	.7935	.7872	.7824	.7778	3.311	3.272	3.226	3.192	3.160
.14	.7984	.7930	.7866	.7817	.7769	3.308	3.267	3.221	3.186	3.154
.17	.7981	.7925	.7858	.7807	.7758	3.303	3.261	3.214	3.178	3.146
.20	.7978	.7920	.7851	.7799	.7749	3.299	3.256	3.207	3.172	3.138
.23	.7976	.7916	.7845	.7791	.7740	3.295	3.251	3.202	3.166	3.132
.26	.7973	.7912	.7840	.7785	.7732	3.291	3.247	3.197	3.160	3.126
.30	.7971	.7908	.7833	.7776	.7722	3.287	3.242	3.191	3.153	3.119
.35	.7968	.7902	.7825	.7767	.7711	3.282	3.236	3.184	3.146	3.111
.40	.7965	.7897	.7818	.7758	.7701	3.277	3.230	3.177	3.139	3.104
.45	.7963	.7893	.7811	.7750	.7691	3.273	3.225	3.172	3.133	3.097
.50	.7960	.7889	.7805	.7742	.7682	3.269	3.220	3.166	3.127	3.091
.60	.7956	.7881	.7794	.7728	.7666	3.261	3.212	3.156	3.117	3.080
.70	.7952	.7874	.7783	.7715	.7650	3.254	3.204	3.148	3.107	3.070
.80	.7949	.7867	.7773	.7702	.7636	3.248	3.197	3.139	3.099	3.061
.90	.7945	.7861	.7763	.7690	.7622	3.242	3.190	3.132	3.091	3.053
1.0	.7941	.7854	.7753	.7679	.7609	3.236	3.183	3.125	3.083	3.046
1.2	.7934	.7842	.7736	.7657	.7584	3.226	3.171	3.112	3.070	3.032
1.4	.7927	.7830	.7719	.7637	.7560	3.216	3.160	3.100	3.057	3.019
1.7	.7916	.7811	.7693	.7606	.7525	3.202	3.145	3.084	3.041	3.002
2.0	.7903	.7792	.7667	.7576	.7491	3.189	3.131	3.069	3.026	2.988
2.3	.7891	.7775	.7643	.7547	.7459	3.177	3.118	3.055	3.012	2.974
2.6	.7879	.7756	.7619	.7519	.7428	3.166	3.106	3.042	2.999	2.961
3.0	.7861	.7731	.7587	.7482	.7386	3.152	3.091	3.027	2.983	2.946
3.5	.7837	.7699	.7546	.7436	.7336	3.135	3.073	3.009	2.966	2.929
4.0	.7812	.7666	.7506	.7391	.7286	3.120	3.057	2.993	2.950	2.914
4.5	.7787	.7633	.7465	.7346	.7237	3.105	3.043	2.979	2.936	2.901
5.0	.7760	.7600	.7425	.7302	.7190	3.092	3.029	2.965	2.924	2.889
6.0	.7707	.7534	.7347	.7215	.7097	3.066	3.002	2.939	2.899	2.866
7.0	.7653	.7468	.7270	.7131	.7007	3.042	2.979	2.918	2.879	2.848
8.0	.7598	.7403	.7195	.7050	.6921	3.021	2.958	2.899	2.862	2.833
9.0	.7543	.7338	.7121	.6971	.6838	3.002	2.940	2.882	2.848	2.821
10	.7489	.7275	.7050	.6895	.6759	2.984	2.924	2.868	2.836	2.811
12	.7380	.7150	.6910	.6747	.6603	2.952	2.893	2.843	2.815	2.795
14	.7275	.7030	.6778	.6608	.6458	2.924	2.869	2.823	2.800	2.784
17	.7123	.6860	.6593	.6415	.6259	2.890	2.841	2.803	2.787	2.778
20	.6978	.6702	.6423	.6239	.6079	2.863	2.820	2.792	2.782	2.781
23	.6835	.6548	.6258	.6068	.5907	2.835	2.801	2.782	2.779	2.785
26	.6699	.6404	.6105	.5911	.5749	2.812	2.787	2.777	2.782	2.795
30	.6529	.6226	.5918	.5720	.5559	2.789	2.776	2.778	2.793	2.815
35	.6332	.6023	.5708	.5508	.5349	2.769	2.771	2.789	2.814	2.848
40	.6151	.5838	.5520	.5319	.5165	2.758	2.774	2.808	2.844	2.887
45	.5983	.5670	.5350	.5151	.5001	2.753	2.783	2.832	2.879	2.932
50	.5827	.5515	.5196	.5000	.4854	2.754	2.798	2.861	2.918	2.980
60	.5522	.5221	.4917	.4731	.4598	2.768	2.838	2.930	3.006	3.085
70	.5261	.4973	.4683	.4510	.4388	2.794	2.889	3.008	3.102	3.196
80	.5036	.4760	.4486	.4324	.4214	2.830	2.948	3.091	3.202	3.311
90	.4841	.4577	.4318	.4168	.4069	2.873	3.012	3.178	3.305	3.427
100	.4672	.4419	.4174	.4035	.3947	2.921	3.080	3.268	3.410	3.544
120	.4416	.4176	.3951	.3831	.3765	3.029	3.219	3.443	3.611	3.770
140	.4216	.3987	.3781	.3679	.3633	3.146	3.364	3.622	3.814	3.996
170	.3983	.3774	.3595	.3517	.3497	3.328	3.588	3.893	4.121	4.334
200	.3805	.3617	.3465	.3409	.3410	3.513	3.814	4.165	4.425	4.668
230	.3652	.3500	.3383	.3349	.3368	3.685	4.027	4.424	4.716	4.986
260	.3528	.3410	.3327	.3314	.3347	3.856	4.240	4.681	5.004	5.300
300	.3398	.3322	.3281	.3292	.3341	4.086	4.522	5.022	5.384	5.713
350	.3277	.3249	.3253	.3289	.3354	4.372	4.873	5.442	5.852	6.220
400	.3192	.3204	.3247	.3304	.3381	4.657	5.220	5.856	6.311	6.717
450	.3133	.3180	.3255	.3330	.3417	4.939	5.562	6.263	6.761	7.203
500	.3094	.3170	.3273	.3362	.3457	5.218	5.899	6.662	7.202	7.679
600	.3062	.3183	.3328	.3438	.3543	5.766	6.556	7.437	8.056	8.600
700	.3071	.3224	.3397	.3521	.3631	6.298	7.191	8.183	8.877	9.482
800	.3109	.3282	.3473	.3604	.3718	6.814	7.804	8.901	9.665	10.33
900	.3167	.3352	.3552	.3687	.3801	7.314	8.397	9.593	10.42	11.14
1000	.3240	.3429	.3632	.3768	.3881	7.800	8.970	10.26	11.16	11.93

E,KEV	DAMAGE KURTOSIS					IONIZATION KURTOSIS				
K/KL=	0.55	0.75	1.0	1.2	1.4	0.55	0.75	1.0	1.2	1.4
.10	4.395	4.338	4.273	4.223	4.177	4.536	4.479	4.411	4.361	4.313
.12	4.387	4.329	4.262	4.212	4.164	4.529	4.470	4.401	4.349	4.301
.14	4.381	4.322	4.253	4.201	4.153	4.523	4.462	4.392	4.339	4.290
.17	4.372	4.311	4.241	4.188	4.139	4.514	4.452	4.380	4.326	4.276
.20	4.364	4.302	4.230	4.176	4.126	4.506	4.443	4.369	4.314	4.263
.23	4.357	4.293	4.220	4.166	4.114	4.499	4.434	4.359	4.304	4.251
.26	4.350	4.286	4.211	4.156	4.104	4.492	4.426	4.350	4.294	4.241
.30	4.342	4.276	4.200	4.143	4.091	4.484	4.417	4.339	4.282	4.228
.35	4.332	4.264	4.187	4.129	4.076	4.475	4.406	4.326	4.268	4.213
.40	4.322	4.254	4.175	4.116	4.062	4.465	4.395	4.315	4.255	4.200
.45	4.313	4.243	4.163	4.104	4.049	4.457	4.385	4.304	4.243	4.187
.50	4.305	4.234	4.153	4.093	4.037	4.449	4.376	4.293	4.232	4.175
.60	4.288	4.216	4.133	4.071	4.015	4.434	4.359	4.274	4.212	4.154
.70	4.273	4.199	4.114	4.052	3.994	4.420	4.343	4.257	4.193	4.134
.80	4.259	4.183	4.097	4.034	3.976	4.407	4.329	4.240	4.176	4.116
.90	4.245	4.169	4.081	4.017	3.958	4.394	4.315	4.225	4.159	4.099
1.0	4.233	4.155	4.066	4.001	3.941	4.382	4.301	4.210	4.144	4.083
1.2	4.211	4.130	4.038	3.971	3.911	4.359	4.276	4.183	4.115	4.053
1.4	4.191	4.107	4.013	3.944	3.883	4.338	4.253	4.157	4.089	4.025
1.7	4.151	4.075	3.977	3.907	3.844	4.309	4.220	4.122	4.052	3.987
2.0	4.134	4.044	3.944	3.872	3.808	4.281	4.190	4.090	4.018	3.953
2.3	4.104	4.013	3.911	3.838	3.774	4.255	4.162	4.060	3.987	3.920
2.6	4.075	3.983	3.880	3.806	3.741	4.230	4.135	4.031	3.957	3.890
3.0	4.039	3.946	3.841	3.767	3.701	4.198	4.101	3.995	3.920	3.852
3.5	3.997	3.902	3.796	3.721	3.655	4.162	4.062	3.953	3.877	3.809
4.0	3.958	3.862	3.754	3.678	3.612	4.127	4.025	3.915	3.838	3.768
4.5	3.922	3.824	3.715	3.639	3.572	4.095	3.990	3.879	3.801	3.731
5.0	3.889	3.789	3.679	3.602	3.535	4.064	3.958	3.844	3.766	3.696
6.0	3.832	3.728	3.614	3.535	3.468	4.006	3.896	3.780	3.701	3.630
7.0	3.779	3.672	3.555	3.475	3.407	3.953	3.840	3.723	3.642	3.571
8.0	3.731	3.621	3.501	3.420	3.352	3.904	3.789	3.670	3.589	3.517
9.0	3.685	3.573	3.452	3.370	3.301	3.858	3.742	3.621	3.540	3.468
10	3.642	3.528	3.405	3.323	3.255	3.816	3.698	3.576	3.494	3.423
12	3.555	3.440	3.317	3.235	3.168	3.739	3.617	3.494	3.412	3.341
14	3.477	3.362	3.239	3.158	3.093	3.669	3.545	3.421	3.339	3.269
17	3.375	3.260	3.138	3.060	2.997	3.577	3.450	3.326	3.245	3.176
20	3.287	3.173	3.053	2.977	2.917	3.495	3.368	3.244	3.165	3.097
23	3.213	3.099	2.981	2.906	2.848	3.417	3.293	3.172	3.093	3.028
26	3.148	3.034	2.918	2.844	2.789	3.346	3.226	3.107	3.031	2.967
30	3.072	2.960	2.845	2.774	2.721	3.263	3.147	3.032	2.957	2.896
35	2.991	2.881	2.770	2.701	2.652	3.171	3.062	2.951	2.879	2.821
40	2.922	2.814	2.706	2.641	2.594	3.092	2.988	2.881	2.813	2.758
45	2.861	2.757	2.653	2.590	2.546	3.023	2.924	2.821	2.755	2.703
50	2.808	2.707	2.607	2.547	2.506	2.962	2.868	2.769	2.705	2.655
60	2.714	2.622	2.531	2.479	2.443	2.862	2.774	2.682	2.623	2.577
70	2.638	2.555	2.474	2.427	2.397	2.780	2.698	2.613	2.558	2.515
80	2.577	2.502	2.430	2.389	2.363	2.714	2.636	2.555	2.504	2.465
90	2.527	2.460	2.395	2.359	2.337	2.658	2.583	2.507	2.459	2.422
100	2.486	2.426	2.368	2.337	2.319	2.611	2.539	2.466	2.420	2.386
120	2.429	2.380	2.335	2.311	2.300	2.543	2.472	2.401	2.359	2.330
140	2.390	2.351	2.316	2.300	2.294	2.491	2.420	2.351	2.312	2.287
170	2.353	2.327	2.306	2.298	2.299	2.432	2.362	2.295	2.259	2.238
200	2.334	2.319	2.309	2.309	2.315	2.387	2.317	2.253	2.220	2.203
230	2.328	2.323	2.323	2.329	2.339	2.342	2.278	2.220	2.191	2.178
260	2.329	2.333	2.342	2.353	2.367	2.305	2.247	2.194	2.169	2.158
300	2.337	2.352	2.372	2.388	2.406	2.263	2.212	2.167	2.147	2.139
350	2.356	2.382	2.413	2.436	2.457	2.220	2.179	2.142	2.126	2.122
400	2.380	2.416	2.455	2.483	2.508	2.187	2.153	2.124	2.112	2.110
450	2.408	2.451	2.498	2.531	2.559	2.160	2.133	2.111	2.102	2.102
500	2.437	2.487	2.541	2.578	2.608	2.139	2.117	2.101	2.095	2.097
600	2.497	2.559	2.625	2.667	2.701	2.108	2.096	2.088	2.088	2.092
700	2.559	2.630	2.704	2.751	2.788	2.088	2.083	2.082	2.085	2.092
800	2.620	2.698	2.779	2.830	2.869	2.077	2.077	2.080	2.086	2.095
900	2.680	2.753	2.849	2.904	2.945	2.072	2.075	2.081	2.089	2.099
1000	2.739	2.826	2.916	2.973	3.016	2.071	2.076	2.085	2.094	2.105

C ON NI, M2/M1=4.92

	REFLECTION COEFFICIENT					SPUTTERING EFFICIENCY					TABLE 7-9
K/KL=	0.55	0.75	1.0	1.2	1.4	0.55	0.75	1.0	1.2	1.4	
E,KEV											
.10	.1763	.1734	.1700	.1674	.1648	.0993	.0931	.0861	.0809	.0762	
.12	.1758	.1729	.1693	.1666	.1640	.0986	.0923	.0851	.0799	.0751	
.14	.1753	.1723	.1688	.1660	.1633	.0980	.0916	.0843	.0790	.0742	
.17	.1747	.1717	.1680	.1651	.1624	.0973	.0907	.0832	.0779	.0730	
.20	.1742	.1710	.1672	.1643	.1615	.0966	.0898	.0823	.0769	.0719	
.23	.1737	.1704	.1666	.1636	.1608	.0959	.0891	.0815	.0760	.0710	
.26	.1732	.1699	.1659	.1629	.1600	.0954	.0884	.0807	.0752	.0701	
.30	.1725	.1692	.1652	.1621	.1591	.0946	.0876	.0798	.0742	.0691	
.35	.1718	.1683	.1642	.1611	.1581	.0938	.0867	.0788	.0731	.0680	
.40	.1711	.1676	.1634	.1602	.1571	.0931	.0859	.0779	.0722	.0670	
.45	.1704	.1668	.1625	.1593	.1562	.0924	.0851	.0770	.0713	.0661	
.50	.1697	.1661	.1618	.1585	.1553	.0918	.0844	.0762	.0705	.0653	
.60	.1685	.1648	.1603	.1569	.1537	.0906	.0831	.0748	.0690	.0638	
.70	.1674	.1635	.1590	.1555	.1522	.0896	.0820	.0736	.0677	.0625	
.80	.1663	.1623	.1577	.1542	.1508	.0886	.0809	.0725	.0666	.0613	
.90	.1652	.1612	.1564	.1529	.1495	.0877	.0800	.0714	.0655	.0603	
1.0	.1642	.1601	.1553	.1517	.1482	.0869	.0791	.0705	.0645	.0593	
1.2	.1622	.1580	.1531	.1494	.1458	.0853	.0774	.0687	.0627	.0575	
1.4	.1604	.1561	.1510	.1472	.1436	.0839	.0759	.0672	.0611	.0559	
1.7	.1578	.1533	.1481	.1442	.1406	.0820	.0739	.0651	.0590	.0538	
2.0	.1554	.1508	.1455	.1415	.1378	.0802	.0720	.0632	.0572	.0520	
2.3	.1531	.1484	.1429	.1389	.1351	.0786	.0703	.0615	.0555	.0503	
2.6	.1509	.1461	.1405	.1364	.1326	.0771	.0688	.0599	.0539	.0488	
3.0	.1482	.1432	.1376	.1334	.1295	.0752	.0669	.0580	.0520	.0470	
3.5	.1450	.1399	.1341	.1298	.1259	.0731	.0647	.0559	.0499	.0449	
4.0	.1421	.1368	.1309	.1266	.1226	.0712	.0628	.0539	.0480	.0431	
4.5	.1393	.1339	.1279	.1235	.1195	.0694	.0610	.0522	.0463	.0415	
5.0	.1367	.1312	.1251	.1207	.1166	.0677	.0593	.0505	.0448	.0400	
6.0	.1318	.1261	.1198	.1153	.1112	.0647	.0563	.0476	.0419	.0373	
7.0	.1274	.1215	.1151	.1105	.1064	.0620	.0536	.0450	.0395	.0350	
8.0	.1233	.1174	.1108	.1062	.1020	.0595	.0512	.0428	.0373	.0330	
9.0	.1196	.1135	.1069	.1023	.0981	.0573	.0491	.0407	.0354	.0312	
10	.1162	.1100	.1033	.0986	.0944	.0553	.0471	.0389	.0337	.0296	
12	.1100	.1036	.0968	.0920	.0879	.0516	.0436	.0356	.0307	.0268	
14	.1046	.0980	.0911	.0863	.0822	.0485	.0406	.0329	.0281	.0245	
17	.0975	.0908	.0838	.0790	.0749	.0444	.0368	.0294	.0249	.0216	
20	.0914	.0846	.0775	.0728	.0688	.0410	.0337	.0266	.0224	.0193	
23	.0861	.0792	.0722	.0674	.0635	.0381	.0310	.0242	.0203	.0174	
26	.0814	.0746	.0675	.0628	.0590	.0355	.0287	.0222	.0185	.0158	
30	.0759	.0691	.0620	.0574	.0538	.0325	.0260	.0199	.0165	.0140	
35	.0700	.0633	.0563	.0517	.0483	.0294	.0233	.0176	.0144	.0122	
40	.0650	.0583	.0514	.0469	.0437	.0268	.0210	.0157	.0128	.0108	
45	.0606	.0540	.0472	.0428	.0397	.0245	.0191	.0141	.0114	.0096	
50	.0567	.0503	.0435	.0392	.0363	.0225	.0174	.0128	.0103	.0086	
60	.0502	.0440	.0375	.0335	.0308	.0194	.0148	.0107	.0086	.0072	
70	.0449	.0389	.0327	.0289	.0265	.0169	.0128	.0092	.0073	.0061	
80	.0404	.0347	.0288	.0252	.0229	.0149	.0112	.0080	.0063	.0053	
90	.0367	.0312	.0255	.0221	.0200	.0132	.0099	.0070	.0055	.0046	
100	.0334	.0282	.0227	.0195	.0176	.0118	.0088	.0062	.0049	.0041	
120	.0281	.0234	.0184	.0155	.0139	.0098	.0072	.0050	.0039	.0033	
140	.0240	.0196	.0151	.0126	.0112	.0083	.0061	.0042	.0033	.0028	
170	.0192	.0154	.0115	.0093	.0082	.0066	.0049	.0033	.0026	.0022	
200	.0156	.0123	.0089	.0070	.0062	.0055	.0040	.0027	.0021	.0018	
230	.0129	.0100	.0070	.0054	.0047	.0046	.0034	.0023	.0018	.0015	
260	.0108	.0083	.0056	.0043	.0037	.0040	.0029	.0020	.0015	.0013	
300	.0086	.0065	.0043	.0032	.0027	.0033	.0024	.0016	.0012	.0010	
350	.0065	.0049	.0031	.0022	.0019	.0028	.0020	.0013	.0010	.0008	
400	.0050	.0037	.0023	.0016	.0013	.0023	.0017	.0011	.0008	.0007	
450	.0039	.0028	.0017	.0011	.0009	.0020	.0014	.0010	.0007	.0006	
500	.0030	.0021	.0012	.0008	.0007	.0018	.0013	.0008	.0006	.0005	
600	.0018	.0013	.0007	.0004	.0004	.0014	.0010	.0007	.0005	.0004	
700	.0011	.0008	.0004	.0003	.0002	.0011	.0008	.0006	.0004	.0003	
800	.0007	.0005	.0002	.0001	.0001	.0010	.0007	.0005	.0004	.0003	
900	.0005	.0003	.0001	.0001	.0001	.0008	.0006	.0004	.0003	.0002	
1000	.0004	.0003	.0001	.0001	.0000	.0007	.0005	.0003	.0002	.0002	

E,KEV	IONIZATION DEFICIENCY K/KL=					SPUTTERING YIELD ALPHA				
	0.55	0.75	1.0	1.2	1.4	0.55	0.75	1.0	1.2	1.4
.10	.0101	.0129	.0161	.0183	.0202	.7559	.7311	.7016	.6793	.6579
.12	.0103	.0132	.0164	.0186	.0205	.7541	.7286	.6983	.6754	.6536
.14	.0105	.0135	.0167	.0189	.0208	.7524	.7263	.6955	.6721	.6499
.17	.0108	.0138	.0171	.0193	.0212	.7504	.7235	.6919	.6679	.6451
.20	.0110	.0141	.0174	.0196	.0215	.7486	.7211	.6888	.6643	.6411
.23	.0112	.0143	.0176	.0198	.0218	.7471	.7190	.6861	.6611	.6376
.26	.0114	.0145	.0178	.0201	.0220	.7458	.7172	.6837	.6584	.6344
.30	.0116	.0148	.0181	.0203	.0223	.7442	.7150	.6809	.6551	.6307
.35	.0118	.0150	.0183	.0206	.0225	.7426	.7127	.6778	.6515	.6267
.40	.0120	.0152	.0186	.0208	.0227	.7412	.7106	.6751	.6484	.6232
.45	.0122	.0154	.0188	.0210	.0229	.7399	.7089	.6728	.6456	.6201
.50	.0123	.0156	.0189	.0211	.0230	.7388	.7073	.6706	.6431	.6173
.60	.0126	.0158	.0192	.0214	.0233	.7370	.7046	.6670	.6388	.6125
.70	.0128	.0160	.0194	.0216	.0234	.7354	.7024	.6639	.6351	.6084
.80	.0129	.0162	.0196	.0218	.0236	.7341	.7005	.6612	.6319	.6048
.90	.0131	.0163	.0197	.0219	.0236	.7329	.6987	.6588	.6290	.6016
1.0	.0132	.0165	.0198	.0220	.0237	.7319	.6972	.6566	.6264	.5987
1.2	.0134	.0167	.0200	.0221	.0238	.7301	.6944	.6527	.6218	.5935
1.4	.0136	.0168	.0201	.0222	.0238	.7286	.6921	.6494	.6179	.5890
1.7	.0138	.0170	.0202	.0222	.0238	.7267	.6890	.6450	.6127	.5832
2.0	.0139	.0171	.0203	.0222	.0237	.7250	.6863	.6412	.6083	.5782
2.3	.0140	.0171	.0203	.0222	.0236	.7238	.6841	.6380	.6044	.5738
2.6	.0141	.0172	.0202	.0221	.0235	.7227	.6822	.6351	.6009	.5699
3.0	.0141	.0172	.0202	.0220	.0233	.7213	.6797	.6316	.5966	.5651
3.5	.0142	.0172	.0201	.0218	.0230	.7196	.6769	.6275	.5917	.5597
4.0	.0142	.0171	.0199	.0215	.0227	.7179	.6742	.6237	.5872	.5548
4.5	.0142	.0170	.0197	.0213	.0223	.7163	.6716	.6201	.5830	.5502
5.0	.0142	.0170	.0196	.0210	.0220	.7146	.6691	.6167	.5790	.5459
6.0	.0141	.0167	.0192	.0205	.0213	.7114	.6643	.6103	.5717	.5379
7.0	.0140	.0165	.0187	.0200	.0207	.7081	.6597	.6044	.5649	.5307
8.0	.0139	.0162	.0183	.0194	.0201	.7048	.6552	.5987	.5586	.5239
9.0	.0137	.0160	.0179	.0189	.0195	.7015	.6509	.5933	.5526	.5176
10	.0136	.0157	.0175	.0184	.0189	.6982	.6466	.5881	.5469	.5117
12	.0133	.0152	.0168	.0175	.0178	.6917	.6384	.5781	.5360	.5005
14	.0130	.0147	.0160	.0166	.0168	.6853	.6305	.5687	.5260	.4902
17	.0126	.0140	.0151	.0155	.0155	.6758	.6193	.5558	.5123	.4765
20	.0121	.0133	.0142	.0144	.0144	.6664	.6086	.5438	.5000	.4643
23	.0117	.0126	.0133	.0135	.0133	.6564	.5976	.5319	.4880	.4528
26	.0112	.0120	.0126	.0126	.0124	.6468	.5872	.5209	.4771	.4423
30	.0107	.0113	.0116	.0116	.0113	.6345	.5742	.5075	.4639	.4298
35	.0101	.0105	.0106	.0105	.0102	.6203	.5595	.4925	.4492	.4161
40	.0095	.0097	.0098	.0096	.0092	.6070	.5461	.4792	.4364	.4040
45	.0090	.0091	.0090	.0088	.0084	.5947	.5338	.4674	.4250	.3934
50	.0085	.0085	.0083	.0081	.0077	.5832	.5226	.4568	.4148	.3839
60	.0077	.0074	.0072	.0070	.0066	.5619	.5029	.4394	.3982	.3683
70	.0070	.0066	.0063	.0061	.0057	.5432	.4858	.4246	.3842	.3552
80	.0063	.0058	.0056	.0053	.0049	.5266	.4707	.4118	.3721	.3440
90	.0058	.0052	.0049	.0047	.0043	.5117	.4574	.4004	.3615	.3343
100	.0053	.0047	.0044	.0042	.0037	.4984	.4454	.3901	.3521	.3257
120	.0046	.0038	.0036	.0034	.0026	.4752	.4242	.3711	.3351	.3106
140	.0040	.0031	.0029	.0029	.0017	.4558	.4065	.3553	.3211	.2984
170	.0032	.0023	.0022	.0022	.0007	.4320	.3850	.3360	.3042	.2838
200	.0027	.0018	.0017	.0018	0.0000	.4127	.3680	.3208	.2911	.2725
230	.0021	.0013	.0013	.0014	0.0000	.3973	.3554	.3104	.2821	.2647
260	.0017	.0010	.0010	.0011	0.0000	.3845	.3452	.3022	.2750	.2585
300	.0012	.0007	.0007	.0008	0.0000	.3702	.3342	.2933	.2674	.2518
350	.0008	.0004	.0004	.0005	0.0000	.3560	.3232	.2848	.2601	.2453
400	.0004	.0002	.0002	.0003	0.0000	.3445	.3145	.2780	.2544	.2401
450	.0002	.0001	.0001	.0001	0.0000	.3352	.3074	.2725	.2497	.2359
500	0.0000	0.0000	0.0000	0.0000	0.0000	.3275	.3015	.2679	.2458	.2322
600	0.0000	0.0000	0.0000	0.0000	0.0000	.3158	.2921	.2604	.2394	.2261
700	0.0000	0.0000	0.0000	0.0000	0.0000	.3074	.2850	.2546	.2342	.2211
800	0.0000	0.0000	0.0000	0.0000	0.0000	.3014	.2794	.2496	.2299	.2167
900	0.0000	0.0000	0.0000	0.0000	0.0000	.2971	.2748	.2454	.2260	.2127
1000	0.0000	0.0000	0.0000	0.0000	0.0000	.2941	.2710	.2416	.2225	.2091

C ON NI, M2/M1=4.92

1000 KEV

100 KEV

10 KEV

1 KEV

.10 KEV

149

TABLE 8- 1

AL ON ZR, M2/M1=3.37

E,KEV	FRACTIONAL DEPOSITED ENERGY					MEAN RANGE,MICROGRAM/SQ.CM.				
K/KL=	0.5	0.65	0.8	1.0	1.3	0.5	0.65	0.8	1.0	1.3
.20	.9071	.8816	.8570	.8258	.7819	.4209	.4188	.4167	.4140	.4100
.23	.9050	.8790	.8540	.8222	.7775	.4656	.4632	.4609	.4577	.4532
.26	.9031	.8767	.8512	.8189	.7736	.5066	.5039	.5013	.4978	.4927
.30	.9009	.8739	.8480	.8151	.7690	.5574	.5543	.5513	.5474	.5417
.35	.8985	.8709	.8444	.8109	.7639	.6166	.6132	.6098	.6054	.5989
.40	.8963	.8682	.8413	.8071	.7595	.6729	.6691	.6653	.6604	.6532
.45	.8944	.8658	.8384	.8038	.7555	.7269	.7227	.7186	.7132	.7053
.50	.8926	.8636	.8358	.8008	.7519	.7793	.7748	.7703	.7644	.7558
.60	.8895	.8597	.8313	.7954	.7455	.8802	.8749	.8697	.8629	.8529
.70	.8868	.8563	.8273	.7907	.7399	.9771	.9710	.9651	.9573	.9458
.80	.8843	.8533	.8238	.7866	.7351	1.071	1.064	1.057	1.048	1.035
.90	.8821	.8506	.8206	.7829	.7307	1.161	1.153	1.146	1.136	1.122
1.0	.8801	.8481	.8177	.7795	.7267	1.249	1.241	1.232	1.222	1.206
1.2	.8766	.8437	.8126	.7735	.7197	1.413	1.403	1.393	1.381	1.363
1.4	.8735	.8399	.8081	.7684	.7136	1.571	1.559	1.549	1.534	1.513
1.7	.8694	.8349	.8024	.7616	.7057	1.800	1.786	1.773	1.756	1.731
2.0	.8659	.8306	.7974	.7558	.6990	2.020	2.005	1.989	1.969	1.940
2.3	.8628	.8269	.7930	.7508	.6931	2.221	2.204	2.187	2.164	2.131
2.6	.8600	.8235	.7891	.7462	.6878	2.418	2.399	2.380	2.355	2.318
3.0	.8567	.8195	.7844	.7408	.6815	2.677	2.654	2.632	2.604	2.562
3.5	.8529	.8150	.7792	.7347	.6746	2.995	2.969	2.943	2.909	2.861
4.0	.8495	.8109	.7745	.7293	.6684	3.308	3.277	3.248	3.209	3.154
4.5	.8464	.8072	.7702	.7243	.6627	3.616	3.581	3.548	3.504	3.441
5.0	.8435	.8038	.7663	.7198	.6575	3.919	3.880	3.842	3.793	3.723
6.0	.8384	.7976	.7592	.7116	.6483	4.475	4.429	4.384	4.326	4.242
7.0	.8338	.7921	.7528	.7044	.6401	5.031	4.976	4.924	4.855	4.757
8.0	.8296	.7870	.7471	.6979	.6328	5.588	5.524	5.462	5.383	5.269
9.0	.8257	.7824	.7419	.6920	.6261	6.144	6.071	5.999	5.908	5.777
10	.8220	.7781	.7370	.6865	.6200	6.699	6.615	6.534	6.430	6.281
12	.8155	.7704	.7282	.6765	.6089	7.713	7.613	7.515	7.390	7.212
14	.8096	.7634	.7203	.6677	.5991	8.752	8.631	8.514	8.365	8.152
17	.8015	.7539	.7097	.6558	.5861	10.36	10.20	10.05	9.853	9.580
20	.7941	.7454	.7001	.6452	.5746	12.00	11.80	11.61	11.36	11.02
23	.7875	.7376	.6914	.6356	.5642	13.44	13.21	12.99	12.71	12.32
26	.7812	.7304	.6834	.6268	.5547	14.93	14.67	14.42	14.10	13.65
30	.7734	.7215	.6735	.6159	.5431	17.01	16.70	16.39	16.00	15.46
35	.7644	.7111	.6622	.6036	.5300	19.75	19.34	18.95	18.46	17.78
40	.7559	.7015	.6517	.5922	.5181	22.61	22.09	21.60	20.99	20.15
45	.7479	.6925	.6419	.5818	.5072	25.56	24.92	24.31	23.57	22.55
50	.7402	.6840	.6327	.5720	.4971	28.57	27.79	27.07	26.18	24.97
60	.7260	.6682	.6157	.5539	.4786	34.08	33.08	32.17	31.05	29.53
70	.7128	.6537	.6003	.5376	.4622	39.99	38.71	37.54	36.13	34.22
80	.7005	.6404	.5861	.5229	.4475	46.23	44.60	43.13	41.36	39.01
90	.6889	.6279	.5731	.5094	.4341	52.73	50.70	48.89	46.72	43.87
100	.6779	.6163	.5609	.4969	.4218	59.41	56.95	54.76	52.15	48.78
120	.6573	.5945	.5384	.4739	.3997	72.04	68.87	66.00	62.59	58.24
140	.6385	.5749	.5183	.4538	.3804	85.46	81.39	77.70	73.33	67.84
170	.6131	.5488	.4919	.4275	.3557	106.8	101.1	95.90	89.81	82.41
200	.5904	.5257	.4688	.4049	.3347	129.2	121.5	114.6	106.5	97.02
230	.5691	.5046	.4480	.3848	.3164	150.6	141.2	132.6	122.8	111.2
260	.5497	.4855	.4293	.3670	.3003	172.8	161.3	150.9	139.1	125.3
300	.5263	.4626	.4072	.3461	.2816	203.2	188.6	175.7	160.9	144.0
350	.5002	.4375	.3832	.3235	.2617	242.2	223.4	206.8	188.0	167.1
400	.4771	.4155	.3622	.3041	.2448	282.0	258.7	238.1	215.0	189.8
450	.4564	.3959	.3438	.2872	.2301	322.3	294.0	269.3	241.7	212.2
500	.4377	.3784	.3274	.2723	.2172	362.7	329.3	300.2	268.1	234.1
600	.4047	.3480	.2994	.2472	.1961	441.8	398.2	360.4	319.1	276.3
700	.3768	.3226	.2762	.2267	.1789	520.8	466.5	419.7	368.7	317.0
800	.3529	.3009	.2566	.2096	.1647	599.4	534.0	477.8	417.0	356.4
900	.3320	.2822	.2399	.1950	.1527	677.2	600.5	534.7	464.0	394.5
1000	.3137	.2658	.2253	.1824	.1424	754.1	665.8	590.4	509.7	431.4
1200	.2829	.2386	.2011	.1618	.1257	904.5	793.1	698.3	597.6	502.0
1400	.2579	.2167	.1820	.1457	.1127	1050.	915.7	801.6	681.2	568.8
1700	.2284	.1911	.1597	.1271	.0979	1260.	1091.	948.8	799.5	662.8
2000	.2056	.1716	.1430	.1133	.0870	1460.	1258.	1088.	910.5	750.5

TABLE 8-2

AL ON ZR, M2/M1=3.37

E,KEV	MEAN DAMAGE DEPTH,MICROGRAM/SQ.CM.					MEAN IONZN.DEPTH,MICROGRAM/SQ.CM.				
K/KL=	0.5	0.65	0.8	1.0	1.3	0.5	0.65	0.8	1.0	1.3
.20	.3775	.3740	.3706	.3662	.3598	.3573	.3536	.3499	.3452	.3383
.23	.4173	.4133	.4094	.4043	.3970	.3949	.3906	.3864	.3810	.3731
.26	.4537	.4493	.4449	.4393	.4312	.4293	.4245	.4199	.4138	.4051
.30	.4989	.4939	.4890	.4827	.4736	.4720	.4666	.4614	.4546	.4449
.35	.5518	.5461	.5406	.5335	.5233	.5220	.5159	.5100	.5023	.4913
.40	.6019	.5956	.5895	.5816	.5702	.5693	.5625	.5560	.5475	.5353
.45	.6500	.6431	.6364	.6277	.6153	.6147	.6073	.6001	.5908	.5775
.50	.6966	.6891	.6818	.6724	.6589	.6588	.6507	.6429	.6328	.6182
.60	.7862	.7775	.7690	.7581	.7425	.7434	.7340	.7249	.7132	.6964
.70	.8720	.8621	.8525	.8401	.8223	.8243	.8137	.8033	.7900	.7709
.80	.9547	.9436	.9328	.9189	.8991	.9023	.8904	.8788	.8639	.8425
.90	1.035	1.022	1.010	.9951	.9731	.9777	.9645	.9517	.9352	.9116
1.0	1.112	1.099	1.086	1.069	1.045	1.051	1.036	1.022	1.004	.9784
1.2	1.257	1.241	1.226	1.206	1.178	1.187	1.170	1.154	1.133	1.103
1.4	1.396	1.378	1.361	1.338	1.307	1.318	1.299	1.280	1.256	1.222
1.7	1.597	1.575	1.555	1.528	1.491	1.507	1.484	1.462	1.434	1.393
2.0	1.790	1.765	1.741	1.711	1.667	1.688	1.662	1.636	1.603	1.557
2.3	1.966	1.938	1.912	1.877	1.829	1.854	1.825	1.796	1.759	1.707
2.6	2.138	2.107	2.078	2.040	1.986	2.016	1.983	1.951	1.911	1.853
3.0	2.363	2.328	2.295	2.251	2.191	2.228	2.190	2.154	2.108	2.042
3.5	2.639	2.598	2.559	2.510	2.440	2.486	2.442	2.401	2.347	2.272
4.0	2.909	2.863	2.818	2.762	2.682	2.738	2.689	2.641	2.581	2.496
4.5	3.173	3.122	3.072	3.008	2.919	2.986	2.930	2.877	2.809	2.714
5.0	3.433	3.376	3.320	3.250	3.151	3.229	3.167	3.107	3.032	2.927
6.0	3.911	3.843	3.778	3.695	3.579	3.676	3.602	3.533	3.444	3.320
7.0	4.386	4.306	4.231	4.134	4.000	4.118	4.033	3.952	3.849	3.706
8.0	4.858	4.767	4.679	4.569	4.415	4.557	4.459	4.366	4.248	4.085
9.0	5.327	5.223	5.124	4.999	4.826	4.993	4.881	4.775	4.642	4.458
10	5.793	5.676	5.565	5.424	5.230	5.424	5.299	5.180	5.031	4.824
12	6.646	6.508	6.375	6.208	5.979	6.219	6.070	5.929	5.751	5.507
14	7.512	7.349	7.192	6.996	6.727	7.021	6.845	6.678	6.469	6.183
17	8.838	8.631	8.434	8.187	7.852	8.240	8.016	7.807	7.546	7.191
20	10.19	9.931	9.689	9.386	8.979	9.471	9.195	8.938	8.620	8.190
23	11.38	11.09	10.81	10.47	10.00	10.57	10.26	9.964	9.601	9.109
26	12.60	12.27	11.95	11.56	11.03	11.69	11.34	11.00	10.59	10.03
30	14.30	13.90	13.52	13.06	12.43	13.23	12.80	12.41	11.92	11.26
35	16.50	16.01	15.54	14.97	14.21	15.21	14.68	14.19	13.60	12.80
40	18.79	18.18	17.62	16.92	16.01	17.24	16.60	16.01	15.29	14.34
45	21.13	20.40	19.73	18.90	17.83	19.31	18.54	17.83	16.99	15.88
50	23.51	22.65	21.86	20.89	19.65	21.40	20.49	19.67	18.69	17.41
60	27.86	26.80	25.80	24.60	23.06	25.29	24.13	23.11	21.90	20.31
70	32.49	31.16	29.93	28.44	26.55	29.33	27.89	26.62	25.14	23.22
80	37.35	35.70	34.19	32.38	30.10	33.51	31.74	30.20	28.41	26.12
90	42.37	40.38	38.55	36.38	33.69	37.77	35.65	33.82	31.69	29.00
100	47.52	45.15	42.99	40.43	37.30	42.09	39.61	37.45	34.96	31.87
120	57.25	54.21	51.45	48.20	44.24	50.38	47.24	44.47	41.28	37.40
140	67.53	63.69	60.21	56.14	51.26	58.90	55.01	51.56	47.59	42.85
170	83.78	78.51	73.78	68.31	61.90	72.02	66.84	62.23	56.96	50.86
200	100.7	93.83	87.68	80.64	72.57	85.37	78.75	72.87	66.20	58.67
230	116.9	108.5	101.0	92.50	82.83	98.34	90.31	83.20	75.16	66.22
260	133.5	123.4	114.5	104.4	93.06	111.4	101.8	93.43	83.97	73.58
300	156.3	143.8	132.8	120.4	106.6	128.9	117.2	106.9	95.46	83.10
350	185.5	169.7	155.8	140.3	123.5	150.7	136.1	123.5	109.4	94.58
400	215.3	195.9	178.9	160.1	140.1	172.4	154.9	139.7	123.0	105.6
450	245.3	222.1	202.0	179.8	156.5	193.8	173.3	155.5	136.1	116.3
500	275.5	248.4	224.9	199.2	172.7	215.0	191.4	171.0	148.9	126.6
600	334.1	299.3	269.3	236.7	203.6	256.3	226.4	200.0	173.3	146.1
700	392.8	350.0	313.2	273.4	233.7	296.4	260.3	229.4	196.4	164.5
800	451.4	409.2	356.4	309.3	263.0	335.4	292.9	256.8	218.5	181.9
900	509.6	449.9	398.9	344.4	291.5	373.1	324.5	283.2	239.5	198.5
1000	567.1	498.8	440.6	378.7	319.3	409.8	355.0	308.6	259.7	214.4
1200	680.1	594.5	521.8	445.0	372.7	480.2	413.3	356.8	297.8	244.1
1400	789.9	687.1	600.0	508.4	423.5	546.9	468.3	402.1	333.3	271.7
1700	948.6	820.2	711.9	598.7	495.6	640.9	545.4	465.3	382.5	309.8
2000	1100.	947.0	818.1	683.9	563.3	728.8	617.2	523.9	427.8	344.7

TABLE 8- 3

AL ON ZR, M2/M1=3.37

E,KEV	RELATIVE RANGE STRAGGLING					RELATIVE DAMAGE STRAGGLING				
K/KL=	0.5	0.65	0.8	1.0	1.3	0.5	0.65	0.8	1.0	1.3
.20	.8713	.8590	.8472	.8320	.8103	.6313	.6253	.6196	.6123	.6019
.23	.8696	.8571	.8451	.8296	.8075	.6306	.6245	.6187	.6112	.6007
.26	.8681	.8554	.8431	.8274	.8050	.6300	.6238	.6179	.6103	.5996
.30	.8663	.8533	.8408	.8248	.8019	.6292	.6229	.6169	.6091	.5983
.35	.8642	.8509	.8382	.8218	.7985	.6284	.6219	.6158	.6079	.5968
.40	.8623	.8488	.8357	.8191	.7954	.6276	.6211	.6148	.6067	.5955
.45	.8604	.8467	.8335	.8166	.7926	.6269	.6202	.6138	.6057	.5943
.50	.8587	.8448	.8314	.8142	.7900	.6262	.6195	.6130	.6047	.5932
.60	.8555	.8412	.8275	.8099	.7851	.6250	.6180	.6114	.6029	.5911
.70	.8525	.8379	.8239	.8060	.7808	.6239	.6167	.6099	.6013	.5893
.80	.8497	.8348	.8205	.8024	.7767	.6228	.6155	.6086	.5999	.5877
.90	.8470	.8319	.8174	.7990	.7730	.6218	.6144	.6074	.5985	.5861
1.0	.8444	.8291	.8144	.7957	.7695	.6208	.6133	.6062	.5972	.5847
1.2	.8396	.8240	.8089	.7898	.7630	.6191	.6114	.6041	.5949	.5821
1.4	.8352	.8191	.8038	.7844	.7571	.6175	.6096	.6021	.5927	.5797
1.7	.8288	.8124	.7967	.7769	.7491	.6152	.6071	.5994	.5898	.5765
2.0	.8229	.8062	.7902	.7699	.7417	.6130	.6048	.5969	.5871	.5736
2.3	.8174	.8003	.7841	.7635	.7349	.6111	.6026	.5947	.5847	.5709
2.6	.8121	.7948	.7783	.7575	.7286	.6092	.6006	.5925	.5823	.5684
3.0	.8054	.7878	.7710	.7499	.7206	.6068	.5981	.5898	.5794	.5653
3.5	.7975	.7796	.7625	.7411	.7114	.6040	.5951	.5866	.5761	.5617
4.0	.7901	.7718	.7545	.7328	.7028	.6013	.5922	.5836	.5729	.5583
4.5	.7830	.7645	.7470	.7250	.6948	.5988	.5895	.5808	.5699	.5551
5.0	.7762	.7575	.7398	.7176	.6872	.5963	.5870	.5781	.5670	.5521
6.0	.7634	.7443	.7262	.7038	.6729	.5918	.5822	.5731	.5618	.5465
7.0	.7515	.7321	.7138	.6910	.6599	.5876	.5777	.5684	.5569	.5414
8.0	.7404	.7207	.7022	.6792	.6479	.5836	.5735	.5640	.5523	.5366
9.0	.7300	.7101	.6914	.6682	.6367	.5798	.5695	.5599	.5479	.5321
10	.7202	.7000	.6811	.6578	.6262	.5761	.5657	.5559	.5438	.5278
12	.7017	.6812	.6620	.6384	.6066	.5692	.5584	.5484	.5360	.5198
14	.6850	.6641	.6447	.6209	.5889	.5627	.5517	.5414	.5289	.5124
17	.6625	.6412	.6215	.5975	.5654	.5538	.5425	.5320	.5191	.5024
20	.6424	.6208	.6009	.5768	.5447	.5457	.5341	.5233	.5102	.4933
23	.6241	.6022	.5821	.5578	.5257	.5382	.5264	.5154	.5020	.4849
26	.6074	.5852	.5650	.5406	.5085	.5313	.5192	.5080	.4944	.4772
30	.5872	.5647	.5443	.5199	.4879	.5228	.5104	.4989	.4851	.4677
35	.5648	.5420	.5215	.4970	.4652	.5131	.5003	.4886	.4746	.4571
40	.5449	.5219	.5012	.4767	.4451	.5042	.4911	.4792	.4650	.4474
45	.5271	.5038	.4830	.4585	.4272	.4959	.4827	.4706	.4563	.4386
50	.5108	.4873	.4665	.4421	.4110	.4883	.4749	.4626	.4482	.4305
60	.4824	.4584	.4375	.4132	.3826	.4741	.4604	.4479	.4333	.4156
70	.4580	.4337	.4128	.3887	.3586	.4615	.4476	.4350	.4203	.4027
80	.4367	.4123	.3914	.3675	.3379	.4503	.4362	.4235	.4087	.3913
90	.4179	.3934	.3725	.3488	.3198	.4402	.4259	.4132	.3984	.3811
100	.4010	.3765	.3557	.3323	.3038	.4310	.4166	.4038	.3890	.3719
120	.3717	.3477	.3272	.3041	.2767	.4147	.4003	.3874	.3726	.3559
140	.3471	.3236	.3034	.2807	.2544	.4008	.3862	.3733	.3587	.3423
170	.3167	.2941	.2744	.2523	.2274	.3829	.3684	.3555	.3411	.3252
200	.2918	.2700	.2509	.2294	.2058	.3679	.3534	.3407	.3265	.3111
230	.2714	.2503	.2317	.2109	.1885	.3547	.3404	.3280	.3141	.2992
260	.2541	.2335	.2154	.1954	.1741	.3432	.3291	.3169	.3033	.2890
300	.2344	.2146	.1973	.1781	.1581	.3298	.3160	.3041	.2910	.2772
350	.2141	.1952	.1786	.1605	.1418	.3155	.3022	.2906	.2780	.2650
400	.1973	.1791	.1633	.1461	.1287	.3033	.2903	.2792	.2670	.2546
450	.1831	.1656	.1505	.1341	.1178	.2927	.2801	.2694	.2576	.2458
500	.1709	.1541	.1396	.1240	.1086	.2834	.2712	.2608	.2494	.2381
600	.1513	.1357	.1224	.1081	.0942	.2680	.2565	.2467	.2361	.2256
700	.1359	.1214	.1090	.0958	.0832	.2555	.2446	.2353	.2254	.2156
800	.1233	.1098	.0983	.0861	.0746	.2450	.2347	.2259	.2165	.2072
900	.1129	.1002	.0895	.0781	.0675	.2361	.2263	.2179	.2089	.2002
1000	.1041	.0922	.0821	.0715	.0617	.2284	.2190	.2110	.2024	.1941
1200	.0901	.0794	.0705	.0611	.0526	.2157	.2071	.1997	.1918	.1841
1400	.0793	.0697	.0617	.0533	.0458	.2057	.1976	.1907	.1833	.1762
1700	.0672	.0589	.0520	.0448	.0384	.1940	.1866	.1802	.1734	.1669
2000	.0583	.0511	.0450	.0387	.0331	.1851	.1781	.1721	.1658	.1596

TABLE 8-4

AL ON ZR, M2/M1=3.37

E,KEV / K/KL=	RELATIVE IONIZATION STRAGGLING					RELATIVE TRANSVERSE RANGE STRAGGLING				
	0.5	0.65	0.8	1.0	1.3	0.5	0.65	0.8	1.0	1.3
.20	.6578	.6526	.6476	.6413	.6324	.8898	.8889	.8881	.8870	.8853
.23	.6572	.6519	.6468	.6403	.6313	.8898	.8889	.8880	.8869	.8852
.26	.6566	.6512	.6461	.6395	.6304	.8898	.8889	.8880	.8868	.8851
.30	.6559	.6504	.6452	.6385	.6292	.8898	.8888	.8879	.8867	.8850
.35	.6552	.6496	.6442	.6374	.6279	.8897	.8888	.8879	.8867	.8849
.40	.6545	.6488	.6433	.6364	.6268	.8897	.8888	.8878	.8866	.8848
.45	.6538	.6480	.6425	.6355	.6258	.8897	.8887	.8878	.8865	.8847
.50	.6532	.6473	.6417	.6346	.6248	.8897	.8887	.8878	.8865	.8847
.60	.6521	.6460	.6403	.6331	.6230	.8897	.8887	.8877	.8864	.8846
.70	.6511	.6449	.6390	.6316	.6214	.8898	.8887	.8877	.8864	.8845
.80	.6501	.6438	.6378	.6303	.6200	.8898	.8888	.8877	.8864	.8845
.90	.6492	.6428	.6367	.6291	.6187	.8899	.8888	.8878	.8864	.8844
1.0	.6483	.6418	.6357	.6280	.6174	.8899	.8889	.8878	.8864	.8844
1.2	.6467	.6401	.6338	.6259	.6152	.8901	.8890	.8879	.8865	.8844
1.4	.6452	.6384	.6320	.6240	.6131	.8902	.8891	.8880	.8866	.8845
1.7	.6431	.6361	.6296	.6214	.6103	.8904	.8893	.8882	.8867	.8846
2.0	.6412	.6340	.6274	.6191	.6078	.8907	.8895	.8884	.8869	.8848
2.3	.6394	.6321	.6253	.6169	.6054	.8909	.8898	.8886	.8871	.8850
2.6	.6376	.6303	.6234	.6148	.6033	.8912	.8900	.8888	.8874	.8852
3.0	.6355	.6279	.6209	.6123	.6006	.8915	.8903	.8892	.8877	.8855
3.5	.6329	.6252	.6181	.6093	.5974	.8920	.8908	.8896	.8881	.8860
4.0	.6304	.6227	.6154	.6065	.5945	.8925	.8913	.8901	.8886	.8864
4.5	.6281	.6202	.6129	.6039	.5918	.8930	.8918	.8906	.8891	.8870
5.0	.6259	.6179	.6105	.6014	.5892	.8935	.8923	.8912	.8897	.8875
6.0	.6217	.6135	.6059	.5967	.5844	.8945	.8933	.8922	.8907	.8885
7.0	.6178	.6095	.6018	.5924	.5800	.8956	.8944	.8932	.8917	.8896
8.0	.6142	.6057	.5979	.5884	.5759	.8966	.8955	.8943	.8929	.8908
9.0	.6107	.6021	.5942	.5846	.5721	.8977	.8966	.8954	.8940	.8920
10	.6074	.5987	.5907	.5811	.5685	.8988	.8977	.8966	.8952	.8932
12	.6013	.5923	.5842	.5744	.5617	.9009	.8999	.8988	.8975	.8956
14	.5956	.5865	.5782	.5684	.5557	.9031	.9021	.9011	.8998	.8980
17	.5878	.5785	.5702	.5602	.5475	.9065	.9055	.9046	.9035	.9018
20	.5809	.5714	.5630	.5530	.5403	.9099	.9090	.9082	.9072	.9057
23	.5744	.5648	.5562	.5462	.5335	.9131	.9124	.9117	.9107	.9095
26	.5684	.5587	.5501	.5400	.5275	.9164	.9158	.9151	.9144	.9132
30	.5611	.5513	.5426	.5326	.5201	.9208	.9203	.9198	.9192	.9183
35	.5530	.5430	.5343	.5243	.5120	.9263	.9260	.9256	.9252	.9246
40	.5456	.5356	.5269	.5170	.5049	.9318	.9316	.9314	.9312	.9309
45	.5390	.5289	.5202	.5104	.4985	.9372	.9372	.9372	.9372	.9372
50	.5329	.5228	.5141	.5044	.4928	.9425	.9427	.9429	.9431	.9433
60	.5221	.5119	.5033	.4938	.4827	.9526	.9532	.9537	.9544	.9552
70	.5127	.5025	.4941	.4849	.4742	.9625	.9635	.9644	.9654	.9668
80	.5045	.4944	.4861	.4772	.4670	.9722	.9736	.9748	.9763	.9781
90	.4972	.4872	.4791	.4705	.4608	.9817	.9835	.9850	.9869	.9892
100	.4906	.4808	.4729	.4646	.4554	.9911	.9932	.9951	.9972	1.0000
120	.4789	.4698	.4624	.4547	.4464	1.010	1.012	1.015	1.018	1.021
140	.4691	.4607	.4538	.4467	.4394	1.027	1.031	1.034	1.037	1.041
170	.4571	.4498	.4437	.4374	.4313	1.052	1.056	1.060	1.065	1.070
200	.4476	.4412	.4358	.4302	.4253	1.076	1.081	1.085	1.090	1.096
230	.4403	.4346	.4299	.4250	.4211	1.098	1.103	1.108	1.114	1.121
260	.4343	.4293	.4251	.4210	.4179	1.118	1.125	1.130	1.137	1.144
300	.4279	.4236	.4202	.4168	.4148	1.144	1.152	1.158	1.165	1.173
350	.4217	.4183	.4156	.4131	.4122	1.174	1.182	1.190	1.197	1.206
400	.4169	.4142	.4122	.4106	.4105	1.202	1.211	1.219	1.227.	1.237
450	.4132	.4112	.4097	.4088	.4095	1.227	1.237	1.246	1.255	1.265
500	.4103	.4089	.4079	.4076	.4090	1.251	1.262	1.271	1.281	1.291
600	.4066	.4061	.4061	.4068	.4093	1.293	1.306	1.317	1.328	1.340
700	.4043	.4047	.4055	.4070	.4103	1.332	1.346	1.358	1.371	1.383
800	.4029	.4041	.4055	.4077	.4117	1.367	1.382	1.395	1.409	1.422
900	.4022	.4040	.4059	.4087	.4132	1.399	1.416	1.430	1.444	1.458
1000	.4019	.4043	.4066	.4099	.4149	1.430	1.447	1.462	1.477	1.491
1200	.4022	.4055	.4086	.4126	.4183	1.486	1.505	1.520	1.537	1.551
1400	.4033	.4072	.4109	.4155	.4216	1.536	1.556	1.573	1.589	1.604
1700	.4055	.4102	.4146	.4197	.4263	1.605	1.625	1.642	1.659	1.673
2000	.4082	.4135	.4182	.4238	.4306	1.666	1.687	1.704	1.721	1.734

AL ON ZR, M2/M1=3.37   TABLE 8- 5

E,KEV K/KL=	RELATIVE TRANSV. DAMAGE STRAGGLING					RELATIVE TRANSV. IONZN. STRAGGLING				
	0.5	0.65	0.8	1.0	1.3	0.5	0.65	0.8	1.0	1.3
.20	.7395	.7358	.7323	.7276	.7207	.6744	.6695	.6648	.6585	.6493
.23	.7391	.7354	.7318	.7270	.7200	.6739	.6689	.6641	.6577	.6483
.26	.7388	.7351	.7313	.7265	.7193	.6735	.6684	.6635	.6569	.6474
.30	.7385	.7346	.7308	.7258	.7185	.6730	.6678	.6627	.6561	.6463
.35	.7381	.7341	.7302	.7251	.7176	.6724	.6671	.6619	.6551	.6451
.40	.7377	.7337	.7297	.7245	.7168	.6719	.6665	.6612	.6542	.6440
.45	.7374	.7333	.7292	.7239	.7161	.6714	.6659	.6605	.6534	.6430
.50	.7371	.7329	.7288	.7234	.7155	.6710	.6654	.6599	.6527	.6421
.60	.7365	.7322	.7280	.7224	.7143	.6702	.6644	.6588	.6513	.6405
.70	.7361	.7316	.7273	.7216	.7132	.6695	.6636	.6578	.6501	.6390
.80	.7356	.7311	.7266	.7208	.7123	.6689	.6628	.6568	.6491	.6377
.90	.7352	.7306	.7260	.7201	.7115	.6683	.6621	.6560	.6481	.6365
1.0	.7348	.7301	.7255	.7195	.7107	.6677	.6614	.6552	.6472	.6354
1.2	.7341	.7293	.7245	.7183	.7093	.6667	.6602	.6538	.6455	.6334
1.4	.7335	.7285	.7236	.7173	.7080	.6657	.6590	.6525	.6440	.6316
1.7	.7326	.7275	.7224	.7159	.7063	.6644	.6575	.6508	.6420	.6292
2.0	.7319	.7266	.7214	.7147	.7049	.6632	.6561	.6492	.6402	.6271
2.3	.7312	.7258	.7204	.7135	.7035	.6622	.6549	.6478	.6385	.6251
2.6	.7305	.7250	.7196	.7125	.7023	.6611	.6537	.6465	.6370	.6233
3.0	.7297	.7240	.7185	.7113	.7008	.6599	.6523	.6448	.6351	.6211
3.5	.7288	.7230	.7173	.7098	.6991	.6584	.6506	.6429	.6329	.6185
4.0	.7280	.7220	.7161	.7085	.6975	.6570	.6490	.6411	.6309	.6162
4.5	.7272	.7211	.7151	.7073	.6961	.6557	.6475	.6394	.6290	.6139
5.0	.7265	.7202	.7141	.7062	.6948	.6545	.6460	.6378	.6272	.6118
6.0	.7251	.7186	.7123	.7041	.6923	.6522	.6434	.6349	.6239	.6080
7.0	.7239	.7172	.7107	.7022	.6901	.6500	.6410	.6322	.6208	.6045
8.0	.7228	.7159	.7092	.7005	.6881	.6480	.6387	.6296	.6179	.6012
9.0	.7218	.7147	.7078	.6989	.6862	.6461	.6365	.6272	.6152	.5981
10	.7209	.7136	.7066	.6974	.6844	.6443	.6345	.6250	.6127	.5951
12	.7192	.7116	.7042	.6947	.6812	.6410	.6307	.6207	.6079	.5897
14	.7177	.7098	.7022	.6923	.6783	.6379	.6272	.6168	.6035	.5847
17	.7157	.7074	.6994	.6891	.6744	.6336	.6223	.6114	.5974	.5777
20	.7140	.7053	.6969	.6861	.6709	.6296	.6177	.6063	.5917	.5712
23	.7126	.7035	.6947	.6835	.6677	.6260	.6136	.6017	.5865	.5652
26	.7113	.7018	.6927	.6811	.6647	.6226	.6097	.5973	.5816	.5596
30	.7097	.6998	.6902	.6781	.6610	.6183	.6047	.5918	.5754	.5525
35	.7080	.6975	.6874	.6746	.6568	.6133	.5989	.5852	.5680	.5441
40	.7065	.6955	.6849	.6714	.6528	.6086	.5934	.5791	.5611	.5363
45	.7051	.6936	.6825	.6685	.6491	.6041	.5882	.5732	.5545	.5289
50	.7039	.6918	.6803	.6657	.6457	.5998	.5832	.5677	.5483	.5218
60	.7021	.6890	.6766	.6609	.6395	.5919	.5739	.5572	.5365	.5085
70	.7005	.6864	.6731	.6564	.6338	.5844	.5651	.5474	.5256	.4962
80	.6989	.6839	.6698	.6521	.6283	.5773	.5569	.5381	.5153	.4848
90	.6973	.6815	.6665	.6480	.6231	.5706	.5491	.5294	.5056	.4741
100	.6957	.6791	.6634	.6440	.6181	.5641	.5416	.5211	.4964	.4639
120	.6928	.6745	.6574	.6363	.6086	.5518	.5275	.5054	.4789	.4448
140	.6898	.6700	.6515	.6290	.5996	.5404	.5144	.4909	.4628	.4274
170	.6851	.6632	.6430	.6185	.5870	.5242	.4963	.4710	.4411	.4042
200	.6802	.6566	.6348	.6086	.5753	.5092	.4797	.4530	.4216	.3838
230	.6752	.6499	.6266	.5988	.5640	.4943	.4634	.4356	.4032	.3649
260	.6701	.6433	.6187	.5895	.5533	.4803	.4483	.4196	.3864	.3479
300	.6633	.6346	.6085	.5777	.5401	.4629	.4297	.4001	.3662	.3276
350	.6548	.6242	.5965	.5640	.5248	.4430	.4087	.3783	.3439	.3054
400	.6464	.6141	.5850	.5511	.5108	.4248	.3898	.3589	.3242	.2861
450	.6380	.6044	.5741	.5390	.4978	.4081	.3726	.3415	.3067	.2692
500	.6298	.5949	.5637	.5277	.4857	.3926	.3569	.3256	.2910	.2541
600	.6131	.5763	.5435	.5060	.4632	.3638	.3284	.2976	.2638	.2286
700	.5973	.5590	.5251	.4866	.4434	.3389	.3041	.2740	.2412	.2076
800	.5824	.5431	.5083	.4691	.4257	.3171	.2830	.2537	.2220	.1901
900	.5684	.5282	.4929	.4532	.4098	.2978	.2646	.2362	.2055	.1751
1000	.5551	.5143	.4786	.4387	.3954	.2807	.2484	.2208	.1912	.1622
1200	.5304	.4891	.4530	.4129	.3702	.2517	.2212	.1952	.1676	.1412
1400	.5081	.4666	.4305	.3907	.3488	.2279	.1992	.1748	.1491	.1249
1700	.4780	.4369	.4013	.3622	.3219	.1996	.1734	.1512	.1280	.1066
2000	.4511	.4109	.3761	.3383	.2996	.1776	.1537	.1335	.1124	.0933

AL ON ZR, M2/M1=3.37

TABLE 8- 6

	RANGE SKEWNESS					DAMAGE SKEWNESS				
K/KL=	0.5	0.65	0.8	1.0	1.3	0.5	0.65	0.8	1.0	1.3
E,KEV										
.20	.3962	.3864	.3769	.3644	.3461	.8460	.8386	.8314	.8220	.8085
.23	.3953	.3854	.3756	.3628	.3441	.8454	.8378	.8304	.8208	.8070
.26	.3945	.3844	.3744	.3614	.3423	.8448	.8371	.8295	.8198	.8057
.30	.3936	.3832	.3730	.3597	.3401	.8441	.8362	.8285	.8185	.8042
.35	.3926	.3820	.3715	.3578	.3378	.8434	.8353	.8274	.8171	.8024
.40	.3917	.3808	.3701	.3561	.3357	.8427	.8344	.8263	.8159	.8009
.45	.3909	.3798	.3689	.3546	.3338	.8421	.8336	.8254	.8148	.7995
.50	.3902	.3788	.3677	.3532	.3320	.8416	.8329	.8246	.8137	.7982
.60	.3888	.3771	.3656	.3506	.3288	.8405	.8316	.8230	.8119	.7959
.70	.3876	.3755	.3637	.3483	.3260	.8396	.8305	.8216	.8102	.7939
.80	.3864	.3741	.3620	.3463	.3234	.8388	.8294	.8204	.8087	.7920
.90	.3854	.3728	.3604	.3444	.3210	.8380	.8285	.8192	.8073	.7903
1.0	.3844	.3716	.3590	.3426	.3188	.8373	.8276	.8181	.8060	.7887
1.2	.3827	.3694	.3563	.3394	.3149	.8361	.8260	.8162	.8037	.7858
1.4	.3811	.3674	.3540	.3365	.3113	.8349	.8245	.8144	.8016	.7832
1.7	.3789	.3646	.3507	.3326	.3065	.8333	.8224	.8120	.7987	.7797
2.0	.3769	.3621	.3477	.3291	.3022	.8317	.8205	.8098	.7960	.7765
2.3	.3751	.3599	.3451	.3259	.2982	.8303	.8188	.8077	.7936	.7736
2.6	.3734	.3578	.3426	.3229	.2946	.8289	.8171	.8058	.7913	.7709
3.0	.3712	.3551	.3394	.3191	.2900	.8272	.8150	.8033	.7884	.7674
3.5	.3687	.3519	.3357	.3147	.2846	.8251	.8125	.8004	.7850	.7634
4.0	.3662	.3489	.3321	.3105	.2796	.8231	.8101	.7977	.7818	.7596
4.5	.3638	.3460	.3287	.3065	.2748	.8212	.8079	.7950	.7787	.7559
5.0	.3615	.3432	.3255	.3027	.2702	.8194	.8057	.7925	.7758	.7524
6.0	.3573	.3381	.3195	.2957	.2618	.8166	.8021	.7882	.7706	.7462
7.0	.3532	.3331	.3138	.2890	.2539	.8138	.7985	.7839	.7655	.7401
8.0	.3492	.3283	.3082	.2826	.2463	.8108	.7948	.7796	.7605	.7343
9.0	.3453	.3236	.3029	.2764	.2390	.8077	.7911	.7753	.7556	.7285
10	.3415	.3191	.2976	.2704	.2319	.8045	.7873	.7711	.7507	.7229
12	.3342	.3105	.2878	.2590	.2187	.7974	.7794	.7623	.7410	.7120
14	.3271	.3021	.2783	.2482	.2062	.7904	.7716	.7537	.7315	.7015
17	.3167	.2899	.2645	.2326	.1882	.7800	.7601	.7412	.7178	.6863
20	.3064	.2780	.2512	.2176	.1712	.7700	.7490	.7292	.7047	.6719
23	.2966	.2666	.2385	.2033	.1549	.7614	.7392	.7183	.6925	.6583
26	.2869	.2555	.2261	.1895	.1394	.7529	.7296	.7077	.6808	.6453
30	.2743	.2411	.2101	.1718	.1195	.7419	.7171	.6939	.6656	.6285
35	.2590	.2237	.1910	.1506	.0959	.7282	.7018	.6773	.6474	.6086
40	.2441	.2069	.1725	.1304	.0735	.7147	.6869	.6611	.6298	.5895
45	.2295	.1906	.1547	.1109	.0521	.7014	.6723	.6454	.6129	.5712
50	.2154	.1748	.1376	.0923	.0317	.6882	.6580	.6301	.5964	.5537
60	.1900	.1456	.1055	.0572	-.0069	.6603	.6284	.5991	.5638	.5194
70	.1649	.1176	.0750	.0242	-.0428	.6341	.6008	.5702	.5336	.4878
80	.1401	.0903	.0458	-.0071	-.0764	.6095	.5749	.5432	.5054	.4584
90	.1155	.0638	.0178	-.0369	-.1082	.5866	.5507	.5180	.4791	.4310
100	.0912	.0381	-.0093	-.0655	-.1384	.5650	.5280	.4943	.4543	.4052
120	.0383	-.0154	-.0637	-.1213	-.1961	.5266	.4872	.4514	.4093	.3580
140	-.0110	-.0648	-.1137	-.1724	-.2486	.4918	.4503	.4127	.3686	.3156
170	-.0779	-.1317	-.1813	-.2413	-.3192	.4447	.4005	.3606	.3141	.2589
200	-.1368	-.1909	-.2413	-.3025	-.3819	.4022	.3558	.3140	.2656	.2086
230	-.1851	-.2410	-.2929	-.3561	-.4373	.3617	.3137	.2706	.2208	.1627
260	-.2282	-.2860	-.3397	-.4046	-.4876	.3244	.2750	.2308	.1799	.1209
300	-.2796	-.3400	-.3959	-.4633	-.5485	.2789	.2280	.1826	.1306	.0707
350	-.3362	-.3999	-.4585	-.5288	-.6168	.2277	.1753	.1287	.0756	.0150
400	-.3863	-.4532	-.5146	-.5876	-.6781	.1816	.1281	.0806	.0266	-.0344
450	-.4315	-.5015	-.5654	-.6410	-.7339	.1398	.0853	.0371	-.0176	-.0788
500	-.4727	-.5457	-.6120	-.6901	-.7851	.1015	.0463	-.0026	-.0577	-.1191
600	-.5434	-.6222	-.6935	-.7767	-.8761	.0337	-.0228	-.0726	-.1284	-.1896
700	-.6060	-.6900	-.7656	-.8532	-.9565	-.0252	-.0826	-.1329	-.1891	-.2499
800	-.6629	-.7513	-.8306	-.9221	-1.029	-.0772	-.1351	-.1859	-.2422	-.3026
900	-.7155	-.8077	-.8903	-.9851	-1.095	-.1237	-.1820	-.2329	-.2893	-.3491
1000	-.7646	-.8602	-.9455	-1.043	-1.156	-.1657	-.2243	-.2753	-.3316	-.3908
1200	-.8551	-.9560	-1.046	-1.148	-1.265	-.2392	-.2979	-.3488	-.4047	-.4628
1400	-.9376	-1.042	-1.136	-1.242	-1.363	-.3019	-.3604	-.4110	-.4664	-.5233
1700	-1.050	-1.159	-1.256	-1.366	-1.491	-.3814	-.4392	-.4891	-.5434	-.5987
2000	-1.153	-1.264	-1.363	-1.476	-1.605	-.4481	-.5049	-.5538	-.6070	-.6608

E,KEV	IONIZATION SKEWNESS 0.5	0.65	0.8	1.0	1.3	RANGE KURTOSIS 0.5	0.65	0.8	1.0	1.3
.20	.9498	.9434	.9372	.9291	.9176	3.373	3.343	3.315	3.279	3.230
.23	.9492	.9427	.9363	.9281	.9164	3.370	3.340	3.311	3.275	3.224
.26	.9487	.9420	.9356	.9272	.9153	3.367	3.337	3.308	3.271	3.219
.30	.9481	.9413	.9347	.9261	.9139	3.364	3.333	3.304	3.266	3.214
.35	.9475	.9405	.9337	.9249	.9125	3.361	3.329	3.299	3.261	3.208
.40	.9469	.9397	.9328	.9239	.9112	3.358	3.326	3.295	3.256	3.202
.45	.9464	.9391	.9320	.9229	.9100	3.355	3.322	3.291	3.252	3.197
.50	.9459	.9384	.9312	.9220	.9089	3.353	3.319	3.288	3.248	3.192
.60	.9450	.9373	.9299	.9204	.9069	3.348	3.314	3.281	3.240	3.184
.70	.9442	.9363	.9287	.9190	.9052	3.344	3.309	3.276	3.234	3.177
.80	.9435	.9354	.9276	.9177	.9036	3.340	3.305	3.271	3.228	3.170
.90	.9428	.9346	.9266	.9165	.9021	3.337	3.300	3.266	3.223	3.164
1.0	.9422	.9338	.9257	.9153	.9008	3.334	3.297	3.262	3.218	3.159
1.2	.9410	.9323	.9240	.9133	.8983	3.328	3.290	3.254	3.210	3.149
1.4	.9400	.9310	.9224	.9115	.8961	3.322	3.284	3.247	3.202	3.140
1.7	.9386	.9293	.9203	.9090	.8931	3.315	3.275	3.238	3.192	3.129
2.0	.9372	.9276	.9184	.9067	.8903	3.308	3.267	3.229	3.182	3.119
2.3	.9361	.9261	.9166	.9046	.8879	3.302	3.260	3.222	3.174	3.109
2.6	.9349	.9247	.9150	.9027	.8856	3.296	3.254	3.215	3.166	3.101
3.0	.9335	.9229	.9129	.9003	.8827	3.289	3.246	3.206	3.156	3.091
3.5	.9317	.9208	.9104	.8973	.8792	3.281	3.237	3.196	3.145	3.079
4.0	.9300	.9187	.9080	.8946	.8760	3.273	3.228	3.186	3.135	3.068
4.5	.9284	.9167	.9057	.8919	.8729	3.265	3.220	3.177	3.126	3.058
5.0	.9267	.9148	.9035	.8894	.8699	3.258	3.212	3.169	3.117	3.049
6.0	.9238	.9112	.8994	.8846	.8644	3.245	3.197	3.153	3.101	3.032
7.0	.9208	.9077	.8954	.8801	.8592	3.233	3.184	3.139	3.086	3.016
8.0	.9179	.9042	.8915	.8757	.8541	3.221	3.171	3.126	3.072	3.002
9.0	.9150	.9009	.8877	.8714	.8493	3.210	3.159	3.114	3.059	2.989
10	.9121	.8976	.8840	.8672	.8446	3.199	3.148	3.102	3.047	2.977
12	.9066	.8912	.8769	.8593	.8357	3.179	3.127	3.080	3.025	2.955
14	.9012	.8850	.8700	.8517	.8272	3.161	3.107	3.060	3.004	2.935
17	.8932	.8759	.8601	.8408	.8152	3.135	3.081	3.033	2.977	2.909
20	.8854	.8672	.8505	.8304	.8038	3.112	3.057	3.009	2.954	2.887
23	.8780	.8586	.8411	.8202	.7927	3.090	3.034	2.986	2.931	2.867
26	.8707	.8502	.8321	.8105	.7822	3.069	3.013	2.965	2.911	2.848
30	.8612	.8395	.8206	.7981	.7690	3.044	2.987	2.940	2.888	2.828
35	.8496	.8267	.8069	.7835	.7535	3.015	2.959	2.913	2.863	2.807
40	.8383	.8145	.7940	.7698	.7390	2.990	2.935	2.890	2.842	2.790
45	.8273	.8029	.7817	.7569	.7255	2.968	2.913	2.869	2.824	2.776
50	.8166	.7918	.7700	.7446	.7127	2.947	2.894	2.852	2.809	2.766
60	.7956	.7711	.7479	.7213	.6885	2.917	2.863	2.823	2.784	2.749
70	.7757	.7517	.7276	.7002	.6667	2.889	2.838	2.800	2.766	2.739
80	.7570	.7335	.7087	.6808	.6470	2.863	2.815	2.782	2.754	2.736
90	.7393	.7163	.6912	.6630	.6291	2.837	2.795	2.767	2.746	2.738
100	.7224	.6999	.6747	.6464	.6126	2.811	2.776	2.755	2.742	2.743
120	.6887	.6666	.6425	.6153	.5824	2.735	2.726	2.726	2.735	2.759
140	.6587	.6370	.6142	.5882	.5565	2.671	2.688	2.709	2.739	2.784
170	.6200	.5987	.5779	.5537	.5240	2.602	2.653	2.702	2.761	2.835
200	.5877	.5667	.5476	.5250	.4972	2.560	2.641	2.713	2.798	2.898
230	.5627	.5418	.5232	.5015	.4754	2.569	2.663	2.748	2.848	2.969
260	.5415	.5206	.5024	.4814	.4571	2.593	2.698	2.793	2.905	3.044
300	.5176	.4969	.4791	.4589	.4367	2.638	2.755	2.861	2.989	3.151
350	.4929	.4725	.4552	.4361	.4163	2.707	2.837	2.956	3.102	3.290
400	.4726	.4526	.4357	.4176	.4000	2.785	2.926	3.057	3.219	3.431
450	.4555	.4359	.4196	.4025	.3870	2.866	3.018	3.161	3.338	3.574
500	.4408	.4219	.4061	.3900	.3765	2.948	3.112	3.266	3.458	3.717
600	.4169	.3996	.3855	.3718	.3621	3.100	3.287	3.465	3.687	3.989
700	.3983	.3827	.3704	.3589	.3524	3.253	3.465	3.666	3.918	4.260
800	.3835	.3697	.3590	.3496	.3460	3.406	3.643	3.867	4.148	4.531
900	.3715	.3594	.3503	.3429	.3417	3.558	3.820	4.068	4.378	4.799
1000	.3617	.3513	.3437	.3381	.3391	3.709	3.995	4.267	4.606	5.064
1200	.3467	.3396	.3350	.3326	.3373	4.002	4.340	4.658	5.053	5.583
1400	.3361	.3322	.3303	.3308	.3383	4.286	4.674	5.039	5.489	6.087
1700	.3258	.3262	.3280	.3323	.3431	4.693	5.156	5.589	6.119	6.814
2000	.3197	.3243	.3294	.3370	.3500	5.079	5.615	6.115	6.722	7.508

E,KEV	DAMAGE KURTOSIS					IONIZATION KURTOSIS				
K/KL=	0.5	0.65	0.8	1.0	1.3	0.5	0.65	0.8	1.0	1.3
.20	4.363	4.321	4.281	4.231	4.159	4.506	4.464	4.423	4.372	4.300
.23	4.359	4.316	4.275	4.224	4.151	4.502	4.459	4.417	4.365	4.291
.26	4.355	4.312	4.270	4.217	4.143	4.498	4.454	4.412	4.359	4.284
.30	4.350	4.306	4.264	4.210	4.134	4.493	4.448	4.405	4.351	4.275
.35	4.345	4.300	4.256	4.202	4.125	4.488	4.442	4.398	4.343	4.265
.40	4.340	4.294	4.250	4.194	4.116	4.483	4.436	4.392	4.335	4.257
.45	4.336	4.289	4.244	4.187	4.108	4.478	4.431	4.386	4.328	4.249
.50	4.331	4.284	4.238	4.181	4.101	4.474	4.426	4.380	4.322	4.241
.60	4.324	4.275	4.228	4.169	4.088	4.466	4.417	4.370	4.311	4.228
.70	4.317	4.267	4.219	4.159	4.076	4.459	4.409	4.361	4.300	4.216
.80	4.310	4.259	4.211	4.150	4.065	4.453	4.401	4.353	4.291	4.206
.90	4.304	4.252	4.203	4.141	4.055	4.447	4.394	4.345	4.282	4.196
1.0	4.298	4.246	4.196	4.133	4.046	4.441	4.388	4.337	4.274	4.187
1.2	4.288	4.234	4.182	4.118	4.029	4.431	4.376	4.324	4.259	4.170
1.4	4.278	4.223	4.170	4.105	4.014	4.421	4.365	4.312	4.246	4.155
1.7	4.265	4.208	4.154	4.086	3.994	4.408	4.350	4.296	4.228	4.135
2.0	4.252	4.193	4.138	4.070	3.975	4.395	4.336	4.280	4.211	4.117
2.3	4.239	4.180	4.124	4.054	3.958	4.384	4.323	4.267	4.196	4.100
2.6	4.227	4.167	4.110	4.039	3.943	4.373	4.311	4.254	4.182	4.085
3.0	4.212	4.151	4.093	4.021	3.923	4.359	4.296	4.237	4.164	4.066
3.5	4.195	4.133	4.074	4.000	3.900	4.343	4.278	4.218	4.144	4.043
4.0	4.179	4.115	4.055	3.980	3.879	4.327	4.262	4.200	4.125	4.023
4.5	4.164	4.099	4.038	3.962	3.859	4.313	4.246	4.183	4.106	4.003
5.0	4.150	4.084	4.022	3.944	3.840	4.299	4.230	4.167	4.089	3.985
6.0	4.127	4.058	3.993	3.913	3.806	4.272	4.202	4.137	4.057	3.951
7.0	4.106	4.034	3.966	3.884	3.775	4.247	4.175	4.108	4.027	3.919
8.0	4.084	4.010	3.941	3.856	3.745	4.224	4.150	4.082	3.999	3.890
9.0	4.063	3.987	3.916	3.830	3.717	4.201	4.126	4.057	3.973	3.863
10	4.042	3.964	3.892	3.804	3.690	4.180	4.103	4.033	3.948	3.837
12	3.996	3.917	3.843	3.754	3.639	4.139	4.060	3.988	3.901	3.788
14	3.952	3.872	3.798	3.708	3.592	4.101	4.020	3.946	3.858	3.743
17	3.892	3.810	3.735	3.644	3.527	4.049	3.964	3.889	3.799	3.683
20	3.837	3.755	3.678	3.586	3.470	4.000	3.914	3.837	3.745	3.628
23	3.792	3.707	3.629	3.535	3.416	3.955	3.866	3.787	3.695	3.577
26	3.751	3.663	3.583	3.487	3.368	3.912	3.822	3.742	3.649	3.530
30	3.700	3.609	3.527	3.429	3.309	3.860	3.767	3.686	3.592	3.473
35	3.641	3.548	3.463	3.363	3.242	3.800	3.705	3.623	3.528	3.410
40	3.587	3.491	3.404	3.303	3.182	3.745	3.649	3.565	3.470	3.352
45	3.536	3.438	3.351	3.249	3.128	3.694	3.597	3.513	3.418	3.300
50	3.487	3.389	3.300	3.198	3.078	3.648	3.549	3.465	3.369	3.253
60	3.388	3.290	3.202	3.102	2.986	3.570	3.466	3.379	3.283	3.167
70	3.300	3.203	3.117	3.019	2.908	3.498	3.392	3.304	3.207	3.094
80	3.222	3.127	3.043	2.947	2.840	3.431	3.324	3.237	3.141	3.030
90	3.154	3.060	2.978	2.884	2.780	3.366	3.261	3.175	3.082	2.973
100	3.092	3.001	2.920	2.828	2.728	3.304	3.203	3.119	3.028	2.923
120	2.993	2.903	2.824	2.736	2.642	3.164	3.081	3.012	2.933	2.835
140	2.911	2.823	2.746	2.662	2.573	3.041	2.976	2.920	2.853	2.762
170	2.810	2.726	2.653	2.573	2.492	2.886	2.846	2.806	2.754	2.673
200	2.729	2.649	2.579	2.505	2.430	2.764	2.742	2.716	2.675	2.602
230	2.657	2.583	2.519	2.450	2.384	2.698	2.677	2.653	2.614	2.546
260	2.597	2.528	2.469	2.406	2.347	2.648	2.626	2.601	2.564	2.499
300	2.530	2.468	2.416	2.361	2.310	2.598	2.573	2.546	2.508	2.448
350	2.463	2.410	2.365	2.318	2.277	2.552	2.522	2.492	2.452	2.396
400	2.411	2.365	2.326	2.288	2.256	2.518	2.482	2.448	2.407	2.355
450	2.369	2.330	2.298	2.266	2.242	2.490	2.450	2.413	2.370	2.320
500	2.337	2.304	2.277	2.252	2.234	2.467	2.422	2.382	2.339	2.292
600	2.297	2.274	2.256	2.240	2.234	2.418	2.371	2.331	2.288	2.247
700	2.273	2.259	2.248	2.241	2.244	2.378	2.331	2.291	2.250	2.214
800	2.261	2.253	2.249	2.249	2.260	2.345	2.298	2.260	2.221	2.188
900	2.255	2.254	2.256	2.262	2.280	2.317	2.271	2.234	2.197	2.168
1000	2.255	2.260	2.266	2.278	2.302	2.292	2.248	2.212	2.177	2.153
1200	2.265	2.280	2.295	2.317	2.350	2.250	2.210	2.178	2.148	2.129
1400	2.284	2.307	2.330	2.359	2.400	2.215	2.180	2.153	2.128	2.114
1700	2.321	2.355	2.387	2.426	2.476	2.172	2.146	2.126	2.108	2.101
2000	2.364	2.407	2.446	2.493	2.550	2.136	2.119	2.106	2.096	2.095

K/KL=	_ REFLECTION COEFFICIENT _					_ SPUTTERING EFFICIENCY _				
E,KEV	0.5	0.65	0.8	1.0	1.3	0.5	0.65	0.8	1.0	1.3
.20	.1316	.1301	.1287	.1268	.1240	.0670	.0642	.0616	.0583	.0538
.23	.1314	.1299	.1284	.1264	.1236	.0668	.0639	.0612	.0579	.0533
.26	.1312	.1297	.1281	.1262	.1233	.0665	.0636	.0609	.0575	.0529
.30	.1310	.1294	.1278	.1258	.1229	.0662	.0633	.0605	.0571	.0524
.35	.1307	.1291	.1275	.1254	.1224	.0659	.0629	.0601	.0566	.0519
.40	.1304	.1288	.1271	.1250	.1220	.0656	.0626	.0597	.0562	.0514
.45	.1302	.1285	.1268	.1247	.1216	.0654	.0623	.0594	.0558	.0509
.50	.1300	.1282	.1265	.1244	.1213	.0651	.0620	.0591	.0554	.0505
.60	.1295	.1277	.1260	.1238	.1206	.0647	.0615	.0585	.0548	.0498
.70	.1291	.1273	.1255	.1232	.1200	.0643	.0610	.0580	.0542	.0492
.80	.1287	.1268	.1251	.1227	.1194	.0639	.0606	.0575	.0537	.0487
.90	.1283	.1264	.1246	.1223	.1189	.0636	.0602	.0571	.0533	.0482
1.0	.1280	.1260	.1242	.1218	.1184	.0633	.0599	.0567	.0529	.0477
1.2	.1273	.1253	.1234	.1210	.1175	.0627	.0592	.0560	.0521	.0469
1.4	.1266	.1246	.1227	.1202	.1166	.0622	.0587	.0554	.0514	.0462
1.7	.1257	.1237	.1217	.1191	.1154	.0615	.0579	.0546	.0506	.0452
2.0	.1249	.1228	.1207	.1181	.1143	.0609	.0572	.0539	.0498	.0444
2.3	.1241	.1219	.1198	.1171	.1133	.0603	.0566	.0532	.0491	.0437
2.6	.1233	.1211	.1189	.1162	.1124	.0598	.0561	.0526	.0485	.0430
3.0	.1223	.1200	.1179	.1151	.1112	.0592	.0554	.0519	.0477	.0422
3.5	.1211	.1188	.1166	.1137	.1097	.0584	.0546	.0511	.0468	.0413
4.0	.1200	.1176	.1154	.1125	.1084	.0577	.0539	.0503	.0460	.0405
4.5	.1189	.1165	.1142	.1113	.1071	.0571	.0532	.0496	.0453	.0398
5.0	.1179	.1154	.1131	.1101	.1059	.0565	.0525	.0489	.0446	.0391
6.0	.1159	.1134	.1110	.1079	.1037	.0554	.0514	.0477	.0434	.0378
7.0	.1140	.1115	.1090	.1059	.1016	.0543	.0503	.0466	.0422	.0367
8.0	.1123	.1097	.1072	.1040	.0996	.0534	.0493	.0456	.0412	.0357
9.0	.1106	.1079	.1054	.1022	.0978	.0525	.0484	.0447	.0403	.0348
10	.1090	.1063	.1037	.1005	.0960	.0517	.0476	.0439	.0395	.0340
12	.1060	.1032	.1006	.0973	.0927	.0502	.0460	.0423	.0379	.0324
14	.1033	.1004	.0977	.0943	.0898	.0488	.0446	.0408	.0365	.0311
17	.0995	.0966	.0938	.0904	.0857	.0469	.0427	.0390	.0346	.0293
20	.0961	.0931	.0903	.0868	.0821	.0452	.0410	.0373	.0330	.0278
23	.0930	.0899	.0870	.0835	.0788	.0437	.0395	.0358	.0316	.0264
26	.0901	.0869	.0840	.0804	.0757	.0424	.0382	.0345	.0302	.0252
30	.0865	.0833	.0803	.0768	.0720	.0407	.0365	.0329	.0287	.0238
35	.0826	.0793	.0763	.0727	.0679	.0389	.0347	.0311	.0270	.0222
40	.0791	.0757	.0726	.0690	.0643	.0372	.0331	.0296	.0255	.0209
45	.0759	.0725	.0694	.0657	.0611	.0358	.0317	.0282	.0242	.0197
50	.0730	.0695	.0664	.0628	.0581	.0344	.0304	.0269	.0230	.0186
60	.0679	.0643	.0612	.0575	.0529	.0320	.0281	.0247	.0209	.0168
70	.0635	.0599	.0567	.0530	.0485	.0299	.0261	.0228	.0192	.0153
80	.0597	.0560	.0529	.0492	.0447	.0282	.0244	.0212	.0178	.0141
90	.0563	.0526	.0495	.0458	.0414	.0266	.0229	.0198	.0165	.0130
100	.0532	.0496	.0465	.0428	.0385	.0252	.0216	.0186	.0154	.0120
120	.0478	.0443	.0413	.0378	.0337	.0228	.0194	.0166	.0136	.0105
140	.0433	.0400	.0371	.0337	.0298	.0208	.0176	.0149	.0121	.0093
170	.0378	.0347	.0319	.0287	.0251	.0184	.0154	.0130	.0104	.0079
200	.0333	.0304	.0278	.0248	.0214	.0165	.0137	.0114	.0091	.0068
230	.0298	.0271	.0246	.0217	.0186	.0149	.0123	.0102	.0081	.0060
260	.0269	.0243	.0219	.0191	.0162	.0135	.0111	.0092	.0072	.0054
300	.0237	.0213	.0190	.0164	.0137	.0120	.0099	.0081	.0064	.0047
350	.0205	.0182	.0161	.0136	.0112	.0105	.0086	.0070	.0055	.0041
400	.0179	.0157	.0137	.0114	.0092	.0093	.0076	.0061	.0048	.0036
450	.0158	.0137	.0118	.0097	.0077	.0083	.0068	.0054	.0043	.0032
500	.0140	.0121	.0103	.0083	.0064	.0075	.0061	.0049	.0039	.0029
600	.0113	.0096	.0079	.0062	.0047	.0064	.0052	.0041	.0032	.0024
700	.0092	.0077	.0063	.0047	.0035	.0055	.0045	.0035	.0027	.0020
800	.0077	.0063	.0050	.0036	.0026	.0049	.0039	.0031	.0024	.0017
900	.0064	.0052	.0040	.0028	.0019	.0044	.0035	.0028	.0021	.0015
1000	.0054	.0043	.0032	.0022	.0015	.0040	.0032	.0025	.0019	.0014
1200	.0039	.0030	.0021	.0014	.0008	.0033	.0026	.0021	.0015	.0011
1400	.0028	.0021	.0014	.0008	.0005	.0028	.0022	.0018	.0013	.0009
1700	.0017	.0013	.0008	.0004	.0002	.0023	.0018	.0014	.0010	.0007
2000	.0011	.0008	.0005	.0003	.0001	.0018	.0014	.0011	.0008	.0006

	IONIZATION DEFICIENCY					SPUTTERING YIELD ALPHA				
K/KL=	0.5	0.65	0.8	1.0	1.3	0.5	0.65	0.8	1.0	1.3
E,KEV										
.20	.0050	.0063	.0075	.0089	.0107	.5322	.5197	.5077	.4921	.4702
.23	.0051	.0064	.0076	.0090	.0109	.5313	.5185	.5062	.4904	.4680
.26	.0052	.0065	.0077	.0092	.0111	.5304	.5174	.5049	.4889	.4661
.30	.0053	.0066	.0079	.0093	.0112	.5295	.5162	.5034	.4871	.4639
.35	.0054	.0068	.0080	.0095	.0114	.5284	.5148	.5017	.4851	.4614
.40	.0055	.0069	.0081	.0096	.0116	.5274	.5136	.5003	.4833	.4592
.45	.0056	.0070	.0083	.0098	.0117	.5266	.5125	.4990	.4817	.4573
.50	.0057	.0071	.0084	.0099	.0118	.5258	.5115	.4978	.4803	.4556
.60	.0059	.0073	.0086	.0101	.0121	.5245	.5098	.4957	.4778	.4525
.70	.0060	.0074	.0087	.0103	.0122	.5233	.5083	.4940	.4756	.4499
.80	.0061	.0075	.0088	.0104	.0124	.5223	.5070	.4924	.4737	.4476
.90	.0062	.0076	.0090	.0105	.0125	.5214	.5059	.4910	.4721	.4456
1.0	.0063	.0077	.0091	.0106	.0126	.5206	.5049	.4898	.4705	.4437
1.2	.0064	.0079	.0092	.0108	.0128	.5193	.5031	.4876	.4679	.4405
1.4	.0065	.0080	.0094	.0110	.0130	.5181	.5015	.4857	.4656	.4377
1.7	.0067	.0082	.0096	.0112	.0132	.5166	.4996	.4833	.4627	.4341
2.0	.0068	.0083	.0097	.0113	.0133	.5154	.4980	.4812	.4603	.4311
2.3	.0069	.0085	.0098	.0114	.0134	.5144	.4966	.4795	.4582	.4285
2.6	.0070	.0085	.0099	.0115	.0135	.5135	.4954	.4780	.4563	.4262
3.0	.0071	.0087	.0100	.0116	.0136	.5125	.4940	.4763	.4541	.4235
3.5	.0072	.0088	.0102	.0118	.0137	.5113	.4924	.4743	.4517	.4206
4.0	.0073	.0088	.0102	.0118	.0137	.5104	.4911	.4727	.4496	.4180
4.5	.0073	.0089	.0103	.0119	.0137	.5095	.4898	.4711	.4477	.4156
5.0	.0074	.0090	.0104	.0119	.0138	.5087	.4887	.4698	.4459	.4135
6.0	.0075	.0091	.0104	.0120	.0138	.5073	.4868	.4673	.4429	.4097
7.0	.0076	.0091	.0105	.0120	.0138	.5060	.4851	.4652	.4402	.4065
8.0	.0076	.0092	.0105	.0120	.0137	.5050	.4835	.4632	.4378	.4035
9.0	.0076	.0092	.0105	.0120	.0137	.5040	.4822	.4615	.4356	.4009
10	.0077	.0092	.0105	.0120	.0136	.5031	.4809	.4599	.4336	.3985
12	.0077	.0092	.0105	.0119	.0134	.5016	.4788	.4572	.4301	.3942
14	.0077	.0092	.0104	.0118	.0133	.5003	.4769	.4547	.4271	.3905
17	.0077	.0091	.0103	.0116	.0130	.4984	.4742	.4513	.4229	.3855
20	.0077	.0091	.0102	.0114	.0127	.4965	.4717	.4482	.4191	.3810
23	.0076	.0090	.0101	.0112	.0124	.4947	.4692	.4452	.4155	.3769
26	.0076	.0089	.0099	.0110	.0121	.4928	.4668	.4423	.4122	.3731
30	.0075	.0087	.0097	.0108	.0117	.4904	.4637	.4387	.4080	.3684
35	.0074	.0085	.0095	.0104	.0113	.4874	.4601	.4344	.4031	.3631
40	.0073	.0084	.0092	.0101	.0109	.4844	.4565	.4304	.3986	.3582
45	.0072	.0082	.0090	.0098	.0105	.4816	.4532	.4266	.3944	.3536
50	.0071	.0080	.0088	.0095	.0101	.4788	.4500	.4230	.3904	.3494
60	.0069	.0077	.0083	.0090	.0094	.4734	.4438	.4161	.3828	.3414
70	.0067	.0074	.0080	.0085	.0088	.4683	.4381	.4098	.3760	.3344
80	.0066	.0071	.0076	.0080	.0083	.4633	.4326	.4040	.3697	.3280
90	.0064	.0068	.0072	.0076	.0078	.4585	.4275	.3985	.3640	.3223
100	.0062	.0065	.0069	.0073	.0074	.4539	.4226	.3934	.3586	.3171
120	.0057	.0060	.0063	.0066	.0066	.4445	.4128	.3833	.3485	.3077
140	.0052	.0055	.0058	.0060	.0060	.4359	.4041	.3745	.3397	.2997
170	.0046	.0048	.0050	.0053	.0051	.4244	.3926	.3631	.3286	.2897
200	.0041	.0042	.0045	.0047	.0045	.4143	.3829	.3535	.3194	.2816
230	.0037	.0038	.0040	.0042	.0040	.4054	.3751	.3463	.3126	.2754
260	.0035	.0035	.0036	.0038	.0036	.3975	.3683	.3401	.3069	.2701
300	.0032	.0031	.0032	.0034	.0031	.3882	.3605	.3329	.3004	.2643
350	.0029	.0027	.0027	.0029	.0026	.3782	.3520	.3253	.2935	.2582
400	.0026	.0023	.0023	.0025	.0022	.3695	.3447	.3188	.2877	.2530
450	.0024	.0020	.0020	.0022	.0019	.3619	.3383	.3131	.2826	.2486
500	.0022	.0018	.0017	.0019	.0016	.3552	.3325	.3080	.2781	.2448
600	.0019	.0012	.0012	.0013	.0011	.3442	.3218	.2982	.2694	.2378
700	.0016	.0009	.0007	.0008	.0007	.3351	.3130	.2902	.2622	.2321
800	.0014	.0005	.0004	.0005	.0004	.3275	.3056	.2834	.2562	.2273
900	.0012	.0002	.0002	.0002	.0002	.3210	.2993	.2777	.2511	.2233
1000	.0011	0.0000	0.0000	0.0000	0.0000	.3154	.2939	.2729	.2469	.2199
1200	.0008	0.0000	0.0000	0.0000	0.0000	.3061	.2853	.2651	.2400	.2145
1400	.0005	0.0000	0.0000	0.0000	0.0000	.2988	.2788	.2594	.2350	.2103
1700	.0002	0.0000	0.0000	0.0000	0.0000	.2903	.2718	.2533	.2298	.2057
2000	0.0000	0.0000	0.0000	0.0000	0.0000	.2838	.2671	.2494	.2265	.2025

AL ON ZR, M2/M1=3.37

2000 KEV

200 KEV

20 KEV

2 KEV

.20 KEV

E,KEV	FRACTIONAL DEPOSITED ENERGY					MEAN RANGE,MICROGRAM/SQ.CM.				
K/KL=	0.4	0.55	0.75	1.0	1.3	0.4	0.55	0.75	1.0	1.3
2.0	.9170	.8884	.8520	.8093	.7619	1.987	1.976	1.960	1.942	1.920
2.3	.9152	.8859	.8489	.8054	.7572	2.202	2.189	2.171	2.150	2.125
2.6	.9135	.8837	.8460	.8019	.7530	2.398	2.383	2.363	2.339	2.311
3.0	.9115	.8811	.8427	.7977	.7480	2.639	2.622	2.600	2.573	2.541
3.5	.9093	.8782	.8389	.7931	.7425	2.920	2.901	2.876	2.846	2.810
4.0	.9073	.8756	.8356	.7890	.7377	3.187	3.166	3.138	3.105	3.065
4.5	.9055	.8733	.8327	.7854	.7334	3.444	3.421	3.390	3.353	3.310
5.0	.9039	.8712	.8299	.7821	.7294	3.694	3.668	3.635	3.595	3.548
6.0	.9011	.8674	.8251	.7762	.7225	4.176	4.146	4.108	4.061	4.006
7.0	.8986	.8641	.8210	.7711	.7165	4.640	4.606	4.562	4.508	4.446
8.0	.8963	.8612	.8173	.7666	.7112	5.090	5.052	5.002	4.941	4.871
9.0	.8943	.8586	.8140	.7626	.7065	5.527	5.484	5.429	5.361	5.283
10	.8925	.8562	.8110	.7589	.7022	5.953	5.906	5.844	5.770	5.684
12	.8892	.8521	.8058	.7525	.6947	6.743	6.688	6.617	6.529	6.429
14	.8864	.8486	.8012	.7469	.6882	7.511	7.448	7.365	7.265	7.149
17	.8827	.8439	.7953	.7396	.6799	8.633	8.556	8.456	8.335	8.196
20	.8794	.8398	.7902	.7333	.6726	9.723	9.632	9.515	9.373	9.209
23	.8765	.8361	.7855	.7277	.6662	10.71	10.61	10.47	10.31	10.13
26	.8738	.8327	.7813	.7226	.6603	11.68	11.57	11.42	11.24	11.03
30	.8706	.8287	.7763	.7165	.6533	12.97	12.84	12.67	12.46	12.23
35	.8670	.8241	.7705	.7097	.6456	14.57	14.41	14.21	13.97	13.69
40	.8637	.8199	.7654	.7035	.6386	16.16	15.98	15.74	15.46	15.14
45	.8607	.8161	.7607	.6980	.6323	17.74	17.53	17.26	16.94	16.57
50	.8579	.8126	.7563	.6928	.6265	19.30	19.07	18.76	18.40	17.98
60	.8529	.8062	.7486	.6836	.6161	22.16	21.87	21.51	21.07	20.58
70	.8483	.8006	.7416	.6754	.6070	25.06	24.72	24.28	23.77	23.18
80	.8441	.7954	.7353	.6680	.5988	28.00	27.60	27.09	26.48	25.79
90	.8402	.7906	.7295	.6612	.5913	30.98	30.50	29.90	29.19	28.39
100	.8366	.7861	.7240	.6549	.5844	33.97	33.42	32.72	31.91	30.99
120	.8300	.7780	.7142	.6435	.5719	39.41	38.75	37.91	36.92	35.81
140	.8239	.7706	.7054	.6333	.5608	45.09	44.28	43.26	42.06	40.72
170	.8156	.7605	.6934	.6197	.5461	54.02	52.93	51.57	49.98	48.23
200	.8079	.7513	.6826	.6074	.5331	63.34	61.91	60.13	58.09	55.86
230	.8008	.7428	.6726	.5962	.5212	71.39	69.75	67.73	65.42	62.84
260	.7941	.7349	.6634	.5859	.5104	79.88	77.96	75.62	72.96	69.97
300	.7857	.7250	.6519	.5732	.4972	91.90	89.50	86.61	83.34	79.70
350	.7759	.7136	.6388	.5588	.4824	107.9	104.8	101.0	96.79	92.17
400	.7666	.7029	.6268	.5457	.4690	124.8	120.7	115.9	110.6	104.9
450	.7578	.6929	.6155	.5336	.4568	142.4	137.3	131.3	124.7	117.8
500	.7493	.6834	.6050	.5223	.4456	160.4	154.2	146.9	139.0	130.8
600	.7335	.6656	.5854	.5016	.4250	193.7	185.8	176.4	166.1	155.5
700	.7187	.6494	.5678	.4831	.4070	229.6	219.5	207.5	194.0	180.7
800	.7050	.6345	.5518	.4666	.3911	267.6	254.7	239.3	222.6	206.3
900	.6920	.6205	.5370	.4516	.3767	307.2	291.2	272.0	251.5	231.9
1000	.6797	.6075	.5234	.4378	.3637	348.0	328.5	305.3	280.7	257.6
1200	.6565	.5833	.4984	.4129	.3405	425.9	400.2	369.6	337.3	307.5
1400	.6354	.5615	.4764	.3913	.3207	508.4	475.0	435.6	394.6	357.3
1700	.6070	.5328	.4477	.3638	.2957	638.9	591.9	537.0	480.8	431.5
2000	.5817	.5076	.4231	.3405	.2749	774.9	712.4	640.0	566.8	504.6
2300	.5582	.4848	.4013	.3204	.2573	907.4	829.9	740.3	650.6	575.6
2600	.5368	.4642	.3821	.3028	.2421	1043.	949.1	841.0	733.6	645.3
3000	.5112	.4399	.3595	.2825	.2247	1228.	1110.	975.2	842.7	736.3
3500	.4829	.4134	.3352	.2610	.2064	1462.	1312.	1142.	976.5	847.0
4000	.4580	.3903	.3144	.2429	.1911	1698.	1515.	1307.	1107.	954.4
4500	.4359	.3700	.2963	.2272	.1781	1935.	1717.	1470.	1235.	1059.
5000	.4161	.3519	.2804	.2136	.1668	2171.	1917.	1631.	1359.	1160.
6000	.3818	.3212	.2539	.1915	.1487	2633.	2306.	1941.	1598.	1353.
7000	.3532	.2958	.2323	.1738	.1343	3088.	2687.	2242.	1826.	1536.
8000	.3288	.2744	.2143	.1593	.1226	3535.	3060.	2533.	2046.	1712.
9000	.3077	.2560	.1990	.1471	.1129	3974.	3424.	2816.	2257.	1880.
10000	.2894	.2401	.1859	.1367	.1046	4403.	3779.	3091.	2460.	2041.
12000	.2588	.2139	.1645	.1199	.0914	5236.	4465.	3617.	2847.	2341.
14000	.2345	.1931	.1478	.1070	.0813	6034.	5119.	4116.	3210.	2633.
17000	.2062	.1692	.1287	.0925	.0701	7172.	6048.	4820.	3718.	3031.
20000	.1847	.1512	.1146	.0820	.0619	8246.	6922.	5478.	4189.	3400.

GE ON TA, M2/M1=2.48   TABLE 9- 2

	MEAN DAMAGE DEPTH,MICROGRAM/SQ.CM.					MEAN IONZN.DEPTH,MICROGRAM/SQ.CM.				
K/KL=	0.4	0.55	0.75	1.0	1.3	0.4	0.55	0.75	1.0	1.3
E,KEV										
2.0	1.754	1.736	1.712	1.684	1.652	1.641	1.622	1.597	1.567	1.533
2.3	1.941	1.920	1.894	1.861	1.824	1.815	1.793	1.765	1.730	1.691
2.6	2.112	2.089	2.059	2.023	1.982	1.975	1.950	1.918	1.880	1.837
3.0	2.323	2.297	2.264	2.223	2.177	2.172	2.144	2.109	2.066	2.017
3.5	2.570	2.541	2.503	2.457	2.406	2.403	2.371	2.331	2.283	2.228
4.0	2.804	2.771	2.729	2.679	2.622	2.621	2.586	2.542	2.488	2.427
4.5	3.029	2.993	2.947	2.892	2.829	2.831	2.793	2.744	2.685	2.618
5.0	3.247	3.208	3.158	3.098	3.030	3.035	2.993	2.940	2.876	2.803
6.0	3.668	3.622	3.564	3.495	3.415	3.427	3.379	3.317	3.243	3.159
7.0	4.072	4.020	3.954	3.875	3.784	3.804	3.749	3.678	3.594	3.498
8.0	4.463	4.404	4.330	4.241	4.140	4.168	4.106	4.027	3.932	3.825
9.0	4.841	4.777	4.694	4.595	4.484	4.521	4.452	4.364	4.259	4.140
10	5.210	5.139	5.048	4.939	4.817	4.864	4.788	4.691	4.576	4.446
12	5.894	5.812	5.706	5.580	5.438	5.501	5.413	5.300	5.167	5.016
14	6.557	6.462	6.341	6.197	6.035	6.117	6.017	5.888	5.735	5.563
17	7.519	7.406	7.261	7.090	6.897	7.012	6.891	6.737	6.555	6.351
20	8.451	8.319	8.150	7.950	7.727	7.876	7.735	7.556	7.345	7.108
23	9.296	9.143	8.959	8.735	8.485	8.662	8.504	8.303	8.066	7.801
26	10.13	9.962	9.752	9.503	9.225	9.433	9.257	9.034	8.770	8.476
30	11.22	11.03	10.79	10.51	10.19	10.45	10.25	9.991	9.691	9.356
35	12.57	12.35	12.07	11.74	11.38	11.70	11.46	11.17	10.82	10.43
40	13.91	13.66	13.34	12.96	12.54	12.93	12.66	12.32	11.92	11.48
45	15.23	14.94	14.58	14.15	13.68	14.15	13.84	13.46	13.01	12.51
50	16.53	16.21	15.80	15.33	14.80	15.35	15.00	14.57	14.07	13.52
60	18.91	18.53	18.05	17.49	16.87	17.54	17.14	16.63	16.03	15.38
70	21.32	20.87	20.30	19.64	18.92	19.75	19.27	18.67	17.98	17.22
80	23.73	23.21	22.55	21.79	20.95	21.97	21.41	20.71	19.91	19.04
90	26.16	25.56	24.80	23.93	22.98	24.18	23.54	22.74	21.83	20.83
100	28.59	27.91	27.05	26.05	24.99	26.39	25.66	24.76	23.72	22.61
120	33.03	32.21	31.18	29.99	28.72	30.46	29.58	28.49	27.26	25.92
140	37.62	36.64	35.40	33.99	32.47	34.62	33.57	32.27	30.80	29.22
170	44.76	43.48	41.89	40.08	38.17	41.04	39.67	38.01	36.15	34.16
200	52.13	50.51	48.51	46.25	43.90	47.61	45.89	43.81	41.51	39.08
230	58.56	56.73	54.44	51.85	49.14	53.46	51.46	49.08	46.46	43.66
260	65.29	63.17	60.54	57.57	54.46	59.51	57.17	54.44	51.45	48.24
300	74.71	72.14	68.96	65.38	61.66	67.88	65.02	61.73	58.17	54.35
350	87.14	83.88	79.87	75.40	70.84	78.77	75.18	71.07	66.66	62.01
400	100.2	96.09	91.13	85.65	80.14	90.04	85.62	80.57	75.20	69.65
450	113.6	108.6	102.6	96.03	89.52	101.6	96.26	90.18	83.75	77.24
500	127.4	121.4	114.3	106.5	98.93	113.3	107.0	99.84	92.29	84.78
600	152.7	145.1	136.1	126.6	116.7	135.4	127.5	118.3	108.6	99.24
700	179.8	170.3	158.8	146.6	134.9	158.4	148.6	137.1	125.0	113.6
800	208.4	196.5	182.3	167.3	153.2	182.2	170.2	156.1	141.3	127.7
900	238.1	223.5	206.3	188.2	171.6	206.5	192.1	175.1	157.6	141.7
1000	268.6	251.2	230.7	209.3	190.0	231.1	214.1	194.2	173.7	155.4
1200	326.9	304.2	277.6	250.2	225.6	278.6	256.7	231.1	204.9	182.0
1400	388.3	359.5	325.9	291.4	261.3	327.4	299.9	267.9	235.5	207.8
1700	485.1	445.7	399.9	353.7	314.6	402.0	365.2	322.7	280.4	245.2
2000	585.7	534.3	475.0	416.0	367.3	477.5	430.6	376.8	323.9	281.1
2300	682.8	619.8	547.6	476.1	418.0	551.8	494.7	429.5	365.9	315.6
2600	782.1	706.7	620.5	535.8	468.1	626.0	558.2	481.3	406.7	348.8
3000	917.3	824.0	718.0	614.7	533.8	724.6	642.1	548.9	459.3	391.3
3500	1089.	972.2	839.7	711.9	614.3	846.7	745.1	630.9	522.3	441.9
4000	1263.	1121.	960.6	807.5	692.8	967.0	846.0	710.4	582.5	490.0
4500	1437.	1269.	1080.	901.2	769.5	1085.	944.6	787.5	640.4	535.9
5000	1611.	1416.	1199.	993.0	844.3	1201.	1041.	862.3	696.0	579.9
6000	1949.	1702.	1427.	1169.	987.1	1426.	1226.	1005.	801.1	662.5
7000	2285.	1984.	1649.	1339.	1124.	1643.	1404.	1141.	899.6	739.5
8000	2616.	2261.	1867.	1503.	1256.	1851.	1574.	1270.	992.4	811.8
9000	2942.	2532.	2079.	1662.	1384.	2051.	1737.	1393.	1080.	880.2
10000	3263.	2798.	2285.	1817.	1507.	2245.	1895.	1510.	1164.	945.1
12000	3888.	3314.	2684.	2112.	1742.	2613.	2193.	1733.	1321.	1066.
14000	4489.	3810.	3065.	2392.	1964.	2960.	2472.	1940.	1466.	1178.
17000	5351.	4517.	3605.	2787.	2276.	3444.	2861.	2226.	1665.	1331.
20000	6169.	5186.	4113.	3155.	2567.	3894.	3220.	2490.	1847.	1470.

164

	RELATIVE RANGE STRAGGLING					RELATIVE DAMAGE STRAGGLING				
K/KL=	0.4	0.55	0.75	1.0	1.3	0.4	0.55	0.75	1.0	1.3
E,KEV										
2.0	.6691	.6600	.6484	.6345	.6187	.5309	.5265	.5208	.5141	.5066
2.3	.6678	.6586	.6467	.6325	.6164	.5304	.5259	.5201	.5133	.5057
2.6	.6666	.6572	.6451	.6307	.6144	.5300	.5254	.5195	.5126	.5048
3.0	.6652	.6556	.6432	.6286	.6119	.5295	.5248	.5188	.5117	.5038
3.5	.6635	.6537	.6411	.6261	.6091	.5289	.5241	.5180	.5107	.5027
4.0	.6620	.6519	.6391	.6238	.6066	.5284	.5234	.5172	.5099	.5017
4.5	.6605	.6503	.6372	.6217	.6042	.5278	.5228	.5165	.5090	.5007
5.0	.6590	.6487	.6354	.6197	.6020	.5274	.5223	.5158	.5083	.4998
6.0	.6563	.6457	.6321	.6160	.5979	.5265	.5212	.5146	.5069	.4982
7.0	.6538	.6429	.6290	.6126	.5942	.5256	.5203	.5135	.5056	.4968
8.0	.6513	.6402	.6261	.6094	.5907	.5248	.5193	.5124	.5044	.4954
9.0	.6490	.6377	.6234	.6064	.5875	.5241	.5185	.5115	.5033	.4942
10	.6468	.6353	.6207	.6036	.5844	.5233	.5177	.5105	.5022	.4930
12	.6425	.6308	.6158	.5983	.5787	.5220	.5162	.5088	.5003	.4909
14	.6385	.6265	.6113	.5934	.5735	.5207	.5148	.5072	.4985	.4889
17	.6328	.6205	.6049	.5866	.5663	.5189	.5128	.5050	.4960	.4862
20	.6275	.6149	.5989	.5803	.5597	.5172	.5109	.5030	.4938	.4838
23	.6225	.6096	.5933	.5744	.5535	.5157	.5092	.5011	.4917	.4815
26	.6176	.6045	.5881	.5689	.5477	.5142	.5076	.4993	.4897	.4794
30	.6115	.5982	.5814	.5619	.5405	.5123	.5055	.4970	.4872	.4767
35	.6043	.5907	.5736	.5538	.5321	.5100	.5031	.4944	.4843	.4736
40	.5976	.5837	.5663	.5462	.5243	.5079	.5008	.4918	.4816	.4707
45	.5912	.5770	.5594	.5391	.5169	.5058	.4985	.4894	.4790	.4679
50	.5850	.5707	.5529	.5324	.5100	.5038	.4964	.4871	.4765	.4653
60	.5735	.5588	.5405	.5197	.4971	.5000	.4923	.4827	.4718	.4603
70	.5629	.5478	.5293	.5081	.4853	.4964	.4885	.4786	.4674	.4557
80	.5530	.5377	.5189	.4975	.4745	.4930	.4849	.4747	.4633	.4514
90	.5439	.5283	.5092	.4876	.4645	.4898	.4815	.4711	.4595	.4474
100	.5353	.5195	.5002	.4784	.4551	.4868	.4782	.4677	.4558	.4436
120	.5193	.5031	.4834	.4613	.4378	.4811	.4722	.4612	.4490	.4365
140	.5050	.4884	.4683	.4460	.4225	.4759	.4666	.4553	.4428	.4300
170	.4860	.4689	.4484	.4258	.4022	.4686	.4590	.4473	.4343	.4212
200	.4693	.4518	.4309	.4081	.3844	.4620	.4520	.4399	.4266	.4133
230	.4542	.4363	.4151	.3922	.3685	.4557	.4454	.4329	.4193	.4059
260	.4407	.4223	.4009	.3779	.3543	.4499	.4393	.4265	.4126	.3990
300	.4245	.4057	.3840	.3609	.3374	.4426	.4317	.4186	.4044	.3907
350	.4066	.3874	.3654	.3422	.3189	.4344	.4231	.4096	.3951	.3813
400	.3909	.3713	.3491	.3259	.3028	.4269	.4153	.4015	.3868	.3729
450	.3769	.3570	.3346	.3114	.2885	.4200	.4082	.3940	.3791	.3653
500	.3643	.3442	.3217	.2984	.2758	.4136	.4015	.3872	.3722	.3583
600	.3423	.3222	.2994	.2760	.2538	.4021	.3896	.3749	.3596	.3458
700	.3237	.3036	.2807	.2572	.2355	.3919	.3791	.3641	.3487	.3350
800	.3075	.2875	.2645	.2410	.2199	.3827	.3698	.3546	.3391	.3255
900	.2933	.2734	.2505	.2270	.2065	.3745	.3613	.3461	.3305	.3170
1000	.2805	.2609	.2380	.2147	.1947	.3669	.3537	.3383	.3228	.3094
1200	.2584	.2391	.2167	.1941	.1751	.3532	.3400	.3246	.3092	.2963
1400	.2400	.2211	.1992	.1773	.1592	.3414	.3282	.3129	.2977	.2852
1700	.2173	.1990	.1779	.1571	.1404	.3264	.3132	.2982	.2833	.2713
2000	.1989	.1811	.1609	.1411	.1255	.3137	.3007	.2860	.2715	.2599
2300	.1840	.1668	.1473	.1285	.1139	.3029	.2902	.2757	.2616	.2505
2600	.1714	.1548	.1360	.1180	.1043	.2934	.2810	.2668	.2531	.2424
3000	.1572	.1414	.1235	.1065	.0937	.2825	.2704	.2567	.2434	.2332
3500	.1427	.1277	.1108	.0949	.0833	.2709	.2592	.2460	.2333	.2236
4000	.1307	.1165	.1006	.0856	.0749	.2611	.2497	.2369	.2247	.2156
4500	.1207	.1071	.0921	.0780	.0680	.2526	.2416	.2292	.2174	.2087
5000	.1121	.0992	.0849	.0716	.0623	.2451	.2344	.2225	.2111	.2027
6000	.0984	.0867	.0737	.0617	.0535	.2328	.2227	.2114	.2007	.1929
7000	.0877	.0770	.0651	.0542	.0469	.2228	.2132	.2025	.1924	.1850
8000	.0792	.0693	.0584	.0484	.0418	.2145	.2053	.1951	.1854	.1784
9000	.0721	.0629	.0528	.0437	.0376	.2074	.1986	.1888	.1795	.1729
10000	.0662	.0577	.0483	.0398	.0342	.2013	.1928	.1834	.1745	.1681
12000	.0569	.0494	.0412	.0338	.0290	.1913	.1833	.1744	.1661	.1601
14000	.0498	.0432	.0359	.0293	.0252	.1833	.1758	.1673	.1594	.1537
17000	.0421	.0364	.0302	.0246	.0211	.1741	.1601	.1589	.1515	.1461
20000	.0366	.0316	.0261	.0212	.0181	.1670	.1601	.1524	.1453	.1402

GE ON TA, M2/M1=2.48   TABLE 9-4
         RELATIVE IONIZATION STRAGGLING   RELATIVE TRANSVERSE RANGE STRAGGLING

K/KL= E,KEV	0.4	0.55	0.75	1.0	1.3	0.4	0.55	0.75	1.0	1.3
2.0	.5708	.5671	.5625	.5570	.5510	.8576	.8568	.8557	.8543	.8528
2.3	.5704	.5666	.5619	.5564	.5503	.8576	.8567	.8556	.8543	.8527
2.6	.5700	.5662	.5614	.5558	.5496	.8576	.8567	.8556	.8542	.8527
3.0	.5696	.5657	.5608	.5551	.5488	.8576	.8567	.8555	.8541	.8526
3.5	.5691	.5651	.5601	.5543	.5479	.8576	.8567	.8555	.8541	.8525
4.0	.5686	.5646	.5595	.5536	.5470	.8576	.8567	.8555	.8540	.8524
4.5	.5682	.5641	.5589	.5529	.5463	.8576	.8566	.8554	.8540	.8523
5.0	.5678	.5636	.5583	.5523	.5456	.8576	.8566	.8554	.8539	.8522
6.0	.5670	.5627	.5573	.5511	.5443	.8576	.8566	.8554	.8539	.8522
7.0	.5663	.5619	.5564	.5500	.5431	.8577	.8567	.8554	.8539	.8521
8.0	.5656	.5611	.5555	.5491	.5421	.8577	.8567	.8554	.8539	.8521
9.0	.5650	.5604	.5547	.5482	.5411	.8578	.8568	.8555	.8539	.8521
10	.5644	.5597	.5539	.5473	.5401	.8578	.8568	.8555	.8540	.8522
12	.5633	.5585	.5525	.5457	.5384	.8580	.8570	.8556	.8541	.8522
14	.5622	.5573	.5512	.5443	.5369	.8581	.8571	.8558	.8542	.8524
17	.5607	.5556	.5494	.5423	.5347	.8584	.8574	.8560	.8544	.8526
20	.5592	.5541	.5477	.5405	.5328	.8587	.8577	.8563	.8547	.8529
23	.5579	.5526	.5461	.5388	.5310	.8590	.8579	.8566	.8550	.8532
26	.5566	.5512	.5446	.5372	.5293	.8592	.8582	.8569	.8553	.8536
30	.5550	.5495	.5427	.5352	.5271	.8597	.8586	.8573	.8558	.8540
35	.5531	.5474	.5405	.5328	.5247	.8602	.8592	.8579	.8564	.8547
40	.5513	.5455	.5384	.5306	.5224	.8608	.8598	.8586	.8571	.8555
45	.5496	.5436	.5365	.5285	.5202	.8614	.8605	.8593	.8579	.8562
50	.5479	.5419	.5346	.5266	.5182	.8621	.8612	.8600	.8586	.8571
60	.5448	.5386	.5311	.5229	.5144	.8633	.8624	.8614	.8601	.8586
70	.5419	.5355	.5278	.5195	.5109	.8646	.8638	.8628	.8616	.8603
80	.5391	.5326	.5248	.5163	.5077	.8660	.8653	.8643	.8632	.8620
90	.5366	.5298	.5219	.5133	.5046	.8674	.8668	.8659	.8649	.8638
100	.5341	.5273	.5192	.5106	.5018	.8689	.8683	.8675	.8666	.8657
120	.5295	.5224	.5141	.5053	.4966	.8717	.8713	.8707	.8700	.8693
140	.5253	.5180	.5095	.5006	.4918	.8747	.8744	.8740	.8735	.8730
170	.5196	.5120	.5034	.4944	.4856	.8793	.8792	.8790	.8789	.8787
200	.5144	.5067	.4979	.4888	.4801	.8839	.8840	.8842	.8844	.8846
230	.5096	.5017	.4927	.4837	.4750	.8886	.8890	.8894	.8898	.8903
260	.5053	.4971	.4881	.4791	.4705	.8933	.8939	.8946	.8953	.8961
300	.5000	.4916	.4825	.4736	.4652	.8996	.9004	.9014	.9025	.9037
350	.4940	.4855	.4764	.4675	.4594	.9072	.9084	.9099	.9115	.9132
400	.4888	.4801	.4710	.4623	.4544	.9147	.9163	.9182	.9203	.9225
450	.4840	.4753	.4662	.4576	.4500	.9221	.9240	.9264	.9289	.9315
500	.4796	.4710	.4620	.4535	.4461	.9292	.9316	.9343	.9374	.9404
600	.4720	.4635	.4546	.4463	.4395	.9420	.9452	.9490	.9531	.9572
700	.4654	.4572	.4485	.4405	.4342	.9546	.9586	.9633	.9685	.9735
800	.4596	.4518	.4433	.4356	.4299	.9672	.9719	.9775	.9835	.9893
900	.4545	.4470	.4389	.4314	.4262	.9798	.9851	.9914	.9982	1.005
1000	.4499	.4428	.4350	.4279	.4232	.9922	.9981	1.005	1.013	1.020
1200	.4417	.4354	.4285	.4223	.4186	1.019	1.025	1.033	1.042	1.050
1400	.4349	.4293	.4233	.4181	.4154	1.044	1.051	1.060	1.070	1.078
1700	.4268	.4223	.4175	.4136	.4120	1.079	1.087	1.098	1.108	1.116
2000	.4206	.4169	.4133	.4105	.4099	1.110	1.120	1.132	1.143	1.152
2300	.4163	.4134	.4106	.4088	.4090	1.139	1.150	1.162	1.174	1.184
2600	.4129	.4107	.4087	.4077	.4086	1.165	1.177	1.190	1.203	1.214
3000	.4095	.4081	.4071	.4070	.4087	1.196	1.210	1.225	1.239	1.250
3500	.4065	.4060	.4059	.4068	.4094	1.233	1.248	1.264	1.280	1.292
4000	.4045	.4047	.4054	.4072	.4104	1.266	1.282	1.300	1.317	1.329
4500	.4031	.4039	.4054	.4079	.4117	1.297	1.314	1.333	1.351	1.364
5000	.4022	.4036	.4057	.4088	.4130	1.326	1.344	1.364	1.383	1.395
6000	.4016	.4039	.4071	.4112	.4161	1.379	1.399	1.421	1.441	1.453
7000	.4018	.4049	.4089	.4137	.4191	1.426	1.447	1.471	1.492	1.505
8000	.4024	.4061	.4109	.4163	.4220	1.468	1.491	1.516	1.538	1.550
9000	.4033	.4076	.4128	.4187	.4248	1.507	1.531	1.557	1.579	1.591
10000	.4044	.4091	.4148	.4211	.4274	1.543	1.567	1.594	1.617	1.629
12000	.4069	.4122	.4187	.4255	.4320	1.606	1.632	1.660	1.683	1.695
14000	.4095	.4153	.4222	.4295	.4361	1.662	1.689	1.717	1.741	1.752
17000	.4135	.4198	.4271	.4346	.4413	1.735	1.762	1.792	1.815	1.825
20000	.4174	.4239	.4314	.4391	.4456	1.797	1.825	1.855	1.878	1.887

RELATIVE TRANSV. DAMAGE STRAGGLING RELATIVE TRANSV. IONZN. STRAGGLING

K/KL= E,KEV	0.4	0.55	0.75	1.0	1.3	0.4	0.55	0.75	1.0	1.3
2.0	.6784	.6748	.6700	.6643	.6575	.5975	.5925	.5861	.5782	.5691
2.3	.6780	.6743	.6695	.6636	.6567	.5970	.5919	.5854	.5773	.5680
2.6	.6777	.6739	.6690	.6630	.6559	.5965	.5914	.5847	.5765	.5670
3.0	.6773	.6734	.6684	.6622	.6550	.5960	.5907	.5839	.5755	.5658
3.5	.6768	.6729	.6677	.6614	.6540	.5954	.5900	.5829	.5744	.5644
4.0	.6764	.6723	.6671	.6606	.6531	.5948	.5893	.5821	.5734	.5632
4.5	.6760	.6719	.6665	.6599	.6523	.5943	.5887	.5813	.5724	.5621
5.0	.6756	.6714	.6659	.6593	.6515	.5938	.5881	.5806	.5716	.5611
6.0	.6749	.6706	.6649	.6581	.6501	.5928	.5869	.5793	.5700	.5592
7.0	.6742	.6698	.6640	.6570	.6488	.5920	.5859	.5781	.5685	.5575
8.0	.6736	.6691	.6632	.6560	.6477	.5911	.5850	.5769	.5672	.5559
9.0	.6730	.6684	.6624	.6551	.6466	.5904	.5841	.5759	.5660	.5545
10	.6724	.6677	.6616	.6542	.6456	.5896	.5832	.5749	.5648	.5532
12	.6714	.6665	.6602	.6526	.6438	.5883	.5817	.5731	.5627	.5507
14	.6704	.6654	.6590	.6511	.6421	.5870	.5802	.5714	.5608	.5485
17	.6690	.6639	.6572	.6491	.6398	.5852	.5782	.5691	.5581	.5455
20	.6678	.6625	.6556	.6473	.6378	.5835	.5763	.5670	.5557	.5427
23	.6666	.6612	.6541	.6456	.6359	.5820	.5746	.5650	.5534	.5402
26	.6655	.6599	.6527	.6441	.6341	.5805	.5729	.5631	.5513	.5378
30	.6641	.6584	.6510	.6421	.6320	.5786	.5709	.5608	.5487	.5349
35	.6624	.6566	.6490	.6399	.6294	.5764	.5684	.5581	.5457	.5315
40	.6609	.6549	.6471	.6378	.6271	.5744	.5661	.5555	.5428	.5283
45	.6595	.6533	.6454	.6358	.6250	.5724	.5640	.5531	.5401	.5253
50	.6581	.6519	.6437	.6340	.6229	.5705	.5619	.5508	.5376	.5225
60	.6556	.6491	.6407	.6306	.6191	.5670	.5581	.5466	.5328	.5172
70	.6533	.6466	.6379	.6275	.6157	.5638	.5545	.5426	.5284	.5124
80	.6512	.6443	.6353	.6246	.6125	.5608	.5512	.5389	.5244	.5079
90	.6493	.6421	.6329	.6220	.6096	.5580	.5481	.5355	.5205	.5037
100	.6475	.6402	.6307	.6195	.6068	.5553	.5451	.5322	.5169	.4997
120	.6442	.6365	.6266	.6149	.6017	.5503	.5397	.5261	.5101	.4922
140	.6414	.6333	.6230	.6107	.5971	.5458	.5346	.5205	.5039	.4854
170	.6377	.6291	.6181	.6052	.5908	.5397	.5278	.5128	.4954	.4760
200	.6345	.6254	.6139	.6003	.5853	.5341	.5216	.5059	.4876	.4675
230	.6319	.6223	.6102	.5960	.5803	.5291	.5159	.4995	.4805	.4596
260	.6295	.6195	.6069	.5920	.5758	.5246	.5107	.4935	.4738	.4522
300	.6268	.6162	.6028	.5872	.5703	.5189	.5041	.4861	.4655	.4431
350	.6238	.6125	.5983	.5818	.5639	.5123	.4966	.4775	.4559	.4326
400	.6212	.6092	.5941	.5767	.5581	.5062	.4897	.4696	.4470	.4229
450	.6189	.6062	.5903	.5721	.5526	.5005	.4832	.4622	.4387	.4139
500	.6168	.6034	.5868	.5677	.5476	.4952	.4770	.4552	.4309	.4054
600	.6133	.5987	.5805	.5599	.5383	.4852	.4657	.4422	.4163	.3896
700	.6102	.5944	.5748	.5528	.5299	.4761	.4554	.4305	.4031	.3754
800	.6074	.5904	.5695	.5461	.5221	.4676	.4459	.4197	.3910	.3625
900	.6047	.5866	.5644	.5398	.5148	.4597	.4370	.4096	.3799	.3508
1000	.6021	.5830	.5597	.5339	.5080	.4523	.4287	.4002	.3695	.3399
1200	.5974	.5763	.5507	.5227	.4951	.4388	.4133	.3828	.3504	.3199
1400	.5929	.5700	.5423	.5124	.4834	.4265	.3994	.3672	.3335	.3025
1700	.5862	.5609	.5306	.4983	.4676	.4097	.3807	.3466	.3114	.2800
2000	.5795	.5522	.5197	.4854	.4534	.3944	.3639	.3284	.2923	.2609
2300	.5726	.5435	.5091	.4731	.4401	.3793	.3480	.3117	.2752	.2442
2600	.5657	.5351	.4991	.4617	.4280	.3653	.3335	.2967	.2601	.2296
3000	.5567	.5244	.4866	.4477	.4132	.3483	.3160	.2789	.2423	.2126
3500	.5459	.5119	.4723	.4318	.3968	.3292	.2967	.2596	.2233	.1946
4000	.5356	.5002	.4591	.4176	.3821	.3122	.2797	.2428	.2070	.1794
4500	.5256	.4891	.4470	.4046	.3689	.2969	.2646	.2281	.1930	.1663
5000	.5160	.4787	.4357	.3927	.3569	.2830	.2510	.2150	.1806	.1550
6000	.4972	.4588	.4147	.3711	.3356	.2585	.2277	.1932	.1605	.1368
7000	.4800	.4410	.3964	.3525	.3174	.2378	.2083	.1753	.1443	.1223
8000	.4643	.4249	.3801	.3363	.3017	.2201	.1918	.1603	.1310	.1105
9000	.4498	.4103	.3654	.3219	.2880	.2048	.1777	.1476	.1198	.1006
10000	.4364	.3969	.3523	.3091	.2758	.1914	.1654	.1367	.1102	.0923
12000	.4123	.3733	.3293	.2870	.2551	.1690	.1452	.1190	.0950	.0791
14000	.3911	.3529	.3099	.2688	.2381	.1513	.1293	.1052	.0833	.0691
17000	.3637	.3270	.2857	.2465	.2176	.1307	.1112	.0898	.0704	.0582
20000	.3401	.3051	.2658	.2286	.2015	.1153	.0978	.0786	.0614	.0505

	RANGE SKEWNESS					DAMAGE SKEWNESS				
K/KL=	0.4	0.55	0.75	1.0	1.3	0.4	0.55	0.75	1.0	1.3
E,KEV										
2.0	.4900	.4780	.4623	.4431	.4210	.9090	.8990	.8861	.8707	.8531
2.3	.4891	.4767	.4606	.4411	.4184	.9082	.8980	.8848	.8691	.8511
2.6	.4882	.4756	.4592	.4392	.4161	.9075	.8971	.8837	.8676	.8493
3.0	.4872	.4743	.4575	.4370	.4134	.9067	.8960	.8823	.8659	.8472
3.5	.4861	.4728	.4555	.4346	.4104	.9058	.8948	.8808	.8639	.8448
4.0	.4850	.4714	.4538	.4324	.4076	.9050	.8938	.8794	.8622	.8427
4.5	.4841	.4702	.4522	.4304	.4051	.9042	.8928	.8781	.8606	.8408
5.0	.4832	.4691	.4507	.4285	.4028	.9035	.8919	.8769	.8591	.8390
6.0	.4816	.4669	.4480	.4251	.3986	.9022	.8902	.8748	.8565	.8357
7.0	.4801	.4650	.4455	.4220	.3948	.9010	.8887	.8728	.8540	.8328
8.0	.4787	.4632	.4432	.4191	.3914	.8999	.8872	.8710	.8518	.8301
9.0	.4774	.4616	.4411	.4165	.3882	.8989	.8859	.8693	.8497	.8276
10	.4761	.4600	.4391	.4140	.3852	.8979	.8847	.8678	.8478	.8253
12	.4739	.4571	.4355	.4095	.3798	.8961	.8824	.8649	.8442	.8210
14	.4718	.4545	.4322	.4054	.3748	.8944	.8803	.8622	.8409	.8172
17	.4688	.4508	.4275	.3997	.3680	.8920	.8773	.8585	.8364	.8118
20	.4660	.4473	.4232	.3945	.3618	.8897	.8745	.8551	.8322	.8069
23	.4635	.4442	.4193	.3898	.3561	.8878	.8721	.8521	.8286	.8025
26	.4611	.4411	.4156	.3853	.3508	.8858	.8697	.8492	.8251	.7984
30	.4579	.4373	.4109	.3795	.3440	.8833	.8667	.8455	.8206	.7931
35	.4541	.4327	.4052	.3728	.3361	.8802	.8629	.8409	.8152	.7869
40	.4505	.4282	.3998	.3663	.3285	.8771	.8592	.8365	.8100	.7808
45	.4469	.4239	.3946	.3601	.3213	.8741	.8556	.8322	.8049	.7750
50	.4434	.4197	.3896	.3541	.3144	.8710	.8520	.8279	.7999	.7693
60	.4369	.4119	.3802	.3430	.3015	.8651	.8449	.8195	.7902	.7585
70	.4306	.4043	.3711	.3324	.2893	.8593	.8380	.8114	.7808	.7481
80	.4244	.3969	.3624	.3223	.2777	.8535	.8313	.8036	.7719	.7382
90	.4183	.3897	.3540	.3124	.2666	.8478	.8247	.7960	.7632	.7286
100	.4123	.3827	.3457	.3029	.2558	.8421	.8183	.7886	.7549	.7195
120	.4010	.3693	.3301	.2849	.2354	.8314	.8062	.7750	.7396	.7024
140	.3899	.3564	.3151	.2677	.2162	.8208	.7945	.7619	.7249	.6862
170	.3738	.3378	.2936	.2433	.1890	.8053	.7775	.7430	.7040	.6632
200	.3583	.3199	.2731	.2203	.1635	.7902	.7610	.7248	.6839	.6414
230	.3432	.3024	.2533	.1982	.1393	.7751	.7443	.7063	.6637	.6199
260	.3286	.2856	.2343	.1772	.1163	.7605	.7282	.6886	.6444	.5994
300	.3098	.2642	.2102	.1506	.0873	.7418	.7077	.6660	.6199	.5736
350	.2874	.2388	.1817	.1192	.0533	.7196	.6834	.6394	.5912	.5434
400	.2660	.2149	.1550	.0898	.0216	.6985	.6605	.6144	.5644	.5153
450	.2456	.1921	.1296	.0620	-.0083	.6785	.6388	.5909	.5392	.4889
500	.2261	.1704	.1056	.0356	-.0366	.6595	.6183	.5687	.5154	.4642
600	.1918	.1322	.0621	-.0133	-.0892	.6247	.5806	.5278	.4716	.4184
700	.1587	.0958	.0215	-.0584	-.1373	.5923	.5458	.4904	.4318	.3771
800	.1264	.0610	-.0169	-.1004	-.1817	.5620	.5135	.4560	.3954	.3394
900	.0948	.0273	-.0534	-.1398	-.2230	.5334	.4833	.4239	.3617	.3046
1000	.0637	-.0054	-.0883	-.1771	-.2618	.5063	.4548	.3939	.3303	.2724
1200	-.0041	-.0741	-.1583	-.2484	-.3342	.4536	.4004	.3376	.2723	.2132
1400	-.0665	-.1371	-.2219	-.3127	-.3991	.4059	.3515	.2874	.2208	.1610
1700	-.1499	-.2211	-.3067	-.3982	-.4852	.3423	.2865	.2210	.1532	.0926
2000	-.2221	-.2942	-.3807	-.4730	-.5607	.2866	.2298	.1631	.0944	.0335
2300	-.2776	-.3525	-.4419	-.5369	-.6264	.2380	.1800	.1121	.0426	-.0185
2600	-.3265	-.4041	-.4966	-.5944	-.6856	.1944	.1354	.0665	-.0038	-.0649
3000	-.3840	-.4653	-.5618	-.6632	-.7568	.1425	.0823	.0122	-.0587	-.1197
3500	-.4469	-.5326	-.6339	-.7396	-.8360	.0854	.0240	-.0471	-.1187	-.1793
4000	-.5025	-.5923	-.6981	-.8079	-.9068	.0351	-.0272	-.0991	-.1712	-.2314
4500	-.5528	-.6463	-.7562	-.8697	-.9710	-.0098	-.0728	-.1454	-.2177	-.2774
5000	-.5988	-.6958	-.8095	-.9264	-1.030	-.0503	-.1139	-.1870	-.2594	-.3186
6000	-.6818	-.7846	-.9046	-1.027	-1.134	-.1215	-.1856	-.2590	-.3312	-.3892
7000	-.7552	-.8630	-.9886	-1.116	-1.226	-.1822	-.2465	-.3199	-.3918	-.4487
8000	-.8215	-.9337	-1.064	-1.196	-1.309	-.2349	-.2993	-.3726	-.4440	-.4999
9000	-.8824	-.9984	-1.133	-1.269	-1.384	-.2814	-.3458	-.4189	-.4898	-.5448
10000	-.9389	-1.058	-1.196	-1.336	-1.453	-.3229	-.3872	-.4601	-.5305	-.5847
12000	-1.042	-1.167	-1.311	-1.456	-1.577	-.3944	-.4583	-.5306	-.6001	-.6529
14000	-1.134	-1.263	-1.413	-1.562	-1.687	-.4543	-.5178	-.5894	-.6580	-.7096
17000	-1.257	-1.392	-1.548	-1.703	-1.831	-.5287	-.5913	-.6619	-.7293	-.7797
20000	-1.368	-1.507	-1.667	-1.826	-1.958	-.5897	-.6515	-.7211	-.7875	-.8369

TABLE 9- 7

GE ON TA, M2/M1=2.48

K/KL=	IONIZATION SKEWNESS					RANGE KURTOSIS				
E,KEV	0.4	0.55	0.75	1.0	1.3	0.4	0.55	0.75	1.0	1.3
2.0	1.004	.9954	.9839	.9703	.9549	3.401	3.366	3.322	3.271	3.215
2.3	1.004	.9945	.9827	.9688	.9531	3.398	3.362	3.318	3.266	3.208
2.6	1.003	.9936	.9817	.9675	.9515	3.395	3.359	3.314	3.261	3.203
3.0	1.002	.9926	.9804	.9659	.9496	3.392	3.355	3.309	3.255	3.196
3.5	1.001	.9915	.9790	.9642	.9475	3.388	3.351	3.303	3.249	3.189
4.0	1.001	.9905	.9777	.9626	.9456	3.385	3.347	3.299	3.243	3.182
4.5	.9998	.9896	.9765	.9611	.9438	3.382	3.343	3.294	3.238	3.176
5.0	.9991	.9887	.9754	.9598	.9422	3.379	3.340	3.290	3.233	3.170
6.0	.9979	.9871	.9734	.9573	.9393	3.374	3.333	3.282	3.224	3.160
7.0	.9967	.9856	.9716	.9551	.9366	3.369	3.328	3.276	3.216	3.152
8.0	.9956	.9843	.9699	.9530	.9342	3.365	3.322	3.270	3.209	3.144
9.0	.9946	.9830	.9683	.9511	.9320	3.361	3.318	3.264	3.202	3.136
10	.9936	.9818	.9668	.9493	.9299	3.357	3.313	3.258	3.196	3.129
12	.9918	.9796	.9641	.9461	.9260	3.350	3.305	3.249	3.185	3.117
14	.9901	.9775	.9616	.9431	.9225	3.343	3.297	3.240	3.175	3.106
17	.9878	.9746	.9581	.9389	.9177	3.334	3.286	3.228	3.162	3.092
20	.9855	.9719	.9548	.9350	.9133	3.326	3.277	3.217	3.150	3.078
23	.9835	.9694	.9518	.9315	.9092	3.318	3.268	3.207	3.138	3.067
26	.9815	.9670	.9489	.9281	.9054	3.310	3.259	3.197	3.128	3.056
30	.9789	.9639	.9453	.9239	.9005	3.301	3.249	3.186	3.115	3.042
35	.9757	.9601	.9409	.9188	.8948	3.290	3.236	3.172	3.101	3.027
40	.9726	.9565	.9366	.9139	.8893	3.279	3.225	3.159	3.087	3.013
45	.9696	.9530	.9325	.9092	.8841	3.269	3.214	3.147	3.074	3.000
50	.9666	.9495	.9285	.9047	.8791	3.260	3.203	3.136	3.062	2.988
60	.9610	.9430	.9210	.8962	.8697	3.242	3.184	3.115	3.040	2.965
70	.9555	.9368	.9139	.8881	.8608	3.225	3.165	3.095	3.020	2.945
80	.9502	.9306	.9069	.8804	.8524	3.209	3.148	3.077	3.002	2.927
90	.9449	.9247	.9002	.8730	.8443	3.194	3.132	3.061	2.985	2.911
100	.9398	.9189	.8937	.8658	.8365	3.180	3.117	3.045	2.969	2.896
120	.9298	.9076	.8811	.8520	.8217	3.153	3.088	3.015	2.940	2.868
140	.9202	.8969	.8692	.8390	.8078	3.128	3.062	2.989	2.914	2.844
170	.9064	.8816	.8524	.8209	.7886	3.094	3.027	2.954	2.881	2.814
200	.8933	.8672	.8367	.8041	.7710	3.064	2.997	2.924	2.853	2.790
230	.8808	.8532	.8215	.7879	.7541	3.037	2.969	2.896	2.828	2.768
260	.8688	.8400	.8071	.7727	.7383	3.013	2.944	2.872	2.806	2.750
300	.8537	.8234	.7893	.7539	.7190	2.983	2.914	2.844	2.782	2.730
350	.8358	.8041	.7687	.7325	.6970	2.950	2.881	2.815	2.758	2.713
400	.8190	.7861	.7498	.7129	.6771	2.919	2.853	2.790	2.739	2.702
450	.8031	.7694	.7323	.6948	.6590	2.890	2.828	2.770	2.725	2.695
500	.7880	.7537	.7160	.6781	.6423	2.864	2.805	2.753	2.715	2.692
600	.7591	.7244	.6858	.6471	.6117	2.807	2.761	2.721	2.696	2.690
700	.7328	.6982	.6592	.6202	.5854	2.760	2.726	2.700	2.688	2.698
800	.7086	.6744	.6355	.5964	.5624	2.720	2.700	2.687	2.689	2.715
900	.6864	.6527	.6141	.5754	.5423	2.688	2.680	2.681	2.698	2.738
1000	.6657	.6328	.5947	.5565	.5245	2.662	2.667	2.681	2.711	2.766
1200	.6257	.5949	.5593	.5237	.4942	2.617	2.648	2.691	2.751	2.833
1400	.5912	.5626	.5295	.4965	.4696	2.591	2.644	2.715	2.803	2.910
1700	.5481	.5224	.4929	.4638	.4404	2.577	2.662	2.769	2.896	3.036
2000	.5133	.4902	.4639	.4382	.4181	2.587	2.699	2.839	2.999	3.171
2300	.4884	.4667	.4423	.4191	.4017	2.627	2.757	2.920	3.106	3.304
2600	.4680	.4476	.4249	.4037	.3888	2.679	2.825	3.008	3.217	3.439
3000	.4460	.4270	.4062	.3874	.3755	2.757	2.923	3.131	3.369	3.621
3500	.4244	.4070	.3883	.3722	.3635	2.864	3.054	3.291	3.563	3.851
4000	.4074	.3914	.3747	.3609	.3550	2.975	3.188	3.454	3.757	4.079
4500	.3937	.3791	.3641	.3525	.3490	3.088	3.324	3.617	3.952	4.306
5000	.3824	.3692	.3559	.3462	.3448	3.201	3.459	3.780	4.145	4.530
6000	.3646	.3546	.3451	.3394	.3415	3.407	3.710	4.086	4.511	4.954
7000	.3518	.3447	.3386	.3362	.3410	3.613	3.961	4.390	4.873	5.371
8000	.3424	.3379	.3347	.3352	.3423	3.818	4.210	4.692	5.230	5.781
9000	.3355	.3333	.3327	.3356	.3446	4.022	4.456	4.989	5.581	6.183
10000	.3304	.3303	.3319	.3370	.3475	4.224	4.700	5.282	5.927	6.578
12000	.3241	.3274	.3329	.3414	.3543	4.619	5.176	5.854	6.600	7.344
14000	.3212	.3273	.3358	.3471	.3615	5.003	5.637	6.407	7.249	8.081
17000	.3209	.3303	.3423	.3565	.3724	5.557	6.301	7.202	8.180	9.137
20000	.3236	.3354	.3499	.3661	.3828	6.087	6.935	7.958	9.064	10.14

K/KL=	DAMAGE KURTOSIS					IONIZATION KURTOSIS				
E,KEV	0.4	0.55	0.75	1.0	1.3	0.4	0.55	0.75	1.0	1.3
2.0	4.248	4.202	4.144	4.075	3.999	4.395	4.349	4.290	4.222	4.146
2.3	4.244	4.197	4.137	4.067	3.990	4.391	4.343	4.284	4.214	4.137
2.6	4.240	4.192	4.131	4.060	3.981	4.387	4.339	4.278	4.207	4.129
3.0	4.236	4.187	4.124	4.052	3.972	4.382	4.333	4.271	4.199	4.119
3.5	4.230	4.180	4.117	4.043	3.961	4.377	4.326	4.263	4.190	4.108
4.0	4.225	4.174	4.110	4.034	3.951	4.372	4.320	4.256	4.181	4.099
4.5	4.221	4.169	4.103	4.027	3.942	4.367	4.315	4.249	4.173	4.090
5.0	4.217	4.164	4.097	4.019	3.934	4.363	4.310	4.243	4.166	4.082
6.0	4.209	4.154	4.086	4.006	3.919	4.355	4.300	4.232	4.153	4.067
7.0	4.201	4.145	4.075	3.995	3.906	4.347	4.292	4.222	4.141	4.054
8.0	4.194	4.137	4.066	3.984	3.894	4.341	4.283	4.212	4.131	4.041
9.0	4.188	4.130	4.057	3.974	3.882	4.334	4.276	4.204	4.120	4.030
10	4.182	4.123	4.049	3.964	3.872	4.328	4.269	4.195	4.111	4.020
12	4.171	4.110	4.034	3.946	3.852	4.316	4.255	4.180	4.094	4.000
14	4.160	4.098	4.020	3.930	3.834	4.305	4.243	4.166	4.078	3.983
17	4.145	4.081	4.000	3.908	3.810	4.290	4.226	4.146	4.056	3.959
20	4.131	4.065	3.982	3.888	3.788	4.276	4.210	4.128	4.036	3.938
23	4.117	4.050	3.966	3.870	3.769	4.263	4.195	4.112	4.018	3.918
26	4.104	4.036	3.951	3.854	3.750	4.250	4.181	4.096	4.001	3.899
30	4.087	4.018	3.931	3.832	3.727	4.234	4.163	4.076	3.979	3.876
35	4.067	3.996	3.908	3.808	3.701	4.214	4.142	4.053	3.954	3.849
40	4.048	3.976	3.887	3.784	3.675	4.196	4.121	4.031	3.930	3.824
45	4.030	3.957	3.866	3.762	3.651	4.179	4.102	4.010	3.908	3.801
50	4.013	3.939	3.846	3.740	3.629	4.162	4.084	3.990	3.887	3.779
60	3.985	3.905	3.807	3.697	3.584	4.130	4.049	3.953	3.847	3.737
70	3.958	3.874	3.771	3.658	3.544	4.099	4.016	3.918	3.810	3.699
80	3.932	3.844	3.737	3.621	3.507	4.071	3.986	3.885	3.776	3.663
90	3.905	3.815	3.705	3.587	3.472	4.044	3.957	3.854	3.743	3.630
100	3.879	3.787	3.675	3.555	3.439	4.018	3.929	3.825	3.713	3.599
120	3.822	3.730	3.618	3.497	3.380	3.968	3.877	3.770	3.656	3.541
140	3.768	3.677	3.566	3.445	3.326	3.923	3.828	3.719	3.604	3.488
170	3.694	3.605	3.495	3.375	3.254	3.860	3.762	3.651	3.533	3.418
200	3.629	3.540	3.432	3.312	3.190	3.803	3.703	3.589	3.471	3.355
230	3.574	3.482	3.371	3.249	3.129	3.755	3.650	3.533	3.414	3.298
260	3.525	3.430	3.315	3.192	3.073	3.710	3.601	3.482	3.361	3.247
300	3.466	3.366	3.248	3.123	3.006	3.653	3.541	3.419	3.298	3.185
350	3.399	3.296	3.173	3.047	2.933	3.586	3.472	3.349	3.228	3.116
400	3.339	3.232	3.107	2.979	2.869	3.523	3.408	3.285	3.166	3.056
450	3.285	3.175	3.048	2.920	2.812	3.463	3.349	3.228	3.110	3.002
500	3.234	3.122	2.994	2.867	2.762	3.406	3.294	3.175	3.060	2.954
600	3.139	3.028	2.901	2.777	2.677	3.278	3.181	3.075	2.969	2.869
700	3.056	2.947	2.823	2.702	2.607	3.167	3.083	2.989	2.893	2.798
800	2.982	2.876	2.756	2.639	2.549	3.070	2.999	2.916	2.827	2.738
900	2.916	2.813	2.697	2.585	2.500	2.986	2.925	2.852	2.770	2.686
1000	2.855	2.757	2.646	2.539	2.458	2.914	2.861	2.796	2.720	2.640
1200	2.744	2.656	2.557	2.462	2.392	2.807	2.763	2.706	2.639	2.566
1400	2.650	2.573	2.486	2.403	2.342	2.724	2.685	2.634	2.574	2.507
1700	2.538	2.475	2.404	2.337	2.288	2.629	2.595	2.550	2.497	2.437
2000	2.451	2.400	2.343	2.290	2.251	2.559	2.526	2.485	2.437	2.384
2300	2.393	2.351	2.303	2.260	2.230	2.509	2.476	2.436	2.390	2.342
2600	2.349	2.313	2.274	2.239	2.217	2.469	2.436	2.396	2.352	2.308
3000	2.305	2.277	2.248	2.223	2.209	2.428	2.394	2.354	2.311	2.271
3500	2.267	2.247	2.228	2.213	2.209	2.388	2.353	2.312	2.271	2.236
4000	2.241	2.229	2.218	2.212	2.216	2.357	2.321	2.280	2.239	2.208
4500	2.225	2.219	2.215	2.216	2.227	2.332	2.295	2.253	2.214	2.186
5000	2.215	2.215	2.217	2.225	2.242	2.311	2.273	2.232	2.194	2.169
6000	2.211	2.220	2.233	2.253	2.279	2.271	2.235	2.197	2.163	2.144
7000	2.217	2.234	2.257	2.285	2.319	2.240	2.207	2.171	2.142	2.127
8000	2.229	2.253	2.284	2.320	2.361	2.215	2.184	2.152	2.126	2.115
9000	2.245	2.276	2.313	2.356	2.402	2.194	2.166	2.137	2.114	2.107
10000	2.263	2.299	2.343	2.393	2.443	2.177	2.151	2.125	2.106	2.101
12000	2.303	2.349	2.404	2.464	2.522	2.149	2.129	2.109	2.095	2.095
14000	2.344	2.398	2.463	2.532	2.597	2.128	2.113	2.098	2.090	2.094
17000	2.405	2.471	2.548	2.628	2.701	2.105	2.096	2.089	2.087	2.096
20000	2.464	2.539	2.626	2.716	2.796	2.089	2.086	2.085	2.089	2.102

E,KEV	\multicolumn{5}{c}{REFLECTION COEFFICIENT}	\multicolumn{5}{c}{SPUTTERING EFFICIENCY}
K/KL=	0.4	0.55	0.75	1.0	1.3	0.4	0.55	0.75	1.0	1.3
2.0	.0936	.0923	.0907	.0887	.0865	.0456	.0435	.0410	.0381	.0349
2.3	.0934	.0921	.0904	.0884	.0861	.0454	.0433	.0408	.0378	.0346
2.6	.0932	.0919	.0902	.0881	.0858	.0453	.0432	.0405	.0375	.0343
3.0	.0930	.0916	.0899	.0878	.0854	.0451	.0429	.0403	.0372	.0339
3.5	.0927	.0913	.0895	.0874	.0850	.0449	.0427	.0399	.0368	.0335
4.0	.0924	.0910	.0892	.0870	.0846	.0447	.0425	.0397	.0365	.0332
4.5	.0922	.0908	.0889	.0867	.0842	.0445	.0422	.0394	.0362	.0329
5.0	.0920	.0905	.0886	.0864	.0838	.0443	.0420	.0392	.0360	.0326
6.0	.0915	.0900	.0881	.0858	.0832	.0440	.0417	.0388	.0355	.0321
7.0	.0911	.0895	.0876	.0852	.0826	.0438	.0414	.0384	.0351	.0316
8.0	.0907	.0891	.0871	.0847	.0820	.0435	.0411	.0381	.0347	.0312
9.0	.0903	.0887	.0866	.0842	.0815	.0433	.0408	.0378	.0344	.0309
10	.0899	.0882	.0862	.0837	.0810	.0431	.0405	.0375	.0341	.0305
12	.0891	.0875	.0853	.0828	.0800	.0427	.0401	.0370	.0335	.0299
14	.0884	.0867	.0846	.0820	.0791	.0423	.0397	.0365	.0330	.0294
17	.0875	.0857	.0835	.0809	.0779	.0418	.0392	.0359	.0324	.0287
20	.0865	.0847	.0824	.0798	.0768	.0414	.0387	.0354	.0318	.0281
23	.0856	.0838	.0815	.0788	.0757	.0409	.0382	.0349	.0313	.0276
26	.0848	.0829	.0805	.0778	.0747	.0405	.0378	.0344	.0308	.0271
30	.0837	.0818	.0794	.0766	.0735	.0400	.0372	.0339	.0302	.0265
35	.0824	.0804	.0780	.0751	.0720	.0394	.0366	.0332	.0295	.0258
40	.0812	.0792	.0767	.0738	.0706	.0389	.0361	.0327	.0289	.0252
45	.0800	.0780	.0755	.0725	.0693	.0384	.0356	.0321	.0284	.0247
50	.0789	.0769	.0743	.0713	.0681	.0380	.0351	.0316	.0279	.0242
60	.0768	.0747	.0721	.0691	.0658	.0372	.0342	.0307	.0269	.0232
70	.0748	.0727	.0700	.0670	.0637	.0364	.0334	.0299	.0261	.0224
80	.0730	.0709	.0682	.0651	.0617	.0357	.0327	.0291	.0253	.0217
90	.0714	.0691	.0664	.0633	.0599	.0351	.0320	.0284	.0246	.0210
100	.0698	.0675	.0648	.0616	.0583	.0345	.0314	.0278	.0240	.0204
120	.0668	.0645	.0617	.0585	.0552	.0333	.0302	.0266	.0228	.0193
140	.0642	.0618	.0590	.0558	.0524	.0322	.0291	.0256	.0218	.0184
170	.0606	.0582	.0553	.0521	.0488	.0307	.0277	.0242	.0205	.0172
200	.0576	.0551	.0522	.0489	.0456	.0295	.0265	.0230	.0194	.0161
230	.0548	.0523	.0493	.0461	.0428	.0284	.0255	.0220	.0184	.0152
260	.0524	.0498	.0468	.0436	.0403	.0275	.0245	.0211	.0176	.0144
300	.0494	.0468	.0438	.0406	.0374	.0264	.0234	.0200	.0165	.0135
350	.0462	.0436	.0406	.0374	.0343	.0252	.0222	.0188	.0154	.0125
400	.0434	.0407	.0377	.0347	.0316	.0242	.0212	.0178	.0144	.0116
450	.0409	.0382	.0353	.0322	.0292	.0232	.0202	.0169	.0136	.0109
500	.0387	.0360	.0331	.0301	.0271	.0224	.0194	.0161	.0128	.0102
600	.0347	.0322	.0293	.0264	.0236	.0208	.0179	.0147	.0116	.0092
700	.0314	.0291	.0263	.0234	.0207	.0195	.0166	.0135	.0106	.0083
800	.0286	.0264	.0237	.0209	.0183	.0183	.0155	.0125	.0097	.0075
900	.0262	.0241	.0215	.0187	.0163	.0172	.0145	.0117	.0090	.0069
1000	.0241	.0222	.0196	.0169	.0146	.0163	.0137	.0109	.0084	.0064
1200	.0207	.0190	.0166	.0141	.0120	.0146	.0122	.0097	.0073	.0056
1400	.0180	.0164	.0142	.0118	.0100	.0132	.0110	.0086	.0065	.0049
1700	.0150	.0135	.0115	.0093	.0077	.0116	.0096	.0074	.0055	.0042
2000	.0126	.0112	.0094	.0075	.0060	.0102	.0085	.0065	.0048	.0036
2300	.0108	.0096	.0079	.0061	.0048	.0093	.0076	.0058	.0042	.0032
2600	.0094	.0082	.0067	.0050	.0039	.0085	.0070	.0053	.0038	.0029
3000	.0079	.0068	.0054	.0039	.0030	.0076	.0062	.0047	.0034	.0025
3500	.0065	.0055	.0042	.0029	.0021	.0068	.0055	.0042	.0030	.0022
4000	.0054	.0045	.0033	.0022	.0015	.0061	.0050	.0037	.0026	.0019
4500	.0045	.0037	.0026	.0016	.0011	.0055	.0045	.0034	.0024	.0017
5000	.0038	.0031	.0021	.0012	.0008	.0051	.0042	.0031	.0022	.0016
6000	.0027	.0022	.0014	.0007	.0004	.0044	.0036	.0026	.0018	.0013
7000	.0020	.0015	.0010	.0004	.0002	.0038	.0031	.0023	.0015	.0011
8000	.0015	.0011	.0007	.0003	.0001	.0034	.0027	.0020	.0013	.0009
9000	.0011	.0008	.0005	.0002	.0001	.0031	.0024	.0017	.0011	.0008
10000	.0098	.0006	.0003	.0001	.0001	.0028	.0022	.0016	.0010	.0007
12000	.0005	.0003	.0002	.0001	.0001	.0023	.0018	.0013	.0008	.0006
14000	.0002	.0002	.0001	.0001	.0001	.0020	.0015	.0011	.0007	.0005
17000	.0001	.0001	.0000	.0000	.0000	.0016	.0012	.0008	.0005	.0004
20000	.0001	.0000	.0000	.0000	.0000	.0013	.0010	.0007	.0004	.0003

GE ON TA, M2/M1=2.48 TABLE 9-10

	IONIZATION DEFICIENCY					SPUTTERING YIELD ALPHA				
K/KL=	0.4	0.55	0.75	1.0	1.3	0.4	0.55	0.75	1.0	1.3
E,KEV										
2.0	.0029	.0038	.0050	.0062	.0075	.3766	.3664	.3534	.3380	.3208
2.3	.0030	.0039	.0051	.0063	.0076	.3761	.3656	.3523	.3367	.3192
2.6	.0030	.0040	.0051	.0064	.0077	.3756	.3649	.3514	.3355	.3177
3.0	.0031	.0041	.0052	.0065	.0078	.3750	.3641	.3503	.3341	.3160
3.5	.0032	.0041	.0053	.0066	.0079	.3743	.3632	.3491	.3325	.3141
4.0	.0032	.0042	.0054	.0067	.0080	.3737	.3624	.3480	.3312	.3125
4.5	.0033	.0043	.0055	.0068	.0081	.3732	.3617	.3470	.3300	.3110
5.0	.0033	.0043	.0056	.0069	.0082	.3727	.3610	.3462	.3289	.3096
6.0	.0034	.0044	.0057	.0070	.0083	.3719	.3599	.3447	.3269	.3073
7.0	.0035	.0045	.0058	.0071	.0084	.3713	.3589	.3434	.3253	.3053
8.0	.0035	.0046	.0059	.0072	.0085	.3707	.3581	.3423	.3239	.3036
9.0	.0036	.0047	.0059	.0073	.0086	.3702	.3574	.3413	.3226	.3021
10	.0036	.0047	.0060	.0074	.0087	.3698	.3568	.3404	.3215	.3007
12	.0037	.0048	.0061	.0075	.0088	.3690	.3557	.3390	.3195	.2983
14	.0038	.0049	.0062	.0075	.0089	.3684	.3548	.3377	.3179	.2963
17	.0038	.0050	.0063	.0076	.0089	.3677	.3538	.3362	.3158	.2938
20	.0039	.0050	.0064	.0077	.0090	.3671	.3528	.3348	.3140	.2916
23	.0040	.0051	.0064	.0078	.0090	.3665	.3520	.3337	.3125	.2897
26	.0040	.0051	.0065	.0078	.0091	.3661	.3513	.3326	.3111	.2880
30	.0041	.0052	.0065	.0079	.0091	.3656	.3505	.3314	.3095	.2860
35	.0041	.0053	.0066	.0079	.0091	.3650	.3496	.3301	.3077	.2838
40	.0042	.0053	.0066	.0079	.0091	.3646	.3488	.3290	.3061	.2819
45	.0042	.0053	.0066	.0080	.0091	.3642	.3482	.3279	.3048	.2802
50	.0042	.0054	.0067	.0080	.0091	.3639	.3475	.3270	.3035	.2787
60	.0043	.0054	.0067	.0079	.0090	.3635	.3466	.3255	.3013	.2760
70	.0043	.0054	.0067	.0079	.0090	.3631	.3458	.3241	.2994	.2737
80	.0043	.0054	.0067	.0079	.0089	.3628	.3451	.3229	.2978	.2716
90	.0043	.0054	.0066	.0078	.0088	.3624	.3444	.3218	.2962	.2698
100	.0043	.0054	.0066	.0078	.0087	.3621	.3437	.3208	.2948	.2681
120	.0043	.0054	.0065	.0076	.0085	.3613	.3425	.3189	.2923	.2651
140	.0043	.0053	.0064	.0075	.0083	.3606	.3413	.3172	.2901	.2624
170	.0043	.0053	.0063	.0073	.0080	.3595	.3396	.3149	.2871	.2589
200	.0042	.0052	.0062	.0071	.0078	.3585	.3381	.3127	.2844	.2558
230	.0042	.0051	.0060	.0069	.0075	.3577	.3368	.3108	.2818	.2530
260	.0042	.0050	.0059	.0067	.0073	.3569	.3355	.3090	.2795	.2504
300	.0041	.0049	.0057	.0065	.0069	.3559	.3339	.3067	.2766	.2473
350	.0041	.0048	.0055	.0062	.0066	.3545	.3319	.3041	.2734	.2438
400	.0040	.0046	.0053	.0059	.0063	.3531	.3300	.3016	.2704	.2407
450	.0039	.0045	.0051	.0057	.0060	.3516	.3282	.2993	.2677	.2379
500	.0038	.0043	.0050	.0055	.0057	.3501	.3264	.2971	.2652	.2354
600	.0035	.0040	.0046	.0050	.0052	.3467	.3226	.2928	.2604	.2306
700	.0032	.0037	.0042	.0047	.0048	.3434	.3190	.2890	.2563	.2265
800	.0030	.0034	.0039	.0043	.0044	.3402	.3158	.2855	.2527	.2232
900	.0028	.0031	.0036	.0040	.0041	.3372	.3127	.2824	.2496	.2203
1000	.0026	.0029	.0033	.0038	.0038	.3342	.3098	.2796	.2469	.2180
1200	.0024	.0026	.0029	.0034	.0033	.3280	.3042	.2745	.2424	.2150
1400	.0022	.0023	.0026	.0030	.0029	.3226	.2993	.2702	.2388	.2128
1700	.0020	.0020	.0022	.0026	.0024	.3156	.2933	.2650	.2345	.2102
2000	.0019	.0017	.0018	.0022	.0020	.3099	.2884	.2608	.2310	.2080
2300	.0017	.0014	.0015	.0018	.0017	.3066	.2853	.2580	.2282	.2054
2600	.0016	.0011	.0012	.0015	.0013	.3038	.2829	.2556	.2258	.2030
3000	.0014	.0008	.0009	.0011	.0010	.3009	.2802	.2531	.2232	.2002
3500	.0013	.0005	.0006	.0007	.0006	.2978	.2774	.2505	.2204	.1971
4000	.0011	.0003	.0003	.0004	.0004	.2951	.2751	.2482	.2180	.1944
4500	.0010	.0001	.0001	.0002	.0002	.2926	.2729	.2462	.2159	.1922
5000	.0009	0.0000	0.0000	0.0000	0.0000	.2903	.2709	.2445	.2141	.1902
6000	.0006	0.0000	0.0000	0.0000	0.0000	.2847	.2664	.2408	.2108	.1874
7000	.0004	0.0000	0.0000	0.0000	0.0000	.2800	.2625	.2376	.2080	.1851
8000	.0002	0.0000	0.0000	0.0000	0.0000	.2759	.2592	.2349	.2056	.1832
9000	.0001	0.0000	0.0000	0.0000	0.0000	.2723	.2563	.2325	.2035	.1815
10000	0.0000	0.0000	0.0000	0.0000	0.0000	.2693	.2537	.2304	.2017	.1800
12000	0.0000	0.0000	0.0000	0.0000	0.0000	.2644	.2495	.2268	.1984	.1773
14000	0.0000	0.0000	0.0000	0.0000	0.0000	.2608	.2463	.2238	.1957	.1750
17000	0.0000	0.0000	0.0000	0.0000	0.0000	.2570	.2426	.2203	.1924	.1720
20000	0.0000	0.0000	0.0000	0.0000	0.0000	.2547	.2400	.2174	.1896	.1693

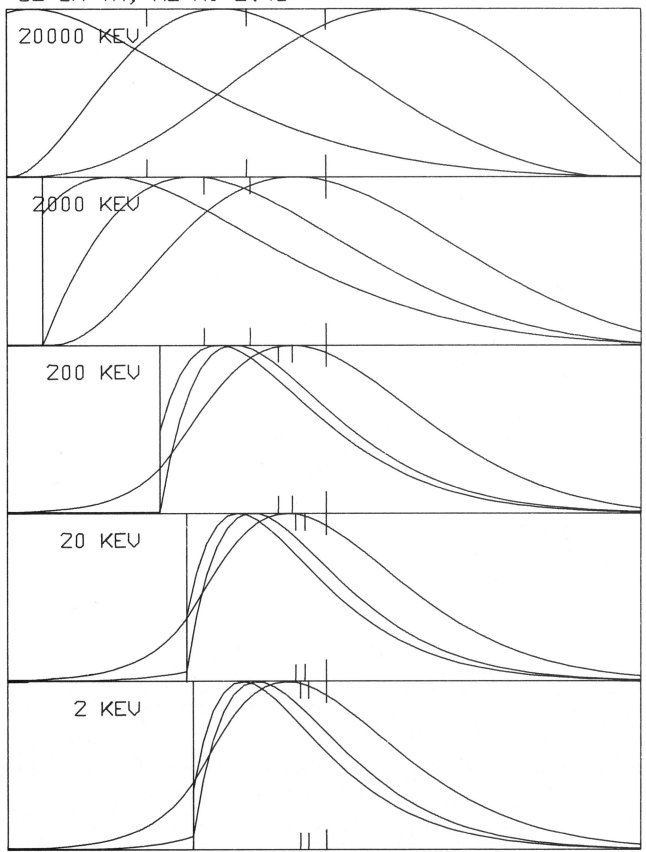

AG ON AU, M2/M1=1.82   TABLE 10- 1

K/KL=	FRACTIONAL DEPOSITED ENERGY					MEAN RANGE,MICROGRAM/SQ.CM.				
E,KEV	0.5	0.65	0.8	1.0	1.2	0.5	0.65	0.8	1.0	1.2
5.0	.8930	.8639	.8360	.8007	.7674	3.233	3.213	3.193	3.168	3.143
6.0	.8898	.8600	.8315	.7954	.7613	3.677	3.653	3.629	3.599	3.569
7.0	.8871	.8566	.8275	.7907	.7561	4.079	4.052	4.026	3.991	3.957
8.0	.8847	.8537	.8240	.7866	.7514	4.457	4.427	4.397	4.359	4.321
9.0	.8826	.8510	.8209	.7829	.7473	4.818	4.785	4.753	4.710	4.668
10	.8806	.8485	.8180	.7796	.7435	5.168	5.132	5.096	5.050	5.004
12	.8771	.8442	.8130	.7737	.7369	5.841	5.799	5.757	5.703	5.651
14	.8741	.8405	.8086	.7685	.7311	6.488	6.440	6.393	6.331	6.271
17	.8701	.8356	.8029	.7619	.7237	7.423	7.366	7.310	7.236	7.164
20	.8667	.8314	.7980	.7562	.7173	8.322	8.255	8.190	8.104	8.021
23	.8637	.8277	.7937	.7512	.7117	9.143	9.068	8.994	8.899	8.805
26	.8609	.8243	.7897	.7466	.7066	9.945	9.861	9.779	9.673	9.569
30	.8577	.8203	.7851	.7413	.7006	10.99	10.90	10.80	10.68	10.56
35	.8540	.8158	.7799	.7353	.6940	12.27	12.16	12.05	11.91	11.78
40	.8507	.8118	.7753	.7300	.6881	13.53	13.40	13.28	13.12	12.96
45	.8478	.8083	.7712	.7252	.6829	14.76	14.61	14.47	14.29	14.12
50	.8451	.8050	.7674	.7209	.6781	15.96	15.80	15.65	15.44	15.25
60	.8403	.7993	.7608	.7133	.6697	18.18	17.99	17.81	17.57	17.34
70	.8360	.7942	.7550	.7066	.6624	20.38	20.16	19.94	19.67	19.40
80	.8322	.7897	.7498	.7006	.6559	22.57	22.31	22.06	21.74	21.43
90	.8287	.7855	.7450	.6951	.6499	24.74	24.45	24.16	23.80	23.44
100	.8254	.7817	.7406	.6901	.6444	26.90	26.56	26.24	25.83	25.43
120	.8195	.7746	.7326	.6810	.6345	30.85	30.45	30.07	29.57	29.10
140	.8142	.7683	.7254	.6729	.6257	34.85	34.38	33.93	33.34	32.78
170	.8071	.7598	.7158	.6621	.6141	40.97	40.37	39.78	39.04	38.33
200	.8006	.7523	.7072	.6525	.6038	47.19	46.43	45.70	44.78	43.90
230	.7948	.7454	.6996	.6439	.5946	52.64	51.78	50.96	49.91	48.91
260	.7895	.7391	.6925	.6360	.5862	58.25	57.27	56.34	55.14	53.99
300	.7828	.7313	.6838	.6264	.5759	66.02	64.84	63.71	62.27	60.90
350	.7750	.7224	.6738	.6155	.5644	76.14	74.65	73.23	71.44	69.74
400	.7678	.7141	.6647	.6055	.5539	86.63	84.78	83.02	80.81	78.74
450	.7611	.7064	.6562	.5963	.5443	97.39	95.12	92.99	90.32	87.84
500	.7547	.6992	.6483	.5877	.5353	108.3	105.6	103.1	99.92	97.00
600	.7429	.6858	.6337	.5719	.5190	128.3	124.9	121.7	117.8	114.1
700	.7321	.6736	.6204	.5577	.5045	149.6	145.2	141.2	136.3	131.8
800	.7219	.6623	.6082	.5448	.4913	172.0	166.5	161.5	155.4	149.8
900	.7123	.6517	.5969	.5329	.4792	195.2	188.5	182.3	174.9	168.2
1000	.7032	.6417	.5863	.5218	.4681	219.1	210.9	203.5	194.6	186.7
1200	.6860	.6231	.5666	.5014	.4477	263.9	253.5	243.9	232.4	222.2
1400	.6702	.6061	.5489	.4832	.4297	311.6	298.3	286.0	271.3	258.5
1700	.6486	.5833	.5253	.4593	.4062	387.5	368.7	351.5	331.2	313.7
2000	.6290	.5629	.5044	.4384	.3860	467.4	442.0	419.0	392.1	369.6
2300	.6105	.5439	.4853	.4195	.3679	543.4	512.2	484.0	451.1	423.7
2600	.5934	.5266	.4679	.4026	.3518	622.0	584.2	550.1	510.6	477.9
3000	.5725	.5056	.4472	.3825	.3329	730.5	682.5	639.6	590.2	550.0
3500	.5490	.4823	.4243	.3606	.3124	870.5	808.1	752.7	689.8	639.5
4000	.5278	.4615	.4041	.3415	.2947	1014.	935.5	866.5	789.1	728.2
4500	.5085	.4428	.3861	.3246	.2791	1159.	1064.	980.3	887.5	815.6
5000	.4909	.4259	.3699	.3095	.2654	1305.	1192.	1094.	984.7	901.7
6000	.4590	.3958	.3417	.2837	.2420	1590.	1442.	1314.	1173.	1068.
7000	.4316	.3702	.3179	.2623	.2228	1877.	1691.	1531.	1357.	1229.
8000	.4076	.3481	.2976	.2441	.2067	2163.	1938.	1745.	1537.	1386.
9000	.3865	.3288	.2799	.2285	.1929	2448.	2182.	1955.	1712.	1538.
10000	.3676	.3116	.2644	.2149	.1809	2730.	2422.	2161.	1882.	1685.
12000	.3355	.2829	.2386	.1926	.1615	3281.	2890.	2558.	2210.	1968.
14000	.3089	.2593	.2178	.1748	.1461	3818.	3342.	2941.	2523.	2236.
17000	.2763	.2308	.1928	.1537	.1280	4598.	3996.	3491.	2968.	2617.
20000	.2503	.2082	.1731	.1373	.1140	5348.	4621.	4014.	3389.	2975.
23000	.2288	.1897	.1572	.1241	.1028	6070.	5220.	4512.	3787.	3314.
26000	.2109	.1744	.1441	.1134	.0937	6764.	5795.	4989.	4167.	3636.
30000	.1912	.1576	.1298	.1017	.0839	7653.	6528.	5595.	4648.	4042.
35000	.1715	.1410	.1157	.0903	.0744	8708.	7396.	6310.	5214.	4519.
40000	.1559	.1279	.1047	.0815	.0670	9710.	8217.	6985.	5745.	4967.
45000	.1434	.1174	.0960	.0745	.0612	10664	8998.	7624.	6247.	5389.
50000	.1333	.1090	.0890	.0690	.0565	11575	9743.	8233.	6724.	5790.

TABLE 10- 2

AG ON AU, M2/M1=1.82

E,KEV	MEAN DAMAGE DEPTH,MICROGRAM/SQ.CM.					MEAN IONZN.DEPTH,MICROGRAM/SQ.CM.				
K/KL=	0.5	0.65	0.8	1.0	1.2	0.5	0.65	0.8	1.0	1.2
5.0	2.797	2.768	2.739	2.703	2.667	2.587	2.556	2.526	2.487	2.450
6.0	3.176	3.142	3.108	3.065	3.023	2.937	2.900	2.865	2.819	2.775
7.0	3.522	3.482	3.444	3.395	3.348	3.255	3.214	3.174	3.122	3.072
8.0	3.846	3.802	3.760	3.705	3.652	3.555	3.509	3.464	3.406	3.350
9.0	4.156	4.108	4.061	4.001	3.943	3.841	3.790	3.741	3.677	3.616
10	4.455	4.403	4.352	4.287	4.224	4.117	4.062	4.008	3.939	3.873
12	5.031	4.970	4.911	4.835	4.762	4.648	4.583	4.521	4.441	4.364
14	5.583	5.513	5.446	5.360	5.277	5.156	5.083	5.012	4.921	4.834
17	6.377	6.295	6.216	6.114	6.016	5.887	5.801	5.717	5.609	5.506
20	7.138	7.043	6.952	6.834	6.721	6.587	6.487	6.391	6.266	6.147
23	7.835	7.729	7.627	7.495	7.369	7.230	7.118	7.009	6.870	6.737
26	8.514	8.396	8.282	8.137	7.997	7.854	7.730	7.610	7.456	7.309
30	9.397	9.263	9.134	8.969	8.812	8.665	8.524	8.388	8.214	8.048
35	10.47	10.32	10.17	9.981	9.800	9.652	9.490	9.333	9.134	8.943
40	11.52	11.35	11.18	10.96	10.76	10.61	10.43	10.25	10.03	9.810
45	12.54	12.35	12.16	11.92	11.69	11.55	11.34	11.15	10.89	10.65
50	13.54	13.33	13.12	12.85	12.60	12.46	12.24	12.02	11.74	11.47
60	15.39	15.14	14.89	14.58	14.29	14.15	13.89	13.63	13.30	12.99
70	17.21	16.91	16.63	16.27	15.93	15.81	15.50	15.21	14.83	14.47
80	19.00	18.66	18.34	17.93	17.55	17.45	17.09	16.76	16.33	15.92
90	20.77	20.39	20.03	19.57	19.13	19.06	18.66	18.28	17.80	17.34
100	22.52	22.10	21.69	21.18	20.69	20.65	20.20	19.78	19.24	18.73
120	25.74	25.24	24.76	24.15	23.58	23.58	23.05	22.55	21.91	21.31
140	28.97	28.39	27.82	27.11	26.45	26.50	25.88	25.30	24.56	23.86
170	33.86	33.13	32.43	31.55	30.73	30.91	30.14	29.41	28.49	27.64
200	38.78	37.89	37.04	35.97	34.98	35.32	34.38	33.49	32.39	31.37
230	43.12	42.11	41.15	39.94	38.82	39.26	38.20	37.20	35.94	34.78
260	47.55	46.40	45.31	43.95	42.68	43.27	42.06	40.92	39.50	38.19
300	53.63	52.27	50.98	49.38	47.89	48.71	47.28	45.94	44.28	42.74
350	61.47	59.81	58.24	56.28	54.49	55.67	53.93	52.30	50.28	48.43
400	69.54	67.53	65.64	63.29	61.16	62.76	60.66	58.71	56.32	54.13
450	77.77	75.37	73.13	70.37	67.87	69.95	67.45	65.15	62.35	59.80
500	86.10	83.29	80.67	77.47	74.59	77.18	74.27	71.60	68.37	65.45
600	101.3	97.87	94.63	90.68	87.13	90.55	86.91	83.65	79.66	76.07
700	117.4	113.1	109.1	104.3	99.97	104.4	99.93	95.95	91.09	86.74
800	134.2	129.0	124.1	118.2	113.0	118.8	113.3	108.4	102.6	97.43
900	151.6	145.2	139.4	132.3	126.2	133.4	126.8	121.0	114.1	108.1
1000	169.3	161.7	154.8	146.6	139.5	148.2	140.5	133.7	125.7	118.7
1200	202.5	192.9	184.2	173.8	164.9	176.4	166.7	157.9	147.9	139.2
1400	237.6	225.5	214.6	201.6	190.6	205.6	193.6	182.4	170.2	159.5
1700	293.1	276.6	261.7	244.4	229.9	250.6	234.6	219.6	203.4	189.6
2000	351.2	329.6	310.2	287.8	269.5	296.6	276.0	256.8	236.4	219.2
2300	406.0	379.8	356.3	329.4	307.5	341.0	316.1	293.1	268.3	247.8
2600	462.6	431.2	403.3	371.4	345.6	385.9	356.3	329.2	299.8	275.9
3000	540.7	501.6	467.0	427.7	396.4	446.3	409.9	376.9	341.0	312.5
3500	641.4	591.6	547.7	498.5	459.8	522.2	476.6	435.9	391.4	356.9
4000	744.6	683.0	629.1	569.2	522.9	598.1	542.7	493.9	440.5	399.9
4500	849.3	775.2	710.7	639.5	585.3	673.7	608.1	550.8	488.3	441.6
5000	954.8	867.6	792.1	709.2	647.0	748.6	672.6	606.5	534.9	482.0
6000	1160.	1047.	949.7	844.0	765.8	894.5	797.6	714.1	624.3	559.2
7000	1367.	1227.	1106.	976.5	882.1	1037.	918.9	817.6	709.5	632.5
8000	1574.	1405.	1261.	1107.	995.8	1177.	1037.	917.4	791.0	702.3
9000	1781.	1583.	1414.	1234.	1107.	1313.	1151.	1014.	869.2	768.9
10000	1987.	1758.	1564.	1359.	1215.	1446.	1262.	1107.	944.4	832.8
12000	2386.	2098.	1855.	1599.	1423.	1703.	1475.	1284.	1086.	953.0
14000	2779.	2430.	2137.	1830.	1622.	1947.	1678.	1452.	1219.	1065.
17000	3354.	2914.	2545.	2163.	1908.	2294.	1963.	1687.	1405.	1221.
20000	3912.	3381.	2937.	2481.	2180.	2621.	2230.	1906.	1577.	1365.
23000	4453.	3832.	3315.	2785.	2439.	2929.	2482.	2112.	1737.	1499.
26000	4977.	4268.	3678.	3077.	2688.	3222.	2720.	2306.	1888.	1625.
30000	5652.	4828.	4144.	3450.	3005.	3591.	3020.	2549.	2076.	1782.
35000	6460.	5496.	4698.	3891.	3379.	4023.	3370.	2832.	2295.	1964.
40000	7231.	6132.	5224.	4309.	3733.	4427.	3696.	3096.	2498.	2133.
45000	7970.	6741.	5726.	4707.	4070.	4808.	4003.	3343.	2688.	2291.
50000	8679.	7324.	6206.	5087.	4390.	5169.	4294.	3577.	2868.	2440.

TABLE 10- 3

AG ON AU, M2/M1=1.82

	RELATIVE RANGE STRAGGLING					RELATIVE DAMAGE STRAGGLING				
K/KL=	0.5	0.65	0.8	1.0	1.2	0.5	0.65	0.8	1.0	1.2
E,KEV										
5.0	.5129	.5067	.5006	.4927	.4852	.4606	.4575	.4544	.4506	.4468
6.0	.5115	.5051	.4988	.4908	.4830	.4601	.4569	.4537	.4498	.4460
7.0	.5102	.5037	.4973	.4890	.4811	.4596	.4563	.4531	.4491	.4452
8.0	.5090	.5023	.4958	.4874	.4794	.4592	.4558	.4526	.4484	.4445
9.0	.5079	.5011	.4945	.4859	.4778	.4588	.4554	.4521	.4479	.4438
10	.5068	.4999	.4932	.4845	.4762	.4584	.4550	.4516	.4473	.4432
12	.5048	.4977	.4908	.4819	.4734	.4578	.4542	.4507	.4463	.4422
14	.5029	.4956	.4886	.4795	.4709	.4571	.4534	.4499	.4454	.4412
17	.5002	.4927	.4855	.4762	.4673	.4562	.4524	.4488	.4442	.4398
20	.4977	.4900	.4826	.4731	.4640	.4554	.4515	.4478	.4431	.4386
23	.4953	.4875	.4799	.4703	.4610	.4547	.4507	.4469	.4420	.4375
26	.4931	.4851	.4774	.4676	.4582	.4540	.4499	.4460	.4411	.4364
30	.4902	.4820	.4742	.4642	.4547	.4531	.4489	.4449	.4399	.4351
35	.4868	.4784	.4704	.4602	.4505	.4520	.4477	.4436	.4384	.4336
40	.4835	.4750	.4669	.4565	.4466	.4510	.4466	.4424	.4371	.4322
45	.4804	.4718	.4635	.4529	.4429	.4500	.4455	.4412	.4359	.4308
50	.4774	.4686	.4602	.4495	.4394	.4490	.4444	.4401	.4346	.4295
60	.4718	.4628	.4541	.4432	.4328	.4472	.4425	.4380	.4324	.4272
70	.4665	.4573	.4484	.4373	.4268	.4456	.4407	.4361	.4303	.4249
80	.4615	.4521	.4431	.4318	.4211	.4440	.4390	.4342	.4283	.4228
90	.4568	.4472	.4381	.4266	.4158	.4424	.4373	.4325	.4264	.4209
100	.4523	.4425	.4333	.4216	.4107	.4410	.4357	.4308	.4246	.4190
120	.4438	.4337	.4242	.4123	.4012	.4383	.4328	.4277	.4213	.4154
140	.4359	.4256	.4159	.4038	.3925	.4357	.4301	.4247	.4181	.4121
170	.4252	.4146	.4046	.3922	.3807	.4321	.4262	.4206	.4138	.4075
200	.4154	.4045	.3944	.3817	.3701	.4287	.4225	.4168	.4097	.4033
230	.4063	.3952	.3848	.3720	.3602	.4254	.4190	.4131	.4058	.3992
260	.3979	.3866	.3760	.3630	.3511	.4222	.4157	.4095	.4021	.3954
300	.3877	.3760	.3653	.3521	.3400	.4182	.4114	.4051	.3975	.3906
350	.3761	.3641	.3531	.3397	.3276	.4136	.4065	.4000	.3921	.3851
400	.3656	.3534	.3422	.3286	.3164	.4092	.4020	.3952	.3871	.3800
450	.3560	.3436	.3322	.3185	.3062	.4051	.3977	.3908	.3825	.3753
500	.3472	.3345	.3231	.3093	.2969	.4013	.3936	.3866	.3782	.3708
600	.3313	.3182	.3065	.2925	.2801	.3943	.3862	.3789	.3701	.3626
700	.3174	.3040	.2921	.2781	.2656	.3879	.3795	.3719	.3628	.3551
800	.3051	.2914	.2794	.2653	.2529	.3820	.3733	.3655	.3562	.3484
900	.2941	.2803	.2681	.2539	.2416	.3765	.3676	.3596	.3502	.3423
1000	.2842	.2702	.2579	.2437	.2314	.3714	.3623	.3541	.3445	.3366
1200	.2667	.2527	.2403	.2259	.2138	.3617	.3523	.3439	.3342	.3262
1400	.2518	.2379	.2253	.2109	.1989	.3529	.3434	.3349	.3251	.3171
1700	.2330	.2194	.2066	.1922	.1805	.3415	.3317	.3231	.3133	.3054
2000	.2174	.2039	.1912	.1769	.1655	.3315	.3217	.3130	.3032	.2953
2300	.2040	.1907	.1781	.1641	.1531	.3228	.3129	.3042	.2944	.2867
2600	.1924	.1792	.1669	.1531	.1426	.3151	.3051	.2964	.2866	.2790
3000	.1790	.1660	.1540	.1408	.1307	.3059	.2960	.2873	.2776	.2701
3500	.1649	.1521	.1406	.1279	.1184	.2959	.2860	.2774	.2678	.2606
4000	.1530	.1405	.1294	.1173	.1083	.2873	.2774	.2688	.2594	.2523
4500	.1428	.1306	.1199	.1083	.0997	.2796	.2698	.2613	.2521	.2452
5000	.1340	.1220	.1117	.1006	.0925	.2727	.2630	.2547	.2456	.2388
6000	.1194	.1082	.0986	.0884	.0810	.2609	.2515	.2434	.2346	.2282
7000	.1077	.0972	.0883	.0788	.0720	.2510	.2418	.2340	.2255	.2194
8000	.0982	.0883	.0799	.0711	.0648	.2425	.2336	.2260	.2179	.2120
9000	.0902	.0809	.0730	.0647	.0589	.2351	.2265	.2191	.2113	.2056
10000	.0834	.0745	.0672	.0594	.0540	.2287	.2203	.2131	.2055	.2000
12000	.0728	.0648	.0581	.0512	.0464	.2180	.2100	.2032	.1960	.1908
14000	.0646	.0574	.0513	.0449	.0407	.2092	.2016	.1951	.1882	.1834
17000	.0554	.0489	.0436	.0380	.0343	.1986	.1915	.1854	.1789	.1744
20000	.0484	.0427	.0378	.0329	.0297	.1901	.1833	.1776	.1715	.1672
23000	.0430	.0378	.0334	.0290	.0261	.1831	.1766	.1711	.1653	.1612
26000	.0387	.0339	.0300	.0259	.0233	.1771	.1709	.1656	.1601	.1562
30000	.0340	.0298	.0263	.0227	.0204	.1704	.1645	.1595	.1542	.1504
35000	.0296	.0259	.0228	.0197	.0177	.1635	.1579	.1531	.1481	.1445
40000	.0261	.0228	.0201	.0173	.0156	.1578	.1524	.1478	.1430	.1396
45000	.0234	.0204	.0180	.0155	.0140	.1530	.1478	.1433	.1386	.1353
50000	.0211	.0185	.0163	.0141	.0126	.1489	.1438	.1394	.1349	.1317

TABLE 10-4

AG ON AU, M2/M1=1.82

E,KEV	RELATIVE IONIZATION STRAGGLING					RELATIVE TRANSVERSE RANGE STRAGGLING				
K/KL=	0.5	0.65	0.8	1.0	1.2	0.5	0.65	0.8	1.0	1.2
5.0	.5095	.5071	.5049	.5020	.4993	.8154	.8147	.8141	.8132	.8124
6.0	.5091	.5067	.5044	.5015	.4987	.8153	.8146	.8140	.8131	.8123
7.0	.5088	.5063	.5039	.5010	.4982	.8152	.8145	.8139	.8130	.8122
8.0	.5085	.5059	.5035	.5005	.4977	.8152	.8145	.8138	.8129	.8121
9.0	.5082	.5056	.5032	.5001	.4972	.8151	.8144	.8137	.8128	.8120
10	.5079	.5053	.5028	.4997	.4968	.8151	.8144	.8137	.8128	.8119
12	.5074	.5047	.5022	.4990	.4960	.8150	.8143	.8136	.8127	.8118
14	.5069	.5042	.5016	.4983	.4953	.8149	.8142	.8135	.8126	.8117
17	.5063	.5035	.5008	.4974	.4943	.8148	.8141	.8134	.8125	.8116
20	.5057	.5028	.5001	.4966	.4935	.8148	.8141	.8134	.8124	.8116
23	.5052	.5022	.4994	.4959	.4927	.8147	.8140	.8133	.8124	.8115
26	.5046	.5016	.4987	.4952	.4919	.8147	.8140	.8133	.8124	.8115
30	.5040	.5009	.4979	.4943	.4910	.8147	.8140	.8133	.8124	.8115
35	.5032	.5000	.4970	.4933	.4899	.8147	.8140	.8133	.8124	.8116
40	.5025	.4992	.4962	.4924	.4890	.8147	.8140	.8133	.8125	.8117
45	.5018	.4984	.4953	.4915	.4880	.8147	.8141	.8134	.8126	.8118
50	.5011	.4977	.4946	.4907	.4872	.8148	.8142	.8135	.8127	.8119
60	.4999	.4964	.4932	.4892	.4856	.8149	.8143	.8137	.8129	.8122
70	.4988	.4952	.4919	.4878	.4841	.8151	.8145	.8140	.8133	.8126
80	.4978	.4941	.4907	.4865	.4828	.8154	.8148	.8143	.8137	.8130
90	.4967	.4929	.4895	.4853	.4815	.8157	.8152	.8147	.8141	.8135
100	.4957	.4919	.4883	.4840	.4802	.8160	.8155	.8151	.8146	.8141
120	.4937	.4897	.4861	.4816	.4777	.8166	.8163	.8159	.8155	.8151
140	.4918	.4877	.4839	.4794	.4754	.8174	.8171	.8169	.8166	.8163
170	.4892	.4849	.4810	.4764	.4723	.8187	.8186	.8185	.8184	.8183
200	.4868	.4823	.4783	.4736	.4694	.8203	.8203	.8203	.8204	.8204
230	.4846	.4800	.4759	.4711	.4669	.8217	.8219	.8221	.8223	.8225
260	.4826	.4778	.4736	.4687	.4645	.8234	.8237	.8240	.8244	.8248
300	.4800	.4752	.4709	.4659	.4616	.8257	.8262	.8267	.8273	.8279
350	.4771	.4721	.4677	.4626	.4583	.8288	.8295	.8303	.8312	.8320
400	.4744	.4692	.4647	.4596	.4553	.8321	.8330	.8340	.8351	.8362
450	.4719	.4666	.4620	.4569	.4525	.8354	.8366	.8378	.8392	.8405
500	.4695	.4641	.4595	.4543	.4500	.8389	.8403	.8417	.8434	.8449
600	.4648	.4592	.4546	.4494	.4452	.8458	.8478	.8495	.8516	.8535
700	.4607	.4550	.4503	.4451	.4410	.8528	.8553	.8573	.8599	.8622
800	.4570	.4512	.4466	.4414	.4373	.8599	.8627	.8652	.8682	.8709
900	.4537	.4479	.4432	.4381	.4342	.8669	.8702	.8730	.8764	.8795
1000	.4507	.4449	.4402	.4352	.4314	.8740	.8775	.8807	.8846	.8881
1200	.4454	.4400	.4351	.4304	.4268	.8879	.8916	.8958	.9007	.9050
1400	.4410	.4359	.4309	.4264	.4231	.9015	.9054	.9104	.9164	.9214
1700	.4355	.4309	.4258	.4217	.4188	.9212	.9255	.9317	.9389	.9449
2000	.4309	.4268	.4218	.4181	.4156	.9401	.9448	.9520	.9603	.9672
2300	.4271	.4232	.4186	.4154	.4132	.9576	.9635	.9716	.9808	.9883
2600	.4238	.4202	.4160	.4131	.4114	.9744	.9815	.9903	1.000	1.008
3000	.4202	.4168	.4133	.4109	.4097	.9960	1.005	1.014	1.025	1.034
3500	.4165	.4135	.4107	.4089	.4082	1.022	1.032	1.043	1.054	1.064
4000	.4137	.4109	.4088	.4075	.4073	1.046	1.058	1.069	1.082	1.092
4500	.4113	.4090	.4075	.4066	.4069	1.070	1.083	1.095	1.108	1.118
5000	.4095	.4075	.4065	.4061	.4067	1.093	1.107	1.119	1.132	1.143
6000	.4070	.4059	.4057	.4062	.4074	1.139	1.154	1.166	1.180	1.190
7000	.4053	.4052	.4057	.4068	.4086	1.181	1.196	1.209	1.222	1.233
8000	.4044	.4050	.4060	.4078	.4099	1.220	1.235	1.247	1.261	1.272
9000	.4039	.4052	.4067	.4089	.4114	1.255	1.270	1.283	1.297	1.307
10000	.4038	.4056	.4075	.4102	.4130	1.288	1.303	1.316	1.330	1.340
12000	.4046	.4072	.4097	.4130	.4163	1.343	1.360	1.374	1.389	1.399
14000	.4059	.4090	.4120	.4158	.4194	1.392	1.410	1.425	1.441	1.451
17000	.4082	.4120	.4155	.4199	.4238	1.457	1.476	1.492	1.508	1.519
20000	.4107	.4150	.4189	.4236	.4277	1.515	1.535	1.551	1.567	1.577
23000	.4132	.4178	.4220	.4270	.4313	1.567	1.587	1.604	1.620	1.629
26000	.4156	.4205	.4249	.4301	.4345	1.614	1.635	1.651	1.667	1.675
30000	.4187	.4238	.4285	.4338	.4383	1.673	1.693	1.708	1.723	1.731
35000	.4222	.4276	.4324	.4379	.4424	1.739	1.758	1.772	1.785	1.791
40000	.4254	.4310	.4359	.4415	.4459	1.800	1.817	1.830	1.841	1.844
45000	.4283	.4340	.4390	.4446	.4490	1.856	1.871	1.882	1.890	1.892
50000	.4310	.4367	.4418	.4474	.4517	1.908	1.921	1.930	1.936	1.935

TABLE 10- 5

AG ON AU, M2/M1=1.82

K/KL= E,KEV	RELATIVE TRANSV. DAMAGE STRAGGLING					RELATIVE TRANSV. IONZN. STRAGGLING				
	0.5	0.65	0.8	1.0	1.2	0.5	0.65	0.8	1.0	1.2
5.0	.6174	.6143	.6111	.6070	.6030	.5237	.5193	.5150	.5093	.5037
6.0	.6168	.6135	.6103	.6061	.6019	.5229	.5184	.5139	.5081	.5023
7.0	.6162	.6128	.6095	.6052	.6010	.5222	.5175	.5129	.5070	.5011
8.0	.6156	.6122	.6088	.6044	.6001	.5215	.5167	.5120	.5059	.5000
9.0	.6150	.6116	.6081	.6037	.5993	.5208	.5160	.5112	.5050	.4989
10	.6145	.6110	.6075	.6030	.5985	.5202	.5153	.5105	.5041	.4980
12	.6136	.6099	.6063	.6017	.5971	.5191	.5140	.5090	.5025	.4962
14	.6127	.6089	.6053	.6005	.5958	.5180	.5128	.5077	.5011	.4946
17	.6114	.6076	.6038	.5989	.5940	.5165	.5111	.5059	.4991	.4924
20	.6102	.6063	.6024	.5974	.5924	.5151	.5096	.5042	.4972	.4904
23	.6091	.6051	.6011	.5960	.5909	.5138	.5082	.5027	.4955	.4886
26	.6081	.6039	.5999	.5947	.5896	.5126	.5069	.5013	.4940	.4869
30	.6067	.6025	.5984	.5930	.5878	.5110	.5052	.4994	.4920	.4848
35	.6051	.6008	.5966	.5911	.5858	.5092	.5032	.4973	.4897	.4823
40	.6036	.5992	.5949	.5893	.5839	.5074	.5013	.4953	.4875	.4800
45	.6022	.5977	.5933	.5876	.5821	.5058	.4995	.4934	.4855	.4778
50	.6008	.5962	.5918	.5860	.5804	.5042	.4978	.4916	.4835	.4757
60	.5982	.5935	.5889	.5830	.5772	.5011	.4945	.4881	.4799	.4718
70	.5958	.5910	.5863	.5802	.5743	.4982	.4915	.4849	.4764	.4682
80	.5935	.5886	.5838	.5776	.5715	.4956	.4887	.4819	.4733	.4649
90	.5913	.5863	.5814	.5751	.5690	.4930	.4860	.4791	.4703	.4617
100	.5893	.5842	.5792	.5728	.5665	.4906	.4834	.4765	.4674	.4588
120	.5854	.5802	.5750	.5684	.5620	.4861	.4787	.4715	.4622	.4533
140	.5819	.5765	.5712	.5644	.5578	.4820	.4744	.4669	.4574	.4482
170	.5771	.5715	.5660	.5590	.5522	.4764	.4684	.4607	.4508	.4413
200	.5729	.5671	.5614	.5541	.5471	.4712	.4630	.4550	.4448	.4350
230	.5690	.5630	.5572	.5496	.5424	.4665	.4579	.4497	.4392	.4292
260	.5655	.5593	.5533	.5456	.5382	.4620	.4532	.4448	.4340	.4237
300	.5613	.5548	.5486	.5406	.5330	.4566	.4475	.4387	.4276	.4170
350	.5566	.5498	.5433	.5350	.5271	.4504	.4409	.4318	.4202	.4093
400	.5524	.5453	.5386	.5299	.5217	.4447	.4349	.4254	.4135	.4023
450	.5486	.5413	.5342	.5253	.5168	.4396	.4293	.4196	.4073	.3958
500	.5451	.5376	.5303	.5210	.5123	.4347	.4241	.4141	.4015	.3897
600	.5391	.5310	.5232	.5134	.5042	.4262	.4149	.4043	.3910	.3786
700	.5339	.5253	.5171	.5067	.4970	.4186	.4066	.3955	.3816	.3687
800	.5294	.5203	.5116	.5007	.4905	.4116	.3991	.3874	.3730	.3597
900	.5254	.5158	.5066	.4952	.4846	.4053	.3921	.3800	.3650	.3514
1000	.5218	.5117	.5021	.4902	.4792	.3993	.3857	.3731	.3577	.3437
1200	.5160	.5049	.4944	.4814	.4696	.3885	.3738	.3603	.3441	.3294
1400	.5109	.4989	.4876	.4737	.4612	.3788	.3632	.3489	.3319	.3168
1700	.5043	.4909	.4785	.4634	.4499	.3659	.3490	.3338	.3159	.3001
2000	.4984	.4839	.4704	.4542	.4398	.3543	.3365	.3204	.3018	.2856
2300	.4929	.4773	.4629	.4456	.4305	.3436	.3250	.3083	.2890	.2725
2600	.4878	.4711	.4559	.4377	.4220	.3338	.3145	.2972	.2774	.2607
3000	.4814	.4635	.4473	.4280	.4116	.3219	.3019	.2840	.2636	.2467
3500	.4739	.4547	.4374	.4170	.3999	.3085	.2878	.2694	.2485	.2314
4000	.4669	.4466	.4283	.4070	.3892	.2965	.2753	.2564	.2351	.2181
4500	.4602	.4389	.4198	.3978	.3795	.2855	.2639	.2448	.2233	.2064
5000	.4538	.4317	.4119	.3892	.3705	.2754	.2536	.2343	.2127	.1959
6000	.4413	.4179	.3970	.3732	.3540	.2570	.2350	.2156	.1942	.1779
7000	.4299	.4053	.3837	.3591	.3396	.2411	.2190	.1998	.1787	.1630
8000	.4193	.3939	.3716	.3465	.3267	.2272	.2052	.1862	.1655	.1503
9000	.4094	.3834	.3606	.3351	.3152	.2148	.1930	.1743	.1542	.1395
10000	.4001	.3737	.3505	.3248	.3048	.2037	.1823	.1639	.1442	.1301
12000	.3828	.3558	.3323	.3063	.2865	.1846	.1640	.1466	.1280	.1149
14000	.3674	.3401	.3164	.2905	.2709	.1687	.1491	.1325	.1150	.1028
17000	.3472	.3198	.2962	.2704	.2513	.1492	.1310	.1157	.0996	.0887
20000	.3297	.3025	.2791	.2537	.2351	.1335	.1167	.1025	.0877	.0778
23000	.3144	.2875	.2644	.2395	.2213	.1207	.1050	.0918	.0782	.0691
26000	.3009	.2743	.2516	.2272	.2096	.1101	.0954	.0831	.0705	.0621
30000	.2850	.2590	.2368	.2131	.1962	.0984	.0849	.0737	.0622	.0547
35000	.2679	.2428	.2213	.1985	.1823	.0869	.0747	.0645	.0542	.0475
40000	.2532	.2289	.2082	.1863	.1708	.0780	.0668	.0575	.0481	.0420
45000	.2404	.2169	.1970	.1759	.1611	.0709	.0606	.0520	.0434	.0378
50000	.2291	.2065	.1873	.1670	.1528	.0653	.0557	.0477	.0397	.0346

TABLE 10- 6

AG ON AU, M2/M1=1.82

E,KEV	RANGE SKEWNESS					DAMAGE SKEWNESS				
K/KL=	0.5	0.65	0.8	1.0	1.2	0.5	0.65	0.8	1.0	1.2
5.0	.5582	.5452	.5324	.5158	.4996	.8964	.8858	.8756	.8623	.8496
6.0	.5564	.5430	.5298	.5127	.4960	.8950	.8841	.8736	.8599	.8468
7.0	.5548	.5410	.5276	.5100	.4929	.8938	.8826	.8718	.8579	.8444
8.0	.5534	.5393	.5255	.5076	.4901	.8927	.8813	.8702	.8560	.8423
9.0	.5521	.5377	.5236	.5053	.4876	.8917	.8801	.8688	.8543	.8403
10	.5508	.5362	.5219	.5033	.4852	.8908	.8789	.8674	.8527	.8385
12	.5486	.5334	.5187	.4995	.4808	.8890	.8768	.8650	.8498	.8352
14	.5465	.5309	.5157	.4960	.4769	.8875	.8749	.8627	.8472	.8323
17	.5436	.5275	.5117	.4914	.4716	.8853	.8723	.8597	.8436	.8282
20	.5410	.5243	.5081	.4871	.4668	.8833	.8699	.8569	.8404	.8246
23	.5386	.5215	.5048	.4833	.4625	.8815	.8677	.8545	.8375	.8213
26	.5363	.5188	.5018	.4798	.4585	.8799	.8657	.8521	.8348	.8183
30	.5334	.5154	.4979	.4753	.4535	.8778	.8632	.8492	.8314	.8145
35	.5301	.5114	.4934	.4701	.4476	.8752	.8601	.8457	.8274	.8101
40	.5268	.5077	.4891	.4652	.4421	.8727	.8572	.8425	.8237	.8059
45	.5237	.5041	.4850	.4605	.4369	.8703	.8545	.8393	.8201	.8019
50	.5208	.5006	.4811	.4561	.4320	.8680	.8517	.8363	.8166	.7981
60	.5153	.4942	.4739	.4479	.4229	.8633	.8465	.8305	.8102	.7910
70	.5100	.4882	.4671	.4402	.4144	.8589	.8416	.8251	.8041	.7844
80	.5049	.4823	.4606	.4328	.4063	.8547	.8369	.8199	.7983	.7780
90	.5000	.4767	.4543	.4258	.3986	.8507	.8324	.8149	.7927	.7719
100	.4953	.4713	.4483	.4190	.3912	.8469	.8281	.8101	.7874	.7661
120	.4864	.4611	.4370	.4063	.3773	.8408	.8207	.8016	.7776	.7554
140	.4778	.4514	.4262	.3943	.3642	.8348	.8135	.7934	.7683	.7451
170	.4654	.4374	.4108	.3772	.3456	.8256	.8029	.7814	.7547	.7304
200	.4535	.4241	.3962	.3611	.3281	.8163	.7922	.7696	.7416	.7162
230	.4423	.4114	.3823	.3458	.3116	.8056	.7807	.7573	.7284	.7023
250	.4314	.3992	.3690	.3311	.2958	.7950	.7694	.7453	.7157	.6889
300	.4174	.3836	.3519	.3124	.2758	.7812	.7547	.7299	.6994	.6717
350	.4005	.3649	.3315	.2903	.2520	.7647	.7373	.7116	.6799	.6513
400	.3844	.3469	.3121	.2692	.2296	.7491	.7207	.6941	.6615	.6320
450	.3688	.3297	.2936	.2491	.2082	.7343	.7049	.6775	.6439	.6137
500	.3537	.3132	.2757	.2298	.1878	.7203	.6899	.6616	.6271	.5962
600	.3250	.2804	.2412	.1933	.1493	.6964	.6634	.6328	.5959	.5634
700	.2980	.2503	.2092	.1593	.1137	.6739	.6384	.6059	.5669	.5330
800	.2724	.2224	.1795	.1275	.0803	.6522	.6147	.5805	.5397	.5047
900	.2481	.1966	.1516	.0976	.0488	.6312	.5920	.5563	.5141	.4781
1000	.2249	.1725	.1255	.0693	.0190	.6108	.5701	.5332	.4898	.4530
1200	.1813	.1318	.0786	.0164	-.0372	.5684	.5261	.4879	.4431	.4054
1400	.1409	.0945	.0358	-.0319	-.0884	.5290	.4857	.4465	.4008	.3624
1700	.0854	.0426	-.0230	-.0974	-.1575	.4756	.4310	.3908	.3440	.3049
2000	.0347	-.0060	-.0767	-.1563	-.2191	.4281	.3825	.3414	.2937	.2541
2300	-.0125	-.0574	-.1295	-.2107	-.2747	.3873	.3402	.2979	.2491	.2088
2600	-.0564	-.1060	-.1786	-.2606	-.3254	.3503	.3018	.2585	.2087	.1679
3000	-.1105	-.1664	-.2391	-.3214	-.3869	.3057	.2557	.2111	.1601	.1188
3500	-.1724	-.2354	-.3077	-.3899	-.4560	.2561	.2044	.1585	.1063	.0644
4000	-.2290	-.2982	-.3697	-.4516	-.5180	.2116	.1586	.1116	.0586	.0163
4500	-.2812	-.3555	-.4263	-.5079	-.5745	.1713	.1171	.0694	.0156	-.0269
5000	-.3297	-.4081	-.4783	-.5595	-.6264	.1343	.0793	.0308	-.0235	-.0662
6000	-.4217	-.5005	-.5703	-.6514	-.7187	.0668	.0109	-.0382	-.0929	-.1354
7000	-.5020	-.5808	-.6506	-.7319	-.7998	.0082	-.0482	-.0976	-.1524	-.1947
8000	-.5730	-.6518	-.7219	-.8038	-.8723	-.0435	-.1002	-.1497	-.2045	-.2464
9000	-.6361	-.7152	-.7860	-.8687	-.9379	-.0896	-.1465	-.1960	-.2506	-.2921
10000	-.6928	-.7724	-.8443	-.9280	-.9980	-.1312	-.1881	-.2375	-.2919	-.3330
12000	-.7834	-.8683	-.9444	-1.032	-1.105	-.2034	-.2600	-.3091	-.3627	-.4030
14000	-.8610	-.9511	-1.031	-1.124	-1.198	-.2648	-.3210	-.3695	-.4224	-.4619
17000	-.9619	-1.059	-1.145	-1.242	-1.320	-.3423	-.3977	-.4454	-.4972	-.5355
20000	-1.050	-1.153	-1.244	-1.346	-1.426	-.4070	-.4616	-.5085	-.5593	-.5966
23000	-1.131	-1.238	-1.332	-1.438	-1.520	-.4624	-.5161	-.5624	-.6122	-.6487
26000	-1.205	-1.316	-1.413	-1.521	-1.604	-.5107	-.5636	-.6092	-.6582	-.6939
30000	-1.297	-1.410	-1.510	-1.620	-1.705	-.5665	-.6185	-.6632	-.7112	-.7462
35000	-1.403	-1.518	-1.620	-1.732	-1.818	-.6258	-.6768	-.7206	-.7676	-.8017
40000	-1.501	-1.618	-1.719	-1.831	-1.917	-.6764	-.7265	-.7695	-.8156	-.8491
45000	-1.594	-1.709	-1.810	-1.922	-2.008	-.7204	-.7696	-.8119	-.8573	-.8903
50000	-1.682	-1.796	-1.895	-2.005	-2.090	-.7590	-.8076	-.8493	-.8941	-.9266

E,KEV	IONIZATION SKEWNESS					RANGE KURTOSIS				
K/KL=	0.5	0.65	0.8	1.0	1.2	0.5	0.65	0.8	1.0	1.2
5.0	.9824	.9730	.9639	.9523	.9412	3.382	3.347	3.314	3.271	3.231
6.0	.9811	.9714	.9621	.9501	.9387	3.377	3.341	3.307	3.263	3.222
7.0	.9799	.9700	.9605	.9483	.9366	3.373	3.336	3.301	3.256	3.215
8.0	.9789	.9688	.9590	.9466	.9347	3.369	3.331	3.296	3.250	3.208
9.0	.9780	.9676	.9577	.9450	.9329	3.365	3.327	3.291	3.245	3.202
10	.9771	.9666	.9565	.9436	.9313	3.362	3.323	3.286	3.240	3.196
12	.9754	.9646	.9542	.9409	.9284	3.356	3.316	3.278	3.230	3.186
14	.9739	.9628	.9521	.9386	.9257	3.350	3.309	3.270	3.222	3.177
17	.9718	.9603	.9493	.9353	.9221	3.342	3.300	3.261	3.211	3.165
20	.9699	.9581	.9467	.9324	.9188	3.335	3.292	3.252	3.201	3.154
23	.9681	.9560	.9444	.9297	.9158	3.329	3.285	3.243	3.192	3.145
26	.9665	.9540	.9421	.9271	.9130	3.322	3.278	3.236	3.184	3.136
30	.9643	.9515	.9394	.9240	.9095	3.315	3.269	3.226	3.174	3.125
35	.9618	.9486	.9361	.9203	.9055	3.306	3.259	3.216	3.162	3.112
40	.9595	.9459	.9331	.9169	.9017	3.297	3.250	3.206	3.151	3.101
45	.9573	.9434	.9302	.9137	.8982	3.290	3.241	3.196	3.141	3.090
50	.9551	.9409	.9275	.9106	.8949	3.282	3.233	3.187	3.131	3.080
60	.9516	.9368	.9229	.9054	.8892	3.268	3.218	3.171	3.114	3.062
70	.9481	.9328	.9184	.9004	.8837	3.255	3.203	3.156	3.098	3.046
80	.9447	.9288	.9140	.8955	.8784	3.242	3.190	3.142	3.083	3.030
90	.9412	.9249	.9096	.8907	.8732	3.231	3.177	3.128	3.069	3.016
100	.9376	.9209	.9053	.8860	.8682	3.220	3.165	3.116	3.056	3.003
120	.9301	.9126	.8963	.8762	.8577	3.199	3.143	3.092	3.032	2.979
140	.9228	.9046	.8876	.8668	.8478	3.179	3.122	3.071	3.010	2.957
170	.9124	.8932	.8754	.8537	.8340	3.152	3.094	3.042	2.981	2.928
200	.9026	.8825	.8641	.8416	.8213	3.127	3.069	3.016	2.955	2.902
230	.8944	.8734	.8543	.8311	.8102	3.104	3.044	2.991	2.930	2.878
260	.8865	.8648	.8451	.8213	.7999	3.082	3.021	2.968	2.908	2.856
300	.8765	.8539	.8334	.8088	.7869	3.055	2.994	2.941	2.881	2.831
350	.8645	.8409	.8196	.7942	.7718	3.025	2.963	2.911	2.852	2.804
400	.8529	.8284	.8064	.7804	.7575	2.997	2.936	2.884	2.827	2.781
450	.8416	.8164	.7938	.7673	.7440	2.972	2.912	2.861	2.805	2.761
500	.8307	.8047	.7817	.7546	.7310	2.950	2.889	2.840	2.786	2.744
600	.8078	.7799	.7565	.7287	.7046	2.907	2.842	2.798	2.752	2.715
700	.7867	.7576	.7337	.7053	.6810	2.871	2.805	2.766	2.725	2.694
800	.7675	.7375	.7130	.6842	.6597	2.840	2.777	2.741	2.705	2.679
900	.7498	.7193	.6942	.6650	.6405	2.814	2.757	2.723	2.690	2.669
1000	.7334	.7029	.6769	.6476	.6231	2.791	2.744	2.710	2.680	2.662
1200	.7047	.6765	.6468	.6175	.5930	2.755	2.749	2.703	2.666	2.654
1400	.6795	.6538	.6209	.5917	.5676	2.727	2.762	2.705	2.663	2.657
1700	.6470	.6245	.5880	.5593	.5357	2.699	2.787	2.719	2.673	2.677
2000	.6191	.5990	.5603	.5324	.5096	2.681	2.810	2.739	2.695	2.710
2300	.5938	.5727	.5357	.5089	.4874	2.658	2.795	2.744	2.721	2.751
2600	.5714	.5491	.5143	.4886	.4685	2.643	2.782	2.754	2.754	2.800
3000	.5454	.5213	.4899	.4657	.4474	2.635	2.773	2.775	2.807	2.872
3500	.5176	.4918	.4646	.4423	.4260	2.640	2.776	2.815	2.884	2.972
4000	.4941	.4671	.4437	.4232	.4089	2.660	2.794	2.866	2.969	3.077
4500	.4739	.4463	.4264	.4076	.3950	2.691	2.824	2.927	3.059	3.187
5000	.4564	.4288	.4118	.3946	.3836	2.731	2.867	2.995	3.154	3.300
6000	.4277	.4043	.3905	.3764	.3681	2.844	3.005	3.162	3.354	3.525
7000	.4054	.3863	.3748	.3635	.3575	2.970	3.157	3.338	3.558	3.752
8000	.3878	.3726	.3631	.3542	.3501	3.103	3.315	3.518	3.763	3.979
9000	.3738	.3621	.3543	.3475	.3451	3.239	3.476	3.698	3.966	4.203
10000	.3627	.3540	.3476	.3426	.3417	3.376	3.636	3.877	4.167	4.423
12000	.3488	.3435	.3397	.3377	.3392	3.638	3.936	4.212	4.546	4.841
14000	.3401	.3373	.3356	.3360	.3393	3.899	4.233	4.542	4.917	5.249
17000	.3327	.3326	.3336	.3368	.3422	4.288	4.673	5.030	5.462	5.845
20000	.3293	.3313	.3343	.3397	.3468	4.670	5.104	5.506	5.993	6.423
23000	.3282	.3319	.3366	.3438	.3520	5.044	5.526	5.972	6.509	6.983
26000	.3285	.3337	.3398	.3484	.3576	5.409	5.938	6.425	7.012	7.526
30000	.3302	.3371	.3448	.3549	.3650	5.882	6.471	7.013	7.662	8.227
35000	.3335	.3423	.3515	.3631	.3740	6.452	7.114	7.721	8.443	9.067
40000	.3375	.3480	.3584	.3712	.3826	6.999	7.732	8.401	9.193	9.871
45000	.3417	.3538	.3653	.3789	.3906	7.525	8.327	9.056	9.914	10.64
50000	.3459	.3596	.3719	.3862	.3980	8.032	8.901	9.687	10.61	11.38

TABLE 10- 8

AG ON AU, M2/M1=1.82

E,KEV	DAMAGE KURTOSIS K/KL= 0.5	0.65	0.8	1.0	1.2	IONIZATION KURTOSIS 0.5	0.65	0.8	1.0	1.2
5.0	4.014	3.972	3.931	3.880	3.831	4.160	4.118	4.079	4.028	3.980
6.0	4.008	3.964	3.923	3.870	3.820	4.154	4.111	4.070	4.018	3.970
7.0	4.002	3.958	3.915	3.862	3.811	4.149	4.105	4.063	4.010	3.960
8.0	3.997	3.952	3.909	3.854	3.802	4.144	4.099	4.056	4.002	3.952
9.0	3.993	3.947	3.903	3.847	3.795	4.139	4.093	4.050	3.996	3.944
10	3.989	3.942	3.897	3.841	3.788	4.135	4.088	4.045	3.989	3.937
12	3.981	3.932	3.887	3.829	3.775	4.127	4.079	4.034	3.977	3.924
14	3.973	3.924	3.877	3.818	3.763	4.120	4.071	4.025	3.967	3.913
17	3.963	3.913	3.864	3.804	3.747	4.110	4.059	4.012	3.953	3.897
20	3.954	3.902	3.853	3.791	3.733	4.100	4.049	4.000	3.940	3.883
23	3.946	3.893	3.842	3.779	3.720	4.092	4.039	3.990	3.928	3.870
26	3.939	3.884	3.833	3.768	3.709	4.083	4.030	3.979	3.917	3.858
30	3.929	3.873	3.820	3.755	3.694	4.073	4.018	3.967	3.903	3.843
35	3.917	3.860	3.806	3.739	3.677	4.061	4.005	3.952	3.887	3.827
40	3.906	3.847	3.792	3.724	3.661	4.050	3.993	3.939	3.872	3.811
45	3.895	3.835	3.779	3.710	3.646	4.040	3.981	3.926	3.859	3.797
50	3.884	3.824	3.767	3.697	3.632	4.030	3.970	3.915	3.846	3.783
60	3.861	3.800	3.742	3.671	3.605	4.014	3.952	3.895	3.825	3.761
70	3.839	3.778	3.720	3.647	3.581	3.998	3.935	3.877	3.805	3.739
80	3.820	3.758	3.699	3.626	3.558	3.983	3.918	3.858	3.785	3.719
90	3.802	3.739	3.679	3.605	3.536	3.967	3.901	3.840	3.766	3.699
100	3.786	3.722	3.661	3.585	3.516	3.952	3.884	3.823	3.747	3.679
120	3.765	3.695	3.630	3.551	3.479	3.917	3.847	3.784	3.707	3.637
140	3.744	3.670	3.602	3.518	3.445	3.884	3.812	3.747	3.669	3.598
170	3.714	3.634	3.561	3.473	3.397	3.838	3.764	3.697	3.617	3.545
200	3.683	3.599	3.522	3.431	3.354	3.797	3.721	3.652	3.570	3.498
230	3.640	3.556	3.479	3.388	3.310	3.764	3.686	3.616	3.533	3.460
260	3.597	3.514	3.438	3.347	3.270	3.734	3.654	3.583	3.499	3.425
300	3.544	3.462	3.387	3.297	3.220	3.697	3.615	3.543	3.458	3.383
350	3.482	3.402	3.328	3.239	3.163	3.653	3.570	3.496	3.410	3.336
400	3.427	3.348	3.275	3.187	3.111	3.612	3.527	3.453	3.366	3.292
450	3.378	3.299	3.227	3.140	3.064	3.573	3.487	3.412	3.325	3.250
500	3.333	3.255	3.182	3.096	3.021	3.536	3.448	3.373	3.286	3.212
600	3.272	3.186	3.109	3.020	2.944	3.456	3.364	3.291	3.204	3.131
700	3.217	3.126	3.045	2.953	2.878	3.385	3.290	3.218	3.132	3.060
800	3.167	3.072	2.988	2.894	2.819	3.321	3.226	3.154	3.068	2.997
900	3.119	3.022	2.936	2.841	2.768	3.264	3.170	3.096	3.012	2.942
1000	3.074	2.975	2.888	2.793	2.721	3.212	3.121	3.045	2.962	2.893
1200	2.972	2.877	2.795	2.706	2.638	3.127	3.056	2.962	2.881	2.814
1400	2.881	2.793	2.716	2.632	2.570	3.055	3.003	2.893	2.814	2.749
1700	2.766	2.686	2.617	2.543	2.487	2.965	2.936	2.807	2.732	2.669
2000	2.671	2.600	2.538	2.472	2.423	2.891	2.877	2.737	2.665	2.605
2300	2.604	2.537	2.480	2.419	2.376	2.825	2.805	2.674	2.608	2.551
2600	2.549	2.487	2.433	2.377	2.338	2.768	2.739	2.621	2.558	2.505
3000	2.490	2.432	2.384	2.333	2.299	2.703	2.660	2.560	2.502	2.454
3500	2.433	2.380	2.336	2.292	2.263	2.634	2.576	2.496	2.445	2.402
4000	2.388	2.340	2.301	2.262	2.238	2.576	2.506	2.445	2.398	2.360
4500	2.352	2.309	2.274	2.240	2.221	2.527	2.448	2.401	2.359	2.324
5000	2.323	2.285	2.254	2.225	2.210	2.484	2.399	2.365	2.326	2.295
6000	2.277	2.249	2.228	2.209	2.201	2.415	2.341	2.313	2.278	2.251
7000	2.247	2.228	2.214	2.204	2.203	2.360	2.300	2.273	2.242	2.218
8000	2.227	2.216	2.210	2.207	2.212	2.317	2.268	2.243	2.213	2.192
9000	2.214	2.211	2.211	2.215	2.224	2.282	2.243	2.218	2.191	2.172
10000	2.208	2.211	2.216	2.226	2.240	2.253	2.223	2.198	2.173	2.156
12000	2.211	2.224	2.238	2.257	2.277	2.214	2.190	2.168	2.146	2.133
14000	2.225	2.246	2.266	2.293	2.318	2.186	2.165	2.145	2.127	2.117
17000	2.254	2.285	2.314	2.349	2.381	2.158	2.139	2.123	2.108	2.103
20000	2.290	2.329	2.364	2.407	2.444	2.138	2.121	2.108	2.097	2.094
23000	2.329	2.374	2.415	2.463	2.504	2.124	2.108	2.098	2.090	2.090
26000	2.368	2.419	2.465	2.518	2.563	2.114	2.099	2.091	2.087	2.089
30000	2.421	2.478	2.529	2.588	2.637	2.104	2.091	2.086	2.085	2.089
35000	2.486	2.549	2.606	2.671	2.723	2.094	2.085	2.084	2.086	2.093
40000	2.548	2.617	2.678	2.748	2.804	2.086	2.082	2.084	2.089	2.098
45000	2.608	2.682	2.747	2.821	2.880	2.080	2.081	2.086	2.094	2.104
50000	2.664	2.743	2.812	2.890	2.951	2.074	2.081	2.089	2.100	2.111

TABLE 10- 9

AG ON AU, M2/M1=1.82

K/KL=	REFLECTION COEFFICIENT					SPUTTERING EFFICIENCY				
E,KEV	0.5	0.65	0.8	1.0	1.2	0.5	0.65	0.8	1.0	1.2
5.0	.0587	.0578	.0569	.0558	.0547	.0309	.0295	.0282	.0266	.0251
6.0	.0584	.0575	.0566	.0554	.0543	.0307	.0293	.0280	.0263	.0248
7.0	.0582	.0573	.0563	.0552	.0540	.0306	.0291	.0278	.0261	.0245
8.0	.0580	.0570	.0561	.0549	.0537	.0304	.0290	.0276	.0259	.0243
9.0	.0578	.0568	.0559	.0547	.0535	.0303	.0288	.0274	.0257	.0241
10	.0576	.0566	.0557	.0544	.0532	.0302	.0287	.0273	.0255	.0239
12	.0573	.0563	.0553	.0540	.0528	.0299	.0284	.0270	.0252	.0236
14	.0569	.0559	.0549	.0536	.0524	.0298	.0282	.0267	.0249	.0233
17	.0565	.0554	.0544	.0530	.0518	.0295	.0279	.0264	.0246	.0229
20	.0560	.0549	.0539	.0525	.0512	.0293	.0276	.0261	.0242	.0226
23	.0556	.0545	.0534	.0520	.0507	.0290	.0274	.0258	.0240	.0223
26	.0552	.0541	.0530	.0516	.0503	.0288	.0272	.0256	.0237	.0220
30	.0547	.0535	.0524	.0510	.0497	.0285	.0269	.0253	.0234	.0217
35	.0541	.0529	.0518	.0503	.0490	.0283	.0266	.0250	.0230	.0213
40	.0535	.0523	.0511	.0497	.0483	.0281	.0263	.0247	.0227	.0210
45	.0529	.0517	.0505	.0491	.0477	.0278	.0260	.0244	.0224	.0207
50	.0524	.0512	.0500	.0485	.0471	.0276	.0258	.0242	.0222	.0204
60	.0514	.0501	.0489	.0474	.0460	.0271	.0253	.0237	.0217	.0199
70	.0504	.0491	.0479	.0464	.0449	.0267	.0249	.0232	.0212	.0195
80	.0495	.0482	.0470	.0454	.0440	.0264	.0245	.0229	.0208	.0191
90	.0487	.0474	.0461	.0445	.0430	.0260	.0242	.0225	.0205	.0187
100	.0479	.0465	.0453	.0437	.0422	.0257	.0239	.0222	.0201	.0184
120	.0463	.0450	.0437	.0421	.0406	.0252	.0233	.0216	.0195	.0178
140	.0449	.0435	.0422	.0406	.0391	.0248	.0228	.0211	.0190	.0172
170	.0430	.0416	.0403	.0386	.0371	.0242	.0222	.0204	.0183	.0165
200	.0413	.0399	.0386	.0369	.0354	.0236	.0216	.0198	.0177	.0159
230	.0397	.0383	.0369	.0353	.0337	.0230	.0210	.0192	.0171	.0153
260	.0383	.0368	.0354	.0338	.0322	.0225	.0204	.0187	.0166	.0148
300	.0365	.0350	.0337	.0320	.0305	.0218	.0198	.0180	.0159	.0142
350	.0345	.0330	.0317	.0300	.0285	.0210	.0190	.0173	.0152	.0136
400	.0328	.0313	.0299	.0283	.0268	.0203	.0184	.0166	.0146	.0130
450	.0312	.0297	.0283	.0267	.0252	.0197	.0178	.0160	.0141	.0124
500	.0298	.0283	.0269	.0253	.0238	.0192	.0172	.0155	.0136	.0120
600	.0273	.0257	.0244	.0228	.0214	.0183	.0164	.0146	.0127	.0111
700	.0252	.0236	.0223	.0207	.0194	.0176	.0156	.0139	.0120	.0104
800	.0233	.0217	.0205	.0190	.0177	.0170	.0149	.0132	.0113	.0098
900	.0217	.0202	.0189	.0174	.0162	.0164	.0143	.0126	.0107	.0093
1000	.0203	.0189	.0176	.0161	.0149	.0158	.0138	.0121	.0102	.0088
1200	.0180	.0169	.0155	.0139	.0128	.0147	.0128	.0111	.0093	.0080
1400	.0161	.0153	.0138	.0122	.0110	.0137	.0119	.0103	.0086	.0073
1700	.0138	.0134	.0118	.0101	.0090	.0125	.0107	.0092	.0076	.0065
2000	.0120	.0119	.0103	.0085	.0074	.0115	.0098	.0084	.0069	.0058
2300	.0104	.0104	.0089	.0072	.0062	.0106	.0091	.0077	.0063	.0053
2600	.0091	.0091	.0077	.0062	.0053	.0100	.0084	.0072	.0058	.0049
3000	.0077	.0076	.0064	.0051	.0043	.0092	.0077	.0065	.0052	.0044
3500	.0063	.0060	.0051	.0040	.0033	.0084	.0070	.0059	.0047	.0039
4000	.0051	.0048	.0040	.0031	.0026	.0077	.0064	.0054	.0042	.0036
4500	.0042	.0038	.0032	.0025	.0020	.0071	.0059	.0049	.0039	.0033
5000	.0035	.0030	.0025	.0020	.0016	.0066	.0055	.0046	.0036	.0030
6000	.0026	.0022	.0018	.0013	.0011	.0059	.0048	.0040	.0031	.0026
7000	.0019	.0016	.0013	.0009	.0007	.0053	.0043	.0036	.0027	.0023
8000	.0015	.0012	.0009	.0007	.0005	.0048	.0039	.0032	.0024	.0020
9000	.0012	.0009	.0007	.0005	.0003	.0044	.0036	.0029	.0022	.0018
10000	.0009	.0007	.0005	.0003	.0002	.0040	.0033	.0027	.0020	.0016
12000	.0006	.0004	.0003	.0002	.0001	.0035	.0028	.0023	.0017	.0014
14000	.0004	.0003	.0002	.0001	.0001	.0030	.0024	.0019	.0014	.0012
17000	.0002	.0001	.0001	.0000	0.0000	.0025	.0020	.0016	.0012	.0009
20000	.0001	.0001	.0000	.0000	0.0000	.0021	.0017	.0013	.0010	.0008
23000	.0001	.0001	.0000	.0000	0.0000	.0018	.0014	.0011	.0008	.0006
26000	.0001	.0001	.0000	.0000	0.0000	.0016	.0012	.0009	.0007	.0005
30000	.0001	.0001	.0001	.0001	0.0000	.0013	.0010	.0008	.0006	.0004
35000	.0001	.0001	.0001	.0001	0.0000	.0011	.0009	.0006	.0004	.0004
40000	.0001	.0001	.0001	.0001	0.0000	.0009	.0007	.0005	.0004	.0003
45000	.0000	.0000	.0000	.0000	0.0000	.0008	.0006	.0005	.0003	.0003
50000	.0000	.0000	.0000	.0000	0.0000	.0007	.0006	.0004	.0003	.0002

AG ON AU, M2/M1=1.82   TABLE 10-10

| E,KEV | IONIZATION DEFICIENCY | | | | | SPUTTERING YIELD ALPHA | | | | |
K/KL=	0.5	0.65	0.8	1.0	1.2	0.5	0.65	0.8	1.0	1.2
5.0	.0025	.0032	.0037	.0044	.0051	.2595	.2522	.2452	.2362	.2276
6.0	.0026	.0032	.0038	.0045	.0052	.2589	.2514	.2441	.2349	.2262
7.0	.0026	.0033	.0039	.0046	.0052	.2583	.2507	.2433	.2339	.2250
8.0	.0027	.0034	.0040	.0047	.0053	.2579	.2500	.2425	.2330	.2239
9.0	.0027	.0034	.0040	.0047	.0054	.2574	.2495	.2418	.2321	.2230
10	.0028	.0035	.0041	.0048	.0054	.2571	.2490	.2412	.2314	.2221
12	.0028	.0035	.0041	.0049	.0055	.2565	.2481	.2402	.2301	.2207
14	.0029	.0036	.0042	.0050	.0056	.2559	.2474	.2393	.2290	.2194
17	.0030	.0037	.0043	.0051	.0057	.2553	.2465	.2382	.2277	.2178
20	.0030	.0037	.0044	.0051	.0058	.2548	.2458	.2373	.2265	.2165
23	.0031	.0038	.0045	.0052	.0059	.2544	.2452	.2365	.2255	.2153
26	.0031	.0039	.0045	.0053	.0059	.2540	.2446	.2358	.2247	.2143
30	.0032	.0039	.0046	.0053	.0060	.2536	.2440	.2350	.2237	.2132
35	.0032	.0040	.0046	.0054	.0060	.2532	.2434	.2342	.2227	.2119
40	.0033	.0040	.0047	.0054	.0061	.2528	.2429	.2335	.2218	.2109
45	.0033	.0041	.0047	.0055	.0061	.2526	.2424	.2329	.2210	.2099
50	.0033	.0041	.0048	.0055	.0062	.2523	.2420	.2324	.2203	.2091
60	.0034	.0042	.0048	.0056	.0062	.2520	.2414	.2315	.2192	.2078
70	.0035	.0042	.0049	.0056	.0062	.2517	.2410	.2309	.2182	.2066
80	.0035	.0042	.0049	.0056	.0063	.2515	.2406	.2303	.2174	.2057
90	.0035	.0043	.0049	.0057	.0063	.2514	.2403	.2298	.2168	.2049
100	.0035	.0043	.0049	.0057	.0063	.2513	.2401	.2294	.2162	.2041
120	.0036	.0043	.0049	.0057	.0062	.2514	.2398	.2288	.2152	.2029
140	.0036	.0043	.0049	.0056	.0062	.2516	.2396	.2283	.2144	.2018
170	.0036	.0043	.0049	.0056	.0061	.2518	.2394	.2277	.2133	.2005
200	.0036	.0043	.0049	.0056	.0061	.2520	.2392	.2272	.2125	.1994
230	.0036	.0043	.0049	.0055	.0060	.2520	.2390	.2268	.2117	.1984
260	.0036	.0043	.0049	.0055	.0059	.2520	.2388	.2263	.2111	.1976
300	.0036	.0043	.0048	.0054	.0058	.2520	.2385	.2258	.2103	.1966
350	.0036	.0042	.0048	.0053	.0057	.2520	.2382	.2252	.2094	.1955
400	.0036	.0042	.0047	.0052	.0056	.2520	.2379	.2247	.2086	.1945
450	.0035	.0041	.0046	.0051	.0055	.2520	.2377	.2243	.2079	.1936
500	.0035	.0041	.0046	.0050	.0054	.2520	.2375	.2238	.2073	.1928
600	.0034	.0040	.0044	.0048	.0051	.2524	.2373	.2232	.2061	.1914
700	.0034	.0038	.0042	.0047	.0049	.2527	.2371	.2225	.2050	.1901
800	.0033	.0037	.0041	.0045	.0047	.2528	.2368	.2219	.2041	.1890
900	.0032	.0036	.0040	.0043	.0045	.2527	.2364	.2213	.2032	.1880
1000	.0032	.0035	.0038	.0042	.0043	.2526	.2360	.2207	.2024	.1870
1200	.0030	.0034	.0036	.0039	.0040	.2515	.2347	.2190	.2005	.1851
1400	.0029	.0033	.0034	.0036	.0037	.2503	.2334	.2176	.1989	.1836
1700	.0027	.0031	.0031	.0033	.0033	.2486	.2317	.2158	.1970	.1817
2000	.0026	.0029	.0029	.0030	.0030	.2471	.2302	.2144	.1956	.1804
2300	.0025	.0027	.0027	.0028	.0028	.2459	.2295	.2142	.1955	.1801
2600	.0023	.0024	.0024	.0026	.0026	.2449	.2288	.2140	.1955	.1800
3000	.0022	.0022	.0022	.0023	.0023	.2437	.2282	.2140	.1956	.1800
3500	.0020	.0019	.0019	.0021	.0020	.2425	.2275	.2140	.1957	.1800
4000	.0018	.0016	.0016	.0018	.0018	.2415	.2270	.2139	.1957	.1800
4500	.0017	.0013	.0014	.0016	.0015	.2406	.2265	.2137	.1956	.1798
5000	.0015	.0011	.0012	.0014	.0013	.2400	.2260	.2135	.1954	.1796
6000	.0013	.0008	.0008	.0010	.0009	.2395	.2252	.2119	.1936	.1781
7000	.0011	.0005	.0005	.0006	.0006	.2390	.2243	.2104	.1918	.1766
8000	.0009	.0003	.0003	.0004	.0003	.2385	.2235	.2090	.1901	.1753
9000	.0008	.0001	.0001	.0002	.0001	.2380	.2227	.2076	.1886	.1740
10000	.0007	0.0000	0.0000	0.0000	0.0000	.2374	.2219	.2064	.1873	.1729
12000	.0004	0.0000	0.0000	0.0000	0.0000	.2353	.2202	.2046	.1855	.1713
14000	.0003	0.0000	0.0000	0.0000	0.0000	.2333	.2185	.2030	.1839	.1700
17000	.0001	0.0000	0.0000	0.0000	0.0000	.2305	.2162	.2009	.1820	.1682
20000	0.0000	0.0000	0.0000	0.0000	0.0000	.2280	.2140	.1990	.1802	.1666
23000	0.0000	0.0000	0.0000	0.0000	0.0000	.2257	.2121	.1973	.1786	.1651
26000	0.0000	0.0000	0.0000	0.0000	0.0000	.2237	.2103	.1958	.1771	.1637
30000	0.0000	0.0000	0.0000	0.0000	0.0000	.2215	.2081	.1938	.1752	.1619
35000	0.0000	0.0000	0.0000	0.0000	0.0000	.2191	.2056	.1916	.1729	.1598
40000	0.0000	0.0000	0.0000	0.0000	0.0000	.2171	.2034	.1895	.1708	.1577
45000	0.0000	0.0000	0.0000	0.0000	0.0000	.2155	.2014	.1875	.1687	.1557
50000	0.0000	0.0000	0.0000	0.0000	0.0000	.2142	.1996	.1856	.1666	.1538

AG ON AU, M2/M1=1.82

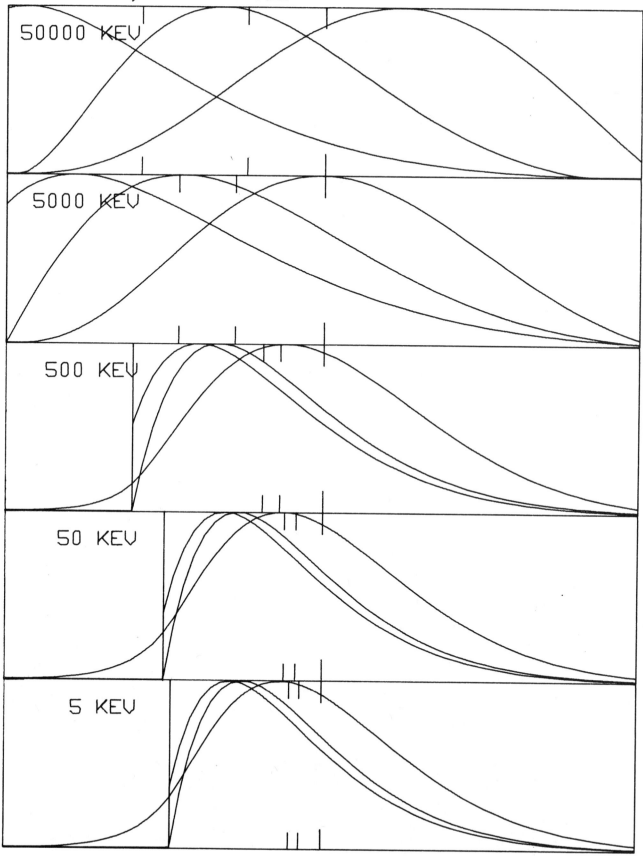

50000 KEV

5000 KEV

500 KEV

50 KEV

5 KEV

TABLE 11- 1

BE ON C, M2/M1=4/3

K/KL= E,KEV	FRACTIONAL DEPOSITED ENERGY					MEAN RANGE,MICROGRAM/SQ.CM.				
	0.6	0.8	1.0	1.3	1.7	0.6	0.8	1.0	1.3	1.7
.010	.9157	.8898	.8647	.8289	.7842	.0298	.0296	.0295	.0293	.0290
.012	.9133	.8866	.8609	.8243	.7785	.0339	.0337	.0335	.0333	.0329
.014	.9111	.8839	.8576	.8202	.7736	.0376	.0374	.0372	.0369	.0365
.017	.9084	.8803	.8534	.8150	.7672	.0427	.0425	.0422	.0419	.0414
.020	.9059	.8772	.8497	.8105	.7618	.0475	.0472	.0470	.0466	.0460
.023	.9038	.8745	.8464	.8065	.7571	.0521	.0518	.0515	.0511	.0505
.026	.9019	.8721	.8435	.8030	.7528	.0565	.0562	.0559	.0554	.0547
.030	.8996	.8692	.8401	.7988	.7478	.0623	.0619	.0615	.0610	.0603
.035	.8971	.8660	.8363	.7942	.7423	.0692	.0688	.0683	.0677	.0669
.040	.8949	.8632	.8329	.7901	.7374	.0759	.0754	.0749	.0742	.0733
.045	.8929	.8606	.8299	.7864	.7330	.0824	.0818	.0813	.0805	.0795
.050	.8911	.8583	.8271	.7831	.7291	.0887	.0881	.0875	.0866	.0855
.060	.8880	.8543	.8224	.7773	.7222	.1004	.0997	.0990	.0979	.0966
.070	.8852	.8508	.8182	.7723	.7162	.1117	.1109	.1101	.1089	.1074
.080	.8828	.8477	.8145	.7678	.7109	.1227	.1217	.1208	.1195	.1178
.090	.8806	.8449	.8111	.7638	.7061	.1334	.1324	.1313	.1299	.1279
.10	.8785	.8423	.8081	.7601	.7018	.1439	.1427	.1416	.1400	.1378
.12	.8748	.8377	.8026	.7535	.6940	.1633	.1619	.1606	.1587	.1562
.14	.8716	.8336	.7977	.7477	.6873	.1823	.1807	.1792	.1770	.1741
.17	.8673	.8282	.7914	.7402	.6785	.2102	.2083	.2065	.2038	.2003
.20	.8635	.8235	.7859	.7336	.6709	.2377	.2354	.2332	.2300	.2258
.23	.8602	.8193	.7810	.7279	.6642	.2622	.2597	.2572	.2536	.2489
.26	.8571	.8155	.7766	.7226	.6582	.2887	.2838	.2811	.2770	.2718
.30	.8534	.8109	.7712	.7163	.6510	.3194	.3161	.3129	.3082	.3022
.35	.8493	.8058	.7652	.7093	.6429	.3605	.3566	.3527	.3471	.3400
.40	.8455	.8011	.7597	.7029	.6357	.4018	.3972	.3926	.3860	.3777
.45	.8420	.7968	.7548	.6971	.6291	.4432	.4377	.4324	.4248	.4151
.50	.8388	.7928	.7501	.6918	.6230	.4844	.4781	.4721	.4633	.4523
.60	.8329	.7856	.7419	.6822	.6122	.5593	.5517	.5444	.5339	.5206
.70	.8277	.7792	.7344	.6736	.6026	.6366	.6274	.6186	.6059	.5900
.80	.8228	.7732	.7276	.6658	.5940	.7159	.7049	.6943	.6792	.6602
.90	.8183	.7677	.7213	.6586	.5860	.7969	.7838	.7712	.7533	.7309
1.0	.8140	.7625	.7154	.6520	.5787	.8789	.8635	.8488	.8279	.8020
1.2	.8062	.7531	.7047	.6398	.5653	1.028	1.009	.9914	.9659	.9342
1.4	.7990	.7444	.6949	.6288	.5534	1.186	1.163	1.140	1.109	1.070
1.7	.7890	.7326	.6816	.6140	.5375	1.437	1.405	1.375	1.332	1.280
2.0	.7798	.7217	.6696	.6008	.5234	1.702	1.659	1.619	1.563	1.495
2.3	.7713	.7116	.6584	.5887	.5105	1.933	1.884	1.838	1.774	1.694
2.6	.7633	.7022	.6481	.5775	.4988	2.179	2.121	2.067	1.993	1.898
3.0	.7532	.6904	.6352	.5638	.4845	2.527	2.454	2.387	2.294	2.176
3.5	.7414	.6769	.6205	.5481	.4684	2.993	2.895	2.806	2.684	2.534
4.0	.7304	.6643	.6069	.5338	.4539	3.485	3.357	3.241	3.085	2.899
4.5	.7199	.6525	.5943	.5205	.4408	3.998	3.834	3.688	3.494	3.268
5.0	.7099	.6415	.5825	.5081	.4287	4.524	4.323	4.144	3.907	3.639
6.0	.6910	.6210	.5605	.4846	.4067	5.510	5.248	5.012	4.700	4.354
7.0	.6736	.6025	.5408	.4639	.3875	6.565	6.223	5.915	5.512	5.079
8.0	.6575	.5855	.5229	.4453	.3706	7.675	7.237	6.844	6.336	5.807
9.0	.6424	.5698	.5065	.4285	.3554	8.824	8.278	7.790	7.166	6.535
10	.6282	.5551	.4914	.4132	.3417	10.00	9.337	8.745	7.998	7.261
12	.6014	.5277	.4637	.3861	.3175	12.30	11.41	10.62	9.628	8.674
14	.5774	.5034	.4396	.3629	.2969	14.69	13.53	12.51	11.25	10.07
17	.5455	.4717	.4085	.3336	.2712	18.39	16.77	15.36	13.65	12.12
20	.5176	.4444	.3821	.3092	.2501	22.17	20.04	18.21	16.01	14.13
23	.4924	.4204	.3594	.2885	.2325	25.89	23.25	20.98	18.30	16.06
26	.4699	.3992	.3395	.2707	.2175	29.63	26.44	23.73	20.54	17.94
30	.4432	.3743	.3165	.2503	.2005	34.63	30.68	27.33	23.44	20.37
35	.4143	.3477	.2921	.2290	.1829	40.85	35.90	31.73	26.95	23.30
40	.3893	.3250	.2714	.2111	.1683	47.00	41.03	36.02	30.32	26.12
45	.3673	.3052	.2536	.1960	.1560	53.08	46.05	40.19	33.58	28.83
50	.3479	.2879	.2382	.1829	.1454	59.06	50.97	44.25	36.72	31.45
60	.3150	.2589	.2126	.1616	.1282	70.71	60.51	52.06	42.71	36.43
70	.2882	.2355	.1923	.1450	.1149	81.96	69.64	59.49	48.35	41.11
80	.2658	.2163	.1758	.1318	.1043	92.80	78.40	66.58	53.69	45.53
90	.2469	.2004	.1623	.1210	.0957	103.3	86.81	73.35	58.76	49.74
100	.2308	.1869	.1510	.1122	.0887	113.4	94.92	79.85	63.59	53.74

BE ON C, M2/M1=4/3

TABLE 11- 2

E,KEV	MEAN DAMAGE DEPTH,MICROGRAM/SQ.CM. K/KL= 0.6	0.8	1.0	1.3	1.7	MEAN IONZN.DEPTH,MICROGRAM/SQ.CM. 0.6	0.8	1.0	1.3	1.7
.010	.0251	.0249	.0247	.0244	.0240	.0232	.0230	.0228	.0224	.0220
.012	.0285	.0282	.0280	.0276	.0272	.0263	.0261	.0258	.0254	.0250
.014	.0316	.0313	.0310	.0306	.0301	.0292	.0289	.0286	.0282	.0276
.017	.0359	.0355	.0352	.0347	.0341	.0332	.0328	.0325	.0320	.0313
.020	.0399	.0395	.0392	.0386	.0379	.0369	.0365	.0361	.0355	.0348
.023	.0437	.0433	.0429	.0423	.0415	.0404	.0400	.0395	.0389	.0381
.026	.0475	.0470	.0465	.0459	.0450	.0439	.0434	.0429	.0422	.0413
.030	.0522	.0517	.0512	.0504	.0495	.0483	.0477	.0472	.0464	.0453
.035	.0580	.0574	.0568	.0560	.0549	.0536	.0529	.0523	.0514	.0503
.040	.0636	.0629	.0622	.0613	.0600	.0587	.0580	.0573	.0563	.0550
.045	.0689	.0682	.0674	.0664	.0650	.0636	.0628	.0621	.0609	.0595
.050	.0741	.0733	.0725	.0713	.0699	.0684	.0676	.0667	.0655	.0639
.060	.0838	.0829	.0819	.0806	.0788	.0773	.0763	.0753	.0739	.0721
.070	.0931	.0920	.0910	.0894	.0875	.0859	.0848	.0836	.0820	.0799
.080	.1022	.1010	.0998	.0980	.0958	.0943	.0929	.0917	.0898	.0875
.090	.1110	.1096	.1083	.1063	.1039	.1023	.1009	.0995	.0974	.0948
.10	.1196	.1181	.1166	.1145	.1118	.1102	.1086	.1070	.1048	.1020
.12	.1355	.1337	.1320	.1295	.1264	.1248	.1230	.1211	.1185	.1152
.14	.1510	.1489	.1470	.1441	.1405	.1390	.1369	.1348	.1318	.1280
.17	.1737	.1712	.1688	.1654	.1611	.1598	.1572	.1547	.1511	.1466
.20	.1958	.1929	.1901	.1861	.1811	.1800	.1770	.1740	.1698	.1645
.23	.2157	.2125	.2093	.2048	.1992	.1983	.1949	.1916	.1868	.1809
.26	.2354	.2318	.2283	.2233	.2170	.2163	.2125	.2088	.2035	.1969
.30	.2616	.2575	.2534	.2477	.2405	.2402	.2358	.2316	.2256	.2180
.35	.2943	.2894	.2847	.2780	.2696	.2700	.2648	.2599	.2528	.2440
.40	.3270	.3213	.3158	.3080	.2983	.2996	.2936	.2878	.2797	.2695
.45	.3594	.3529	.3466	.3378	.3268	.3289	.3221	.3155	.3062	.2947
.50	.3917	.3843	.3772	.3672	.3548	.3580	.3503	.3429	.3324	.3195
.60	.4505	.4417	.4332	.4213	.4066	.4113	.4021	.3932	.3808	.3654
.70	.5106	.5001	.4901	.4759	.4586	.4654	.4544	.4440	.4292	.4112
.80	.5719	.5594	.5476	.5311	.5108	.5202	.5073	.4950	.4778	.4568
.90	.6339	.6194	.6057	.5865	.5631	.5756	.5605	.5462	.5263	.5022
1.0	.6965	.6798	.6640	.6420	.6153	.6311	.6138	.5975	.5748	.5473
1.2	.8106	.7903	.7713	.7446	.7124	.7335	.7125	.6927	.6652	.6320
1.4	.9301	.9054	.8822	.8501	.8113	.8394	.8139	.7900	.7568	.7171
1.7	1.119	1.086	1.055	1.013	.9629	1.004	.9707	.9393	.8964	.8454
2.0	1.316	1.273	1.234	1.180	1.117	1.175	1.131	1.092	1.037	.9740
2.3	1.489	1.439	1.394	1.333	1.259	1.329	1.279	1.232	1.169	1.095
2.6	1.671	1.613	1.561	1.491	1.404	1.488	1.430	1.376	1.302	1.216
3.0	1.928	1.856	1.793	1.707	1.601	1.709	1.638	1.572	1.482	1.378
3.5	2.269	2.176	2.094	1.985	1.854	1.998	1.905	1.822	1.710	1.582
4.0	2.627	2.510	2.406	2.270	2.111	2.296	2.180	2.076	1.940	1.785
4.5	2.999	2.854	2.726	2.560	2.370	2.602	2.459	2.334	2.170	1.986
5.0	3.380	3.205	3.051	2.852	2.630	2.912	2.741	2.592	2.400	2.187
6.0	4.093	3.868	3.668	3.409	3.129	3.502	3.281	3.088	2.843	2.574
7.0	4.853	4.565	4.308	3.979	3.635	4.115	3.832	3.590	3.285	2.955
8.0	5.650	5.287	4.966	4.557	4.145	4.743	4.392	4.094	3.724	3.330
9.0	6.473	6.029	5.636	5.140	4.655	5.382	4.955	4.598	4.159	3.699
10	7.316	6.782	6.312	5.725	5.164	6.028	5.519	5.099	4.589	4.060
12	8.949	8.245	7.628	6.864	6.151	7.293	6.617	6.080	5.438	4.760
14	10.65	9.747	8.963	8.003	7.132	8.570	7.713	7.045	6.259	5.434
17	13.28	12.05	10.98	9.702	8.585	10.49	9.345	8.461	7.439	6.401
20	15.98	14.38	13.01	11.38	10.01	12.40	10.95	9.835	8.558	7.321
23	18.61	16.65	14.97	13.00	11.39	14.28	12.53	11.16	9.606	8.197
26	21.27	18.92	16.92	14.60	12.73	16.12	14.08	12.45	10.61	9.034
30	24.83	21.94	19.50	16.68	14.49	18.54	16.09	14.11	11.88	10.10
35	29.29	25.70	22.67	19.22	16.62	21.49	18.52	16.10	13.39	11.36
40	33.73	29.40	25.78	21.68	18.68	24.35	20.87	18.01	14.82	12.55
45	38.13	33.06	28.83	24.07	20.68	27.13	23.13	19.83	16.18	13.68
50	42.47	36.65	31.81	26.40	22.63	29.82	25.31	21.58	17.47	14.76
60	50.98	43.64	37.57	30.86	26.35	34.99	29.46	24.90	19.91	16.78
70	59.22	50.38	43.09	35.09	29.89	39.90	33.37	28.00	22.18	18.65
80	67.21	56.87	48.39	39.13	33.25	44.56	37.06	30.92	24.30	20.39
90	74.94	63.14	53.48	42.98	36.46	49.01	40.56	33.67	26.30	22.03
100	82.42	69.19	58.38	46.68	39.54	53.27	43.90	36.29	28.19	23.57

TABLE 11- 3

BE ON C, M2/M1=4/3

	RELATIVE RANGE STRAGGLING					RELATIVE DAMAGE STRAGGLING				
K/KL=	0.6	0.8	1.0	1.3	1.7	0.6	0.8	1.0	1.3	1.7
E,KEV										
.010	.4076	.4039	.4003	.3950	.3883	.4247	.4227	.4208	.4180	.4145
.012	.4069	.4031	.3994	.3940	.3870	.4244	.4224	.4204	.4176	.4139
.014	.4063	.4024	.3986	.3931	.3860	.4241	.4221	.4201	.4172	.4135
.017	.4054	.4014	.3975	.3919	.3845	.4238	.4217	.4196	.4166	.4128
.020	.4047	.4006	.3966	.3907	.3833	.4235	.4214	.4192	.4162	.4123
.023	.4040	.3998	.3957	.3897	.3821	.4233	.4210	.4189	.4157	.4118
.026	.4033	.3990	.3948	.3888	.3810	.4230	.4208	.4185	.4153	.4113
.030	.4024	.3981	.3938	.3876	.3797	.4227	.4204	.4181	.4149	.4107
.035	.4014	.3969	.3926	.3862	.3781	.4224	.4200	.4177	.4143	.4101
.040	.4004	.3958	.3914	.3849	.3767	.4220	.4196	.4172	.4138	.4095
.045	.3995	.3948	.3903	.3837	.3753	.4217	.4192	.4168	.4133	.4089
.050	.3986	.3938	.3892	.3825	.3740	.4214	.4189	.4164	.4129	.4084
.060	.3969	.3920	.3873	.3804	.3717	.4208	.4182	.4157	.4120	.4075
.070	.3953	.3903	.3854	.3784	.3695	.4203	.4176	.4150	.4113	.4066
.080	.3938	.3886	.3837	.3765	.3674	.4198	.4170	.4144	.4105	.4058
.090	.3923	.3871	.3820	.3747	.3655	.4193	.4165	.4138	.4099	.4050
.10	.3909	.3855	.3804	.3730	.3636	.4189	.4160	.4132	.4092	.4043
.12	.3882	.3827	.3774	.3698	.3601	.4180	.4150	.4122	.4081	.4030
.14	.3856	.3800	.3746	.3668	.3569	.4172	.4141	.4112	.4070	.4017
.17	.3820	.3762	.3706	.3625	.3524	.4161	.4129	.4098	.4054	.4000
.20	.3786	.3726	.3669	.3586	.3482	.4150	.4116	.4085	.4040	.3984
.23	.3754	.3693	.3633	.3549	.3443	.4139	.4105	.4072	.4026	.3969
.26	.3723	.3660	.3600	.3514	.3406	.4129	.4094	.4060	.4013	.3954
.30	.3684	.3620	.3558	.3470	.3360	.4116	.4080	.4045	.3996	.3936
.35	.3639	.3572	.3508	.3418	.3306	.4101	.4063	.4027	.3976	.3914
.40	.3596	.3527	.3462	.3370	.3255	.4086	.4047	.4009	.3957	.3894
.45	.3555	.3485	.3418	.3324	.3208	.4072	.4031	.3993	.3939	.3874
.50	.3516	.3445	.3377	.3281	.3163	.4058	.4016	.3977	.3922	.3855
.60	.3443	.3369	.3299	.3200	.3079	.4033	.3989	.3947	.3890	.3821
.70	.3377	.3300	.3227	.3126	.3002	.4009	.3963	.3920	.3860	.3788
.80	.3315	.3236	.3161	.3058	.2932	.3986	.3938	.3893	.3831	.3758
.90	.3258	.3176	.3100	.2994	.2866	.3965	.3914	.3868	.3804	.3728
1.0	.3204	.3121	.3043	.2935	.2806	.3943	.3891	.3843	.3778	.3700
1.2	.3105	.3018	.2937	.2826	.2693	.3902	.3847	.3796	.3727	.3647
1.4	.3016	.2925	.2842	.2728	.2593	.3864	.3805	.3752	.3680	.3597
1.7	.2897	.2803	.2716	.2598	.2461	.3810	.3748	.3691	.3616	.3529
2.0	.2793	.2695	.2606	.2485	.2346	.3761	.3695	.3635	.3557	.3467
2.3	.2700	.2598	.2506	.2384	.2242	.3715	.3645	.3584	.3502	.3410
2.6	.2616	.2510	.2416	.2292	.2149	.3672	.3599	.3535	.3452	.3358
3.0	.2515	.2406	.2309	.2183	.2040	.3618	.3542	.3476	.3390	.3293
3.5	.2403	.2290	.2192	.2064	.1920	.3557	.3477	.3408	.3319	.3221
4.0	.2304	.2189	.2089	.1960	.1816	.3501	.3417	.3346	.3255	.3156
4.5	.2216	.2098	.1997	.1867	.1725	.3448	.3362	.3289	.3197	.3096
5.0	.2136	.2017	.1914	.1784	.1643	.3399	.3311	.3237	.3143	.3042
6.0	.1994	.1874	.1770	.1640	.1505	.3306	.3216	.3139	.3042	.2942
7.0	.1873	.1753	.1649	.1519	.1390	.3224	.3132	.3053	.2955	.2856
8.0	.1769	.1649	.1545	.1416	.1292	.3150	.3058	.2977	.2878	.2781
9.0	.1678	.1558	.1455	.1327	.1208	.3084	.2991	.2909	.2809	.2714
10	.1597	.1478	.1375	.1249	.1135	.3023	.2930	.2848	.2747	.2653
12	.1462	.1345	.1243	.1119	.1012	.2919	.2824	.2741	.2640	.2550
14	.1351	.1235	.1135	.1015	.0914	.2829	.2733	.2650	.2550	.2462
17	.1216	.1102	.1005	.0890	.0796	.2716	.2619	.2536	.2437	.2354
20	.1107	.0996	.0902	.0792	.0705	.2620	.2524	.2442	.2345	.2264
23	.1015	.0909	.0819	.0715	.0635	.2537	.2442	.2362	.2267	.2190
26	.0938	.0836	.0750	.0652	.0577	.2465	.2372	.2292	.2200	.2125
30	.0851	.0755	.0674	.0582	.0515	.2381	.2290	.2213	.2123	.2052
35	.0763	.0673	.0598	.0514	.0453	.2292	.2204	.2129	.2043	.1975
40	.0690	.0606	.0537	.0459	.0405	.2217	.2131	.2059	.1976	.1911
45	.0630	.0552	.0487	.0414	.0365	.2152	.2069	.1999	.1918	.1856
50	.0580	.0506	.0445	.0377	.0333	.2096	.2015	.1946	.1867	.1807
60	.0499	.0433	.0379	.0320	.0282	.2002	.1924	.1859	.1784	.1727
70	.0439	.0379	.0331	.0278	.0245	.1928	.1852	.1789	.1716	.1663
80	.0393	.0339	.0294	.0246	.0217	.1867	.1793	.1731	.1661	.1609
90	.0357	.0307	.0266	.0222	.0196	.1816	.1744	.1683	.1614	.1564
100	.0330	.0283	.0244	.0203	.0179	.1774	.1702	.1642	.1575	.1526

TABLE 11- 4

BE ON C, M2/M1=4/3

K/KL= E,KEV	RELATIVE IONIZATION STRAGGLING					RELATIVE TRANSVERSE RANGE STRAGGLING				
	0.6	0.8	1.0	1.3	1.7	0.6	0.8	1.0	1.3	1.7
.010	.4739	.4726	.4713	.4695	.4672	.7632	.7628	.7625	.7620	.7613
.012	.4737	.4724	.4711	.4692	.4669	.7631	.7627	.7624	.7619	.7612
.014	.4736	.4722	.4708	.4690	.4666	.7630	.7626	.7623	.7617	.7611
.017	.4734	.4719	.4706	.4686	.4662	.7629	.7625	.7621	.7616	.7609
.020	.4732	.4717	.4703	.4683	.4659	.7627	.7624	.7620	.7615	.7608
.023	.4730	.4715	.4701	.4681	.4656	.7626	.7622	.7619	.7613	.7606
.026	.4729	.4714	.4699	.4678	.4653	.7625	.7621	.7618	.7612	.7605
.030	.4727	.4711	.4696	.4675	.4649	.7624	.7620	.7616	.7611	.7604
.035	.4725	.4709	.4693	.4672	.4645	.7622	.7618	.7614	.7609	.7602
.040	.4723	.4706	.4691	.4669	.4642	.7621	.7617	.7613	.7608	.7601
.045	.4721	.4704	.4688	.4666	.4638	.7619	.7615	.7612	.7606	.7599
.050	.4719	.4702	.4686	.4663	.4635	.7618	.7614	.7610	.7605	.7598
.060	.4716	.4698	.4681	.4658	.4630	.7615	.7611	.7608	.7602	.7596
.070	.4712	.4694	.4677	.4653	.4624	.7612	.7609	.7605	.7600	.7594
.080	.4709	.4691	.4674	.4649	.4620	.7610	.7607	.7603	.7598	.7592
.090	.4707	.4688	.4670	.4645	.4615	.7608	.7605	.7601	.7597	.7591
.10	.4704	.4685	.4667	.4641	.4611	.7606	.7603	.7600	.7595	.7590
.12	.4699	.4679	.4660	.4634	.4603	.7602	.7599	.7596	.7592	.7587
.14	.4694	.4674	.4654	.4628	.4596	.7598	.7596	.7593	.7590	.7585
.17	.4688	.4666	.4646	.4618	.4585	.7594	.7592	.7590	.7587	.7584
.20	.4681	.4659	.4638	.4610	.4576	.7590	.7589	.7587	.7585	.7583
.23	.4675	.4652	.4631	.4602	.4567	.7587	.7586	.7585	.7583	.7582
.26	.4670	.4646	.4624	.4594	.4559	.7583	.7583	.7582	.7582	.7581
.30	.4663	.4638	.4615	.4585	.4548	.7580	.7580	.7581	.7581	.7582
.35	.4654	.4629	.4605	.4573	.4536	.7577	.7578	.7580	.7581	.7583
.40	.4646	.4620	.4595	.4562	.4525	.7576	.7578	.7580	.7583	.7587
.45	.4638	.4611	.4586	.4552	.4514	.7575	.7578	.7581	.7585	.7591
.50	.4631	.4603	.4577	.4543	.4503	.7575	.7579	.7583	.7589	.7596
.60	.4616	.4587	.4560	.4524	.4483	.7576	.7582	.7587	.7595	.7605
.70	.4603	.4572	.4544	.4507	.4465	.7579	.7587	.7594	.7604	.7617
.80	.4591	.4558	.4529	.4491	.4448	.7585	.7594	.7603	.7615	.7631
.90	.4579	.4545	.4515	.4476	.4433	.7592	.7603	.7613	.7628	.7647
1.0	.4567	.4533	.4502	.4462	.4418	.7601	.7613	.7625	.7643	.7664
1.2	.4546	.4509	.4477	.4435	.4390	.7620	.7635	.7651	.7672	.7699
1.4	.4526	.4487	.4454	.4411	.4365	.7642	.7661	.7680	.7705	.7737
1.7	.4499	.4458	.4423	.4378	.4332	.7682	.7706	.7728	.7760	.7798
2.0	.4474	.4431	.4395	.4350	.4303	.7726	.7754	.7781	.7818	.7863
2.3	.4451	.4406	.4369	.4323	.4277	.7771	.7804	.7835	.7878	.7930
2.6	.4429	.4383	.4345	.4299	.4253	.7819	.7856	.7891	.7939	.7998
3.0	.4403	.4356	.4317	.4271	.4226	.7885	.7927	.7967	.8022	.8089
3.5	.4374	.4326	.4286	.4240	.4197	.7969	.8018	.8064	.8126	.8202
4.0	.4348	.4299	.4260	.4214	.4172	.8054	.8110	.8161	.8230	.8313
4.5	.4325	.4275	.4236	.4192	.4152	.8140	.8201	.8257	.8333	.8423
5.0	.4304	.4254	.4215	.4172	.4134	.8225	.8292	.8353	.8435	.8530
6.0	.4266	.4216	.4179	.4140	.4106	.8395	.8472	.8542	.8634	.8735
7.0	.4234	.4186	.4150	.4114	.4085	.8560	.8647	.8725	.8827	.8933
8.0	.4207	.4160	.4127	.4094	.4070	.8720	.8816	.8903	.9012	.9124
9.0	.4184	.4139	.4108	.4078	.4059	.8874	.8980	.9074	.9192	.9309
10	.4164	.4122	.4093	.4066	.4050	.9024	.9138	.9239	.9365	.9487
12	.4133	.4095	.4071	.4051	.4043	.9309	.9440	.9556	.9701	.9838
14	.4108	.4075	.4056	.4043	.4041	.9577	.9724	.9853	1.001	1.017
17	.4082	.4056	.4044	.4038	.4045	.9951	1.012	1.027	1.045	1.062
20	.4064	.4045	.4038	.4040	.4054	1.030	1.048	1.065	1.084	1.103
23	.4053	.4041	.4039	.4047	.4068	1.063	1.084	1.101	1.122	1.141
26	.4047	.4040	.4043	.4056	.4083	1.095	1.116	1.135	1.157	1.176
30	.4042	.4043	.4052	.4071	.4105	1.133	1.157	1.176	1.200	1.219
35	.4042	.4051	.4065	.4091	.4131	1.177	1.202	1.223	1.248	1.267
40	.4045	.4061	.4080	.4111	.4157	1.217	1.243	1.266	1.291	1.311
45	.4051	.4073	.4096	.4132	.4181	1.253	1.281	1.304	1.331	1.350
50	.4059	.4085	.4112	.4152	.4205	1.286	1.315	1.339	1.366	1.385
60	.4077	.4112	.4144	.4190	.4248	1.345	1.376	1.402	1.430	1.449
70	.4099	.4139	.4176	.4227	.4287	1.396	1.429	1.456	1.485	1.503
80	.4122	.4167	.4207	.4260	.4322	1.442	1.475	1.503	1.533	1.551
90	.4146	.4193	.4236	.4292	.4353	1.482	1.517	1.545	1.576	1.594
100	.4170	.4219	.4264	.4322	.4382	1.519	1.554	1.583	1.614	1.632

K/KL= E,KEV	RELATIVE TRANSV. DAMAGE STRAGGLING					RELATIVE TRANSV. IONZN. STRAGGLING				
	0.6	0.8	1.0	1.3	1.7	0.6	0.8	1.0	1.3	1.7
.010	.5526	.5502	.5478	.5443	.5398	.4629	.4596	.4564	.4516	.4454
.012	.5520	.5496	.5472	.5436	.5389	.4623	.4589	.4556	.4507	.4444
.014	.5516	.5491	.5466	.5429	.5381	.4617	.4583	.4549	.4499	.4434
.017	.5509	.5483	.5458	.5420	.5371	.4610	.4575	.4540	.4488	.4421
.020	.5504	.5477	.5451	.5412	.5362	.4604	.4567	.4531	.4479	.4410
.023	.5498	.5471	.5444	.5405	.5353	.4597	.4560	.4524	.4470	.4399
.026	.5493	.5465	.5438	.5397	.5345	.4591	.4553	.4516	.4461	.4390
.030	.5486	.5457	.5430	.5389	.5335	.4584	.4545	.4507	.4451	.4378
.035	.5478	.5449	.5420	.5378	.5324	.4575	.4535	.4496	.4439	.4364
.040	.5470	.5440	.5411	.5368	.5313	.4566	.4526	.4486	.4427	.4352
.045	.5462	.5432	.5403	.5359	.5303	.4558	.4517	.4476	.4417	.4340
.050	.5455	.5425	.5394	.5350	.5293	.4550	.4509	.4467	.4407	.4328
.060	.5442	.5410	.5379	.5334	.5275	.4536	.4493	.4450	.4388	.4308
.070	.5429	.5397	.5365	.5319	.5259	.4522	.4478	.4435	.4371	.4289
.080	.5416	.5383	.5351	.5304	.5243	.4509	.4464	.4420	.4355	.4271
.090	.5404	.5371	.5338	.5290	.5228	.4496	.4450	.4405	.4339	.4254
.10	.5393	.5359	.5326	.5277	.5214	.4484	.4437	.4392	.4325	.4238
.12	.5371	.5336	.5302	.5252	.5188	.4461	.4413	.4366	.4297	.4209
.14	.5350	.5315	.5280	.5229	.5163	.4439	.4390	.4342	.4272	.4182
.17	.5321	.5285	.5249	.5197	.5129	.4409	.4358	.4309	.4237	.4144
.20	.5294	.5257	.5220	.5167	.5098	.4380	.4328	.4278	.4204	.4109
.23	.5268	.5230	.5193	.5138	.5068	.4353	.4300	.4248	.4173	.4076
.26	.5243	.5204	.5167	.5111	.5040	.4327	.4273	.4220	.4144	.4046
.30	.5212	.5173	.5134	.5078	.5006	.4294	.4239	.4186	.4107	.4007
.35	.5176	.5136	.5097	.5039	.4965	.4256	.4200	.4145	.4065	.3963
.40	.5143	.5102	.5062	.5003	.4928	.4221	.4163	.4107	.4026	.3921
.45	.5112	.5070	.5029	.4970	.4893	.4187	.4129	.4072	.3989	.3883
.50	.5082	.5040	.4998	.4938	.4860	.4156	.4096	.4038	.3954	.3846
.60	.5027	.4983	.4940	.4878	.4799	.4097	.4035	.3975	.3889	.3778
.70	.4977	.4932	.4888	.4825	.4744	.4044	.3980	.3919	.3829	.3716
.80	.4932	.4886	.4841	.4776	.4693	.3995	.3930	.3866	.3775	.3660
.90	.4892	.4844	.4798	.4732	.4647	.3950	.3883	.3818	.3725	.3607
1.0	.4854	.4806	.4759	.4691	.4604	.3908	.3840	.3773	.3678	.3558
1.2	.4788	.4737	.4688	.4618	.4528	.3833	.3761	.3692	.3593	.3469
1.4	.4731	.4678	.4627	.4553	.4461	.3766	.3691	.3619	.3517	.3389
1.7	.4658	.4601	.4547	.4470	.4372	.3678	.3599	.3523	.3416	.3283
2.0	.4596	.4537	.4480	.4398	.4296	.3601	.3518	.3439	.3327	.3189
2.3	.4544	.4482	.4422	.4336	.4229	.3534	.3447	.3364	.3248	.3105
2.6	.4500	.4433	.4370	.4281	.4170	.3474	.3383	.3296	.3176	.3029
3.0	.4447	.4377	.4310	.4215	.4098	.3403	.3306	.3215	.3089	.2937
3.5	.4392	.4315	.4243	.4143	.4019	.3323	.3220	.3124	.2992	.2834
4.0	.4344	.4262	.4185	.4079	.3949	.3252	.3143	.3042	.2905	.2742
4.5	.4302	.4215	.4134	.4021	.3885	.3188	.3073	.2968	.2825	.2659
5.0	.4265	.4173	.4087	.3969	.3827	.3129	.3009	.2900	.2753	.2583
6.0	.4207	.4103	.4007	.3877	.3726	.3024	.2894	.2777	.2622	.2445
7.0	.4157	.4043	.3938	.3797	.3637	.2932	.2793	.2670	.2508	.2325
8.0	.4112	.3989	.3876	.3725	.3557	.2850	.2703	.2574	.2406	.2219
9.0	.4071	.3939	.3819	.3659	.3484	.2774	.2620	.2487	.2314	.2124
10	.4033	.3893	.3766	.3598	.3417	.2705	.2545	.2407	.2231	.2039
12	.3961	.3807	.3667	.3486	.3295	.2577	.2406	.2262	.2082	.1887
14	.3895	.3729	.3579	.3386	.3188	.2464	.2285	.2137	.1954	.1759
17	.3804	.3623	.3461	.3255	.3048	.2317	.2129	.1976	.1791	.1599
20	.3721	.3528	.3356	.3139	.2926	.2189	.1996	.1841	.1655	.1466
23	.3641	.3437	.3257	.3032	.2816	.2074	.1880	.1723	.1537	.1356
26	.3566	.3353	.3167	.2935	.2717	.1971	.1777	.1620	.1435	.1261
30	.3473	.3251	.3057	.2819	.2600	.1850	.1658	.1501	.1318	.1154
35	.3366	.3135	.2935	.2691	.2473	.1719	.1530	.1375	.1196	.1043
40	.3268	.3031	.2826	.2579	.2361	.1606	.1421	.1269	.1093	.0951
45	.3177	.2935	.2727	.2478	.2263	.1507	.1327	.1177	.1006	.0873
50	.3092	.2848	.2638	.2388	.2175	.1419	.1244	.1098	.0931	.0807
60	.2938	.2691	.2480	.2231	.2025	.1271	.1106	.0967	.0809	.0700
70	.2800	.2554	.2345	.2099	.1899	.1151	.0995	.0863	.0716	.0618
80	.2675	.2432	.2226	.1986	.1793	.1050	.0903	.0780	.0642	.0553
90	.2561	.2323	.2121	.1886	.1701	.0966	.0827	.0711	.0583	.0501
100	.2455	.2223	.2027	.1799	.1621	.0893	.0763	.0655	.0536	.0459

TABLE 11- 6

BE ON C, M2/M1=4/3

E,KEV	RANGE SKEWNESS K/KL= 0.6	0.8	1.0	1.3	1.7	DAMAGE SKEWNESS 0.6	0.8	1.0	1.3	1.7
.010	.6276	.6168	.6062	.5906	.5703	.8822	.8739	.8658	.8539	.8387
.012	.6262	.6151	.6042	.5881	.5673	.8812	.8726	.8643	.8521	.8365
.014	.6250	.6136	.6024	.5859	.5646	.8803	.8715	.8630	.8505	.8345
.017	.6234	.6116	.6001	.5831	.5611	.8791	.8701	.8612	.8484	.8319
.020	.6220	.6099	.5980	.5805	.5580	.8781	.8688	.8597	.8465	.8297
.023	.6207	.6083	.5961	.5783	.5552	.8772	.8676	.8584	.8449	.8276
.026	.6195	.6068	.5944	.5761	.5526	.8763	.8666	.8571	.8433	.8258
.030	.6180	.6050	.5922	.5736	.5495	.8752	.8652	.8555	.8414	.8235
.035	.6162	.6028	.5898	.5706	.5459	.8740	.8637	.8537	.8393	.8209
.040	.6146	.6009	.5875	.5679	.5426	.8728	.8623	.8521	.8373	.8185
.045	.6130	.5990	.5853	.5653	.5395	.8717	.8609	.8505	.8354	.8163
.050	.6115	.5973	.5833	.5629	.5367	.8706	.8596	.8490	.8337	.8142
.060	.6088	.5941	.5796	.5586	.5315	.8686	.8573	.8463	.8305	.8104
.070	.6063	.5911	.5762	.5545	.5268	.8668	.8551	.8438	.8275	.8069
.080	.6039	.5882	.5730	.5508	.5223	.8651	.8531	.8415	.8248	.8037
.090	.6016	.5856	.5700	.5472	.5182	.8634	.8511	.8393	.8222	.8007
.10	.5994	.5830	.5671	.5439	.5142	.8619	.8493	.8372	.8197	.7978
.12	.5953	.5783	.5617	.5377	.5070	.8590	.8460	.8334	.8153	.7926
.14	.5914	.5738	.5567	.5319	.5003	.8563	.8428	.8298	.8112	.7878
.17	.5858	.5674	.5496	.5237	.4909	.8524	.8383	.8247	.8053	.7810
.20	.5805	.5614	.5429	.5161	.4821	.8486	.8339	.8198	.7997	.7745
.23	.5756	.5558	.5367	.5090	.4741	.8449	.8297	.8151	.7943	.7685
.26	.5709	.5505	.5307	.5023	.4664	.8413	.8256	.8105	.7891	.7626
.30	.5648	.5436	.5231	.4937	.4566	.8366	.8202	.8046	.7825	.7551
.35	.5574	.5353	.5139	.4834	.4449	.8308	.8137	.7975	.7745	.7461
.40	.5503	.5273	.5052	.4735	.4339	.8252	.8075	.7906	.7668	.7374
.45	.5434	.5196	.4967	.4640	.4232	.8198	.8014	.7839	.7593	.7291
.50	.5367	.5121	.4885	.4549	.4130	.8145	.7954	.7774	.7521	.7212
.60	.5241	.4980	.4731	.4377	.3938	.8047	.7844	.7654	.7388	.7063
.70	.5120	.4845	.4584	.4214	.3757	.7951	.7738	.7538	.7260	.6922
.80	.5004	.4716	.4443	.4059	.3585	.7857	.7634	.7426	.7136	.6787
.90	.4891	.4592	.4308	.3909	.3420	.7765	.7533	.7316	.7016	.6656
1.0	.4783	.4471	.4178	.3766	.3263	.7675	.7434	.7209	.6900	.6529
1.2	.4576	.4243	.3930	.3493	.2964	.7498	.7239	.7000	.6673	.6283
1.4	.4380	.4026	.3696	.3237	.2684	.7327	.7054	.6802	.6458	.6052
1.7	.4103	.3722	.3367	.2880	.2295	.7085	.6791	.6521	.6157	.5729
2.0	.3844	.3437	.3061	.2547	.1936	.6857	.6544	.6259	.5876	.5431
2.3	.3604	.3169	.2771	.2230	.1588	.6647	.6313	.6013	.5613	.5151
2.6	.3376	.2915	.2497	.1931	.1263	.6448	.6095	.5781	.5366	.4889
3.0	.3086	.2596	.2154	.1560	.0861	.6194	.5820	.5490	.5057	.4563
3.5	.2744	.2222	.1755	.1131	.0403	.5895	.5498	.5151	.4699	.4188
4.0	.2420	.1872	.1384	.0736	-.0014	.5611	.5197	.4836	.4368	.3843
4.5	.2112	.1542	.1036	.0370	-.0396	.5342	.4913	.4540	.4060	.3524
5.0	.1817	.1229	.0710	.0027	-.0749	.5084	.4644	.4262	.3770	.3225
6.0	.1214	.0622	.0097	-.0588	-.1345	.4559	.4116	.3726	.3222	.2668
7.0	.0675	.0077	-.0455	-.1144	-.1884	.4089	.3643	.3248	.2733	.2173
8.0	.0190	-.0417	-.0958	-.1654	-.2382	.3667	.3219	.2817	.2293	.1730
9.0	-.0246	-.0866	-.1419	-.2127	-.2849	.3288	.2835	.2426	.1894	.1329
10	-.0641	-.1278	-.1846	-.2568	-.3291	.2945	.2486	.2070	.1530	.0964
12	-.1268	-.1975	-.2603	-.3394	-.4167	.2386	.1893	.1452	.0888	.0321
14	-.1813	-.2589	-.3275	-.4134	-.4957	.1907	.1382	.0917	.0332	-.0233
17	-.2535	-.3403	-.4166	-.5114	-.6008	.1288	.0723	.0228	-.0381	-.0943
20	-.3181	-.4126	-.4953	-.5974	-.6926	.0753	.0156	-.0361	-.0988	-.1545
23	-.3832	-.4819	-.5679	-.6735	-.7702	.0240	-.0365	-.0888	-.1518	-.2068
26	-.4435	-.5454	-.6340	-.7421	-.8396	-.0221	-.0831	-.1356	-.1985	-.2528
30	-.5173	-.6226	-.7138	-.8243	-.9225	-.0770	-.1381	-.1906	-.2532	-.3066
35	-.6005	-.7090	-.8027	-.9155	-1.014	-.1369	-.1979	-.2501	-.3122	-.3644
40	-.6753	-.7863	-.8820	-.9965	-1.095	-.1890	-.2497	-.3015	-.3630	-.4143
45	-.7432	-.8563	-.9535	-1.069	-1.168	-.2349	-.2952	-.3467	-.4075	-.4580
50	-.8052	-.9201	-1.019	-1.136	-1.235	-.2756	-.3356	-.3868	-.4471	-.4968
60	-.9151	-1.033	-1.134	-1.253	-1.354	-.3448	-.4044	-.4551	-.5146	-.5630
70	-1.010	-1.130	-1.233	-1.355	-1.458	-.4016	-.4610	-.5115	-.5704	-.6179
80	-1.093	-1.216	-1.321	-1.445	-1.550	-.4491	-.5086	-.5591	-.6178	-.6644
90	-1.167	-1.292	-1.399	-1.526	-1.633	-.4893	-.5492	-.5999	-.6586	-.7046
100	-1.233	-1.360	-1.469	-1.599	-1.710	-.5237	-.5842	-.6353	-.6942	-.7398

BE ON C, M2/M1=4/3

TABLE 11- 7

K/KL= E,KEV	IONIZATION SKEWNESS					RANGE KURTOSIS				
	0.6	0.8	1.0	1.3	1.7	0.6	0.8	1.0	1.3	1.7
.010	.9523	.9451	.9380	.9277	.9146	3.420	3.391	3.363	3.322	3.271
.012	.9514	.9439	.9366	.9261	.9127	3.417	3.386	3.358	3.316	3.264
.014	.9506	.9429	.9355	.9247	.9110	3.413	3.383	3.353	3.310	3.257
.017	.9496	.9416	.9339	.9228	.9087	3.409	3.377	3.347	3.303	3.249
.020	.9486	.9405	.9326	.9212	.9068	3.405	3.373	3.342	3.297	3.241
.023	.9478	.9394	.9314	.9197	.9050	3.402	3.369	3.337	3.291	3.235
.026	.9470	.9385	.9302	.9184	.9034	3.399	3.365	3.332	3.286	3.229
.030	.9460	.9373	.9288	.9167	.9014	3.395	3.360	3.327	3.280	3.221
.035	.9448	.9359	.9272	.9148	.8991	3.390	3.355	3.321	3.273	3.213
.040	.9437	.9346	.9257	.9130	.8970	3.386	3.350	3.315	3.266	3.206
.045	.9427	.9334	.9243	.9114	.8951	3.382	3.345	3.310	3.260	3.199
.050	.9418	.9322	.9230	.9098	.8933	3.378	3.340	3.305	3.254	3.192
.060	.9400	.9301	.9206	.9070	.8900	3.371	3.332	3.296	3.244	3.181
.070	.9383	.9281	.9184	.9044	.8870	3.364	3.325	3.288	3.235	3.170
.080	.9367	.9263	.9163	.9020	.8842	3.358	3.318	3.280	3.226	3.161
.090	.9352	.9245	.9143	.8997	.8816	3.352	3.312	3.273	3.218	3.152
.10	.9338	.9228	.9124	.8975	.8791	3.347	3.305	3.266	3.211	3.144
.12	.9310	.9197	.9089	.8935	.8745	3.337	3.294	3.254	3.197	3.129
.14	.9285	.9168	.9056	.8897	.8702	3.327	3.283	3.242	3.184	3.115
.17	.9248	.9125	.9009	.8844	.8642	3.314	3.268	3.226	3.167	3.096
.20	.9212	.9085	.8964	.8794	.8585	3.301	3.255	3.211	3.151	3.079
.23	.9179	.9048	.8923	.8748	.8533	3.289	3.242	3.197	3.136	3.063
.26	.9147	.9011	.8883	.8703	.8483	3.278	3.230	3.185	3.122	3.049
.30	.9105	.8965	.8832	.8646	.8420	3.264	3.214	3.168	3.105	3.031
.35	.9055	.8908	.8770	.8578	.8345	3.247	3.196	3.149	3.085	3.010
.40	.9006	.8854	.8711	.8512	.8273	3.232	3.180	3.132	3.067	2.991
.45	.8958	.8801	.8654	.8449	.8204	3.217	3.164	3.115	3.050	2.974
.50	.8912	.8750	.8598	.8389	.8138	3.203	3.149	3.100	3.034	2.958
.60	.8823	.8652	.8492	.8273	.8013	3.176	3.120	3.070	3.003	2.927
.70	.8738	.8558	.8392	.8164	.7895	3.151	3.094	3.044	2.976	2.900
.80	.8656	.8469	.8296	.8061	.7784	3.128	3.071	3.019	2.952	2.876
.90	.8577	.8383	.8204	.7962	.7679	3.107	3.049	2.997	2.929	2.855
1.0	.8501	.8300	.8116	.7868	.7578	3.087	3.028	2.976	2.909	2.835
1.2	.8354	.8140	.7946	.7686	.7386	3.049	2.989	2.937	2.870	2.799
1.4	.8216	.7991	.7789	.7519	.7211	3.015	2.955	2.903	2.838	2.770
1.7	.8024	.7785	.7572	.7290	.6973	2.971	2.911	2.860	2.797	2.734
2.0	.7847	.7596	.7374	.7084	.6760	2.933	2.874	2.824	2.765	2.707
2.3	.7678	.7416	.7186	.6888	.6561	2.901	2.841	2.792	2.734	2.680
2.6	.7520	.7249	.7012	.6708	.6381	2.873	2.812	2.764	2.709	2.659
3.0	.7326	.7044	.6800	.6491	.6164	2.840	2.780	2.734	2.684	2.640
3.5	.7107	.6814	.6563	.6249	.5924	2.804	2.748	2.706	2.662	2.628
4.0	.6908	.6607	.6352	.6034	.5713	2.773	2.722	2.685	2.649	2.626
4.5	.6727	.6419	.6161	.5843	.5526	2.746	2.701	2.670	2.644	2.633
5.0	.6561	.6248	.5987	.5670	.5357	2.722	2.684	2.661	2.645	2.646
6.0	.6258	.5938	.5676	.5363	.5060	2.656	2.646	2.647	2.663	2.702
7.0	.5997	.5673	.5413	.5107	.4814	2.610	2.625	2.648	2.693	2.765
8.0	.5769	.5445	.5188	.4890	.4609	2.582	2.619	2.660	2.729	2.830
9.0	.5568	.5245	.4992	.4704	.4434	2.570	2.624	2.681	2.770	2.895
10	.5389	.5068	.4821	.4543	.4286	2.570	2.638	2.708	2.814	2.960
12	.5082	.4772	.4539	.4285	.4057	2.633	2.706	2.781	2.898	3.062
14	.4829	.4532	.4314	.4083	.3883	2.713	2.788	2.866	2.990	3.168
17	.4526	.4249	.4054	.3855	.3694	2.842	2.921	3.005	3.139	3.334
20	.4287	.4033	.3858	.3688	.3562	2.971	3.057	3.149	3.296	3.510
23	.4104	.3874	.3718	.3571	.3479	3.059	3.169	3.282	3.457	3.701
26	.3955	.3748	.3610	.3485	.3422	3.147	3.282	3.417	3.621	3.896
30	.3797	.3620	.3502	.3402	.3373	3.263	3.432	3.599	3.843	4.156
35	.3647	.3503	.3408	.3336	.3343	3.410	3.623	3.828	4.120	4.481
40	.3534	.3419	.3345	.3297	.3333	3.559	3.815	4.057	4.396	4.802
45	.3449	.3360	.3304	.3277	.3337	3.711	4.007	4.285	4.669	5.118
50	.3385	.3319	.3280	.3270	.3350	3.865	4.199	4.511	4.938	5.427
60	.3302	.3274	.3263	.3284	.3393	4.174	4.582	4.957	5.462	6.025
70	.3260	.3262	.3274	.3321	.3450	4.486	4.959	5.392	5.968	6.597
80	.3246	.3273	.3304	.3372	.3512	4.796	5.329	5.815	6.455	7.144
90	.3252	.3297	.3346	.3432	.3577	5.103	5.691	6.225	6.925	7.668
100	.3272	.3332	.3395	.3496	.3642	5.408	6.046	6.624	7.378	8.172

193

	DAMAGE KURTOSIS					IONIZATION KURTOSIS				
K/KL=	0.6	0.8	1.0	1.3	1.7	0.6	0.8	1.0	1.3	1.7
E,KEV										
.010	3.888	3.857	3.827	3.784	3.729	4.014	3.984	3.955	3.913	3.860
.012	3.884	3.852	3.821	3.777	3.721	4.011	3.979	3.949	3.906	3.852
.014	3.881	3.848	3.816	3.771	3.714	4.007	3.975	3.945	3.901	3.845
.017	3.876	3.842	3.810	3.763	3.705	4.003	3.970	3.938	3.893	3.836
.020	3.872	3.837	3.804	3.756	3.696	3.999	3.965	3.933	3.886	3.829
.023	3.868	3.833	3.799	3.750	3.689	3.995	3.961	3.928	3.880	3.822
.026	3.865	3.829	3.794	3.744	3.683	3.992	3.956	3.923	3.875	3.815
.030	3.860	3.824	3.788	3.738	3.674	3.987	3.951	3.917	3.868	3.807
.035	3.855	3.818	3.781	3.730	3.665	3.982	3.946	3.910	3.860	3.798
.040	3.851	3.812	3.775	3.722	3.657	3.978	3.940	3.904	3.853	3.790
.045	3.846	3.807	3.769	3.716	3.649	3.973	3.935	3.898	3.847	3.782
.050	3.842	3.802	3.764	3.709	3.642	3.969	3.930	3.893	3.840	3.775
.060	3.835	3.793	3.754	3.698	3.628	3.962	3.921	3.883	3.829	3.762
.070	3.827	3.785	3.745	3.687	3.616	3.955	3.913	3.874	3.819	3.751
.080	3.821	3.777	3.736	3.677	3.605	3.948	3.906	3.866	3.809	3.740
.090	3.814	3.770	3.728	3.668	3.595	3.941	3.898	3.858	3.800	3.730
.10	3.808	3.763	3.720	3.659	3.585	3.935	3.891	3.850	3.791	3.720
.12	3.796	3.750	3.705	3.643	3.567	3.924	3.879	3.836	3.776	3.702
.14	3.786	3.738	3.692	3.628	3.550	3.913	3.866	3.822	3.761	3.686
.17	3.770	3.720	3.673	3.607	3.527	3.898	3.849	3.804	3.740	3.664
.20	3.755	3.704	3.655	3.588	3.506	3.883	3.833	3.786	3.721	3.643
.23	3.742	3.689	3.639	3.570	3.486	3.869	3.818	3.770	3.703	3.623
.26	3.729	3.675	3.624	3.553	3.468	3.856	3.804	3.754	3.687	3.605
.30	3.713	3.657	3.604	3.532	3.445	3.839	3.785	3.735	3.665	3.582
.35	3.692	3.635	3.581	3.506	3.418	3.819	3.763	3.711	3.640	3.555
.40	3.673	3.613	3.558	3.482	3.392	3.800	3.742	3.689	3.616	3.530
.45	3.654	3.593	3.536	3.459	3.367	3.781	3.722	3.668	3.594	3.506
.50	3.636	3.573	3.515	3.437	3.344	3.764	3.703	3.648	3.572	3.484
.60	3.598	3.533	3.473	3.392	3.297	3.729	3.666	3.609	3.531	3.441
.70	3.563	3.496	3.434	3.351	3.254	3.697	3.632	3.573	3.493	3.401
.80	3.531	3.461	3.398	3.313	3.215	3.667	3.600	3.539	3.458	3.365
.90	3.500	3.429	3.365	3.278	3.179	3.638	3.570	3.508	3.425	3.331
1.0	3.472	3.399	3.333	3.246	3.146	3.611	3.541	3.478	3.394	3.299
1.2	3.421	3.345	3.277	3.188	3.087	3.559	3.486	3.421	3.335	3.238
1.4	3.374	3.296	3.226	3.136	3.034	3.511	3.436	3.369	3.282	3.185
1.7	3.310	3.229	3.158	3.067	2.965	3.446	3.368	3.300	3.212	3.114
2.0	3.253	3.170	3.098	3.005	2.904	3.388	3.308	3.238	3.150	3.053
2.3	3.199	3.113	3.039	2.946	2.846	3.334	3.252	3.182	3.093	3.002
2.6	3.149	3.060	2.985	2.892	2.794	3.285	3.201	3.131	3.042	2.957
3.0	3.087	2.996	2.921	2.828	2.732	3.225	3.141	3.069	2.981	2.904
3.5	3.016	2.925	2.849	2.758	2.666	3.159	3.074	3.002	2.915	2.846
4.0	2.951	2.860	2.786	2.698	2.608	3.100	3.014	2.943	2.857	2.795
4.5	2.891	2.802	2.730	2.644	2.558	3.048	2.962	2.891	2.806	2.749
5.0	2.835	2.749	2.680	2.597	2.515	3.001	2.915	2.845	2.761	2.708
6.0	2.713	2.644	2.587	2.517	2.442	2.918	2.833	2.765	2.684	2.629
7.0	2.612	2.558	2.512	2.452	2.385	2.848	2.764	2.698	2.621	2.563
8.0	2.530	2.488	2.451	2.400	2.339	2.788	2.706	2.641	2.568	2.507
9.0	2.463	2.431	2.400	2.357	2.303	2.736	2.655	2.593	2.522	2.460
10	2.410	2.384	2.359	2.322	2.274	2.691	2.611	2.551	2.483	2.419
12	2.361	2.333	2.307	2.273	2.236	2.614	2.539	2.482	2.420	2.358
14	2.334	2.301	2.272	2.239	2.213	2.553	2.481	2.428	2.370	2.311
17	2.313	2.273	2.240	2.208	2.195	2.479	2.412	2.364	2.312	2.258
20	2.303	2.257	2.223	2.192	2.190	2.421	2.359	2.315	2.268	2.220
23	2.280	2.240	2.211	2.187	2.193	2.376	2.319	2.277	2.234	2.192
26	2.261	2.228	2.206	2.189	2.201	2.340	2.286	2.247	2.206	2.170
30	2.241	2.219	2.205	2.198	2.217	2.300	2.251	2.214	2.178	2.149
35	2.225	2.215	2.211	2.216	2.242	2.261	2.216	2.184	2.151	2.129
40	2.216	2.218	2.223	2.238	2.269	2.230	2.190	2.160	2.131	2.115
45	2.213	2.225	2.238	2.262	2.298	2.205	2.169	2.142	2.115	2.105
50	2.215	2.235	2.256	2.287	2.328	2.185	2.152	2.127	2.104	2.098
60	2.228	2.263	2.296	2.339	2.387	2.154	2.128	2.107	2.088	2.089
70	2.252	2.297	2.338	2.391	2.445	2.132	2.111	2.094	2.080	2.085
80	2.283	2.336	2.382	2.442	2.501	2.116	2.099	2.086	2.076	2.083
90	2.320	2.376	2.427	2.491	2.555	2.104	2.091	2.081	2.075	2.084
100	2.360	2.418	2.471	2.539	2.607	2.095	2.085	2.079	2.076	2.085

	REFLECTION COEFFICIENT					SPUTTERING EFFICIENCY				
K/KL=	0.6	0.8	1.0	1.3	1.7	0.6	0.8	1.0	1.3	1.7
E,KEV										
.010	.0336	.0331	.0326	.0319	.0311	.0252	.0243	.0234	.0221	.0206
.012	.0335	.0330	.0325	.0318	.0309	.0251	.0241	.0232	.0220	.0204
.014	.0334	.0329	.0324	.0317	.0307	.0250	.0240	.0231	.0218	.0202
.017	.0333	.0327	.0322	.0315	.0306	.0249	.0239	.0229	.0216	.0200
.020	.0331	.0326	.0321	.0313	.0304	.0248	.0238	.0228	.0215	.0198
.023	.0330	.0325	.0320	.0312	.0302	.0247	.0237	.0227	.0213	.0196
.026	.0329	.0324	.0318	.0310	.0301	.0246	.0236	.0226	.0212	.0195
.030	.0328	.0322	.0317	.0309	.0299	.0245	.0234	.0224	.0210	.0193
.035	.0326	.0320	.0315	.0307	.0296	.0244	.0233	.0223	.0208	.0191
.040	.0325	.0319	.0313	.0305	.0294	.0243	.0232	.0221	.0207	.0189
.045	.0323	.0317	.0311	.0303	.0292	.0242	.0231	.0220	.0205	.0187
.050	.0322	.0316	.0310	.0301	.0291	.0241	.0230	.0219	.0204	.0186
.060	.0319	.0313	.0307	.0298	.0287	.0240	.0228	.0217	.0202	.0183
.070	.0317	.0310	.0304	.0295	.0284	.0238	.0226	.0215	.0200	.0181
.080	.0314	.0308	.0301	.0292	.0281	.0237	.0225	.0213	.0198	.0179
.090	.0312	.0305	.0299	.0290	.0278	.0236	.0223	.0212	.0196	.0177
.10	.0310	.0303	.0297	.0287	.0276	.0234	.0222	.0210	.0194	.0175
.12	.0305	.0299	.0292	.0282	.0271	.0232	.0220	.0208	.0191	.0172
.14	.0301	.0294	.0288	.0278	.0266	.0231	.0218	.0205	.0189	.0169
.17	.0296	.0289	.0282	.0272	.0259	.0228	.0215	.0202	.0185	.0166
.20	.0290	.0283	.0276	.0266	.0254	.0226	.0212	.0199	.0182	.0163
.23	.0285	.0278	.0271	.0261	.0248	.0224	.0210	.0197	.0180	.0160
.26	.0280	.0273	.0266	.0255	.0243	.0222	.0208	.0195	.0177	.0157
.30	.0274	.0267	.0259	.0249	.0236	.0219	.0205	.0192	.0174	.0154
.35	.0267	.0260	.0252	.0242	.0229	.0217	.0202	.0189	.0171	.0151
.40	.0261	.0253	.0245	.0235	.0222	.0214	.0199	.0186	.0168	.0147
.45	.0255	.0247	.0239	.0228	.0215	.0212	.0197	.0183	.0165	.0145
.50	.0249	.0241	.0233	.0222	.0209	.0209	.0194	.0181	.0162	.0142
.60	.0238	.0230	.0222	.0211	.0198	.0205	.0189	.0176	.0157	.0137
.70	.0228	.0220	.0212	.0201	.0188	.0201	.0185	.0171	.0153	.0132
.80	.0220	.0211	.0203	.0192	.0179	.0197	.0181	.0167	.0149	.0128
.90	.0211	.0203	.0195	.0184	.0171	.0194	.0178	.0164	.0145	.0125
1.0	.0204	.0195	.0187	.0176	.0163	.0191	.0175	.0161	.0142	.0121
1.2	.0191	.0182	.0174	.0163	.0149	.0186	.0169	.0155	.0136	.0116
1.4	.0179	.0170	.0162	.0151	.0138	.0181	.0165	.0150	.0132	.0111
1.7	.0164	.0155	.0147	.0136	.0123	.0175	.0159	.0144	.0125	.0105
2.0	.0151	.0142	.0134	.0123	.0111	.0170	.0153	.0138	.0120	.0100
2.3	.0141	.0132	.0123	.0113	.0101	.0165	.0148	.0133	.0115	.0095
2.6	.0131	.0122	.0114	.0104	.0092	.0161	.0143	.0128	.0110	.0090
3.0	.0121	.0111	.0103	.0093	.0082	.0155	.0137	.0122	.0104	.0085
3.5	.0109	.0100	.0092	.0082	.0072	.0149	.0131	.0116	.0098	.0079
4.0	.0099	.0090	.0082	.0073	.0063	.0143	.0125	.0110	.0092	.0074
4.5	.0091	.0082	.0074	.0065	.0056	.0137	.0119	.0105	.0087	.0070
5.0	.0083	.0075	.0067	.0059	.0051	.0132	.0114	.0100	.0083	.0066
6.0	.0069	.0063	.0057	.0049	.0043	.0120	.0104	.0091	.0075	.0060
7.0	.0059	.0053	.0048	.0042	.0037	.0110	.0096	.0084	.0069	.0054
8.0	.0050	.0046	.0041	.0036	.0032	.0102	.0089	.0078	.0063	.0050
9.0	.0043	.0040	.0036	.0031	.0028	.0095	.0083	.0072	.0058	.0046
10	.0037	.0035	.0031	.0027	.0025	.0089	.0078	.0068	.0054	.0042
12	.0031	.0028	.0025	.0020	.0018	.0082	.0071	.0061	.0048	.0038
14	.0028	.0024	.0020	.0015	.0013	.0078	.0066	.0056	.0044	.0034
17	.0024	.0019	.0014	.0010	.0008	.0073	.0060	.0050	.0038	.0030
20	.0022	.0016	.0011	.0006	.0004	.0069	.0055	.0045	.0034	.0027
23	.0019	.0013	.0008	.0004	.0002	.0063	.0050	.0041	.0031	.0024
26	.0015	.0010	.0006	.0003	.0001	.0059	.0046	.0037	.0028	.0021
30	.0012	.0008	.0004	.0001	.0000	.0053	.0042	.0033	.0024	.0019
35	.0008	.0005	.0003	.0001	0.0000	.0047	.0037	.0029	.0021	.0016
40	.0006	.0003	.0002	.0000	0.0000	.0042	.0033	.0026	.0019	.0014
45	.0003	.0002	.0001	.0000	0.0000	.0038	.0029	.0023	.0016	.0012
50	.0002	.0001	.0000	.0000	0.0000	.0034	.0026	.0021	.0015	.0011
60	0.0000	0.0000	0.0000	.0000	0.0000	.0028	.0022	.0017	.0012	.0009
70	0.0000	0.0000	0.0000	.0000	0.0000	.0024	.0019	.0015	.0010	.0007
80	0.0000	0.0000	0.0000	.0000	0.0000	.0021	.0016	.0013	.0008	.0006
90	0.0000	0.0000	0.0000	.0000	0.0000	.0019	.0014	.0011	.0007	.0005
100	0.0000	0.0000	0.0000	.0000	0.0000	.0017	.0013	.0010	.0006	.0005

TABLE 11-10

BE ON C, M2/M1=4/3

K/KL= E,KEV	IONIZATION DEFICIENCY					SPUTTERING YIELD ALPHA				
	0.6	0.8	1.0	1.3	1.7	0.6	0.8	1.0	1.3	1.7
.010	.0016	.0021	.0025	.0031	.0038	.2021	.1972	.1924	.1856	.1771
.012	.0016	.0021	.0026	.0032	.0039	.2018	.1967	.1918	.1848	.1761
.014	.0017	.0022	.0026	.0032	.0040	.2015	.1963	.1913	.1842	.1752
.017	.0017	.0022	.0027	.0033	.0041	.2011	.1958	.1906	.1833	.1741
.020	.0018	.0023	.0028	.0034	.0042	.2008	.1953	.1900	.1826	.1732
.023	.0018	.0023	.0028	.0035	.0042	.2005	.1949	.1896	.1820	.1724
.026	.0018	.0024	.0029	.0035	.0043	.2003	.1946	.1891	.1814	.1717
.030	.0019	.0024	.0029	.0036	.0043	.2000	.1942	.1887	.1808	.1709
.035	.0019	.0025	.0030	.0036	.0044	.1998	.1938	.1882	.1801	.1700
.040	.0020	.0025	.0030	.0037	.0045	.1996	.1935	.1877	.1795	.1692
.045	.0020	.0025	.0031	.0037	.0045	.1994	.1932	.1874	.1790	.1686
.050	.0020	.0026	.0031	.0038	.0046	.1993	.1930	.1871	.1785	.1680
.060	.0021	.0026	.0032	.0039	.0047	.1991	.1926	.1865	.1778	.1670
.070	.0021	.0027	.0032	.0039	.0047	.1989	.1924	.1861	.1772	.1662
.080	.0021	.0027	.0033	.0040	.0048	.1989	.1922	.1858	.1767	.1655
.090	.0022	.0028	.0033	.0040	.0048	.1988	.1920	.1855	.1762	.1649
.10	.0022	.0028	.0033	.0041	.0049	.1988	.1919	.1852	.1758	.1644
.12	.0022	.0029	.0034	.0041	.0049	.1988	.1917	.1848	.1752	.1635
.14	.0023	.0029	.0035	.0042	.0050	.1989	.1915	.1845	.1747	.1628
.17	.0023	.0030	.0035	.0043	.0051	.1990	.1914	.1842	.1741	.1619
.20	.0024	.0030	.0036	.0043	.0051	.1993	.1914	.1840	.1737	.1612
.23	.0024	.0031	.0036	.0043	.0051	.1996	.1915	.1840	.1734	.1607
.26	.0024	.0031	.0037	.0044	.0052	.1999	.1917	.1840	.1732	.1603
.30	.0025	.0031	.0037	.0044	.0052	.2003	.1919	.1840	.1730	.1599
.35	.0025	.0032	.0037	.0044	.0052	.2009	.1922	.1842	.1729	.1594
.40	.0025	.0032	.0038	.0045	.0052	.2014	.1926	.1843	.1728	.1591
.45	.0026	.0032	.0038	.0045	.0052	.2020	.1929	.1845	.1728	.1588
.50	.0026	.0032	.0038	.0045	.0052	.2025	.1932	.1847	.1728	.1585
.60	.0026	.0033	.0038	.0045	.0052	.2034	.1938	.1850	.1727	.1581
.70	.0026	.0033	.0038	.0045	.0052	.2043	.1945	.1854	.1728	.1578
.80	.0026	.0033	.0038	.0045	.0051	.2052	.1951	.1858	.1728	.1576
.90	.0027	.0033	.0038	.0045	.0051	.2061	.1957	.1861	.1730	.1575
1.0	.0027	.0033	.0038	.0044	.0051	.2069	.1962	.1865	.1731	.1574
1.2	.0027	.0033	.0038	.0044	.0050	.2086	.1976	.1875	.1737	.1575
1.4	.0027	.0033	.0037	.0043	.0049	.2102	.1988	.1884	.1742	.1577
1.7	.0027	.0032	.0037	.0042	.0047	.2123	.2004	.1895	.1750	.1579
2.0	.0026	.0032	.0036	.0041	.0046	.2140	.2017	.1905	.1755	.1581
2.3	.0026	.0032	.0036	.0040	.0045	.2155	.2027	.1911	.1758	.1581
2.6	.0026	.0031	.0035	.0039	.0044	.2168	.2035	.1916	.1760	.1580
3.0	.0026	.0030	.0034	.0038	.0043	.2181	.2044	.1922	.1762	.1579
3.5	.0025	.0030	.0033	.0036	.0041	.2194	.2051	.1926	.1764	.1577
4.0	.0025	.0029	.0032	.0035	.0039	.2202	.2057	.1930	.1765	.1577
4.5	.0024	.0028	.0031	.0034	.0038	.2208	.2060	.1932	.1765	.1576
5.0	.0024	.0027	.0030	.0033	.0037	.2211	.2063	.1934	.1765	.1576
6.0	.0023	.0026	.0028	.0030	.0033	.2196	.2055	.1930	.1762	.1578
7.0	.0022	.0026	.0026	.0029	.0030	.2183	.2049	.1927	.1760	.1581
8.0	.0021	.0023	.0025	.0027	.0027	.2175	.2046	.1926	.1759	.1584
9.0	.0020	.0022	.0023	.0025	.0025	.2170	.2046	.1928	.1759	.1587
10	.0019	.0020	.0022	.0024	.0023	.2169	.2049	.1931	.1760	.1590
12	.0016	.0018	.0020	.0020	.0019	.2199	.2080	.1953	.1771	.1600
14	.0014	.0017	.0018	.0017	.0016	.2230	.2111	.1975	.1781	.1608
17	.0013	.0014	.0015	.0014	.0012	.2272	.2149	.2002	.1793	.1618
20	.0013	.0012	.0013	.0014	.0010	.2305	.2178	.2022	.1801	.1623
23	.0022	.0010	.0011	.0022	.0007	.2311	.2180	.2021	.1797	.1621
26	.0030	.0008	.0009	.0031	.0006	.2312	.2177	.2017	.1792	.1617
30	.0040	.0006	.0007	.0041	.0004	.2308	.2169	.2008	.1784	.1611
35	.0051	.0004	.0004	.0051	.0003	.2301	.2156	.1996	.1773	.1603
40	.0058	.0002	.0002	.0059	.0001	.2292	.2142	.1983	.1762	.1594
45	.0062	.0001	.0001	.0063	.0001	.2282	.2129	.1971	.1751	.1586
50	.0064	0.0000	0.0000	.0065	0.0000	.2272	.2118	.1960	.1742	.1579
60	.0062	0.0000	0.0000	.0063	0.0000	.2256	.2099	.1941	.1724	.1566
70	.0053	0.0000	0.0000	.0054	0.0000	.2242	.2085	.1927	.1710	.1555
80	.0039	0.0000	0.0000	.0040	0.0000	.2232	.2077	.1917	.1698	.1546
90	.0021	0.0000	0.0000	.0021	0.0000	.2224	.2073	.1911	.1689	.1539
100	0.0000	0.0000	0.0000	0.0000	0.0000	.2219	.2072	.1908	.1682	.1533

BE ON C, M2/M1=4/3

100 KEV

10 KEV

1 KEV

.10 KEV

.01 KEV

TABLE 12- 1

GE ON GE,M2=M1

E,KEV	FRACTIONAL DEPOSITED ENERGY K/KL=					MEAN RANGE,MICROGRAM/SQ.CM.				
	0.6	0.8	1.0	1.2	1.4	0.6	0.8	1.0	1.2	1.4
1.0	.8898	.8567	.8251	.7951	.7665	.7412	.7363	.7314	.7266	.7219
1.2	.8866	.8527	.8203	.7896	.7604	.8426	.8367	.8309	.8252	.8196
1.4	.8839	.8492	.8162	.7849	.7552	.9344	.9277	.9211	.9146	.9083
1.7	.8803	.8447	.8109	.7788	.7485	1.062	1.054	1.046	1.039	1.031
2.0	.8772	.8408	.8063	.7736	.7427	1.181	1.172	1.163	1.155	1.146
2.3	.8745	.8374	.8022	.7690	.7376	1.295	1.285	1.276	1.266	1.256
2.6	.8721	.8343	.7986	.7649	.7331	1.406	1.395	1.384	1.373	1.363
3.0	.8692	.8307	.7944	.7601	.7277	1.548	1.536	1.524	1.512	1.500
3.5	.8660	.8267	.7896	.7548	.7218	1.720	1.706	1.692	1.678	1.665
4.0	.8632	.8231	.7855	.7501	.7167	1.886	1.870	1.854	1.839	1.824
4.5	.8607	.8199	.7817	.7458	.7121	2.047	2.029	2.012	1.995	1.978
5.0	.8583	.8171	.7784	.7420	.7079	2.203	2.184	2.164	2.146	2.127
6.0	.8543	.8120	.7724	.7354	.7007	2.493	2.470	2.448	2.426	2.404
7.0	.8508	.8077	.7673	.7296	.6945	2.774	2.747	2.722	2.696	2.672
8.0	.8477	.8038	.7628	.7245	.6890	3.047	3.017	2.988	2.959	2.931
9.0	.8449	.8003	.7586	.7199	.6840	3.313	3.279	3.247	3.215	3.183
10	.8423	.7970	.7549	.7157	.6795	3.573	3.536	3.500	3.464	3.430
12	.8376	.7913	.7481	.7082	.6714	4.054	4.011	3.968	3.927	3.887
14	.8335	.7862	.7423	.7016	.6643	4.526	4.475	4.426	4.378	4.332
17	.8281	.7795	.7345	.6930	.6550	5.220	5.158	5.098	5.040	4.984
20	.8233	.7737	.7278	.6856	.6470	5.900	5.827	5.756	5.687	5.620
23	.8191	.7686	.7219	.6790	.6399	6.510	6.428	6.348	6.270	6.195
26	.8153	.7639	.7166	.6731	.6336	7.117	7.025	6.936	6.849	6.764
30	.8107	.7583	.7101	.6660	.6259	7.929	7.822	7.719	7.618	7.520
35	.8054	.7520	.7029	.6580	.6174	8.949	8.821	8.698	8.578	8.462
40	.8007	.7463	.6964	.6509	.6098	9.972	9.821	9.676	9.536	9.400
45	.7963	.7410	.6904	.6444	.6029	11.00	10.82	10.65	10.49	10.33
50	.7923	.7362	.6849	.6384	.5966	12.02	11.82	11.62	11.44	11.26
60	.7850	.7275	.6750	.6276	.5852	13.87	13.63	13.40	13.18	12.96
70	.7784	.7197	.6662	.6181	.5752	15.79	15.49	15.21	14.95	14.69
80	.7724	.7125	.6582	.6094	.5661	17.75	17.40	17.06	16.74	16.43
90	.7668	.7059	.6508	.6015	.5578	19.75	19.33	18.93	18.55	18.20
100	.7615	.6997	.6439	.5941	.5502	21.77	21.28	20.82	20.38	19.96
120	.7519	.6885	.6314	.5807	.5363	25.47	24.87	24.30	23.77	23.26
140	.7431	.6783	.6202	.5687	.5239	29.36	28.62	27.92	27.26	26.63
170	.7310	.6644	.6050	.5527	.5075	35.52	34.51	33.56	32.68	31.84
200	.7200	.6519	.5914	.5385	.4930	41.99	40.64	39.39	38.24	37.16
230	.7097	.6403	.5789	.5254	.4798	47.74	46.18	44.73	43.38	42.12
260	.7002	.6295	.5673	.5134	.4677	53.78	51.96	50.25	48.66	47.17
300	.6882	.6163	.5532	.4988	.4531	62.29	60.00	57.87	55.89	54.06
350	.6744	.6011	.5371	.4824	.4369	73.55	70.51	67.73	65.19	62.87
400	.6615	.5871	.5225	.4676	.4223	85.35	81.42	77.89	74.68	71.81
450	.6494	.5742	.5091	.4541	.4090	97.55	92.60	88.23	84.30	80.83
500	.6381	.5621	.4967	.4416	.3969	110.0	104.0	98.70	93.98	89.80
600	.6167	.5396	.4737	.4189	.3750	133.7	125.7	118.8	112.6	107.4
700	.5973	.5196	.4535	.3990	.3561	158.6	148.3	139.4	131.6	124.9
800	.5795	.5015	.4354	.3814	.3394	184.4	171.4	160.3	150.6	142.5
900	.5632	.4850	.4191	.3657	.3246	210.9	194.8	181.4	169.7	160.1
1000	.5480	.4699	.4043	.3514	.3113	237.9	218.4	202.5	188.7	177.5
1200	.5198	.4424	.3778	.3263	.2880	290.7	264.5	243.7	225.7	211.4
1400	.4950	.4185	.3551	.3051	.2683	344.7	311.0	284.8	262.3	244.8
1700	.4628	.3879	.3264	.2785	.2440	426.7	381.0	345.9	316.1	293.5
2000	.4351	.3621	.3026	.2566	.2241	509.2	450.7	405.8	368.5	340.8
2300	.4108	.3399	.2825	.2384	.2077	590.1	519.1	463.9	419.0	386.2
2600	.3894	.3206	.2651	.2228	.1938	670.5	586.6	520.7	468.1	430.2
3000	.3644	.2984	.2453	.2052	.1781	776.6	675.0	594.6	531.5	487.0
3500	.3377	.2748	.2246	.1870	.1620	906.9	782.8	683.9	607.8	555.2
4000	.3149	.2550	.2073	.1718	.1486	1035.	887.4	770.1	680.9	620.4
4500	.2952	.2380	.1926	.1591	.1374	1159.	989.1	853.4	751.3	683.1
5000	.2779	.2232	.1799	.1481	.1278	1281.	1088.	933.9	819.4	743.4
6000	.2491	.1988	.1592	.1304	.1124	1516.	1277.	1088.	948.0	857.9
7000	.2261	.1795	.1431	.1167	.1004	1740.	1457.	1233.	1069.	965.1
8000	.2073	.1640	.1301	.1058	.0910	1955.	1628.	1370.	1183.	1066.
9000	.1916	.1512	.1197	.0971	.0834	2161.	1791.	1501.	1291.	1162.
10000	.1786	.1406	.1111	.0899	.0772	2359.	1947.	1626.	1394.	1253.

	MEAN DAMAGE DEPTH,MICROGRAM/SQ.CM.					MEAN IONZN.DEPTH,MICROGRAM/SQ.CM.				
K/KL=	0.6	0.8	1.0	1.2	1.4	0.6	0.8	1.0	1.2	1.4
E,KEV										
1.0	.6180	.6112	.6046	.5982	.5920	.5675	.5604	.5534	.5467	.5401
1.2	.7017	.6937	.6860	.6784	.6710	.6443	.6358	.6276	.6196	.6118
1.4	.7777	.7687	.7598	.7513	.7429	.7139	.7043	.6950	.6859	.6771
1.7	.8832	.8726	.8623	.8523	.8426	.8105	.7993	.7885	.7779	.7676
2.0	.9821	.9700	.9583	.9470	.9360	.9012	.8884	.8761	.8641	.8524
2.3	1.077	1.063	1.050	1.037	1.025	.9877	.9735	.9597	.9463	.9333
2.6	1.168	1.153	1.138	1.124	1.111	1.071	1.056	1.040	1.025	1.011
3.0	1.285	1.269	1.252	1.237	1.221	1.179	1.161	1.144	1.127	1.111
3.5	1.427	1.408	1.389	1.371	1.354	1.308	1.288	1.268	1.249	1.231
4.0	1.563	1.542	1.521	1.501	1.481	1.433	1.410	1.388	1.367	1.346
4.5	1.695	1.671	1.648	1.626	1.604	1.553	1.528	1.504	1.480	1.458
5.0	1.823	1.797	1.772	1.747	1.724	1.670	1.643	1.616	1.590	1.565
6.0	2.061	2.030	2.001	1.972	1.945	1.887	1.855	1.824	1.794	1.765
7.0	2.290	2.255	2.221	2.189	2.158	2.096	2.060	2.024	1.990	1.957
8.0	2.512	2.473	2.435	2.399	2.364	2.299	2.257	2.218	2.179	2.142
9.0	2.728	2.685	2.643	2.602	2.563	2.496	2.450	2.405	2.363	2.322
10	2.939	2.891	2.845	2.800	2.757	2.687	2.637	2.588	2.541	2.496
12	3.330	3.274	3.220	3.168	3.118	3.043	2.984	2.927	2.873	2.820
14	3.710	3.646	3.584	3.524	3.467	3.389	3.321	3.256	3.193	3.134
17	4.268	4.190	4.116	4.044	3.976	3.894	3.812	3.734	3.660	3.588
20	4.811	4.720	4.632	4.549	4.469	4.385	4.290	4.198	4.111	4.027
23	5.300	5.198	5.100	5.006	4.916	4.830	4.722	4.620	4.522	4.428
26	5.785	5.671	5.562	5.457	5.357	5.269	5.149	5.034	4.925	4.821
30	6.429	6.298	6.172	6.053	5.938	5.850	5.712	5.581	5.456	5.337
35	7.232	7.078	6.931	6.790	6.657	6.573	6.411	6.257	6.111	5.972
40	8.033	7.853	7.683	7.522	7.368	7.290	7.103	6.925	6.757	6.598
45	8.830	8.624	8.430	8.245	8.071	8.002	7.788	7.586	7.395	7.214
50	9.621	9.389	9.169	8.961	8.765	8.707	8.465	8.238	8.023	7.820
60	11.06	10.79	10.53	10.28	10.05	10.00	9.712	9.441	9.185	8.944
70	12.54	12.21	11.90	11.61	11.33	11.31	10.97	10.65	10.35	10.06
80	14.04	13.65	13.28	12.94	12.62	12.64	12.23	11.86	11.51	11.18
90	15.55	15.10	14.68	14.28	13.92	13.97	13.50	13.07	12.66	12.29
100	17.08	16.56	16.08	15.63	15.21	15.31	14.77	14.28	13.82	13.39
120	19.88	19.26	18.67	18.12	17.61	17.79	17.13	16.53	15.98	15.46
140	22.80	22.04	21.33	20.67	20.06	20.33	19.54	18.82	18.15	17.53
170	27.39	26.37	25.44	24.58	23.80	24.28	23.24	22.30	21.44	20.65
200	32.15	30.85	29.66	28.57	27.60	28.33	27.01	25.82	24.75	23.77
230	36.40	34.90	33.52	32.26	31.12	32.03	30.51	29.12	27.87	26.71
260	40.84	39.09	37.48	36.02	34.69	35.84	34.07	32.44	31.00	29.66
300	47.05	44.90	42.93	41.15	39.55	41.07	38.92	36.94	35.20	33.58
350	55.22	52.45	49.95	47.72	45.74	47.83	45.10	42.62	40.47	38.48
400	63.72	60.26	57.15	54.40	52.01	54.76	51.38	48.33	45.72	43.35
450	72.47	68.24	64.46	61.16	58.33	61.79	57.71	54.06	50.96	48.18
500	81.40	76.33	71.84	67.95	64.67	68.90	64.06	59.77	56.15	52.95
600	98.23	91.69	85.89	80.93	76.82	82.55	76.28	70.82	66.17	62.17
700	115.9	107.6	100.3	94.11	89.08	96.49	88.59	81.81	76.06	71.20
800	134.2	123.9	114.9	107.4	101.4	110.6	100.9	92.71	85.78	80.05
900	153.0	140.6	129.7	120.7	113.7	124.8	113.2	103.5	95.33	88.72
1000	172.1	157.3	144.5	134.0	125.9	139.0	125.4	114.1	104.7	97.20
1200	209.4	190.0	173.4	159.9	149.5	166.8	149.3	134.7	122.8	113.6
1400	247.5	223.1	202.4	185.7	172.9	194.4	172.8	154.8	140.3	129.3
1700	305.7	273.0	245.6	223.7	207.4	235.3	207.1	183.8	165.3	151.7
2000	364.5	322.8	288.3	261.1	241.1	275.3	240.3	211.6	189.1	173.0
2300	421.7	371.1	329.5	297.0	273.4	314.3	272.4	238.2	211.8	193.1
2600	478.8	419.0	370.1	332.1	305.0	352.4	303.5	263.8	233.4	212.3
3000	554.5	482.0	423.1	377.8	346.1	401.7	343.5	296.6	261.0	236.7
3500	648.1	559.4	487.8	433.3	395.7	461.1	391.3	335.4	293.4	265.4
4000	740.3	635.2	550.7	486.9	443.7	518.2	437.0	372.3	324.1	292.5
4500	830.9	709.2	611.9	538.9	490.2	573.1	480.7	407.4	353.2	318.1
5000	919.8	781.6	671.4	589.3	535.2	626.1	522.6	440.9	380.9	342.5
6000	1092.	921.4	785.8	685.8	621.2	726.7	602.0	504.0	432.7	388.2
7000	1258.	1055.	894.6	777.1	702.5	821.1	676.0	562.5	480.6	430.3
8000	1418.	1183.	998.4	864.0	779.7	910.2	745.5	617.2	525.2	469.3
9000	1571.	1306.	1098.	946.9	853.4	994.7	811.2	668.7	567.0	506.3
10000	1719.	1424.	1193.	1026.	923.8	1075.	873.6	717.4	606.5	541.0

K/KL=	RELATIVE RANGE STRAGGLING					RELATIVE DAMAGE STRAGGLING				
E,KEV	0.6	0.8	1.0	1.2	1.4	0.6	0.8	1.0	1.2	1.4
1.0	.3280	.3245	.3210	.3176	.3143	.3995	.3974	.3953	.3932	.3912
1.2	.3275	.3238	.3202	.3167	.3133	.3992	.3970	.3948	.3927	.3907
1.4	.3269	.3232	.3195	.3159	.3124	.3990	.3967	.3945	.3923	.3902
1.7	.3262	.3224	.3186	.3149	.3113	.3986	.3963	.3940	.3918	.3896
2.0	.3256	.3216	.3178	.3140	.3103	.3983	.3959	.3936	.3913	.3891
2.3	.3250	.3210	.3170	.3132	.3094	.3980	.3956	.3932	.3909	.3886
2.6	.3245	.3203	.3163	.3124	.3086	.3978	.3952	.3928	.3905	.3882
3.0	.3238	.3195	.3154	.3114	.3075	.3974	.3948	.3924	.3900	.3877
3.5	.3229	.3186	.3144	.3103	.3063	.3970	.3944	.3918	.3894	.3870
4.0	.3222	.3177	.3134	.3092	.3052	.3967	.3940	.3914	.3889	.3865
4.5	.3214	.3169	.3125	.3082	.3041	.3963	.3936	.3909	.3884	.3859
5.0	.3207	.3161	.3116	.3073	.3031	.3960	.3932	.3905	.3879	.3854
6.0	.3194	.3146	.3100	.3056	.3013	.3954	.3925	.3897	.3870	.3845
7.0	.3181	.3132	.3085	.3040	.2996	.3948	.3918	.3890	.3862	.3836
8.0	.3169	.3119	.3071	.3025	.2980	.3942	.3912	.3883	.3855	.3828
9.0	.3158	.3107	.3058	.3010	.2965	.3937	.3906	.3876	.3848	.3821
10	.3147	.3095	.3045	.2997	.2950	.3932	.3901	.3870	.3841	.3814
12	.3126	.3072	.3021	.2971	.2924	.3923	.3890	.3859	.3829	.3801
14	.3106	.3051	.2998	.2948	.2899	.3914	.3881	.3849	.3818	.3789
17	.3078	.3021	.2967	.2914	.2864	.3902	.3867	.3834	.3802	.3772
20	.3052	.2993	.2937	.2883	.2832	.3891	.3854	.3820	.3787	.3757
23	.3027	.2967	.2909	.2854	.2802	.3880	.3842	.3807	.3773	.3742
26	.3004	.2942	.2883	.2827	.2773	.3869	.3831	.3795	.3760	.3728
30	.2974	.2910	.2849	.2792	.2737	.3856	.3816	.3779	.3743	.3710
35	.2938	.2873	.2810	.2751	.2695	.3840	.3798	.3760	.3723	.3689
40	.2905	.2837	.2774	.2713	.2656	.3824	.3782	.3742	.3704	.3669
45	.2873	.2804	.2739	.2677	.2619	.3809	.3765	.3725	.3686	.3650
50	.2843	.2772	.2706	.2643	.2583	.3795	.3750	.3708	.3669	.3632
60	.2787	.2713	.2643	.2578	.2517	.3768	.3720	.3677	.3636	.3598
70	.2734	.2658	.2586	.2519	.2457	.3742	.3693	.3647	.3605	.3567
80	.2686	.2607	.2533	.2465	.2401	.3717	.3667	.3619	.3576	.3538
90	.2640	.2559	.2484	.2414	.2349	.3694	.3642	.3592	.3548	.3510
100	.2598	.2514	.2438	.2367	.2301	.3672	.3618	.3567	.3522	.3483
120	.2518	.2431	.2351	.2278	.2210	.3630	.3573	.3520	.3473	.3432
140	.2446	.2356	.2274	.2199	.2130	.3592	.3532	.3477	.3428	.3385
170	.2350	.2256	.2171	.2094	.2023	.3539	.3476	.3418	.3366	.3320
200	.2265	.2167	.2080	.2001	.1929	.3491	.3424	.3364	.3310	.3262
230	.2188	.2087	.1997	.1916	.1845	.3445	.3375	.3313	.3258	.3209
260	.2117	.2013	.1922	.1840	.1769	.3402	.3330	.3266	.3209	.3160
300	.2032	.1926	.1833	.1749	.1678	.3349	.3274	.3208	.3150	.3100
350	.1939	.1829	.1735	.1649	.1579	.3289	.3211	.3143	.3083	.3033
400	.1855	.1744	.1648	.1561	.1493	.3234	.3154	.3084	.3024	.2973
450	.1781	.1667	.1571	.1484	.1417	.3184	.3102	.3031	.2969	.2918
500	.1713	.1598	.1502	.1414	.1349	.3138	.3054	.2982	.2920	.2868
600	.1593	.1477	.1382	.1294	.1231	.3058	.2970	.2898	.2834	.2778
700	.1492	.1375	.1280	.1194	.1134	.2987	.2895	.2824	.2758	.2699
800	.1404	.1287	.1194	.1109	.1051	.2923	.2829	.2758	.2692	.2631
900	.1327	.1211	.1119	.1036	.0980	.2864	.2769	.2699	.2632	.2570
1000	.1258	.1144	.1053	.0972	.0918	.2811	.2714	.2644	.2578	.2515
1200	.1143	.1031	.0945	.0869	.0817	.2710	.2615	.2545	.2480	.2421
1400	.1048	.0940	.0857	.0785	.0737	.2623	.2530	.2459	.2396	.2342
1700	.0932	.0829	.0752	.0687	.0641	.2511	.2422	.2350	.2291	.2243
2000	.0840	.0742	.0670	.0610	.0568	.2418	.2332	.2260	.2204	.2162
2300	.0765	.0673	.0605	.0549	.0510	.2343	.2258	.2188	.2134	.2095
2600	.0702	.0616	.0551	.0500	.0463	.2277	.2195	.2127	.2074	.2036
3000	.0633	.0553	.0493	.0445	.0412	.2203	.2123	.2056	.2005	.1969
3500	.0563	.0490	.0434	.0391	.0362	.2124	.2047	.1983	.1934	.1899
4000	.0506	.0439	.0388	.0349	.0322	.2058	.1983	.1921	.1873	.1840
4500	.0460	.0398	.0350	.0314	.0290	.2002	.1928	.1868	.1822	.1789
5000	.0421	.0364	.0319	.0285	.0264	.1952	.1880	.1822	.1777	.1745
6000	.0359	.0310	.0270	.0241	.0223	.1869	.1800	.1745	.1701	.1671
7000	.0314	.0270	.0234	.0209	.0193	.1803	.1736	.1682	.1640	.1610
8000	.0280	.0240	.0207	.0184	.0171	.1748	.1682	.1630	.1589	.1560
9000	.0253	.0217	.0187	.0166	.0153	.1701	.1637	.1586	.1546	.1518
10000	.0233	.0198	.0171	.0152	.0140	.1661	.1598	.1547	.1508	.1481

K/KL = E,KEV	RELATIVE IONIZATION STRAGGLING					RELATIVE TRANSVERSE RANGE STRAGGLING				
	0.6	0.8	1.0	1.2	1.4	0.6	0.8	1.0	1.2	1.4
1.0	.4525	.4512	.4499	.4488	.4477	.7070	.7068	.7066	.7064	.7062
1.2	.4523	.4510	.4497	.4485	.4474	.7068	.7066	.7064	.7063	.7061
1.4	.4522	.4508	.4495	.4483	.4471	.7067	.7065	.7063	.7061	.7060
1.7	.4520	.4506	.4492	.4480	.4468	.7065	.7063	.7061	.7059	.7058
2.0	.4518	.4503	.4490	.4477	.4465	.7063	.7061	.7059	.7058	.7056
2.3	.4516	.4501	.4488	.4475	.4463	.7061	.7059	.7058	.7056	.7055
2.6	.4515	.4500	.4486	.4473	.4460	.7060	.7058	.7056	.7055	.7053
3.0	.4513	.4497	.4483	.4470	.4457	.7057	.7056	.7054	.7053	.7051
3.5	.4511	.4495	.4480	.4467	.4454	.7055	.7053	.7052	.7050	.7049
4.0	.4509	.4493	.4478	.4464	.4451	.7052	.7051	.7049	.7048	.7047
4.5	.4507	.4490	.4475	.4461	.4448	.7050	.7048	.7047	.7046	.7045
5.0	.4505	.4488	.4473	.4459	.4445	.7047	.7046	.7045	.7044	.7043
6.0	.4501	.4484	.4469	.4454	.4440	.7043	.7042	.7041	.7040	.7040
7.0	.4498	.4481	.4465	.4450	.4436	.7039	.7038	.7037	.7037	.7037
8.0	.4495	.4477	.4461	.4446	.4432	.7035	.7034	.7034	.7034	.7034
9.0	.4492	.4474	.4457	.4442	.4428	.7031	.7031	.7031	.7031	.7031
10	.4490	.4471	.4454	.4438	.4424	.7027	.7027	.7028	.7028	.7029
12	.4485	.4466	.4448	.4432	.4417	.7020	.7021	.7022	.7023	.7024
14	.4480	.4460	.4442	.4426	.4411	.7013	.7015	.7016	.7018	.7019
17	.4473	.4453	.4434	.4417	.4402	.7005	.7007	.7009	.7012	.7014
20	.4467	.4446	.4427	.4409	.4394	.6997	.7000	.7003	.7007	.7010
23	.4461	.4440	.4420	.4402	.4386	.6989	.6993	.6997	.7002	.7006
26	.4456	.4433	.4413	.4395	.4379	.6982	.6987	.6992	.6997	.7002
30	.4449	.4426	.4405	.4386	.4370	.6974	.6980	.6987	.6993	.6998
35	.4440	.4416	.4395	.4376	.4359	.6966	.6974	.6981	.6989	.6996
40	.4432	.4408	.4386	.4366	.4349	.6959	.6968	.6977	.6986	.6995
45	.4425	.4400	.4377	.4357	.4340	.6954	.6965	.6975	.6985	.6995
50	.4418	.4392	.4369	.4348	.4331	.6950	.6962	.6973	.6985	.6996
60	.4404	.4377	.4353	.4332	.4314	.6941	.6956	.6971	.6985	.6998
70	.4392	.4363	.4338	.4317	.4298	.6937	.6955	.6972	.6988	.7003
80	.4380	.4350	.4325	.4303	.4284	.6936	.6956	.6975	.6994	.7012
90	.4369	.4339	.4313	.4290	.4271	.6937	.6960	.6982	.7002	.7022
100	.4359	.4327	.4301	.4278	.4259	.6940	.6965	.6989	.7012	.7034
120	.4339	.4306	.4279	.4255	.4236	.6949	.6979	.7007	.7034	.7060
140	.4321	.4287	.4259	.4235	.4216	.6963	.6997	.7030	.7060	.7090
170	.4297	.4262	.4233	.4209	.4190	.6991	.7032	.7071	.7107	.7142
200	.4276	.4240	.4210	.4186	.4167	.7026	.7074	.7118	.7159	.7198
230	.4256	.4218	.4188	.4165	.4144	.7064	.7118	.7167	.7215	.7257
260	.4238	.4199	.4169	.4146	.4124	.7105	.7165	.7219	.7273	.7318
300	.4216	.4177	.4148	.4125	.4102	.7164	.7230	.7292	.7354	.7401
350	.4193	.4154	.4125	.4103	.4078	.7241	.7316	.7384	.7456	.7506
400	.4174	.4134	.4106	.4086	.4059	.7320	.7403	.7479	.7560	.7611
450	.4157	.4117	.4090	.4071	.4044	.7401	.7491	.7573	.7663	.7716
500	.4142	.4102	.4077	.4059	.4032	.7482	.7580	.7668	.7766	.7820
600	.4120	.4079	.4060	.4044	.4019	.7644	.7756	.7855	.7976	.8026
700	.4103	.4062	.4049	.4034	.4010	.7804	.7929	.8038	.8176	.8224
800	.4089	.4049	.4040	.4027	.4006	.7961	.8097	.8215	.8367	.8414
900	.4078	.4039	.4034	.4023	.4005	.8113	.8260	.8387	.8549	.8598
1000	.4068	.4032	.4030	.4021	.4006	.8262	.8418	.8553	.8723	.8774
1200	.4051	.4022	.4023	.4019	.4012	.8550	.8724	.8872	.9043	.9112
1400	.4038	.4018	.4020	.4021	.4022	.8822	.9012	.9171	.9339	.9425
1700	.4026	.4017	.4021	.4029	.4039	.9204	.9413	.9586	.9749	.9857
2000	.4020	.4021	.4027	.4040	.4058	.9556	.9783	.9967	1.012	1.025
2300	.4021	.4029	.4040	.4057	.4079	.9900	1.014	1.033	1.049	1.062
2600	.4025	.4039	.4054	.4074	.4100	1.022	1.047	1.067	1.083	1.097
3000	.4033	.4053	.4074	.4098	.4126	1.061	1.087	1.109	1.125	1.139
3500	.4045	.4073	.4099	.4128	.4158	1.106	1.133	1.155	1.172	1.186
4000	.4059	.4093	.4124	.4156	.4188	1.146	1.174	1.197	1.215	1.228
4500	.4074	.4112	.4148	.4182	.4215	1.183	1.212	1.236	1.254	1.266
5000	.4090	.4132	.4171	.4207	.4241	1.216	1.246	1.270	1.289	1.301
6000	.4121	.4169	.4213	.4252	.4287	1.276	1.307	1.332	1.351	1.363
7000	.4152	.4204	.4251	.4292	.4327	1.327	1.359	1.385	1.404	1.416
8000	.4192	.4236	.4286	.4328	.4363	1.372	1.405	1.432	1.451	1.463
9000	.4211	.4267	.4317	.4360	.4395	1.412	1.446	1.473	1.493	1.504
10000	.4238	.4296	.4346	.4389	.4425	1.448	1.483	1.510	1.530	1.541

TABLE 12- 5

GE ON GE,M2=M1

K/KL= E,KEV	RELATIVE TRANSV. DAMAGE STRAGGLING					RELATIVE TRANSV. IONZN. STRAGGLING				
	0.6	0.8	1.0	1.2	1.4	0.6	0.8	1.0	1.2	1.4
1.0	.5292	.5263	.5234	.5206	.5178	.4341	.4301	.4261	.4223	.4184
1.2	.5285	.5255	.5226	.5197	.5169	.4334	.4293	.4252	.4212	.4173
1.4	.5280	.5249	.5219	.5189	.5160	.4327	.4285	.4243	.4203	.4162
1.7	.5272	.5240	.5209	.5179	.5149	.4319	.4275	.4232	.4190	.4149
2.0	.5264	.5232	.5201	.5169	.5139	.4310	.4266	.4222	.4179	.4136
2.3	.5258	.5225	.5192	.5161	.5129	.4303	.4257	.4212	.4168	.4125
2.6	.5251	.5218	.5185	.5152	.5121	.4296	.4249	.4203	.4159	.4115
3.0	.5243	.5208	.5175	.5142	.5109	.4286	.4239	.4192	.4147	.4102
3.5	.5233	.5198	.5163	.5130	.5097	.4276	.4227	.4179	.4132	.4087
4.0	.5223	.5187	.5152	.5118	.5084	.4265	.4216	.4167	.4119	.4073
4.5	.5214	.5178	.5142	.5107	.5073	.4255	.4205	.4155	.4107	.4060
5.0	.5205	.5168	.5132	.5097	.5062	.4246	.4195	.4144	.4095	.4047
6.0	.5188	.5151	.5114	.5077	.5042	.4228	.4176	.4124	.4074	.4024
7.0	.5173	.5134	.5096	.5059	.5023	.4211	.4158	.4105	.4054	.4003
8.0	.5157	.5118	.5080	.5042	.5005	.4196	.4141	.4087	.4035	.3984
9.0	.5143	.5103	.5064	.5026	.4988	.4180	.4124	.4070	.4017	.3965
10	.5129	.5088	.5049	.5010	.4972	.4166	.4109	.4054	.4000	.3947
12	.5102	.5060	.5020	.4980	.4941	.4138	.4080	.4023	.3968	.3914
14	.5076	.5034	.4993	.4952	.4913	.4112	.4052	.3995	.3938	.3883
17	.5041	.4997	.4955	.4914	.4873	.4075	.4014	.3955	.3897	.3841
20	.5007	.4963	.4920	.4878	.4837	.4040	.3978	.3918	.3859	.3801
23	.4975	.4930	.4886	.4844	.4802	.4008	.3944	.3883	.3823	.3765
26	.4945	.4899	.4855	.4812	.4770	.3977	.3912	.3850	.3789	.3730
30	.4906	.4860	.4815	.4771	.4729	.3938	.3872	.3808	.3747	.3687
35	.4862	.4815	.4769	.4725	.4682	.3892	.3825	.3761	.3698	.3637
40	.4821	.4773	.4727	.4682	.4638	.3850	.3782	.3716	.3652	.3590
45	.4783	.4734	.4687	.4641	.4598	.3810	.3741	.3674	.3609	.3547
50	.4747	.4698	.4650	.4604	.4559	.3773	.3703	.3635	.3569	.3506
60	.4679	.4629	.4580	.4533	.4487	.3703	.3631	.3561	.3494	.3429
70	.4619	.4567	.4518	.4469	.4421	.3639	.3566	.3495	.3426	.3360
80	.4564	.4511	.4461	.4411	.4362	.3582	.3506	.3434	.3364	.3297
90	.4514	.4460	.4409	.4358	.4308	.3528	.3452	.3378	.3307	.3239
100	.4467	.4413	.4361	.4309	.4258	.3479	.3401	.3326	.3254	.3185
120	.4384	.4327	.4273	.4220	.4168	.3390	.3309	.3231	.3157	.3087
140	.4311	.4253	.4197	.4142	.4090	.3311	.3227	.3148	.3072	.2999
170	.4217	.4156	.4097	.4041	.3987	.3207	.3120	.3037	.2959	.2884
200	.4138	.4074	.4012	.3953	.3899	.3118	.3027	.2941	.2861	.2784
230	.4071	.4004	.3939	.3878	.3821	.2735	.2664	.2595	.2527	.2459
260	.4013	.3942	.3875	.3811	.3752	.2417	.2362	.2306	.2248	.2187
300	.3944	.3869	.3798	.3732	.3669	.2120	.2076	.2030	.1981	.1926
350	.3870	.3790	.3715	.3644	.3579	.1952	.1910	.1865	.1816	.1762
400	.3806	.3721	.3641	.3567	.3499	.1991	.1935	.1879	.1822	.1761
450	.3749	.3659	.3575	.3498	.3428	.2209	.2126	.2048	.1974	.1902
500	.3697	.3604	.3516	.3435	.3363	.2576	.2456	.2348	.2250	.2162
600	.3608	.3507	.3409	.3324	.3251	.4856	.4549	.4289	.4073	.3902
700	.3532	.3424	.3317	.3228	.3154	.6867	.6395	.5998	.5677	.5431
800	.3465	.3351	.3237	.3143	.3069	.8524	.7911	.7400	.6990	.6680
900	.3406	.3286	.3165	.3068	.2992	.9821	.9095	.8490	.8008	.7646
1000	.3354	.3227	.3101	.3000	.2923	1.078	.9967	.9289	.8749	.8347
1200	.3265	.3126	.2992	.2882	.2798	1.075	.9898	.9186	.8615	.8186
1400	.3189	.3038	.2897	.2782	.2691	1.028	.9423	.8705	.8128	.7690
1700	.3089	.2923	.2776	.2653	.2554	.9275	.8444	.7745	.7179	.6745
2000	.3000	.2824	.2671	.2542	.2437	.8228	.7435	.6766	.6222	.5802
2300	.2913	.2730	.2573	.2441	.2333	.7711	.6929	.6272	.5740	.5333
2600	.2834	.2645	.2484	.2350	.2242	.7261	.6491	.5847	.5329	.4937
3000	.2737	.2543	.2379	.2243	.2135	.6743	.5990	.5363	.4865	.4493
3500	.2629	.2430	.2263	.2126	.2019	.6196	.5464	.4860	.4385	.4039
4000	.2532	.2331	.2162	.2025	.1920	.5733	.5023	.4442	.3990	.3666
4500	.2445	.2242	.2073	.1937	.1833	.5335	.4647	.4088	.3657	.3355
5000	.2365	.2162	.1993	.1858	.1757	.4989	.4323	.3784	.3374	.3091
6000	.2225	.2023	.1856	.1724	.1626	.4415	.3791	.3292	.2917	.2667
7000	.2104	.1906	.1742	.1613	.1519	.3957	.3374	.2911	.2568	.2344
8000	.1999	.1805	.1645	.1519	.1430	.3583	.3040	.2611	.2294	.2091
9000	.1905	.1716	.1561	.1440	.1353	.3272	.2768	.2370	.2077	.1890
10000	.1821	.1637	.1487	.1370	.1287	.3011	.2544	.2175	.1903	.1728

K/KL=	RANGE SKEWNESS					DAMAGE SKEWNESS				
E,KEV	0.6	0.8	1.0	1.2	1.4	0.6	0.8	1.0	1.2	1.4
1.0	.6422	.6286	.6153	.6022	.5894	.8135	.8034	.7936	.7840	.7747
1.2	.6405	.6265	.6127	.5993	.5861	.8122	.8018	.7917	.7819	.7723
1.4	.6390	.6246	.6105	.5967	.5832	.8111	.8004	.7901	.7800	.7703
1.7	.6369	.6221	.6076	.5933	.5794	.8096	.7986	.7880	.7776	.7676
2.0	.6351	.6199	.6049	.5904	.5761	.8083	.7970	.7861	.7755	.7652
2.3	.6334	.6178	.6026	.5877	.5731	.8071	.7956	.7844	.7736	.7631
2.6	.6319	.6159	.6004	.5852	.5703	.8060	.7942	.7828	.7718	.7611
3.0	.6300	.6136	.5977	.5821	.5669	.8047	.7926	.7809	.7696	.7587
3.5	.6277	.6109	.5945	.5786	.5630	.8031	.7906	.7787	.7671	.7559
4.0	.6256	.6084	.5917	.5753	.5594	.8015	.7888	.7766	.7648	.7533
4.5	.6236	.6060	.5889	.5723	.5560	.8001	.7872	.7747	.7626	.7510
5.0	.6217	.6038	.5864	.5694	.5529	.7988	.7856	.7728	.7606	.7487
6.0	.6183	.5998	.5818	.5642	.5472	.7962	.7826	.7695	.7568	.7446
7.0	.6150	.5960	.5775	.5595	.5420	.7939	.7799	.7663	.7533	.7408
8.0	.6120	.5924	.5734	.5550	.5371	.7917	.7773	.7634	.7501	.7373
9.0	.6090	.5890	.5696	.5508	.5325	.7896	.7749	.7607	.7471	.7340
10	.6062	.5858	.5659	.5467	.5281	.7875	.7725	.7580	.7442	.7309
12	.6010	.5797	.5592	.5393	.5201	.7839	.7683	.7533	.7390	.7253
14	.5960	.5741	.5529	.5324	.5126	.7805	.7643	.7489	.7341	.7201
17	.5890	.5661	.5440	.5227	.5021	.7755	.7586	.7426	.7273	.7127
20	.5823	.5585	.5356	.5136	.4923	.7708	.7532	.7366	.7208	.7057
23	.5761	.5515	.5279	.5051	.4832	.7663	.7481	.7310	.7146	.6990
26	.5701	.5447	.5204	.4971	.4746	.7620	.7433	.7256	.7086	.6925
30	.5624	.5361	.5109	.4868	.4636	.7563	.7370	.7186	.7010	.6841
35	.5531	.5257	.4996	.4746	.4506	.7494	.7293	.7102	.6918	.6743
40	.5442	.5158	.4888	.4629	.4382	.7427	.7219	.7020	.6830	.6648
45	.5357	.5063	.4783	.4517	.4263	.7361	.7146	.6941	.6746	.6559
50	.5274	.4971	.4683	.4410	.4149	.7297	.7075	.6865	.6664	.6473
60	.5117	.4797	.4494	.4207	.3935	.7172	.6937	.6716	.6509	.6316
70	.4968	.4632	.4315	.4016	.3733	.7053	.6805	.6575	.6362	.6168
80	.4825	.4474	.4145	.3835	.3542	.6938	.6680	.6440	.6222	.6025
90	.4688	.4323	.3982	.3662	.3359	.6829	.6560	.6312	.6088	.5887
100	.4556	.4178	.3826	.3496	.3185	.6723	.6445	.6189	.5959	.5754
120	.4306	.3903	.3529	.3184	.2855	.6526	.6232	.5959	.5712	.5491
140	.4070	.3644	.3252	.2892	.2548	.6341	.6032	.5744	.5481	.5245
170	.3740	.3283	.2866	.2484	.2123	.6081	.5752	.5445	.5161	.4903
200	.3432	.2948	.2508	.2105	.1732	.5838	.5492	.5168	.4867	.4590
230	.3142	.2632	.2172	.1735	.1365	.5606	.5243	.4903	.4589	.4301
260	.2868	.2335	.1856	.1388	.1022	.5386	.5008	.4656	.4330	.4032
300	.2526	.1964	.1462	.0958	.0596	.5112	.4714	.4348	.4010	.3700
350	.2127	.1534	.1008	.0463	.0107	.4795	.4374	.3994	.3643	.3320
400	.1758	.1135	.0588	.0011	-.0343	.4502	.4060	.3670	.3307	.2973
450	.1411	.0764	.0197	-.0406	-.0760	.4229	.3768	.3370	.2998	.2653
500	.1086	.0415	-.0169	-.0793	-.1149	.3973	.3494	.3090	.2711	.2355
600	.0476	-.0235	-.0848	-.1486	-.1866	.3511	.2988	.2587	.2193	.1804
700	-.0073	-.0820	-.1458	-.2101	-.2506	.3094	.2534	.2137	.1731	.1316
800	-.0574	-.1351	-.2010	-.2658	-.3086	.2710	.2123	.1728	.1314	.0881
900	-.1035	-.1839	-.2515	-.3166	-.3615	.2355	.1747	.1351	.0932	.0488
1000	-.1463	-.2289	-.2982	-.3635	-.4102	.2023	.1400	.1002	.0579	.0130
1200	-.2227	-.3093	-.3810	-.4475	-.4969	.1390	.0766	.0345	-.0076	-.0499
1400	-.2907	-.3805	-.4542	-.5218	-.5733	.0829	.0212	-.0231	-.0648	-.1040
1700	-.3808	-.4746	-.5508	-.6198	-.6736	.0100	-.0502	-.0974	-.1384	-.1731
2000	-.4602	-.5572	-.6355	-.7057	-.7611	-.0523	-.1110	-.1604	-.2006	-.2315
2300	-.5329	-.6323	-.7129	-.7837	-.8396	-.1047	-.1628	-.2125	-.2522	-.2818
2600	-.5989	-.7004	-.7830	-.8544	-.9105	-.1506	-.2084	-.2580	-.2972	-.3260
3000	-.6785	-.7823	-.8673	-.9394	-.9956	-.2042	-.2615	-.3107	-.3494	-.3775
3500	-.7670	-.8734	-.9610	-1.034	-1.090	-.2616	-.3185	-.3670	-.4051	-.4328
4000	-.8459	-.9544	-1.044	-1.118	-1.174	-.3109	-.3675	-.4153	-.4529	-.4804
4500	-.9171	-1.027	-1.119	-1.193	-1.250	-.3541	-.4104	-.4575	-.4948	-.5221
5000	-.9819	-1.094	-1.187	-1.262	-1.319	-.3923	-.4484	-.4950	-.5318	-.5590
6000	-1.096	-1.212	-1.308	-1.384	-1.441	-.4574	-.5133	-.5588	-.5951	-.6221
7000	-1.195	-1.313	-1.412	-1.489	-1.547	-.5112	-.5669	-.6119	-.6478	-.6745
8000	-1.282	-1.402	-1.503	-1.582	-1.640	-.5568	-.6123	-.6571	-.6927	-.7189
9000	-1.360	-1.482	-1.584	-1.665	-1.724	-.5962	-.6515	-.6964	-.7317	-.7574
10000	-1.430	-1.554	-1.658	-1.739	-1.800	-.6305	-.6858	-.7310	-.7661	-.7911

TABLE 12- 7

GE ON GE,M2=M1

K/KL=	IONIZATION SKEWNESS					RANGE KURTOSIS				
E,KEV	0.6	0.8	1.0	1.2	1.4	0.6	0.8	1.0	1.2	1.4
1.0	.8875	.8789	.8706	.8627	.8550	3.383	3.348	3.314	3.282	3.251
1.2	.8864	.8775	.8691	.8609	.8530	3.379	3.343	3.308	3.275	3.243
1.4	.8854	.8764	.8677	.8593	.8512	3.375	3.338	3.303	3.269	3.236
1.7	.8841	.8748	.8659	.8573	.8490	3.370	3.332	3.295	3.261	3.228
2.0	.8830	.8734	.8643	.8555	.8470	3.365	3.326	3.289	3.254	3.220
2.3	.8819	.8722	.8628	.8538	.8452	3.361	3.321	3.283	3.247	3.213
2.6	.8810	.8710	.8615	.8523	.8435	3.357	3.317	3.278	3.242	3.207
3.0	.8797	.8696	.8598	.8505	.8415	3.352	3.311	3.272	3.235	3.199
3.5	.8783	.8679	.8579	.8483	.8392	3.347	3.305	3.265	3.227	3.191
4.0	.8770	.8663	.8561	.8464	.8370	3.342	3.299	3.258	3.220	3.183
4.5	.8758	.8649	.8545	.8445	.8350	3.337	3.293	3.252	3.213	3.176
5.0	.8746	.8635	.8529	.8428	.8332	3.333	3.288	3.246	3.207	3.169
6.0	.8724	.8610	.8501	.8397	.8298	3.325	3.279	3.236	3.195	3.157
7.0	.8704	.8586	.8474	.8368	.8266	3.317	3.270	3.226	3.185	3.146
8.0	.8685	.8564	.8449	.8341	.8237	3.310	3.262	3.217	3.175	3.136
9.0	.8666	.8543	.8426	.8315	.8210	3.303	3.255	3.209	3.167	3.127
10	.8648	.8523	.8404	.8291	.8183	3.297	3.247	3.201	3.158	3.118
12	.8616	.8485	.8362	.8246	.8135	3.285	3.234	3.187	3.143	3.102
14	.8584	.8450	.8323	.8203	.8090	3.274	3.222	3.174	3.129	3.088
17	.8540	.8400	.8268	.8144	.8027	3.258	3.205	3.156	3.110	3.068
20	.8498	.8352	.8216	.8089	.7969	3.244	3.190	3.139	3.093	3.050
23	.8458	.8308	.8168	.8037	.7914	3.230	3.175	3.124	3.077	3.033
26	.8420	.8266	.8122	.7988	.7862	3.218	3.161	3.109	3.062	3.018
30	.8371	.8211	.8063	.7925	.7796	3.202	3.144	3.091	3.043	2.999
35	.8312	.8146	.7993	.7850	.7717	3.183	3.124	3.071	3.022	2.978
40	.8255	.8084	.7926	.7779	.7643	3.165	3.105	3.051	3.002	2.958
45	.8200	.8024	.7861	.7711	.7573	3.149	3.088	3.034	2.984	2.940
50	.8147	.7966	.7800	.7646	.7505	3.133	3.072	3.017	2.968	2.923
60	.8046	.7856	.7682	.7523	.7376	3.103	3.041	2.985	2.936	2.892
70	.7951	.7752	.7572	.7407	.7256	3.076	3.013	2.957	2.908	2.864
80	.7860	.7654	.7468	.7298	.7144	3.052	2.988	2.932	2.883	2.840
90	.7774	.7561	.7369	.7196	.7039	3.029	2.964	2.909	2.860	2.818
100	.7691	.7471	.7275	.7098	.6939	3.008	2.943	2.888	2.840	2.799
120	.7532	.7302	.7096	.6912	.6752	2.968	2.903	2.849	2.803	2.763
140	.7385	.7145	.6931	.6743	.6581	2.933	2.869	2.816	2.772	2.735
170	.7183	.6931	.6708	.6514	.6348	2.889	2.826	2.776	2.735	2.701
200	.7000	.6736	.6509	.6311	.6138	2.851	2.791	2.744	2.707	2.677
230	.6824	.6547	.6318	.6118	.5922	2.817	2.760	2.716	2.678	2.657
250	.6661	.6373	.6145	.5943	.5726	2.788	2.734	2.694	2.656	2.643
300	.6465	.6163	.5937	.5734	.5492	2.756	2.706	2.671	2.634	2.632
350	.6247	.5930	.5709	.5507	.5238	2.725	2.681	2.653	2.619	2.627
400	.6054	.5724	.5511	.5310	.5021	2.702	2.665	2.644	2.614	2.631
450	.5883	.5541	.5336	.5138	.4835	2.685	2.655	2.640	2.616	2.642
500	.5728	.5377	.5182	.4986	.4674	2.673	2.650	2.642	2.626	2.657
600	.5466	.5094	.4929	.4740	.4438	2.655	2.646	2.653	2.664	2.695
700	.5246	.4859	.4721	.4540	.4254	2.651	2.657	2.678	2.714	2.744
800	.5057	.4663	.4546	.4372	.4106	2.658	2.679	2.713	2.770	2.803
900	.4891	.4497	.4395	.4230	.3984	2.673	2.708	2.755	2.832	2.867
1000	.4745	.4355	.4264	.4106	.3882	2.695	2.744	2.804	2.896	2.936
1200	.4492	.4138	.4037	.3895	.3720	2.756	2.832	2.916	3.021	3.081
1400	.4286	.3973	.3858	.3731	.3600	2.829	2.932	3.038	3.152	3.234
1700	.4041	.3786	.3654	.3548	.3474	2.953	3.092	3.229	3.355	3.469
2000	.3852	.3650	.3506	.3420	.3391	3.085	3.258	3.424	3.563	3.705
2300	.3708	.3543	.3416	.3350	.3344	3.214	3.417	3.608	3.769	3.932
2600	.3595	.3461	.3354	.3305	.3316	3.346	3.576	3.792	3.975	4.158
3000	.3478	.3379	.3299	.3272	.3299	3.524	3.790	4.037	4.249	4.454
3500	.3372	.3310	.3262	.3258	.3299	3.748	4.056	4.339	4.586	4.818
4000	.3299	.3266	.3247	.3263	.3313	3.971	4.320	4.637	4.918	5.174
4500	.3249	.3240	.3247	.3279	.3336	4.193	4.580	4.930	5.242	5.521
5000	.3216	.3229	.3257	.3303	.3365	4.412	4.836	5.218	5.559	5.859
6000	.3189	.3234	.3296	.3363	.3433	4.840	5.334	5.777	6.173	6.512
7000	.3196	.3265	.3350	.3431	.3507	5.255	5.815	6.315	6.759	7.135
8000	.3226	.3312	.3411	.3501	.3582	5.655	6.279	6.832	7.320	7.731
9000	.3272	.3371	.3475	.3571	.3658	6.042	6.726	7.331	7.858	8.302
10000	.3328	.3438	.3541	.3639	.3731	6.416	7.159	7.812	8.375	8.850

E,KEV	DAMAGE KURTOSIS 0.6	0.8	1.0	1.2	1.4	IONIZATION KURTOSIS 0.6	0.8	1.0	1.2	1.4
1.0	3.664	3.630	3.597	3.565	3.534	3.787	3.754	3.722	3.692	3.662
1.2	3.660	3.625	3.591	3.558	3.527	3.783	3.749	3.716	3.685	3.655
1.4	3.656	3.620	3.585	3.552	3.520	3.779	3.744	3.711	3.679	3.649
1.7	3.651	3.614	3.578	3.544	3.511	3.774	3.738	3.704	3.671	3.640
2.0	3.647	3.608	3.572	3.537	3.503	3.770	3.733	3.698	3.665	3.633
2.3	3.642	3.603	3.566	3.531	3.496	3.766	3.728	3.692	3.658	3.626
2.6	3.639	3.599	3.561	3.525	3.490	3.762	3.724	3.687	3.653	3.620
3.0	3.634	3.593	3.554	3.518	3.482	3.757	3.718	3.681	3.646	3.612
3.5	3.628	3.587	3.547	3.509	3.473	3.752	3.712	3.674	3.638	3.604
4.0	3.623	3.581	3.540	3.502	3.465	3.747	3.706	3.667	3.631	3.596
4.5	3.618	3.575	3.534	3.495	3.458	3.742	3.701	3.661	3.624	3.589
5.0	3.614	3.570	3.528	3.488	3.451	3.738	3.695	3.655	3.618	3.582
6.0	3.605	3.560	3.517	3.476	3.438	3.729	3.686	3.645	3.606	3.569
7.0	3.597	3.551	3.507	3.465	3.426	3.722	3.677	3.635	3.595	3.558
8.0	3.589	3.543	3.497	3.455	3.415	3.714	3.669	3.626	3.586	3.548
9.0	3.582	3.535	3.488	3.445	3.405	3.707	3.661	3.617	3.576	3.538
10	3.576	3.527	3.480	3.436	3.395	3.701	3.653	3.609	3.568	3.529
12	3.563	3.513	3.465	3.420	3.378	3.688	3.640	3.594	3.552	3.512
14	3.552	3.499	3.451	3.406	3.363	3.677	3.627	3.580	3.537	3.496
17	3.535	3.480	3.431	3.385	3.341	3.660	3.609	3.561	3.516	3.474
20	3.520	3.462	3.413	3.366	3.321	3.645	3.592	3.542	3.497	3.454
23	3.505	3.447	3.395	3.346	3.301	3.630	3.576	3.525	3.479	3.435
26	3.490	3.433	3.378	3.328	3.282	3.616	3.560	3.509	3.462	3.418
30	3.472	3.414	3.357	3.305	3.258	3.599	3.541	3.489	3.440	3.396
35	3.450	3.392	3.332	3.278	3.230	3.578	3.519	3.465	3.415	3.370
40	3.429	3.371	3.308	3.253	3.205	3.558	3.497	3.442	3.392	3.346
45	3.409	3.351	3.286	3.229	3.181	3.539	3.477	3.421	3.370	3.323
50	3.390	3.331	3.265	3.207	3.158	3.520	3.457	3.400	3.349	3.301
60	3.355	3.288	3.225	3.169	3.118	3.486	3.420	3.362	3.309	3.261
70	3.322	3.249	3.189	3.133	3.081	3.453	3.386	3.326	3.272	3.223
80	3.291	3.214	3.155	3.100	3.047	3.423	3.354	3.293	3.238	3.189
90	3.262	3.182	3.124	3.069	3.015	3.395	3.324	3.262	3.207	3.157
100	3.235	3.152	3.095	3.039	2.985	3.368	3.296	3.233	3.177	3.128
120	3.183	3.103	3.039	2.980	2.926	3.317	3.244	3.179	3.122	3.073
140	3.137	3.059	2.990	2.928	2.872	3.270	3.196	3.130	3.073	3.025
170	3.074	3.002	2.925	2.858	2.802	3.208	3.132	3.065	3.007	2.959
200	3.019	2.951	2.868	2.799	2.742	3.153	3.074	3.007	2.950	2.901
230	2.969	2.900	2.818	2.748	2.690	3.102	3.016	2.955	2.898	2.840
260	2.923	2.854	2.773	2.703	2.644	3.055	2.963	2.908	2.851	2.785
300	2.869	2.798	2.720	2.651	2.592	3.000	2.900	2.853	2.797	2.720
350	2.809	2.735	2.662	2.596	2.536	2.939	2.831	2.792	2.737	2.651
400	2.757	2.681	2.613	2.549	2.489	2.886	2.771	2.740	2.687	2.592
450	2.710	2.632	2.569	2.508	2.448	2.839	2.719	2.695	2.643	2.543
500	2.667	2.588	2.530	2.472	2.414	2.797	2.674	2.655	2.604	2.501
600	2.588	2.508	2.461	2.410	2.356	2.726	2.601	2.590	2.545	2.448
700	2.523	2.443	2.405	2.361	2.311	2.667	2.543	2.537	2.496	2.408
800	2.469	2.390	2.360	2.322	2.277	2.618	2.495	2.493	2.455	2.375
900	2.424	2.347	2.323	2.290	2.250	2.576	2.456	2.455	2.420	2.348
1000	2.396	2.312	2.293	2.265	2.229	2.540	2.424	2.421	2.389	2.325
1200	2.331	2.269	2.252	2.231	2.204	2.485	2.380	2.366	2.333	2.283
1400	2.292	2.241	2.225	2.209	2.191	2.441	2.347	2.322	2.289	2.250
1700	2.252	2.217	2.202	2.192	2.186	2.389	2.309	2.270	2.237	2.210
2000	2.227	2.206	2.192	2.187	2.190	2.347	2.280	2.230	2.197	2.181
2300	2.213	2.201	2.193	2.193	2.201	2.307	2.249	2.200	2.170	2.159
2600	2.206	2.202	2.200	2.204	2.216	2.273	2.222	2.176	2.150	2.141
3000	2.204	2.209	2.214	2.224	2.239	2.234	2.191	2.152	2.129	2.123
3500	2.209	2.224	2.237	2.252	2.271	2.195	2.160	2.128	2.111	2.107
4000	2.220	2.242	2.262	2.283	2.305	2.164	2.136	2.111	2.098	2.096
4500	2.234	2.264	2.290	2.315	2.339	2.140	2.117	2.099	2.089	2.088
5000	2.251	2.286	2.318	2.347	2.373	2.120	2.102	2.089	2.083	2.082
6000	2.289	2.333	2.374	2.410	2.439	2.093	2.081	2.077	2.076	2.076
7000	2.329	2.381	2.429	2.470	2.502	2.078	2.070	2.072	2.073	2.075
8000	2.371	2.429	2.482	2.526	2.561	2.070	2.066	2.070	2.074	2.077
9000	2.412	2.476	2.533	2.580	2.618	2.069	2.067	2.071	2.076	2.081
10000	2.454	2.522	2.582	2.631	2.672	2.072	2.072	2.074	2.079	2.087

GE ON GE,M2=M1   TABLE 12- 9

K/KL=	REFLECTION COEFFICIENT					SPUTTERING EFFICIENCY				
E,KEV	0.6	0.8	1.0	1.2	1.4	0.6	0.8	1.0	1.2	1.4
1.0	.0179	.0175	.0171	.0168	.0164	.0217	.0207	.0197	.0188	.0180
1.2	.0178	.0174	.0171	.0167	.0163	.0216	.0206	.0196	.0186	.0178
1.4	.0178	.0174	.0170	.0166	.0162	.0215	.0204	.0194	.0185	.0176
1.7	.0177	.0173	.0169	.0165	.0161	.0214	.0203	.0193	.0183	.0174
2.0	.0176	.0172	.0168	.0164	.0160	.0212	.0202	.0191	.0181	.0172
2.3	.0176	.0171	.0167	.0163	.0159	.0212	.0200	.0190	.0180	.0171
2.6	.0175	.0170	.0166	.0162	.0158	.0211	.0199	.0189	.0179	.0169
3.0	.0174	.0170	.0165	.0161	.0157	.0210	.0198	.0187	.0177	.0167
3.5	.0173	.0168	.0164	.0160	.0156	.0209	.0197	.0186	.0175	.0166
4.0	.0172	.0167	.0163	.0159	.0155	.0208	.0195	.0184	.0174	.0164
4.5	.0171	.0166	.0162	.0158	.0153	.0207	.0194	.0183	.0172	.0162
5.0	.0170	.0166	.0161	.0157	.0152	.0206	.0193	.0182	.0171	.0161
6.0	.0169	.0164	.0159	.0155	.0150	.0204	.0191	.0179	.0169	.0159
7.0	.0167	.0162	.0157	.0153	.0148	.0203	.0190	.0178	.0167	.0157
8.0	.0166	.0161	.0156	.0151	.0147	.0201	.0188	.0176	.0165	.0155
9.0	.0165	.0159	.0154	.0150	.0145	.0200	.0187	.0174	.0163	.0153
10	.0163	.0158	.0153	.0148	.0144	.0199	.0185	.0173	.0162	.0152
12	.0161	.0155	.0150	.0145	.0141	.0197	.0183	.0170	.0159	.0149
14	.0159	.0153	.0148	.0143	.0138	.0195	.0181	.0168	.0157	.0146
17	.0155	.0150	.0144	.0139	.0134	.0192	.0178	.0165	.0154	.0143
20	.0152	.0147	.0141	.0136	.0131	.0190	.0175	.0162	.0151	.0140
23	.0149	.0144	.0138	.0133	.0128	.0188	.0173	.0160	.0148	.0138
26	.0147	.0141	.0135	.0130	.0125	.0186	.0171	.0158	.0146	.0135
30	.0143	.0137	.0132	.0126	.0121	.0184	.0169	.0155	.0143	.0133
35	.0139	.0133	.0127	.0122	.0117	.0181	.0166	.0152	.0140	.0129
40	.0136	.0129	.0124	.0118	.0113	.0179	.0164	.0149	.0137	.0126
45	.0132	.0126	.0120	.0115	.0109	.0176	.0161	.0147	.0134	.0124
50	.0129	.0123	.0117	.0111	.0106	.0174	.0159	.0144	.0132	.0122
60	.0123	.0117	.0111	.0105	.0100	.0171	.0155	.0140	.0128	.0117
70	.0118	.0111	.0105	.0100	.0095	.0167	.0151	.0137	.0124	.0114
80	.0113	.0106	.0100	.0095	.0090	.0164	.0147	.0134	.0121	.0110
90	.0108	.0102	.0096	.0090	.0085	.0161	.0144	.0131	.0118	.0107
100	.0104	.0098	.0092	.0086	.0081	.0159	.0141	.0128	.0115	.0104
120	.0097	.0090	.0084	.0079	.0074	.0154	.0137	.0123	.0110	.0099
140	.0091	.0084	.0078	.0072	.0067	.0149	.0133	.0118	.0106	.0095
170	.0082	.0076	.0070	.0064	.0060	.0144	.0127	.0112	.0100	.0089
200	.0075	.0069	.0063	.0058	.0053	.0138	.0123	.0107	.0094	.0084
230	.0069	.0063	.0057	.0052	.0048	.0134	.0118	.0103	.0090	.0080
260	.0064	.0058	.0052	.0047	.0043	.0130	.0114	.0099	.0086	.0076
300	.0058	.0052	.0046	.0042	.0038	.0125	.0109	.0094	.0082	.0071
350	.0052	.0046	.0041	.0036	.0033	.0120	.0103	.0089	.0077	.0067
400	.0047	.0041	.0036	.0031	.0029	.0115	.0098	.0084	.0072	.0062
450	.0042	.0036	.0032	.0028	.0025	.0110	.0094	.0080	.0069	.0059
500	.0038	.0033	.0028	.0024	.0022	.0106	.0089	.0076	.0065	.0056
600	.0032	.0027	.0023	.0020	.0018	.0098	.0082	.0069	.0059	.0050
700	.0028	.0022	.0019	.0016	.0014	.0091	.0075	.0063	.0054	.0046
800	.0024	.0019	.0016	.0013	.0011	.0085	.0069	.0059	.0049	.0042
900	.0021	.0016	.0013	.0011	.0009	.0080	.0065	.0054	.0046	.0039
1000	.0018	.0014	.0011	.0009	.0007	.0075	.0061	.0051	.0042	.0036
1200	.0014	.0010	.0008	.0006	.0005	.0069	.0055	.0045	.0038	.0032
1400	.0011	.0008	.0006	.0004	.0003	.0063	.0051	.0041	.0034	.0029
1700	.0008	.0005	.0004	.0003	.0002	.0057	.0045	.0036	.0030	.0026
2000	.0006	.0004	.0002	.0001	.0001	.0052	.0041	.0032	.0027	.0023
2300	.0004	.0003	.0002	.0001	.0000	.0047	.0037	.0029	.0024	.0020
2600	.0003	.0002	.0001	.0000	.0000	.0043	.0034	.0026	.0021	.0018
3000	.0002	.0001	.0001	.0000	0.0000	.0039	.0030	.0023	.0019	.0016
3500	.0001	.0001	.0000	0.0000	0.0000	.0034	.0027	.0020	.0016	.0014
4000	.0001	.0000	.0000	0.0000	0.0000	.0030	.0023	.0018	.0014	.0012
4500	.0000	.0000	.0000	0.0000	0.0000	.0027	.0021	.0016	.0012	.0010
5000	.0000	.0000	.0000	0.0000	0.0000	.0024	.0019	.0014	.0011	.0009
6000	0.0000	.0000	.0000	0.0000	0.0000	.0020	.0015	.0011	.0009	.0007
7000	0.0000	.0000	.0000	0.0000	0.0000	.0017	.0013	.0009	.0007	.0006
8000	0.0000	.0000	.0000	0.0000	0.0000	.0015	.0011	.0008	.0006	.0005
9000	0.0000	.0000	.0000	0.0000	0.0000	.0013	.0010	.0007	.0005	.0004
10000	0.0000	.0000	.0000	0.0000	0.0000	.0012	.0009	.0006	.0005	.0004

E,KEV / K/KL=	IONIZATION DEFICIENCY					SPUTTERING YIELD ALPHA				
	0.6	0.8	1.0	1.2	1.4	0.6	0.8	1.0	1.2	1.4
1.0	.0018	.0024	.0028	.0033	.0037	.1637	.1586	.1536	.1488	.1443
1.2	.0019	.0024	.0029	.0033	.0037	.1633	.1580	.1529	.1481	.1434
1.4	.0019	.0025	.0030	.0034	.0038	.1629	.1576	.1524	.1474	.1427
1.7	.0020	.0025	.0030	.0035	.0039	.1625	.1570	.1517	.1466	.1417
2.0	.0020	.0026	.0031	.0035	.0040	.1622	.1565	.1511	.1459	.1409
2.3	.0021	.0026	.0031	.0036	.0040	.1619	.1561	.1506	.1453	.1402
2.6	.0021	.0027	.0032	.0037	.0041	.1617	.1558	.1501	.1447	.1396
3.0	.0021	.0027	.0032	.0037	.0041	.1614	.1554	.1496	.1441	.1389
3.5	.0022	.0028	.0033	.0038	.0042	.1611	.1549	.1491	.1435	.1382
4.0	.0022	.0028	.0034	.0038	.0043	.1609	.1546	.1486	.1429	.1375
4.5	.0023	.0029	.0034	.0039	.0043	.1607	.1543	.1482	.1424	.1370
5.0	.0023	.0029	.0034	.0039	.0044	.1606	.1540	.1478	.1420	.1365
6.0	.0023	.0030	.0035	.0040	.0044	.1603	.1536	.1473	.1413	.1357
7.0	.0024	.0030	.0036	.0041	.0045	.1602	.1533	.1468	.1407	.1351
8.0	.0024	.0031	.0036	.0041	.0045	.1601	.1531	.1464	.1402	.1345
9.0	.0025	.0031	.0037	.0041	.0046	.1600	.1529	.1461	.1398	.1341
10	.0025	.0031	.0037	.0042	.0046	.1599	.1527	.1459	.1395	.1337
12	.0025	.0032	.0038	.0042	.0047	.1599	.1524	.1454	.1389	.1330
14	.0026	.0032	.0038	.0043	.0047	.1599	.1522	.1451	.1385	.1324
17	.0026	.0033	.0039	.0044	.0048	.1600	.1520	.1448	.1380	.1317
20	.0027	.0033	.0039	.0044	.0048	.1601	.1520	.1445	.1376	.1312
23	.0027	.0034	.0040	.0044	.0048	.1603	.1521	.1444	.1373	.1308
26	.0028	.0034	.0040	.0045	.0049	.1606	.1522	.1443	.1371	.1305
30	.0028	.0034	.0040	.0045	.0049	.1609	.1524	.1443	.1369	.1302
35	.0028	.0035	.0040	.0045	.0049	.1614	.1527	.1443	.1367	.1299
40	.0028	.0035	.0041	.0045	.0049	.1618	.1530	.1444	.1366	.1297
45	.0029	.0035	.0041	.0045	.0049	.1623	.1533	.1445	.1366	.1296
50	.0029	.0035	.0041	.0045	.0049	.1628	.1536	.1446	.1366	.1295
60	.0029	.0036	.0041	.0045	.0049	.1638	.1541	.1451	.1368	.1294
70	.0029	.0036	.0041	.0045	.0049	.1648	.1547	.1455	.1371	.1295
80	.0029	.0036	.0041	.0045	.0048	.1657	.1552	.1460	.1374	.1295
90	.0029	.0036	.0041	.0045	.0048	.1666	.1558	.1464	.1376	.1296
100	.0029	.0035	.0040	.0045	.0047	.1674	.1563	.1468	.1379	.1297
120	.0029	.0035	.0040	.0044	.0047	.1689	.1576	.1476	.1383	.1300
140	.0029	.0035	.0040	.0043	.0046	.1702	.1588	.1483	.1387	.1302
170	.0029	.0034	.0039	.0042	.0044	.1719	.1604	.1492	.1392	.1306
200	.0029	.0034	.0038	.0041	.0043	.1734	.1617	.1500	.1397	.1310
230	.0028	.0033	.0037	.0040	.0041	.1748	.1627	.1507	.1402	.1311
260	.0028	.0032	.0036	.0039	.0039	.1759	.1635	.1513	.1406	.1313
300	.0028	.0031	.0035	.0037	.0037	.1773	.1645	.1520	.1412	.1316
350	.0027	.0029	.0034	.0035	.0035	.1787	.1655	.1529	.1418	.1320
400	.0026	.0028	.0032	.0034	.0033	.1799	.1663	.1536	.1424	.1326
450	.0026	.0027	.0031	.0033	.0031	.1809	.1670	.1543	.1431	.1331
500	.0025	.0026	.0030	.0031	.0029	.1818	.1677	.1549	.1436	.1338
600	.0024	.0024	.0028	.0029	.0027	.1827	.1685	.1561	.1450	.1359
700	.0022	.0022	.0026	.0027	.0025	.1835	.1692	.1570	.1462	.1378
800	.0021	.0020	.0024	.0026	.0023	.1844	.1700	.1579	.1472	.1393
900	.0020	.0019	.0023	.0024	.0022	.1853	.1708	.1588	.1481	.1406
1000	.0019	.0018	.0021	.0022	.0020	.1862	.1716	.1595	.1488	.1415
1200	.0018	.0016	.0019	.0020	.0017	.1893	.1741	.1611	.1498	.1421
1400	.0017	.0015	.0017	.0017	.0015	.1921	.1764	.1625	.1505	.1423
1700	.0016	.0013	.0014	.0014	.0012	.1955	.1791	.1640	.1512	.1421
2000	.0014	.0011	.0012	.0012	.0009	.1980	.1811	.1651	.1516	.1418
2300	.0012	.0010	.0010	.0010	.0007	.1984	.1815	.1653	.1516	.1417
2600	.0010	.0008	.0008	.0008	.0006	.1983	.1816	.1652	.1516	.1415
3000	.0008	.0005	.0006	.0006	.0004	.1979	.1814	.1650	.1513	.1412
3500	.0005	.0004	.0004	.0004	.0003	.1971	.1809	.1646	.1509	.1409
4000	.0003	.0002	.0002	.0002	.0001	.1962	.1802	.1640	.1505	.1405
4500	.0001	.0001	.0001	.0001	.0001	.1954	.1796	.1635	.1500	.1401
5000	0.0000	0.0000	0.0000	0.0000	0.0000	.1946	.1789	.1630	.1495	.1397
6000	0.0000	0.0000	0.0000	0.0000	0.0000	.1932	.1777	.1620	.1485	.1389
7000	0.0000	0.0000	0.0000	0.0000	0.0000	.1923	.1767	.1612	.1475	.1382
8000	0.0000	0.0000	0.0000	0.0000	0.0000	.1917	.1760	.1605	.1466	.1374
9000	0.0000	0.0000	0.0000	0.0000	0.0000	.1915	.1755	.1599	.1457	.1367
10000	0.0000	0.0000	0.0000	0.0000	0.0000	.1915	.1751	.1595	.1449	.1360

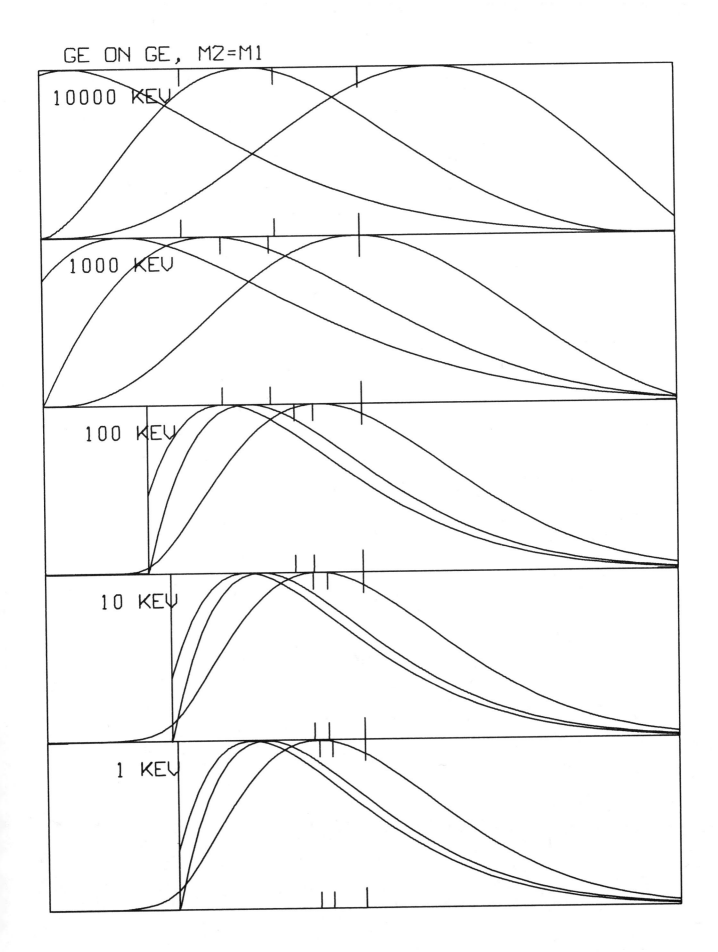

GE ON GE, M2=M1

10000 KEV

1000 KEV

100 KEV

10 KEV

1 KEV

TABLE 13- 1

C ON BE, M2/M1=0.75

K/KL= E,KEV	FRACTIONAL DEPOSITED ENERGY 0.8	1.0	1.3	1.7	2.0	MEAN RANGE,MICROGRAM/SQ.CM. 0.8	1.0	1.3	1.7	2.0
.020	.8832	.8568	.8192	.7723	.7395	.0368	.0366	.0363	.0359	.0356
.023	.8806	.8538	.8154	.7678	.7345	.0407	.0405	.0402	.0397	.0394
.026	.8783	.8510	.8121	.7637	.7300	.0443	.0441	.0437	.0432	.0429
.030	.8756	.8477	.8081	.7589	.7246	.0488	.0485	.0481	.0475	.0471
.035	.8726	.8441	.8036	.7536	.7187	.0540	.0536	.0532	.0526	.0521
.040	.8699	.8409	.7997	.7489	.7135	.0589	.0585	.0580	.0573	.0568
.045	.8675	.8380	.7962	.7447	.7089	.0636	.0632	.0627	.0619	.0614
.050	.8653	.8354	.7930	.7408	.7046	.0682	.0678	.0672	.0664	.0658
.060	.8613	.8307	.7874	.7341	.6972	.0771	.0766	.0759	.0750	.0743
.070	.8579	.8266	.7825	.7282	.6907	.0857	.0851	.0843	.0832	.0824
.080	.8549	.8230	.7781	.7231	.6851	.0939	.0933	.0924	.0912	.0903
.090	.8522	.8198	.7742	.7185	.6801	.1020	.1013	.1002	.0989	.0979
.10	.8497	.8169	.7707	.7143	.6755	.1098	.1090	.1079	.1064	.1053
.12	.8454	.8117	.7645	.7070	.6677	.1243	.1234	.1221	.1204	.1191
.14	.8416	.8073	.7591	.7006	.6609	.1385	.1374	.1359	.1339	.1324
.17	.8366	.8014	.7520	.6924	.6521	.1590	.1578	.1559	.1535	.1518
.20	.8323	.7964	.7459	.6852	.6446	.1790	.1775	.1753	.1725	.1705
.23	.8284	.7919	.7405	.6789	.6378	.1971	.1954	.1930	.1898	.1875
.26	.8249	.7877	.7356	.6732	.6317	.2149	.2130	.2103	.2068	.2042
.30	.8207	.7827	.7297	.6663	.6243	.2384	.2363	.2332	.2291	.2262
.35	.8159	.7771	.7230	.6586	.6161	.2677	.2652	.2615	.2567	.2533
.40	.8116	.7721	.7171	.6518	.6088	.2967	.2938	.2895	.2840	.2801
.45	.8077	.7675	.7117	.6455	.6022	.3254	.3221	.3172	.3109	.3064
.50	.8040	.7632	.7067	.6398	.5961	.3539	.3501	.3446	.3375	.3324
.60	.7975	.7556	.6978	.6297	.5854	.4059	.4013	.3947	.3863	.3803
.70	.7916	.7489	.6898	.6207	.5759	.4586	.4532	.4453	.4354	.4282
.80	.7863	.7427	.6827	.6126	.5675	.5119	.5055	.4963	.4846	.4762
.90	.7813	.7370	.6761	.6052	.5598	.5657	.5582	.5475	.5339	.5242
1.0	.7767	.7317	.6700	.5984	.5527	.6196	.6110	.5987	.5831	.5720
1.2	.7684	.7222	.6590	.5861	.5399	.7182	.7079	.6929	.6741	.6607
1.4	.7608	.7136	.6491	.5751	.5287	.8205	.8079	.7898	.7670	.7509
1.7	.7506	.7019	.6358	.5606	.5138	.9807	.9638	.9398	.9099	.8889
2.0	.7412	.6914	.6240	.5477	.5007	1.147	1.125	1.094	1.056	1.029
2.3	.7327	.6818	.6131	.5360	.4888	1.293	1.268	1.233	1.189	1.158
2.6	.7247	.6728	.6031	.5252	.4781	1.445	1.416	1.376	1.325	1.289
3.0	.7147	.6617	.5908	.5121	.4650	1.659	1.624	1.574	1.512	1.467
3.5	.7032	.6490	.5768	.4974	.4504	1.943	1.896	1.832	1.752	1.696
4.0	.6925	.6373	.5640	.4840	.4373	2.239	2.179	2.097	1.998	1.928
4.5	.6824	.6263	.5522	.4718	.4254	2.546	2.471	2.368	2.246	2.163
5.0	.6729	.6161	.5411	.4605	.4144	2.859	2.768	2.643	2.497	2.400
6.0	.6551	.5969	.5207	.4398	.3945	3.444	3.325	3.164	2.974	2.850
7.0	.6389	.5797	.5025	.4216	.3772	4.066	3.912	3.703	3.462	3.308
8.0	.6239	.5639	.4861	.4053	.3619	4.718	4.522	4.257	3.957	3.771
9.0	.6100	.5494	.4711	.3906	.3481	5.392	5.147	4.819	4.455	4.235
10	.5970	.5359	.4574	.3773	.3357	6.081	5.784	5.388	4.954	4.700
12	.5727	.5110	.4323	.3534	.3137	7.418	7.023	6.500	5.933	5.605
14	.5510	.4889	.4104	.3328	.2950	8.807	8.296	7.624	6.908	6.505
17	.5222	.4601	.3823	.3068	.2714	10.96	10.25	9.318	8.354	7.838
20	.4969	.4353	.3583	.2850	.2518	13.17	12.22	11.00	9.774	9.144
23	.4740	.4131	.3375	.2664	.2353	15.34	14.16	12.65	11.15	10.41
26	.4534	.3934	.3193	.2504	.2211	17.53	16.09	14.28	12.50	11.64
30	.4290	.3702	.2981	.2319	.2049	20.45	18.66	16.42	14.25	13.25
35	.4025	.3452	.2755	.2126	.1879	24.11	21.84	19.03	16.37	15.19
40	.3794	.3238	.2564	.1964	.1737	27.73	24.98	21.57	18.41	17.06
45	.3590	.3051	.2399	.1825	.1617	31.32	28.05	24.04	20.37	18.87
50	.3410	.2886	.2255	.1706	.1513	34.86	31.07	26.45	22.27	20.62
60	.3104	.2611	.2021	.1515	.1346	41.74	36.91	31.06	25.88	23.96
70	.2852	.2387	.1833	.1364	.1214	48.41	42.52	35.44	29.28	27.11
80	.2640	.2200	.1679	.1241	.1107	54.88	47.94	39.63	32.50	30.10
90	.2458	.2042	.1549	.1139	.1018	61.15	53.16	43.64	35.56	32.96
100	.2301	.1905	.1438	.1053	.0943	67.23	58.21	47.50	38.49	35.69
120	.2043	.1683	.1260	.0916	.0822	78.87	67.82	54.78	43.99	40.85
140	.1840	.1509	.1123	.0811	.0731	89.89	76.87	61.59	49.09	45.64
170	.1607	.1313	.0969	.0696	.0629	105.4	89.56	71.05	56.14	52.29
200	.1434	.1168	.0858	.0613	.0555	119.9	101.3	79.77	62.59	58.39

TABLE 13- 2

C ON BE, M2/M1=0.75

E,KEV	MEAN DAMAGE DEPTH,MICROGRAM/SQ.CM. K/KL= 0.8	1.0	1.3	1.7	2.0	MEAN IONZN.DEPTH,MICROGRAM/SQ.CM. 0.8	1.0	1.3	1.7	2.0
.020	.0304	.0301	.0297	.0292	.0289	.0278	.0275	.0271	.0266	.0262
.023	.0336	.0333	.0329	.0323	.0319	.0308	.0304	.0300	.0294	.0289
.026	.0366	.0362	.0357	.0351	.0347	.0335	.0331	.0326	.0319	.0314
.030	.0402	.0398	.0393	.0386	.0381	.0368	.0364	.0358	.0351	.0345
.035	.0445	.0441	.0434	.0426	.0421	.0407	.0402	.0396	.0387	.0381
.040	.0485	.0481	.0474	.0465	.0458	.0444	.0439	.0432	.0422	.0416
.045	.0524	.0519	.0511	.0502	.0495	.0479	.0474	.0466	.0456	.0448
.050	.0562	.0556	.0548	.0538	.0530	.0514	.0508	.0499	.0488	.0480
.050	.0635	.0628	.0619	.0606	.0598	.0580	.0573	.0563	.0550	.0541
.070	.0705	.0697	.0686	.0672	.0662	.0644	.0636	.0624	.0610	.0599
.080	.0772	.0764	.0751	.0736	.0725	.0705	.0697	.0684	.0667	.0655
.090	.0838	.0828	.0815	.0797	.0785	.0765	.0755	.0741	.0723	.0709
.10	.0901	.0891	.0876	.0857	.0844	.0823	.0812	.0796	.0776	.0762
.12	.1020	.1008	.0990	.0968	.0953	.0931	.0918	.0900	.0876	.0860
.14	.1134	.1121	.1101	.1076	.1058	.1035	.1020	.0999	.0973	.0954
.17	.1301	.1285	.1261	.1231	.1209	.1186	.1169	.1144	.1112	.1089
.20	.1462	.1443	.1415	.1380	.1356	.1332	.1312	.1283	.1246	.1220
.23	.1609	.1587	.1556	.1517	.1489	.1465	.1442	.1410	.1368	.1339
.26	.1753	.1729	.1694	.1650	.1620	.1595	.1570	.1534	.1488	.1455
.30	.1943	.1915	.1875	.1826	.1791	.1767	.1738	.1696	.1644	.1607
.35	.2177	.2145	.2099	.2041	.2000	.1978	.1944	.1896	.1835	.1793
.40	.2408	.2372	.2319	.2252	.2206	.2186	.2148	.2092	.2023	.1974
.45	.2637	.2595	.2535	.2460	.2408	.2391	.2348	.2285	.2207	.2152
.50	.2863	.2816	.2749	.2665	.2607	.2594	.2544	.2474	.2387	.2326
.60	.3276	.3221	.3141	.3042	.2973	.2965	.2907	.2824	.2721	.2649
.70	.3693	.3627	.3533	.3417	.3337	.3338	.3269	.3172	.3051	.2967
.80	.4112	.4035	.3926	.3791	.3699	.3710	.3631	.3518	.3378	.3282
.90	.4532	.4443	.4318	.4164	.4059	.4083	.3991	.3862	.3703	.3593
1.0	.4952	.4851	.4709	.4535	.4416	.4454	.4350	.4203	.4024	.3900
1.2	.5724	.5603	.5432	.5223	.5081	.5140	.5015	.4839	.4624	.4476
1.4	.6517	.6372	.6167	.5919	.5750	.5839	.5689	.5479	.5224	.5048
1.7	.7747	.7559	.7295	.6977	.6764	.6910	.6717	.6449	.6126	.5905
2.0	.9011	.8774	.8442	.8047	.7786	.8002	.7759	.7425	.7027	.6758
2.3	1.013	.9859	.9480	.9025	.8723	.8988	.8710	.8325	.7865	.7554
2.6	1.129	1.098	1.054	1.002	.9672	.9999	.9679	.9236	.8706	.8349
3.0	1.291	1.253	1.200	1.137	1.096	1.139	1.100	1.047	.9834	.9410
3.5	1.502	1.454	1.388	1.310	1.259	1.318	1.269	1.203	1.125	1.074
4.0	1.722	1.662	1.580	1.485	1.425	1.501	1.441	1.361	1.267	1.206
4.5	1.948	1.875	1.776	1.662	1.591	1.688	1.616	1.519	1.409	1.337
5.0	2.178	2.091	1.973	1.840	1.758	1.876	1.791	1.678	1.550	1.467
6.0	2.606	2.496	2.346	2.178	2.075	2.234	2.124	1.983	1.820	1.717
7.0	3.059	2.919	2.731	2.522	2.397	2.602	2.465	2.290	2.089	1.963
8.0	3.530	3.356	3.124	2.870	2.721	2.978	2.809	2.596	2.356	2.207
9.0	4.016	3.803	3.522	3.219	3.046	3.358	3.155	2.900	2.620	2.446
10	4.511	4.257	3.924	3.569	3.370	3.742	3.502	3.202	2.880	2.681
12	5.457	5.136	4.704	4.249	3.999	4.486	4.176	3.785	3.393	3.136
14	6.458	6.037	5.493	4.928	4.625	5.237	4.850	4.359	3.889	3.577
17	7.994	7.419	6.683	5.938	5.557	6.368	5.851	5.198	4.600	4.213
20	9.565	8.819	7.873	6.935	6.475	7.493	6.838	6.012	5.276	4.822
23	11.09	10.18	9.025	7.898	7.358	8.597	7.801	6.804	5.910	5.404
26	12.63	11.54	10.17	8.844	8.227	9.688	8.745	7.571	6.516	5.963
30	14.70	13.35	11.68	10.08	9.364	11.12	9.973	8.559	7.286	6.678
35	17.30	15.61	13.53	11.59	10.75	12.86	11.46	9.741	8.198	7.525
40	19.89	17.85	15.35	13.05	12.10	14.56	12.90	10.87	9.061	8.329
45	22.46	20.06	17.13	14.47	13.41	16.21	14.29	11.95	9.882	9.095
50	25.00	22.23	18.87	15.85	14.68	17.82	15.63	12.99	10.67	9.826
60	29.93	26.43	22.20	18.47	17.11	20.90	18.19	14.94	12.14	11.20
70	34.75	30.50	25.41	20.98	19.44	23.83	20.61	16.77	13.51	12.47
80	39.46	34.46	28.50	23.38	21.67	26.63	22.90	18.49	14.79	13.66
90	44.05	38.32	31.48	25.69	23.82	29.30	25.08	20.11	15.99	14.79
100	48.54	42.06	34.37	27.91	25.89	31.85	27.17	21.65	17.13	15.86
120	57.19	49.26	39.89	32.12	29.84	36.68	31.08	24.53	19.24	17.85
140	65.45	56.10	45.10	36.08	33.55	41.18	34.72	27.18	21.17	19.68
170	77.18	65.77	52.42	41.61	38.75	47.41	39.73	30.82	23.80	22.18
200	88.21	74.84	59.24	46.74	43.59	53.15	44.33	34.13	26.18	24.46

TABLE 13- 3

C ON BE, M2/M1=0.75

K/KL= E,KEV	RELATIVE RANGE STRAGGLING					RELATIVE DAMAGE STRAGGLING				
	0.8	1.0	1.3	1.7	2.0	0.8	1.0	1.3	1.7	2.0
.020	.2671	.2649	.2618	.2576	.2546	.3893	.3876	.3852	.3821	.3798
.023	.2668	.2645	.2613	.2571	.2540	.3890	.3873	.3848	.3817	.3794
.026	.2664	.2642	.2608	.2565	.2534	.3888	.3871	.3845	.3813	.3790
.030	.2660	.2637	.2603	.2559	.2527	.3886	.3868	.3842	.3809	.3785
.035	.2656	.2632	.2597	.2552	.2520	.3883	.3864	.3838	.3804	.3780
.040	.2651	.2627	.2591	.2546	.2512	.3880	.3861	.3834	.3800	.3775
.045	.2647	.2623	.2586	.2540	.2506	.3877	.3858	.3831	.3796	.3771
.050	.2644	.2618	.2581	.2534	.2500	.3875	.3856	.3828	.3792	.3767
.060	.2636	.2610	.2572	.2524	.2488	.3871	.3851	.3822	.3785	.3760
.070	.2630	.2603	.2564	.2514	.2478	.3867	.3846	.3817	.3779	.3753
.080	.2623	.2596	.2556	.2505	.2468	.3863	.3842	.3812	.3773	.3746
.090	.2617	.2589	.2548	.2496	.2459	.3859	.3838	.3807	.3768	.3741
.10	.2611	.2583	.2541	.2488	.2450	.3856	.3834	.3802	.3763	.3735
.12	.2600	.2571	.2528	.2474	.2434	.3849	.3826	.3794	.3753	.3725
.14	.2590	.2560	.2516	.2460	.2420	.3843	.3820	.3786	.3745	.3715
.17	.2575	.2544	.2498	.2441	.2399	.3834	.3810	.3776	.3732	.3702
.20	.2561	.2529	.2482	.2423	.2380	.3826	.3801	.3765	.3721	.3690
.23	.2548	.2515	.2467	.2406	.2363	.3819	.3793	.3756	.3711	.3679
.26	.2536	.2502	.2453	.2390	.2346	.3812	.3786	.3748	.3701	.3669
.30	.2520	.2485	.2434	.2370	.2325	.3803	.3776	.3737	.3689	.3656
.35	.2501	.2465	.2413	.2347	.2300	.3793	.3764	.3724	.3674	.3640
.40	.2483	.2446	.2392	.2325	.2277	.3782	.3753	.3711	.3661	.3626
.45	.2466	.2427	.2372	.2303	.2255	.3772	.3742	.3699	.3647	.3612
.50	.2449	.2410	.2354	.2283	.2234	.3763	.3732	.3688	.3634	.3598
.60	.2418	.2377	.2318	.2245	.2194	.3743	.3711	.3665	.3610	.3572
.70	.2388	.2346	.2285	.2210	.2157	.3725	.3691	.3644	.3587	.3548
.80	.2360	.2316	.2254	.2176	.2123	.3708	.3673	.3624	.3565	.3525
.90	.2334	.2288	.2224	.2145	.2090	.3691	.3655	.3604	.3544	.3504
1.0	.2309	.2262	.2196	.2115	.2059	.3676	.3638	.3586	.3524	.3483
1.2	.2261	.2212	.2143	.2059	.2001	.3647	.3607	.3552	.3487	.3445
1.4	.2217	.2166	.2094	.2007	.1948	.3620	.3578	.3521	.3453	.3409
1.7	.2156	.2102	.2027	.1937	.1876	.3582	.3537	.3476	.3405	.3360
2.0	.2101	.2045	.1967	.1874	.1811	.3546	.3499	.3435	.3361	.3315
2.3	.2050	.1991	.1911	.1815	.1751	.3510	.3461	.3395	.3319	.3271
2.6	.2002	.1941	.1859	.1761	.1695	.3476	.3426	.3357	.3279	.3231
3.0	.1944	.1880	.1795	.1695	.1628	.3434	.3381	.3310	.3230	.3181
3.5	.1877	.1811	.1723	.1621	.1553	.3385	.3330	.3255	.3173	.3124
4.0	.1817	.1749	.1658	.1555	.1486	.3340	.3282	.3206	.3122	.3072
4.5	.1762	.1692	.1599	.1494	.1426	.3298	.3238	.3160	.3074	.3025
5.0	.1711	.1639	.1545	.1439	.1371	.3258	.3197	.3117	.3030	.2981
6.0	.1618	.1544	.1446	.1339	.1272	.3186	.3122	.3039	.2950	.2901
7.0	.1537	.1460	.1361	.1252	.1188	.3121	.3055	.2970	.2880	.2831
8.0	.1465	.1387	.1286	.1178	.1116	.3063	.2995	.2908	.2818	.2770
9.0	.1400	.1322	.1219	.1112	.1052	.3009	.2941	.2852	.2762	.2714
10	.1342	.1263	.1160	.1053	.0996	.2960	.2891	.2801	.2711	.2663
12	.1239	.1160	.1059	.0956	.0902	.2871	.2800	.2711	.2620	.2574
14	.1151	.1074	.0975	.0876	.0825	.2793	.2722	.2632	.2543	.2498
17	.1041	.0967	.0872	.0778	.0733	.2692	.2621	.2532	.2445	.2402
20	.0951	.0879	.0789	.0700	.0658	.2607	.2537	.2449	.2363	.2322
23	.0877	.0807	.0720	.0637	.0598	.2534	.2464	.2378	.2294	.2255
26	.0814	.0746	.0663	.0583	.0548	.2470	.2401	.2316	.2234	.2196
30	.0742	.0678	.0599	.0524	.0492	.2395	.2328	.2244	.2165	.2129
35	.0669	.0608	.0533	.0465	.0436	.2316	.2250	.2168	.2091	.2057
40	.0609	.0551	.0480	.0416	.0391	.2248	.2183	.2104	.2029	.1997
45	.0558	.0503	.0436	.0376	.0354	.2189	.2126	.2048	.1975	.1944
50	.0515	.0463	.0399	.0343	.0322	.2136	.2075	.1998	.1928	.1898
60	.0449	.0401	.0343	.0294	.0276	.2049	.1990	.1916	.1849	.1821
70	.0397	.0353	.0301	.0256	.0241	.1978	.1920	.1849	.1784	.1758
80	.0356	.0316	.0268	.0227	.0214	.1917	.1861	.1792	.1729	.1704
90	.0323	.0285	.0241	.0204	.0192	.1865	.1810	.1743	.1683	.1658
100	.0295	.0260	.0219	.0185	.0175	.1820	.1766	.1701	.1642	.1618
120	.0251	.0220	.0185	.0155	.0147	.1744	.1692	.1629	.1573	.1551
140	.0218	.0191	.0159	.0134	.0126	.1682	.1632	.1571	.1517	.1496
170	.0181	.0158	.0131	.0110	.0104	.1609	.1560	.1501	.1449	.1429
200	.0155	.0135	.0111	.0092	.0088	.1550	.1503	.1446	.1395	.1375

TABLE 13- 4

C ON RE, M2/M1=0.75

E,KEV / K/KL=	RELATIVE IONIZATION STRAGGLING					RELATIVE TRANSVERSE RANGE STRAGGLING				
	0.8	1.0	1.3	1.7	2.0	0.8	1.0	1.3	1.7	2.0
.020	.4442	.4433	.4419	.4402	.4391	.6505	.6505	.6507	.6508	.6509
.023	.4441	.4431	.4417	.4400	.4388	.6503	.6504	.6505	.6507	.6508
.026	.4440	.4430	.4416	.4398	.4386	.6502	.6502	.6504	.6506	.6507
.030	.4439	.4428	.4414	.4396	.4384	.6500	.6501	.6502	.6504	.6507
.035	.4437	.4427	.4412	.4394	.4382	.6497	.6498	.6500	.6502	.6504
.040	.4436	.4425	.4410	.4392	.4379	.6495	.6496	.6498	.6500	.6502
.045	.4435	.4424	.4408	.4390	.4377	.6493	.6494	.6496	.6498	.6500
.050	.4433	.4422	.4407	.4388	.4375	.6490	.6492	.6494	.6496	.6498
.060	.4431	.4420	.4404	.4385	.4371	.6486	.6487	.6490	.6493	.6495
.070	.4429	.4418	.4401	.4381	.4368	.6482	.6483	.6486	.6489	.6492
.080	.4427	.4415	.4399	.4379	.4365	.6477	.6479	.6482	.6486	.6489
.090	.4426	.4413	.4396	.4376	.4362	.6473	.6476	.6479	.6483	.6486
.10	.4424	.4412	.4394	.4373	.4359	.6469	.6472	.6475	.6480	.6483
.12	.4421	.4408	.4390	.4369	.4355	.6462	.6465	.6469	.6474	.6478
.14	.4418	.4405	.4386	.4365	.4350	.6455	.6458	.6462	.6468	.6473
.17	.4414	.4400	.4381	.4359	.4344	.6445	.6448	.6454	.6461	.6466
.20	.4410	.4396	.4377	.4353	.4338	.6435	.6439	.6445	.6454	.6460
.23	.4407	.4392	.4372	.4349	.4333	.6426	.6430	.6437	.6446	.6453
.26	.4404	.4389	.4368	.4344	.4328	.6417	.6422	.6430	.6440	.6447
.30	.4400	.4384	.4363	.4338	.4322	.6406	.6412	.6421	.6432	.6440
.35	.4395	.4378	.4356	.4331	.4314	.6394	.6401	.6411	.6424	.6433
.40	.4390	.4373	.4350	.4324	.4307	.6382	.6390	.6401	.6416	.6427
.45	.4385	.4368	.4345	.4318	.4301	.6372	.6381	.6393	.6409	.6421
.50	.4381	.4363	.4339	.4312	.4295	.6363	.6372	.6386	.6404	.6416
.60	.4373	.4354	.4329	.4301	.4283	.6344	.6355	.6372	.6392	.6407
.70	.4365	.4346	.4320	.4290	.4272	.6329	.6342	.6360	.6384	.6400
.80	.4358	.4337	.4311	.4281	.4262	.6316	.6331	.6351	.6377	.6396
.90	.4351	.4330	.4302	.4271	.4252	.6306	.6321	.6344	.6373	.6394
1.0	.4344	.4322	.4294	.4263	.4243	.6297	.6314	.6339	.6370	.6393
1.2	.4331	.4308	.4278	.4246	.4226	.6281	.6301	.6330	.6366	.6392
1.4	.4319	.4295	.4264	.4231	.4211	.6270	.6293	.6326	.6367	.6396
1.7	.4303	.4277	.4244	.4210	.4190	.6262	.6289	.6327	.6375	.6409
2.0	.4287	.4260	.4227	.4192	.4172	.6261	.6291	.6335	.6390	.6428
2.3	.4273	.4244	.4210	.4174	.4154	.6263	.6298	.6346	.6406	.6449
2.6	.4259	.4230	.4194	.4159	.4139	.6270	.6308	.6361	.6427	.6473
3.0	.4242	.4212	.4176	.4140	.4121	.6285	.6327	.6386	.6459	.6510
3.5	.4223	.4192	.4155	.4120	.4102	.6310	.6357	.6423	.6504	.6561
4.0	.4206	.4174	.4137	.4103	.4085	.6340	.6392	.6465	.6554	.6616
4.5	.4191	.4158	.4122	.4088	.4072	.6374	.6431	.6510	.6607	.6674
5.0	.4177	.4144	.4108	.4075	.4060	.6411	.6473	.6558	.6662	.6733
6.0	.4151	.4119	.4084	.4054	.4040	.6488	.6561	.6662	.6782	.6862
7.0	.4129	.4098	.4064	.4037	.4025	.6570	.6653	.6767	.6901	.6989
8.0	.4111	.4080	.4048	.4024	.4015	.6656	.6748	.6873	.7019	.7113
9.0	.4094	.4065	.4035	.4014	.4007	.6745	.6844	.6979	.7135	.7234
10	.4080	.4052	.4024	.4006	.4001	.6834	.6939	.7083	.7247	.7351
12	.4053	.4030	.4008	.3995	.3997	.7019	.7131	.7283	.7457	.7566
14	.4032	.4013	.3997	.3989	.3997	.7200	.7318	.7478	.7659	.7770
17	.4009	.3997	.3988	.3986	.4003	.7462	.7588	.7756	.7946	.8061
20	.3994	.3988	.3985	.3989	.4012	.7712	.7844	.8021	.8218	.8335
23	.3988	.3987	.3990	.4000	.4026	.7951	.8093	.8282	.8489	.8609
26	.3987	.3989	.3997	.4013	.4041	.8178	.8330	.8530	.8746	.8870
30	.3988	.3995	.4009	.4031	.4062	.8464	.8627	.8841	.9068	.9196
35	.3994	.4005	.4026	.4056	.4088	.8798	.8975	.9204	.9443	.9573
40	.4003	.4019	.4044	.4080	.4113	.9109	.9297	.9540	.9789	.9920
45	.4014	.4032	.4062	.4103	.4137	.9400	.9599	.9854	1.011	1.024
50	.4025	.4047	.4080	.4125	.4160	.9674	.9883	1.015	1.041	1.054
60	.4050	.4077	.4117	.4167	.4203	1.019	1.042	1.070	1.097	1.110
70	.4075	.4106	.4151	.4205	.4242	1.066	1.089	1.119	1.147	1.159
80	.4099	.4134	.4182	.4240	.4277	1.108	1.132	1.163	1.191	1.203
90	.4123	.4160	.4211	.4271	.4308	1.146	1.171	1.203	1.231	1.242
100	.4145	.4185	.4238	.4299	.4337	1.180	1.207	1.239	1.267	1.278
120	.4186	.4229	.4286	.4349	.4386	1.242	1.270	1.303	1.332	1.341
140	.4223	.4268	.4328	.4392	.4428	1.295	1.324	1.358	1.386	1.395
170	.4272	.4319	.4381	.4445	.4480	1.363	1.393	1.427	1.456	1.464
200	.4314	.4362	.4425	.4488	.4521	1.420	1.451	1.486	1.514	1.521

TABLE 13- 5

C ON BE, M2/M1=0.75

E,KEV	RELATIVE TRANSV. DAMAGE STRAGGLING					RELATIVE TRANSV. IONZN. STRAGGLING				
K/KL=	0.8	1.0	1.3	1.7	2.0	0.8	1.0	1.3	1.7	2.0
.020	.5198	.5175	.5141	.5097	.5064	.4240	.4208	.4161	.4100	.4055
.023	.5192	.5169	.5134	.5089	.5055	.4234	.4201	.4153	.4091	.4045
.026	.5187	.5163	.5127	.5081	.5047	.4228	.4194	.4146	.4082	.4035
.030	.5180	.5156	.5119	.5072	.5038	.4220	.4186	.4136	.4071	.4024
.035	.5172	.5147	.5110	.5062	.5027	.4212	.4177	.4126	.4059	.4011
.040	.5165	.5139	.5102	.5053	.5017	.4204	.4168	.4116	.4048	.3999
.045	.5158	.5132	.5094	.5044	.5007	.4196	.4160	.4107	.4038	.3988
.050	.5151	.5125	.5086	.5035	.4998	.4189	.4152	.4098	.4028	.3977
.060	.5138	.5111	.5071	.5019	.4981	.4175	.4137	.4082	.4010	.3958
.070	.5126	.5098	.5057	.5004	.4965	.4162	.4123	.4067	.3993	.3940
.080	.5114	.5086	.5044	.4990	.4951	.4150	.4110	.4052	.3978	.3923
.090	.5103	.5074	.5032	.4977	.4937	.4138	.4098	.4039	.3963	.3908
.10	.5092	.5063	.5020	.4964	.4923	.4126	.4086	.4026	.3949	.3893
.12	.5071	.5041	.4997	.4940	.4898	.4105	.4063	.4002	.3923	.3865
.14	.5051	.5021	.4976	.4918	.4875	.4084	.4042	.3979	.3898	.3840
.17	.5024	.4992	.4946	.4886	.4843	.4056	.4012	.3947	.3865	.3805
.20	.4997	.4965	.4918	.4857	.4813	.4029	.3984	.3918	.3833	.3772
.23	.4972	.4939	.4891	.4829	.4784	.4003	.3957	.3890	.3804	.3742
.26	.4948	.4914	.4866	.4803	.4757	.3979	.3932	.3864	.3776	.3713
.30	.4917	.4883	.4834	.4770	.4723	.3948	.3900	.3831	.3742	.3677
.35	.4881	.4846	.4796	.4731	.4684	.3911	.3863	.3792	.3701	.3636
.40	.4847	.4812	.4761	.4694	.4647	.3877	.3828	.3756	.3663	.3597
.45	.4815	.4779	.4727	.4660	.4612	.3845	.3794	.3721	.3627	.3560
.50	.4785	.4749	.4696	.4628	.4579	.3814	.3763	.3689	.3594	.3525
.60	.4728	.4691	.4637	.4567	.4517	.3756	.3703	.3627	.3530	.3460
.70	.4675	.4638	.4582	.4511	.4460	.3702	.3648	.3571	.3472	.3401
.80	.4627	.4588	.4532	.4460	.4408	.3653	.3598	.3519	.3418	.3346
.90	.4582	.4543	.4486	.4412	.4360	.3606	.3551	.3470	.3368	.3295
1.0	.4540	.4500	.4442	.4367	.4314	.3563	.3506	.3424	.3321	.3247
1.2	.4461	.4419	.4360	.4284	.4229	.3483	.3424	.3340	.3233	.3158
1.4	.4390	.4348	.4286	.4208	.4153	.3410	.3350	.3264	.3155	.3078
1.7	.4297	.4253	.4189	.4108	.4051	.3314	.3252	.3162	.3051	.2972
2.0	.4215	.4170	.4104	.4021	.3962	.3229	.3165	.3073	.2958	.2878
2.3	.4144	.4097	.4029	.3943	.3883	.3153	.3087	.2992	.2875	.2793
2.6	.4081	.4032	.3961	.3873	.3812	.3084	.3016	.2919	.2799	.2715
3.0	.4005	.3954	.3881	.3790	.3726	.3002	.2931	.2830	.2708	.2623
3.5	.3923	.3869	.3792	.3697	.3631	.2910	.2836	.2732	.2606	.2520
4.0	.3851	.3794	.3714	.3615	.3547	.2829	.2752	.2645	.2516	.2428
4.5	.3786	.3727	.3644	.3542	.3472	.2755	.2676	.2567	.2435	.2346
5.0	.3728	.3667	.3580	.3475	.3403	.2689	.2607	.2495	.2361	.2271
6.0	.3628	.3562	.3469	.3357	.3282	.2571	.2485	.2369	.2230	.2138
7.0	.3542	.3472	.3373	.3256	.3178	.2470	.2380	.2260	.2118	.2025
8.0	.3468	.3393	.3289	.3166	.3086	.2382	.2288	.2164	.2020	.1926
9.0	.3401	.3323	.3214	.3087	.3004	.2302	.2206	.2079	.1933	.1838
10	.3342	.3259	.3146	.3014	.2931	.2231	.2132	.2002	.1855	.1760
12	.3240	.3150	.3027	.2888	.2801	.2106	.2002	.1866	.1720	.1624
14	.3152	.3055	.2925	.2780	.2691	.2000	.1892	.1751	.1607	.1510
17	.3040	.2935	.2795	.2641	.2551	.1865	.1753	.1607	.1466	.1369
20	.2943	.2832	.2684	.2524	.2433	.1752	.1636	.1488	.1349	.1255
23	.2857	.2740	.2586	.2422	.2330	.1652	.1535	.1386	.1249	.1159
26	.2779	.2658	.2499	.2331	.2239	.1564	.1447	.1299	.1163	.1078
30	.2687	.2560	.2396	.2225	.2134	.1461	.1345	.1199	.1065	.0987
35	.2586	.2454	.2285	.2112	.2021	.1352	.1238	.1094	.0962	.0892
40	.2496	.2361	.2188	.2013	.1924	.1260	.1147	.1007	.0877	.0814
45	.2416	.2278	.2103	.1927	.1839	.1179	.1069	.0932	.0806	.0749
50	.2344	.2204	.2026	.1851	.1764	.1109	.1001	.0868	.0744	.0693
60	.2214	.2072	.1894	.1720	.1637	.0992	.0890	.0764	.0650	.0604
70	.2102	.1961	.1783	.1611	.1532	.0898	.0801	.0683	.0576	.0536
80	.2006	.1864	.1688	.1519	.1443	.0820	.0728	.0616	.0517	.0481
90	.1921	.1780	.1606	.1441	.1367	.0754	.0666	.0561	.0468	.0436
100	.1845	.1706	.1534	.1372	.1301	.0698	.0614	.0515	.0428	.0398
120	.1715	.1579	.1412	.1257	.1191	.0607	.0531	.0441	.0364	.0338
140	.1607	.1475	.1314	.1165	.1103	.0536	.0467	.0385	.0316	.0294
170	.1473	.1349	.1196	.1056	.0999	.0457	.0396	.0324	.0264	.0246
200	.1366	.1247	.1103	.0972	.0919	.0400	.0345	.0281	.0227	.0212

TABLE 13- 6

C ON RE, M2/M1=0.75

K/KL = E,KEV	RANGE SKEWNESS					DAMAGE SKEWNESS				
	0.8	1.0	1.3	1.7	2.0	0.8	1.0	1.3	1.7	2.0
.020	.6418	.6316	.6167	.5973	.5832	.7868	.7788	.7670	.7519	.7410
.023	.6403	.6299	.6146	.5948	.5804	.7858	.7775	.7655	.7501	.7390
.026	.6390	.6284	.6128	.5926	.5778	.7848	.7764	.7641	.7484	.7371
.030	.6373	.6265	.6106	.5899	.5748	.7837	.7751	.7625	.7465	.7349
.035	.6355	.6244	.6080	.5868	.5714	.7825	.7736	.7608	.7443	.7325
.040	.6338	.6225	.6057	.5841	.5683	.7814	.7723	.7592	.7424	.7303
.045	.6323	.6207	.6036	.5815	.5654	.7804	.7711	.7577	.7406	.7283
.050	.6308	.6189	.6016	.5791	.5628	.7794	.7700	.7563	.7389	.7264
.060	.6280	.6158	.5979	.5747	.5579	.7776	.7679	.7538	.7358	.7229
.070	.6254	.6128	.5944	.5707	.5534	.7760	.7659	.7514	.7330	.7198
.080	.6230	.6101	.5912	.5669	.5493	.7744	.7641	.7492	.7304	.7169
.090	.6206	.6075	.5882	.5634	.5454	.7729	.7624	.7472	.7279	.7141
.10	.6184	.6050	.5853	.5600	.5417	.7714	.7607	.7452	.7256	.7116
.12	.6143	.6004	.5800	.5539	.5350	.7686	.7575	.7415	.7212	.7068
.14	.6104	.5961	.5751	.5482	.5288	.7660	.7546	.7380	.7172	.7025
.17	.6049	.5900	.5682	.5403	.5202	.7624	.7504	.7332	.7116	.6964
.20	.5997	.5842	.5617	.5330	.5122	.7590	.7466	.7288	.7065	.6908
.23	.5949	.5789	.5558	.5262	.5049	.7564	.7436	.7251	.7020	.6859
.26	.5902	.5738	.5501	.5197	.4979	.7539	.7406	.7216	.6978	.6812
.30	.5843	.5673	.5428	.5115	.4891	.7506	.7368	.7170	.6924	.6751
.35	.5771	.5595	.5341	.5018	.4786	.7466	.7321	.7114	.6858	.6680
.40	.5703	.5521	.5258	.4925	.4687	.7425	.7274	.7060	.6794	.6610
.45	.5636	.5449	.5178	.4836	.4592	.7384	.7228	.7006	.6732	.6543
.50	.5572	.5379	.5101	.4750	.4500	.7342	.7181	.6953	.6671	.6478
.60	.5451	.5248	.4957	.4590	.4330	.7253	.7083	.6843	.6550	.6351
.70	.5336	.5123	.4820	.4438	.4169	.7167	.6988	.6738	.6435	.6231
.80	.5225	.5004	.4689	.4294	.4015	.7084	.6899	.6638	.6325	.6116
.90	.5118	.4889	.4563	.4156	.3869	.7006	.6813	.6543	.6221	.6007
1.0	.5015	.4778	.4442	.4023	.3729	.6931	.6731	.6453	.6121	.5902
1.2	.4819	.4568	.4212	.3771	.3463	.6803	.6590	.6294	.5941	.5709
1.4	.4634	.4369	.3995	.3535	.3215	.6680	.6456	.6143	.5772	.5527
1.7	.4373	.4089	.3691	.3205	.2868	.6503	.6262	.5927	.5531	.5271
2.0	.4128	.3827	.3408	.2897	.2546	.6330	.6075	.5720	.5303	.5031
2.3	.3896	.3579	.3140	.2608	.2243	.6147	.5877	.5505	.5071	.4793
2.6	.3676	.3344	.2886	.2335	.1957	.5970	.5687	.5299	.4851	.4567
3.0	.3399	.3048	.2568	.1994	.1601	.5744	.5444	.5037	.4573	.4285
3.5	.3074	.2702	.2196	.1596	.1187	.5477	.5159	.4731	.4251	.3957
4.0	.2770	.2378	.1849	.1225	.0804	.5226	.4893	.4447	.3952	.3655
4.5	.2483	.2074	.1523	.0878	.0445	.4991	.4644	.4181	.3673	.3374
5.0	.2211	.1786	.1216	.0551	.0108	.4769	.4409	.3932	.3413	.3112
6.0	.1714	.1249	.0626	-.0087	-.0546	.4367	.3983	.3478	.2937	.2632
7.0	.1254	.0758	.0095	-.0655	-.1124	.4000	.3597	.3069	.2512	.2202
8.0	.0824	.0307	-.0385	-.1162	-.1639	.3661	.3242	.2697	.2126	.1815
9.0	.0419	-.0112	-.0823	-.1618	-.2101	.3344	.2914	.2355	.1774	.1461
10	.0034	-.0503	-.1226	-.2032	-.2518	.3047	.2607	.2038	.1449	.1135
12	-.0714	-.1222	-.1912	-.2694	-.3177	.2479	.2032	.1456	.0860	.0545
14	-.1391	-.1867	-.2520	-.3274	-.3752	.1972	.1523	.0943	.0346	.0030
17	-.2292	-.2726	-.3331	-.4047	-.4516	.1305	.0855	.0275	-.0322	-.0636
20	-.3079	-.3486	-.4056	-.4745	-.5206	.0727	.0278	-.0300	-.0892	-.1204
23	-.3741	-.4165	-.4754	-.5454	-.5914	.0229	-.0221	-.0799	-.1388	-.1694
26	-.4335	-.4782	-.5398	-.6114	-.6574	-.0213	-.0664	-.1242	-.1827	-.2127
30	-.5046	-.5529	-.6184	-.6927	-.7386	-.0734	-.1186	-.1762	-.2342	-.2635
35	-.5836	-.6366	-.7071	-.7848	-.8308	-.1299	-.1751	-.2326	-.2898	-.3182
40	-.6543	-.7117	-.7872	-.8682	-.9141	-.1790	-.2242	-.2814	-.3379	-.3655
45	-.7184	-.7800	-.8601	-.9441	-.9901	-.2223	-.2675	-.3244	-.3803	-.4071
50	-.7774	-.8427	-.9271	-1.014	-1.060	-.2610	-.3061	-.3628	-.4180	-.4441
60	-.8867	-.9580	-1.049	-1.140	-1.185	-.3271	-.3719	-.4279	-.4818	-.5068
70	-.9819	-1.058	-1.154	-1.248	-1.294	-.3826	-.4270	-.4824	-.5353	-.5592
80	-1.066	-1.146	-1.247	-1.344	-1.388	-.4302	-.4743	-.5291	-.5811	-.6042
90	-1.141	-1.225	-1.329	-1.428	-1.473	-.4718	-.5156	-.5700	-.6212	-.6436
100	-1.209	-1.295	-1.403	-1.503	-1.548	-.5086	-.5522	-.6061	-.6567	-.6785
120	-1.327	-1.418	-1.529	-1.633	-1.677	-.5714	-.6146	-.6678	-.7173	-.7382
140	-1.427	-1.521	-1.636	-1.741	-1.784	-.6234	-.6663	-.7190	-.7677	-.7879
170	-1.553	-1.649	-1.766	-1.873	-1.916	-.6873	-.7299	-.7821	-.8300	-.8496
200	-1.658	-1.754	-1.872	-1.980	-2.023	-.7393	-.7817	-.8336	-.8811	-.9003

K/KL= E,KEV	IONIZATION SKEWNESS					RANGE KURTOSIS				
	0.8	1.0	1.3	1.7	2.0	0.8	1.0	1.3	1.7	2.0
.020	.8650	.8582	.8484	.8359	.8271	3.359	3.334	3.297	3.251	3.218
.023	.8641	.8572	.8471	.8344	.8254	3.356	3.330	3.292	3.245	3.212
.026	.8633	.8562	.8460	.8331	.8239	3.353	3.326	3.288	3.240	3.206
.030	.8624	.8551	.8447	.8315	.8221	3.349	3.322	3.283	3.234	3.200
.035	.8613	.8539	.8432	.8297	.8201	3.344	3.317	3.277	3.227	3.192
.040	.8604	.8528	.8418	.8280	.8183	3.340	3.312	3.272	3.221	3.185
.045	.8595	.8517	.8406	.8265	.8166	3.337	3.308	3.267	3.216	3.179
.050	.8586	.8508	.8394	.8252	.8150	3.333	3.304	3.262	3.210	3.174
.060	.8571	.8489	.8373	.8226	.8122	3.327	3.297	3.254	3.201	3.163
.070	.8557	.8473	.8353	.8202	.8096	3.321	3.290	3.246	3.192	3.154
.080	.8543	.8457	.8335	.8181	.8073	3.315	3.284	3.239	3.184	3.146
.090	.8531	.8443	.8317	.8161	.8050	3.310	3.278	3.233	3.177	3.138
.10	.8518	.8429	.8301	.8142	.8030	3.305	3.273	3.227	3.170	3.130
.12	.8496	.8404	.8271	.8107	.7992	3.296	3.263	3.216	3.157	3.117
.14	.8476	.8380	.8244	.8075	.7956	3.287	3.253	3.205	3.146	3.105
.17	.8446	.8346	.8205	.8030	.7908	3.275	3.240	3.191	3.130	3.089
.20	.8419	.8315	.8169	.7988	.7863	3.264	3.228	3.178	3.116	3.074
.23	.8393	.8286	.8136	.7950	.7822	3.254	3.217	3.166	3.103	3.060
.26	.8369	.8259	.8104	.7914	.7782	3.244	3.207	3.154	3.091	3.048
.30	.8337	.8223	.8063	.7868	.7733	3.232	3.193	3.140	3.076	3.032
.35	.8300	.8181	.8015	.7813	.7674	3.217	3.178	3.124	3.058	3.014
.40	.8263	.8140	.7969	.7761	.7618	3.203	3.163	3.108	3.042	2.998
.45	.8228	.8101	.7924	.7711	.7565	3.190	3.150	3.094	3.027	2.982
.50	.8193	.8063	.7881	.7663	.7514	3.178	3.137	3.080	3.013	2.968
.60	.8129	.7991	.7800	.7572	.7418	3.154	3.112	3.055	2.987	2.941
.70	.8066	.7922	.7723	.7487	.7328	3.133	3.090	3.031	2.963	2.918
.80	.8006	.7856	.7650	.7406	.7243	3.113	3.069	3.010	2.941	2.896
.90	.7948	.7792	.7579	.7329	.7162	3.094	3.050	2.990	2.921	2.877
1.0	.7892	.7730	.7511	.7254	.7084	3.076	3.032	2.972	2.903	2.859
1.2	.7783	.7612	.7381	.7112	.6936	3.043	2.997	2.937	2.869	2.826
1.4	.7680	.7500	.7258	.6980	.6799	3.013	2.967	2.906	2.839	2.797
1.7	.7535	.7343	.7088	.6798	.6611	2.973	2.927	2.867	2.802	2.762
2.0	.7399	.7197	.6931	.6632	.6441	2.939	2.893	2.835	2.772	2.735
2.3	.7268	.7054	.6778	.6472	.6278	2.909	2.861	2.801	2.740	2.706
2.6	.7144	.6921	.6637	.6324	.6128	2.883	2.833	2.773	2.714	2.683
3.0	.6988	.6756	.6462	.6143	.5945	2.852	2.801	2.742	2.687	2.659
3.5	.6809	.6566	.6264	.5940	.5742	2.818	2.768	2.712	2.663	2.641
4.0	.6643	.6393	.6085	.5758	.5561	2.788	2.742	2.692	2.649	2.632
4.5	.6489	.6234	.5921	.5594	.5398	2.762	2.722	2.678	2.644	2.631
5.0	.6344	.6087	.5771	.5445	.5251	2.738	2.705	2.670	2.644	2.636
6.0	.6075	.5819	.5498	.5179	.4987	2.687	2.678	2.672	2.668	2.666
7.0	.5835	.5584	.5263	.4953	.4766	2.647	2.663	2.685	2.703	2.707
8.0	.5620	.5375	.5058	.4757	.4578	2.619	2.658	2.705	2.745	2.755
9.0	.5425	.5188	.4877	.4584	.4416	2.599	2.659	2.731	2.790	2.807
10	.5246	.5018	.4716	.4431	.4276	2.586	2.666	2.760	2.839	2.862
12	.4909	.4704	.4434	.4157	.4050	2.576	2.689	2.824	2.938	2.975
14	.4625	.4443	.4205	.3937	.3874	2.585	2.726	2.894	3.040	3.091
17	.4281	.4128	.3935	.3682	.3676	2.624	2.798	3.008	3.194	3.264
20	.4013	.3886	.3731	.3495	.3534	2.685	2.883	3.126	3.346	3.434
23	.3837	.3726	.3594	.3392	.3444	2.765	2.973	3.231	3.473	3.579
26	.3702	.3603	.3490	.3322	.3380	2.854	3.069	3.339	3.600	3.722
30	.3564	.3481	.3389	.3262	.3324	2.983	3.204	3.487	3.771	3.914
35	.3441	.3373	.3302	.3222	.3284	3.152	3.381	3.678	3.987	4.153
40	.3354	.3298	.3246	.3206	.3266	3.326	3.563	3.874	4.206	4.393
45	.3292	.3247	.3211	.3205	.3263	3.502	3.747	4.073	4.427	4.633
50	.3247	.3213	.3191	.3213	.3270	3.677	3.933	4.273	4.649	4.873
60	.3195	.3182	.3187	.3241	.3309	3.982	4.266	4.645	5.066	5.317
70	.3172	.3178	.3206	.3282	.3360	4.301	4.614	5.032	5.495	5.771
80	.3168	.3189	.3236	.3329	.3416	4.630	4.973	5.430	5.933	6.231
90	.3176	.3210	.3273	.3380	.3473	4.968	5.340	5.836	6.378	6.697
100	.3191	.3237	.3313	.3430	.3530	5.312	5.713	6.247	6.827	7.165
120	.3234	.3299	.3398	.3531	.3637	6.010	6.470	7.076	7.729	8.102
140	.3286	.3366	.3482	.3626	.3736	6.715	7.231	7.908	8.628	9.032
170	.3371	.3467	.3601	.3759	.3867	7.769	8.365	9.143	9.959	10.41
200	.3455	.3564	.3711	.3877	.3980	8.808	9.481	10.35	11.26	11.75

C ON BE, M2/M1=0.75   TABLE 13- 8

K/KL=	DAMAGE KURTOSIS					IONIZATION KURTOSIS				
E,KEV	0.8	1.0	1.3	1.7	2.0	0.8	1.0	1.3	1.7	2.0
.020	3.626	3.598	3.559	3.510	3.475	3.747	3.720	3.682	3.635	3.601
.023	3.622	3.594	3.554	3.504	3.468	3.743	3.716	3.678	3.629	3.595
.026	3.619	3.590	3.550	3.498	3.462	3.740	3.713	3.673	3.624	3.589
.030	3.615	3.586	3.544	3.492	3.455	3.737	3.709	3.668	3.618	3.583
.035	3.611	3.581	3.538	3.485	3.447	3.733	3.704	3.663	3.611	3.575
.040	3.607	3.577	3.533	3.479	3.441	3.729	3.700	3.658	3.605	3.569
.045	3.604	3.573	3.528	3.473	3.434	3.726	3.696	3.653	3.600	3.562
.050	3.601	3.569	3.524	3.468	3.428	3.722	3.692	3.648	3.595	3.557
.060	3.595	3.562	3.516	3.458	3.418	3.716	3.685	3.640	3.585	3.546
.070	3.589	3.556	3.508	3.449	3.408	3.711	3.679	3.633	3.577	3.537
.080	3.584	3.550	3.501	3.441	3.399	3.706	3.673	3.626	3.569	3.529
.090	3.579	3.544	3.495	3.433	3.391	3.701	3.668	3.620	3.561	3.521
.10	3.574	3.539	3.488	3.426	3.383	3.697	3.662	3.614	3.554	3.513
.12	3.564	3.527	3.476	3.412	3.368	3.688	3.653	3.603	3.542	3.500
.14	3.555	3.517	3.465	3.400	3.355	3.681	3.644	3.593	3.530	3.487
.17	3.542	3.504	3.449	3.383	3.337	3.670	3.632	3.579	3.515	3.470
.20	3.531	3.491	3.435	3.367	3.321	3.660	3.621	3.566	3.500	3.455
.23	3.524	3.483	3.425	3.355	3.307	3.650	3.610	3.554	3.487	3.440
.26	3.517	3.475	3.415	3.343	3.294	3.641	3.600	3.543	3.474	3.427
.30	3.508	3.465	3.403	3.328	3.277	3.630	3.588	3.529	3.458	3.410
.35	3.497	3.452	3.388	3.310	3.258	3.617	3.573	3.512	3.440	3.391
.40	3.486	3.439	3.373	3.293	3.239	3.604	3.559	3.496	3.422	3.372
.45	3.475	3.426	3.358	3.276	3.221	3.591	3.545	3.481	3.406	3.355
.50	3.463	3.413	3.343	3.260	3.204	3.580	3.532	3.467	3.390	3.338
.60	3.435	3.382	3.309	3.224	3.170	3.557	3.507	3.440	3.360	3.307
.70	3.408	3.352	3.277	3.191	3.138	3.536	3.484	3.414	3.333	3.279
.80	3.384	3.325	3.247	3.161	3.109	3.516	3.462	3.390	3.307	3.252
.90	3.361	3.301	3.221	3.133	3.082	3.497	3.441	3.368	3.282	3.227
1.0	3.339	3.278	3.196	3.108	3.056	3.478	3.421	3.346	3.259	3.203
1.2	3.304	3.242	3.159	3.067	3.011	3.443	3.383	3.305	3.215	3.158
1.4	3.272	3.210	3.126	3.030	2.971	3.410	3.348	3.267	3.175	3.117
1.7	3.227	3.165	3.080	2.981	2.917	3.364	3.299	3.215	3.121	3.062
2.0	3.186	3.124	3.037	2.935	2.869	3.321	3.254	3.167	3.073	3.013
2.3	3.143	3.077	2.987	2.885	2.821	3.282	3.211	3.123	3.027	2.968
2.6	3.101	3.033	2.940	2.838	2.777	3.246	3.171	3.082	2.985	2.926
3.0	3.051	2.978	2.883	2.781	2.724	3.200	3.123	3.032	2.935	2.876
3.5	2.993	2.917	2.819	2.719	2.667	3.148	3.068	2.976	2.879	2.821
4.0	2.941	2.862	2.762	2.664	2.617	3.099	3.018	2.926	2.830	2.773
4.5	2.894	2.813	2.712	2.616	2.573	3.053	2.973	2.880	2.786	2.730
5.0	2.852	2.769	2.668	2.574	2.535	3.009	2.932	2.839	2.746	2.691
6.0	2.779	2.697	2.597	2.507	2.472	2.922	2.859	2.767	2.678	2.624
7.0	2.716	2.636	2.539	2.453	2.421	2.845	2.796	2.705	2.620	2.567
8.0	2.661	2.584	2.491	2.409	2.380	2.778	2.740	2.651	2.569	2.519
9.0	2.613	2.539	2.450	2.373	2.345	2.718	2.689	2.604	2.525	2.478
10	2.569	2.499	2.415	2.343	2.317	2.665	2.644	2.562	2.485	2.442
12	2.488	2.429	2.357	2.295	2.274	2.574	2.558	2.490	2.410	2.382
14	2.422	2.372	2.312	2.260	2.243	2.501	2.486	2.431	2.348	2.335
17	2.344	2.307	2.263	2.225	2.213	2.415	2.400	2.361	2.275	2.280
20	2.287	2.261	2.230	2.204	2.196	2.349	2.333	2.307	2.220	2.239
23	2.252	2.234	2.212	2.195	2.190	2.306	2.290	2.268	2.190	2.208
26	2.228	2.216	2.201	2.191	2.190	2.273	2.256	2.237	2.167	2.184
30	2.207	2.201	2.195	2.193	2.197	2.239	2.223	2.204	2.146	2.160
35	2.193	2.194	2.196	2.203	2.210	2.208	2.192	2.173	2.129	2.137
40	2.188	2.194	2.203	2.217	2.229	2.186	2.169	2.150	2.117	2.120
45	2.189	2.199	2.214	2.234	2.249	2.168	2.151	2.132	2.108	2.107
50	2.194	2.208	2.228	2.253	2.271	2.154	2.137	2.118	2.102	2.098
60	2.214	2.234	2.262	2.296	2.317	2.130	2.115	2.098	2.088	2.086
70	2.239	2.265	2.299	2.339	2.364	2.113	2.100	2.085	2.079	2.080
80	2.267	2.297	2.337	2.383	2.411	2.101	2.089	2.077	2.073	2.077
90	2.296	2.330	2.375	2.425	2.455	2.092	2.081	2.071	2.070	2.076
100	2.325	2.363	2.412	2.467	2.499	2.085	2.076	2.068	2.068	2.076
120	2.383	2.426	2.483	2.545	2.580	2.076	2.070	2.066	2.069	2.080
140	2.438	2.487	2.551	2.618	2.656	2.070	2.067	2.067	2.073	2.086
170	2.514	2.571	2.644	2.719	2.760	2.067	2.068	2.072	2.082	2.096
200	2.584	2.648	2.729	2.811	2.853	2.066	2.071	2.079	2.094	2.107

C ON BE, M2/M1=0.75   TABLE 13- 9

| K/KL= | REFLECTION COEFFICIENT | | | | | SPUTTERING EFFICIENCY | | | | |
E,KEV	0.8	1.0	1.3	1.7	2.0	0.8	1.0	1.3	1.7	2.0
.020	.0089	.0087	.0084	.0081	.0079	.0209	.0201	.0190	.0177	.0167
.023	.0088	.0087	.0084	.0081	.0078	.0208	.0200	.0189	.0175	.0166
.026	.0088	.0086	.0084	.0080	.0078	.0207	.0199	.0188	.0174	.0164
.030	.0088	.0086	.0083	.0080	.0077	.0206	.0198	.0187	.0173	.0163
.035	.0087	.0086	.0083	.0079	.0077	.0205	.0197	.0185	.0171	.0161
.040	.0087	.0085	.0082	.0079	.0076	.0205	.0196	.0184	.0170	.0160
.045	.0087	.0085	.0082	.0078	.0076	.0204	.0195	.0183	.0168	.0158
.050	.0086	.0085	.0082	.0078	.0075	.0203	.0194	.0182	.0167	.0157
.060	.0086	.0084	.0081	.0077	.0075	.0202	.0193	.0180	.0165	.0155
.070	.0085	.0083	.0080	.0076	.0074	.0201	.0191	.0179	.0163	.0153
.080	.0085	.0083	.0080	.0076	.0073	.0200	.0190	.0177	.0161	.0151
.090	.0084	.0082	.0079	.0075	.0072	.0199	.0189	.0176	.0160	.0149
.10	.0084	.0082	.0079	.0075	.0072	.0198	.0188	.0175	.0159	.0148
.12	.0083	.0081	.0077	.0073	.0071	.0196	.0186	.0172	.0156	.0145
.14	.0082	.0080	.0076	.0072	.0069	.0194	.0184	.0170	.0154	.0143
.17	.0081	.0078	.0075	.0071	.0068	.0192	.0182	.0168	.0151	.0140
.20	.0080	.0077	.0074	.0070	.0067	.0190	.0180	.0166	.0149	.0138
.23	.0079	.0076	.0073	.0068	.0065	.0189	.0178	.0164	.0147	.0136
.26	.0078	.0075	.0072	.0067	.0064	.0188	.0177	.0162	.0145	.0134
.30	.0077	.0074	.0070	.0066	.0063	.0186	.0175	.0160	.0143	.0131
.35	.0075	.0072	.0069	.0064	.0061	.0184	.0173	.0158	.0140	.0129
.40	.0074	.0071	.0067	.0063	.0060	.0182	.0171	.0155	.0138	.0126
.45	.0073	.0070	.0066	.0061	.0058	.0181	.0169	.0153	.0136	.0124
.50	.0071	.0069	.0065	.0060	.0057	.0179	.0167	.0151	.0134	.0122
.60	.0069	.0066	.0062	.0057	.0054	.0176	.0164	.0148	.0130	.0118
.70	.0067	.0064	.0060	.0055	.0052	.0173	.0161	.0144	.0126	.0115
.80	.0065	.0062	.0058	.0053	.0050	.0170	.0158	.0141	.0123	.0112
.90	.0063	.0060	.0056	.0051	.0048	.0168	.0155	.0138	.0120	.0109
1.0	.0061	.0058	.0054	.0049	.0046	.0165	.0153	.0136	.0118	.0107
1.2	.0058	.0055	.0051	.0046	.0043	.0161	.0149	.0132	.0114	.0103
1.4	.0055	.0052	.0048	.0043	.0040	.0158	.0145	.0128	.0110	.0099
1.7	.0052	.0049	.0044	.0039	.0036	.0153	.0141	.0124	.0105	.0094
2.0	.0048	.0045	.0041	.0036	.0033	.0149	.0137	.0120	.0101	.0090
2.3	.0046	.0042	.0038	.0033	.0030	.0146	.0133	.0115	.0097	.0086
2.6	.0043	.0040	.0035	.0031	.0028	.0142	.0129	.0112	.0093	.0083
3.0	.0040	.0037	.0032	.0028	.0025	.0138	.0124	.0107	.0089	.0079
3.5	.0037	.0033	.0029	.0024	.0022	.0133	.0119	.0102	.0083	.0074
4.0	.0034	.0031	.0026	.0022	.0019	.0129	.0115	.0097	.0079	.0070
4.5	.0031	.0028	.0024	.0020	.0017	.0125	.0111	.0093	.0075	.0067
5.0	.0029	.0026	.0022	.0018	.0016	.0121	.0107	.0089	.0072	.0063
6.0	.0024	.0022	.0019	.0016	.0014	.0115	.0100	.0083	.0066	.0058
7.0	.0020	.0019	.0017	.0014	.0012	.0109	.0095	.0077	.0061	.0053
8.0	.0017	.0017	.0016	.0013	.0011	.0104	.0090	.0073	.0057	.0049
9.0	.0014	.0015	.0014	.0012	.0010	.0099	.0085	.0069	.0053	.0046
10	.0012	.0013	.0013	.0011	.0009	.0095	.0081	.0065	.0050	.0043
12	.0009	.0011	.0011	.0009	.0007	.0086	.0074	.0059	.0045	.0039
14	.0007	.0008	.0009	.0007	.0006	.0079	.0067	.0054	.0041	.0035
17	.0005	.0006	.0006	.0005	.0004	.0070	.0059	.0047	.0036	.0031
20	.0004	.0004	.0005	.0004	.0003	.0063	.0053	.0042	.0032	.0028
23	.0003	.0003	.0004	.0003	.0002	.0058	.0048	.0038	.0029	.0025
26	.0002	.0002	.0003	.0002	.0002	.0054	.0045	.0035	.0026	.0023
30	.0001	.0002	.0002	.0002	.0001	.0049	.0041	.0031	.0023	.0020
35	.0001	.0001	.0001	.0001	.0001	.0044	.0036	.0028	.0021	.0018
40	.0001	.0001	.0001	.0000	.0000	.0040	.0033	.0025	.0018	.0016
45	.0000	.0000	.0000	.0000	.0000	.0037	.0030	.0022	.0016	.0014
50	.0000	.0000	.0000	.0000	.0000	.0034	.0028	.0020	.0015	.0013
60	.0000	0.0000	0.0000	.0000	0.0000	.0029	.0024	.0017	.0012	.0010
70	.0000	0.0000	0.0000	0.0000	0.0000	.0026	.0021	.0015	.0010	.0009
80	0.0000	0.0000	0.0000	0.0000	0.0000	.0023	.0018	.0013	.0009	.0008
90	0.0000	0.0000	0.0000	0.0000	0.0000	.0020	.0016	.0011	.0008	.0007
100	0.0000	0.0000	0.0000	0.0000	0.0000	.0018	.0014	.0010	.0007	.0006
120	0.0000	0.0000	0.0000	0.0000	0.0000	.0015	.0011	.0008	.0005	.0005
140	0.0000	0.0000	0.0000	0.0000	0.0000	.0013	.0009	.0006	.0004	.0004
170	0.0000	0.0000	0.0000	0.0000	0.0000	.0010	.0008	.0005	.0003	.0003
200	0.0000	0.0000	0.0000	0.0000	0.0000	.0008	.0006	.0004	.0003	.0002

TABLE 13-10

C ON BE, M2/M1=0.75

E,KEV	IONIZATION DEFICIENCY K/KL= 0.8	1.0	1.3	1.7	2.0	SPUTTERING YIELD ALPHA 0.8	1.0	1.3	1.7	2.0
.020	.0019	.0023	.0029	.0035	.0039	.1447	.1410	.1358	.1292	.1245
.023	.0019	.0023	.0029	.0036	.0040	.1444	.1407	.1354	.1286	.1239
.026	.0020	.0024	.0030	.0036	.0041	.1442	.1404	.1350	.1281	.1233
.030	.0020	.0024	.0030	.0037	.0041	.1439	.1401	.1345	.1276	.1227
.035	.0021	.0025	.0031	.0037	.0042	.1437	.1397	.1340	.1269	.1220
.040	.0021	.0025	.0031	.0038	.0042	.1434	.1394	.1336	.1264	.1213
.045	.0021	.0026	.0032	.0038	.0043	.1432	.1391	.1332	.1259	.1208
.050	.0022	.0026	.0032	.0039	.0043	.1431	.1389	.1329	.1255	.1203
.060	.0022	.0027	.0033	.0040	.0044	.1428	.1385	.1324	.1247	.1194
.070	.0023	.0027	.0033	.0040	.0045	.1426	.1382	.1319	.1241	.1187
.080	.0023	.0028	.0034	.0041	.0045	.1424	.1379	.1315	.1236	.1181
.090	.0023	.0028	.0034	.0041	.0046	.1423	.1377	.1312	.1232	.1176
.10	.0024	.0028	.0035	.0042	.0046	.1422	.1375	.1309	.1228	.1172
.12	.0024	.0029	.0035	.0042	.0047	.1420	.1372	.1304	.1221	.1165
.14	.0025	.0029	.0036	.0043	.0048	.1419	.1370	.1301	.1216	.1159
.17	.0025	.0030	.0037	.0044	.0048	.1418	.1368	.1297	.1210	.1152
.20	.0026	.0031	.0037	.0044	.0049	.1418	.1367	.1294	.1205	.1146
.23	.0026	.0031	.0038	.0045	.0049	.1419	.1366	.1292	.1202	.1142
.26	.0026	.0031	.0038	.0045	.0049	.1420	.1366	.1290	.1199	.1138
.30	.0027	.0032	.0038	.0046	.0050	.1422	.1367	.1289	.1196	.1134
.35	.0027	.0032	.0039	.0046	.0050	.1425	.1368	.1289	.1193	.1129
.40	.0028	.0033	.0039	.0046	.0050	.1428	.1370	.1289	.1191	.1126
.45	.0028	.0033	.0040	.0047	.0051	.1431	.1372	.1289	.1189	.1123
.50	.0028	.0033	.0040	.0047	.0051	.1434	.1374	.1289	.1188	.1121
.60	.0029	.0034	.0040	.0047	.0051	.1440	.1377	.1290	.1186	.1118
.70	.0029	.0034	.0040	.0047	.0051	.1446	.1381	.1291	.1185	.1117
.80	.0029	.0034	.0041	.0047	.0051	.1452	.1385	.1293	.1185	.1115
.90	.0030	.0034	.0041	.0047	.0051	.1458	.1390	.1295	.1185	.1115
1.0	.0030	.0035	.0041	.0047	.0050	.1463	.1394	.1297	.1185	.1115
1.2	.0030	.0035	.0041	.0047	.0050	.1475	.1404	.1304	.1189	.1116
1.4	.0030	.0035	.0041	.0047	.0049	.1486	.1413	.1311	.1192	.1117
1.7	.0030	.0035	.0041	.0046	.0049	.1501	.1426	.1321	.1198	.1120
2.0	.0030	.0035	.0040	.0045	.0048	.1515	.1438	.1329	.1203	.1123
2.3	.0030	.0035	.0040	.0045	.0047	.1528	.1447	.1335	.1205	.1125
2.6	.0030	.0034	.0039	.0044	.0046	.1540	.1456	.1341	.1208	.1127
3.0	.0030	.0034	.0039	.0043	.0044	.1554	.1467	.1347	.1211	.1130
3.5	.0030	.0033	.0038	.0042	.0043	.1570	.1478	.1354	.1214	.1134
4.0	.0030	.0033	.0037	.0041	.0041	.1583	.1489	.1361	.1218	.1139
4.5	.0029	.0032	.0036	.0040	.0040	.1595	.1498	.1367	.1222	.1143
5.0	.0029	.0032	.0036	.0039	.0039	.1606	.1507	.1372	.1225	.1147
6.0	.0027	.0031	.0034	.0037	.0036	.1623	.1521	.1382	.1233	.1155
7.0	.0026	.0030	.0033	.0035	.0034	.1638	.1533	.1392	.1241	.1164
8.0	.0025	.0029	.0031	.0033	.0032	.1650	.1545	.1401	.1248	.1171
9.0	.0023	.0027	.0030	.0031	.0030	.1661	.1555	.1411	.1255	.1179
10	.0022	.0026	.0029	.0030	.0028	.1671	.1566	.1420	.1263	.1186
12	.0020	.0024	.0027	.0026	.0025	.1689	.1588	.1442	.1279	.1202
14	.0019	.0022	.0025	.0023	.0022	.1705	.1607	.1462	.1294	.1215
17	.0017	.0019	.0023	.0019	.0019	.1724	.1631	.1486	.1312	.1232
20	.0015	.0017	.0021	.0016	.0016	.1739	.1649	.1504	.1324	.1244
23	.0014	.0015	.0019	.0013	.0013	.1752	.1659	.1510	.1328	.1249
26	.0013	.0013	.0018	.0011	.0011	.1762	.1667	.1514	.1330	.1252
30	.0011	.0011	.0016	.0010	.0008	.1774	.1673	.1515	.1330	.1254
35	.0010	.0010	.0014	.0009	.0005	.1785	.1679	.1514	.1328	.1254
40	.0008	.0008	.0012	.0010	.0003	.1793	.1682	.1511	.1325	.1254
45	.0007	.0007	.0011	.0011	.0001	.1799	.1683	.1508	.1321	.1253
50	.0006	.0006	.0010	.0013	0.0000	.1803	.1684	.1505	.1317	.1251
60	.0004	.0004	.0007	.0026	0.0000	.1807	.1684	.1500	.1311	.1249
70	.0003	.0003	.0004	.0036	0.0000	.1808	.1682	.1496	.1305	.1246
80	.0002	.0001	.0003	.0045	0.0000	.1807	.1679	.1490	.1299	.1243
90	.0001	.0001	.0001	.0052	0.0000	.1804	.1676	.1485	.1293	.1239
100	0.0000	0.0000	0.0000	.0057	0.0000	.1801	.1672	.1480	.1287	.1235
120	0.0000	0.0000	0.0000	.0062	0.0000	.1793	.1664	.1469	.1276	.1227
140	0.0000	0.0000	0.0000	.0062	0.0000	.1784	.1655	.1459	.1265	.1219
170	0.0000	0.0000	0.0000	.0055	0.0000	.1770	.1642	.1443	.1250	.1206
200	0.0000	0.0000	0.0000	.0042	0.0000	.1755	.1630	.1429	.1237	.1193

C ON BE, M2/M1=0.75

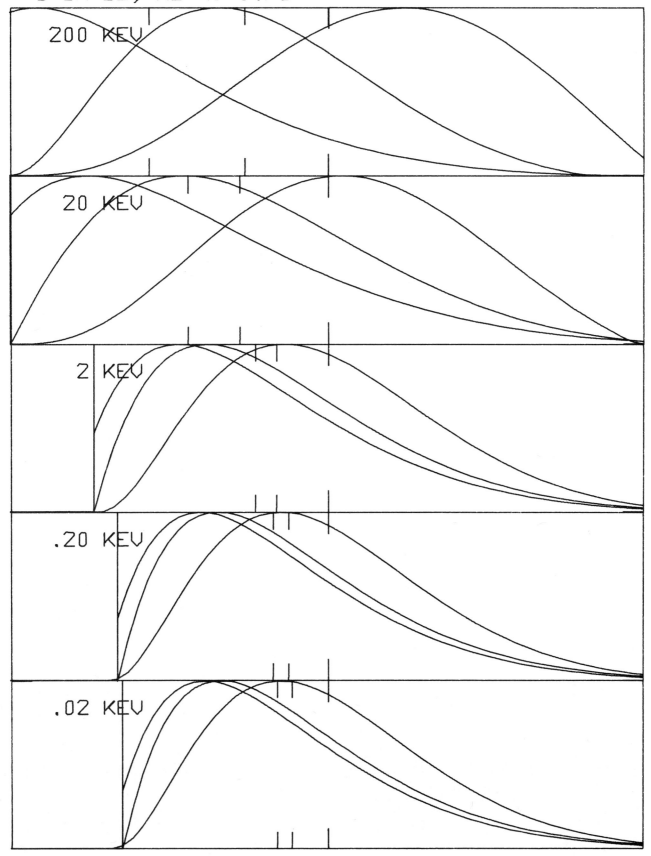

200 KEV

20 KEV

2 KEV

.20 KEV

.02 KEV

AU ON AG, M2/M1=0.55   TABLE 14- 1
      FRACTIONAL DEPOSITED ENERGY   MEAN RANGE,MICROGRAM/SQ.CM.
K/KL= 0.5   0.65   0.8   1.0   1.2   0.5   0.65   0.8   1.0   1.2
E,KEV

E,KEV	0.5	0.65	0.8	1.0	1.2	0.5	0.65	0.8	1.0	1.2
10	.8886	.8584	.8295	.7930	.7585	3.078	3.061	3.044	3.021	2.999
12	.8853	.8544	.8248	.7874	.7522	3.503	3.482	3.462	3.435	3.409
14	.8825	.8509	.8207	.7826	.7468	3.886	3.863	3.840	3.809	3.779
17	.8789	.8464	.8154	.7764	.7398	4.417	4.389	4.362	4.326	4.291
20	.8758	.8425	.8109	.7711	.7339	4.915	4.883	4.853	4.812	4.772
23	.8730	.8391	.8069	.7665	.7286	5.392	5.357	5.322	5.277	5.233
26	.8705	.8361	.8033	.7623	.7240	5.855	5.816	5.777	5.727	5.678
30	.8676	.8324	.7991	.7573	.7184	6.453	6.409	6.365	6.308	6.253
35	.8643	.8284	.7944	.7519	.7123	7.176	7.126	7.076	7.011	6.947
40	.8614	.8248	.7902	.7470	.7069	7.877	7.820	7.764	7.690	7.618
45	.8588	.8216	.7865	.7427	.7021	8.558	8.494	8.431	8.349	8.269
50	.8563	.8186	.7830	.7387	.6976	9.220	9.149	9.080	8.990	8.901
60	.8520	.8133	.7769	.7317	.6898	10.45	10.36	10.28	10.18	10.07
70	.8483	.8087	.7716	.7256	.6830	11.64	11.54	11.45	11.33	11.21
80	.8449	.8047	.7669	.7201	.6771	12.80	12.70	12.59	12.45	12.32
90	.8419	.8010	.7627	.7153	.6717	13.95	13.82	13.71	13.55	13.40
100	.8391	.7976	.7588	.7108	.6668	15.06	14.93	14.80	14.63	14.46
120	.8344	.7918	.7521	.7031	.6583	17.13	16.97	16.82	16.62	16.42
140	.8302	.7867	.7462	.6964	.6509	19.17	18.99	18.81	18.57	18.34
170	.8247	.7800	.7385	.6876	.6413	22.20	21.97	21.75	21.46	21.18
200	.8198	.7741	.7317	.6799	.6328	25.20	24.92	24.66	24.31	23.98
230	.8153	.7687	.7256	.6729	.6253	27.86	27.55	27.25	26.87	26.49
260	.8111	.7637	.7199	.6666	.6184	30.54	30.19	29.86	29.42	29.00
300	.8061	.7576	.7130	.6589	.6100	34.15	33.75	33.35	32.85	32.35
350	.8002	.7507	.7052	.6502	.6007	38.74	38.25	37.78	37.17	36.57
400	.7949	.7444	.6982	.6423	.5923	43.39	42.81	42.24	41.52	40.81
450	.7900	.7386	.6916	.6351	.5846	48.09	47.40	46.73	45.88	45.06
500	.7854	.7332	.6856	.6284	.5775	52.80	52.00	51.22	50.24	49.29
600	.7770	.7234	.6747	.6164	.5648	61.33	60.36	59.42	58.23	57.08
700	.7694	.7146	.6649	.6057	.5535	70.25	69.06	67.92	66.46	65.06
800	.7624	.7064	.6559	.5960	.5432	79.50	78.05	76.65	74.88	73.21
900	.7559	.6989	.6476	.5870	.5339	89.00	87.25	85.57	83.46	81.46
1000	.7497	.6918	.6399	.5787	.5253	98.69	96.60	94.62	92.13	89.79
1200	.7384	.6787	.6257	.5635	.5096	116.4	113.8	111.4	108.3	105.4
1400	.7280	.6669	.6129	.5500	.4956	135.3	132.0	129.0	125.1	121.5
1700	.7137	.6508	.5956	.5319	.4772	165.5	160.9	156.7	151.4	146.5
2000	.7007	.6364	.5803	.5158	.4610	197.5	191.3	185.5	178.5	172.1
2300	.6884	.6232	.5662	.5011	.4463	226.2	219.0	212.2	203.9	196.2
2600	.6770	.6110	.5534	.4876	.4329	256.5	248.0	239.9	230.0	220.9
3000	.6628	.5961	.5377	.4714	.4169	299.4	288.4	278.2	265.7	254.4
3500	.6464	.5790	.5200	.4532	.3991	356.2	341.4	327.9	311.6	297.1
4000	.6312	.5635	.5040	.4369	.3833	415.8	396.5	379.1	358.5	340.3
4500	.6172	.5491	.4893	.4222	.3691	477.5	452.9	431.3	405.9	383.8
5000	.6040	.5358	.4758	.4087	.3562	540.6	510.4	484.0	453.5	427.3
6000	.5794	.5109	.4509	.3844	.3333	661.7	620.3	585.6	546.1	511.8
7000	.5573	.4889	.4292	.3635	.3137	788.5	734.2	689.4	639.4	596.2
8000	.5374	.4692	.4100	.3452	.2966	919.3	850.8	794.7	732.6	680.0
9000	.5193	.4514	.3927	.3290	.2817	1053.	969.2	900.6	825.5	763.0
10000	.5026	.4352	.3772	.3144	.2684	1188.	1088.	1007.	917.6	844.9
12000	.4723	.4064	.3499	.2894	.2457	1455.	1324.	1216.	1097.	1004.
14000	.4460	.3817	.3269	.2685	.2270	1724.	1561.	1423.	1273.	1159.
17000	.4124	.3505	.2981	.2427	.2041	2129.	1913.	1728.	1530.	1383.
20000	.3841	.3246	.2744	.2218	.1857	2531.	2259.	2026.	1777.	1598.
23000	.3598	.3028	.2548	.2048	.1708	2925.	2595.	2314.	2015.	1803.
26000	.3387	.2839	.2380	.1904	.1583	3313.	2924.	2593.	2244.	2000.
30000	.3143	.2625	.2190	.1742	.1444	3820.	3351.	2955.	2539.	2253.
35000	.2887	.2400	.1994	.1577	.1303	4437.	3868.	3390.	2892.	2554.
40000	.2671	.2213	.1831	.1441	.1188	5034.	4367.	3807.	3228.	2841.
45000	.2487	.2054	.1694	.1328	.1092	5614.	4849.	4209.	3550.	3114.
50000	.2327	.1917	.1577	.1232	.1011	6177.	5315.	4596.	3859.	3377.
60000	.2065	.1694	.1386	.1077	.0881	7253.	6204.	5332.	4444.	3871.
70000	.1859	.1519	.1240	.0959	.0783	8273.	7042.	6023.	4990.	4332.
80000	.1693	.1381	.1123	.0866	.0706	9241.	7836.	6675.	5504.	4765.
90000	.1559	.1269	.1030	.0792	.0644	10164	8592.	7294.	5990.	5173.
100000	.1448	.1177	.0954	.0732	.0595	11047	9313.	7883.	6451.	5560.

MEAN DAMAGE DEPTH,MICROGRAM/SQ.CM. MEAN IONZN.DEPTH,MICROGRAM/SQ.CM.

K/KL= E,KEV	0.5	0.65	0.8	1.0	1.2	0.5	0.65	0.8	1.0	1.2
10	2.412	2.390	2.368	2.340	2.312	2.222	2.198	2.175	2.145	2.115
12	2.744	2.717	2.691	2.658	2.626	2.526	2.498	2.471	2.435	2.400
14	3.043	3.013	2.984	2.946	2.909	2.802	2.769	2.738	2.697	2.658
17	3.457	3.422	3.388	3.344	3.301	3.182	3.145	3.108	3.061	3.015
20	3.846	3.806	3.767	3.717	3.668	3.540	3.497	3.455	3.402	3.349
23	4.219	4.174	4.130	4.074	4.020	3.882	3.834	3.788	3.727	3.669
26	4.579	4.530	4.482	4.419	4.359	4.213	4.160	4.109	4.042	3.978
30	5.046	4.990	4.935	4.865	4.797	4.641	4.581	4.523	4.448	4.376
35	5.610	5.546	5.483	5.403	5.326	5.158	5.090	5.023	4.937	4.855
40	6.156	6.083	6.013	5.923	5.836	5.658	5.581	5.506	5.410	5.317
45	6.686	6.605	6.527	6.426	6.330	6.143	6.057	5.974	5.867	5.764
50	7.201	7.112	7.026	6.916	6.809	6.615	6.520	6.429	6.311	6.197
60	8.156	8.052	7.952	7.823	7.699	7.489	7.379	7.272	7.134	7.002
70	9.084	8.965	8.850	8.702	8.560	8.338	8.211	8.088	7.930	7.779
80	9.990	9.855	9.725	9.557	9.397	9.166	9.022	8.883	8.704	8.534
90	10.88	10.73	10.58	10.39	10.21	9.974	9.813	9.658	9.459	9.269
100	11.75	11.58	11.42	11.21	11.01	10.77	10.59	10.42	10.20	9.986
120	13.35	13.15	12.96	12.72	12.49	12.23	12.02	11.82	11.56	11.32
140	14.93	14.71	14.49	14.21	13.94	13.67	13.43	13.20	12.90	12.62
170	17.28	17.00	16.73	16.39	16.07	15.80	15.51	15.22	14.86	14.52
200	19.60	19.26	18.94	18.54	18.16	17.90	17.55	17.21	16.78	16.38
230	21.66	21.29	20.93	20.48	20.04	19.78	19.38	19.00	18.52	18.07
260	23.74	23.32	22.92	22.40	21.92	21.66	21.22	20.79	20.25	19.74
300	26.53	26.04	25.57	24.98	24.42	24.19	23.67	23.17	22.55	21.96
350	30.06	29.48	28.92	28.21	27.55	27.36	26.74	26.15	25.41	24.71
400	33.63	32.93	32.28	31.45	30.68	30.55	29.82	29.13	28.26	27.45
450	37.21	36.40	35.64	34.69	33.79	33.75	32.90	32.10	31.10	30.17
500	40.80	39.87	39.00	37.91	36.90	36.94	35.97	35.06	33.92	32.86
600	47.34	46.21	45.15	43.84	42.61	42.80	41.62	40.52	39.14	37.86
700	54.12	52.76	51.48	49.89	48.43	48.82	47.40	46.07	44.41	42.89
800	61.11	59.47	57.95	56.06	54.33	54.97	53.27	51.69	49.73	47.94
900	68.25	66.31	64.51	62.30	60.28	61.22	59.21	57.35	55.07	52.98
1000	75.51	73.24	71.15	68.59	66.26	67.53	65.19	63.03	60.40	58.01
1200	88.85	86.04	83.47	80.34	77.47	79.28	76.39	73.73	70.48	67.55
1400	102.9	99.45	96.29	92.46	88.96	91.47	87.91	84.65	80.70	77.13
1700	125.3	120.6	116.3	111.2	106.6	110.5	105.7	101.4	96.22	91.60
2000	148.7	142.6	137.1	130.5	124.7	130.2	124.0	118.4	111.9	106.1
2300	169.9	162.7	156.2	148.4	141.6	148.4	141.1	134.5	126.8	119.9
2600	192.0	183.6	175.9	166.8	158.8	167.2	158.7	150.8	141.7	133.8
3000	223.1	212.6	203.2	191.9	182.1	193.0	182.5	172.9	161.8	152.2
3500	264.0	250.5	238.4	224.0	211.8	226.3	212.8	200.7	186.9	175.0
4000	306.7	289.8	274.5	256.7	241.8	260.3	243.5	228.6	211.9	197.6
4500	350.7	329.9	311.4	289.8	272.0	294.9	274.3	256.6	236.7	220.0
5000	395.5	370.7	348.5	323.0	302.2	329.7	305.1	284.5	261.3	242.0
6000	481.0	448.7	419.9	387.0	360.3	397.1	363.9	338.0	309.0	284.7
7000	570.4	529.3	492.8	451.5	418.6	465.4	423.2	391.3	355.8	326.2
8000	662.5	611.6	566.8	516.4	476.8	534.1	482.6	444.1	401.5	366.5
9000	756.5	695.0	641.2	581.1	534.6	602.8	541.9	496.3	446.2	405.7
10000	851.7	779.1	715.7	645.6	591.8	671.2	600.8	547.9	489.9	443.9
12000	1038.	943.4	861.1	770.8	702.8	805.2	717.7	648.9	574.2	517.0
14000	1227.	1109.	1006.	894.4	811.7	937.1	832.0	746.7	654.8	586.6
17000	1513.	1356.	1221.	1076.	970.8	1130.	998.0	887.4	769.7	685.2
20000	1799.	1601.	1433.	1253.	1125.	1318.	1157.	1022.	878.1	777.7
23000	2077.	1839.	1637.	1423.	1272.	1499.	1309.	1149.	980.5	864.7
26000	2352.	2073.	1837.	1588.	1415.	1674.	1456.	1271.	1078.	947.3
30000	2714.	2380.	2098.	1802.	1600.	1900.	1643.	1426.	1202.	1051.
35000	3159.	2755.	2415.	2061.	1822.	2170.	1865.	1609.	1347.	1174.
40000	3593.	3119.	2721.	2310.	2035.	2427.	2076.	1783.	1483.	1288.
45000	4018.	3474.	3019.	2551.	2241.	2672.	2277.	1947.	1612.	1396.
50000	4432.	3819.	3308.	2784.	2439.	2908.	2469.	2104.	1735.	1499.
60000	5232.	4484.	3862.	3228.	2817.	3352.	2830.	2399.	1964.	1690.
70000	5995.	5116.	4386.	3647.	3173.	3766.	3166.	2671.	2175.	1865.
80000	6725.	5719.	4886.	4044.	3509.	4154.	3480.	2925.	2371.	2028.
90000	7426.	6296.	5362.	4422.	3829.	4521.	3775.	3163.	2554.	2181.
100000	8099.	6850.	5819.	4783.	4134.	4868.	4055.	3389.	2727.	2324.

	RELATIVE RANGE STRAGGLING					RELATIVE DAMAGE STRAGGLING				
K/KL=	0.5	0.65	0.8	1.0	1.2	0.5	0.65	0.8	1.0	1.2
E,KEV										
10	.2130	.2112	.2095	.2072	.2049	.3784	.3766	.3750	.3728	.3707
12	.2127	.2109	.2091	.2067	.2044	.3781	.3763	.3746	.3724	.3703
14	.2124	.2105	.2087	.2063	.2039	.3779	.3761	.3743	.3720	.3699
17	.2121	.2101	.2082	.2057	.2033	.3776	.3757	.3739	.3716	.3693
20	.2117	.2097	.2078	.2052	.2027	.3774	.3754	.3735	.3711	.3688
23	.2114	.2094	.2073	.2047	.2022	.3771	.3751	.3732	.3708	.3684
26	.2111	.2090	.2070	.2043	.2017	.3769	.3749	.3729	.3704	.3680
30	.2108	.2086	.2065	.2037	.2011	.3766	.3745	.3725	.3700	.3675
35	.2103	.2081	.2059	.2031	.2004	.3763	.3742	.3721	.3694	.3669
40	.2099	.2076	.2054	.2025	.1997	.3760	.3738	.3717	.3690	.3664
45	.2095	.2072	.2049	.2020	.1991	.3757	.3735	.3713	.3685	.3659
50	.2091	.2068	.2044	.2014	.1985	.3754	.3731	.3709	.3681	.3655
60	.2084	.2060	.2036	.2005	.1975	.3750	.3726	.3703	.3674	.3646
70	.2078	.2052	.2027	.1995	.1965	.3745	.3720	.3696	.3666	.3638
80	.2071	.2045	.2020	.1987	.1955	.3740	.3715	.3691	.3660	.3631
90	.2065	.2038	.2012	.1979	.1946	.3736	.3710	.3685	.3653	.3624
100	.2059	.2032	.2005	.1971	.1938	.3732	.3705	.3679	.3647	.3617
120	.2048	.2019	.1992	.1956	.1922	.3724	.3696	.3669	.3636	.3604
140	.2037	.2008	.1979	.1942	.1907	.3717	.3688	.3660	.3625	.3593
170	.2022	.1991	.1961	.1923	.1886	.3706	.3675	.3646	.3610	.3576
200	.2008	.1975	.1944	.1905	.1867	.3696	.3664	.3633	.3595	.3560
230	.1994	.1961	.1928	.1887	.1848	.3687	.3653	.3622	.3582	.3546
260	.1981	.1946	.1913	.1871	.1831	.3679	.3643	.3610	.3570	.3532
300	.1964	.1928	.1894	.1850	.1809	.3667	.3630	.3596	.3554	.3515
350	.1944	.1907	.1871	.1825	.1782	.3653	.3615	.3578	.3534	.3494
400	.1925	.1886	.1849	.1802	.1758	.3640	.3599	.3562	.3516	.3474
450	.1907	.1867	.1828	.1780	.1734	.3627	.3584	.3545	.3498	.3455
500	.1890	.1848	.1808	.1758	.1712	.3614	.3569	.3529	.3480	.3436
600	.1856	.1812	.1770	.1718	.1669	.3588	.3541	.3498	.3446	.3400
700	.1825	.1779	.1735	.1680	.1629	.3563	.3513	.3468	.3414	.3366
800	.1796	.1747	.1701	.1645	.1593	.3540	.3487	.3440	.3384	.3334
900	.1768	.1717	.1670	.1612	.1558	.3517	.3462	.3413	.3355	.3304
1000	.1742	.1689	.1640	.1581	.1526	.3495	.3438	.3387	.3327	.3275
1200	.1692	.1636	.1584	.1521	.1465	.3453	.3392	.3338	.3276	.3222
1400	.1646	.1587	.1533	.1468	.1409	.3414	.3349	.3293	.3229	.3173
1700	.1584	.1521	.1464	.1396	.1336	.3359	.3291	.3231	.3164	.3107
2000	.1527	.1462	.1403	.1333	.1271	.3309	.3237	.3175	.3106	.3047
2300	.1475	.1406	.1346	.1274	.1211	.3258	.3184	.3122	.3051	.2991
2600	.1427	.1356	.1294	.1221	.1158	.3211	.3136	.3072	.3000	.2940
3000	.1368	.1295	.1231	.1157	.1094	.3153	.3077	.3012	.2938	.2878
3500	.1303	.1227	.1162	.1088	.1024	.3087	.3010	.2944	.2869	.2810
4000	.1244	.1167	.1101	.1026	.0963	.3028	.2950	.2883	.2808	.2748
4500	.1192	.1113	.1047	.0972	.0909	.2974	.2896	.2828	.2753	.2694
5000	.1144	.1064	.0998	.0923	.0861	.2924	.2847	.2778	.2703	.2644
6000	.1058	.0978	.0912	.0839	.0779	.2837	.2759	.2691	.2616	.2558
7000	.0985	.0905	.0840	.0769	.0711	.2762	.2684	.2616	.2541	.2484
8000	.0922	.0843	.0778	.0710	.0654	.2696	.2618	.2551	.2476	.2420
9000	.0867	.0788	.0725	.0659	.0605	.2637	.2560	.2493	.2419	.2364
10000	.0818	.0740	.0679	.0614	.0563	.2584	.2507	.2441	.2368	.2314
12000	.0735	.0661	.0603	.0542	.0495	.2491	.2416	.2351	.2280	.2228
14000	.0667	.0597	.0542	.0485	.0441	.2413	.2339	.2276	.2207	.2156
17000	.0585	.0520	.0470	.0417	.0379	.2315	.2244	.2182	.2116	.2067
20000	.0520	.0460	.0414	.0365	.0331	.2234	.2165	.2105	.2041	.1994
23000	.0469	.0414	.0371	.0326	.0294	.2167	.2099	.2041	.1979	.1933
26000	.0427	.0375	.0336	.0294	.0265	.2109	.2043	.1986	.1925	.1881
30000	.0381	.0334	.0298	.0260	.0234	.2042	.1978	.1923	.1864	.1821
35000	.0334	.0293	.0261	.0226	.0204	.1971	.1909	.1856	.1799	.1759
40000	.0298	.0261	.0231	.0201	.0180	.1912	.1851	.1800	.1745	.1706
45000	.0268	.0235	.0208	.0180	.0161	.1860	.1801	.1751	.1698	.1660
50000	.0243	.0213	.0188	.0163	.0146	.1815	.1758	.1709	.1656	.1619
60000	.0204	.0180	.0158	.0137	.0122	.1739	.1684	.1637	.1587	.1551
70000	.0176	.0155	.0136	.0118	.0105	.1676	.1623	.1578	.1530	.1496
80000	.0156	.0137	.0120	.0103	.0093	.1624	.1572	.1528	.1482	.1449
90000	.0140	.0122	.0107	.0092	.0083	.1579	.1529	.1486	.1440	.1408
100000	.0128	.0111	.0098	.0084	.0075	.1540	.1491	.1449	.1404	.1373

TABLE 14- 4

AU ON AG, M2/M1=0.55

E,KEV	RELATIVE IONIZATION STRAGGLING					RELATIVE TRANSVERSE RANGE STRAGGLING				
K/KL=	0.5	0.65	0.8	1.0	1.2	0.5	0.65	0.8	1.0	1.2
10	.4301	.4291	.4282	.4271	.4260	.5968	.5971	.5974	.5978	.5983
12	.4300	.4290	.4281	.4269	.4258	.5965	.5969	.5972	.5976	.5981
14	.4299	.4289	.4279	.4267	.4256	.5963	.5966	.5969	.5974	.5979
17	.4298	.4287	.4277	.4265	.4254	.5959	.5962	.5966	.5971	.5976
20	.4297	.4286	.4276	.4263	.4252	.5955	.5959	.5963	.5968	.5973
23	.4296	.4285	.4275	.4262	.4250	.5951	.5955	.5959	.5964	.5970
26	.4295	.4284	.4273	.4260	.4248	.5947	.5952	.5956	.5961	.5967
30	.4294	.4283	.4272	.4258	.4246	.5942	.5947	.5951	.5957	.5963
35	.4293	.4281	.4270	.4256	.4244	.5936	.5941	.5946	.5952	.5959
40	.4292	.4280	.4268	.4254	.4242	.5931	.5936	.5941	.5947	.5954
45	.4291	.4279	.4267	.4252	.4240	.5925	.5930	.5936	.5943	.5950
50	.4290	.4277	.4265	.4251	.4238	.5919	.5925	.5931	.5938	.5945
60	.4289	.4275	.4263	.4248	.4234	.5909	.5915	.5921	.5929	.5937
70	.4287	.4273	.4261	.4245	.4231	.5898	.5905	.5912	.5921	.5929
80	.4286	.4272	.4258	.4242	.4228	.5889	.5896	.5903	.5912	.5922
90	.4285	.4270	.4256	.4240	.4225	.5879	.5887	.5895	.5905	.5915
100	.4284	.4268	.4254	.4238	.4223	.5870	.5878	.5887	.5897	.5908
120	.4282	.4266	.4251	.4233	.4218	.5853	.5862	.5871	.5883	.5894
140	.4280	.4263	.4248	.4229	.4213	.5836	.5846	.5856	.5869	.5882
170	.4277	.4259	.4243	.4224	.4207	.5814	.5825	.5837	.5851	.5866
200	.4275	.4256	.4238	.4218	.4201	.5793	.5806	.5819	.5836	.5852
230	.4272	.4252	.4234	.4213	.4195	.5774	.5788	.5802	.5820	.5837
260	.4270	.4249	.4230	.4208	.4189	.5755	.5771	.5786	.5806	.5825
300	.4267	.4244	.4225	.4202	.4182	.5734	.5751	.5767	.5789	.5810
350	.4263	.4239	.4218	.4194	.4174	.5710	.5729	.5748	.5772	.5795
400	.4259	.4234	.4212	.4187	.4166	.5689	.5710	.5731	.5757	.5783
450	.4256	.4229	.4207	.4181	.4159	.5671	.5694	.5717	.5745	.5773
500	.4252	.4225	.4201	.4174	.4152	.5655	.5680	.5704	.5735	.5765
600	.4246	.4217	.4191	.4163	.4140	.5626	.5655	.5682	.5718	.5752
700	.4240	.4208	.4181	.4152	.4128	.5604	.5636	.5667	.5706	.5744
800	.4234	.4200	.4172	.4141	.4117	.5587	.5623	.5657	.5700	.5742
900	.4227	.4192	.4163	.4131	.4107	.5575	.5614	.5651	.5698	.5743
1000	.4221	.4184	.4154	.4122	.4097	.5567	.5608	.5648	.5699	.5747
1200	.4205	.4166	.4134	.4101	.4076	.5556	.5603	.5649	.5706	.5761
1400	.4190	.4148	.4116	.4082	.4057	.5555	.5608	.5658	.5722	.5782
1700	.4169	.4125	.4092	.4058	.4034	.5565	.5626	.5683	.5756	.5824
2000	.4150	.4105	.4071	.4038	.4015	.5586	.5654	.5718	.5799	.5874
2300	.4133	.4088	.4054	.4022	.4000	.5613	.5689	.5760	.5848	.5930
2600	.4118	.4072	.4039	.4008	.3988	.5646	.5729	.5806	.5901	.5990
3000	.4100	.4055	.4023	.3994	.3975	.5696	.5787	.5872	.5976	.6072
3500	.4081	.4037	.4006	.3980	.3964	.5765	.5866	.5959	.6074	.6179
4000	.4065	.4022	.3993	.3969	.3956	.5838	.5948	.6050	.6174	.6287
4500	.4051	.4009	.3983	.3961	.3950	.5914	.6033	.6142	.6274	.6395
5000	.4039	.3999	.3974	.3956	.3947	.5992	.6119	.6235	.6375	.6502
6000	.4018	.3983	.3962	.3949	.3945	.6148	.6289	.6419	.6575	.6715
7000	.4002	.3972	.3955	.3947	.3948	.6304	.6458	.6602	.6771	.6922
8000	.3991	.3964	.3952	.3949	.3953	.6459	.6626	.6780	.6961	.7120
9000	.3982	.3960	.3952	.3953	.3961	.6611	.6792	.6954	.7144	.7311
10000	.3976	.3958	.3953	.3958	.3969	.6761	.6955	.7124	.7322	.7495
12000	.3970	.3962	.3964	.3975	.3991	.7056	.7284	.7463	.7671	.7851
14000	.3969	.3969	.3977	.3994	.4015	.7336	.7595	.7779	.7995	.8179
17000	.3973	.3985	.3999	.4023	.4049	.7730	.8025	.8216	.8440	.8629
20000	.3982	.4002	.4023	.4052	.4083	.8096	.8416	.8613	.8842	.9034
23000	.3997	.4023	.4048	.4081	.4114	.8447	.8770	.8975	.9212	.9408
26000	.4013	.4043	.4071	.4108	.4144	.8775	.9096	.9308	.9552	.9750
30000	.4035	.4069	.4102	.4142	.4180	.9180	.9494	.9715	.9965	1.017
35000	.4062	.4101	.4137	.4181	.4221	.9640	.9945	1.017	1.043	1.063
40000	.4088	.4130	.4169	.4216	.4257	1.006	1.035	1.059	1.085	1.105
45000	.4113	.4158	.4199	.4248	.4290	1.044	1.072	1.096	1.122	1.142
50000	.4136	.4183	.4226	.4276	.4319	1.079	1.107	1.131	1.157	1.177
60000	.4179	.4229	.4275	.4327	.4371	1.142	1.168	1.192	1.218	1.237
70000	.4218	.4270	.4317	.4370	.4414	1.196	1.222	1.246	1.271	1.289
80000	.4253	.4306	.4353	.4407	.4451	1.245	1.271	1.294	1.318	1.335
90000	.4284	.4338	.4385	.4440	.4482	1.288	1.314	1.337	1.360	1.376
100000	.4312	.4366	.4414	.4468	.4510	1.327	1.354	1.376	1.398	1.412

K/KL= E,KEV	RELATIVE TRANSV. DAMAGE STRAGGLING					RELATIVE TRANSV. IONZN. STRAGGLING				
	0.5	0.65	0.8	1.0	1.2	0.5	0.65	0.8	1.0	1.2
10	.4728	.4699	.4670	.4633	.4596	.3915	.3878	.3841	.3794	.3747
12	.4720	.4690	.4661	.4622	.4585	.3906	.3868	.3831	.3782	.3734
14	.4713	.4682	.4652	.4613	.4575	.3899	.3859	.3821	.3771	.3722
17	.4703	.4671	.4640	.4600	.4561	.3888	.3848	.3808	.3757	.3707
20	.4693	.4661	.4629	.4588	.4548	.3878	.3837	.3797	.3744	.3693
23	.4685	.4652	.4619	.4577	.4536	.3869	.3827	.3786	.3732	.3680
26	.4676	.4643	.4610	.4567	.4525	.3861	.3818	.3776	.3721	.3668
30	.4665	.4631	.4598	.4554	.4511	.3850	.3806	.3763	.3707	.3653
35	.4652	.4617	.4583	.4538	.4495	.3836	.3792	.3748	.3691	.3635
40	.4640	.4604	.4569	.4524	.4480	.3824	.3778	.3733	.3675	.3619
45	.4628	.4592	.4556	.4510	.4465	.3812	.3765	.3720	.3661	.3603
50	.4617	.4580	.4544	.4497	.4451	.3800	.3753	.3707	.3647	.3589
60	.4595	.4557	.4520	.4472	.4426	.3779	.3730	.3683	.3621	.3562
70	.4575	.4536	.4498	.4449	.4402	.3759	.3709	.3660	.3597	.3537
80	.4555	.4516	.4477	.4427	.4379	.3739	.3688	.3639	.3575	.3513
90	.4536	.4496	.4457	.4406	.4357	.3721	.3669	.3619	.3554	.3491
100	.4518	.4478	.4438	.4387	.4337	.3703	.3650	.3599	.3533	.3470
120	.4484	.4443	.4402	.4349	.4298	.3669	.3615	.3563	.3495	.3430
140	.4452	.4410	.4368	.4314	.4263	.3637	.3582	.3529	.3460	.3393
170	.4408	.4364	.4321	.4266	.4213	.3593	.3537	.3482	.3411	.3343
200	.4366	.4321	.4277	.4221	.4167	.3553	.3494	.3438	.3366	.3296
230	.4326	.4280	.4236	.4179	.4124	.3514	.3454	.3397	.3323	.3253
260	.4289	.4242	.4197	.4139	.4083	.3478	.3417	.3359	.3284	.3212
300	.4243	.4195	.4149	.4089	.4033	.3432	.3370	.3311	.3234	.3161
350	.4189	.4140	.4093	.4032	.3975	.3380	.3317	.3255	.3177	.3103
400	.4140	.4090	.4042	.3980	.3921	.3332	.3267	.3204	.3125	.3049
450	.4094	.4043	.3994	.3932	.3872	.3287	.3220	.3157	.3075	.2999
500	.4052	.4000	.3950	.3886	.3826	.3245	.3177	.3112	.3029	.2951
600	.3974	.3921	.3869	.3803	.3741	.3166	.3095	.3028	.2944	.2864
700	.3905	.3850	.3796	.3729	.3665	.3095	.3022	.2953	.2867	.2785
800	.3843	.3786	.3731	.3662	.3597	.3031	.2956	.2886	.2797	.2714
900	.3786	.3727	.3671	.3601	.3534	.2973	.2896	.2824	.2733	.2649
1000	.3734	.3674	.3616	.3544	.3477	.2919	.2840	.2766	.2674	.2589
1200	.3640	.3577	.3517	.3442	.3372	.2824	.2741	.2664	.2569	.2481
1400	.3558	.3492	.3430	.3352	.3281	.2740	.2654	.2575	.2477	.2387
1700	.3453	.3382	.3317	.3236	.3162	.2631	.2541	.2458	.2356	.2264
2000	.3363	.3289	.3220	.3136	.3060	.2536	.2442	.2356	.2252	.2159
2300	.3287	.3209	.3138	.3050	.2971	.2452	.2354	.2265	.2159	.2064
2600	.3220	.3139	.3064	.2974	.2893	.2376	.2275	.2184	.2076	.1980
3000	.3141	.3056	.2978	.2884	.2800	.2286	.2181	.2087	.1977	.1880
3500	.3055	.2966	.2884	.2786	.2699	.2187	.2078	.1982	.1870	.1772
4000	.2980	.2887	.2801	.2700	.2612	.2100	.1988	.1890	.1776	.1678
4500	.2913	.2816	.2728	.2624	.2534	.2023	.1907	.1808	.1694	.1596
5000	.2852	.2753	.2662	.2555	.2464	.1953	.1835	.1735	.1620	.1522
6000	.2746	.2642	.2547	.2435	.2342	.1830	.1709	.1607	.1493	.1397
7000	.2656	.2547	.2448	.2334	.2238	.1726	.1603	.1501	.1387	.1293
8000	.2576	.2464	.2363	.2246	.2149	.1636	.1511	.1409	.1297	.1204
9000	.2506	.2390	.2287	.2168	.2071	.1556	.1431	.1330	.1219	.1128
10000	.2442	.2324	.2219	.2099	.2001	.1486	.1361	.1260	.1150	.1061
12000	.2331	.2210	.2102	.1979	.1881	.1364	.1243	.1143	.1036	.0951
14000	.2237	.2112	.2003	.1879	.1780	.1264	.1146	.1048	.0943	.0862
17000	.2117	.1990	.1878	.1754	.1656	.1140	.1028	.0933	.0832	.0757
20000	.2016	.1888	.1775	.1651	.1554	.1039	.0933	.0842	.0745	.0674
23000	.1928	.1799	.1687	.1563	.1468	.0957	.0855	.0768	.0675	.0609
26000	.1851	.1722	.1611	.1488	.1394	.0886	.0789	.0706	.0618	.0555
30000	.1761	.1633	.1523	.1401	.1310	.0808	.0716	.0638	.0555	.0497
35000	.1665	.1539	.1430	.1311	.1222	.0727	.0641	.0568	.0492	.0439
40000	.1582	.1458	.1351	.1234	.1148	.0661	.0580	.0512	.0441	.0392
45000	.1510	.1388	.1283	.1169	.1085	.0605	.0530	.0466	.0399	.0355
50000	.1447	.1327	.1223	.1112	.1031	.0558	.0487	.0427	.0365	.0323
60000	.1339	.1223	.1124	.1017	.0940	.0482	.0418	.0365	.0310	.0274
70000	.1251	.1139	.1043	.0941	.0868	.0425	.0366	.0318	.0270	.0237
80000	.1176	.1069	.0977	.0879	.0809	.0379	.0326	.0283	.0238	.0209
90000	.1113	.1009	.0920	.0827	.0760	.0343	.0295	.0255	.0214	.0187
100000	.1058	.0957	.0872	.0782	.0718	.0314	.0269	.0232	.0194	.0170

TABLE 14- 6

AU ON AG, M2/M1=0.55

	RANGE SKEWNESS					DAMAGE SKEWNESS				
K/KL=	0.5	0.65	0.8	1.0	1.2	0.5	0.65	0.8	1.0	1.2
E,KEV										
10	.6302	.6201	.6102	.5973	.5846	.7627	.7537	.7449	.7336	.7227
12	.6284	.6181	.6079	.5945	.5815	.7616	.7523	.7432	.7316	.7204
14	.6269	.6162	.6058	.5921	.5787	.7606	.7511	.7418	.7298	.7183
17	.6248	.6138	.6030	.5889	.5750	.7594	.7495	.7399	.7275	.7157
20	.6229	.6116	.6005	.5860	.5718	.7583	.7481	.7382	.7255	.7133
23	.6212	.6096	.5982	.5833	.5688	.7573	.7468	.7367	.7237	.7112
26	.6195	.6076	.5960	.5808	.5660	.7563	.7456	.7353	.7220	.7092
30	.6174	.6052	.5933	.5777	.5625	.7552	.7442	.7335	.7199	.7068
35	.6149	.6024	.5901	.5741	.5585	.7539	.7425	.7315	.7175	.7040
40	.6125	.5997	.5871	.5707	.5547	.7526	.7409	.7296	.7152	.7014
45	.6103	.5971	.5842	.5675	.5512	.7515	.7395	.7279	.7131	.6990
50	.6081	.5947	.5815	.5644	.5478	.7504	.7381	.7262	.7111	.6967
60	.6041	.5901	.5765	.5588	.5416	.7484	.7356	.7232	.7076	.6926
70	.6002	.5858	.5718	.5536	.5359	.7466	.7332	.7205	.7042	.6888
80	.5965	.5818	.5673	.5486	.5304	.7448	.7310	.7178	.7010	.6852
90	.5930	.5778	.5630	.5438	.5252	.7431	.7288	.7152	.6980	.6817
100	.5896	.5740	.5588	.5392	.5202	.7414	.7267	.7127	.6951	.6784
120	.5831	.5669	.5511	.5307	.5109	.7382	.7227	.7080	.6895	.6721
140	.5769	.5601	.5437	.5226	.5022	.7351	.7189	.7035	.6842	.6661
170	.5681	.5504	.5332	.5111	.4898	.7307	.7134	.6971	.6767	.6577
200	.5596	.5412	.5233	.5002	.4781	.7264	.7082	.6910	.6696	.6497
230	.5517	.5325	.5139	.4900	.4672	.7230	.7039	.6859	.6635	.6428
260	.5440	.5241	.5049	.4803	.4567	.7197	.6996	.6808	.6576	.6361
300	.5341	.5133	.4933	.4677	.4433	.7152	.6939	.6742	.6498	.6273
350	.5222	.5004	.4795	.4528	.4273	.7094	.6868	.6659	.6402	.6167
400	.5108	.4881	.4662	.4385	.4121	.7035	.6796	.6576	.6307	.6062
450	.4998	.4762	.4535	.4248	.3975	.6975	.6724	.6494	.6214	.5959
500	.4892	.4647	.4412	.4116	.3835	.6914	.6651	.6411	.6121	.5858
600	.4690	.4429	.4180	.3865	.3570	.6784	.6499	.6241	.5930	.5652
700	.4500	.4223	.3960	.3630	.3321	.6656	.6351	.6076	.5748	.5455
800	.4318	.4027	.3752	.3407	.3085	.6532	.6207	.5918	.5574	.5269
900	.4145	.3840	.3553	.3195	.2861	.6410	.6069	.5766	.5408	.5093
1000	.3978	.3661	.3363	.2992	.2648	.6293	.5936	.5620	.5249	.4925
1200	.3663	.3321	.3003	.2608	.2245	.6077	.5689	.5350	.4957	.4616
1400	.3370	.3006	.2668	.2253	.1872	.5870	.5454	.5097	.4685	.4330
1700	.2962	.2568	.2205	.1762	.1359	.5570	.5123	.4741	.4307	.3936
2000	.2586	.2165	.1780	.1314	.0892	.5282	.4810	.4410	.3958	.3575
2300	.2235	.1788	.1383	.0895	.0456	.4976	.4490	.4076	.3611	.3222
2600	.1905	.1436	.1012	.0505	.0052	.4684	.4187	.3763	.3287	.2894
3000	.1496	.0999	.0554	.0024	-.0446	.4318	.3811	.3376	.2888	.2491
3500	.1025	.0497	.0028	-.0526	-.1015	.3896	.3381	.2936	.2437	.2036
4000	.0590	.0035	-.0454	-.1029	-.1533	.3511	.2990	.2536	.2029	.1627
4500	.0186	-.0393	-.0901	-.1493	-.2010	.3157	.2632	.2172	.1659	.1257
5000	-.0192	-.0792	-.1316	-.1924	-.2453	.2832	.2303	.1839	.1321	.0918
6000	-.0882	-.1501	-.2063	-.2706	-.3257	.2262	.1725	.1254	.0730	.0327
7000	-.1504	-.2145	-.2735	-.3403	-.3971	.1762	.1219	.0744	.0218	-.0185
8000	-.2072	-.2742	-.3350	-.4034	-.4613	.1317	.0770	.0294	-.0233	-.0635
9000	-.2597	-.3300	-.3919	-.4612	-.5197	.0915	.0367	-.0111	-.0637	-.1037
10000	-.3086	-.3827	-.4449	-.5146	-.5734	.0548	-.0000	-.0477	-.1002	-.1400
12000	-.3986	-.4878	-.5462	-.6128	-.6700	-.0108	-.0652	-.1124	-.1643	-.2035
14000	-.4788	-.5814	-.6356	-.6989	-.7544	-.0674	-.1211	-.1678	-.2190	-.2576
17000	-.5848	-.7029	-.7517	-.8107	-.8636	-.1395	-.1922	-.2380	-.2881	-.3258
20000	-.6776	-.8059	-.8506	-.9062	-.9572	-.2001	-.2519	-.2968	-.3460	-.3828
23000	-.7625	-.8854	-.9306	-.9867	-1.038	-.2513	-.3024	-.3466	-.3949	-.4310
26000	-.8398	-.9545	-1.001	-1.058	-1.109	-.2961	-.3464	-.3900	-.4375	-.4730
30000	-.9295	-1.035	-1.083	-1.142	-1.194	-.3480	-.3976	-.4404	-.4870	-.5216
35000	-1.029	-1.121	-1.172	-1.233	-1.286	-.4035	-.4522	-.4942	-.5398	-.5736
40000	-1.115	-1.197	-1.250	-1.314	-1.367	-.4512	-.4991	-.5404	-.5851	-.6182
45000	-1.192	-1.264	-1.320	-1.385	-1.440	-.4928	-.5400	-.5807	-.6248	-.6572
50000	-1.260	-1.325	-1.383	-1.450	-1.505	-.5297	-.5764	-.6165	-.6599	-.6918
60000	-1.376	-1.433	-1.495	-1.564	-1.619	-.5928	-.6384	-.6777	-.7200	-.7511
70000	-1.473	-1.529	-1.591	-1.662	-1.717	-.6451	-.6900	-.7285	-.7701	-.8005
80000	-1.555	-1.612	-1.677	-1.749	-1.803	-.6896	-.7339	-.7719	-.8128	-.8427
90000	-1.625	-1.690	-1.755	-1.826	-1.879	-.7283	-.7720	-.8096	-.8500	-.8795
100000	-1.686	-1.761	-1.826	-1.897	-1.948	-.7623	-.8056	-.8428	-.8829	-.9121

K/KL=	IONIZATION SKEWNESS					RANGE KURTOSIS				
	0.5	0.65	0.8	1.0	1.2	0.5	0.65	0.8	1.0	1.2
E,KEV										
10	.8370	.8294	.8220	.8126	.8037	3.347	3.322	3.298	3.268	3.239
12	.8361	.8282	.8207	.8110	.8018	3.343	3.317	3.293	3.262	3.232
14	.8353	.8272	.8195	.8095	.8001	3.339	3.313	3.288	3.257	3.226
17	.8343	.8259	.8179	.8077	.7979	3.334	3.308	3.282	3.249	3.218
20	.8334	.8248	.8165	.8060	.7960	3.330	3.303	3.276	3.243	3.211
23	.8325	.8237	.8153	.8045	.7943	3.326	3.298	3.271	3.237	3.205
26	.8318	.8228	.8141	.8031	.7927	3.322	3.294	3.267	3.232	3.199
30	.8309	.8216	.8127	.8014	.7908	3.317	3.288	3.261	3.225	3.191
35	.8298	.8202	.8110	.7994	.7885	3.312	3.282	3.254	3.217	3.183
40	.8288	.8189	.8095	.7976	.7864	3.307	3.276	3.247	3.210	3.175
45	.8279	.8177	.8081	.7959	.7845	3.302	3.271	3.241	3.204	3.168
50	.8270	.8166	.8067	.7943	.7827	3.297	3.265	3.235	3.197	3.162
60	.8254	.8146	.8043	.7914	.7793	3.288	3.256	3.225	3.186	3.149
70	.8239	.8127	.8021	.7887	.7763	3.280	3.247	3.215	3.175	3.138
80	.8225	.8109	.7999	.7862	.7734	3.272	3.238	3.206	3.166	3.128
90	.8212	.8092	.7979	.7838	.7706	3.265	3.230	3.197	3.156	3.118
100	.8199	.8075	.7959	.7815	.7680	3.257	3.222	3.189	3.147	3.108
120	.8176	.8045	.7923	.7772	.7632	3.244	3.208	3.174	3.131	3.091
140	.8153	.8017	.7889	.7732	.7586	3.231	3.194	3.159	3.116	3.075
170	.8121	.7976	.7840	.7674	.7522	3.214	3.175	3.140	3.095	3.054
200	.8089	.7936	.7794	.7620	.7461	3.197	3.158	3.121	3.076	3.034
230	.8059	.7898	.7749	.7567	.7402	3.182	3.141	3.104	3.058	3.016
260	.8030	.7861	.7706	.7517	.7346	3.167	3.126	3.088	3.041	2.998
300	.7992	.7813	.7650	.7452	.7275	3.148	3.106	3.067	3.020	2.977
350	.7946	.7756	.7583	.7375	.7190	3.127	3.084	3.044	2.996	2.953
400	.7901	.7700	.7519	.7302	.7110	3.107	3.063	3.023	2.975	2.932
450	.7857	.7647	.7457	.7233	.7034	3.088	3.044	3.003	2.955	2.912
500	.7815	.7595	.7398	.7166	.6962	3.070	3.026	2.985	2.937	2.894
600	.7744	.7506	.7294	.7048	.6833	3.037	2.991	2.950	2.902	2.860
700	.7672	.7417	.7193	.6934	.6711	3.007	2.961	2.920	2.872	2.830
800	.7597	.7327	.7093	.6823	.6593	2.980	2.934	2.893	2.846	2.805
900	.7521	.7238	.6993	.6715	.6479	2.956	2.910	2.869	2.823	2.784
1000	.7443	.7148	.6895	.6609	.6369	2.934	2.888	2.848	2.803	2.765
1200	.7266	.6948	.6681	.6382	.6134	2.893	2.847	2.809	2.766	2.732
1400	.7095	.6760	.6480	.6172	.5920	2.859	2.814	2.777	2.738	2.707
1700	.6857	.6500	.6207	.5891	.5636	2.817	2.775	2.741	2.707	2.682
2000	.6640	.6268	.5967	.5646	.5391	2.784	2.746	2.716	2.687	2.667
2300	.6446	.6064	.5759	.5438	.5187	2.756	2.721	2.695	2.672	2.659
2600	.6269	.5880	.5574	.5256	.5009	2.733	2.702	2.681	2.664	2.657
3000	.6055	.5661	.5356	.5042	.4803	2.711	2.685	2.670	2.662	2.662
3500	.5816	.5421	.5119	.4814	.4585	2.693	2.675	2.668	2.670	2.680
4000	.5604	.5211	.4914	.4619	.4402	2.684	2.674	2.675	2.686	2.706
4500	.5414	.5024	.4734	.4451	.4244	2.682	2.680	2.689	2.710	2.739
5000	.5242	.4857	.4575	.4303	.4108	2.685	2.691	2.708	2.739	2.776
6000	.4930	.4555	.4293	.4047	.3876	2.701	2.728	2.758	2.806	2.860
7000	.4668	.4308	.4067	.3845	.3695	2.729	2.775	2.818	2.884	2.954
8000	.4446	.4103	.3882	.3684	.3554	2.767	2.829	2.886	2.969	3.054
9000	.4256	.3933	.3732	.3554	.3442	2.811	2.886	2.959	3.058	3.157
10000	.4093	.3791	.3608	.3450	.3354	2.859	2.947	3.035	3.151	3.263
12000	.3831	.3587	.3439	.3316	.3250	2.961	3.056	3.184	3.341	3.477
14000	.3630	.3440	.3324	.3231	.3188	3.073	3.175	3.341	3.533	3.692
17000	.3409	.3289	.3213	.3157	.3144	3.254	3.368	3.583	3.824	4.012
20000	.3256	.3192	.3149	.3124	.3134	3.444	3.575	3.830	4.111	4.325
23000	.3174	.3141	.3121	.3118	.3146	3.641	3.810	4.088	4.396	4.630
26000	.3121	.3113	.3111	.3128	.3169	3.840	4.047	4.344	4.673	4.927
30000	.3083	.3098	.3116	.3154	.3210	4.105	4.360	4.678	5.032	5.308
35000	.3066	.3103	.3141	.3198	.3268	4.431	4.742	5.082	5.463	5.763
40000	.3069	.3123	.3176	.3250	.3330	4.751	5.112	5.470	5.875	6.196
45000	.3085	.3151	.3216	.3304	.3393	5.064	5.468	5.843	6.269	6.609
50000	.3109	.3185	.3260	.3358	.3453	5.369	5.810	6.202	6.646	7.003
60000	.3167	.3259	.3350	.3464	.3568	5.955	6.457	6.879	7.359	7.745
70000	.3232	.3337	.3438	.3563	.3673	6.511	7.058	7.509	8.022	8.432
80000	.3297	.3412	.3522	.3654	.3768	7.041	7.619	8.100	8.643	9.074
90000	.3361	.3485	.3601	.3738	.3854	7.546	8.146	8.656	9.227	9.677
100000	.3421	.3554	.3674	.3815	.3932	8.029	8.643	9.181	9.781	10.25

K/KL=	DAMAGE KURTOSIS					IONIZATION KURTOSIS				
E,KEV	0.5	0.65	0.8	1.0	1.2	0.5	0.65	0.8	1.0	1.2
10	3.591	3.560	3.531	3.493	3.457	3.698	3.668	3.639	3.603	3.568
12	3.588	3.556	3.525	3.486	3.449	3.695	3.664	3.634	3.597	3.561
14	3.585	3.552	3.520	3.481	3.443	3.692	3.660	3.630	3.591	3.555
17	3.580	3.547	3.514	3.473	3.434	3.688	3.655	3.624	3.584	3.547
20	3.577	3.542	3.509	3.467	3.427	3.684	3.650	3.618	3.578	3.540
23	3.573	3.538	3.504	3.461	3.420	3.681	3.647	3.614	3.572	3.533
26	3.570	3.534	3.499	3.455	3.414	3.678	3.643	3.609	3.567	3.527
30	3.567	3.529	3.494	3.449	3.407	3.675	3.638	3.604	3.561	3.520
35	3.562	3.524	3.487	3.441	3.398	3.671	3.633	3.598	3.553	3.512
40	3.558	3.519	3.481	3.434	3.390	3.667	3.629	3.592	3.547	3.504
45	3.555	3.514	3.476	3.428	3.383	3.664	3.624	3.587	3.541	3.497
50	3.551	3.510	3.471	3.422	3.376	3.661	3.620	3.582	3.535	3.491
60	3.545	3.502	3.462	3.411	3.364	3.655	3.613	3.573	3.524	3.479
70	3.540	3.495	3.453	3.401	3.353	3.649	3.606	3.565	3.515	3.468
80	3.535	3.488	3.445	3.392	3.342	3.645	3.600	3.558	3.506	3.458
90	3.529	3.482	3.437	3.383	3.332	3.640	3.594	3.550	3.497	3.449
100	3.524	3.476	3.430	3.374	3.322	3.635	3.588	3.544	3.489	3.440
120	3.514	3.463	3.415	3.357	3.303	3.627	3.577	3.531	3.475	3.423
140	3.505	3.451	3.402	3.341	3.286	3.620	3.568	3.520	3.461	3.408
170	3.492	3.435	3.383	3.319	3.262	3.609	3.554	3.503	3.442	3.386
200	3.480	3.420	3.366	3.300	3.240	3.599	3.541	3.488	3.424	3.366
230	3.472	3.410	3.353	3.284	3.223	3.589	3.528	3.472	3.406	3.347
260	3.465	3.400	3.341	3.270	3.207	3.579	3.516	3.458	3.389	3.328
300	3.456	3.387	3.325	3.252	3.187	3.567	3.500	3.439	3.368	3.305
350	3.443	3.371	3.306	3.229	3.162	3.553	3.482	3.418	3.343	3.278
400	3.430	3.354	3.286	3.207	3.138	3.539	3.464	3.398	3.321	3.254
450	3.416	3.337	3.267	3.185	3.114	3.526	3.448	3.379	3.299	3.231
500	3.402	3.320	3.247	3.163	3.091	3.514	3.433	3.361	3.279	3.209
600	3.367	3.278	3.202	3.114	3.039	3.497	3.409	3.333	3.247	3.174
700	3.333	3.239	3.159	3.068	2.991	3.479	3.385	3.306	3.216	3.141
800	3.301	3.203	3.120	3.026	2.948	3.459	3.361	3.278	3.186	3.109
900	3.271	3.169	3.083	2.988	2.908	3.438	3.335	3.250	3.156	3.078
1000	3.243	3.137	3.049	2.952	2.873	3.415	3.309	3.222	3.125	3.047
1200	3.196	3.083	2.992	2.893	2.813	3.359	3.246	3.155	3.056	2.977
1400	3.152	3.035	2.941	2.841	2.762	3.304	3.186	3.092	2.992	2.913
1700	3.091	2.969	2.874	2.774	2.696	3.227	3.104	3.008	2.907	2.828
2000	3.033	2.910	2.814	2.716	2.639	3.159	3.032	2.934	2.834	2.757
2300	2.967	2.846	2.753	2.657	2.584	3.102	2.973	2.877	2.778	2.703
2600	2.904	2.788	2.697	2.604	2.535	3.050	2.922	2.827	2.730	2.658
3000	2.829	2.718	2.631	2.542	2.479	2.989	2.861	2.768	2.675	2.606
3500	2.746	2.643	2.561	2.478	2.420	2.921	2.795	2.706	2.616	2.551
4000	2.675	2.579	2.502	2.424	2.372	2.861	2.738	2.652	2.567	2.505
4500	2.614	2.525	2.452	2.379	2.332	2.808	2.688	2.604	2.523	2.465
5000	2.561	2.479	2.411	2.342	2.300	2.759	2.643	2.562	2.485	2.430
6000	2.488	2.413	2.352	2.292	2.257	2.669	2.562	2.483	2.414	2.365
7000	2.431	2.364	2.309	2.257	2.228	2.594	2.496	2.419	2.358	2.313
8000	2.386	2.326	2.277	2.232	2.208	2.530	2.441	2.367	2.311	2.271
9000	2.350	2.296	2.253	2.214	2.195	2.475	2.394	2.323	2.273	2.237
10000	2.320	2.272	2.235	2.202	2.187	2.427	2.354	2.286	2.241	2.208
12000	2.272	2.237	2.211	2.189	2.182	2.352	2.292	2.236	2.197	2.170
14000	2.239	2.215	2.199	2.187	2.186	2.293	2.244	2.199	2.166	2.144
17000	2.207	2.199	2.194	2.194	2.202	2.227	2.191	2.160	2.134	2.117
20000	2.191	2.195	2.200	2.210	2.224	2.178	2.153	2.133	2.112	2.099
23000	2.191	2.203	2.215	2.232	2.250	2.148	2.128	2.113	2.096	2.087
26000	2.198	2.215	2.233	2.256	2.278	2.126	2.110	2.098	2.085	2.078
30000	2.213	2.237	2.261	2.290	2.316	2.105	2.093	2.084	2.074	2.070
35000	2.237	2.269	2.298	2.333	2.364	2.087	2.078	2.072	2.066	2.064
40000	2.265	2.303	2.337	2.377	2.411	2.076	2.069	2.065	2.061	2.062
45000	2.295	2.337	2.375	2.420	2.457	2.068	2.063	2.061	2.059	2.061
50000	2.325	2.371	2.413	2.461	2.501	2.063	2.060	2.058	2.059	2.063
60000	2.386	2.439	2.486	2.540	2.584	2.058	2.057	2.057	2.061	2.067
70000	2.444	2.503	2.555	2.614	2.662	2.056	2.057	2.059	2.066	2.074
80000	2.500	2.563	2.619	2.683	2.733	2.056	2.059	2.063	2.072	2.081
90000	2.552	2.620	2.680	2.747	2.801	2.056	2.062	2.068	2.078	2.089
100000	2.602	2.674	2.737	2.808	2.863	2.056	2.065	2.074	2.086	2.098

TABLE 14- 9

AU ON AG, M2/M1=0.55

E,KEV	REFLECTION COEFFICIENT					SPUTTERING EFFICIENCY				
K/KL=	0.5	0.65	0.8	1.0	1.2	0.5	0.65	0.8	1.0	1.2
10	.0036	.0035	.0034	.0033	.0032	.0202	.0194	.0186	.0175	.0166
12	.0036	.0035	.0034	.0033	.0032	.0201	.0193	.0184	.0174	.0164
14	.0036	.0035	.0034	.0033	.0031	.0200	.0192	.0183	.0172	.0163
17	.0036	.0035	.0034	.0032	.0031	.0199	.0190	.0181	.0171	.0161
20	.0035	.0034	.0033	.0032	.0031	.0198	.0189	.0180	.0169	.0159
23	.0035	.0034	.0033	.0032	.0031	.0197	.0188	.0179	.0168	.0157
26	.0035	.0034	.0033	.0032	.0030	.0197	.0187	.0178	.0166	.0156
30	.0035	.0034	.0033	.0031	.0030	.0196	.0186	.0177	.0165	.0154
35	.0035	.0034	.0033	.0031	.0030	.0195	.0185	.0175	.0163	.0153
40	.0035	.0033	.0032	.0031	.0030	.0194	.0183	.0174	.0162	.0151
45	.0034	.0033	.0032	.0031	.0029	.0193	.0182	.0173	.0161	.0150
50	.0034	.0033	.0032	.0030	.0029	.0192	.0181	.0172	.0159	.0148
60	.0034	.0033	.0032	.0030	.0029	.0191	.0180	.0170	.0157	.0146
70	.0034	.0032	.0031	.0030	.0028	.0190	.0178	.0168	.0155	.0144
80	.0033	.0032	.0031	.0029	.0028	.0188	.0177	.0166	.0154	.0142
90	.0033	.0032	.0031	.0029	.0028	.0187	.0176	.0165	.0152	.0141
100	.0033	.0031	.0030	.0029	.0027	.0186	.0175	.0164	.0151	.0139
120	.0032	.0031	.0030	.0028	.0027	.0184	.0172	.0161	.0148	.0137
140	.0032	.0030	.0029	.0027	.0026	.0183	.0171	.0159	.0146	.0134
170	.0031	.0030	.0028	.0027	.0025	.0181	.0168	.0157	.0143	.0131
200	.0031	.0029	.0028	.0026	.0024	.0179	.0166	.0154	.0141	.0128
230	.0030	.0029	.0027	.0025	.0024	.0177	.0164	.0152	.0138	.0126
260	.0030	.0028	.0027	.0025	.0023	.0176	.0163	.0150	.0136	.0124
300	.0029	.0027	.0026	.0024	.0022	.0174	.0161	.0148	.0134	.0121
350	.0028	.0027	.0025	.0023	.0021	.0172	.0158	.0146	.0131	.0119
400	.0028	.0026	.0024	.0022	.0021	.0170	.0156	.0143	.0129	.0116
450	.0027	.0025	.0024	.0022	.0020	.0168	.0154	.0141	.0127	.0114
500	.0026	.0024	.0023	.0021	.0019	.0167	.0152	.0139	.0124	.0112
600	.0025	.0023	.0022	.0020	.0018	.0163	.0148	.0135	.0120	.0107
700	.0024	.0022	.0020	.0018	.0017	.0160	.0145	.0131	.0116	.0103
800	.0023	.0021	.0019	.0017	.0016	.0157	.0141	.0128	.0113	.0100
900	.0022	.0020	.0018	.0016	.0015	.0155	.0139	.0125	.0110	.0097
1000	.0021	.0019	.0018	.0016	.0014	.0152	.0136	.0122	.0107	.0094
1200	.0020	.0018	.0016	.0014	.0012	.0148	.0131	.0117	.0102	.0089
1400	.0018	.0015	.0015	.0013	.0011	.0144	.0127	.0113	.0098	.0085
1700	.0017	.0015	.0013	.0011	.0009	.0139	.0122	.0108	.0092	.0080
2000	.0015	.0013	.0012	.0010	.0008	.0134	.0117	.0103	.0088	.0075
2300	.0014	.0012	.0010	.0009	.0007	.0129	.0112	.0098	.0083	.0071
2600	.0013	.0011	.0009	.0008	.0006	.0124	.0107	.0093	.0078	.0067
3000	.0011	.0010	.0008	.0007	.0005	.0119	.0102	.0088	.0073	.0062
3500	.0010	.0008	.0007	.0005	.0004	.0112	.0096	.0082	.0068	.0057
4000	.0009	.0007	.0006	.0005	.0004	.0106	.0090	.0077	.0063	.0053
4500	.0008	.0006	.0005	.0004	.0003	.0101	.0086	.0073	.0059	.0050
5000	.0007	.0006	.0004	.0003	.0002	.0096	.0081	.0069	.0055	.0047
6000	.0006	.0004	.0003	.0002	.0002	.0089	.0074	.0063	.0050	.0042
7000	.0004	.0003	.0002	.0002	.0001	.0082	.0068	.0058	.0046	.0038
8000	.0004	.0003	.0002	.0001	.0001	.0077	.0064	.0053	.0042	.0035
9000	.0003	.0002	.0001	.0001	.0000	.0072	.0059	.0050	.0039	.0033
10000	.0002	.0001	.0001	.0000	.0000	.0068	.0056	.0047	.0037	.0030
12000	.0002	.0001	.0000	.0000	.0000	.0061	.0050	.0041	.0032	.0027
14000	.0001	.0000	.0000	.0000	.0000	.0055	.0045	.0037	.0029	.0023
17000	.0000	.0000	.0000	.0000	0.0000	.0048	.0039	.0032	.0024	.0020
20000	.0000	.0000	.0000	.0000	0.0000	.0042	.0035	.0028	.0021	.0017
23000	.0000	0.0000	0.0000	0.0000	0.0000	.0038	.0031	.0025	.0018	.0015
26000	0.0000	0.0000	0.0000	0.0000	0.0000	.0034	.0028	.0022	.0016	.0013
30000	0.0000	0.0000	0.0000	0.0000	0.0000	.0030	.0024	.0019	.0014	.0011
35000	0.0000	0.0000	0.0000	0.0000	0.0000	.0026	.0021	.0016	.0012	.0010
40000	0.0000	0.0000	0.0000	0.0000	0.0000	.0023	.0018	.0014	.0010	.0008
45000	0.0000	0.0000	0.0000	0.0000	0.0000	.0020	.0016	.0012	.0009	.0007
50000	0.0000	0.0000	0.0000	0.0000	0.0000	.0018	.0014	.0011	.0008	.0006
60000	0.0000	0.0000	0.0000	0.0000	0.0000	.0015	.0011	.0009	.0006	.0005
70000	0.0000	0.0000	0.0000	0.0000	0.0000	.0012	.0009	.0007	.0005	.0004
80000	0.0000	0.0000	0.0000	0.0000	0.0000	.0011	.0008	.0006	.0004	.0003
90000	0.0000	0.0000	0.0000	0.0000	0.0000	.0009	.0007	.0005	.0004	.0003
100000	0.0000	0.0000	0.0000	0.0000	0.0000	.0008	.0006	.0005	.0003	.0003

AU ON AG, M2/M1=0.55   TABLE 14-10

E,KEV	IONIZATION DEFICIENCY K/KL=					SPUTTERING YIELD ALPHA				
	0.5	0.65	0.8	1.0	1.2	0.5	0.65	0.8	1.0	1.2
10	.0018	.0023	.0027	.0032	.0037	.1357	.1318	.1279	.1231	.1185
12	.0019	.0023	.0028	.0033	.0038	.1354	.1314	.1274	.1225	.1178
14	.0019	.0024	.0028	.0034	.0038	.1352	.1310	.1270	.1219	.1171
17	.0020	.0024	.0029	.0034	.0039	.1349	.1306	.1265	.1213	.1164
20	.0020	.0025	.0030	.0035	.0040	.1346	.1303	.1261	.1207	.1157
23	.0020	.0026	.0030	.0036	.0041	.1345	.1300	.1257	.1203	.1152
26	.0021	.0026	.0031	.0036	.0041	.1343	.1298	.1254	.1199	.1147
30	.0021	.0026	.0031	.0037	.0042	.1342	.1295	.1250	.1194	.1141
35	.0022	.0027	.0032	.0037	.0043	.1340	.1292	.1247	.1189	.1136
40	.0022	.0027	.0032	.0038	.0043	.1339	.1290	.1244	.1185	.1131
45	.0022	.0028	.0033	.0039	.0044	.1338	.1288	.1241	.1182	.1127
50	.0023	.0028	.0033	.0039	.0044	.1338	.1287	.1239	.1179	.1123
60	.0023	.0029	.0034	.0040	.0045	.1337	.1285	.1236	.1174	.1117
70	.0024	.0029	.0034	.0040	.0045	.1337	.1284	.1233	.1170	.1112
80	.0024	.0030	.0035	.0041	.0046	.1337	.1283	.1231	.1167	.1108
90	.0025	.0030	.0035	.0041	.0046	.1338	.1282	.1230	.1165	.1105
100	.0025	.0031	.0036	.0042	.0047	.1339	.1282	.1229	.1163	.1102
120	.0025	.0031	.0036	.0042	.0047	.1341	.1283	.1228	.1160	.1098
140	.0026	.0032	.0037	.0043	.0048	.1344	.1284	.1228	.1159	.1095
170	.0026	.0032	.0037	.0043	.0048	.1349	.1287	.1229	.1157	.1092
200	.0027	.0033	.0038	.0044	.0049	.1354	.1290	.1230	.1157	.1090
230	.0027	.0033	.0038	.0044	.0049	.1360	.1294	.1233	.1158	.1089
260	.0028	.0034	.0039	.0044	.0049	.1365	.1298	.1235	.1159	.1089
300	.0028	.0034	.0039	.0045	.0049	.1373	.1304	.1239	.1161	.1090
350	.0028	.0034	.0039	.0045	.0050	.1382	.1311	.1244	.1164	.1091
400	.0029	.0035	.0040	.0045	.0050	.1391	.1317	.1249	.1167	.1092
450	.0029	.0035	.0040	.0045	.0050	.1399	.1324	.1254	.1170	.1094
500	.0029	.0035	.0040	.0046	.0050	.1407	.1330	.1258	.1173	.1095
600	.0030	.0036	.0041	.0046	.0050	.1421	.1341	.1267	.1178	.1098
700	.0030	.0036	.0041	.0046	.0050	.1435	.1351	.1275	.1183	.1101
800	.0031	.0036	.0041	.0046	.0049	.1447	.1360	.1282	.1188	.1104
900	.0031	.0036	.0041	.0046	.0049	.1459	.1369	.1289	.1193	.1108
1000	.0031	.0036	.0041	.0045	.0049	.1469	.1378	.1296	.1198	.1111
1200	.0031	.0036	.0040	.0044	.0047	.1490	.1394	.1309	.1208	.1118
1400	.0031	.0036	.0040	.0043	.0046	.1508	.1409	.1321	.1217	.1125
1700	.0031	.0035	.0039	.0042	.0044	.1532	.1428	.1336	.1229	.1135
2000	.0030	.0035	.0038	.0040	.0042	.1551	.1444	.1350	.1240	.1144
2300	.0030	.0034	.0037	.0039	.0041	.1564	.1455	.1359	.1248	.1152
2600	.0030	.0033	.0036	.0038	.0039	.1575	.1465	.1368	.1255	.1160
3000	.0029	.0032	.0035	.0037	.0038	.1587	.1477	.1379	.1264	.1169
3500	.0029	.0031	.0033	.0035	.0036	.1601	.1490	.1391	.1274	.1179
4000	.0028	.0030	.0032	.0033	.0034	.1614	.1503	.1403	.1283	.1189
4500	.0028	.0029	.0030	.0032	.0032	.1626	.1516	.1414	.1292	.1198
5000	.0027	.0028	.0029	.0030	.0030	.1638	.1529	.1425	.1301	.1206
6000	.0025	.0026	.0026	.0027	.0024	.1668	.1563	.1455	.1324	.1223
7000	.0024	.0024	.0024	.0024	.0019	.1693	.1593	.1481	.1343	.1238
8000	.0022	.0023	.0021	.0022	.0016	.1714	.1617	.1501	.1359	.1250
9000	.0021	.0021	.0019	.0020	.0016	.1732	.1636	.1518	.1372	.1259
10000	.0019	.0020	.0017	.0018	.0017	.1746	.1651	.1531	.1381	.1267
12000	.0017	.0016	.0014	.0015	.0031	.1763	.1662	.1541	.1389	.1273
14000	.0015	.0013	.0011	.0012	.0045	.1773	.1667	.1545	.1393	.1276
17000	.0012	.0009	.0008	.0010	.0063	.1783	.1668	.1545	.1393	.1278
20000	.0009	.0006	.0005	.0008	.0075	.1788	.1666	.1543	.1392	.1277
23000	.0007	.0004	.0004	.0006	.0069	.1792	.1668	.1544	.1391	.1277
26000	.0006	.0003	.0003	.0005	.0060	.1794	.1669	.1545	.1390	.1276
30000	.0004	.0002	.0002	.0003	.0048	.1794	.1668	.1544	.1388	.1274
35000	.0003	.0001	.0001	.0001	.0033	.1793	.1666	.1542	.1384	.1270
40000	.0002	.0000	.0000	.0001	.0020	.1789	.1663	.1539	.1379	.1266
45000	.0001	.0000	.0000	.0000	.0009	.1785	.1660	.1535	.1375	.1260
50000	0.0000	0.0000	0.0000	0.0000	0.0000	.1780	.1655	.1531	.1370	.1255
60000	0.0000	0.0000	0.0000	0.0000	0.0000	.1769	.1645	.1520	.1359	.1243
70000	0.0000	0.0000	0.0000	0.0000	0.0000	.1756	.1633	.1509	.1349	.1231
80000	0.0000	0.0000	0.0000	0.0000	0.0000	.1743	.1621	.1497	.1339	.1218
90000	0.0000	0.0000	0.0000	0.0000	0.0000	.1730	.1609	.1484	.1329	.1205
100000	0.0000	0.0000	0.0000	0.0000	0.0000	.1716	.1596	.1472	.1319	.1193

AU ON AG, M2/M1=0.55

100000 KEV

10000 KEV

1000 KEV

100 KEV

10 KEV

	FRACTIONAL DEPOSITED ENERGY					MEAN RANGE,MICROGRAM/SQ.CM.				
K/KL=	0.5	0.65	0.8	1.0	1.3	0.5	0.65	0.8	1.0	1.3
E,KEV										
2.0	.9100	.8851	.8611	.8304	.7870	.9252	.9212	.9172	.9119	.9042
2.3	.9080	.8826	.8581	.8269	.7827	1.022	1.017	1.013	1.007	.9979
2.6	.9062	.8804	.8555	.8237	.7789	1.111	1.106	1.101	1.094	1.084
3.0	.9041	.8777	.8523	.8199	.7743	1.222	1.216	1.210	1.203	1.192
3.5	.9017	.8748	.8488	.8158	.7694	1.351	1.345	1.338	1.330	1.317
4.0	.8997	.8722	.8458	.8122	.7650	1.474	1.467	1.460	1.450	1.437
4.5	.8978	.8698	.8430	.8089	.7611	1.592	1.584	1.576	1.566	1.551
5.0	.8961	.8677	.8405	.8059	.7575	1.706	1.698	1.689	1.678	1.661
6.0	.8931	.8639	.8360	.8007	.7512	1.926	1.915	1.906	1.893	1.873
7.0	.8904	.8607	.8322	.7961	.7458	2.135	2.123	2.112	2.097	2.075
8.0	.8881	.8579	.8288	.7921	.7410	2.336	2.323	2.311	2.294	2.270
9.0	.8860	.8552	.8257	.7885	.7367	2.530	2.516	2.502	2.484	2.457
10	.8841	.8528	.8230	.7853	.7329	2.718	2.703	2.687	2.667	2.638
12	.8807	.8486	.8180	.7795	.7260	3.069	3.051	3.033	3.010	2.976
14	.8778	.8450	.8138	.7745	.7201	3.405	3.385	3.364	3.338	3.299
17	.8739	.8402	.8082	.7680	.7125	3.888	3.863	3.839	3.808	3.761
20	.8707	.8362	.8035	.7625	.7060	4.349	4.321	4.293	4.256	4.203
23	.8678	.8327	.7994	.7577	.7004	4.773	4.741	4.710	4.669	4.609
26	.8652	.8296	.7958	.7535	.6955	5.184	5.149	5.114	5.069	5.003
30	.8621	.8258	.7915	.7484	.6896	5.717	5.678	5.639	5.587	5.512
35	.8587	.8217	.7866	.7428	.6831	6.365	6.319	6.274	6.215	6.129
40	.8556	.8180	.7824	.7378	.6774	6.993	6.941	6.890	6.824	6.726
45	.8528	.8147	.7785	.7333	.6722	7.605	7.547	7.490	7.415	7.307
50	.8503	.8116	.7749	.7292	.6675	8.201	8.137	8.074	7.991	7.871
60	.8457	.8060	.7685	.7218	.6589	9.303	9.228	9.154	9.058	8.917
70	.8417	.8011	.7628	.7153	.6515	10.38	10.29	10.21	10.10	9.935
80	.8380	.7967	.7578	.7095	.6448	11.44	11.34	11.24	11.12	10.93
90	.8347	.7927	.7531	.7042	.6388	12.48	12.37	12.26	12.11	11.91
100	.8316	.7890	.7489	.6993	.6333	13.50	13.38	13.25	13.09	12.86
120	.8261	.7824	.7414	.6907	.6235	15.38	15.24	15.09	14.91	14.63
140	.8212	.7766	.7347	.6831	.6150	17.26	17.08	16.92	16.70	16.38
170	.8147	.7688	.7258	.6731	.6038	20.06	19.85	19.64	19.37	18.98
200	.8089	.7619	.7180	.6643	.5940	22.86	22.60	22.35	22.02	21.55
230	.8037	.7558	.7111	.6565	.5853	25.33	25.04	24.75	24.38	23.85
260	.7989	.7501	.7047	.6494	.5775	27.83	27.50	27.18	26.76	26.17
300	.7930	.7432	.6970	.6407	.5680	31.24	30.85	30.47	29.98	29.28
350	.7863	.7354	.6882	.6309	.5573	35.60	35.13	34.67	34.07	33.22
400	.7801	.7282	.6802	.6221	.5478	40.07	39.49	38.93	38.22	37.20
450	.7743	.7216	.6728	.6139	.5390	44.60	43.91	43.25	42.40	41.19
500	.7689	.7153	.6659	.6064	.5310	49.17	48.36	47.58	46.59	45.20
600	.7590	.7040	.6534	.5927	.5164	57.43	56.45	55.50	54.29	52.58
700	.7500	.6937	.6421	.5805	.5036	66.17	64.95	63.77	62.28	60.18
800	.7416	.6843	.6318	.5694	.4920	75.30	73.78	72.34	70.51	67.96
900	.7338	.6755	.6223	.5592	.4815	84.73	82.88	81.12	78.91	75.85
1000	.7264	.6673	.6134	.5497	.4718	94.39	92.17	90.06	87.44	83.82
1200	.7128	.6522	.5972	.5325	.4542	112.1	109.3	106.7	103.5	98.91
1400	.7003	.6385	.5826	.5172	.4388	131.1	127.6	124.3	120.1	114.4
1700	.6833	.6200	.5631	.4969	.4186	161.8	156.6	151.9	146.2	138.4
2000	.6679	.6035	.5458	.4791	.4012	194.3	187.3	180.9	173.1	162.9
2300	.6534	.5882	.5299	.4629	.3855	223.9	215.6	208.0	198.7	186.3
2600	.6400	.5741	.5153	.4482	.3714	255.2	245.3	236.0	224.8	210.0
3000	.6235	.5569	.4978	.4306	.3548	299.2	286.5	274.7	260.5	242.0
3500	.6047	.5375	.4782	.4111	.3365	357.4	340.4	324.8	306.1	282.4
4000	.5875	.5200	.4606	.3939	.3206	418.2	396.2	376.1	352.3	322.9
4500	.5716	.5041	.4448	.3785	.3064	480.9	453.2	428.1	398.8	363.4
5000	.5569	.4894	.4303	.3645	.2937	544.9	511.1	480.7	445.4	403.6
6000	.5297	.4626	.4042	.3397	.2717	669.0	623.5	583.0	536.2	481.8
7000	.5057	.4393	.3818	.3186	.2532	797.5	738.5	686.3	626.8	558.8
8000	.4843	.4188	.3622	.3004	.2373	928.8	854.8	790.0	716.6	634.5
9000	.4650	.4005	.3448	.2845	.2236	1062.	971.8	893.4	805.4	708.7
10000	.4475	.3839	.3293	.2703	.2115	1196.	1089.	996.2	892.9	781.4
12000	.4166	.3552	.3026	.2463	.1913	1463.	1321.	1199.	1064.	922.3
14000	.3901	.3309	.2804	.2266	.1749	1729.	1550.	1397.	1229.	1057.
17000	.3565	.3006	.2530	.2027	.1554	2119.	1884.	1684.	1467.	1250.
20000	.3283	.2755	.2308	.1837	.1401	2498.	2207.	1959.	1694.	1432.

TA ON GE, M2/M1=0.403   TABLE 15- 2

	MEAN DAMAGE DEPTH,MICROGRAM/SQ.CM.					MEAN IONZN.DEPTH,MICROGRAM/SQ.CM.				
K/KL=	0.5	0.65	0.8	1.0	1.3	0.5	0.65	0.8	1.0	1.3
E,KEV										
2.0	.6954	.6905	.6857	.6794	.6702	.6425	.6371	.6319	.6251	.6152
2.3	.7681	.7625	.7569	.7497	.7392	.7095	.7034	.6974	.6897	.6784
2.6	.8350	.8287	.8226	.8145	.8029	.7712	.7644	.7578	.7491	.7366
3.0	.9181	.9110	.9041	.8951	.8820	.8478	.8402	.8328	.8230	.8090
3.5	1.015	1.007	.9996	.9894	.9747	.9375	.9289	.9205	.9096	.8937
4.0	1.108	1.099	1.090	1.079	1.062	1.023	1.013	1.004	.9916	.9739
4.5	1.196	1.185	1.177	1.165	1.147	1.104	1.094	1.084	1.070	1.051
5.0	1.282	1.271	1.261	1.247	1.228	1.183	1.172	1.161	1.146	1.125
6.0	1.446	1.434	1.422	1.406	1.384	1.335	1.322	1.309	1.292	1.268
7.0	1.603	1.589	1.576	1.558	1.532	1.480	1.465	1.450	1.431	1.403
8.0	1.754	1.739	1.723	1.704	1.675	1.619	1.602	1.586	1.564	1.533
9.0	1.900	1.883	1.866	1.844	1.812	1.753	1.735	1.716	1.693	1.659
10	2.042	2.023	2.004	1.980	1.945	1.883	1.863	1.843	1.817	1.780
12	2.305	2.283	2.262	2.234	2.194	2.126	2.102	2.079	2.049	2.006
14	2.558	2.533	2.508	2.476	2.431	2.359	2.332	2.305	2.271	2.222
17	2.921	2.891	2.862	2.825	2.771	2.693	2.661	2.629	2.589	2.531
20	3.268	3.234	3.200	3.157	3.095	3.012	2.975	2.939	2.892	2.826
23	3.587	3.549	3.511	3.463	3.393	3.306	3.264	3.224	3.172	3.097
26	3.897	3.854	3.813	3.759	3.682	3.591	3.545	3.500	3.442	3.359
30	4.300	4.251	4.204	4.143	4.056	3.961	3.908	3.858	3.792	3.699
35	4.789	4.733	4.679	4.609	4.509	4.410	4.350	4.291	4.216	4.109
40	5.264	5.201	5.140	5.061	4.948	4.846	4.778	4.712	4.627	4.506
45	5.728	5.657	5.589	5.500	5.374	5.271	5.195	5.121	5.027	4.892
50	6.181	6.102	6.026	5.928	5.790	5.686	5.602	5.520	5.415	5.266
60	7.017	6.925	6.836	6.721	6.559	6.454	6.354	6.259	6.136	5.961
70	7.837	7.730	7.627	7.495	7.308	7.205	7.090	6.979	6.837	6.637
80	8.643	8.521	8.403	8.253	8.040	7.942	7.811	7.685	7.524	7.296
90	9.436	9.299	9.166	8.997	8.758	8.668	8.520	8.377	8.196	7.940
100	10.22	10.06	9.916	9.727	9.462	9.382	9.216	9.058	8.856	8.572
120	11.66	11.48	11.30	11.08	10.77	10.70	10.51	10.32	10.08	9.745
140	13.09	12.88	12.68	12.42	12.05	12.01	11.78	11.56	11.29	10.90
170	15.25	14.98	14.73	14.41	13.96	13.97	13.69	13.42	13.08	12.60
200	17.40	17.08	16.77	16.38	15.85	15.93	15.58	15.25	14.84	14.27
230	19.31	18.95	18.60	18.16	17.55	17.68	17.28	16.91	16.44	15.79
260	21.25	20.84	20.44	19.95	19.26	19.44	19.00	18.58	18.05	17.31
300	23.89	23.40	22.94	22.35	21.55	21.83	21.31	20.81	20.19	19.33
350	27.25	26.66	26.10	25.39	24.43	24.87	24.23	23.63	22.88	21.86
400	30.69	29.98	29.30	28.46	27.33	27.95	27.18	26.47	25.58	24.38
450	34.16	33.32	32.53	31.55	30.22	31.06	30.15	29.31	28.28	26.88
500	37.66	36.69	35.77	34.63	33.12	34.18	33.13	32.16	30.97	29.37
600	44.04	42.84	41.71	40.32	38.47	39.93	38.62	37.43	35.98	34.03
700	50.72	49.25	47.87	46.18	43.94	45.89	44.28	42.82	41.07	38.72
800	57.65	55.87	54.20	52.16	49.49	52.01	50.06	48.31	46.20	43.42
900	64.77	62.64	60.66	58.24	55.11	58.25	55.93	53.85	51.37	48.12
1000	72.03	69.52	67.20	64.38	60.76	64.57	61.86	59.43	56.54	52.80
1200	85.44	82.32	79.42	75.92	71.42	76.42	73.06	70.01	66.38	61.74
1400	99.65	95.76	92.16	87.84	82.33	88.73	84.58	80.81	76.33	70.68
1700	122.2	117.0	112.1	106.3	99.07	107.9	102.4	97.34	91.40	84.09
2000	146.0	139.1	132.7	125.3	116.1	127.7	120.6	114.1	106.4	97.42
2300	167.6	159.4	152.0	143.1	132.2	146.2	137.8	130.1	121.1	110.3
2600	190.3	180.6	171.8	161.3	148.5	165.3	155.2	146.2	135.6	123.0
3000	222.0	209.9	199.0	186.1	170.6	191.2	178.9	167.8	154.9	139.8
3500	263.5	248.0	234.0	217.7	198.3	224.5	208.9	195.0	179.0	160.4
4000	306.7	287.3	269.9	249.8	226.3	258.3	239.2	222.2	202.8	180.7
4500	351.1	327.4	306.3	282.0	254.2	292.5	269.6	249.3	226.3	200.6
5000	396.2	368.0	342.9	314.3	282.0	326.8	299.9	276.2	249.5	220.0
6000	483.2	446.3	413.6	376.8	335.7	393.4	358.8	328.5	294.5	257.6
7000	573.1	526.4	485.3	439.4	389.0	460.2	417.4	379.9	338.3	293.7
8000	664.9	607.7	557.5	501.7	441.5	526.8	475.2	430.3	380.7	328.6
9000	757.8	689.5	629.7	563.7	493.3	593.0	532.3	479.7	422.0	362.1
10000	851.3	771.3	701.7	625.0	544.3	658.5	588.5	528.0	462.1	394.6
12000	1038.	934.3	844.2	745.4	643.5	787.0	698.1	621.7	539.0	456.4
14000	1224.	1095.	984.0	862.6	739.2	912.0	804.0	711.4	612.1	514.7
17000	1497.	1330.	1188.	1032.	876.5	1093.	955.9	839.3	715.3	596.4
20000	1763.	1558.	1384.	1194.	1007.	1265.	1100.	960.1	811.9	672.3

TA ON GE, M2/M1=0.403

TABLE 15- 3

E, KEV / (K/KL=)	RELATIVE RANGE STRAGGLING 0.5	0.65	0.8	1.0	1.3	RELATIVE DAMAGE STRAGGLING 0.5	0.65	0.8	1.0	1.3
2.0	.1691	.1680	.1670	.1656	.1636	.3699	.3685	.3672	.3655	.3630
2.3	.1690	.1679	.1668	.1654	.1633	.3698	.3684	.3670	.3653	.3627
2.6	.1689	.1678	.1667	.1653	.1631	.3697	.3683	.3669	.3651	.3625
3.0	.1688	.1677	.1665	.1651	.1629	.3696	.3682	.3667	.3649	.3622
3.5	.1687	.1675	.1664	.1649	.1626	.3695	.3680	.3665	.3646	.3619
4.0	.1686	.1674	.1662	.1647	.1624	.3694	.3679	.3664	.3644	.3617
4.5	.1685	.1673	.1661	.1645	.1622	.3693	.3677	.3662	.3643	.3615
5.0	.1684	.1672	.1659	.1643	.1620	.3692	.3676	.3661	.3641	.3612
6.0	.1682	.1670	.1657	.1640	.1616	.3691	.3674	.3658	.3638	.3609
7.0	.1681	.1668	.1655	.1638	.1613	.3690	.3673	.3656	.3635	.3605
8.0	.1680	.1666	.1653	.1636	.1610	.3689	.3671	.3654	.3633	.3602
9.0	.1678	.1665	.1651	.1633	.1608	.3688	.3670	.3653	.3631	.3600
10	.1677	.1663	.1649	.1632	.1605	.3687	.3669	.3651	.3629	.3597
12	.1675	.1661	.1646	.1628	.1601	.3685	.3667	.3649	.3626	.3593
14	.1673	.1658	.1644	.1625	.1597	.3684	.3665	.3646	.3623	.3589
17	.1670	.1655	.1640	.1620	.1592	.3683	.3662	.3643	.3619	.3584
20	.1668	.1652	.1637	.1617	.1587	.3681	.3660	.3641	.3615	.3580
23	.1666	.1650	.1634	.1613	.1583	.3680	.3659	.3638	.3612	.3576
26	.1664	.1647	.1631	.1610	.1579	.3679	.3657	.3636	.3609	.3572
30	.1661	.1644	.1627	.1606	.1574	.3678	.3655	.3633	.3606	.3568
35	.1658	.1640	.1623	.1601	.1569	.3676	.3653	.3630	.3602	.3562
40	.1655	.1637	.1619	.1597	.1564	.3675	.3651	.3628	.3598	.3558
45	.1652	.1634	.1616	.1592	.1559	.3674	.3649	.3625	.3595	.3553
50	.1650	.1631	.1612	.1588	.1554	.3674	.3648	.3623	.3592	.3549
60	.1645	.1625	.1606	.1581	.1546	.3673	.3646	.3620	.3587	.3542
70	.1640	.1620	.1600	.1574	.1538	.3672	.3644	.3616	.3582	.3536
80	.1635	.1614	.1594	.1568	.1530	.3671	.3642	.3613	.3578	.3530
90	.1631	.1609	.1589	.1562	.1523	.3671	.3640	.3610	.3573	.3524
100	.1627	.1605	.1583	.1556	.1516	.3670	.3638	.3607	.3569	.3518
120	.1619	.1596	.1573	.1544	.1503	.3669	.3634	.3602	.3561	.3507
140	.1611	.1587	.1564	.1534	.1491	.3668	.3631	.3596	.3554	.3497
170	.1600	.1575	.1550	.1519	.1474	.3666	.3626	.3588	.3542	.3482
200	.1590	.1563	.1537	.1504	.1458	.3663	.3620	.3580	.3531	.3467
230	.1580	.1552	.1525	.1491	.1443	.3662	.3616	.3573	.3521	.3454
260	.1570	.1541	.1513	.1478	.1428	.3660	.3611	.3566	.3511	.3441
300	.1558	.1527	.1498	.1461	.1410	.3656	.3604	.3556	.3498	.3424
350	.1542	.1510	.1480	.1441	.1388	.3650	.3594	.3543	.3481	.3404
400	.1528	.1494	.1462	.1422	.1367	.3644	.3584	.3529	.3464	.3383
450	.1514	.1479	.1446	.1404	.1347	.3636	.3573	.3515	.3447	.3363
500	.1500	.1464	.1429	.1387	.1328	.3627	.3561	.3501	.3430	.3343
600	.1474	.1435	.1398	.1353	.1291	.3610	.3537	.3471	.3394	.3303
700	.1449	.1408	.1369	.1321	.1257	.3590	.3512	.3441	.3360	.3265
800	.1425	.1382	.1341	.1292	.1225	.3569	.3486	.3412	.3327	.3228
900	.1402	.1357	.1315	.1264	.1195	.3548	.3460	.3383	.3295	.3194
1000	.1381	.1333	.1290	.1237	.1167	.3525	.3435	.3355	.3264	.3161
1200	.1339	.1288	.1242	.1187	.1114	.3477	.3382	.3298	.3203	.3099
1400	.1300	.1247	.1199	.1141	.1066	.3429	.3331	.3244	.3147	.3042
1700	.1247	.1190	.1139	.1079	.1002	.3361	.3259	.3170	.3071	.2965
2000	.1199	.1140	.1087	.1025	.0946	.3296	.3193	.3103	.3003	.2898
2300	.1154	.1092	.1038	.0975	.0896	.3233	.3129	.3039	.2940	.2836
2600	.1112	.1048	.0993	.0929	.0850	.3173	.3070	.2980	.2882	.2780
3000	.1062	.0996	.0940	.0875	.0797	.3100	.2998	.2910	.2813	.2714
3500	.1006	.0938	.0881	.0816	.0738	.3018	.2918	.2831	.2737	.2640
4000	.0956	.0887	.0829	.0765	.0688	.2945	.2847	.2762	.2670	.2576
4500	.0911	.0841	.0783	.0719	.0644	.2879	.2783	.2701	.2611	.2520
5000	.0870	.0800	.0742	.0679	.0606	.2820	.2726	.2645	.2557	.2469
6000	.0797	.0728	.0671	.0611	.0541	.2719	.2629	.2551	.2467	.2383
7000	.0736	.0668	.0613	.0555	.0488	.2634	.2547	.2473	.2392	.2311
8000	.0684	.0616	.0563	.0507	.0445	.2561	.2477	.2405	.2327	.2249
9000	.0638	.0572	.0521	.0467	.0408	.2497	.2416	.2346	.2271	.2195
10000	.0598	.0533	.0484	.0432	.0376	.2441	.2362	.2294	.2221	.2147
12000	.0530	.0469	.0423	.0376	.0325	.2346	.2271	.2206	.2137	.2067
14000	.0475	.0418	.0375	.0332	.0285	.2269	.2197	.2135	.2068	.2000
17000	.0411	.0359	.0320	.0282	.0241	.2177	.2108	.2048	.1984	.1920
20000	.0362	.0314	.0280	.0246	.0210	.2105	.2038	.1980	.1917	.1855

K/KL= E,KEV	RELATIVE IONIZATION STRAGGLING					RELATIVE TRANSVERSE RANGE STRAGGLING				
	0.5	0.65	0.8	1.0	1.3	0.5	0.65	0.8	1.0	1.3
2.0	.4205	.4197	.4189	.4180	.4167	.5556	.5560	.5563	.5568	.5575
2.3	.4204	.4196	.4189	.4179	.4166	.5555	.5559	.5563	.5568	.5575
2.6	.4204	.4196	.4188	.4178	.4165	.5555	.5559	.5562	.5567	.5575
3.0	.4204	.4195	.4187	.4177	.4163	.5554	.5558	.5562	.5567	.5575
3.5	.4203	.4195	.4187	.4176	.4162	.5553	.5557	.5561	.5566	.5574
4.0	.4203	.4194	.4186	.4175	.4161	.5552	.5556	.5560	.5566	.5574
4.5	.4203	.4194	.4185	.4175	.4160	.5551	.5555	.5559	.5565	.5574
5.0	.4203	.4194	.4185	.4174	.4159	.5550	.5554	.5559	.5564	.5573
6.0	.4203	.4193	.4184	.4173	.4158	.5548	.5552	.5557	.5563	.5572
7.0	.4202	.4193	.4183	.4172	.4157	.5546	.5550	.5555	.5561	.5571
8.0	.4202	.4192	.4183	.4171	.4156	.5543	.5548	.5553	.5560	.5569
9.0	.4202	.4192	.4183	.4171	.4155	.5541	.5546	.5551	.5558	.5568
10	.4202	.4192	.4182	.4170	.4154	.5539	.5544	.5550	.5556	.5567
12	.4203	.4192	.4182	.4169	.4153	.5535	.5541	.5546	.5553	.5564
14	.4203	.4192	.4181	.4169	.4151	.5531	.5537	.5542	.5550	.5561
17	.4204	.4192	.4181	.4168	.4150	.5525	.5531	.5537	.5545	.5557
20	.4204	.4192	.4181	.4167	.4149	.5519	.5526	.5532	.5540	.5552
23	.4205	.4193	.4181	.4167	.4148	.5514	.5520	.5527	.5535	.5548
26	.4206	.4193	.4181	.4167	.4147	.5508	.5515	.5522	.5531	.5544
30	.4207	.4194	.4182	.4166	.4146	.5501	.5508	.5515	.5524	.5538
35	.4209	.4195	.4182	.4166	.4146	.5492	.5499	.5507	.5517	.5532
40	.4211	.4196	.4183	.4166	.4145	.5483	.5491	.5499	.5510	.5525
45	.4213	.4197	.4183	.4166	.4144	.5475	.5483	.5492	.5503	.5519
50	.4214	.4198	.4184	.4166	.4144	.5467	.5475	.5484	.5496	.5512
60	.4218	.4201	.4185	.4167	.4143	.5451	.5460	.5470	.5482	.5500
70	.4222	.4203	.4187	.4167	.4142	.5436	.5446	.5456	.5469	.5489
80	.4225	.4206	.4189	.4168	.4142	.5421	.5432	.5443	.5457	.5478
90	.4229	.4208	.4190	.4168	.4141	.5407	.5419	.5431	.5446	.5468
100	.4232	.4211	.4192	.4169	.4141	.5394	.5406	.5419	.5435	.5458
120	.4239	.4216	.4195	.4170	.4140	.5368	.5382	.5396	.5413	.5439
140	.4246	.4220	.4198	.4171	.4139	.5345	.5360	.5374	.5393	.5421
170	.4255	.4226	.4201	.4172	.4137	.5313	.5330	.5346	.5367	.5398
200	.4262	.4231	.4204	.4173	.4135	.5284	.5303	.5321	.5344	.5378
230	.4271	.4237	.4207	.4173	.4133	.5256	.5276	.5296	.5322	.5359
260	.4278	.4241	.4209	.4174	.4131	.5231	.5253	.5274	.5302	.5341
300	.4287	.4246	.4212	.4173	.4127	.5201	.5225	.5248	.5278	.5322
350	.4295	.4251	.4213	.4171	.4123	.5168	.5195	.5221	.5254	.5302
400	.4302	.4254	.4213	.4169	.4118	.5140	.5169	.5198	.5234	.5286
450	.4307	.4255	.4212	.4165	.4112	.5116	.5148	.5178	.5217	.5273
500	.4312	.4256	.4211	.4162	.4107	.5095	.5129	.5162	.5204	.5264
600	.4319	.4256	.4205	.4152	.4094	.5058	.5096	.5133	.5180	.5248
700	.4322	.4253	.4198	.4142	.4082	.5030	.5072	.5113	.5166	.5240
800	.4322	.4248	.4190	.4132	.4070	.5009	.5056	.5100	.5157	.5238
900	.4319	.4241	.4182	.4121	.4059	.4994	.5044	.5093	.5154	.5241
1000	.4314	.4234	.4172	.4111	.4048	.4983	.5038	.5089	.5155	.5248
1200	.4293	.4212	.4149	.4086	.4025	.4972	.5033	.5092	.5166	.5269
1400	.4270	.4190	.4126	.4063	.4005	.4971	.5039	.5104	.5186	.5300
1700	.4236	.4158	.4096	.4033	.3980	.4985	.5062	.5136	.5228	.5356
2000	.4205	.4130	.4069	.4008	.3961	.5010	.5097	.5179	.5281	.5421
2300	.4175	.4104	.4045	.3987	.3946	.5044	.5140	.5229	.5340	.5493
2600	.4149	.4080	.4025	.3971	.3935	.5084	.5188	.5285	.5404	.5567
3000	.4118	.4054	.4002	.3953	.3924	.5143	.5257	.5363	.5493	.5670
3500	.4085	.4026	.3980	.3937	.3916	.5224	.5350	.5466	.5608	.5799
4000	.4058	.4005	.3963	.3926	.3911	.5308	.5446	.5571	.5724	.5929
4500	.4036	.3988	.3950	.3918	.3910	.5396	.5543	.5678	.5841	.6057
5000	.4019	.3974	.3941	.3914	.3912	.5484	.5641	.5784	.5956	.6183
6000	.3997	.3961	.3935	.3917	.3923	.5660	.5837	.5995	.6184	.6431
7000	.3983	.3955	.3936	.3924	.3938	.5836	.6029	.6201	.6405	.6668
8000	.3975	.3953	.3940	.3935	.3954	.6009	.6217	.6401	.6617	.6894
9000	.3970	.3954	.3946	.3947	.3971	.6178	.6400	.6595	.6822	.7110
10000	.3968	.3958	.3954	.3959	.3989	.6344	.6578	.6782	.7019	.7318
12000	.3969	.3968	.3973	.3986	.4023	.6664	.6920	.7140	.7392	.7708
14000	.3975	.3982	.3994	.4014	.4056	.6969	.7243	.7478	.7742	.8070
17000	.3989	.4007	.4026	.4054	.4103	.7400	.7698	.7949	.8227	.8568
20000	.4007	.4033	.4059	.4093	.4146	.7805	.8122	.8387	.8674	.9024

RELATIVE TRANSV. DAMAGE STRAGGLING / RELATIVE TRANSV. IONZN. STRAGGLING

K/KL= E,KEV	0.5	0.65	0.8	1.0	1.3	0.5	0.65	0.8	1.0	1.3
2.0	.4558	.4532	.4507	.4473	.4424	.3820	.3788	.3757	.3717	.3657
2.3	.4555	.4529	.4503	.4468	.4418	.3817	.3784	.3753	.3711	.3650
2.6	.4553	.4525	.4499	.4464	.4413	.3814	.3781	.3748	.3706	.3644
3.0	.4549	.4522	.4494	.4459	.4407	.3810	.3776	.3743	.3700	.3636
3.5	.4545	.4517	.4489	.4453	.4400	.3806	.3771	.3737	.3693	.3628
4.0	.4542	.4513	.4485	.4448	.4393	.3802	.3767	.3732	.3687	.3621
4.5	.4539	.4509	.4480	.4443	.4387	.3799	.3763	.3727	.3681	.3614
5.0	.4536	.4506	.4476	.4438	.4382	.3795	.3759	.3723	.3676	.3608
6.0	.4530	.4499	.4469	.4430	.4372	.3789	.3752	.3715	.3667	.3597
7.0	.4525	.4493	.4462	.4422	.4363	.3784	.3745	.3708	.3658	.3587
8.0	.4520	.4488	.4456	.4415	.4355	.3779	.3739	.3701	.3651	.3578
9.0	.4516	.4483	.4451	.4409	.4348	.3774	.3734	.3695	.3643	.3569
10	.4511	.4478	.4445	.4402	.4341	.3769	.3729	.3689	.3637	.3561
12	.4503	.4469	.4435	.4391	.4328	.3761	.3719	.3678	.3625	.3547
14	.4496	.4460	.4426	.4381	.4316	.3753	.3710	.3668	.3614	.3534
17	.4485	.4449	.4413	.4367	.4300	.3743	.3698	.3655	.3599	.3517
20	.4475	.4438	.4402	.4354	.4286	.3733	.3687	.3643	.3585	.3502
23	.4466	.4428	.4391	.4343	.4273	.3723	.3677	.3631	.3573	.3488
26	.4457	.4419	.4381	.4332	.4261	.3715	.3667	.3621	.3561	.3474
30	.4446	.4407	.4368	.4318	.4245	.3704	.3655	.3608	.3546	.3458
35	.4433	.4392	.4353	.4301	.4227	.3690	.3641	.3592	.3529	.3439
40	.4420	.4379	.4338	.4286	.4210	.3678	.3627	.3578	.3514	.3422
45	.4408	.4366	.4324	.4271	.4194	.3666	.3614	.3564	.3499	.3405
50	.4396	.4353	.4311	.4257	.4179	.3655	.3602	.3551	.3484	.3389
60	.4374	.4329	.4286	.4231	.4151	.3633	.3579	.3526	.3458	.3360
70	.4352	.4307	.4263	.4206	.4124	.3613	.3557	.3503	.3433	.3334
80	.4332	.4286	.4241	.4183	.4099	.3594	.3537	.3481	.3410	.3309
90	.4313	.4266	.4220	.4160	.4076	.3576	.3517	.3461	.3388	.3285
100	.4294	.4246	.4200	.4139	.4053	.3558	.3499	.3441	.3368	.3262
120	.4259	.4210	.4162	.4100	.4012	.3525	.3464	.3405	.3329	.3221
140	.4227	.4176	.4127	.4064	.3973	.3494	.3431	.3370	.3292	.3182
170	.4182	.4129	.4078	.4013	.3920	.3451	.3386	.3323	.3242	.3128
200	.4140	.4086	.4033	.3966	.3871	.3412	.3344	.3279	.3196	.3079
230	.4100	.4045	.3991	.3922	.3824	.3374	.3305	.3238	.3153	.3033
260	.4063	.4006	.3951	.3880	.3781	.3339	.3268	.3200	.3113	.2990
300	.4017	.3958	.3901	.3828	.3726	.3296	.3222	.3152	.3062	.2937
350	.3964	.3903	.3844	.3769	.3664	.3246	.3169	.3096	.3004	.2876
400	.3915	.3852	.3791	.3714	.3607	.3199	.3120	.3045	.2950	.2819
450	.3869	.3804	.3742	.3663	.3554	.3156	.3075	.2997	.2900	.2766
500	.3827	.3760	.3697	.3616	.3504	.3116	.3032	.2953	.2853	.2716
600	.3750	.3680	.3614	.3529	.3413	.3040	.2952	.2869	.2766	.2624
700	.3682	.3609	.3539	.3452	.3332	.2973	.2881	.2794	.2687	.2542
800	.3620	.3544	.3471	.3381	.3258	.2912	.2815	.2726	.2615	.2466
900	.3563	.3484	.3409	.3316	.3190	.2856	.2756	.2663	.2549	.2397
1000	.3510	.3429	.3352	.3256	.3127	.2804	.2700	.2605	.2488	.2333
1200	.3416	.3328	.3246	.3145	.3012	.2711	.2601	.2499	.2376	.2217
1400	.3332	.3240	.3153	.3048	.2910	.2629	.2513	.2406	.2278	.2114
1700	.3223	.3123	.3031	.2920	.2778	.2520	.2396	.2283	.2148	.1981
2000	.3127	.3022	.2926	.2810	.2665	.2424	.2294	.2175	.2036	.1866
2300	.3042	.2932	.2832	.2713	.2565	.2333	.2198	.2076	.1934	.1764
2600	.2966	.2852	.2748	.2626	.2477	.2250	.2112	.1987	.1843	.1674
3000	.2874	.2756	.2649	.2524	.2373	.2149	.2008	.1881	.1736	.1569
3500	.2774	.2651	.2541	.2412	.2261	.2038	.1894	.1766	.1620	.1455
4000	.2685	.2559	.2446	.2316	.2164	.1939	.1794	.1665	.1520	.1358
4500	.2605	.2477	.2362	.2230	.2079	.1850	.1704	.1576	.1432	.1274
5000	.2533	.2403	.2287	.2154	.2003	.1770	.1624	.1496	.1354	.1199
6000	.2405	.2273	.2156	.2024	.1875	.1626	.1484	.1360	.1222	.1076
7000	.2296	.2164	.2047	.1915	.1769	.1504	.1366	.1246	.1114	.0977
8000	.2201	.2069	.1953	.1822	.1678	.1400	.1267	.1151	.1024	.0894
9000	.2118	.1986	.1871	.1741	.1600	.1310	.1181	.1069	.0948	.0824
10000	.2043	.1913	.1798	.1670	.1531	.1231	.1106	.0998	.0881	.0764
12000	.1915	.1787	.1675	.1550	.1416	.1100	.0982	.0882	.0774	.0667
14000	.1809	.1683	.1573	.1451	.1322	.0995	.0885	.0791	.0690	.0592
17000	.1678	.1556	.1450	.1332	.1209	.0873	.0772	.0686	.0595	.0507
20000	.1571	.1453	.1350	.1238	.1119	.0781	.0688	.0609	.0526	.0446

TABLE 15- 6

TA ON GE, M2/M1=0.403

E,KEV	RANGE SKEWNESS					DAMAGE SKEWNESS				
K/KL=	0.5	0.65	0.8	1.0	1.3	0.5	0.65	0.8	1.0	1.3
2.0	.6101	.6031	.5962	.5870	.5735	.7484	.7407	.7332	.7234	.7094
2.3	.6094	.6022	.5951	.5858	.5720	.7480	.7401	.7324	.7224	.7080
2.6	.6088	.6014	.5942	.5846	.5706	.7476	.7395	.7316	.7215	.7069
3.0	.6080	.6005	.5931	.5833	.5689	.7471	.7388	.7308	.7204	.7055
3.5	.6072	.5994	.5918	.5818	.5671	.7466	.7381	.7299	.7192	.7039
4.0	.6064	.5985	.5907	.5805	.5654	.7462	.7375	.7291	.7182	.7026
4.5	.6056	.5976	.5897	.5793	.5639	.7458	.7370	.7284	.7173	.7014
5.0	.6050	.5968	.5887	.5781	.5625	.7455	.7365	.7277	.7164	.7003
6.0	.6037	.5953	.5870	.5760	.5600	.7450	.7357	.7266	.7150	.6983
7.0	.6026	.5939	.5854	.5742	.5578	.7446	.7350	.7257	.7137	.6967
8.0	.6015	.5927	.5839	.5725	.5557	.7443	.7344	.7249	.7127	.6952
9.0	.6005	.5915	.5826	.5709	.5538	.7440	.7339	.7242	.7117	.6939
10	.5996	.5904	.5813	.5695	.5521	.7438	.7335	.7236	.7108	.6927
12	.5979	.5884	.5791	.5668	.5489	.7435	.7328	.7225	.7093	.6906
14	.5963	.5866	.5770	.5644	.5460	.7433	.7323	.7216	.7080	.6887
17	.5941	.5840	.5741	.5611	.5421	.7432	.7316	.7205	.7064	.6863
20	.5921	.5817	.5714	.5580	.5385	.7431	.7312	.7197	.7050	.6843
23	.5902	.5795	.5690	.5553	.5353	.7430	.7307	.7188	.7037	.6824
26	.5885	.5775	.5668	.5527	.5323	.7430	.7303	.7181	.7026	.6807
30	.5862	.5749	.5639	.5495	.5285	.7430	.7299	.7173	.7012	.6787
35	.5835	.5719	.5605	.5457	.5241	.7433	.7296	.7165	.6998	.6764
40	.5809	.5690	.5573	.5421	.5200	.7436	.7294	.7158	.6985	.6744
45	.5784	.5662	.5542	.5387	.5160	.7441	.7294	.7153	.6974	.6725
50	.5760	.5635	.5513	.5354	.5123	.7447	.7295	.7149	.6964	.6708
60	.5715	.5586	.5458	.5293	.5054	.7463	.7301	.7146	.6951	.6680
70	.5673	.5538	.5407	.5236	.4989	.7480	.7308	.7145	.6938	.6654
80	.5631	.5493	.5357	.5181	.4927	.7496	.7315	.7143	.6926	.6629
90	.5591	.5448	.5309	.5129	.4868	.7512	.7321	.7140	.6913	.6604
100	.5552	.5406	.5263	.5078	.4811	.7527	.7327	.7137	.6901	.6580
120	.5479	.5325	.5176	.4982	.4704	.7558	.7339	.7132	.6876	.6534
140	.5409	.5248	.5093	.4892	.4603	.7585	.7347	.7124	.6849	.6487
170	.5307	.5138	.4974	.4762	.4460	.7619	.7355	.7108	.6808	.6418
200	.5210	.5032	.4860	.4639	.4324	.7645	.7357	.7089	.6765	.6350
230	.5118	.4932	.4753	.4523	.4195	.7669	.7361	.7076	.6732	.6291
260	.5029	.4836	.4649	.4411	.4072	.7686	.7361	.7059	.6696	.6231
300	.4914	.4712	.4517	.4268	.3915	.7702	.7354	.7031	.6643	.6149
350	.4777	.4563	.4358	.4096	.3728	.7712	.7335	.6988	.6572	.6045
400	.4645	.4420	.4205	.3933	.3550	.7713	.7309	.6938	.6494	.5938
450	.4517	.4283	.4059	.3776	.3379	.7707	.7276	.6881	.6412	.5829
500	.4395	.4150	.3918	.3624	.3215	.7695	.7238	.6820	.6326	.5720
600	.4161	.3898	.3650	.3337	.2904	.7691	.7166	.6693	.6143	.5491
700	.3941	.3661	.3398	.3068	.2613	.7658	.7075	.6554	.5956	.5267
800	.3731	.3436	.3159	.2813	.2339	.7600	.6968	.6406	.5768	.5049
900	.3532	.3222	.2932	.2572	.2080	.7523	.6848	.6253	.5581	.4838
1000	.3341	.3017	.2715	.2342	.1834	.7429	.6719	.6095	.5397	.4633
1200	.2981	.2630	.2306	.1907	.1370	.7166	.6414	.5755	.5022	.4233
1400	.2647	.2272	.1927	.1507	.0944	.6883	.6103	.5422	.4666	.3861
1700	.2185	.1778	.1407	.0958	.0363	.6452	.5645	.4943	.4168	.3351
2000	.1763	.1327	.0933	.0459	-.0162	.6026	.5206	.4493	.3709	.2889
2300	.1369	.0906	.0491	-.0004	-.0648	.5588	.4765	.4050	.3267	.2456
2600	.1001	.0515	.0081	-.0432	-.1096	.5172	.4340	.3637	.2859	.2059
3000	.0547	.0033	-.0423	-.0958	-.1644	.4653	.3836	.3130	.2362	.1579
3500	.0027	-.0518	-.0997	-.1555	-.2263	.4061	.3255	.2560	.1807	.1046
4000	-.0449	-.1022	-.1520	-.2097	-.2823	.3527	.2734	.2052	.1313	.0574
4500	-.0890	-.1486	-.2001	-.2594	-.3335	.3044	.2264	.1594	.0871	.0152
5000	-.1301	-.1918	-.2447	-.3053	-.3807	.2604	.1838	.1181	.0473	-.0228
6000	-.2046	-.2698	-.3252	-.3878	-.4650	.1852	.1112	.0477	-.0205	-.0878
7000	-.2714	-.3394	-.3966	-.4608	-.5393	.1209	.0494	-.0120	-.0780	-.1429
8000	-.3320	-.4023	-.4610	-.5264	-.6057	.0652	-.0042	-.0637	-.1277	-.1906
9000	-.3876	-.4599	-.5198	-.5861	-.6660	.0163	-.0512	-.1091	-.1713	-.2324
10000	-.4391	-.5130	-.5738	-.6409	-.7212	-.0272	-.0930	-.1494	-.2100	-.2697
12000	-.5323	-.6086	-.6709	-.7389	-.8196	-.1014	-.1643	-.2182	-.2762	-.3335
14000	-.6151	-.6932	-.7563	-.8248	-.9056	-.1626	-.2232	-.2751	-.3311	-.3866
17000	-.7246	-.8044	-.8683	-.9370	-1.017	-.2370	-.2949	-.3447	-.3985	-.4520
20000	-.8210	-.9017	-.9658	-1.034	-1.114	-.2963	-.3524	-.4008	-.4530	-.5052

K/KL=	IONIZATION SKEWNESS					RANGE KURTOSIS				
	0.5	0.65	0.8	1.0	1.3	0.5	0.65	0.8	1.0	1.3
E,KEV										
2.0	.8252	.8186	.8122	.8039	.7921	3.342	3.325	3.309	3.288	3.257
2.3	.8249	.8181	.8115	.8031	.7910	3.340	3.323	3.306	3.285	3.254
2.6	.8246	.8176	.8109	.8023	.7901	3.339	3.321	3.304	3.282	3.251
3.0	.8242	.8171	.8103	.8014	.7889	3.337	3.319	3.302	3.279	3.247
3.5	.8238	.8165	.8095	.8005	.7877	3.335	3.317	3.299	3.276	3.243
4.0	.8235	.8161	.8089	.7997	.7866	3.333	3.314	3.296	3.273	3.239
4.5	.8232	.8156	.8083	.7989	.7856	3.331	3.312	3.294	3.270	3.236
5.0	.8230	.8153	.8078	.7983	.7848	3.330	3.310	3.292	3.268	3.233
6.0	.8226	.8146	.8069	.7971	.7832	3.327	3.307	3.288	3.263	3.228
7.0	.8224	.8142	.8062	.7961	.7819	3.324	3.304	3.284	3.259	3.223
8.0	.8222	.8137	.8056	.7953	.7808	3.322	3.301	3.281	3.255	3.218
9.0	.8221	.8134	.8051	.7946	.7797	3.319	3.299	3.278	3.252	3.214
10	.8220	.8131	.8047	.7939	.7788	3.317	3.296	3.275	3.249	3.211
12	.8218	.8127	.8039	.7928	.7772	3.314	3.292	3.270	3.243	3.204
14	.8218	.8123	.8033	.7918	.7758	3.310	3.287	3.266	3.238	3.198
17	.8219	.8120	.8026	.7906	.7740	3.305	3.282	3.259	3.231	3.190
20	.8221	.8118	.8020	.7897	.7725	3.301	3.277	3.254	3.224	3.183
23	.8224	.8117	.8016	.7888	.7712	3.296	3.272	3.249	3.219	3.176
26	.8227	.8117	.8013	.7881	.7700	3.292	3.268	3.244	3.213	3.170
30	.8232	.8118	.8009	.7873	.7686	3.287	3.262	3.238	3.206	3.163
35	.8239	.8120	.8006	.7865	.7670	3.282	3.256	3.231	3.199	3.154
40	.8247	.8122	.8004	.7858	.7656	3.276	3.249	3.224	3.191	3.146
45	.8255	.8125	.8003	.7851	.7644	3.271	3.244	3.218	3.185	3.138
50	.8264	.8129	.8002	.7845	.7632	3.266	3.238	3.212	3.178	3.131
60	.8281	.8137	.8003	.7837	.7612	3.256	3.228	3.200	3.166	3.118
70	.8299	.8146	.8004	.7829	.7593	3.247	3.218	3.190	3.155	3.106
80	.8317	.8155	.8006	.7822	.7576	3.239	3.209	3.180	3.144	3.095
90	.8334	.8164	.8007	.7815	.7559	3.231	3.200	3.171	3.135	3.084
100	.8352	.8173	.8009	.7809	.7543	3.223	3.192	3.162	3.125	3.074
120	.8388	.8194	.8016	.7800	.7515	3.208	3.176	3.145	3.107	3.055
140	.8422	.8212	.8020	.7789	.7487	3.194	3.161	3.130	3.091	3.038
170	.8468	.8234	.8023	.7771	.7445	3.174	3.140	3.108	3.069	3.015
200	.8508	.8252	.8022	.7751	.7402	3.156	3.121	3.089	3.048	2.994
230	.8551	.8273	.8025	.7733	.7362	3.139	3.103	3.070	3.028	2.974
260	.8589	.8289	.8023	.7712	.7321	3.123	3.086	3.052	3.010	2.955
300	.8631	.8303	.8014	.7680	.7264	3.102	3.065	3.030	2.988	2.932
350	.8671	.8310	.7995	.7635	.7192	3.079	3.040	3.005	2.963	2.907
400	.8698	.8307	.7969	.7585	.7118	3.057	3.018	2.983	2.940	2.885
450	.8716	.8296	.7937	.7532	.7044	3.037	2.998	2.962	2.919	2.865
500	.8726	.8279	.7900	.7476	.6970	3.019	2.979	2.943	2.901	2.847
600	.8757	.8245	.7822	.7360	.6818	2.983	2.943	2.907	2.865	2.813
700	.8753	.8189	.7731	.7237	.6669	2.952	2.912	2.876	2.835	2.786
800	.8718	.8114	.7628	.7111	.6524	2.925	2.885	2.850	2.810	2.763
900	.8659	.8025	.7518	.6983	.6383	2.901	2.861	2.827	2.789	2.744
1000	.8580	.7924	.7402	.6854	.6246	2.879	2.840	2.807	2.770	2.729
1200	.8301	.7647	.7118	.6561	.5961	2.840	2.802	2.771	2.739	2.704
1400	.8012	.7371	.6845	.6287	.5702	2.808	2.773	2.745	2.716	2.688
1700	.7598	.6983	.6467	.5917	.5361	2.772	2.740	2.717	2.696	2.678
2000	.7221	.6634	.6132	.5594	.5069	2.746	2.719	2.701	2.687	2.680
2300	.6911	.6335	.5842	.5318	.4823	2.725	2.704	2.691	2.684	2.688
2600	.6634	.6068	.5585	.5076	.4612	2.711	2.695	2.688	2.689	2.704
3000	.6311	.5757	.5287	.4799	.4372	2.700	2.692	2.693	2.704	2.732
3500	.5964	.5426	.4974	.4511	.4126	2.697	2.699	2.709	2.732	2.778
4000	.5670	.5146	.4712	.4274	.3927	2.703	2.715	2.735	2.769	2.830
4500	.5416	.4909	.4492	.4078	.3764	2.716	2.738	2.767	2.813	2.888
5000	.5197	.4706	.4305	.3914	.3630	2.734	2.766	2.804	2.860	2.950
6000	.4854	.4402	.4037	.3685	.3449	2.779	2.831	2.887	2.964	3.078
7000	.4582	.4167	.3834	.3519	.3322	2.835	2.907	2.980	3.077	3.213
8000	.4361	.3980	.3678	.3396	.3233	2.901	2.990	3.079	3.194	3.351
9000	.4176	.3829	.3555	.3304	.3170	2.972	3.079	3.182	3.313	3.490
10000	.4019	.3704	.3457	.3233	.3126	3.047	3.171	3.288	3.434	3.629
12000	.3769	.3511	.3313	.3140	.3077	3.205	3.360	3.502	3.677	3.905
14000	.3578	.3373	.3217	.3089	.3062	3.369	3.553	3.717	3.917	4.174
17000	.3365	.3230	.3132	.3061	.3079	3.619	3.841	4.036	4.268	4.566
20000	.3213	.3139	.3091	.3069	.3123	3.868	4.125	4.346	4.608	4.940

TABLE 15- 8

TA ON GE, M2/M1=0.403

	DAMAGE KURTOSIS					IONIZATION KURTOSIS				
K/KL=	0.5	0.65	0.8	1.0	1.3	0.5	0.65	0.8	1.0	1.3
E,KEV										
2.0	3.629	3.601	3.575	3.540	3.492	3.733	3.706	3.680	3.646	3.599
2.3	3.628	3.599	3.572	3.537	3.487	3.732	3.704	3.677	3.643	3.594
2.6	3.626	3.597	3.569	3.534	3.483	3.731	3.702	3.675	3.640	3.590
3.0	3.625	3.595	3.567	3.530	3.479	3.729	3.700	3.672	3.636	3.586
3.5	3.623	3.593	3.563	3.526	3.474	3.728	3.698	3.669	3.632	3.581
4.0	3.622	3.591	3.561	3.523	3.469	3.726	3.696	3.666	3.629	3.577
4.5	3.621	3.589	3.558	3.520	3.465	3.725	3.694	3.664	3.626	3.573
5.0	3.620	3.587	3.556	3.517	3.462	3.724	3.693	3.662	3.624	3.570
6.0	3.618	3.585	3.553	3.512	3.455	3.723	3.690	3.659	3.619	3.564
7.0	3.617	3.583	3.550	3.508	3.450	3.722	3.688	3.656	3.615	3.558
8.0	3.616	3.581	3.547	3.505	3.445	3.721	3.687	3.654	3.612	3.554
9.0	3.615	3.579	3.545	3.502	3.441	3.721	3.685	3.652	3.609	3.550
10	3.615	3.578	3.543	3.499	3.437	3.721	3.684	3.650	3.607	3.547
12	3.614	3.576	3.540	3.494	3.431	3.720	3.683	3.647	3.602	3.540
14	3.614	3.575	3.537	3.490	3.425	3.720	3.682	3.645	3.599	3.535
17	3.615	3.573	3.534	3.485	3.418	3.721	3.681	3.642	3.594	3.528
20	3.615	3.572	3.532	3.481	3.412	3.722	3.680	3.640	3.591	3.523
23	3.614	3.571	3.529	3.477	3.406	3.724	3.680	3.639	3.588	3.518
26	3.614	3.569	3.527	3.474	3.401	3.725	3.680	3.638	3.585	3.514
30	3.614	3.568	3.525	3.470	3.395	3.728	3.681	3.637	3.583	3.509
35	3.615	3.567	3.522	3.466	3.389	3.731	3.682	3.636	3.580	3.503
40	3.617	3.568	3.521	3.463	3.383	3.735	3.684	3.636	3.577	3.498
45	3.619	3.569	3.520	3.460	3.378	3.739	3.686	3.636	3.575	3.494
50	3.623	3.570	3.520	3.458	3.374	3.743	3.688	3.636	3.574	3.490
60	3.632	3.576	3.523	3.457	3.368	3.752	3.692	3.638	3.572	3.483
70	3.641	3.582	3.526	3.456	3.362	3.760	3.697	3.640	3.570	3.478
80	3.651	3.588	3.529	3.455	3.356	3.769	3.702	3.641	3.568	3.472
90	3.662	3.594	3.531	3.454	3.351	3.777	3.707	3.643	3.567	3.467
100	3.672	3.600	3.534	3.452	3.346	3.786	3.712	3.645	3.566	3.462
120	3.695	3.613	3.538	3.449	3.335	3.804	3.723	3.651	3.564	3.454
140	3.715	3.625	3.542	3.444	3.325	3.821	3.733	3.655	3.563	3.446
170	3.742	3.639	3.546	3.437	3.309	3.845	3.747	3.660	3.560	3.434
200	3.762	3.649	3.547	3.430	3.294	3.866	3.758	3.664	3.556	3.421
230	3.771	3.653	3.547	3.425	3.281	3.889	3.771	3.669	3.552	3.410
260	3.776	3.655	3.546	3.419	3.268	3.909	3.782	3.672	3.548	3.398
300	3.782	3.656	3.543	3.410	3.251	3.933	3.793	3.674	3.541	3.381
350	3.789	3.656	3.537	3.397	3.229	3.957	3.803	3.673	3.530	3.360
400	3.795	3.655	3.529	3.382	3.206	3.975	3.808	3.669	3.516	3.339
450	3.801	3.653	3.520	3.366	3.184	3.990	3.810	3.662	3.502	3.317
500	3.808	3.651	3.510	3.349	3.161	4.001	3.809	3.653	3.486	3.295
600	3.856	3.664	3.497	3.310	3.111	4.034	3.812	3.636	3.453	3.250
700	3.889	3.668	3.477	3.271	3.063	4.048	3.802	3.612	3.417	3.205
800	3.908	3.663	3.454	3.231	3.018	4.044	3.782	3.582	3.379	3.162
900	3.912	3.649	3.426	3.192	2.975	4.026	3.753	3.547	3.339	3.120
1000	3.905	3.628	3.396	3.153	2.934	3.994	3.717	3.508	3.298	3.079
1200	3.821	3.544	3.312	3.070	2.855	3.849	3.594	3.398	3.200	2.992
1400	3.727	3.456	3.229	2.993	2.785	3.700	3.473	3.293	3.107	2.913
1700	3.583	3.327	3.113	2.891	2.695	3.496	3.307	3.150	2.984	2.809
2000	3.448	3.209	3.009	2.802	2.619	3.321	3.164	3.028	2.878	2.721
2300	3.331	3.106	2.917	2.723	2.554	3.222	3.072	2.941	2.797	2.653
2600	3.227	3.013	2.835	2.654	2.499	3.143	2.996	2.867	2.728	2.594
3000	3.104	2.906	2.741	2.575	2.438	3.058	2.911	2.785	2.651	2.528
3500	2.973	2.792	2.643	2.494	2.377	2.974	2.827	2.701	2.571	2.460
4000	2.864	2.698	2.563	2.429	2.329	2.906	2.758	2.633	2.507	2.405
4500	2.772	2.621	2.497	2.377	2.291	2.850	2.700	2.576	2.453	2.360
5000	2.694	2.555	2.442	2.334	2.261	2.801	2.651	2.528	2.408	2.322
6000	2.587	2.469	2.372	2.282	2.227	2.713	2.570	2.455	2.343	2.267
7000	2.508	2.405	2.324	2.249	2.208	2.640	2.506	2.398	2.294	2.226
8000	2.447	2.359	2.290	2.227	2.198	2.579	2.453	2.352	2.256	2.194
9000	2.400	2.324	2.266	2.214	2.194	2.526	2.409	2.315	2.225	2.169
10000	2.363	2.299	2.249	2.207	2.195	2.480	2.370	2.283	2.200	2.149
12000	2.311	2.265	2.231	2.205	2.206	2.400	2.306	2.232	2.161	2.119
14000	2.278	2.247	2.226	2.213	2.224	2.332	2.254	2.191	2.133	2.098
17000	2.251	2.240	2.234	2.236	2.258	2.247	2.190	2.144	2.102	2.077
20000	2.241	2.245	2.252	2.266	2.296	2.175	2.137	2.107	2.079	2.064

TABLE 15- 9

TA ON GE, M2/M1=0.403

E,KEV	REFLECTION COEFFICIENT					SPUTTERING EFFICIENCY				
K/KL=	0.5	0.65	0.8	1.0	1.3	0.5	0.65	0.8	1.0	1.3
2.0	.0012	.0012	.0012	.0011	.0011	.0206	.0199	.0192	.0183	.0171
2.3	.0012	.0012	.0012	.0011	.0011	.0205	.0198	.0191	.0182	.0170
2.6	.0012	.0012	.0012	.0011	.0011	.0205	.0197	.0190	.0181	.0169
3.0	.0012	.0012	.0012	.0011	.0011	.0204	.0197	.0190	.0180	.0168
3.5	.0012	.0012	.0012	.0011	.0011	.0203	.0196	.0189	.0179	.0166
4.0	.0012	.0012	.0011	.0011	.0011	.0203	.0195	.0188	.0178	.0165
4.5	.0012	.0012	.0011	.0011	.0010	.0202	.0194	.0187	.0177	.0164
5.0	.0012	.0012	.0011	.0011	.0010	.0202	.0194	.0186	.0177	.0163
6.0	.0012	.0012	.0011	.0011	.0010	.0201	.0193	.0185	.0175	.0162
7.0	.0012	.0012	.0011	.0011	.0010	.0200	.0192	.0184	.0174	.0160
8.0	.0012	.0012	.0011	.0011	.0010	.0200	.0191	.0183	.0173	.0159
9.0	.0012	.0012	.0011	.0011	.0010	.0199	.0190	.0182	.0172	.0158
10	.0012	.0012	.0011	.0011	.0010	.0199	.0190	.0181	.0171	.0157
12	.0012	.0011	.0011	.0011	.0010	.0198	.0189	.0180	.0169	.0155
14	.0012	.0011	.0011	.0011	.0010	.0197	.0188	.0179	.0168	.0153
17	.0012	.0011	.0011	.0010	.0010	.0196	.0186	.0177	.0166	.0151
20	.0012	.0011	.0011	.0010	.0010	.0195	.0185	.0176	.0165	.0149
23	.0012	.0011	.0011	.0010	.0010	.0194	.0184	.0175	.0163	.0148
26	.0012	.0011	.0011	.0010	.0009	.0194	.0183	.0174	.0162	.0147
30	.0012	.0011	.0011	.0010	.0009	.0193	.0182	.0173	.0161	.0145
35	.0011	.0011	.0011	.0010	.0009	.0192	.0181	.0172	.0159	.0143
40	.0011	.0011	.0010	.0010	.0009	.0191	.0180	.0170	.0158	.0142
45	.0011	.0011	.0010	.0010	.0009	.0190	.0179	.0169	.0157	.0141
50	.0011	.0011	.0010	.0010	.0009	.0190	.0179	.0168	.0156	.0139
60	.0011	.0011	.0010	.0010	.0009	.0189	.0177	.0167	.0154	.0137
70	.0011	.0011	.0010	.0009	.0009	.0188	.0176	.0166	.0152	.0135
80	.0011	.0010	.0010	.0009	.0009	.0187	.0175	.0164	.0151	.0133
90	.0011	.0010	.0010	.0009	.0008	.0187	.0175	.0163	.0150	.0132
100	.0011	.0010	.0010	.0009	.0008	.0186	.0174	.0162	.0149	.0130
120	.0011	.0010	.0010	.0009	.0008	.0185	.0173	.0161	.0146	.0128
140	.0011	.0010	.0009	.0009	.0008	.0186	.0172	.0159	.0144	.0126
170	.0010	.0010	.0009	.0009	.0008	.0185	.0171	.0157	.0142	.0123
200	.0010	.0010	.0009	.0008	.0007	.0184	.0169	.0155	.0139	.0120
230	.0010	.0009	.0009	.0008	.0007	.0183	.0167	.0154	.0137	.0118
260	.0010	.0009	.0009	.0008	.0007	.0181	.0166	.0152	.0136	.0116
300	.0010	.0009	.0008	.0008	.0007	.0179	.0164	.0150	.0133	.0113
350	.0009	.0009	.0008	.0007	.0006	.0177	.0162	.0147	.0131	.0110
400	.0009	.0008	.0008	.0007	.0006	.0175	.0160	.0145	.0128	.0108
450	.0009	.0008	.0008	.0007	.0006	.0174	.0158	.0143	.0126	.0105
500	.0009	.0008	.0007	.0007	.0005	.0172	.0156	.0141	.0124	.0103
600	.0008	.0008	.0007	.0006	.0005	.0172	.0154	.0138	.0120	.0099
700	.0008	.0007	.0006	.0006	.0005	.0171	.0152	.0136	.0116	.0095
800	.0008	.0007	.0006	.0005	.0004	.0170	.0150	.0133	.0113	.0092
900	.0007	.0006	.0006	.0005	.0004	.0169	.0148	.0130	.0110	.0089
1000	.0007	.0006	.0005	.0005	.0004	.0167	.0146	.0128	.0107	.0086
1200	.0007	.0006	.0005	.0004	.0003	.0161	.0140	.0122	.0102	.0081
1400	.0006	.0005	.0004	.0004	.0003	.0155	.0135	.0117	.0097	.0076
1700	.0005	.0004	.0004	.0003	.0002	.0146	.0127	.0109	.0090	.0070
2000	.0005	.0004	.0003	.0002	.0002	.0138	.0120	.0103	.0084	.0065
2300	.0004	.0003	.0003	.0002	.0001	.0132	.0114	.0097	.0079	.0061
2600	.0004	.0003	.0002	.0002	.0001	.0126	.0108	.0092	.0074	.0057
3000	.0003	.0003	.0002	.0001	.0001	.0120	.0102	.0086	.0069	.0052
3500	.0003	.0002	.0002	.0001	.0001	.0112	.0095	.0079	.0063	.0047
4000	.0002	.0002	.0001	.0001	.0000	.0106	.0089	.0073	.0058	.0043
4500	.0002	.0001	.0001	.0001	.0000	.0100	.0083	.0068	.0054	.0040
5000	.0002	.0001	.0001	.0000	.0000	.0095	.0079	.0064	.0050	.0037
6000	.0001	.0001	.0001	.0000	.0000	.0087	.0071	.0057	.0044	.0033
7000	.0001	.0001	.0000	.0000	.0000	.0080	.0065	.0052	.0040	.0029
8000	.0001	.0000	.0000	.0000	.0000	.0074	.0060	.0048	.0036	.0026
9000	.0000	.0000	.0000	.0000	.0000	.0068	.0055	.0044	.0033	.0024
10000	.0000	.0000	.0000	.0000	.0000	.0064	.0051	.0041	.0031	.0022
12000	0.0000	0.0000	0.0000	0.0000	.0000	.0056	.0045	.0035	.0026	.0019
14000	0.0000	0.0000	0.0000	0.0000	.0000	.0050	.0039	.0031	.0023	.0016
17000	0.0000	0.0000	0.0000	0.0000	.0000	.0042	.0033	.0026	.0019	.0014
20000	0.0000	0.0000	0.0000	0.0000	.0000	.0036	.0028	.0022	.0016	.0011

TABLE 15-10

TA ON GE, M2/M1=0.403

K/KL=	IONIZATION DEFICIENCY					SPUTTERING YIELD ALPHA				
E,KEV	0.5	0.65	0.8	1.0	1.3	0.5	0.65	0.8	1.0	1.3
2.0	.0015	.0019	.0023	.0027	.0033	.1315	.1283	.1253	.1214	.1158
2.3	.0015	.0019	.0023	.0028	.0034	.1313	.1280	.1250	.1210	.1153
2.6	.0016	.0020	.0023	.0028	.0035	.1311	.1278	.1246	.1206	.1148
3.0	.0016	.0020	.0024	.0029	.0035	.1308	.1275	.1243	.1201	.1143
3.5	.0016	.0021	.0024	.0029	.0036	.1306	.1271	.1239	.1196	.1137
4.0	.0017	.0021	.0025	.0030	.0036	.1304	.1268	.1235	.1192	.1132
4.5	.0017	.0021	.0025	.0030	.0037	.1302	.1266	.1232	.1188	.1127
5.0	.0017	.0022	.0026	.0031	.0037	.1300	.1264	.1229	.1185	.1123
6.0	.0018	.0022	.0026	.0031	.0038	.1297	.1260	.1224	.1179	.1115
7.0	.0018	.0023	.0027	.0032	.0039	.1294	.1256	.1220	.1174	.1109
8.0	.0018	.0023	.0027	.0033	.0040	.1292	.1254	.1217	.1170	.1104
9.0	.0019	.0024	.0028	.0033	.0040	.1290	.1251	.1214	.1166	.1099
10	.0019	.0024	.0028	.0034	.0041	.1289	.1249	.1211	.1162	.1095
12	.0020	.0025	.0029	.0034	.0042	.1286	.1245	.1206	.1156	.1087
14	.0020	.0025	.0030	.0035	.0042	.1284	.1242	.1202	.1151	.1081
17	.0021	.0026	.0030	.0036	.0043	.1282	.1238	.1197	.1145	.1073
20	.0021	.0026	.0031	.0037	.0044	.1280	.1235	.1193	.1140	.1067
23	.0022	.0027	.0032	.0037	.0045	.1278	.1233	.1190	.1136	.1061
26	.0022	.0027	.0032	.0038	.0045	.1277	.1231	.1187	.1132	.1057
30	.0022	.0028	.0033	.0038	.0046	.1275	.1228	.1184	.1128	.1051
35	.0023	.0028	.0033	.0039	.0046	.1274	.1226	.1181	.1124	.1046
40	.0023	.0029	.0034	.0040	.0047	.1274	.1225	.1179	.1121	.1042
45	.0024	.0029	.0034	.0040	.0048	.1273	.1224	.1177	.1118	.1038
50	.0024	.0030	.0035	.0041	.0048	.1273	.1223	.1176	.1116	.1034
60	.0025	.0030	.0036	.0042	.0049	.1274	.1223	.1174	.1112	.1028
70	.0025	.0031	.0036	.0042	.0050	.1276	.1223	.1172	.1109	.1024
80	.0026	.0032	.0037	.0043	.0050	.1278	.1223	.1172	.1107	.1020
90	.0026	.0032	.0037	.0044	.0051	.1280	.1225	.1172	.1106	.1017
100	.0027	.0033	.0038	.0044	.0051	.1282	.1226	.1172	.1105	.1015
120	.0028	.0034	.0039	.0045	.0052	.1289	.1229	.1173	.1104	.1011
140	.0028	.0034	.0040	.0046	.0053	.1295	.1233	.1175	.1103	.1008
170	.0029	.0035	.0040	.0046	.0053	.1304	.1239	.1178	.1103	.1006
200	.0030	.0036	.0041	.0047	.0054	.1313	.1245	.1182	.1105	.1005
230	.0031	.0037	.0042	.0048	.0054	.1319	.1251	.1186	.1106	.1004
260	.0031	.0037	.0042	.0048	.0055	.1325	.1255	.1189	.1109	.1004
300	.0032	.0038	.0043	.0049	.0055	.1333	.1262	.1194	.1112	.1005
350	.0033	.0039	.0044	.0049	.0055	.1343	.1270	.1201	.1116	.1006
400	.0033	.0039	.0044	.0050	.0055	.1353	.1278	.1207	.1120	.1008
450	.0034	.0040	.0045	.0050	.0055	.1363	.1286	.1213	.1124	.1010
500	.0034	.0040	.0045	.0050	.0055	.1372	.1294	.1219	.1128	.1012
600	.0035	.0041	.0045	.0050	.0055	.1398	.1313	.1233	.1136	.1016
700	.0036	.0041	.0046	.0050	.0054	.1419	.1330	.1246	.1144	.1020
800	.0037	.0042	.0046	.0050	.0054	.1438	.1345	.1256	.1151	.1024
900	.0037	.0042	.0046	.0050	.0053	.1455	.1357	.1266	.1157	.1027
1000	.0037	.0041	.0045	.0049	.0052	.1468	.1368	.1274	.1162	.1031
1200	.0036	.0040	.0044	.0048	.0050	.1484	.1382	.1286	.1171	.1036
1400	.0034	.0039	.0043	.0046	.0049	.1495	.1392	.1295	.1178	.1042
1700	.0032	.0037	.0040	.0044	.0046	.1508	.1405	.1306	.1188	.1050
2000	.0031	.0035	.0038	.0042	.0044	.1519	.1415	.1316	.1197	.1058
2300	.0030	.0034	.0037	.0040	.0041	.1528	.1425	.1326	.1208	.1070
2600	.0030	.0033	.0036	.0038	.0039	.1536	.1433	.1336	.1218	.1082
3000	.0029	.0032	.0034	.0036	.0036	.1546	.1444	.1347	.1231	.1096
3500	.0029	.0031	.0033	.0034	.0034	.1558	.1457	.1360	.1245	.1111
4000	.0029	.0030	.0031	.0032	.0031	.1571	.1470	.1372	.1257	.1123
4500	.0029	.0029	.0030	.0030	.0029	.1583	.1481	.1383	.1267	.1133
5000	.0028	.0028	.0028	.0028	.0027	.1595	.1493	.1394	.1276	.1142
6000	.0027	.0026	.0026	.0025	.0023	.1630	.1522	.1417	.1289	.1148
7000	.0026	.0024	.0023	.0023	.0020	.1660	.1547	.1435	.1300	.1152
8000	.0025	.0022	.0021	.0021	.0018	.1684	.1567	.1451	.1307	.1154
9000	.0024	.0020	.0019	.0018	.0016	.1704	.1584	.1463	.1314	.1156
10000	.0023	.0018	.0017	.0016	.0014	.1720	.1596	.1472	.1318	.1157
12000	.0020	.0014	.0013	.0012	.0010	.1740	.1614	.1485	.1325	.1159
14000	.0018	.0011	.0009	.0009	.0007	.1749	.1622	.1491	.1328	.1160
17000	.0014	.0005	.0004	.0004	.0003	.1748	.1622	.1491	.1329	.1163
20000	.0010	0.0000	0.0000	0.0000	0.0000	.1733	.1612	.1484	.1327	.1167

TA ON GE, M2/M1=0.403

20000 KEV

2000 KEV

200 KEV

20 KEV

2 KEV

K/KL =	FRACTIONAL DEPOSITED ENERGY					MEAN RANGE, MICROGRAM/SQ.CM.				
E,KEV	0.5	0.65	0.8	1.0	1.3	0.5	0.65	0.8	1.0	1.3
1.0	.9026	.8758	.8501	.8173	.7711	.4543	.4522	.4501	.4474	.4433
1.2	.8997	.8723	.8459	.8123	.7652	.5161	.5136	.5112	.5079	.5032
1.4	.8972	.8692	.8422	.8080	.7600	.5723	.5694	.5666	.5629	.5575
1.7	.8940	.8651	.8375	.8024	.7533	.6501	.6468	.6436	.6393	.6329
2.0	.8912	.8617	.8334	.7976	.7475	.7232	.7195	.7157	.7109	.7037
2.3	.8888	.8586	.8298	.7933	.7424	.7931	.7889	.7847	.7793	.7712
2.6	.8866	.8559	.8266	.7895	.7379	.8606	.8559	.8513	.8453	.8364
3.0	.8839	.8526	.8227	.7850	.7326	.9476	.9423	.9371	.9303	.9202
3.5	.8810	.8490	.8185	.7801	.7267	1.052	1.046	1.040	1.033	1.021
4.0	.8785	.8458	.8148	.7757	.7215	1.153	1.147	1.140	1.131	1.118
4.5	.8761	.8429	.8114	.7717	.7169	1.251	1.244	1.236	1.226	1.212
5.0	.8740	.8403	.8084	.7682	.7127	1.346	1.338	1.329	1.319	1.303
6.0	.8703	.8358	.8031	.7620	.7055	1.522	1.512	1.503	1.490	1.472
7.0	.8670	.8319	.7985	.7567	.6993	1.692	1.681	1.670	1.656	1.635
8.0	.8641	.8284	.7944	.7519	.6938	1.857	1.844	1.832	1.816	1.792
9.0	.8614	.8252	.7907	.7475	.6888	2.017	2.003	1.989	1.971	1.945
10	.8590	.8222	.7873	.7436	.6842	2.173	2.158	2.143	2.123	2.094
12	.8546	.8169	.7811	.7364	.6758	2.463	2.445	2.427	2.404	2.370
14	.8507	.8121	.7756	.7300	.6685	2.745	2.725	2.704	2.678	2.639
17	.8455	.8058	.7683	.7217	.6589	3.159	3.134	3.109	3.077	3.030
20	.8409	.8003	.7620	.7144	.6506	3.562	3.532	3.503	3.465	3.410
23	.8369	.7954	.7563	.7079	.6432	3.924	3.891	3.859	3.816	3.754
26	.8332	.7910	.7512	.7021	.6366	4.284	4.247	4.210	4.163	4.094
30	.8286	.7855	.7450	.6950	.6286	4.761	4.719	4.677	4.622	4.543
35	.8235	.7794	.7380	.6870	.6196	5.358	5.307	5.258	5.193	5.100
40	.8188	.7738	.7317	.6799	.6116	5.952	5.893	5.836	5.761	5.652
45	.8144	.7687	.7258	.6733	.6043	6.545	6.477	6.410	6.324	6.199
50	.8104	.7639	.7204	.6672	.5976	7.133	7.056	6.980	6.882	6.741
60	.8030	.7552	.7107	.6563	.5856	8.204	8.111	8.021	7.903	7.735
70	.7964	.7475	.7019	.6466	.5749	9.299	9.188	9.080	8.940	8.739
80	.7903	.7403	.6940	.6378	.5653	10.42	10.28	10.16	9.990	9.753
90	.7846	.7337	.6866	.6297	.5566	11.55	11.39	11.24	11.05	10.77
100	.7792	.7276	.6798	.6221	.5485	12.70	12.51	12.34	12.11	11.79
120	.7694	.7163	.6673	.6085	.5339	14.77	14.56	14.34	14.07	13.68
140	.7604	.7061	.6561	.5963	.5211	16.96	16.69	16.43	16.10	15.63
170	.7482	.6923	.6411	.5801	.5041	20.42	20.05	19.70	19.26	18.63
200	.7371	.6799	.6277	.5657	.4892	24.05	23.57	23.10	22.52	21.70
230	.7268	.6685	.6154	.5527	.4759	27.20	26.66	26.14	25.47	24.52
260	.7173	.6580	.6041	.5408	.4638	30.54	29.91	29.30	28.52	27.42
300	.7055	.6451	.5904	.5264	.4493	35.28	34.49	33.73	32.75	31.39
350	.6920	.6305	.5749	.5102	.4331	41.61	40.55	39.54	38.26	36.51
400	.6795	.6171	.5609	.4957	.4188	48.30	46.90	45.59	43.96	41.75
450	.6679	.6047	.5480	.4825	.4059	55.27	53.48	51.82	49.79	47.06
500	.6570	.5932	.5361	.4704	.3941	62.43	60.21	58.17	55.70	52.42
600	.6369	.5722	.5144	.4485	.3731	75.78	72.88	70.22	66.99	62.72
700	.6188	.5534	.4953	.4294	.3550	90.12	86.30	82.83	78.66	73.22
800	.6023	.5365	.4782	.4125	.3391	105.2	100.3	95.85	90.58	83.82
900	.5871	.5210	.4628	.3973	.3250	121.0	114.7	109.2	102.7	94.48
1000	.5730	.5068	.4486	.3835	.3123	137.1	129.4	122.7	114.9	105.1
1200	.5470	.4810	.4232	.3590	.2902	168.6	157.9	148.9	138.8	126.0
1400	.5240	.4584	.4012	.3381	.2715	201.4	187.4	175.7	162.8	146.6
1700	.4939	.4292	.3731	.3117	.2482	252.5	232.7	216.5	198.7	177.2
2000	.4679	.4044	.3495	.2898	.2291	304.9	278.8	257.5	234.2	207.1
2300	.4448	.3826	.3291	.2712	.2133	356.9	324.6	298.0	268.9	236.0
2600	.4242	.3635	.3113	.2552	.1997	409.2	370.5	338.2	303.0	264.2
3000	.4000	.3411	.2908	.2369	.1844	479.1	431.3	391.2	347.5	300.8
3500	.3739	.3173	.2690	.2177	.1685	566.1	506.4	456.2	401.5	344.9
4000	.3513	.2969	.2506	.2016	.1553	652.2	580.3	519.7	453.9	387.3
4500	.3316	.2792	.2348	.1879	.1441	737.1	652.8	581.6	504.5	428.1
5000	.3142	.2637	.2209	.1760	.1345	820.6	723.8	641.9	553.8	467.6
6000	.2848	.2377	.1980	.1565	.1190	983.2	861.3	758.0	647.5	542.6
7000	.2607	.2167	.1797	.1412	.1068	1140.	993.0	868.6	736.3	613.2
8000	.2407	.1994	.1648	.1289	.0971	1291.	1119.	974.1	820.6	679.9
9000	.2238	.1849	.1524	.1188	.0893	1436.	1241.	1075.	901.0	743.3
10000	.2093	.1727	.1420	.1104	.0828	1576.	1357.	1172.	977.7	803.7

E,KEV	MEAN DAMAGE DEPTH,MICROGRAM/SQ.CM.					MEAN IONZN.DEPTH,MICROGRAM/SQ.CM.				
K/KL=	0.5	0.65	0.8	1.0	1.3	0.5	0.65	0.8	1.0	1.3
1.0	.3239	.3215	.3192	.3161	.3117	.2994	.2968	.2943	.2910	.2862
1.2	.3683	.3655	.3627	.3590	.3538	.3404	.3373	.3343	.3304	.3247
1.4	.4085	.4052	.4020	.3979	.3920	.3775	.3740	.3705	.3661	.3596
1.7	.4641	.4603	.4566	.4518	.4448	.4289	.4247	.4207	.4155	.4080
2.0	.5163	.5120	.5078	.5023	.4944	.4771	.4724	.4678	.4619	.4533
2.3	.5663	.5615	.5568	.5507	.5419	.5232	.5180	.5129	.5062	.4966
2.6	.6147	.6093	.6041	.5974	.5876	.5678	.5621	.5564	.5491	.5384
3.0	.6772	.6711	.6653	.6576	.6466	.6255	.6189	.6126	.6043	.5923
3.5	.7526	.7457	.7390	.7302	.7177	.6951	.6876	.6803	.6708	.6571
4.0	.8256	.8178	.8102	.8003	.7862	.7624	.7539	.7456	.7349	.7196
4.5	.8963	.8876	.8792	.8682	.8525	.8276	.8181	.8090	.7971	.7800
5.0	.9651	.9555	.9461	.9341	.9168	.8910	.8805	.8704	.8573	.8385
6.0	1.093	1.081	1.070	1.056	1.036	1.009	.9963	.9845	.9692	.9473
7.0	1.216	1.203	1.191	1.175	1.151	1.122	1.108	1.095	1.077	1.052
8.0	1.337	1.322	1.308	1.289	1.263	1.233	1.217	1.202	1.182	1.154
9.0	1.454	1.438	1.422	1.401	1.372	1.342	1.324	1.307	1.284	1.253
10	1.569	1.551	1.533	1.511	1.478	1.448	1.428	1.409	1.384	1.349
12	1.782	1.761	1.740	1.714	1.676	1.644	1.621	1.598	1.569	1.528
14	1.991	1.966	1.942	1.912	1.868	1.836	1.809	1.783	1.750	1.703
17	2.299	2.269	2.240	2.202	2.149	2.120	2.087	2.055	2.014	1.957
20	2.602	2.565	2.530	2.486	2.423	2.398	2.359	2.321	2.272	2.205
23	2.874	2.833	2.794	2.743	2.672	2.649	2.605	2.562	2.507	2.430
26	3.145	3.099	3.055	2.998	2.919	2.899	2.849	2.801	2.739	2.653
30	3.508	3.454	3.403	3.337	3.245	3.233	3.174	3.118	3.047	2.947
35	3.963	3.899	3.837	3.759	3.650	3.651	3.581	3.514	3.430	3.312
40	4.419	4.343	4.271	4.179	4.052	4.070	3.987	3.908	3.809	3.672
45	4.875	4.787	4.703	4.597	4.451	4.489	4.392	4.301	4.186	4.028
50	5.330	5.228	5.132	5.011	4.846	4.906	4.794	4.690	4.559	4.380
60	6.162	6.039	5.922	5.776	5.576	5.672	5.537	5.409	5.251	5.034
70	7.015	6.867	6.726	6.550	6.311	6.456	6.291	6.137	5.947	5.688
80	7.885	7.708	7.540	7.331	7.051	7.252	7.055	6.871	6.646	6.342
90	8.767	8.558	8.361	8.117	7.792	8.058	7.825	7.609	7.346	6.994
100	9.657	9.414	9.186	8.905	8.532	8.869	8.597	8.348	8.045	7.642
120	11.29	10.99	10.71	10.37	9.910	10.38	10.04	9.727	9.356	8.861
140	13.00	12.63	12.29	11.86	11.31	11.94	11.51	11.13	10.68	10.09
170	15.67	15.18	14.73	14.18	13.46	14.36	13.79	13.29	12.70	11.94
200	18.45	17.82	17.24	16.54	15.65	16.85	16.13	15.49	14.74	13.79
230	20.91	20.17	19.50	18.69	17.65	19.13	18.28	17.53	16.65	15.53
260	23.47	22.62	21.83	20.89	19.70	21.48	20.49	19.60	18.57	17.27
300	27.07	26.03	25.07	23.92	22.48	24.71	23.50	22.42	21.16	19.61
350	31.81	30.48	29.27	27.83	26.05	28.88	27.37	26.02	24.44	22.53
400	36.76	35.12	33.62	31.85	29.69	33.17	31.32	29.66	27.74	25.46
450	41.87	39.88	38.07	35.94	33.37	37.52	35.31	33.34	31.05	28.37
500	47.09	44.72	42.58	40.07	37.07	41.92	39.33	37.02	34.35	31.26
600	56.80	53.79	51.08	47.91	44.13	50.22	46.95	44.02	40.67	36.81
700	67.11	63.34	59.93	55.98	51.31	58.79	54.73	51.12	47.00	42.31
800	77.90	73.24	69.05	64.22	58.57	67.54	62.62	58.25	53.30	47.74
900	89.04	83.40	78.35	72.56	65.88	76.41	70.56	65.40	59.57	53.09
1000	100.4	93.73	87.77	80.96	73.18	85.35	78.53	72.52	65.77	58.35
1200	122.4	113.7	106.0	97.28	87.36	102.8	94.09	86.44	77.88	68.58
1400	145.3	134.3	124.7	113.7	101.5	120.4	109.6	100.2	89.73	78.48
1700	180.8	166.0	153.0	138.5	122.6	146.9	132.8	120.5	107.0	92.76
2000	217.2	198.2	181.6	163.1	143.3	173.2	155.6	140.3	123.6	106.4
2300	252.9	229.7	209.4	187.0	163.3	199.0	177.8	159.6	139.7	119.4
2600	288.9	261.2	237.1	210.7	183.0	224.4	199.6	178.3	155.2	131.9
3000	337.3	303.3	273.9	241.8	208.7	257.7	228.0	202.5	175.0	147.9
3500	397.9	355.7	319.3	279.9	240.0	298.4	262.4	231.6	198.7	166.7
4000	458.3	407.6	364.0	317.1	270.4	337.8	295.6	259.6	221.3	184.6
4500	518.3	458.3	408.0	353.4	299.9	376.1	327.7	286.5	242.8	201.6
5000	577.6	509.3	451.0	388.0	328.5	413.4	358.7	312.4	263.5	217.8
6000	693.8	607.8	534.6	457.0	383.5	484.8	418.0	361.5	302.3	248.1
7000	806.6	702.9	614.9	522.2	435.6	552.6	473.9	407.6	338.5	276.3
8000	915.9	794.7	692.1	584.5	485.3	617.1	526.9	451.1	372.4	302.5
9000	1022.	883.3	766.4	644.2	532.8	678.6	577.3	492.3	404.4	327.2
10000	1125.	969.1	838.1	701.6	578.3	737.6	625.4	531.5	434.7	350.5

K/KL=	RELATIVE RANGE STRAGGLING					RELATIVE DAMAGE STRAGGLING				
E,KEV	0.5	0.65	0.8	1.0	1.3	0.5	0.65	0.8	1.0	1.3
1.0	.1311	.1303	.1294	.1283	.1266	.3563	.3549	.3535	.3517	.3492
1.2	.1310	.1301	.1292	.1281	.1264	.3563	.3548	.3534	.3515	.3489
1.4	.1309	.1300	.1291	.1279	.1262	.3563	.3547	.3532	.3513	.3486
1.7	.1308	.1299	.1289	.1277	.1259	.3562	.3546	.3531	.3511	.3483
2.0	.1307	.1297	.1288	.1275	.1257	.3562	.3546	.3530	.3510	.3481
2.3	.1306	.1296	.1286	.1274	.1255	.3562	.3545	.3529	.3508	.3479
2.6	.1305	.1295	.1285	.1272	.1253	.3563	.3545	.3528	.3507	.3477
3.0	.1304	.1294	.1284	.1270	.1251	.3563	.3545	.3528	.3506	.3475
3.5	.1303	.1292	.1282	.1268	.1248	.3564	.3545	.3527	.3504	.3472
4.0	.1302	.1291	.1280	.1266	.1246	.3564	.3545	.3527	.3503	.3470
4.5	.1301	.1290	.1279	.1264	.1243	.3565	.3545	.3526	.3502	.3468
5.0	.1300	.1289	.1277	.1263	.1241	.3566	.3546	.3526	.3501	.3467
6.0	.1298	.1287	.1275	.1260	.1238	.3568	.3547	.3526	.3500	.3464
7.0	.1297	.1285	.1273	.1257	.1234	.3571	.3548	.3526	.3499	.3462
8.0	.1295	.1283	.1271	.1255	.1231	.3573	.3550	.3527	.3499	.3460
9.0	.1294	.1281	.1269	.1252	.1228	.3576	.3551	.3528	.3498	.3458
10	.1293	.1280	.1267	.1250	.1226	.3579	.3553	.3529	.3498	.3456
12	.1290	.1277	.1263	.1246	.1221	.3584	.3557	.3531	.3498	.3454
14	.1288	.1274	.1260	.1242	.1216	.3590	.3561	.3533	.3499	.3452
17	.1285	.1270	.1256	.1237	.1210	.3599	.3567	.3537	.3500	.3450
20	.1282	.1266	.1251	.1232	.1204	.3608	.3573	.3541	.3501	.3448
23	.1279	.1263	.1247	.1227	.1198	.3617	.3580	.3545	.3503	.3446
26	.1276	.1260	.1244	.1223	.1193	.3627	.3587	.3550	.3504	.3444
30	.1273	.1256	.1239	.1217	.1186	.3638	.3595	.3555	.3506	.3442
35	.1269	.1251	.1233	.1210	.1178	.3652	.3604	.3560	.3507	.3438
40	.1264	.1246	.1227	.1204	.1171	.3664	.3612	.3564	.3507	.3434
45	.1260	.1241	.1222	.1198	.1163	.3675	.3619	.3568	.3507	.3430
50	.1256	.1236	.1216	.1191	.1156	.3685	.3625	.3571	.3506	.3425
60	.1248	.1227	.1206	.1180	.1142	.3705	.3638	.3577	.3505	.3417
70	.1241	.1218	.1196	.1168	.1129	.3721	.3647	.3580	.3502	.3406
80	.1233	.1209	.1186	.1157	.1116	.3733	.3653	.3580	.3496	.3395
90	.1225	.1200	.1176	.1146	.1104	.3742	.3656	.3579	.3489	.3382
100	.1218	.1192	.1167	.1136	.1092	.3749	.3657	.3575	.3480	.3369
120	.1203	.1175	.1149	.1115	.1069	.3759	.3655	.3563	.3458	.3339
140	.1189	.1159	.1131	.1096	.1047	.3761	.3647	.3547	.3434	.3309
170	.1168	.1136	.1105	.1068	.1016	.3755	.3629	.3518	.3395	.3263
200	.1148	.1114	.1081	.1041	.0988	.3740	.3604	.3486	.3356	.3219
230	.1128	.1091	.1057	.1015	.0959	.3712	.3570	.3447	.3312	.3173
260	.1109	.1070	.1034	.0991	.0933	.3681	.3534	.3407	.3269	.3130
300	.1085	.1043	.1006	.0960	.0900	.3638	.3486	.3355	.3215	.3075
350	.1056	.1012	.0972	.0925	.0863	.3581	.3426	.3293	.3151	.3013
400	.1028	.0982	.0941	.0892	.0828	.3525	.3368	.3234	.3092	.2955
450	.1003	.0955	.0912	.0861	.0797	.3470	.3312	.3178	.3037	.2903
500	.0978	.0929	.0885	.0833	.0768	.3416	.3259	.3126	.2986	.2855
600	.0931	.0879	.0833	.0780	.0714	.3304	.3154	.3026	.2892	.2768
700	.0889	.0834	.0787	.0733	.0667	.3204	.3060	.2939	.2811	.2694
800	.0851	.0794	.0746	.0692	.0626	.3114	.2977	.2861	.2740	.2628
900	.0816	.0758	.0709	.0655	.0589	.3034	.2903	.2792	.2676	.2570
1000	.0783	.0725	.0676	.0622	.0557	.2961	.2836	.2731	.2620	.2519
1200	.0725	.0666	.0617	.0564	.0501	.2843	.2727	.2630	.2527	.2433
1400	.0674	.0616	.0567	.0515	.0455	.2745	.2637	.2546	.2450	.2361
1700	.0610	.0552	.0505	.0456	.0399	.2624	.2525	.2442	.2354	.2272
2000	.0557	.0500	.0455	.0407	.0355	.2525	.2435	.2358	.2276	.2199
2300	.0513	.0456	.0414	.0369	.0320	.2446	.2361	.2288	.2211	.2137
2600	.0475	.0418	.0380	.0337	.0291	.2378	.2297	.2228	.2155	.2084
3000	.0431	.0376	.0342	.0302	.0260	.2301	.2226	.2161	.2091	.2024
3500	.0386	.0332	.0303	.0266	.0228	.2222	.2151	.2090	.2025	.1960
4000	.0349	.0297	.0272	.0238	.0203	.2157	.2089	.2031	.1968	.1906
4500	.0318	.0268	.0245	.0214	.0182	.2101	.2036	.1980	.1920	.1860
5000	.0291	.0243	.0224	.0195	.0165	.2053	.1990	.1936	.1877	.1819
6000	.0249	.0205	.0189	.0164	.0139	.1974	.1914	.1862	.1806	.1750
7000	.0216	.0178	.0163	.0141	.0120	.1911	.1852	.1802	.1748	.1694
8000	.0191	.0158	.0143	.0124	.0105	.1859	.1802	.1753	.1700	.1646
9000	.0171	.0144	.0128	.0111	.0094	.1816	.1759	.1710	.1658	.1606
10000	.0156	.0134	.0116	.0101	.0086	.1779	.1722	.1674	.1622	.1571

K/KL= E,KEV	RELATIVE IONIZATION STRAGGLING					RELATIVE TRANSVERSE RANGE STRAGGLING				
	0.5	0.65	0.8	1.0	1.3	0.5	0.65	0.8	1.0	1.3
1.0	.4064	.4055	.4048	.4038	.4024	.5220	.5226	.5231	.5238	.5248
1.2	.4065	.4056	.4048	.4038	.4024	.5218	.5223	.5229	.5236	.5247
1.4	.4066	.4057	.4048	.4038	.4024	.5216	.5222	.5227	.5235	.5246
1.7	.4067	.4057	.4049	.4038	.4023	.5213	.5219	.5225	.5232	.5244
2.0	.4068	.4058	.4049	.4038	.4023	.5210	.5216	.5222	.5230	.5242
2.3	.4069	.4059	.4050	.4038	.4023	.5207	.5213	.5219	.5228	.5240
2.6	.4071	.4061	.4051	.4039	.4023	.5204	.5210	.5217	.5225	.5238
3.0	.4073	.4062	.4052	.4040	.4023	.5200	.5207	.5213	.5222	.5235
3.5	.4076	.4064	.4054	.4041	.4024	.5195	.5202	.5209	.5218	.5231
4.0	.4078	.4066	.4056	.4042	.4025	.5190	.5197	.5205	.5214	.5228
4.5	.4081	.4069	.4057	.4044	.4025	.5185	.5193	.5200	.5210	.5224
5.0	.4084	.4071	.4059	.4045	.4026	.5181	.5188	.5196	.5206	.5221
6.0	.4089	.4076	.4063	.4048	.4028	.5171	.5179	.5187	.5198	.5214
7.0	.4095	.4080	.4067	.4051	.4030	.5162	.5171	.5179	.5190	.5207
8.0	.4100	.4085	.4071	.4054	.4032	.5154	.5162	.5171	.5183	.5200
9.0	.4106	.4090	.4075	.4057	.4034	.5145	.5154	.5163	.5175	.5193
10	.4112	.4095	.4079	.4060	.4036	.5137	.5146	.5156	.5168	.5186
12	.4123	.4104	.4087	.4066	.4040	.5120	.5130	.5141	.5154	.5174
14	.4134	.4113	.4095	.4073	.4045	.5105	.5116	.5126	.5140	.5161
17	.4150	.4127	.4106	.4082	.4051	.5082	.5094	.5106	.5121	.5144
20	.4167	.4141	.4117	.4090	.4056	.5061	.5074	.5087	.5103	.5127
23	.4183	.4155	.4129	.4099	.4062	.5041	.5054	.5068	.5085	.5111
26	.4199	.4168	.4140	.4108	.4068	.5021	.5035	.5050	.5068	.5095
30	.4219	.4185	.4154	.4118	.4075	.4996	.5012	.5027	.5047	.5076
35	.4243	.4204	.4170	.4130	.4082	.4968	.4985	.5001	.5023	.5054
40	.4265	.4222	.4184	.4140	.4088	.4942	.4960	.4977	.5001	.5034
45	.4286	.4239	.4196	.4150	.4094	.4917	.4936	.4955	.4980	.5016
50	.4305	.4253	.4208	.4158	.4098	.4894	.4915	.4935	.4961	.4999
60	.4344	.4283	.4231	.4173	.4106	.4850	.4873	.4895	.4925	.4967
70	.4377	.4308	.4249	.4185	.4111	.4812	.4837	.4861	.4893	.4939
80	.4406	.4328	.4264	.4194	.4115	.4778	.4805	.4832	.4866	.4916
90	.4431	.4346	.4276	.4201	.4116	.4747	.4777	.4805	.4843	.4896
100	.4452	.4360	.4285	.4205	.4116	.4720	.4752	.4782	.4822	.4879
120	.4487	.4380	.4296	.4208	.4111	.4672	.4707	.4741	.4786	.4849
140	.4513	.4394	.4301	.4206	.4104	.4632	.4671	.4709	.4758	.4828
170	.4539	.4405	.4302	.4199	.4091	.4586	.4630	.4673	.4729	.4807
200	.4553	.4407	.4297	.4188	.4077	.4552	.4602	.4650	.4711	.4798
230	.4552	.4400	.4284	.4170	.4058	.4527	.4581	.4633	.4700	.4794
260	.4546	.4390	.4270	.4151	.4041	.4508	.4567	.4624	.4696	.4798
300	.4533	.4373	.4249	.4127	.4019	.4493	.4557	.4619	.4698	.4810
350	.4510	.4349	.4222	.4098	.3994	.4484	.4556	.4625	.4712	.4833
400	.4483	.4324	.4196	.4071	.3972	.4485	.4564	.4639	.4733	.4865
450	.4455	.4297	.4170	.4046	.3953	.4493	.4578	.4659	.4760	.4901
500	.4425	.4271	.4146	.4024	.3937	.4506	.4598	.4684	.4792	.4942
600	.4350	.4208	.4092	.3982	.3909	.4544	.4647	.4744	.4865	.5031
700	.4283	.4152	.4047	.3948	.3889	.4592	.4706	.4813	.4946	.5127
800	.4225	.4105	.4010	.3922	.3875	.4647	.4772	.4888	.5031	.5226
900	.4174	.4065	.3979	.3902	.3866	.4707	.4842	.4966	.5119	.5326
1000	.4131	.4031	.3954	.3887	.3860	.4770	.4914	.5046	.5209	.5427
1200	.4074	.3991	.3929	.3876	.3864	.4900	.5062	.5210	.5390	.5629
1400	.4032	.3965	.3915	.3875	.3873	.5034	.5212	.5374	.5569	.5826
1700	.3990	.3941	.3906	.3882	.3893	.5235	.5436	.5616	.5831	.6110
2000	.3963	.3929	.3907	.3895	.3915	.5435	.5655	.5851	.6082	.6380
2300	.3948	.3925	.3913	.3911	.3940	.5632	.5871	.6082	.6327	.6641
2600	.3939	.3927	.3923	.3929	.3964	.5824	.6080	.6303	.6562	.6888
3000	.3935	.3934	.3939	.3954	.3995	.6071	.6347	.6585	.6858	.7198
3500	.3937	.3947	.3960	.3984	.4032	.6366	.6664	.6917	.7205	.7560
4000	.3945	.3963	.3983	.4014	.4067	.6646	.6962	.7229	.7528	.7895
4500	.3956	.3981	.4007	.4042	.4099	.6913	.7245	.7522	.7832	.8207
5000	.3969	.3999	.4030	.4070	.4129	.7168	.7513	.7800	.8118	.8501
6000	.3998	.4037	.4074	.4120	.4183	.7645	.8013	.8315	.8645	.9038
7000	.4030	.4074	.4115	.4165	.4230	.8087	.8471	.8784	.9122	.9523
8000	.4063	.4110	.4154	.4206	.4272	.8498	.8895	.9216	.9560	.9965
9000	.4096	.4145	.4189	.4243	.4309	.8883	.9290	.9618	.9964	1.037
10000	.4128	.4178	.4223	.4276	.4342	.9246	.9660	.9993	1.034	1.075

ZR ON AL, M2/M1=0.297

E,KEV	RELATIVE TRANSV. DAMAGE STRAGGLING					RELATIVE TRANSV. IONZN. STRAGGLING				
K/KL=	0.5	0.65	0.8	1.0	1.3	0.5	0.65	0.8	1.0	1.3
1.0	.4288	.4257	.4226	.4186	.4128	.3626	.3590	.3554	.3508	.3440
1.2	.4284	.4251	.4220	.4179	.4119	.3622	.3584	.3547	.3500	.3430
1.4	.4280	.4247	.4215	.4173	.4111	.3618	.3579	.3542	.3493	.3422
1.7	.4275	.4241	.4208	.4164	.4101	.3613	.3573	.3534	.3484	.3411
2.0	.4271	.4236	.4202	.4157	.4093	.3608	.3568	.3528	.3476	.3401
2.3	.4267	.4231	.4196	.4151	.4085	.3604	.3563	.3522	.3469	.3392
2.6	.4263	.4227	.4191	.4145	.4078	.3601	.3558	.3517	.3463	.3385
3.0	.4259	.4222	.4185	.4138	.4069	.3596	.3553	.3510	.3455	.3375
3.5	.4254	.4215	.4178	.4129	.4059	.3591	.3546	.3503	.3446	.3364
4.0	.4249	.4209	.4171	.4122	.4050	.3586	.3540	.3496	.3438	.3355
4.5	.4244	.4204	.4165	.4114	.4041	.3581	.3535	.3489	.3430	.3346
5.0	.4239	.4199	.4159	.4108	.4033	.3577	.3530	.3483	.3423	.3337
6.0	.4231	.4190	.4149	.4096	.4019	.3569	.3520	.3473	.3411	.3322
7.0	.4224	.4181	.4139	.4084	.4006	.3562	.3512	.3463	.3399	.3308
8.0	.4216	.4172	.4130	.4074	.3994	.3555	.3504	.3453	.3389	.3296
9.0	.4209	.4164	.4121	.4064	.3983	.3548	.3496	.3445	.3379	.3284
10	.4202	.4157	.4112	.4054	.3972	.3542	.3489	.3436	.3369	.3273
12	.4189	.4142	.4096	.4037	.3951	.3530	.3475	.3421	.3352	.3252
14	.4177	.4128	.4081	.4020	.3932	.3519	.3462	.3407	.3335	.3233
17	.4159	.4108	.4059	.3996	.3906	.3503	.3444	.3386	.3313	.3207
20	.4141	.4089	.4039	.3974	.3881	.3488	.3427	.3367	.3291	.3183
23	.4125	.4071	.4020	.3953	.3858	.3473	.3410	.3350	.3271	.3160
26	.4108	.4054	.4001	.3933	.3836	.3459	.3395	.3332	.3252	.3139
30	.4088	.4032	.3978	.3908	.3808	.3441	.3375	.3311	.3228	.3112
35	.4063	.4005	.3950	.3878	.3775	.3419	.3351	.3285	.3200	.3080
40	.4039	.3980	.3923	.3849	.3744	.3399	.3328	.3260	.3173	.3050
45	.4016	.3956	.3897	.3822	.3714	.3379	.3306	.3236	.3147	.3021
50	.3994	.3932	.3872	.3795	.3686	.3359	.3285	.3213	.3122	.2993
60	.3952	.3887	.3825	.3745	.3631	.3322	.3245	.3170	.3075	.2941
70	.3912	.3845	.3780	.3697	.3580	.3287	.3207	.3129	.3031	.2893
80	.3875	.3806	.3739	.3653	.3533	.3254	.3171	.3091	.2989	.2847
90	.3840	.3768	.3699	.3611	.3488	.3223	.3137	.3054	.2950	.2804
100	.3807	.3733	.3662	.3572	.3445	.3193	.3105	.3020	.2912	.2763
120	.3745	.3668	.3593	.3498	.3366	.3138	.3044	.2954	.2841	.2685
140	.3689	.3608	.3530	.3431	.3293	.3086	.2988	.2894	.2775	.2613
170	.3612	.3525	.3443	.3338	.3194	.3015	.2910	.2810	.2685	.2515
200	.3542	.3451	.3364	.3254	.3104	.2950	.2838	.2733	.2602	.2426
230	.3479	.3382	.3289	.3174	.3018	.2890	.2771	.2660	.2522	.2340
260	.3420	.3317	.3221	.3100	.2940	.2833	.2708	.2591	.2449	.2262
300	.3348	.3239	.3136	.3010	.2844	.2763	.2630	.2507	.2358	.2166
350	.3266	.3148	.3039	.2907	.2736	.2682	.2541	.2411	.2255	.2058
400	.3190	.3066	.2952	.2814	.2639	.2607	.2459	.2323	.2162	.1962
450	.3120	.2991	.2871	.2729	.2551	.2537	.2382	.2242	.2077	.1875
500	.3055	.2920	.2797	.2651	.2471	.2471	.2312	.2167	.1998	.1795
600	.2934	.2790	.2660	.2508	.2328	.2349	.2180	.2028	.1854	.1652
700	.2825	.2675	.2540	.2385	.2204	.2239	.2063	.1907	.1730	.1531
800	.2728	.2573	.2434	.2277	.2097	.2138	.1958	.1799	.1621	.1425
900	.2639	.2480	.2340	.2181	.2003	.2046	.1863	.1703	.1525	.1333
1000	.2557	.2396	.2255	.2095	.1920	.1961	.1777	.1616	.1439	.1252
1200	.2407	.2247	.2106	.1949	.1780	.1800	.1619	.1463	.1292	.1118
1400	.2278	.2119	.1981	.1828	.1665	.1660	.1485	.1334	.1171	.1008
1700	.2113	.1959	.1826	.1679	.1525	.1484	.1318	.1176	.1025	.0877
2000	.1977	.1828	.1699	.1558	.1413	.1338	.1182	.1049	.0909	.0775
2300	.1866	.1722	.1597	.1461	.1323	.1223	.1076	.0952	.0820	.0697
2600	.1771	.1632	.1511	.1380	.1247	.1126	.0988	.0871	.0748	.0633
3000	.1663	.1530	.1414	.1289	.1162	.1018	.0890	.0782	.0669	.0564
3500	.1551	.1424	.1314	.1195	.1076	.0908	.0791	.0693	.0590	.0497
4000	.1457	.1336	.1231	.1118	.1005	.0818	.0711	.0622	.0528	.0443
4500	.1378	.1261	.1162	.1053	.0945	.0743	.0646	.0563	.0477	.0400
5000	.1309	.1197	.1101	.0998	.0894	.0681	.0590	.0514	.0435	.0364
6000	.1195	.1092	.1003	.0907	.0811	.0581	.0503	.0438	.0369	.0308
7000	.1105	.1009	.0926	.0836	.0746	.0507	.0438	.0381	.0320	.0267
8000	.1032	.0941	.0863	.0778	.0693	.0451	.0389	.0337	.0283	.0235
9000	.0972	.0885	.0811	.0730	.0650	.0407	.0351	.0303	.0254	.0211
10000	.0921	.0838	.0767	.0690	.0614	.0373	.0320	.0277	.0231	.0191

TABLE 16- 6

7R ON AL, M2/M1=0.297

K/KL=	RANGE SKEWNESS					DAMAGE SKEWNESS				
	0.5	0.65	0.8	1.0	1.3	0.5	0.65	0.8	1.0	1.3
E,KEV										
1.0	.5671	.5605	.5539	.5453	.5326	.7070	.6980	.6892	.6780	.6620
1.2	.5661	.5593	.5525	.5436	.5305	.7073	.6979	.6889	.6772	.6606
1.4	.5652	.5582	.5512	.5421	.5287	.7076	.6979	.6886	.6766	.6595
1.7	.5640	.5567	.5495	.5401	.5263	.7081	.6981	.6883	.6759	.6582
2.0	.5629	.5554	.5480	.5384	.5241	.7088	.6983	.6883	.6754	.6571
2.3	.5619	.5542	.5467	.5368	.5222	.7095	.6987	.6883	.6750	.6563
2.6	.5609	.5531	.5454	.5353	.5204	.7103	.6992	.6885	.6748	.6556
3.0	.5597	.5517	.5438	.5334	.5182	.7115	.6999	.6888	.6747	.6548
3.5	.5583	.5501	.5419	.5313	.5156	.7130	.7010	.6894	.6747	.6541
4.0	.5570	.5485	.5402	.5293	.5133	.7146	.7021	.6901	.6749	.6536
4.5	.5557	.5471	.5386	.5274	.5111	.7163	.7033	.6909	.6752	.6532
5.0	.5545	.5457	.5370	.5256	.5090	.7180	.7046	.6918	.6755	.6529
6.0	.5523	.5432	.5342	.5224	.5052	.7212	.7070	.6934	.6763	.6524
7.0	.5502	.5408	.5315	.5194	.5017	.7245	.7095	.6952	.6772	.6523
8.0	.5482	.5385	.5290	.5165	.4984	.7279	.7122	.6972	.6783	.6523
9.0	.5462	.5363	.5266	.5138	.4952	.7314	.7149	.6992	.6795	.6524
10	.5443	.5342	.5242	.5112	.4923	.7349	.7177	.7013	.6808	.6527
12	.5408	.5303	.5199	.5064	.4867	.7419	.7234	.7058	.6838	.6536
14	.5375	.5266	.5158	.5019	.4815	.7489	.7291	.7103	.6868	.6547
17	.5326	.5212	.5100	.4954	.4742	.7595	.7376	.7169	.6911	.6562
20	.5280	.5161	.5044	.4892	.4673	.7699	.7459	.7233	.6952	.6576
23	.5236	.5113	.4992	.4835	.4608	.7816	.7551	.7303	.6997	.6593
26	.5194	.5067	.4942	.4780	.4547	.7929	.7639	.7369	.7039	.6608
30	.5139	.5007	.4877	.4709	.4467	.8070	.7748	.7449	.7088	.6624
35	.5073	.4934	.4799	.4624	.4372	.8232	.7871	.7539	.7140	.6636
40	.5008	.4863	.4723	.4541	.4280	.8378	.7981	.7618	.7184	.6643
45	.4945	.4795	.4649	.4461	.4192	.8510	.8080	.7687	.7220	.6644
50	.4884	.4728	.4578	.4384	.4106	.8631	.8169	.7748	.7250	.6641
60	.4767	.4602	.4442	.4237	.3944	.8841	.8329	.7862	.7312	.6638
70	.4655	.4481	.4312	.4096	.3790	.9018	.8458	.7948	.7348	.6617
80	.4547	.4363	.4187	.3961	.3641	.9168	.8561	.8010	.7363	.6581
90	.4442	.4250	.4066	.3831	.3499	.9297	.8644	.8053	.7361	.6532
100	.4340	.4140	.3948	.3705	.3362	.9409	.8710	.8079	.7345	.6474
120	.4145	.3930	.3724	.3464	.3099	.9639	.8829	.8106	.7277	.6324
140	.3959	.3730	.3512	.3236	.2852	.9805	.8895	.8090	.7178	.6157
170	.3697	.3447	.3211	.2915	.2506	.9952	.8914	.8004	.6988	.5888
200	.3450	.3182	.2930	.2616	.2183	1.000	.8861	.7866	.6768	.5610
230	.3215	.2930	.2662	.2330	.1876	.9928	.8709	.7652	.6496	.5305
260	.2993	.2691	.2409	.2061	.1588	.9806	.8524	.7418	.6218	.5005
300	.2712	.2389	.2090	.1723	.1227	.9594	.8248	.7093	.5850	.4620
350	.2384	.2038	.1719	.1330	.0809	.9277	.7877	.6680	.5402	.4167
400	.2077	.1709	.1373	.0965	.0423	.8928	.7493	.6271	.4975	.3744
450	.1788	.1401	.1049	.0624	-.0063	.8560	.7105	.5871	.4569	.3350
500	.1515	.1110	.0743	.0303	-.0275	.8183	.6721	.5483	.4183	.2982
600	.1004	.0565	.0172	-.0295	-.0901	.7315	.5891	.4690	.3435	.2298
700	.0539	.0071	-.0344	-.0834	-.1463	.6507	.5132	.3975	.2773	.1698
800	.0111	-.0382	-.0817	-.1326	-.1974	.5768	.4446	.3335	.2185	.1169
900	-.0285	-.0801	-.1253	-.1777	-.2441	.5097	.3827	.2762	.1662	.0698
1000	-.0656	-.1192	-.1658	-.2196	-.2872	.4488	.3268	.2247	.1194	.0278
1200	-.1337	-.1907	-.2397	-.2956	-.3651	.3484	.2351	.1404	.0427	-.0421
1400	-.1946	-.2544	-.3053	-.3628	-.4335	.2639	.1583	.0700	-.0211	-.1006
1700	-.2755	-.3387	-.3916	-.4509	-.5229	.1588	.0631	-.0171	-.1001	-.1731
2000	-.3469	-.4125	-.4670	-.5274	-.6000	.0728	-.0147	-.0883	-.1647	-.2328
2300	-.4104	-.4778	-.5333	-.5943	-.6670	.0022	-.0790	-.1474	-.2187	-.2830
2600	-.4680	-.5367	-.5929	-.6542	-.7268	-.0579	-.1338	-.1979	-.2650	-.3263
3000	-.5374	-.6075	-.6643	-.7258	-.7980	-.1259	-.1960	-.2554	-.3180	-.3761
3500	-.6149	-.6861	-.7433	-.8048	-.8764	-.1961	-.2606	-.3154	-.3736	-.4286
4000	-.6842	-.7562	-.8136	-.8748	-.9457	-.2543	-.3144	-.3658	-.4205	-.4732
4500	-.7471	-.8195	-.8769	-.9378	-1.008	-.3035	-.3602	-.4088	-.4609	-.5119
5000	-.8046	-.8774	-.9346	-.9952	-1.065	-.3456	-.3998	-.4463	-.4964	-.5459
6000	-.9073	-.9802	-1.037	-1.097	-1.164	-.4143	-.4650	-.5086	-.5559	-.6035
7000	-.9970	-1.070	-1.126	-1.184	-1.251	-.4678	-.5166	-.5588	-.6045	-.6509
8000	-1.077	-1.149	-1.204	-1.262	-1.328	-.5107	-.5588	-.6002	-.6453	-.6909
9000	-1.149	-1.221	-1.275	-1.332	-1.396	-.5457	-.5938	-.6353	-.6802	-.7255
10000	-1.215	-1.286	-1.340	-1.395	-1.459	-.5747	-.6234	-.6654	-.7107	-.7558

	IONIZATION SKEWNESS					RANGE KURTOSIS				
K/KL=	0.5	0.65	0.8	1.0	1.3	0.5	0.65	0.8	1.0	1.3
E,KEV										
1.0	.7896	.7818	.7743	.7647	.7512	3.302	3.287	3.273	3.254	3.227
1.2	.7901	.7820	.7742	.7642	.7502	3.300	3.285	3.270	3.251	3.223
1.4	.7906	.7822	.7741	.7639	.7495	3.298	3.282	3.267	3.248	3.219
1.7	.7913	.7825	.7741	.7635	.7486	3.295	3.279	3.263	3.243	3.214
2.0	.7921	.7830	.7743	.7633	.7479	3.293	3.276	3.260	3.240	3.210
2.3	.7929	.7835	.7746	.7632	.7474	3.290	3.274	3.257	3.236	3.206
2.6	.7938	.7842	.7749	.7632	.7470	3.288	3.271	3.255	3.233	3.202
3.0	.7951	.7851	.7755	.7634	.7466	3.286	3.268	3.251	3.229	3.198
3.5	.7969	.7864	.7764	.7638	.7463	3.282	3.265	3.247	3.225	3.193
4.0	.7987	.7877	.7773	.7643	.7462	3.280	3.261	3.243	3.221	3.188
4.5	.8005	.7892	.7784	.7649	.7462	3.277	3.258	3.240	3.217	3.183
5.0	.8024	.7906	.7795	.7655	.7463	3.274	3.255	3.237	3.213	3.179
6.0	.8061	.7935	.7817	.7669	.7466	3.269	3.250	3.231	3.206	3.172
7.0	.8099	.7966	.7840	.7684	.7472	3.265	3.245	3.225	3.200	3.165
8.0	.8137	.7996	.7864	.7700	.7478	3.260	3.240	3.220	3.195	3.159
9.0	.8176	.8028	.7889	.7717	.7485	3.256	3.235	3.215	3.189	3.152
10	.8215	.8059	.7914	.7734	.7492	3.252	3.231	3.210	3.184	3.147
12	.8292	.8122	.7964	.7770	.7510	3.245	3.223	3.201	3.174	3.136
14	.8369	.8185	.8014	.7806	.7528	3.238	3.215	3.193	3.165	3.126
17	.8481	.8276	.8087	.7858	.7554	3.228	3.204	3.182	3.153	3.113
20	.8591	.8365	.8157	.7907	.7578	3.218	3.194	3.171	3.141	3.101
23	.8706	.8458	.8232	.7961	.7607	3.209	3.184	3.161	3.131	3.089
26	.8815	.8547	.8303	.8011	.7634	3.201	3.175	3.151	3.120	3.078
30	.8954	.8657	.8390	.8072	.7664	3.190	3.163	3.139	3.107	3.064
35	.9115	.8784	.8488	.8139	.7694	3.177	3.150	3.124	3.092	3.048
40	.9264	.8900	.8576	.8196	.7717	3.164	3.137	3.111	3.078	3.033
45	.9401	.9005	.8654	.8245	.7734	3.153	3.124	3.098	3.065	3.019
50	.9530	.9101	.8724	.8288	.7746	3.141	3.113	3.085	3.052	3.006
60	.9784	.9291	.8861	.8369	.7767	3.120	3.090	3.062	3.028	2.981
70	1.000	.9446	.8968	.8426	.7770	3.100	3.069	3.041	3.006	2.959
80	1.018	.9572	.9049	.8462	.7758	3.081	3.050	3.021	2.986	2.939
90	1.034	.9673	.9109	.8480	.7736	3.064	3.032	3.003	2.967	2.921
100	1.047	.9753	.9151	.8485	.7703	3.048	3.016	2.986	2.950	2.904
120	1.068	.9865	.9191	.8458	.7609	3.017	2.984	2.954	2.918	2.872
140	1.083	.9927	.9191	.8401	.7500	2.990	2.956	2.926	2.890	2.845
170	1.096	.9950	.9137	.8279	.7319	2.954	2.920	2.890	2.855	2.812
200	1.101	.9916	.9042	.8129	.7131	2.923	2.889	2.860	2.827	2.787
230	1.098	.9822	.8896	.7931	.6913	2.894	2.861	2.832	2.801	2.764
260	1.090	.9702	.8734	.7729	.6700	2.869	2.836	2.809	2.779	2.746
300	1.076	.9518	.8508	.7462	.6431	2.841	2.809	2.784	2.757	2.728
350	1.054	.9265	.8220	.7140	.6120	2.812	2.783	2.759	2.736	2.714
400	1.029	.8999	.7934	.6836	.5836	2.789	2.762	2.742	2.723	2.707
450	1.002	.8727	.7653	.6550	.5578	2.771	2.747	2.730	2.715	2.706
500	.9739	.8454	.7381	.6282	.5342	2.757	2.736	2.722	2.712	2.709
600	.9057	.7818	.6790	.5752	.4910	2.734	2.720	2.713	2.713	2.723
700	.8431	.7212	.6273	.5302	.4553	2.722	2.715	2.716	2.724	2.748
800	.7870	.6746	.5826	.4922	.4258	2.719	2.719	2.727	2.745	2.781
900	.7373	.6307	.5442	.4601	.4013	2.721	2.729	2.744	2.771	2.819
1000	.6933	.5925	.5112	.4331	.3810	2.729	2.744	2.766	2.801	2.861
1200	.6271	.5374	.4655	.3974	.3545	2.753	2.784	2.820	2.872	2.953
1400	.5744	.4948	.4313	.3717	.3361	2.788	2.834	2.884	2.952	3.052
1700	.5130	.4465	.3938	.3450	.3179	2.856	2.923	2.992	3.081	3.207
2000	.4664	.4108	.3671	.3272	.3067	2.935	3.022	3.108	3.216	3.365
2300	.4309	.3841	.3475	.3147	.2997	3.017	3.124	3.223	3.349	3.522
2600	.4028	.3634	.3329	.3061	.2956	3.104	3.230	3.342	3.484	3.677
3000	.3738	.3426	.3188	.2986	.2930	3.227	3.375	3.502	3.663	3.882
3500	.3471	.3241	.3070	.2934	.2927	3.386	3.559	3.705	3.886	4.130
4000	.3279	.3115	.2998	.2914	.2943	3.550	3.744	3.907	4.104	4.371
4500	.3142	.3030	.2955	.2913	.2972	3.715	3.928	4.108	4.319	4.603
5000	.3044	.2975	.2935	.2926	.3008	3.881	4.110	4.307	4.529	4.828
6000	.2934	.2927	.2936	.2976	.3092	4.212	4.468	4.695	4.936	5.257
7000	.2899	.2932	.2975	.3045	.3182	4.537	4.815	5.071	5.324	5.661
8000	.2915	.2972	.3034	.3124	.3273	4.855	5.151	5.435	5.696	6.042
9000	.2964	.3035	.3107	.3207	.3362	5.166	5.476	5.786	6.053	6.404
10000	.3037	.3112	.3188	.3291	.3448	5.468	5.790	6.126	6.396	6.749

E,KEV	DAMAGE KURTOSIS K/KL= 0.5	0.65	0.8	1.0	1.3	IONIZATION KURTOSIS 0.5	0.65	0.8	1.0	1.3
1.0	3.585	3.552	3.520	3.480	3.424	3.689	3.656	3.625	3.585	3.530
1.2	3.586	3.552	3.519	3.478	3.420	3.691	3.657	3.624	3.583	3.526
1.4	3.588	3.552	3.519	3.476	3.416	3.693	3.658	3.624	3.582	3.523
1.7	3.591	3.554	3.518	3.474	3.413	3.696	3.659	3.624	3.580	3.519
2.0	3.594	3.555	3.519	3.473	3.410	3.700	3.661	3.625	3.580	3.517
2.3	3.597	3.558	3.520	3.473	3.407	3.703	3.664	3.626	3.579	3.515
2.6	3.601	3.560	3.521	3.473	3.406	3.707	3.666	3.628	3.580	3.513
3.0	3.607	3.564	3.524	3.473	3.404	3.713	3.671	3.630	3.580	3.512
3.5	3.614	3.569	3.527	3.474	3.403	3.720	3.676	3.634	3.582	3.511
4.0	3.621	3.574	3.530	3.476	3.402	3.728	3.682	3.638	3.584	3.511
4.5	3.628	3.580	3.534	3.478	3.401	3.736	3.688	3.643	3.587	3.511
5.0	3.635	3.585	3.538	3.480	3.401	3.745	3.695	3.648	3.590	3.512
6.0	3.648	3.595	3.546	3.484	3.402	3.761	3.707	3.657	3.596	3.513
7.0	3.662	3.606	3.554	3.489	3.403	3.777	3.720	3.667	3.603	3.516
8.0	3.675	3.617	3.563	3.495	3.405	3.794	3.734	3.678	3.610	3.519
9.0	3.690	3.629	3.572	3.501	3.407	3.811	3.747	3.689	3.617	3.522
10	3.705	3.641	3.581	3.508	3.410	3.828	3.761	3.699	3.625	3.526
12	3.733	3.665	3.601	3.522	3.417	3.862	3.789	3.721	3.640	3.534
14	3.762	3.690	3.622	3.537	3.425	3.896	3.816	3.743	3.656	3.542
17	3.808	3.728	3.652	3.559	3.436	3.946	3.856	3.775	3.678	3.554
20	3.855	3.766	3.682	3.580	3.446	3.996	3.896	3.806	3.700	3.565
23	3.917	3.813	3.717	3.602	3.457	4.047	3.937	3.839	3.724	3.578
26	3.977	3.858	3.750	3.623	3.466	4.097	3.977	3.870	3.746	3.590
30	4.052	3.915	3.791	3.647	3.478	4.160	4.027	3.909	3.773	3.604
35	4.136	3.978	3.837	3.675	3.490	4.234	4.084	3.953	3.803	3.618
40	4.211	4.034	3.877	3.699	3.499	4.302	4.136	3.993	3.829	3.630
45	4.277	4.083	3.912	3.719	3.506	4.366	4.184	4.028	3.852	3.640
50	4.335	4.127	3.943	3.737	3.512	4.426	4.229	4.060	3.872	3.647
60	4.410	4.188	3.992	3.771	3.523	4.546	4.316	4.123	3.910	3.659
70	4.475	4.240	4.032	3.795	3.528	4.648	4.389	4.173	3.938	3.665
80	4.534	4.285	4.063	3.812	3.527	4.735	4.448	4.212	3.957	3.665
90	4.589	4.324	4.089	3.822	3.523	4.809	4.497	4.242	3.969	3.661
100	4.643	4.360	4.110	3.828	3.514	4.872	4.536	4.264	3.976	3.653
120	4.793	4.451	4.155	3.828	3.485	4.979	4.593	4.289	3.972	3.625
140	4.917	4.522	4.182	3.815	3.450	5.054	4.627	4.296	3.957	3.591
170	5.055	4.591	4.198	3.782	3.392	5.123	4.647	4.284	3.919	3.533
200	5.140	4.623	4.189	3.737	3.331	5.153	4.642	4.253	3.869	3.472
230	5.131	4.588	4.134	3.665	3.259	5.133	4.605	4.199	3.798	3.400
260	5.097	4.536	4.069	3.592	3.189	5.095	4.557	4.139	3.726	3.329
300	5.027	4.453	3.977	3.496	3.101	5.028	4.483	4.055	3.630	3.241
350	4.916	4.337	3.860	3.381	3.001	4.928	4.383	3.948	3.516	3.138
400	4.792	4.217	3.745	3.275	2.912	4.819	4.279	3.843	3.409	3.046
450	4.663	4.097	3.634	3.177	2.832	4.705	4.174	3.741	3.310	2.963
500	4.532	3.980	3.529	3.087	2.761	4.590	4.071	3.644	3.218	2.887
600	4.235	3.730	3.320	2.922	2.641	4.328	3.839	3.439	3.044	2.755
700	3.969	3.511	3.141	2.785	2.545	4.092	3.634	3.261	2.899	2.647
800	3.735	3.322	2.990	2.673	2.468	3.882	3.455	3.109	2.777	2.557
900	3.533	3.160	2.862	2.580	2.405	3.698	3.299	2.979	2.674	2.483
1000	3.359	3.022	2.754	2.503	2.355	3.537	3.165	2.868	2.589	2.421
1200	3.119	2.837	2.614	2.407	2.294	3.302	2.977	2.719	2.478	2.339
1400	2.939	2.701	2.514	2.343	2.256	3.118	2.835	2.609	2.400	2.281
1700	2.742	2.557	2.412	2.283	2.224	2.907	2.673	2.489	2.317	2.220
2000	2.603	2.458	2.347	2.249	2.210	2.746	2.554	2.402	2.259	2.178
2300	2.505	2.392	2.305	2.231	2.207	2.623	2.460	2.332	2.212	2.145
2600	2.434	2.345	2.278	2.223	2.211	2.524	2.386	2.277	2.176	2.120
3000	2.366	2.304	2.258	2.222	2.223	2.421	2.309	2.220	2.139	2.095
3500	2.312	2.274	2.248	2.232	2.244	2.324	2.237	2.168	2.105	2.073
4000	2.279	2.261	2.250	2.247	2.269	2.252	2.183	2.130	2.082	2.058
4500	2.260	2.257	2.258	2.267	2.295	2.197	2.144	2.102	2.065	2.048
5000	2.252	2.260	2.271	2.288	2.322	2.155	2.113	2.081	2.053	2.042
6000	2.254	2.279	2.303	2.334	2.376	2.099	2.074	2.054	2.038	2.035
7000	2.272	2.308	2.340	2.379	2.428	2.068	2.052	2.041	2.032	2.035
8000	2.298	2.340	2.379	2.424	2.477	2.053	2.042	2.035	2.032	2.037
9000	2.329	2.375	2.417	2.466	2.524	2.048	2.040	2.036	2.034	2.041
10000	2.363	2.411	2.455	2.506	2.567	2.051	2.044	2.040	2.039	2.047

K/KL=	REFLECTION COEFFICIENT					SPUTTERING EFFICIENCY				
E,KEV	0.5	0.65	0.8	1.0	1.3	0.5	0.65	0.8	1.0	1.3
1.0	.0003	.0003	.0003	.0003	.0003	.0198	.0190	.0183	.0174	.0162
1.2	.0003	.0003	.0003	.0003	.0003	.0197	.0189	.0182	.0173	.0160
1.4	.0003	.0003	.0003	.0003	.0003	.0197	.0189	.0181	.0172	.0159
1.7	.0003	.0003	.0003	.0003	.0002	.0196	.0188	.0180	.0171	.0157
2.0	.0003	.0003	.0003	.0003	.0002	.0195	.0187	.0179	.0169	.0156
2.3	.0003	.0003	.0003	.0003	.0002	.0195	.0186	.0178	.0168	.0155
2.6	.0003	.0003	.0003	.0003	.0002	.0194	.0186	.0178	.0168	.0154
3.0	.0003	.0003	.0003	.0003	.0002	.0194	.0185	.0177	.0166	.0152
3.5	.0003	.0003	.0003	.0003	.0002	.0194	.0184	.0176	.0165	.0151
4.0	.0003	.0003	.0003	.0003	.0002	.0193	.0184	.0175	.0164	.0150
4.5	.0003	.0003	.0003	.0003	.0002	.0193	.0183	.0174	.0164	.0149
5.0	.0003	.0003	.0003	.0003	.0002	.0192	.0183	.0174	.0163	.0148
6.0	.0003	.0003	.0003	.0003	.0002	.0192	.0182	.0173	.0161	.0146
7.0	.0003	.0003	.0003	.0002	.0002	.0191	.0181	.0172	.0160	.0145
8.0	.0003	.0003	.0003	.0002	.0002	.0191	.0181	.0181	.0159	.0143
9.0	.0003	.0003	.0003	.0002	.0002	.0191	.0180	.0170	.0158	.0142
10	.0003	.0003	.0003	.0002	.0002	.0190	.0180	.0170	.0158	.0141
12	.0003	.0003	.0003	.0002	.0002	.0190	.0179	.0169	.0156	.0140
14	.0003	.0003	.0003	.0002	.0002	.0190	.0179	.0168	.0155	.0138
17	.0003	.0003	.0003	.0002	.0002	.0190	.0178	.0167	.0154	.0136
20	.0003	.0003	.0003	.0002	.0002	.0190	.0178	.0167	.0153	.0135
23	.0003	.0003	.0002	.0002	.0002	.0191	.0178	.0166	.0152	.0133
26	.0003	.0003	.0002	.0002	.0002	.0192	.0178	.0166	.0151	.0132
30	.0003	.0003	.0002	.0002	.0002	.0193	.0179	.0165	.0150	.0131
35	.0003	.0003	.0002	.0002	.0002	.0194	.0179	.0165	.0149	.0129
40	.0003	.0003	.0002	.0002	.0002	.0195	.0179	.0164	.0147	.0127
45	.0003	.0003	.0002	.0002	.0002	.0195	.0179	.0164	.0146	.0126
50	.0003	.0002	.0002	.0002	.0002	.0195	.0178	.0163	.0145	.0124
60	.0003	.0002	.0002	.0002	.0002	.0194	.0177	.0161	.0143	.0122
70	.0003	.0002	.0002	.0002	.0002	.0192	.0175	.0159	.0141	.0120
80	.0003	.0002	.0002	.0002	.0002	.0191	.0174	.0158	.0139	.0117
90	.0003	.0002	.0002	.0002	.0002	.0190	.0172	.0156	.0138	.0115
100	.0002	.0002	.0002	.0002	.0001	.0189	.0171	.0155	.0136	.0113
120	.0002	.0002	.0002	.0002	.0001	.0188	.0170	.0152	.0133	.0110
140	.0002	.0002	.0002	.0002	.0001	.0188	.0168	.0150	.0130	.0106
170	.0002	.0002	.0002	.0001	.0001	.0187	.0166	.0147	.0125	.0102
200	.0002	.0002	.0002	.0001	.0001	.0186	.0164	.0144	.0121	.0097
230	.0002	.0002	.0001	.0001	.0001	.0182	.0160	.0140	.0117	.0093
260	.0002	.0002	.0001	.0001	.0001	.0178	.0156	.0136	.0113	.0090
300	.0002	.0001	.0001	.0001	.0001	.0173	.0152	.0132	.0109	.0085
350	.0002	.0001	.0001	.0001	.0001	.0167	.0146	.0126	.0103	.0080
400	.0002	.0001	.0001	.0001	.0000	.0162	.0140	.0121	.0098	.0076
450	.0001	.0001	.0001	.0001	.0000	.0156	.0135	.0116	.0094	.0072
500	.0001	.0001	.0001	.0001	.0000	.0151	.0131	.0112	.0089	.0068
600	.0001	.0001	.0001	.0000	.0000	.0142	.0122	.0103	.0082	.0061
700	.0001	.0001	.0000	.0000	.0000	.0134	.0115	.0096	.0075	.0056
800	.0001	.0001	.0000	.0000	.0000	.0127	.0108	.0090	.0069	.0051
900	.0001	.0000	.0000	.0000	.0000	.0121	.0102	.0084	.0064	.0047
1000	.0001	.0000	.0000	.0000	.0000	.0115	.0097	.0079	.0059	.0043
1200	.0000	.0000	.0000	.0000	.0000	.0106	.0088	.0071	.0053	.0038
1400	.0000	.0000	.0000	.0000	.0000	.0097	.0080	.0064	.0047	.0034
1700	.0000	.0000	.0000	.0000	.0000	.0087	.0070	.0056	.0041	.0030
2000	.0000	.0000	.0000	.0000	.0000	.0078	.0062	.0049	.0036	.0026
2300	.0000	.0000	0.0000	0.0000	0.0000	.0071	.0056	.0044	.0032	.0023
2600	.0000	0.0000	0.0000	0.0000	0.0000	.0064	.0050	.0039	.0029	.0021
3000	0.0000	0.0000	0.0000	0.0000	0.0000	.0057	.0044	.0035	.0025	.0018
3500	0.0000	0.0000	0.0000	0.0000	0.0000	.0049	.0038	.0030	.0022	.0016
4000	0.0000	0.0000	0.0000	0.0000	0.0000	.0043	.0033	.0026	.0019	.0014
4500	0.0000	0.0000	0.0000	0.0000	0.0000	.0038	.0029	.0023	.0017	.0012
5000	0.0000	0.0000	0.0000	0.0000	0.0000	.0033	.0026	.0020	.0015	.0011
6000	0.0000	0.0000	0.0000	0.0000	0.0000	.0027	.0021	.0016	.0012	.0009
7000	0.0000	0.0000	0.0000	0.0000	0.0000	.0022	.0017	.0014	.0010	.0007
8000	0.0000	0.0000	0.0000	0.0000	0.0000	.0019	.0015	.0012	.0009	.0006
9000	0.0000	0.0000	0.0000	0.0000	0.0000	.0017	.0014	.0011	.0008	.0005
10000	0.0000	0.0000	0.0000	0.0000	0.0000	.0016	.0013	.0010	.0007	.0005

	IONIZATION DEFICIENCY					SPUTTERING YIELD ALPHA				
K/KL=	0.5	0.65	0.8	1.0	1.3	0.5	0.65	0.8	1.0	1.3
E,KEV										
1.0	.0016	.0020	.0024	.0029	.0035	.1265	.1231	.1200	.1159	.1101
1.2	.0017	.0021	.0025	.0030	.0036	.1262	.1228	.1195	.1153	.1094
1.4	.0017	.0021	.0025	.0030	.0037	.1260	.1225	.1191	.1148	.1088
1.7	.0017	.0022	.0026	.0031	.0038	.1257	.1221	.1186	.1142	.1080
2.0	.0018	.0022	.0027	.0032	.0039	.1254	.1217	.1182	.1137	.1074
2.3	.0018	.0023	.0027	.0032	.0039	.1252	.1215	.1179	.1133	.1069
2.6	.0019	.0023	.0028	.0033	.0040	.1251	.1212	.1176	.1129	.1064
3.0	.0019	.0024	.0028	.0034	.0041	.1249	.1209	.1172	.1125	.1058
3.5	.0020	.0024	.0029	.0034	.0041	.1247	.1207	.1168	.1120	.1052
4.0	.0020	.0025	.0029	.0035	.0042	.1246	.1204	.1165	.1116	.1047
4.5	.0020	.0025	.0030	.0035	.0043	.1244	.1202	.1162	.1112	.1042
5.0	.0021	.0026	.0030	.0036	.0043	.1243	.1201	.1160	.1109	.1038
6.0	.0021	.0026	.0031	.0037	.0044	.1242	.1198	.1156	.1104	.1032
7.0	.0022	.0027	.0032	.0037	.0045	.1240	.1196	.1153	.1100	.1027
8.0	.0022	.0028	.0032	.0038	.0045	.1240	.1194	.1151	.1097	.1022
9.0	.0023	.0028	.0033	.0039	.0046	.1239	.1193	.1149	.1094	.1018
10	.0023	.0029	.0033	.0039	.0047	.1239	.1192	.1148	.1091	.1015
12	.0024	.0029	.0034	.0040	.0048	.1239	.1191	.1145	.1088	.1009
14	.0025	.0030	.0035	.0041	.0049	.1240	.1191	.1144	.1085	.1005
17	.0026	.0031	.0036	.0042	.0050	.1243	.1192	.1143	.1082	.0999
20	.0026	.0032	.0037	.0043	.0051	.1246	.1193	.1143	.1080	.0995
23	.0027	.0033	.0038	.0044	.0052	.1251	.1196	.1144	.1079	.0992
26	.0028	.0034	.0039	.0045	.0052	.1257	.1200	.1146	.1079	.0990
30	.0029	.0035	.0040	.0046	.0053	.1264	.1204	.1148	.1078	.0988
35	.0030	.0036	.0041	.0047	.0054	.1273	.1210	.1151	.1079	.0986
40	.0030	.0037	.0042	.0048	.0055	.1281	.1216	.1155	.1080	.0985
45	.0031	.0037	.0043	.0049	.0056	.1288	.1221	.1158	.1081	.0984
50	.0032	.0038	.0044	.0050	.0056	.1295	.1226	.1161	.1083	.0983
60	.0033	.0039	.0045	.0051	.0057	.1304	.1233	.1167	.1086	.0984
70	.0034	.0041	.0046	.0052	.0058	.1312	.1240	.1173	.1090	.0984
80	.0035	.0041	.0047	.0052	.0058	.1321	.1248	.1178	.1093	.0986
90	.0036	.0042	.0047	.0053	.0059	.1329	.1255	.1184	.1097	.0987
100	.0037	.0043	.0048	.0054	.0059	.1338	.1262	.1189	.1101	.0988
120	.0038	.0044	.0049	.0054	.0059	.1361	.1279	.1202	.1108	.0992
140	.0039	.0045	.0049	.0055	.0059	.1381	.1295	.1214	.1115	.0995
170	.0040	.0045	.0050	.0055	.0059	.1408	.1316	.1229	.1124	.0999
200	.0041	.0046	.0050	.0055	.0058	.1430	.1334	.1242	.1131	.1004
230	.0041	.0046	.0050	.0054	.0057	.1445	.1345	.1250	.1136	.1006
260	.0041	.0046	.0050	.0054	.0056	.1456	.1355	.1258	.1140	.1008
300	.0041	.0046	.0049	.0053	.0055	.1468	.1365	.1265	.1145	.1012
350	.0041	.0045	.0049	.0051	.0053	.1480	.1375	.1274	.1151	.1016
400	.0040	.0045	.0048	.0050	.0051	.1490	.1384	.1280	.1156	.1021
450	.0040	.0044	.0047	.0049	.0050	.1497	.1390	.1286	.1161	.1026
500	.0040	.0043	.0046	.0048	.0048	.1503	.1396	.1291	.1166	.1031
600	.0039	.0042	.0044	.0045	.0044	.1509	.1401	.1298	.1176	.1045
700	.0038	.0040	.0042	.0042	.0041	.1514	.1406	.1305	.1185	.1059
800	.0037	.0039	.0040	.0040	.0038	.1519	.1411	.1312	.1194	.1070
900	.0036	.0038	.0039	.0038	.0036	.1524	.1418	.1320	.1202	.1080
1000	.0036	.0036	.0037	.0036	.0033	.1530	.1425	.1327	.1210	.1088
1200	.0034	.0034	.0034	.0032	.0030	.1549	.1453	.1351	.1228	.1099
1400	.0033	.0032	.0031	.0029	.0026	.1567	.1478	.1373	.1243	.1107
1700	.0031	.0028	.0027	.0025	.0022	.1590	.1511	.1399	.1261	.1115
2000	.0029	.0025	.0024	.0022	.0019	.1609	.1536	.1420	.1274	.1120
2300	.0026	.0021	.0020	.0018	.0016	.1622	.1546	.1428	.1280	.1124
2600	.0024	.0018	.0016	.0015	.0013	.1632	.1551	.1433	.1284	.1126
3000	.0021	.0013	.0012	.0011	.0009	.1643	.1555	.1436	.1286	.1128
3500	.0018	.0009	.0008	.0007	.0006	.1652	.1556	.1437	.1287	.1129
4000	.0015	.0005	.0005	.0004	.0003	.1659	.1556	.1437	.1287	.1130
4500	.0013	.0002	.0002	.0002	.0001	.1663	.1555	.1435	.1285	.1130
5000	.0010	0.0000	0.0000	0.0000	0.0000	.1666	.1553	.1433	.1284	.1129
6000	.0007	0.0000	0.0000	0.0000	0.0000	.1668	.1549	.1430	.1280	.1127
7000	.0004	0.0000	0.0000	0.0000	0.0000	.1667	.1546	.1427	.1277	.1125
8000	.0003	0.0000	0.0000	0.0000	0.0000	.1664	.1544	.1424	.1274	.1123
9000	.0001	0.0000	0.0000	0.0000	0.0000	.1659	.1543	.1423	.1271	.1120
10000	0.0000	0.0000	0.0000	0.0000	0.0000	.1653	.1543	.1422	.1269	.1118

ZR ON AL, M2/M1=0.297

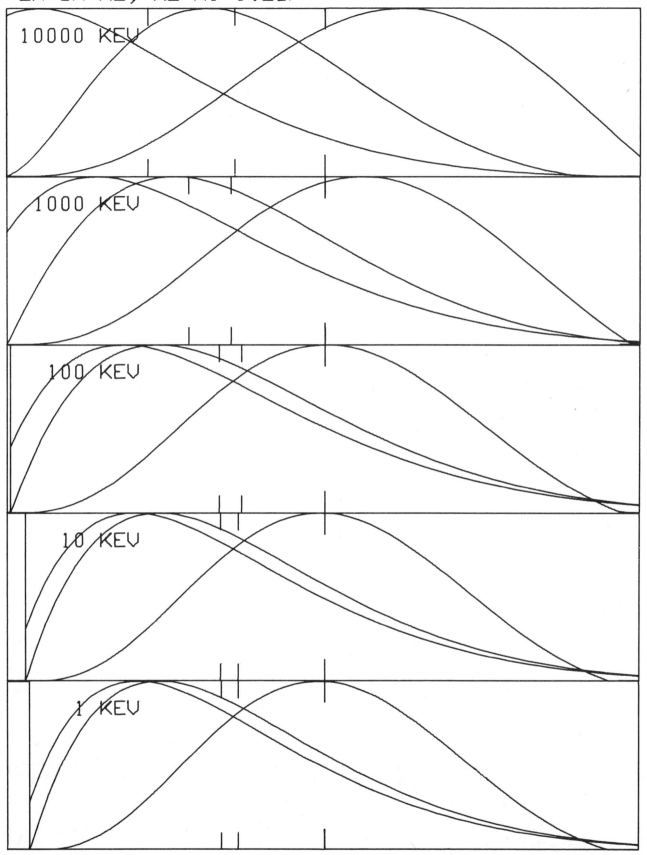

10000 KEV

1000 KEV

100 KEV

10 KEV

1 KEV

257

	FRACTIONAL DEPOSITED ENERGY					MEAN RANGE,MICROGRAM/SQ.CM.				
K/KL=	0.55	0.75	1.0	1.2	1.4	0.55	0.75	1.0	1.2	1.4
E,KEV										
.50	.8917	.8561	.8143	.7827	.7528	.2367	.2352	.2335	.2321	.2307
.60	.8884	.8520	.8091	.7769	.7463	.2688	.2670	.2649	.2633	.2616
.70	.8857	.8484	.8047	.7718	.7407	.2980	.2961	.2937	.2918	.2899
.80	.8832	.8452	.8007	.7673	.7357	.3255	.3233	.3207	.3186	.3165
.90	.8810	.8423	.7971	.7633	.7313	.3518	.3494	.3465	.3442	.3419
1.0	.8789	.8397	.7939	.7596	.7272	.3772	.3746	.3714	.3689	.3664
1.2	.8753	.8351	.7882	.7531	.7201	.4259	.4229	.4192	.4163	.4134
1.4	.8721	.8310	.7832	.7475	.7139	.4726	.4692	.4649	.4616	.4583
1.7	.8680	.8257	.7767	.7401	.7059	.5397	.5356	.5306	.5267	.5228
2.0	.8643	.8211	.7710	.7338	.6989	.6038	.5991	.5933	.5888	.5843
2.3	.8611	.8170	.7660	.7282	.6929	.6628	.6575	.6510	.6459	.6409
2.6	.8582	.8134	.7616	.7232	.6874	.7200	.7141	.7069	.7013	.6957
3.0	.8547	.8089	.7562	.7172	.6808	.7943	.7876	.7794	.7730	.7668
3.5	.8508	.8040	.7502	.7105	.6736	.8844	.8768	.8674	.8600	.8528
4.0	.8473	.7995	.7448	.7045	.6671	.9721	.9634	.9528	.9444	.9363
4.5	.8440	.7955	.7399	.6991	.6613	1.057	1.048	1.036	1.027	1.017
5.0	.8410	.7917	.7354	.6941	.6559	1.141	1.130	1.117	1.107	1.096
6.0	.8357	.7850	.7274	.6853	.6464	1.295	1.282	1.267	1.254	1.243
7.0	.8309	.7791	.7203	.6774	.6380	1.445	1.431	1.413	1.399	1.385
8.0	.8265	.7737	.7139	.6704	.6305	1.593	1.577	1.556	1.541	1.525
9.0	.8225	.7687	.7080	.6640	.6236	1.739	1.720	1.697	1.680	1.662
10	.8188	.7641	.7026	.6581	.6174	1.883	1.862	1.836	1.816	1.797
12	.8120	.7558	.6928	.6475	.6061	2.147	2.122	2.092	2.068	2.046
14	.8058	.7483	.6841	.6381	.5961	2.410	2.381	2.346	2.319	2.292
17	.7976	.7383	.6725	.6256	.5831	2.806	2.770	2.726	2.692	2.659
20	.7902	.7294	.6623	.6147	.5717	3.202	3.158	3.104	3.063	3.023
23	.7834	.7215	.6532	.6049	.5615	3.550	3.500	3.441	3.394	3.349
26	.7772	.7142	.6448	.5959	.5524	3.903	3.848	3.781	3.728	3.678
30	.7695	.7053	.6346	.5851	.5413	4.387	4.321	4.242	4.181	4.121
35	.7606	.6952	.6232	.5730	.5290	5.009	4.929	4.831	4.756	4.684
40	.7525	.6859	.6128	.5620	.5180	5.648	5.549	5.431	5.341	5.254
45	.7449	.6774	.6032	.5520	.5079	6.297	6.179	6.038	5.931	5.828
50	.7378	.6694	.5944	.5428	.4987	6.954	6.815	6.649	6.524	6.403
60	.7246	.6546	.5782	.5260	.4819	8.139	7.970	7.768	7.614	7.466
70	.7127	.6413	.5638	.5113	.4672	9.399	9.187	8.936	8.746	8.564
80	.7018	.6293	.5509	.4981	.4541	10.72	10.46	10.15	9.912	9.690
90	.6917	.6182	.5392	.4862	.4423	12.09	11.76	11.39	11.10	10.84
100	.6822	.6080	.5284	.4753	.4315	13.49	13.10	12.65	12.31	11.99
120	.6650	.5894	.5090	.4558	.4124	16.08	15.59	15.02	14.59	14.19
140	.6495	.5730	.4920	.4388	.3960	18.85	18.23	17.50	16.96	16.46
170	.6289	.5514	.4699	.4169	.3748	23.33	22.43	21.40	20.65	19.96
200	.6106	.5325	.4507	.3981	.3568	28.10	26.84	25.44	24.44	23.53
230	.5940	.5154	.4337	.3815	.3410	32.48	30.95	29.30	28.08	26.98
260	.5788	.5001	.4185	.3667	.3270	37.10	35.23	33.25	31.79	30.46
300	.5605	.4817	.4005	.3494	.3106	43.60	41.17	38.65	36.81	35.15
350	.5401	.4613	.3808	.3305	.2930	52.17	48.90	45.56	43.17	41.05
400	.5217	.4433	.3635	.3142	.2777	61.12	56.88	52.58	49.59	46.96
450	.5051	.4272	.3482	.2997	.2644	70.32	65.02	59.65	56.01	52.84
500	.4900	.4126	.3345	.2869	.2525	79.71	73.26	66.74	62.41	58.68
600	.4626	.3866	.3105	.2646	.2321	98.04	89.44	80.60	75.03	70.07
700	.4389	.3645	.2903	.2460	.2153	116.9	105.8	94.40	87.43	81.24
800	.4182	.3452	.2730	.2303	.2010	136.1	122.3	108.1	99.58	92.16
900	.3997	.3283	.2579	.2166	.1888	155.4	138.7	121.6	111.5	102.8
1000	.3830	.3132	.2446	.2047	.1781	174.8	155.1	134.9	123.1	113.2
1200	.3540	.2873	.2223	.1848	.1604	213.1	187.1	150.8	145.2	133.2
1400	.3296	.2659	.2040	.1688	.1462	251.0	218.6	185.8	166.5	152.4
1700	.2993	.2396	.1820	.1496	.1293	307.0	264.4	221.9	196.9	179.7
2000	.2745	.2184	.1646	.1346	.1161	361.5	308.6	256.3	225.7	205.5
2300	.2539	.2009	.1504	.1224	.1055	414.6	351.3	289.1	253.1	230.1
2600	.2364	.1862	.1385	.1123	.0967	466.1	392.5	320.6	279.3	253.5
3000	.2167	.1699	.1256	.1014	.0873	532.6	445.2	360.7	312.6	283.1
3500	.1967	.1535	.1127	.0906	.0779	612.2	508.1	408.0	351.8	318.0
4000	.1804	.1403	.1025	.0822	.0706	688.3	567.9	452.8	388.8	350.8
4500	.1670	.1295	.0943	.0755	.0648	761.2	624.9	495.2	423.9	381.9
5000	.1557	.1207	.0877	.0700	.0601	831.2	679.5	535.6	457.2	411.4

	MEAN DAMAGE DEPTH,MICROGRAM/SQ.CM.					MEAN IONZN.DEPTH,MICROGRAM/SQ.CM.				
$K/K_L=$	0.55	0.75	1.0	1.2	1.4	0.55	0.75	1.0	1.2	1.4
E,KEV										
.50	.1579	.1564	.1546	.1532	.1518	.1459	.1443	.1423	.1408	.1393
.60	.1797	.1779	.1757	.1741	.1725	.1660	.1641	.1617	.1599	.1581
.70	.1994	.1974	.1949	.1930	.1912	.1843	.1820	.1793	.1772	.1752
.80	.2180	.2157	.2130	.2108	.2088	.2014	.1989	.1959	.1935	.1913
.90	.2357	.2332	.2302	.2278	.2256	.2177	.2150	.2117	.2091	.2066
1.0	.2528	.2501	.2468	.2443	.2418	.2335	.2306	.2269	.2242	.2215
1.2	.2859	.2827	.2789	.2760	.2731	.2641	.2606	.2564	.2531	.2500
1.4	.3178	.3142	.3098	.3064	.3031	.2935	.2895	.2847	.2810	.2774
1.7	.3640	.3596	.3544	.3503	.3464	.3361	.3313	.3256	.3211	.3168
2.0	.4084	.4033	.3972	.3925	.3879	.3772	.3716	.3648	.3597	.3547
2.3	.4491	.4433	.4364	.4312	.4260	.4147	.4084	.4008	.3950	.3894
2.6	.4888	.4824	.4747	.4689	.4632	.4514	.4444	.4359	.4295	.4232
3.0	.5409	.5335	.5247	.5180	.5115	.4994	.4914	.4817	.4744	.4672
3.5	.6047	.5961	.5858	.5780	.5705	.5584	.5490	.5377	.5291	.5209
4.0	.6672	.6573	.6456	.6366	.6280	.6163	.6054	.5925	.5826	.5731
4.5	.7287	.7174	.7041	.6939	.6842	.6732	.6608	.6461	.6349	.6242
5.0	.7891	.7764	.7614	.7500	.7392	.7292	.7151	.6986	.6861	.6742
6.0	.9005	.8853	.8675	.8540	.8411	.8322	.8156	.7959	.7810	.7668
7.0	1.011	.9932	.9722	.9563	.9412	.9348	.9150	.8918	.8744	.8578
8.0	1.121	1.100	1.076	1.057	1.040	1.037	1.014	.9867	.9665	.9473
9.0	1.230	1.206	1.178	1.157	1.137	1.138	1.112	1.081	1.057	1.036
10	1.339	1.311	1.279	1.256	1.233	1.239	1.209	1.173	1.147	1.122
12	1.539	1.506	1.467	1.439	1.412	1.426	1.389	1.346	1.314	1.285
14	1.741	1.701	1.655	1.621	1.589	1.615	1.570	1.519	1.481	1.445
17	2.048	1.996	1.937	1.894	1.854	1.902	1.843	1.777	1.729	1.685
20	2.357	2.292	2.219	2.166	2.118	2.191	2.117	2.035	1.976	1.921
23	2.636	2.560	2.475	2.413	2.357	2.453	2.367	2.271	2.202	2.139
26	2.919	2.831	2.733	2.662	2.597	2.720	2.619	2.508	2.429	2.356
30	3.305	3.199	3.081	2.996	2.920	3.084	2.961	2.827	2.733	2.646
35	3.800	3.669	3.523	3.420	3.328	3.549	3.396	3.229	3.114	3.009
40	4.305	4.147	3.970	3.847	3.739	4.023	3.836	3.633	3.496	3.371
45	4.818	4.629	4.421	4.276	4.150	4.503	4.280	4.039	3.877	3.733
50	5.334	5.114	4.872	4.705	4.561	4.986	4.724	4.443	4.256	4.091
60	6.281	6.005	5.704	5.497	5.322	5.892	5.559	5.203	4.966	4.765
70	7.271	6.931	6.562	6.311	6.100	6.825	6.411	5.972	5.682	5.443
80	8.294	7.883	7.440	7.141	6.891	7.777	7.277	6.748	6.402	6.121
90	9.342	8.856	8.333	7.982	7.692	8.742	8.151	7.528	7.122	6.798
100	10.41	9.842	9.235	8.831	8.498	9.715	9.028	8.307	7.841	7.472
120	12.38	11.67	10.92	10.43	10.02	11.55	10.68	9.784	9.207	8.756
140	14.45	13.59	12.68	12.07	11.58	13.43	12.37	11.28	10.58	10.04
170	17.74	16.50	15.40	14.61	13.97	16.32	14.95	13.54	12.64	11.96
200	21.19	19.73	18.20	17.20	16.41	19.28	17.57	15.80	14.70	13.87
230	24.32	22.63	20.84	19.66	18.72	22.05	20.04	17.98	16.68	15.71
260	27.60	25.63	23.54	22.17	21.07	24.87	22.54	20.15	18.65	17.53
300	32.17	29.76	27.22	25.56	24.23	28.70	25.91	23.04	21.26	19.93
350	38.16	35.12	31.93	29.86	28.23	33.57	30.14	26.65	24.48	22.88
400	44.38	40.63	36.72	34.20	32.24	38.50	34.39	30.22	27.65	25.77
450	50.76	46.24	41.55	38.56	36.25	43.45	38.63	33.75	30.77	28.60
500	57.25	51.91	46.39	42.90	40.24	48.41	42.85	37.24	33.83	31.37
600	69.75	62.87	55.79	51.33	47.96	58.04	51.04	44.02	39.77	36.73
700	82.65	74.04	65.21	59.71	55.60	67.64	59.13	50.61	45.48	41.87
800	95.82	85.31	74.62	68.01	63.14	77.18	67.07	57.00	51.00	46.81
900	109.2	96.63	83.96	76.20	70.57	86.62	74.88	63.22	56.32	51.57
1000	122.6	107.9	93.21	84.28	77.87	95.94	82.52	69.26	61.46	56.16
1200	148.9	130.1	111.2	99.90	91.96	114.2	97.38	80.88	71.28	64.88
1400	175.3	152.0	128.8	115.1	105.6	131.9	111.6	91.90	80.52	73.08
1700	214.5	184.2	154.4	137.0	125.3	157.3	132.0	107.4	93.47	84.55
2000	253.1	215.7	179.1	158.0	144.1	181.6	151.2	122.0	105.5	95.18
2300	290.9	246.4	203.0	178.2	162.1	204.7	169.5	135.6	116.7	105.1
2600	327.9	276.2	226.0	197.6	179.5	226.9	186.8	148.5	127.3	114.5
3000	375.8	314.6	255.6	222.4	201.6	255.0	208.8	164.8	140.6	126.2
3500	433.6	360.7	290.8	251.8	227.9	288.3	234.6	183.7	155.9	139.7
4000	489.2	404.9	324.3	279.7	252.8	319.6	258.8	201.3	170.2	152.3
4500	542.7	447.3	356.3	306.3	276.4	349.3	281.6	217.9	183.6	164.1
5000	594.3	488.0	387.0	331.7	299.0	377.6	303.3	233.5	196.1	175.2

NI ON C, M2/M1=0.203

TABLE 17- 3

	RELATIVE RANGE STRAGGLING					RELATIVE DAMAGE STRAGGLING				
K/KL=	0.55	0.75	1.0	1.2	1.4	0.55	0.75	1.0	1.2	1.4
E,KEV										
.50	.0943	.0935	.0925	.0918	.0910	.3370	.3353	.3332	.3316	.3301
.60	.0943	.0934	.0924	.0916	.0908	.3372	.3353	.3332	.3315	.3299
.70	.0942	.0934	.0923	.0915	.0907	.3373	.3354	.3331	.3314	.3298
.80	.0942	.0933	.0922	.0914	.0905	.3375	.3355	.3331	.3314	.3297
.90	.0941	.0932	.0921	.0913	.0904	.3376	.3356	.3331	.3313	.3296
1.0	.0941	.0932	.0920	.0912	.0903	.3378	.3357	.3332	.3313	.3295
1.2	.0940	.0931	.0919	.0910	.0901	.3381	.3359	.3333	.3313	.3295
1.4	.0939	.0930	.0918	.0909	.0900	.3385	.3361	.3334	.3313	.3294
1.7	.0939	.0928	.0916	.0907	.0897	.3391	.3366	.3336	.3315	.3294
2.0	.0938	.0927	.0915	.0905	.0895	.3398	.3371	.3339	.3316	.3295
2.3	.0937	.0926	.0914	.0903	.0894	.3404	.3375	.3342	.3318	.3295
2.6	.0937	.0926	.0912	.0902	.0892	.3411	.3380	.3346	.3320	.3296
3.0	.0936	.0925	.0911	.0900	.0890	.3420	.3387	.3350	.3323	.3298
3.5	.0935	.0923	.0909	.0898	.0888	.3432	.3397	.3357	.3327	.3301
4.0	.0934	.0922	.0908	.0897	.0886	.3444	.3406	.3363	.3332	.3303
4.5	.0934	.0921	.0906	.0895	.0884	.3456	.3415	.3369	.3336	.3306
5.0	.0933	.0920	.0905	.0893	.0882	.3469	.3425	.3376	.3341	.3309
6.0	.0932	.0919	.0903	.0890	.0878	.3494	.3445	.3390	.3351	.3316
7.0	.0931	.0917	.0900	.0888	.0875	.3519	.3464	.3403	.3361	.3322
8.0	.0929	.0915	.0898	.0885	.0872	.3544	.3482	.3416	.3370	.3328
9.0	.0928	.0914	.0896	.0883	.0870	.3567	.3500	.3428	.3378	.3334
10	.0927	.0912	.0894	.0880	.0867	.3590	.3517	.3439	.3386	.3338
12	.0925	.0910	.0891	.0876	.0862	.3638	.3552	.3463	.3402	.3349
14	.0924	.0907	.0887	.0872	.0857	.3681	.3583	.3483	.3416	.3357
17	.0921	.0903	.0882	.0866	.0850	.3738	.3624	.3508	.3431	.3365
20	.0918	.0899	.0877	.0860	.0844	.3787	.3657	.3526	.3441	.3369
23	.0915	.0895	.0872	.0854	.0837	.3835	.3689	.3542	.3448	.3370
26	.0912	.0892	.0867	.0849	.0831	.3876	.3715	.3554	.3451	.3369
30	.0908	.0886	.0861	.0841	.0823	.3921	.3743	.3564	.3452	.3364
35	.0903	.0880	.0853	.0832	.0813	.3965	.3767	.3569	.3446	.3354
40	.0898	.0874	.0845	.0823	.0803	.3997	.3782	.3568	.3436	.3340
45	.0893	.0867	.0837	.0815	.0794	.4019	.3790	.3561	.3423	.3324
50	.0888	.0861	.0829	.0806	.0784	.4034	.3792	.3551	.3407	.3306
60	.0877	.0848	.0814	.0789	.0766	.4041	.3776	.3515	.3363	.3260
70	.0866	.0835	.0799	.0773	.0748	.4032	.3750	.3474	.3317	.3214
80	.0856	.0822	.0785	.0757	.0732	.4012	.3717	.3432	.3271	.3169
90	.0845	.0810	.0770	.0742	.0716	.3985	.3681	.3389	.3226	.3126
100	.0834	.0798	.0757	.0727	.0700	.3952	.3642	.3346	.3183	.3085
120	.0813	.0773	.0730	.0698	.0670	.3862	.3551	.3258	.3098	.3006
140	.0793	.0750	.0704	.0672	.0642	.3769	.3463	.3175	.3021	.2935
170	.0764	.0718	.0669	.0635	.0605	.3634	.3339	.3065	.2920	.2841
200	.0737	.0689	.0638	.0603	.0572	.3509	.3228	.2968	.2833	.2761
230	.0711	.0660	.0607	.0572	.0541	.3396	.3132	.2887	.2760	.2694
260	.0686	.0634	.0580	.0544	.0513	.3293	.3045	.2816	.2697	.2635
300	.0655	.0601	.0547	.0511	.0480	.3171	.2943	.2733	.2623	.2567
350	.0620	.0565	.0510	.0474	.0443	.3039	.2834	.2644	.2545	.2494
400	.0589	.0533	.0478	.0442	.0412	.2926	.2741	.2569	.2479	.2431
450	.0561	.0504	.0449	.0414	.0385	.2829	.2661	.2504	.2422	.2377
500	.0535	.0478	.0423	.0389	.0360	.2745	.2592	.2448	.2371	.2330
600	.0489	.0432	.0380	.0347	.0320	.2621	.2486	.2359	.2290	.2252
700	.0449	.0394	.0343	.0312	.0287	.2523	.2402	.2287	.2224	.2187
800	.0415	.0361	.0313	.0284	.0260	.2443	.2333	.2227	.2169	.2133
900	.0386	.0333	.0287	.0259	.0237	.2376	.2274	.2176	.2121	.2086
1000	.0360	.0309	.0264	.0238	.0217	.2319	.2223	.2131	.2078	.2045
1200	.0317	.0270	.0229	.0206	.0187	.2229	.2141	.2056	.2007	.1975
1400	.0283	.0239	.0202	.0181	.0164	.2158	.2076	.1995	.1949	.1918
1700	.0242	.0203	.0170	.0152	.0137	.2074	.1997	.1921	.1877	.1847
2000	.0211	.0175	.0146	.0130	.0118	.2008	.1934	.1861	.1818	.1789
2300	.0185	.0154	.0128	.0114	.0103	.1954	.1881	.1810	.1768	.1739
2600	.0165	.0136	.0113	.0101	.0091	.1907	.1836	.1766	.1725	.1697
3000	.0144	.0118	.0098	.0087	.0079	.1853	.1783	.1715	.1675	.1648
3500	.0123	.0101	.0084	.0075	.0067	.1795	.1727	.1661	.1622	.1596
4000	.0108	.0089	.0074	.0065	.0059	.1744	.1679	.1614	.1576	.1551
4500	.0097	.0079	.0066	.0059	.0053	.1699	.1636	.1573	.1536	.1511
5000	.0088	.0073	.0060	.0054	.0049	.1658	.1597	.1536	.1500	.1476

E,KEV	| K/KL=	RELATIVE IONIZATION STRAGGLING				RELATIVE TRANSVERSE RANGE STRAGGLING				
	0.55	0.75	1.0	1.2	1.4	0.55	0.75	1.0	1.2	1.4
.50	.3862	.3852	.3840	.3832	.3825	.4918	.4926	.4937	.4945	.4953
.60	.3865	.3854	.3843	.3834	.3826	.4915	.4923	.4934	.4943	.4951
.70	.3868	.3857	.3845	.3836	.3828	.4912	.4921	.4932	.4941	.4950
.80	.3871	.3860	.3847	.3837	.3829	.4909	.4918	.4930	.4939	.4948
.90	.3874	.3862	.3849	.3839	.3830	.4906	.4916	.4928	.4937	.4946
1.0	.3877	.3865	.3851	.3841	.3832	.4903	.4913	.4925	.4935	.4944
1.2	.3884	.3870	.3855	.3844	.3835	.4898	.4908	.4921	.4931	.4941
1.4	.3890	.3876	.3859	.3848	.3838	.4893	.4903	.4916	.4927	.4937
1.7	.3901	.3884	.3866	.3854	.3843	.4884	.4896	.4909	.4920	.4931
2.0	.3911	.3893	.3874	.3860	.3848	.4876	.4888	.4902	.4914	.4925
2.3	.3921	.3902	.3881	.3866	.3853	.4869	.4881	.4896	.4907	.4919
2.6	.3932	.3911	.3888	.3872	.3858	.4861	.4874	.4889	.4901	.4913
3.0	.3946	.3922	.3897	.3880	.3865	.4851	.4864	.4880	.4893	.4905
3.5	.3963	.3937	.3910	.3890	.3874	.4839	.4853	.4870	.4883	.4896
4.0	.3981	.3952	.3922	.3901	.3882	.4827	.4842	.4859	.4873	.4887
4.5	.3999	.3967	.3934	.3911	.3891	.4816	.4831	.4849	.4863	.4877
5.0	.4017	.3982	.3945	.3920	.3899	.4805	.4820	.4839	.4854	.4869
6.0	.4053	.4012	.3969	.3940	.3916	.4783	.4799	.4820	.4836	.4851
7.0	.4088	.4041	.3992	.3959	.3931	.4762	.4780	.4801	.4818	.4835
8.0	.4122	.4069	.4014	.3977	.3946	.4742	.4761	.4784	.4802	.4819
9.0	.4156	.4096	.4034	.3994	.3960	.4724	.4743	.4767	.4786	.4804
10	.4198	.4121	.4054	.4010	.3972	.4705	.4726	.4751	.4770	.4790
12	.4254	.4174	.4094	.4041	.3998	.4670	.4693	.4720	.4741	.4762
14	.4315	.4221	.4128	.4069	.4020	.4638	.4662	.4691	.4714	.4736
17	.4396	.4283	.4173	.4103	.4046	.4594	.4620	.4653	.4678	.4702
20	.4466	.4335	.4209	.4130	.4067	.4554	.4583	.4618	.4645	.4672
23	.4532	.4381	.4239	.4152	.4082	.4516	.4547	.4585	.4614	.4643
26	.4591	.4421	.4263	.4169	.4094	.4481	.4514	.4555	.4587	.4617
30	.4657	.4466	.4289	.4186	.4105	.4439	.4476	.4520	.4554	.4587
35	.4725	.4510	.4313	.4200	.4113	.4394	.4434	.4482	.4519	.4555
40	.4780	.4543	.4329	.4207	.4117	.4354	.4397	.4449	.4490	.4529
45	.4824	.4569	.4339	.4210	.4117	.4319	.4366	.4422	.4465	.4507
50	.4858	.4588	.4344	.4210	.4114	.4289	.4338	.4398	.4444	.4488
60	.4909	.4612	.4342	.4195	.4099	.4234	.4290	.4357	.4408	.4457
70	.4937	.4621	.4331	.4175	.4080	.4192	.4253	.4326	.4382	.4436
80	.4949	.4619	.4314	.4153	.4060	.4158	.4225	.4304	.4365	.4424
90	.4948	.4608	.4294	.4130	.4040	.4131	.4203	.4289	.4354	.4417
100	.4937	.4591	.4272	.4106	.4020	.4110	.4188	.4279	.4349	.4416
120	.4869	.4525	.4210	.4051	.3975	.4081	.4168	.4271	.4348	.4423
140	.4792	.4456	.4151	.4001	.3935	.4066	.4162	.4275	.4360	.4441
170	.4676	.4357	.4072	.3937	.3884	.4061	.4170	.4297	.4393	.4484
200	.4568	.4268	.4006	.3885	.3845	.4071	.4193	.4333	.4439	.4538
230	.4474	.4197	.3958	.3852	.3821	.4092	.4225	.4379	.4493	.4601
260	.4390	.4136	.3920	.3826	.3804	.4120	.4264	.4430	.4553	.4668
300	.4293	.4068	.3879	.3801	.3789	.4166	.4324	.4505	.4638	.4763
350	.4192	.4000	.3842	.3781	.3778	.4232	.4407	.4605	.4750	.4884
400	.4110	.3947	.3816	.3770	.3775	.4304	.4494	.4708	.4864	.5007
450	.4043	.3906	.3799	.3764	.3776	.4380	.4585	.4813	.4978	.5130
500	.3990	.3874	.3788	.3764	.3780	.4458	.4677	.4919	.5093	.5252
600	.3928	.3846	.3788	.3779	.3802	.4620	.4865	.5133	.5323	.5496
700	.3889	.3832	.3798	.3800	.3827	.4782	.5051	.5341	.5545	.5729
800	.3865	.3829	.3813	.3823	.3854	.4943	.5233	.5542	.5758	.5952
900	.3851	.3831	.3830	.3847	.3880	.5101	.5410	.5737	.5963	.6165
1000	.3843	.3838	.3849	.3871	.3906	.5256	.5582	.5924	.6159	.6368
1200	.3846	.3860	.3888	.3918	.3956	.5566	.5923	.6291	.6541	.6762
1400	.3859	.3887	.3926	.3962	.4001	.5857	.6240	.6629	.6891	.7121
1700	.3888	.3929	.3980	.4021	.4062	.6258	.6673	.7087	.7361	.7601
2000	.3920	.3970	.4030	.4074	.4115	.6622	.7062	.7494	.7778	.8023
2300	.3953	.4010	.4074	.4121	.4162	.6953	.7413	.7859	.8150	.8399
2600	.3984	.4046	.4114	.4162	.4204	.7257	.7734	.8190	.8485	.8736
3000	.4024	.4090	.4162	.4210	.4252	.7627	.8120	.8586	.8885	.9135
3500	.4070	.4139	.4213	.4262	.4303	.8042	.8550	.9022	.9323	.9572
4000	.4110	.4181	.4256	.4306	.4347	.8414	.8932	.9407	.9708	.9952
4500	.4146	.4218	.4294	.4344	.4384	.8751	.9275	.9750	1.005	1.029
5000	.4178	.4251	.4327	.4377	.4416	.9058	.9587	1.006	1.036	1.059

TABLE 17- 5

NI ON C, M2/M1=0.203

E,KEV K/KL=	RELATIVE TRANSV. DAMAGE STRAGGLING					RELATIVE TRANSV. IONZN. STRAGGLING				
	0.55	0.75	1.0	1.2	1.4	0.55	0.75	1.0	1.2	1.4
.50	.3931	.3884	.3827	.3782	.3739	.3359	.3307	.3244	.3194	.3147
.60	.3930	.3882	.3823	.3777	.3733	.3358	.3304	.3239	.3189	.3139
.70	.3930	.3880	.3820	.3773	.3728	.3358	.3303	.3236	.3184	.3134
.80	.3931	.3879	.3818	.3770	.3723	.3358	.3302	.3233	.3180	.3129
.90	.3931	.3879	.3816	.3767	.3720	.3359	.3301	.3231	.3177	.3124
1.0	.3932	.3878	.3814	.3764	.3716	.3359	.3300	.3229	.3174	.3120
1.2	.3933	.3878	.3812	.3760	.3711	.3361	.3299	.3226	.3169	.3114
1.4	.3935	.3878	.3810	.3757	.3706	.3362	.3299	.3223	.3165	.3109
1.7	.3937	.3878	.3807	.3753	.3700	.3365	.3299	.3221	.3160	.3102
2.0	.3939	.3879	.3805	.3749	.3695	.3367	.3300	.3218	.3156	.3096
2.3	.3942	.3879	.3804	.3747	.3691	.3370	.3301	.3217	.3153	.3092
2.6	.3944	.3880	.3803	.3744	.3687	.3373	.3301	.3216	.3151	.3088
3.0	.3947	.3881	.3802	.3741	.3683	.3376	.3303	.3215	.3147	.3083
3.5	.3949	.3881	.3800	.3738	.3677	.3379	.3304	.3213	.3144	.3077
4.0	.3951	.3882	.3798	.3734	.3672	.3382	.3304	.3211	.3140	.3071
4.5	.3953	.3881	.3796	.3730	.3667	.3385	.3305	.3209	.3136	.3066
5.0	.3953	.3880	.3793	.3726	.3662	.3386	.3305	.3207	.3133	.3061
6.0	.3954	.3879	.3788	.3719	.3652	.3390	.3305	.3204	.3126	.3052
7.0	.3954	.3876	.3783	.3711	.3642	.3391	.3304	.3199	.3119	.3043
8.0	.3952	.3872	.3776	.3703	.3632	.3392	.3302	.3194	.3112	.3034
9.0	.3948	.3867	.3769	.3694	.3622	.3391	.3299	.3189	.3105	.3024
10	.3945	.3862	.3762	.3685	.3612	.3389	.3295	.3183	.3097	.3015
12	.3934	.3849	.3746	.3667	.3591	.3383	.3286	.3170	.3081	.2996
14	.3923	.3835	.3730	.3648	.3570	.3376	.3276	.3156	.3064	.2976
17	.3903	.3813	.3704	.3619	.3538	.3362	.3259	.3134	.3039	.2947
20	.3883	.3790	.3677	.3590	.3506	.3347	.3241	.3112	.3013	.2918
23	.3861	.3767	.3651	.3561	.3474	.3332	.3223	.3090	.2988	.2889
26	.3840	.3743	.3625	.3532	.3443	.3316	.3205	.3068	.2963	.2861
30	.3811	.3712	.3590	.3495	.3402	.3294	.3181	.3038	.2929	.2824
35	.3777	.3674	.3547	.3448	.3352	.3266	.3149	.3001	.2888	.2778
40	.3744	.3637	.3506	.3404	.3304	.3238	.3118	.2964	.2847	.2733
45	.3712	.3601	.3466	.3360	.3257	.3211	.3086	.2927	.2806	.2689
50	.3682	.3567	.3427	.3318	.3211	.3183	.3055	.2891	.2767	.2646
60	.3627	.3503	.3353	.3237	.3124	.3128	.2990	.2818	.2687	.2560
70	.3574	.3442	.3283	.3160	.3041	.3075	.2929	.2748	.2611	.2479
80	.3524	.3384	.3215	.3087	.2964	.3026	.2870	.2682	.2539	.2403
90	.3476	.3327	.3150	.3017	.2890	.2979	.2814	.2618	.2471	.2331
100	.3430	.3273	.3088	.2950	.2820	.2935	.2761	.2558	.2405	.2263
120	.3341	.3165	.2964	.2817	.2683	.2861	.2665	.2442	.2279	.2132
140	.3258	.3065	.2849	.2696	.2559	.2792	.2576	.2335	.2165	.2015
170	.3141	.2928	.2694	.2533	.2396	.2693	.2453	.2191	.2012	.1860
200	.3031	.2802	.2556	.2391	.2254	.2598	.2339	.2062	.1877	.1725
230	.2919	.2680	.2428	.2262	.2129	.2492	.2221	.1936	.1751	.1604
260	.2814	.2569	.2313	.2148	.2019	.2390	.2111	.1822	.1638	.1497
300	.2685	.2434	.2176	.2013	.1890	.2264	.1978	.1686	.1505	.1372
350	.2539	.2285	.2027	.1869	.1753	.2119	.1829	.1538	.1363	.1239
400	.2409	.2155	.1899	.1745	.1636	.1988	.1698	.1410	.1241	.1126
450	.2293	.2040	.1788	.1639	.1536	.1869	.1581	.1300	.1137	.1031
500	.2187	.1937	.1691	.1547	.1450	.1761	.1477	.1203	.1047	.0948
600	.2001	.1765	.1535	.1401	.1313	.1566	.1302	.1049	.0908	.0822
700	.1845	.1624	.1409	.1285	.1205	.1402	.1158	.0926	.0799	.0723
800	.1713	.1506	.1305	.1190	.1116	.1263	.1039	.0826	.0711	.0643
900	.1600	.1406	.1218	.1111	.1042	.1145	.0939	.0744	.0639	.0579
1000	.1502	.1320	.1143	.1043	.0979	.1044	.0853	.0674	.0578	.0525
1200	.1351	.1187	.1028	.0938	.0880	.0893	.0727	.0572	.0490	.0444
1400	.1233	.1084	.0939	.0856	.0804	.0778	.0632	.0496	.0424	.0384
1700	.1096	.0965	.0837	.0764	.0717	.0648	.0526	.0413	.0352	.0320
2000	.0993	.0875	.0760	.0694	.0651	.0552	.0449	.0352	.0301	.0273
2300	.0911	.0804	.0699	.0638	.0599	.0479	.0390	.0307	.0263	.0238
2600	.0845	.0746	.0650	.0594	.0556	.0422	.0345	.0272	.0233	.0211
3000	.0775	.0685	.0597	.0545	.0511	.0364	.0298	.0236	.0203	.0183
3500	.0707	.0626	.0545	.0498	.0466	.0310	.0255	.0203	.0174	.0157
4000	.0656	.0580	.0504	.0460	.0431	.0272	.0224	.0177	.0152	.0138
4500	.0616	.0543	.0471	.0430	.0402	.0245	.0200	.0158	.0135	.0122
5000	.0584	.0513	.0444	.0404	.0377	.0225	.0182	.0143	.0121	.0110

K/KL= E,KEV	RANGE SKEWNESS 0.55	0.75	1.0	1.2	1.4	DAMAGE SKEWNESS 0.55	0.75	1.0	1.2	1.4
.50	.5059	.4984	.4892	.4819	.4748	.6276	.6146	.5993	.5878	.5767
.60	.5049	.4972	.4877	.4802	.4728	.6301	.6164	.6004	.5882	.5767
.70	.5040	.4961	.4863	.4787	.4711	.6324	.6181	.6014	.5887	.5767
.80	.5032	.4951	.4851	.4773	.4695	.6347	.6198	.6024	.5893	.5769
.90	.5025	.4942	.4840	.4760	.4681	.6370	.6216	.6036	.5900	.5772
1.0	.5018	.4934	.4830	.4748	.4668	.6393	.6234	.6048	.5909	.5777
1.2	.5005	.4918	.4811	.4727	.4644	.6442	.6272	.6075	.5927	.5788
1.4	.4993	.4903	.4793	.4707	.4622	.6492	.6312	.6104	.5948	.5802
1.7	.4977	.4883	.4769	.4680	.4592	.6572	.6376	.6150	.5983	.5826
2.0	.4961	.4865	.4747	.4655	.4564	.6653	.6442	.6199	.6020	.5853
2.3	.4947	.4848	.4727	.4632	.4539	.6733	.6507	.6249	.6058	.5881
2.6	.4933	.4832	.4708	.4611	.4516	.6814	.6573	.6299	.6098	.5911
3.0	.4916	.4811	.4684	.4584	.4487	.6923	.6662	.6367	.6151	.5952
3.5	.4895	.4787	.4655	.4553	.4452	.7060	.6775	.6452	.6219	.6003
4.0	.4875	.4764	.4628	.4523	.4419	.7198	.6887	.6538	.6286	.6055
4.5	.4855	.4741	.4602	.4494	.4388	.7335	.6998	.6622	.6352	.6106
5.0	.4836	.4719	.4577	.4466	.4358	.7472	.7109	.6706	.6418	.6156
6.0	.4801	.4679	.4531	.4416	.4303	.7749	.7334	.6878	.6555	.6263
7.0	.4767	.4641	.4487	.4367	.4250	.8015	.7549	.7041	.6683	.6362
8.0	.4734	.4603	.4444	.4321	.4200	.8270	.7753	.7193	.6802	.6452
9.0	.4702	.4567	.4403	.4276	.4152	.8514	.7947	.7336	.6912	.6536
10	.4670	.4531	.4363	.4232	.4105	.8748	.8131	.7471	.7016	.6613
12	.4611	.4464	.4287	.4150	.4017	.9223	.8501	.7742	.7223	.6767
14	.4553	.4400	.4214	.4071	.3932	.9649	.8830	.7977	.7399	.6894
17	.4469	.4305	.4109	.3957	.3811	1.021	.9252	.8269	.7611	.7040
20	.4387	.4214	.4007	.3848	.3694	1.068	.9604	.8501	.7770	.7144
23	.4309	.4128	.3911	.3744	.3584	1.110	.9919	.8702	.7900	.7222
26	.4234	.4044	.3817	.3644	.3477	1.147	1.018	.8862	.7994	.7271
30	.4135	.3935	.3696	.3514	.3340	1.187	1.047	.9019	.8074	.7301
35	.4015	.3802	.3550	.3358	.3175	1.227	1.074	.9146	.8116	.7291
40	.3899	.3675	.3409	.3208	.3017	1.259	1.093	.9211	.8109	.7242
45	.3787	.3551	.3273	.3064	.2864	1.283	1.106	.9228	.8064	.7163
50	.3677	.3431	.3141	.2924	.2717	1.301	1.114	.9207	.7989	.7060
60	.3468	.3200	.2889	.2656	.2435	1.326	1.117	.9045	.7730	.6761
70	.3268	.2982	.2650	.2403	.2171	1.337	1.110	.8819	.7432	.6439
80	.3078	.2773	.2423	.2164	.1921	1.338	1.096	.8555	.7115	.6109
90	.2895	.2574	.2207	.1936	.1683	1.330	1.077	.8267	.6790	.5781
100	.2721	.2384	.2000	.1719	.1457	1.316	1.054	.7965	.6464	.5459
120	.2397	.2020	.1607	.1306	.1028	1.265	.9923	.7278	.5770	.4804
140	.2077	.1683	.1243	.0926	.0635	1.206	.9280	.6612	.5120	.4198
170	.1648	.1218	.0744	.0406	.0098	1.113	.8338	.5685	.4236	.3386
200	.1254	.0793	.0291	-.0064	-.0386	1.022	.7451	.4851	.3458	.2675
230	.0885	.0395	-.0132	-.0502	-.0834	.9266	.6601	.4110	.2791	.2069
260	.0541	.0026	-.0523	-.0904	-.1246	.8371	.5819	.3446	.2200	.1534
300	.0117	-.0428	-.1001	-.1396	-.1747	.7276	.4880	.2663	.1509	.0908
350	-.0368	-.0944	-.1543	-.1951	-.2311	.6060	.3853	.1820	.0772	.0239
400	-.0812	-.1414	-.2033	-.2451	-.2818	.4997	.2966	.1099	.0144	-.0333
450	-.1220	-.1845	-.2480	-.2906	-.3278	.4068	.2195	.0478	-.0396	-.0827
500	-.1500	-.2243	-.2892	-.3324	-.3699	.3254	.1524	-.0061	-.0866	-.1260
600	-.2293	-.2966	-.3633	-.4072	-.4449	.2018	.0507	-.0882	-.1593	-.1948
700	-.2907	-.3602	-.4282	-.4724	-.5101	.1028	-.0307	-.1540	-.2179	-.2509
800	-.3460	-.4171	-.4858	-.5301	-.5678	.0215	-.0976	-.2085	-.2668	-.2980
900	-.3961	-.4685	-.5377	-.5820	-.6194	-.0463	-.1539	-.2548	-.3086	-.3385
1000	-.4422	-.5154	-.5848	-.6290	-.6662	-.1038	-.2020	-.2947	-.3449	-.3739
1200	-.5231	-.5971	-.6663	-.7098	-.7463	-.1904	-.2772	-.3601	-.4058	-.4334
1400	-.5941	-.6684	-.7369	-.7798	-.8156	-.2570	-.3364	-.4128	-.4556	-.4822
1700	-.6871	-.7612	-.8286	-.8703	-.9050	-.3340	-.4062	-.4763	-.5162	-.5419
2000	-.7684	-.8418	-.9079	-.9486	-.9823	-.3935	-.4612	-.5274	-.5656	-.5907
2300	-.8409	-.9134	-.9782	-1.018	-1.051	-.4420	-.5067	-.5703	-.6073	-.6319
2600	-.9065	-.9781	-1.042	-1.080	-1.112	-.4829	-.5456	-.6073	-.6434	-.6676
3000	-.9857	-1.056	-1.118	-1.156	-1.187	-.5295	-.5901	-.6499	-.6850	-.7088
3500	-1.074	-1.143	-1.203	-1.239	-1.269	-.5790	-.6376	-.6955	-.7296	-.7528
4000	-1.154	-1.221	-1.279	-1.314	-1.344	-.6218	-.6787	-.7349	-.7681	-.7907
4500	-1.226	-1.291	-1.348	-1.383	-1.411	-.6602	-.7153	-.7699	-.8021	-.8240
5000	-1.293	-1.357	-1.412	-1.446	-1.474	-.6952	-.7485	-.8014	-.8325	-.8538

TABLE 17- 7

NI ON C, M2/M1=0.203

K/KL=	IONIZATION SKEWNESS					RANGE KURTOSIS				
E,KEV	0.55	0.75	1.0	1.2	1.4	0.55	0.75	1.0	1.2	1.4
.50	.7206	.7093	.6962	.6863	.6770	3.249	3.234	3.216	3.202	3.189
.60	.7234	.7114	.6976	.6872	.6775	3.247	3.232	3.213	3.199	3.185
.70	.7259	.7134	.6989	.6881	.6780	3.245	3.229	3.211	3.196	3.182
.80	.7284	.7154	.7003	.6891	.6786	3.244	3.228	3.208	3.193	3.179
.90	.7309	.7173	.7017	.6901	.6793	3.242	3.226	3.206	3.191	3.176
1.0	.7334	.7194	.7032	.6912	.6801	3.241	3.224	3.204	3.188	3.173
1.2	.7386	.7236	.7064	.6937	.6818	3.238	3.221	3.200	3.184	3.169
1.4	.7440	.7280	.7097	.6963	.6838	3.235	3.218	3.197	3.180	3.165
1.7	.7524	.7349	.7150	.7004	.6870	3.232	3.214	3.192	3.175	3.159
2.0	.7609	.7419	.7204	.7048	.6904	3.229	3.210	3.188	3.170	3.154
2.3	.7693	.7489	.7259	.7092	.6939	3.226	3.206	3.184	3.166	3.149
2.6	.7777	.7559	.7314	.7137	.6975	3.223	3.203	3.180	3.162	3.145
3.0	.7891	.7653	.7388	.7197	.7023	3.219	3.199	3.175	3.157	3.139
3.5	.8033	.7771	.7480	.7272	.7083	3.215	3.194	3.170	3.151	3.133
4.0	.8174	.7888	.7571	.7346	.7142	3.211	3.190	3.164	3.145	3.127
4.5	.8315	.8003	.7661	.7418	.7200	3.207	3.185	3.160	3.140	3.121
5.0	.8454	.8117	.7749	.7489	.7257	3.203	3.181	3.155	3.135	3.116
6.0	.8737	.8350	.7930	.7637	.7375	3.196	3.173	3.146	3.126	3.106
7.0	.9007	.8570	.8099	.7773	.7484	3.189	3.165	3.138	3.117	3.097
8.0	.9263	.8777	.8258	.7900	.7585	3.182	3.158	3.130	3.109	3.088
9.0	.9508	.8973	.8406	.8017	.7677	3.176	3.151	3.122	3.101	3.080
10	.9741	.9159	.8545	.8127	.7762	3.170	3.145	3.115	3.093	3.072
12	1.021	.9533	.8824	.8346	.7930	3.158	3.132	3.102	3.079	3.058
14	1.063	.9862	.9064	.8532	.8070	3.147	3.120	3.089	3.066	3.044
17	1.117	1.028	.9361	.8754	.8234	3.132	3.104	3.071	3.047	3.025
20	1.162	1.062	.9593	.8922	.8352	3.117	3.088	3.055	3.031	3.008
23	1.200	1.088	.9764	.9034	.8429	3.103	3.073	3.039	3.014	2.991
26	1.232	1.109	.9896	.9114	.8478	3.089	3.059	3.024	2.999	2.976
30	1.268	1.132	1.003	.9182	.8511	3.073	3.041	3.006	2.980	2.957
35	1.304	1.153	1.013	.9221	.8515	3.053	3.021	2.985	2.959	2.936
40	1.331	1.169	1.019	.9223	.8488	3.035	3.002	2.966	2.940	2.917
45	1.353	1.180	1.022	.9197	.8439	3.019	2.985	2.949	2.923	2.900
50	1.370	1.188	1.021	.9152	.8373	3.003	2.970	2.933	2.908	2.885
60	1.400	1.202	1.018	.9023	.8195	2.973	2.939	2.903	2.878	2.856
70	1.416	1.207	1.007	.8853	.7992	2.948	2.913	2.877	2.853	2.833
80	1.421	1.203	.9925	.8657	.7776	2.925	2.891	2.856	2.833	2.814
90	1.418	1.194	.9743	.8443	.7553	2.905	2.871	2.837	2.816	2.798
100	1.408	1.179	.9536	.8216	.7328	2.887	2.854	2.822	2.802	2.786
120	1.360	1.125	.8958	.7653	.6815	2.854	2.823	2.795	2.778	2.766
140	1.305	1.068	.8390	.7119	.6341	2.828	2.800	2.775	2.762	2.753
170	1.220	.9838	.7605	.6403	.5718	2.799	2.776	2.758	2.750	2.746
200	1.139	.9071	.6918	.5791	.5195	2.779	2.761	2.750	2.748	2.750
230	1.063	.8407	.6364	.5320	.4801	2.764	2.752	2.749	2.753	2.760
260	.9936	.7815	.5887	.4923	.4473	2.755	2.749	2.753	2.763	2.777
300	.9107	.7127	.5348	.4483	.4114	2.749	2.752	2.767	2.784	2.804
350	.8211	.6401	.4799	.4045	.3761	2.751	2.765	2.792	2.818	2.846
400	.7448	.5799	.4358	.3702	.3487	2.761	2.784	2.823	2.857	2.893
450	.6797	.5297	.4002	.3431	.3273	2.776	2.810	2.859	2.901	2.943
500	.6242	.4877	.3714	.3217	.3106	2.795	2.839	2.899	2.948	2.996
600	.5441	.4313	.3366	.2976	.2915	2.842	2.906	2.987	3.049	3.109
700	.4835	.3905	.3134	.2829	.2801	2.898	2.981	3.080	3.154	3.223
800	.4366	.3602	.2976	.2738	.2735	2.960	3.059	3.174	3.259	3.336
900	.3998	.3373	.2867	.2685	.2699	3.025	3.140	3.270	3.363	3.448
1000	.3707	.3198	.2793	.2656	.2684	3.093	3.222	3.365	3.466	3.557
1200	.3335	.2981	.2714	.2640	.2689	3.228	3.382	3.547	3.661	3.762
1400	.3098	.2858	.2689	.2659	.2724	3.367	3.543	3.727	3.851	3.961
1700	.2893	.2770	.2705	.2722	.2800	3.580	3.783	3.990	4.127	4.246
2000	.2790	.2749	.2753	.2801	.2888	3.794	4.020	4.245	4.392	4.519
2300	.2746	.2762	.2815	.2884	.2978	4.009	4.252	4.492	4.647	4.780
2600	.2737	.2794	.2883	.2968	.3066	4.221	4.480	4.732	4.893	5.031
3000	.2757	.2852	.2974	.3074	.3176	4.459	4.775	5.040	5.207	5.351
3500	.2809	.2935	.3084	.3196	.3302	4.838	5.131	5.409	5.582	5.739
4000	.2874	.3019	.3185	.3305	.3414	5.168	5.474	5.761	5.938	6.090
4500	.2944	.3100	.3277	.3402	.3513	5.487	5.803	6.098	6.278	6.432
5000	.3013	.3176	.3358	.3487	.3601	5.797	6.121	6.421	6.603	6.760

	DAMAGE KURTOSIS					IONIZATION KURTOSIS				
K/KL=	0.55	0.75	1.0	1.2	1.4	0.55	0.75	1.0	1.2	1.4
E,KEV										
.50	3.443	3.396	3.341	3.300	3.262	3.551	3.503	3.448	3.407	3.369
.60	3.454	3.404	3.346	3.303	3.264	3.563	3.512	3.454	3.411	3.371
.70	3.464	3.412	3.351	3.307	3.265	3.574	3.521	3.459	3.415	3.373
.80	3.474	3.419	3.356	3.310	3.267	3.584	3.529	3.465	3.419	3.376
.90	3.484	3.427	3.362	3.314	3.270	3.595	3.537	3.471	3.423	3.378
1.0	3.494	3.435	3.367	3.318	3.272	3.606	3.546	3.477	3.428	3.382
1.2	3.515	3.451	3.379	3.327	3.278	3.628	3.564	3.491	3.438	3.389
1.4	3.536	3.468	3.392	3.337	3.286	3.651	3.582	3.505	3.449	3.397
1.7	3.570	3.496	3.412	3.352	3.297	3.687	3.611	3.527	3.466	3.410
2.0	3.605	3.524	3.434	3.369	3.310	3.724	3.641	3.550	3.484	3.424
2.3	3.639	3.551	3.455	3.386	3.323	3.760	3.671	3.572	3.502	3.439
2.6	3.673	3.580	3.476	3.403	3.336	3.797	3.701	3.596	3.521	3.454
3.0	3.720	3.617	3.505	3.426	3.354	3.846	3.741	3.626	3.546	3.473
3.5	3.779	3.665	3.541	3.455	3.377	3.908	3.791	3.665	3.577	3.498
4.0	3.839	3.713	3.578	3.483	3.400	3.970	3.842	3.703	3.607	3.522
4.5	3.899	3.761	3.614	3.512	3.422	4.032	3.891	3.741	3.637	3.546
5.0	3.959	3.809	3.650	3.540	3.444	4.094	3.941	3.778	3.667	3.569
6.0	4.081	3.906	3.723	3.598	3.489	4.219	4.041	3.855	3.728	3.617
7.0	4.199	4.000	3.792	3.653	3.532	4.339	4.137	3.926	3.784	3.662
8.0	4.313	4.089	3.858	3.704	3.571	4.454	4.227	3.993	3.837	3.703
9.0	4.423	4.174	3.920	3.751	3.607	4.563	4.313	4.056	3.885	3.740
10	4.529	4.256	3.978	3.796	3.641	4.668	4.393	4.114	3.931	3.775
12	4.749	4.421	4.096	3.886	3.708	4.881	4.559	4.233	4.022	3.843
14	4.948	4.569	4.200	3.963	3.765	5.071	4.703	4.335	4.099	3.900
17	5.208	4.761	4.329	4.057	3.832	5.316	4.884	4.460	4.191	3.967
20	5.430	4.921	4.434	4.130	3.883	5.519	5.030	4.556	4.259	4.015
23	5.620	5.062	4.522	4.187	3.921	5.681	5.130	4.620	4.297	4.042
26	5.784	5.181	4.593	4.231	3.948	5.817	5.209	4.667	4.323	4.059
30	5.969	5.311	4.666	4.271	3.971	5.966	5.292	4.712	4.342	4.070
35	6.156	5.438	4.729	4.299	3.982	6.115	5.370	4.747	4.350	4.070
40	6.305	5.532	4.768	4.308	3.979	6.232	5.427	4.765	4.346	4.059
45	6.424	5.602	4.788	4.304	3.965	6.325	5.470	4.772	4.333	4.041
50	6.520	5.652	4.793	4.288	3.943	6.398	5.502	4.771	4.314	4.017
60	6.684	5.707	4.756	4.212	3.862	6.544	5.588	4.768	4.274	3.958
70	6.779	5.716	4.694	4.124	3.774	6.626	5.629	4.741	4.219	3.890
80	6.822	5.692	4.617	4.030	3.685	6.658	5.635	4.694	4.153	3.817
90	6.824	5.643	4.532	3.935	3.597	6.651	5.610	4.632	4.080	3.742
100	6.792	5.576	4.440	3.839	3.512	6.611	5.561	4.558	4.002	3.665
120	6.583	5.352	4.218	3.636	3.341	6.365	5.324	4.332	3.796	3.486
140	6.338	5.118	4.007	3.450	3.188	6.092	5.072	4.109	3.600	3.321
170	5.957	4.779	3.722	3.210	2.993	5.686	4.709	3.803	3.339	3.105
200	5.589	4.470	3.478	3.011	2.834	5.311	4.382	3.539	3.119	2.925
230	5.233	4.197	3.287	2.866	2.719	5.003	4.130	3.346	2.963	2.799
260	4.907	3.955	3.125	2.748	2.626	4.730	3.912	3.183	2.835	2.695
300	4.520	3.676	2.945	2.621	2.527	4.411	3.663	3.003	2.695	2.583
350	4.107	3.385	2.767	2.499	2.432	4.075	3.406	2.822	2.558	2.474
400	3.763	3.148	2.627	2.407	2.362	3.795	3.196	2.679	2.452	2.389
450	3.477	2.955	2.517	2.337	2.309	3.559	3.023	2.565	2.368	2.322
500	3.240	2.798	2.431	2.284	2.270	3.361	2.880	2.472	2.301	2.268
600	2.955	2.617	2.340	2.234	2.232	3.082	2.689	2.357	2.221	2.199
700	2.755	2.497	2.287	2.210	2.215	2.875	2.551	2.279	2.169	2.153
800	2.612	2.415	2.257	2.202	2.210	2.715	2.448	2.224	2.133	2.120
900	2.508	2.358	2.240	2.202	2.212	2.589	2.369	2.183	2.108	2.097
1000	2.431	2.319	2.233	2.207	2.219	2.489	2.306	2.152	2.089	2.079
1200	2.340	2.276	2.230	2.222	2.238	2.351	2.217	2.104	2.059	2.052
1400	2.295	2.261	2.241	2.244	2.263	2.258	2.157	2.073	2.040	2.036
1700	2.270	2.265	2.270	2.283	2.305	2.168	2.101	2.045	2.024	2.023
2000	2.274	2.285	2.305	2.325	2.349	2.113	2.067	2.030	2.016	2.017
2300	2.290	2.312	2.342	2.366	2.392	2.079	2.047	2.021	2.013	2.015
2600	2.313	2.342	2.379	2.407	2.434	2.056	2.034	2.017	2.013	2.017
3000	2.345	2.383	2.426	2.458	2.487	2.038	2.025	2.016	2.015	2.020
3500	2.386	2.431	2.482	2.518	2.550	2.026	2.020	2.018	2.020	2.027
4000	2.421	2.475	2.533	2.573	2.607	2.020	2.019	2.021	2.026	2.035
4500	2.452	2.514	2.580	2.624	2.660	2.018	2.019	2.025	2.033	2.042
5000	2.476	2.548	2.622	2.671	2.709	2.016	2.021	2.029	2.039	2.050

K/KL=	REFLECTION COEFFICIENT					SPUTTERING EFFICIENCY				
E,KEV	0.55	0.75	1.0	1.2	1.4	0.55	0.75	1.0	1.2	1.4
.50	.0000	.0000	.0000	.0000	.0000	.0186	.0176	.0165	.0157	.0149
.60	.0000	.0000	.0000	.0000	.0000	.0185	.0175	.0164	.0156	.0148
.70	.0000	.0000	.0000	.0000	.0000	.0185	.0175	.0163	.0155	.0147
.80	.0000	.0000	.0000	.0000	.0000	.0184	.0174	.0162	.0154	.0146
.90	.0000	.0000	.0000	.0000	.0000	.0184	.0174	.0162	.0153	.0145
1.0	.0000	.0000	.0000	.0000	.0000	.0184	.0174	.0161	.0152	.0144
1.2	.0000	.0000	.0000	.0000	.0000	.0184	.0173	.0160	.0151	.0143
1.4	.0000	.0000	.0000	.0000	.0000	.0184	.0173	.0160	.0150	.0142
1.7	.0000	.0000	.0000	.0000	.0000	.0184	.0172	.0159	.0149	.0140
2.0	.0000	.0000	.0000	.0000	.0000	.0184	.0172	.0158	.0148	.0139
2.3	.0000	.0000	.0000	.0000	.0000	.0184	.0172	.0158	.0147	.0138
2.6	.0000	.0000	.0000	.0000	.0000	.0185	.0172	.0157	.0147	.0137
3.0	.0000	.0000	.0000	.0000	.0000	.0185	.0172	.0157	.0146	.0136
3.5	.0000	.0000	.0000	.0000	.0000	.0186	.0172	.0156	.0145	.0135
4.0	.0000	.0000	.0000	.0000	.0000	.0187	.0172	.0156	.0145	.0135
4.5	.0000	.0000	.0000	.0000	.0000	.0187	.0172	.0156	.0144	.0134
5.0	.0000	.0000	.0000	.0000	.0000	.0188	.0172	.0156	.0144	.0133
6.0	.0000	.0000	.0000	.0000	.0000	.0189	.0173	.0155	.0143	.0132
7.0	.0000	.0000	.0000	.0000	.0000	.0190	.0173	.0155	.0142	.0131
8.0	.0000	.0000	.0000	.0000	.0000	.0191	.0174	.0155	.0142	.0130
9.0	.0000	.0000	.0000	.0000	.0000	.0192	.0174	.0155	.0141	.0130
10	.0000	.0000	.0000	.0000	.0000	.0193	.0175	.0155	.0141	.0129
12	.0000	.0000	.0000	.0000	.0000	.0195	.0175	.0154	.0140	.0127
14	.0000	.0000	.0000	.0000	.0000	.0196	.0175	.0153	.0139	.0126
17	.0000	.0000	.0000	.0000	.0000	.0197	.0175	.0152	.0137	.0124
20	.0000	.0000	.0000	.0000	.0000	.0198	.0175	.0151	.0136	.0122
23	.0000	.0000	.0000	.0000	.0000	.0198	.0174	.0150	.0134	.0120
26	.0000	.0000	.0000	.0000	.0000	.0197	.0173	.0148	.0132	.0118
30	.0000	.0000	.0000	.0000	.0000	.0197	.0172	.0146	.0129	.0116
35	.0000	.0000	.0000	.0000	.0000	.0196	.0170	.0144	.0127	.0113
40	.0000	.0000	.0000	.0000	.0000	.0194	.0168	.0141	.0124	.0111
45	.0000	.0000	.0000	.0000	.0000	.0193	.0167	.0139	.0121	.0108
50	.0000	.0000	.0000	.0000	.0000	.0191	.0165	.0136	.0119	.0106
60	.0000	.0000	.0000	.0000	.0000	.0188	.0161	.0132	.0114	.0101
70	.0000	.0000	.0000	.0000	.0000	.0184	.0157	.0128	.0109	.0097
80	.0000	.0000	.0000	.0000	.0000	.0181	.0154	.0124	.0105	.0093
90	.0000	.0000	.0000	.0000	.0000	.0178	.0150	.0120	.0101	.0090
100	.0000	.0000	.0000	.0000	.0000	.0174	.0147	.0116	.0098	.0086
120	.0000	.0000	.0000	.0000	.0000	.0167	.0140	.0110	.0091	.0081
140	.0000	.0000	.0000	.0000	.0000	.0160	.0133	.0104	.0086	.0075
170	.0000	.0000	.0000	.0000	.0000	.0150	.0125	.0097	.0079	.0069
200	.0000	.0000	.0000	.0000	.0000	.0142	.0118	.0090	.0072	.0063
230	.0000	.0000	.0000	.0000	.0000	.0135	.0111	.0084	.0067	.0058
260	.0000	.0000	.0000	.0000	.0000	.0129	.0106	.0078	.0062	.0054
300	.0000	.0000	.0000	.0000	.0000	.0122	.0099	.0072	.0056	.0049
350	.0000	.0000	.0000	.0000	.0000	.0114	.0092	.0065	.0050	.0044
400	.0000	.0000	.0000	.0000	.0000	.0107	.0085	.0059	.0045	.0039
450	.0000	.0000	.0000	.0000	.0000	.0101	.0079	.0055	.0041	.0036
500	.0000	.0000	.0000	.0000	.0000	.0095	.0074	.0050	.0037	.0033
600	.0000	.0000	0.0000	0.0000	0.0000	.0085	.0065	.0043	.0032	.0028
700	.0000	.0000	0.0000	0.0000	0.0000	.0077	.0058	.0038	.0029	.0025
800	.0000	.0000	0.0000	0.0000	0.0000	.0069	.0051	.0034	.0026	.0022
900	.0000	0.0000	0.0000	0.0000	0.0000	.0063	.0046	.0030	.0023	.0020
1000	0.0000	0.0000	0.0000	0.0000	0.0000	.0058	.0042	.0027	.0021	.0018
1200	0.0000	0.0000	0.0000	0.0000	0.0000	.0049	.0035	.0023	.0018	.0015
1400	0.0000	0.0000	0.0000	0.0000	0.0000	.0042	.0030	.0020	.0015	.0013
1700	0.0000	0.0000	0.0000	0.0000	0.0000	.0035	.0025	.0016	.0013	.0011
2000	0.0000	0.0000	0.0000	0.0000	0.0000	.0029	.0021	.0014	.0011	.0009
2300	0.0000	0.0000	0.0000	0.0000	0.0000	.0024	.0018	.0012	.0009	.0008
2600	0.0000	0.0000	0.0000	0.0000	0.0000	.0021	.0016	.0011	.0008	.0007
3000	0.0000	0.0000	0.0000	0.0000	0.0000	.0017	.0013	.0009	.0007	.0006
3500	0.0000	0.0000	0.0000	0.0000	0.0000	.0014	.0011	.0008	.0006	.0005
4000	0.0000	0.0000	0.0000	0.0000	0.0000	.0012	.0010	.0007	.0005	.0004
4500	0.0000	0.0000	0.0000	0.0000	0.0000	.0011	.0008	.0006	.0004	.0003
5000	0.0000	0.0000	0.0000	0.0000	0.0000	.0010	.0008	.0005	.0004	.0003

NI ON C, M2/M1=0.203   TABLE 17-10

K/KL=	IONIZATION DEFICIENCY					SPUTTERING YIELD ALPHA				
E,KEV	0.55	0.75	1.0	1.2	1.4	0.55	0.75	1.0	1.2	1.4
.50	.0017	.0023	.0028	.0033	.0037	.1218	.1174	.1123	.1085	.1048
.60	.0018	.0023	.0029	.0034	.0038	.1215	.1170	.1118	.1078	.1040
.70	.0018	.0024	.0030	.0034	.0038	.1212	.1167	.1113	.1072	.1034
.80	.0019	.0024	.0030	.0035	.0039	.1210	.1163	.1108	.1068	.1028
.90	.0019	.0025	.0031	.0036	.0040	.1208	.1161	.1105	.1063	.1024
1.0	.0019	.0025	.0032	.0036	.0040	.1206	.1158	.1101	.1059	.1019
1.2	.0020	.0026	.0032	.0037	.0041	.1203	.1154	.1096	.1052	.1011
1.4	.0021	.0027	.0033	.0038	.0042	.1201	.1150	.1091	.1047	.1005
1.7	.0021	.0027	.0034	.0039	.0043	.1198	.1146	.1085	.1039	.0997
2.0	.0022	.0028	.0035	.0040	.0044	.1196	.1143	.1080	.1033	.0990
2.3	.0023	.0029	.0036	.0041	.0045	.1195	.1140	.1076	.1028	.0984
2.6	.0023	.0030	.0037	.0042	.0046	.1194	.1137	.1072	.1024	.0979
3.0	.0024	.0030	.0038	.0043	.0047	.1193	.1135	.1068	.1019	.0973
3.5	.0025	.0031	.0039	.0044	.0048	.1192	.1133	.1064	.1014	.0967
4.0	.0025	.0032	.0040	.0045	.0049	.1192	.1131	.1061	.1010	.0962
4.5	.0026	.0033	.0040	.0046	.0050	.1193	.1130	.1058	.1006	.0958
5.0	.0027	.0034	.0041	.0046	.0051	.1193	.1129	.1056	.1003	.0954
6.0	.0028	.0035	.0043	.0048	.0052	.1195	.1128	.1053	.0998	.0948
7.0	.0029	.0036	.0044	.0049	.0053	.1197	.1128	.1051	.0994	.0943
8.0	.0030	.0038	.0045	.0050	.0055	.1200	.1129	.1049	.0991	.0939
9.0	.0031	.0039	.0046	.0051	.0056	.1203	.1130	.1048	.0989	.0935
10	.0032	.0039	.0047	.0052	.0056	.1206	.1131	.1047	.0987	.0932
12	.0033	.0041	.0049	.0054	.0058	.1213	.1134	.1047	.0985	.0928
14	.0035	.0043	.0051	.0055	.0059	.1219	.1137	.1047	.0983	.0925
17	.0036	.0045	.0052	.0057	.0061	.1229	.1142	.1048	.0982	.0922
20	.0038	.0046	.0054	.0058	.0062	.1238	.1147	.1049	.0981	.0919
23	.0039	.0047	.0055	.0059	.0063	.1246	.1153	.1051	.0980	.0918
26	.0040	.0048	.0056	.0060	.0064	.1254	.1158	.1053	.0980	.0917
30	.0042	.0049	.0057	.0061	.0064	.1263	.1165	.1055	.0980	.0916
35	.0043	.0050	.0057	.0061	.0065	.1274	.1172	.1058	.0980	.0915
40	.0044	.0051	.0058	.0062	.0065	.1283	.1179	.1061	.0980	.0915
45	.0045	.0052	.0058	.0062	.0065	.1292	.1186	.1063	.0980	.0915
50	.0046	.0053	.0059	.0062	.0065	.1300	.1191	.1066	.0981	.0915
60	.0048	.0054	.0059	.0062	.0065	.1316	.1203	.1070	.0981	.0915
70	.0049	.0055	.0060	.0062	.0064	.1330	.1212	.1074	.0982	.0915
80	.0049	.0056	.0060	.0061	.0064	.1341	.1220	.1077	.0982	.0916
90	.0050	.0056	.0060	.0061	.0063	.1350	.1226	.1079	.0982	.0916
100	.0050	.0056	.0059	.0060	.0062	.1357	.1231	.1081	.0983	.0917
120	.0050	.0055	.0058	.0058	.0060	.1365	.1237	.1082	.0983	.0918
140	.0049	.0054	.0056	.0057	.0058	.1371	.1242	.1084	.0983	.0920
170	.0047	.0052	.0054	.0054	.0055	.1377	.1246	.1086	.0985	.0924
200	.0046	.0050	.0052	.0051	.0052	.1380	.1250	.1090	.0988	.0928
230	.0045	.0048	.0049	.0048	.0049	.1379	.1252	.1100	.0998	.0937
260	.0044	.0047	.0048	.0046	.0046	.1378	.1254	.1110	.1007	.0946
300	.0043	.0045	.0045	.0043	.0042	.1377	.1257	.1124	.1019	.0958
350	.0041	.0043	.0042	.0039	.0039	.1377	.1261	.1138	.1033	.0970
400	.0040	.0041	.0040	.0036	.0035	.1380	.1267	.1151	.1045	.0981
450	.0039	.0039	.0037	.0033	.0032	.1384	.1273	.1162	.1055	.0990
500	.0037	.0037	.0035	.0031	.0030	.1390	.1281	.1172	.1063	.0997
600	.0036	.0034	.0027	.0024	.0025	.1419	.1307	.1183	.1071	.1005
700	.0034	.0031	.0022	.0019	.0022	.1445	.1330	.1191	.1077	.1010
800	.0032	.0028	.0019	.0016	.0019	.1468	.1349	.1196	.1081	.1014
900	.0030	.0025	.0018	.0015	.0016	.1487	.1365	.1201	.1084	.1017
1000	.0028	.0022	.0018	.0016	.0014	.1503	.1378	.1204	.1087	.1019
1200	.0024	.0016	.0031	.0029	.0009	.1517	.1387	.1209	.1091	.1022
1400	.0020	.0011	.0044	.0042	.0006	.1525	.1391	.1212	.1093	.1024
1700	.0015	.0005	.0060	.0057	.0003	.1531	.1392	.1213	.1095	.1026
2000	.0011	0.0000	.0071	.0068	0.0000	.1532	.1391	.1213	.1096	.1026
2300	.0008	0.0000	.0077	.0074	0.0000	.1530	.1387	.1212	.1095	.1024
2600	.0006	0.0000	.0079	.0076	0.0000	.1528	.1384	.1210	.1094	.1022
3000	.0003	0.0000	.0076	.0073	0.0000	.1524	.1379	.1207	.1091	.1019
3500	.0001	0.0000	.0065	.0063	0.0000	.1518	.1374	.1202	.1087	.1014
4000	.0000	0.0000	.0048	.0046	0.0000	.1511	.1370	.1197	.1083	.1008
4500	0.0000	0.0000	.0026	.0025	0.0000	.1505	.1367	.1192	.1078	.1002
5000	0.0000	0.0000	0.0000	0.0000	0.0000	.1500	.1365	.1186	.1073	.0995

NI ON C, M2/M1=0.203

5000 KEV

500 KEV

50 KEV

5 KEV

.50 KEV

	FRACTIONAL DEPOSITED ENERGY					MEAN RANGE, MICROGRAM/SQ.CM.				
K/KL=	0.5	0.65	0.8	1.0	1.3	0.5	0.65	0.8	1.0	1.3
E,KEV										
5.0	.8868	.8562	.8269	.7899	.7384	1.259	1.253	1.246	1.238	1.225
6.0	.8835	.8520	.8221	.7842	.7316	1.430	1.422	1.414	1.404	1.390
7.0	.8805	.8484	.8178	.7792	.7258	1.585	1.576	1.568	1.556	1.540
8.0	.8779	.8452	.8140	.7748	.7206	1.731	1.722	1.712	1.699	1.681
9.0	.8756	.8423	.8106	.7709	.7159	1.871	1.860	1.850	1.836	1.815
10	.8734	.8396	.8075	.7672	.7116	2.006	1.994	1.983	1.968	1.945
12	.8696	.8349	.8020	.7608	.7041	2.265	2.252	2.238	2.220	2.195
14	.8662	.8308	.7972	.7552	.6975	2.513	2.498	2.482	2.462	2.433
17	.8618	.8254	.7910	.7480	.6891	2.870	2.851	2.833	2.809	2.775
20	.8580	.8207	.7855	.7417	.6818	3.210	3.189	3.168	3.141	3.100
23	.8548	.8167	.7809	.7363	.6756	3.523	3.500	3.476	3.445	3.400
26	.8518	.8131	.7767	.7315	.6699	3.827	3.801	3.775	3.741	3.691
30	.8483	.8088	.7716	.7256	.6632	4.221	4.191	4.162	4.123	4.067
35	.8443	.8039	.7660	.7191	.6557	4.700	4.665	4.632	4.587	4.522
40	.8407	.7995	.7609	.7133	.6491	5.165	5.126	5.087	5.037	4.964
45	.8373	.7954	.7562	.7080	.6430	5.617	5.574	5.531	5.475	5.393
50	.8342	.7916	.7519	.7030	.6374	6.059	6.010	5.963	5.901	5.811
60	.8285	.7847	.7440	.6941	.6273	6.874	6.818	6.762	6.689	6.584
70	.8233	.7785	.7370	.6861	.6184	7.672	7.607	7.543	7.459	7.337
80	.8185	.7728	.7305	.6789	.6104	8.456	8.382	8.309	8.213	8.074
90	.8142	.7676	.7247	.6724	.6031	9.228	9.143	9.061	8.954	8.797
100	.8100	.7628	.7192	.6663	.5964	9.986	9.893	9.801	9.681	9.507
120	.8026	.7540	.7093	.6553	.5844	11.38	11.27	11.16	11.02	10.82
140	.7959	.7460	.7005	.6456	.5739	12.77	12.64	12.52	12.35	12.12
170	.7867	.7354	.6887	.6327	.5600	14.86	14.70	14.54	14.34	14.05
200	.7785	.7259	.6783	.6214	.5479	16.95	16.75	16.56	16.31	15.96
230	.7710	.7173	.6688	.6111	.5370	18.78	18.56	18.35	18.07	17.67
260	.7641	.7093	.6601	.6018	.5271	20.65	20.40	20.16	19.85	19.40
300	.7555	.6995	.6495	.5905	.5153	23.19	22.90	22.62	22.25	21.72
350	.7457	.6885	.6376	.5778	.5022	26.46	25.10	25.76	25.31	24.67
400	.7366	.6784	.6267	.5664	.4905	29.81	29.38	28.96	28.42	27.65
450	.7283	.6691	.6168	.5561	.4800	33.22	32.70	32.20	31.56	30.64
500	.7205	.6605	.6077	.5465	.4704	36.66	36.05	35.46	34.71	33.65
600	.7061	.6447	.5911	.5294	.4533	42.88	42.14	41.42	40.50	39.20
700	.6931	.6306	.5764	.5143	.4384	49.47	48.55	47.66	46.52	44.92
800	.6814	.6179	.5632	.5009	.4253	56.38	55.23	54.13	52.73	50.78
900	.6706	.6064	.5512	.4888	.4135	63.54	62.13	60.78	59.08	56.73
1000	.6606	.5957	.5402	.4778	.4028	70.89	69.17	67.56	65.53	62.75
1200	.6428	.5763	.5204	.4582	.3838	84.38	82.26	80.24	77.70	74.19
1400	.6269	.5593	.5032	.4412	.3676	98.88	96.16	93.59	90.36	85.94
1700	.6059	.5372	.4808	.4193	.3468	122.3	118.3	114.7	110.1	104.1
2000	.5874	.5180	.4617	.4006	.3292	147.1	141.6	136.6	130.5	122.5
2300	.5705	.5015	.4451	.3841	.3140	170.0	163.3	157.3	150.1	140.3
2600	.5550	.4867	.4304	.3696	.3007	194.1	185.9	178.6	170.0	158.3
3000	.5365	.4691	.4131	.3524	.2851	228.0	217.3	208.0	197.1	182.5
3500	.5159	.4497	.3942	.3339	.2684	272.6	258.3	245.9	231.6	212.9
4000	.4975	.4326	.3776	.3178	.2541	319.2	300.7	284.7	266.6	243.3
4500	.4809	.4173	.3629	.3036	.2416	367.1	343.9	324.0	301.5	273.5
5000	.4659	.4034	.3496	.2909	.2305	416.0	387.8	363.7	336.5	303.4
6000	.4391	.3783	.3258	.2690	.2116	511.4	473.5	441.5	404.8	361.8
7000	.4161	.3568	.3057	.2507	.1960	609.5	560.7	519.7	472.6	418.9
8000	.3961	.3381	.2884	.2351	.1829	709.2	648.6	597.7	539.4	474.7
9000	.3782	.3216	.2733	.2216	.1716	809.9	736.6	675.1	605.1	529.3
10000	.3623	.3070	.2599	.2098	.1617	910.6	824.3	751.7	669.7	582.5
12000	.3345	.2819	.2372	.1900	.1455	1110.	996.5	900.7	794.5	684.5
14000	.3112	.2611	.2186	.1740	.1325	1307.	1166.	1046.	915.0	782.1
17000	.2824	.2356	.1961	.1548	.1171	1598.	1414.	1257.	1088.	921.4
20000	.2590	.2150	.1781	.1396	.1051	1881.	1653.	1459.	1253.	1053.
23000	.2394	.1980	.1633	.1273	.0954	2157.	1885.	1654.	1410.	1178.
26000	.2228	.1837	.1510	.1171	.0875	2424.	2109.	1841.	1560.	1297.
30000	.2043	.1678	.1374	.1060	.0789	2769.	2396.	2081.	1751.	1449.
35000	.1854	.1518	.1238	.0950	.0704	3182.	2739.	2366.	1977.	1626.
40000	.1700	.1388	.1129	.0863	.0638	3577.	3066.	2636.	2191.	1794.
45000	.1573	.1283	.1041	.0793	.0585	3955.	3378.	2893.	2393.	1952.
50000	.1467	.1195	.0968	.0737	.0543	4318.	3676.	3139.	2585.	2103.

E,KEV	MEAN DAMAGE DEPTH,MICROGRAM/SQ.CM.					MEAN IONZN.DEPTH,MICROGRAM/SQ.CM.				
K/KL=	0.5	0.65	0.8	1.0	1.3	0.5	0.65	0.8	1.0	1.3
5.0	.7816	.7758	.7701	.7626	.7519	.7217	.7152	.7089	.7006	.6886
6.0	.8900	.8829	.8760	.8671	.8543	.8218	.8139	.8063	.7963	.7820
7.0	.9881	.9800	.9721	.9619	.9473	.9123	.9033	.8945	.8832	.8669
8.0	1.080	1.071	1.062	1.051	1.035	.9971	.9871	.9773	.9647	.9465
9.0	1.168	1.158	1.148	1.136	1.118	1.078	1.067	1.056	1.043	1.023
10	1.253	1.242	1.232	1.218	1.199	1.157	1.145	1.133	1.118	1.096
12	1.418	1.405	1.393	1.377	1.354	1.309	1.295	1.281	1.263	1.238
14	1.577	1.562	1.548	1.530	1.504	1.455	1.439	1.423	1.403	1.374
17	1.807	1.789	1.772	1.750	1.719	1.668	1.648	1.629	1.605	1.570
20	2.029	2.008	1.988	1.963	1.926	1.873	1.850	1.828	1.799	1.758
23	2.232	2.209	2.186	2.157	2.116	2.060	2.034	2.009	1.977	1.931
26	2.430	2.404	2.379	2.347	2.302	2.243	2.214	2.186	2.151	2.100
30	2.691	2.661	2.632	2.595	2.543	2.483	2.450	2.418	2.378	2.319
35	3.010	2.976	2.942	2.899	2.838	2.779	2.740	2.703	2.655	2.587
40	3.325	3.284	3.245	3.196	3.126	3.070	3.025	2.982	2.926	2.848
45	3.634	3.587	3.543	3.487	3.407	3.357	3.305	3.255	3.192	3.104
50	3.938	3.885	3.835	3.772	3.683	3.639	3.580	3.524	3.453	3.354
60	4.499	4.436	4.376	4.300	4.194	4.158	4.088	4.021	3.936	3.818
70	5.057	4.982	4.911	4.821	4.697	4.677	4.593	4.513	4.413	4.274
80	5.613	5.525	5.442	5.337	5.193	5.194	5.095	5.002	4.884	4.723
90	6.165	6.063	5.967	5.847	5.681	5.710	5.594	5.486	5.351	5.165
100	6.714	6.597	6.487	6.350	6.164	6.222	6.089	5.965	5.811	5.601
120	7.732	7.588	7.454	7.288	7.061	7.174	7.011	6.859	6.672	6.417
140	8.758	8.584	8.422	8.223	7.954	8.138	7.939	7.755	7.529	7.224
170	10.32	10.09	9.881	9.627	9.287	9.604	9.342	9.102	8.812	8.427
200	11.89	11.60	11.34	11.03	10.61	11.09	10.75	10.45	10.09	9.616
230	13.30	12.97	12.66	12.30	11.82	12.44	12.04	11.69	11.26	10.71
260	14.75	14.35	14.00	13.58	13.03	13.81	13.35	12.93	12.44	11.80
300	16.71	16.24	15.81	15.31	14.65	15.69	15.11	14.61	14.02	13.26
350	19.23	18.64	18.11	17.50	16.70	18.08	17.36	16.73	16.00	15.08
400	21.80	21.08	20.45	19.71	18.77	20.51	19.63	18.87	18.00	16.91
450	24.40	23.55	22.80	21.94	20.85	22.97	21.91	21.01	19.99	18.72
500	27.02	26.02	25.16	24.17	22.92	25.44	24.20	23.15	21.98	20.53
600	31.82	30.57	29.50	28.27	26.75	30.07	28.47	27.16	25.69	23.90
700	36.84	35.30	33.99	32.50	30.67	34.82	32.85	31.23	29.45	27.30
800	42.03	40.17	38.60	36.84	34.66	39.66	37.29	35.36	33.24	30.71
900	47.34	45.15	43.30	41.24	38.69	44.57	41.78	39.51	37.05	34.12
1000	52.75	50.20	48.07	45.69	42.76	49.50	46.29	43.68	40.85	37.51
1200	62.74	59.52	56.92	54.03	50.42	58.69	54.80	51.55	48.06	43.97
1400	73.29	69.32	66.16	62.66	58.27	68.16	63.49	59.55	55.34	50.44
1700	90.02	84.81	80.66	76.08	70.38	82.82	76.83	71.74	66.36	60.14
2000	107.6	101.0	95.72	89.89	82.72	97.86	90.38	84.04	77.42	69.80
2300	123.5	116.0	109.8	102.9	94.51	112.1	103.1	95.56	88.05	79.14
2600	140.1	131.6	124.4	116.2	106.4	126.6	116.0	107.2	98.68	88.41
3000	163.4	153.2	144.4	134.4	122.6	146.3	133.4	122.8	112.8	100.6
3500	194.0	181.3	170.2	157.6	142.9	171.4	155.5	142.5	130.4	115.7
4000	225.8	210.3	196.6	181.1	163.4	196.8	177.8	162.3	147.8	130.4
4500	258.5	239.8	223.8	204.7	183.8	222.4	200.1	182.1	164.9	144.8
5000	291.8	269.8	250.3	228.4	204.2	248.0	222.4	201.8	181.9	159.0
6000	356.2	327.5	302.6	274.3	243.6	297.5	265.7	240.7	214.8	186.3
7000	422.7	386.5	355.5	320.3	282.6	347.0	308.8	278.8	246.7	212.5
8000	490.6	446.4	408.6	366.0	321.0	396.3	351.3	316.1	277.5	237.7
9000	559.4	506.6	461.6	411.3	358.8	445.0	393.2	352.5	307.5	261.9
10000	628.5	566.8	514.4	456.1	395.9	493.3	434.5	388.1	336.5	285.3
12000	764.5	685.1	617.4	542.9	467.3	588.0	515.4	456.9	392.2	329.6
14000	900.5	802.5	718.8	627.6	536.4	679.8	593.2	522.2	444.8	371.3
17000	1103.	975.9	867.4	750.6	636.1	812.1	704.5	615.0	518.8	429.4
20000	1302.	1145.	1012.	868.9	731.4	938.0	809.7	702.0	587.8	483.3
23000	1497.	1310.	1151.	982.7	822.7	1058.	909.5	784.1	652.5	533.6
26000	1688.	1471.	1287.	1093.	910.5	1173.	1004.	861.9	713.6	580.9
30000	1935.	1679.	1461.	1233.	1023.	1319.	1124.	959.7	790.1	640.1
35000	2234.	1928.	1670.	1400.	1155.	1491.	1265.	1074.	879.1	708.6
40000	2521.	2167.	1869.	1559.	1281.	1653.	1397.	1181.	961.9	772.3
45000	2797.	2396.	2060.	1711.	1401.	1807.	1522.	1281.	1040.	831.8
50000	3063.	2616.	2243.	1856.	1516.	1953.	1640.	1376.	1113.	887.9

TABLE 18- 3

TA ON AL, M2/M1=0.149

E,KEV	RELATIVE RANGE STRAGGLING					RELATIVE DAMAGE STRAGGLING				
K/KL=	0.5	0.65	0.8	1.0	1.3	0.5	0.65	0.8	1.0	1.3
5.0	.0713	.0708	.0703	.0696	.0687	.3186	.3174	.3163	.3150	.3131
6.0	.0713	.0707	.0702	.0695	.0685	.3186	.3174	.3163	.3149	.3129
7.0	.0712	.0707	.0701	.0694	.0684	.3187	.3174	.3162	.3147	.3127
8.0	.0712	.0706	.0701	.0694	.0683	.3187	.3174	.3162	.3147	.3125
9.0	.0712	.0706	.0700	.0693	.0682	.3188	.3174	.3162	.3146	.3124
10	.0711	.0706	.0700	.0692	.0682	.3189	.3175	.3161	.3145	.3123
12	.0711	.0705	.0699	.0691	.0680	.3191	.3176	.3162	.3144	.3120
14	.0710	.0704	.0698	.0690	.0679	.3193	.3177	.3162	.3144	.3119
17	.0710	.0704	.0697	.0689	.0677	.3197	.3180	.3164	.3144	.3117
20	.0709	.0703	.0697	.0688	.0676	.3202	.3183	.3166	.3144	.3116
23	.0709	.0702	.0696	.0687	.0675	.3207	.3187	.3168	.3145	.3115
26	.0709	.0702	.0695	.0686	.0674	.3212	.3191	.3171	.3147	.3114
30	.0708	.0701	.0694	.0685	.0672	.3220	.3196	.3175	.3149	.3114
35	.0708	.0701	.0694	.0684	.0671	.3230	.3204	.3180	.3152	.3114
40	.0708	.0700	.0693	.0683	.0669	.3241	.3213	.3187	.3155	.3115
45	.0707	.0700	.0692	.0682	.0668	.3253	.3221	.3193	.3159	.3116
50	.0707	.0699	.0691	.0681	.0667	.3265	.3230	.3200	.3163	.3117
60	.0707	.0698	.0690	.0680	.0664	.3289	.3250	.3214	.3173	.3121
70	.0706	.0697	.0689	.0678	.0662	.3314	.3269	.3229	.3183	.3125
80	.0706	.0697	.0688	.0677	.0660	.3339	.3288	.3243	.3192	.3128
90	.0705	.0696	.0687	.0675	.0659	.3363	.3306	.3257	.3200	.3132
100	.0705	.0695	.0686	.0674	.0657	.3387	.3324	.3270	.3209	.3135
120	.0704	.0694	.0684	.0672	.0654	.3437	.3362	.3298	.3227	.3142
140	.0703	.0693	.0682	.0669	.0651	.3483	.3396	.3323	.3242	.3147
170	.0702	.0691	.0680	.0666	.0646	.3543	.3439	.3353	.3260	.3152
200	.0701	.0689	.0677	.0662	.0641	.3595	.3475	.3377	.3272	.3154
230	.0699	.0687	.0675	.0659	.0637	.3644	.3506	.3396	.3282	.3153
260	.0698	.0685	.0672	.0656	.0633	.3686	.3531	.3411	.3287	.3149
300	.0696	.0682	.0669	.0652	.0628	.3732	.3557	.3424	.3290	.3142
350	.0693	.0678	.0664	.0646	.0621	.3775	.3579	.3433	.3288	.3130
400	.0690	.0675	.0660	.0641	.0614	.3806	.3593	.3436	.3280	.3115
450	.0687	.0671	.0655	.0635	.0608	.3826	.3600	.3433	.3269	.3098
500	.0684	.0667	.0650	.0629	.0601	.3838	.3602	.3427	.3256	.3080
600	.0678	.0659	.0641	.0618	.0588	.3835	.3589	.3400	.3214	.3036
700	.0671	.0650	.0631	.0608	.0575	.3817	.3566	.3368	.3171	.2992
800	.0664	.0642	.0622	.0597	.0563	.3790	.3536	.3332	.3128	.2951
900	.0657	.0634	.0613	.0586	.0552	.3756	.3503	.3295	.3086	.2912
1000	.0650	.0626	.0603	.0576	.0540	.3718	.3466	.3257	.3046	.2875
1200	.0635	.0609	.0585	.0556	.0518	.3621	.3377	.3173	.2969	.2808
1400	.0621	.0592	.0567	.0536	.0497	.3525	.3290	.3095	.2899	.2749
1700	.0600	.0569	.0542	.0510	.0469	.3389	.3169	.2988	.2808	.2672
2000	.0580	.0547	.0519	.0485	.0443	.3267	.3061	.2895	.2730	.2606
2300	.0560	.0526	.0496	.0462	.0420	.3162	.2972	.2819	.2666	.2552
2600	.0541	.0506	.0475	.0441	.0398	.3069	.2894	.2752	.2611	.2504
3000	.0518	.0481	.0450	.0415	.0372	.2961	.2802	.2674	.2546	.2449
3500	.0491	.0453	.0421	.0386	.0344	.2845	.2705	.2592	.2478	.2390
4000	.0467	.0428	.0396	.0361	.0320	.2748	.2624	.2522	.2420	.2339
4500	.0445	.0406	.0373	.0339	.0298	.2665	.2554	.2462	.2370	.2295
5000	.0425	.0385	.0353	.0319	.0279	.2594	.2493	.2410	.2326	.2255
6000	.0388	.0349	.0318	.0285	.0248	.2490	.2403	.2330	.2255	.2188
7000	.0357	.0318	.0288	.0257	.0222	.2409	.2331	.2265	.2196	.2133
8000	.0329	.0292	.0263	.0233	.0201	.2344	.2272	.2210	.2147	.2085
9000	.0306	.0269	.0242	.0213	.0183	.2289	.2222	.2164	.2103	.2043
10000	.0285	.0250	.0223	.0196	.0167	.2242	.2178	.2124	.2066	.2006
12000	.0251	.0218	.0194	.0169	.0144	.2167	.2107	.2056	.2001	.1944
14000	.0223	.0193	.0171	.0149	.0126	.2106	.2049	.2000	.1947	.1892
17000	.0191	.0164	.0144	.0125	.0105	.2034	.1979	.1931	.1880	.1827
20000	.0166	.0141	.0124	.0107	.0090	.1976	.1922	.1875	.1824	.1773
23000	.0146	.0124	.0108	.0093	.0078	.1928	.1873	.1827	.1777	.1727
26000	.0130	.0110	.0096	.0083	.0069	.1885	.1831	.1786	.1736	.1687
30000	.0113	.0095	.0083	.0071	.0060	.1836	.1782	.1737	.1688	.1640
35000	.0098	.0081	.0071	.0061	.0051	.1782	.1729	.1685	.1637	.1590
40000	.0087	.0071	.0062	.0053	.0045	.1734	.1683	.1639	.1592	.1547
45000	.0079	.0064	.0056	.0048	.0040	.1691	.1641	.1599	.1553	.1508
50000	.0074	.0058	.0051	.0044	.0037	.1651	.1603	.1562	.1518	.1474

TA ON AL, M2/M1=0.149   TABLE 18- 4

	RELATIVE IONIZATION STRAGGLING					RELATIVE TRANSVERSE RANGE STRAGGLING				
K/KL=	0.5	0.65	0.8	1.0	1.3	0.5	0.65	0.8	1.0	1.3
E,KEV										
5.0	.3645	.3640	.3636	.3631	.3625	.4739	.4747	.4755	.4765	.4780
6.0	.3648	.3642	.3638	.3632	.3626	.4736	.4744	.4752	.4763	.4779
7.0	.3650	.3644	.3639	.3633	.3627	.4733	.4742	.4750	.4761	.4777
8.0	.3652	.3646	.3641	.3634	.3627	.4730	.4739	.4748	.4759	.4776
9.0	.3655	.3648	.3642	.3636	.3628	.4728	.4737	.4745	.4757	.4774
10	.3657	.3650	.3644	.3637	.3629	.4725	.4734	.4743	.4755	.4772
12	.3662	.3654	.3647	.3640	.3631	.4719	.4729	.4738	.4750	.4769
14	.3667	.3658	.3651	.3643	.3633	.4714	.4724	.4733	.4746	.4765
17	.3676	.3666	.3657	.3647	.3636	.4706	.4716	.4726	.4739	.4759
20	.3685	.3674	.3664	.3653	.3640	.4697	.4708	.4719	.4732	.4753
23	.3694	.3681	.3670	.3658	.3643	.4689	.4701	.4711	.4726	.4747
26	.3703	.3689	.3677	.3663	.3647	.4682	.4693	.4704	.4719	.4741
30	.3716	.3700	.3686	.3671	.3652	.4672	.4683	.4695	.4711	.4733
35	.3733	.3714	.3698	.3680	.3659	.4659	.4672	.4684	.4700	.4724
40	.3750	.3729	.3710	.3690	.3666	.4647	.4660	.4673	.4690	.4714
45	.3768	.3744	.3723	.3699	.3673	.4635	.4649	.4662	.4679	.4705
50	.3786	.3759	.3735	.3709	.3679	.4623	.4637	.4651	.4669	.4696
60	.3823	.3790	.3761	.3730	.3694	.4601	.4616	.4631	.4650	.4678
70	.3859	.3820	.3786	.3749	.3707	.4580	.4595	.4611	.4631	.4661
80	.3895	.3849	.3810	.3768	.3720	.4559	.4576	.4592	.4614	.4645
90	.3930	.3877	.3833	.3785	.3732	.4540	.4557	.4574	.4597	.4630
100	.3964	.3904	.3855	.3802	.3742	.4521	.4539	.4557	.4580	.4615
120	.4034	.3960	.3900	.3835	.3764	.4484	.4504	.4523	.4549	.4586
140	.4099	.4011	.3940	.3865	.3783	.4450	.4472	.4492	.4520	.4559
170	.4184	.4077	.3991	.3901	.3806	.4404	.4428	.4450	.4480	.4524
200	.4258	.4132	.4033	.3930	.3823	.4362	.4388	.4413	.4445	.4492
230	.4328	.4182	.4068	.3953	.3834	.4322	.4350	.4376	.4411	.4461
260	.4389	.4224	.4098	.3972	.3843	.4286	.4315	.4343	.4380	.4434
300	.4457	.4270	.4129	.3990	.3850	.4241	.4273	.4304	.4344	.4402
350	.4524	.4314	.4157	.4005	.3855	.4193	.4227	.4261	.4305	.4368
400	.4577	.4346	.4177	.4014	.3856	.4150	.4187	.4224	.4271	.4339
450	.4617	.4369	.4189	.4018	.3854	.4113	.4153	.4192	.4242	.4315
500	.4646	.4384	.4195	.4018	.3851	.4079	.4122	.4164	.4217	.4294
600	.4677	.4387	.4186	.4004	.3834	.4020	.4068	.4114	.4174	.4259
700	.4686	.4378	.4169	.3985	.3817	.3972	.4025	.4076	.4141	.4235
800	.4681	.4361	.4148	.3964	.3799	.3934	.3991	.4047	.4118	.4219
900	.4665	.4339	.4125	.3941	.3781	.3904	.3965	.4025	.4101	.4209
1000	.4642	.4315	.4100	.3919	.3765	.3879	.3945	.4009	.4090	.4205
1200	.4561	.4252	.4044	.3866	.3730	.3843	.3917	.3988	.4079	.4207
1400	.4476	.4190	.3991	.3818	.3700	.3821	.3903	.3981	.4081	.4221
1700	.4354	.4105	.3921	.3758	.3665	.3808	.3901	.3989	.4101	.4258
2000	.4245	.4030	.3863	.3711	.3640	.3810	.3914	.4012	.4135	.4307
2300	.4155	.3967	.3818	.3684	.3631	.3824	.3938	.4045	.4180	.4365
2600	.4078	.3912	.3781	.3664	.3626	.3846	.3969	.4085	.4230	.4429
3000	.3992	.3852	.3742	.3647	.3626	.3884	.4019	.4146	.4304	.4519
3500	.3904	.3793	.3707	.3636	.3632	.3942	.4091	.4231	.4403	.4636
4000	.3836	.3748	.3683	.3632	.3643	.4006	.4169	.4320	.4505	.4755
4500	.3784	.3715	.3667	.3634	.3656	.4076	.4252	.4413	.4610	.4874
5000	.3743	.3691	.3658	.3640	.3672	.4149	.4336	.4508	.4716	.4993
6000	.3708	.3679	.3665	.3665	.3709	.4299	.4510	.4700	.4930	.5231
7000	.3691	.3681	.3681	.3695	.3746	.4452	.4684	.4891	.5138	.5460
8000	.3687	.3690	.3701	.3725	.3783	.4606	.4855	.5077	.5340	.5679
9000	.3690	.3704	.3723	.3755	.3817	.4757	.5023	.5258	.5534	.5888
10000	.3697	.3720	.3746	.3784	.3850	.4906	.5187	.5434	.5722	.6088
12000	.3722	.3757	.3792	.3839	.3909	.5210	.5516	.5781	.6085	.6471
14000	.3752	.3794	.3836	.3888	.3962	.5494	.5823	.6102	.6422	.6821
17000	.3799	.3849	.3897	.3955	.4031	.5886	.6243	.6543	.6881	.7298
20000	.3845	.3899	.3952	.4013	.4090	.6238	.6620	.6939	.7294	.7725
23000	.3888	.3946	.4001	.4063	.4141	.6557	.6962	.7297	.7669	.8111
26000	.3928	.3988	.4044	.4108	.4186	.6847	.7272	.7624	.8012	.8465
30000	.3977	.4038	.4096	.4160	.4238	.7197	.7647	.8020	.8428	.8893
35000	.4030	.4093	.4151	.4216	.4293	.7584	.8063	.8461	.8894	.9372
40000	.4076	.4139	.4198	.4263	.4339	.7926	.8432	.8855	.9312	.9802
45000	.4116	.4180	.4238	.4303	.4378	.8232	.8763	.9209	.9691	1.019
50000	.4150	.4215	.4273	.4338	.4411	.8509	.9063	.9532	1.004	1.055

TA ON AL, M2/M1=0.149   TABLE 18- 5
RELATIVE TRANSV. DAMAGE STRAGGLING   RELATIVE TRANSV. IONZN. STRAGGLING

K/KL= E,KEV	0.5	0.65	0.8	1.0	1.3	0.5	0.65	0.8	1.0	1.3
5.0	.3346	.3303	.3262	.3208	.3131	.2897	.2851	.2807	.2750	.2668
6.0	.3350	.3306	.3263	.3207	.3128	.2900	.2853	.2807	.2748	.2664
7.0	.3354	.3309	.3264	.3207	.3126	.2904	.2855	.2808	.2747	.2661
8.0	.3359	.3312	.3266	.3208	.3124	.2908	.2858	.2810	.2747	.2659
9.0	.3364	.3315	.3269	.3209	.3123	.2913	.2861	.2812	.2748	.2657
10	.3368	.3319	.3271	.3210	.3123	.2917	.2865	.2814	.2749	.2656
12	.3377	.3326	.3277	.3213	.3123	.2926	.2871	.2818	.2751	.2655
14	.3386	.3333	.3282	.3217	.3123	.2934	.2878	.2823	.2754	.2655
17	.3399	.3344	.3290	.3222	.3125	.2947	.2888	.2831	.2758	.2655
20	.3411	.3354	.3298	.3227	.3127	.2959	.2897	.2838	.2763	.2656
23	.3423	.3364	.3306	.3233	.3130	.2971	.2907	.2846	.2768	.2658
26	.3435	.3373	.3314	.3239	.3132	.2982	.2917	.2854	.2773	.2660
30	.3449	.3385	.3324	.3246	.3136	.2996	.2928	.2863	.2780	.2663
35	.3465	.3398	.3335	.3254	.3140	.3012	.2941	.2874	.2787	.2666
40	.3479	.3410	.3344	.3260	.3143	.3027	.2953	.2883	.2794	.2669
45	.3491	.3420	.3352	.3266	.3145	.3040	.2964	.2891	.2800	.2671
50	.3502	.3430	.3360	.3271	.3147	.3051	.2974	.2899	.2805	.2673
60	.3523	.3447	.3374	.3281	.3151	.3073	.2992	.2914	.2815	.2677
70	.3540	.3460	.3385	.3288	.3153	.3091	.3006	.2925	.2822	.2679
80	.3553	.3471	.3393	.3293	.3154	.3106	.3018	.2934	.2827	.2679
90	.3563	.3479	.3399	.3296	.3153	.3118	.3028	.2941	.2831	.2678
100	.3572	.3486	.3403	.3297	.3150	.3128	.3035	.2946	.2833	.2677
120	.3583	.3494	.3407	.3297	.3144	.3142	.3045	.2953	.2835	.2671
140	.3589	.3497	.3408	.3294	.3135	.3151	.3051	.2955	.2833	.2663
170	.3591	.3496	.3402	.3283	.3116	.3157	.3054	.2953	.2825	.2647
200	.3588	.3489	.3392	.3268	.3095	.3158	.3051	.2947	.2813	.2628
230	.3580	.3478	.3379	.3250	.3071	.3154	.3045	.2937	.2799	.2606
260	.3569	.3465	.3363	.3230	.3045	.3148	.3036	.2925	.2783	.2583
300	.3553	.3446	.3339	.3202	.3010	.3137	.3022	.2907	.2759	.2551
350	.3530	.3419	.3307	.3164	.2965	.3119	.3000	.2881	.2726	.2510
400	.3505	.3390	.3274	.3125	.2919	.3100	.2976	.2852	.2692	.2469
450	.3480	.3360	.3240	.3086	.2874	.3078	.2951	.2822	.2657	.2428
500	.3454	.3330	.3206	.3047	.2829	.3056	.2924	.2792	.2621	.2387
600	.3406	.3271	.3138	.2969	.2737	.3009	.2866	.2725	.2547	.2307
700	.3358	.3212	.3071	.2893	.2650	.2962	.2808	.2660	.2475	.2230
800	.3309	.3153	.3004	.2819	.2568	.2916	.2752	.2597	.2405	.2156
900	.3261	.3096	.2940	.2748	.2491	.2871	.2698	.2536	.2338	.2085
1000	.3212	.3039	.2877	.2680	.2418	.2828	.2645	.2477	.2273	.2017
1200	.3115	.2925	.2750	.2542	.2277	.2753	.2549	.2364	.2146	.1881
1400	.3022	.2817	.2632	.2415	.2152	.2680	.2459	.2261	.2030	.1758
1700	.2890	.2669	.2471	.2247	.1988	.2574	.2332	.2119	.1874	.1597
2000	.2766	.2534	.2329	.2100	.1848	.2471	.2215	.1990	.1737	.1459
2300	.2638	.2404	.2198	.1971	.1728	.2353	.2091	.1864	.1611	.1342
2600	.2520	.2285	.2080	.1857	.1624	.2240	.1976	.1749	.1499	.1241
3000	.2375	.2142	.1941	.1724	.1505	.2101	.1837	.1611	.1367	.1125
3500	.2215	.1987	.1791	.1583	.1379	.1942	.1682	.1461	.1227	.1003
4000	.2073	.1852	.1663	.1464	.1274	.1800	.1546	.1332	.1108	.0903
4500	.1948	.1734	.1552	.1363	.1186	.1673	.1427	.1220	.1007	.0818
5000	.1836	.1630	.1456	.1275	.1110	.1559	.1321	.1123	.0920	.0746
6000	.1650	.1463	.1305	.1142	.0995	.1362	.1148	.0970	.0790	.0639
7000	.1499	.1329	.1186	.1038	.0906	.1201	.1009	.0849	.0689	.0556
8000	.1373	.1219	.1088	.0954	.0834	.1066	.0894	.0752	.0609	.0492
9000	.1267	.1126	.1007	.0884	.0775	.0953	.0799	.0671	.0543	.0440
10000	.1177	.1049	.0939	.0826	.0725	.0858	.0719	.0604	.0489	.0397
12000	.1044	.0932	.0836	.0737	.0648	.0722	.0604	.0507	.0410	.0333
14000	.0943	.0843	.0757	.0669	.0589	.0621	.0520	.0436	.0353	.0286
17000	.0829	.0743	.0669	.0593	.0523	.0510	.0427	.0360	.0291	.0237
20000	.0746	.0670	.0604	.0536	.0473	.0431	.0362	.0305	.0248	.0202
23000	.0681	.0613	.0554	.0493	.0435	.0371	.0313	.0265	.0216	.0176
26000	.0630	.0568	.0513	.0457	.0404	.0326	.0276	.0234	.0192	.0156
30000	.0576	.0520	.0471	.0420	.0370	.0280	.0238	.0203	.0166	.0136
35000	.0524	.0473	.0429	.0383	.0337	.0238	.0203	.0173	.0143	.0116
40000	.0484	.0437	.0396	.0353	.0311	.0208	.0177	.0152	.0125	.0102
45000	.0453	.0408	.0369	.0329	.0289	.0186	.0158	.0135	.0111	.0090
50000	.0427	.0384	.0347	.0308	.0270	.0169	.0143	.0121	.0098	.0080

TA ON AL, M2/M1=0.149   TABLE 18- 6

	RANGE SKEWNESS					DAMAGE SKEWNESS				
K/KL=	0.5	0.65	0.8	1.0	1.3	0.5	0.65	0.8	1.0	1.3
E,KEV										
5.0	.4541	.4486	.4431	.4360	.4255	.5110	.5017	.4928	.4817	.4662
6.0	.4532	.4476	.4420	.4346	.4237	.5133	.5034	.4939	.4821	.4658
7.0	.4525	.4467	.4409	.4334	.4222	.5153	.5048	.4949	.4825	.4654
8.0	.4519	.4459	.4400	.4322	.4209	.5172	.5062	.4959	.4830	.4652
9.0	.4512	.4451	.4391	.4312	.4196	.5192	.5077	.4969	.4835	.4652
10	.4507	.4444	.4383	.4303	.4184	.5212	.5093	.4981	.4842	.4652
12	.4496	.4432	.4368	.4285	.4163	.5255	.5127	.5006	.4857	.4655
14	.4486	.4420	.4355	.4269	.4144	.5302	.5164	.5035	.4876	.4661
17	.4472	.4403	.4336	.4247	.4117	.5378	.5225	.5083	.4909	.4675
20	.4459	.4388	.4318	.4227	.4093	.5459	.5291	.5135	.4945	.4693
23	.4447	.4374	.4303	.4209	.4072	.5539	.5357	.5189	.4984	.4714
26	.4436	.4361	.4288	.4192	.4051	.5622	.5425	.5244	.5025	.4736
30	.4421	.4345	.4269	.4170	.4026	.5737	.5519	.5321	.5081	.4768
35	.4404	.4325	.4247	.4145	.3996	.5883	.5639	.5418	.5153	.4809
40	.4387	.4306	.4225	.4120	.3967	.6031	.5760	.5516	.5225	.4850
45	.4371	.4287	.4205	.4097	.3940	.6181	.5882	.5614	.5296	.4891
50	.4356	.4270	.4185	.4075	.3914	.6330	.6003	.5711	.5368	.4932
60	.4327	.4237	.4149	.4034	.3867	.6636	.6252	.5912	.5515	.5018
70	.4299	.4205	.4114	.3995	.3821	.6932	.6492	.6104	.5655	.5098
80	.4271	.4175	.4080	.3957	.3778	.7218	.6720	.6286	.5786	.5172
90	.4245	.4145	.4047	.3920	.3736	.7493	.6939	.6459	.5910	.5240
100	.4218	.4115	.4015	.3884	.3695	.7757	.7149	.6623	.6026	.5303
120	.4169	.4061	.3954	.3817	.3618	.8300	.7578	.6960	.6265	.5435
140	.4121	.4007	.3896	.3752	.3545	.8788	.7960	.7256	.6471	.5543
170	.4050	.3929	.3811	.3658	.3439	.9426	.8449	.7627	.6721	.5664
200	.3981	.3852	.3728	.3567	.3337	.9964	.8853	.7926	.6913	.5744
230	.3915	.3780	.3649	.3481	.3240	1.045	.9204	.8178	.7071	.5800
250	.3851	.3709	.3572	.3397	.3146	1.087	.9494	.8379	.7188	.5828
300	.3766	.3617	.3472	.3287	.3025	1.133	.9804	.8583	.7292	.5831
350	.3664	.3504	.3351	.3155	.2879	1.177	1.009	.8755	.7356	.5791
400	.3563	.3395	.3233	.3027	.2738	1.210	1.029	.8854	.7361	.5715
450	.3465	.3288	.3118	.2903	.2601	1.235	1.042	.8897	.7322	.5612
500	.3370	.3184	.3007	.2783	.2469	1.252	1.049	.8896	.7248	.5487
600	.3186	.2984	.2792	.2551	.2216	1.266	1.049	.8756	.6955	.5130
700	.3010	.2793	.2588	.2331	.1977	1.266	1.039	.8540	.6624	.4765
800	.2841	.2610	.2393	.2122	.1751	1.254	1.020	.8278	.6278	.4406
900	.2678	.2434	.2206	.1922	.1536	1.236	.9968	.7987	.5928	.4061
1000	.2522	.2266	.2027	.1731	.1330	1.212	.9696	.7678	.5582	.3732
1200	.2222	.1942	.1683	.1366	.0939	1.141	.8980	.6955	.4874	.3111
1400	.1942	.1641	.1364	.1027	.0579	1.066	.8253	.6253	.4219	.2553
1700	.1553	.1223	.0924	.0563	.0087	.9554	.7210	.5279	.3342	.1822
2000	.1194	.0841	.0522	.0141	-.0356	.8505	.6253	.4408	.2582	.1195
2300	.0857	.0481	.0145	-.0253	-.0767	.7494	.5381	.3653	.1949	.0675
2600	.0542	.0146	-.0205	-.0617	-.1144	.6568	.4593	.2981	.1395	.0220
3000	.0153	-.0266	-.0634	-.1062	-.1603	.5459	.3662	.2195	.0754	-.0308
3500	-.0292	-.0735	-.1120	-.1564	-.2119	.4258	.2659	.1355	.0075	-.0869
4000	-.0699	-.1162	-.1561	-.2017	-.2581	.3230	.1807	.0644	-.0499	-.1346
4500	-.1074	-.1554	-.1964	-.2429	-.3000	.2350	.1077	.0036	-.0990	-.1759
5000	-.1422	-.1917	-.2335	-.2807	-.3383	.1594	.0449	-.0488	-.1416	-.2121
6000	-.2058	-.2574	-.3005	-.3484	-.4061	.0522	-.0459	-.1267	-.2074	-.2707
7000	-.2620	-.3152	-.3591	-.4073	-.4649	-.0312	-.1172	-.1885	-.2603	-.3188
8000	-.3125	-.3668	-.4111	-.4595	-.5167	-.0984	-.1751	-.2391	-.3044	-.3595
9000	-.3583	-.4133	-.4579	-.5062	-.5629	-.1538	-.2234	-.2819	-.3421	-.3947
10000	-.4001	-.4557	-.5004	-.5484	-.6046	-.2004	-.2645	-.3186	-.3749	-.4257
12000	-.4735	-.5294	-.5737	-.6210	-.6757	-.2723	-.3300	-.3790	-.4306	-.4786
14000	-.5376	-.5933	-.6371	-.6835	-.7367	-.3279	-.3817	-.4276	-.4762	-.5226
17000	-.6212	-.6763	-.7191	-.7640	-.8151	-.3924	-.4429	-.4861	-.5322	-.5770
20000	-.6937	-.7479	-.7897	-.8332	-.8825	-.4429	-.4917	-.5334	-.5781	-.6218
23000	-.7580	-.8113	-.8520	-.8943	-.9419	-.4845	-.5323	-.5732	-.6170	-.6600
26000	-.8160	-.8682	-.9080	-.9491	-.9952	-.5203	-.5674	-.6077	-.6508	-.6932
30000	-.8855	-.9365	-.9750	-1.015	-1.059	-.5617	-.6081	-.6477	-.6900	-.7318
35000	-.9629	-1.012	-1.049	-1.088	-1.130	-.6068	-.6521	-.6908	-.7323	-.7731
40000	-1.032	-1.080	-1.116	-1.153	-1.194	-.6467	-.6908	-.7285	-.7689	-.8088
45000	-1.095	-1.141	-1.176	-1.212	-1.252	-.6832	-.7257	-.7622	-.8014	-.8402
50000	-1.152	-1.197	-1.231	-1.266	-1.305	-.7171	-.7579	-.7929	-.8307	-.8684

TABLE 18- 7

TA ON AL, M2/M1=0.149

K/KL=	IONIZATION SKEWNESS					RANGE KURTOSIS				
	0.5	0.65	0.8	1.0	1.3	0.5	0.65	0.8	1.0	1.3
E,KEV										
5.0	.6108	.6029	.5955	.5863	.5739	3.207	3.197	3.188	3.175	3.157
6.0	.6136	.6051	.5971	.5873	.5741	3.206	3.196	3.186	3.173	3.154
7.0	.6159	.6069	.5985	.5882	.5744	3.204	3.194	3.184	3.170	3.151
8.0	.6182	.6088	.6000	.5892	.5747	3.203	3.192	3.182	3.168	3.149
9.0	.6205	.6106	.6014	.5902	.5752	3.202	3.191	3.180	3.167	3.147
10	.6229	.6125	.6030	.5913	.5757	3.201	3.190	3.179	3.165	3.145
12	.6278	.6166	.6063	.5936	.5769	3.199	3.187	3.176	3.162	3.141
14	.6332	.6210	.6098	.5963	.5784	3.197	3.185	3.174	3.159	3.138
17	.6416	.6280	.6156	.6006	.5809	3.194	3.182	3.170	3.155	3.133
20	.6505	.6354	.6217	.6051	.5837	3.191	3.179	3.167	3.151	3.129
23	.6593	.6428	.6278	.6099	.5867	3.189	3.176	3.164	3.148	3.125
26	.6683	.6503	.6341	.6147	.5898	3.187	3.174	3.161	3.145	3.122
30	.6806	.6606	.6426	.6213	.5941	3.184	3.170	3.158	3.141	3.117
35	.6962	.6736	.6534	.6296	.5995	3.180	3.167	3.153	3.136	3.112
40	.7121	.6868	.6643	.6379	.6048	3.177	3.163	3.149	3.132	3.107
45	.7279	.6999	.6751	.6462	.6102	3.174	3.160	3.146	3.128	3.103
50	.7438	.7130	.6858	.6544	.6155	3.171	3.156	3.142	3.124	3.098
60	.7767	.7403	.7084	.6717	.6268	3.165	3.150	3.135	3.117	3.090
70	.8082	.7662	.7296	.6879	.6373	3.160	3.144	3.129	3.110	3.083
80	.8382	.7907	.7496	.7030	.6469	3.155	3.139	3.123	3.103	3.076
90	.8667	.8138	.7683	.7170	.6557	3.150	3.133	3.117	3.097	3.069
100	.8938	.8357	.7859	.7300	.6637	3.145	3.128	3.112	3.091	3.063
120	.9483	.8794	.8210	.7559	.6796	3.135	3.118	3.101	3.080	3.051
140	.9966	.9178	.8513	.7779	.6926	3.127	3.108	3.091	3.069	3.039
170	1.059	.9663	.8891	.8045	.7074	3.114	3.095	3.077	3.054	3.024
200	1.110	1.006	.9191	.8249	.7178	3.102	3.082	3.064	3.041	3.010
230	1.153	1.037	.9420	.8397	.7238	3.091	3.070	3.051	3.027	2.996
260	1.189	1.063	.9601	.8506	.7271	3.079	3.058	3.039	3.015	2.983
300	1.228	1.090	.9784	.8605	.7285	3.066	3.044	3.024	2.999	2.967
350	1.265	1.115	.9940	.8671	.7266	3.049	3.027	3.007	2.982	2.949
400	1.294	1.133	1.004	.8690	.7218	3.035	3.012	2.991	2.966	2.933
450	1.315	1.145	1.009	.8674	.7149	3.021	2.997	2.976	2.951	2.919
500	1.330	1.152	1.010	.8630	.7065	3.007	2.984	2.963	2.938	2.906
600	1.351	1.162	1.007	.8468	.6838	2.982	2.958	2.937	2.912	2.881
700	1.358	1.160	.9962	.8261	.6599	2.960	2.936	2.914	2.890	2.861
800	1.354	1.150	.9798	.8029	.6358	2.940	2.916	2.895	2.872	2.844
900	1.342	1.134	.9595	.7782	.6120	2.923	2.899	2.879	2.856	2.831
1000	1.324	1.113	.9364	.7526	.5888	2.907	2.884	2.864	2.843	2.820
1200	1.257	1.047	.8713	.6915	.5396	2.878	2.856	2.838	2.820	2.801
1400	1.185	.9790	.8077	.6347	.4956	2.855	2.835	2.819	2.804	2.790
1700	1.081	.8832	.7206	.5598	.4396	2.829	2.812	2.800	2.790	2.784
2000	.9861	.7979	.6452	.4972	.3942	2.811	2.797	2.789	2.784	2.786
2300	.9037	.7287	.5876	.4524	.3630	2.797	2.788	2.784	2.785	2.795
2600	.8305	.6686	.5387	.4155	.3381	2.788	2.783	2.784	2.791	2.809
3000	.7455	.6002	.4842	.3758	.3121	2.782	2.784	2.790	2.805	2.832
3500	.6565	.5299	.4296	.3374	.2877	2.783	2.792	2.806	2.828	2.867
4000	.5830	.4730	.3865	.3082	.2699	2.790	2.806	2.827	2.858	2.907
4500	.5223	.4267	.3521	.2858	.2568	2.801	2.825	2.852	2.891	2.949
5000	.4720	.3890	.3248	.2686	.2471	2.817	2.847	2.881	2.926	2.993
6000	.4081	.3430	.2931	.2506	.2379	2.855	2.900	2.945	3.004	3.087
7000	.3629	.3117	.2729	.2408	.2342	2.901	2.959	3.015	3.085	3.181
8000	.3300	.2898	.2598	.2357	.2336	2.951	3.021	3.086	3.167	3.274
9000	.3055	.2742	.2513	.2337	.2348	3.005	3.085	3.159	3.248	3.365
10000	.2872	.2632	.2460	.2335	.2372	3.061	3.151	3.232	3.329	3.454
12000	.2640	.2498	.2404	.2353	.2433	3.172	3.279	3.372	3.482	3.622
14000	.2511	.2437	.2396	.2396	.2506	3.288	3.408	3.512	3.631	3.783
17000	.2424	.2417	.2431	.2483	.2624	3.464	3.602	3.718	3.849	4.013
20000	.2406	.2445	.2493	.2579	.2740	3.642	3.794	3.919	4.060	4.232
23000	.2425	.2494	.2567	.2675	.2850	3.819	3.982	4.115	4.263	4.441
26000	.2463	.2554	.2644	.2769	.2953	3.995	4.167	4.306	4.460	4.641
30000	.2529	.2640	.2747	.2887	.3080	4.225	4.407	4.553	4.713	4.895
35000	.2619	.2748	.2869	.3021	.3222	4.505	4.698	4.850	5.015	5.196
40000	.2708	.2850	.2981	.3142	.3347	4.777	4.977	5.134	5.303	5.480
45000	.2791	.2943	.3082	.3249	.3458	5.041	5.247	5.407	5.578	5.750
50000	.2867	.3027	.3173	.3345	.3556	5.296	5.507	5.669	5.842	6.008

K/KL= E,KEV	DAMAGE KURTOSIS					IONIZATION KURTOSIS				
	0.5	0.65	0.8	1.0	1.3	0.5	0.65	0.8	1.0	1.3
5.0	3.114	3.082	3.052	3.015	2.966	3.218	3.185	3.155	3.117	3.067
6.0	3.125	3.091	3.059	3.019	2.967	3.230	3.195	3.162	3.122	3.069
7.0	3.135	3.098	3.064	3.023	2.968	3.241	3.203	3.168	3.126	3.070
8.0	3.143	3.105	3.070	3.026	2.969	3.250	3.211	3.175	3.131	3.072
9.0	3.152	3.112	3.075	3.030	2.971	3.260	3.219	3.181	3.135	3.075
10	3.161	3.119	3.081	3.034	2.973	3.271	3.227	3.188	3.140	3.077
12	3.180	3.135	3.093	3.043	2.977	3.292	3.245	3.202	3.150	3.083
14	3.201	3.152	3.107	3.053	2.983	3.315	3.264	3.217	3.162	3.090
17	3.234	3.179	3.129	3.069	2.992	3.351	3.294	3.242	3.180	3.101
20	3.269	3.208	3.152	3.087	3.003	3.389	3.325	3.268	3.200	3.113
23	3.302	3.236	3.176	3.105	3.014	3.427	3.356	3.293	3.220	3.126
26	3.336	3.266	3.200	3.124	3.026	3.465	3.388	3.320	3.240	3.139
30	3.383	3.306	3.234	3.149	3.043	3.518	3.432	3.356	3.267	3.157
35	3.446	3.358	3.276	3.181	3.063	3.585	3.487	3.401	3.301	3.179
40	3.511	3.410	3.318	3.212	3.083	3.653	3.543	3.446	3.336	3.201
45	3.578	3.462	3.360	3.243	3.102	3.722	3.598	3.491	3.370	3.223
50	3.646	3.515	3.402	3.274	3.121	3.790	3.654	3.536	3.403	3.244
60	3.796	3.622	3.487	3.337	3.160	3.933	3.770	3.630	3.474	3.290
70	3.940	3.724	3.569	3.397	3.196	4.071	3.880	3.719	3.540	3.332
80	4.077	3.823	3.646	3.452	3.229	4.201	3.984	3.802	3.601	3.370
90	4.207	3.917	3.719	3.504	3.259	4.325	4.082	3.879	3.658	3.405
100	4.330	4.007	3.789	3.553	3.287	4.444	4.175	3.952	3.711	3.436
120	4.569	4.192	3.929	3.651	3.341	4.683	4.361	4.097	3.815	3.497
140	4.780	4.356	4.052	3.735	3.386	4.894	4.523	4.222	3.903	3.547
170	5.053	4.568	4.209	3.840	3.439	5.163	4.727	4.377	4.009	3.605
200	5.291	4.745	4.339	3.921	3.478	5.385	4.892	4.499	4.089	3.646
230	5.476	4.898	4.453	3.990	3.507	5.562	5.015	4.586	4.144	3.676
260	5.642	5.026	4.547	4.044	3.527	5.708	5.115	4.654	4.183	3.696
300	5.826	5.166	4.647	4.096	3.542	5.866	5.219	4.721	4.218	3.710
350	6.007	5.300	4.738	4.135	3.546	6.018	5.315	4.778	4.241	3.713
400	6.148	5.398	4.799	4.154	3.538	6.131	5.383	4.813	4.247	3.703
450	6.256	5.469	4.836	4.155	3.520	6.215	5.430	4.832	4.240	3.684
500	6.339	5.518	4.855	4.143	3.495	6.276	5.461	4.839	4.224	3.657
600	6.471	5.564	4.829	4.060	3.405	6.376	5.516	4.841	4.170	3.572
700	6.533	5.561	4.772	3.964	3.313	6.412	5.523	4.812	4.100	3.482
800	6.541	5.523	4.696	3.864	3.224	6.398	5.493	4.758	4.019	3.393
900	6.508	5.460	4.608	3.764	3.140	6.347	5.435	4.687	3.933	3.306
1000	6.443	5.379	4.512	3.666	3.062	6.265	5.355	4.602	3.842	3.223
1200	6.147	5.112	4.273	3.466	2.921	5.931	5.056	4.338	3.617	3.057
1400	5.829	4.841	4.045	3.286	2.800	5.585	4.754	4.079	3.408	2.911
1700	5.363	4.460	3.736	3.058	2.652	5.094	4.334	3.728	3.135	2.729
2000	4.938	4.122	3.472	2.871	2.536	4.661	3.971	3.430	2.910	2.582
2300	4.579	3.851	3.272	2.741	2.458	4.344	3.718	3.231	2.766	2.488
2600	4.264	3.617	3.104	2.637	2.397	4.074	3.508	3.068	2.652	2.414
3000	3.904	3.354	2.920	2.528	2.335	3.773	3.276	2.892	2.532	2.337
3500	3.532	3.087	2.738	2.425	2.280	3.467	3.046	2.721	2.419	2.265
4000	3.234	2.876	2.597	2.349	2.242	3.223	2.865	2.588	2.334	2.211
4500	2.993	2.708	2.487	2.292	2.215	3.025	2.721	2.484	2.269	2.170
5000	2.801	2.576	2.402	2.251	2.197	2.865	2.605	2.401	2.219	2.138
6000	2.602	2.441	2.318	2.213	2.183	2.668	2.461	2.299	2.156	2.095
7000	2.477	2.359	2.271	2.197	2.183	2.531	2.363	2.232	2.116	2.067
8000	2.397	2.309	2.246	2.194	2.189	2.432	2.294	2.185	2.089	2.048
9000	2.344	2.279	2.234	2.197	2.200	2.358	2.242	2.151	2.070	2.034
10000	2.310	2.262	2.230	2.205	2.213	2.301	2.202	2.125	2.056	2.025
12000	2.264	2.241	2.228	2.222	2.240	2.208	2.136	2.080	2.030	2.008
14000	2.245	2.239	2.239	2.244	2.269	2.145	2.091	2.050	2.013	1.999
17000	2.245	2.255	2.265	2.282	2.314	2.083	2.048	2.021	1.998	1.991
20000	2.262	2.280	2.298	2.322	2.359	2.045	2.022	2.004	1.990	1.989
23000	2.285	2.310	2.332	2.361	2.402	2.021	2.006	1.995	1.987	1.990
26000	2.311	2.340	2.366	2.399	2.444	2.005	1.996	1.990	1.987	1.993
30000	2.347	2.381	2.411	2.448	2.496	1.994	1.989	1.988	1.989	1.999
35000	2.388	2.428	2.463	2.505	2.558	1.987	1.987	1.989	1.994	2.008
40000	2.425	2.470	2.510	2.557	2.615	1.986	1.989	1.994	2.002	2.018
45000	2.457	2.508	2.554	2.606	2.668	1.987	1.993	2.000	2.010	2.028
50000	2.485	2.542	2.594	2.652	2.718	1.990	1.998	2.006	2.018	2.039

TA ON AL, M2/M1=0.149

TABLE 18- 9

E,KEV	REFLECTION COEFFICIENT					SPUTTERING EFFICIENCY				
K/KL=	0.5	0.65	0.8	1.0	1.3	0.5	0.65	0.8	1.0	1.3
5.0	.0000	.0000	.0000	.0000	.0000	.0168	.0161	.0154	.0145	.0133
6.0	.0000	.0000	.0000	.0000	.0000	.0168	.0160	.0153	.0144	.0132
7.0	.0000	.0000	.0000	.0000	.0000	.0168	.0160	.0152	.0143	.0131
8.0	.0000	.0000	.0000	.0000	.0000	.0167	.0159	.0152	.0142	.0130
9.0	.0000	.0000	.0000	.0000	.0000	.0167	.0159	.0151	.0142	.0129
10	.0000	.0000	.0000	.0000	.0000	.0167	.0159	.0151	.0141	.0128
12	.0000	.0000	.0000	.0000	.0000	.0167	.0158	.0150	.0140	.0127
14	.0000	.0000	.0000	.0000	.0000	.0167	.0158	.0150	.0140	.0126
17	.0000	.0000	.0000	.0000	.0000	.0168	.0158	.0149	.0139	.0125
20	.0000	.0000	.0000	.0000	.0000	.0168	.0158	.0149	.0138	.0124
23	.0000	.0000	.0000	.0000	.0000	.0169	.0158	.0149	.0138	.0123
26	.0000	.0000	.0000	.0000	.0000	.0169	.0159	.0149	.0137	.0122
30	.0000	.0000	.0000	.0000	.0000	.0170	.0159	.0149	.0137	.0121
35	.0000	.0000	.0000	.0000	.0000	.0171	.0160	.0149	.0137	.0121
40	.0000	.0000	.0000	.0000	.0000	.0172	.0160	.0149	.0137	.0120
45	.0000	.0000	.0000	.0000	.0000	.0173	.0161	.0149	.0136	.0119
50	.0000	.0000	.0000	.0000	.0000	.0174	.0161	.0150	.0136	.0119
60	.0000	.0000	.0000	.0000	.0000	.0177	.0162	.0150	.0136	.0118
70	.0000	.0000	.0000	.0000	.0000	.0179	.0163	.0151	.0136	.0117
80	.0000	.0000	.0000	.0000	.0000	.0181	.0164	.0151	.0136	.0116
90	.0000	.0000	.0000	.0000	.0000	.0183	.0165	.0151	.0135	.0116
100	.0000	.0000	.0000	.0000	.0000	.0184	.0165	.0151	.0135	.0115
120	.0000	.0000	.0000	.0000	.0000	.0186	.0166	.0151	.0134	.0114
140	.0000	.0000	.0000	.0000	.0000	.0186	.0166	.0151	.0133	.0112
170	.0000	.0000	.0000	.0000	.0000	.0186	.0166	.0150	.0132	.0110
200	.0000	.0000	.0000	.0000	.0000	.0186	.0165	.0149	.0131	.0108
230	.0000	.0000	.0000	.0000	.0000	.0184	.0164	.0148	.0129	.0106
260	.0000	.0000	.0000	.0000	.0000	.0183	.0163	.0147	.0127	.0105
300	.0000	.0000	.0000	.0000	.0000	.0181	.0162	.0145	.0125	.0102
350	.0000	.0000	.0000	.0000	.0000	.0178	.0160	.0143	.0123	.0100
400	.0000	.0000	.0000	.0000	.0000	.0176	.0158	.0141	.0120	.0097
450	.0000	.0000	.0000	.0000	.0000	.0173	.0156	.0139	.0118	.0095
500	.0000	.0000	.0000	.0000	.0000	.0171	.0154	.0137	.0115	.0093
600	.0000	.0000	.0000	.0000	.0000	.0169	.0151	.0133	.0111	.0088
700	.0000	.0000	.0000	.0000	.0000	.0167	.0148	.0129	.0107	.0084
800	.0000	.0000	.0000	.0000	.0000	.0164	.0145	.0126	.0103	.0081
900	.0000	.0000	.0000	.0000	.0000	.0161	.0142	.0122	.0100	.0078
1000	.0000	.0000	.0000	.0000	.0000	.0159	.0139	.0119	.0096	.0075
1200	.0000	.0000	.0000	.0000	.0000	.0152	.0132	.0113	.0090	.0069
1400	.0000	.0000	.0000	.0000	.0000	.0145	.0126	.0107	.0085	.0064
1700	.0000	.0000	.0000	.0000	.0000	.0136	.0117	.0100	.0078	.0057
2000	.0000	.0000	.0000	.0000	.0000	.0128	.0110	.0093	.0072	.0052
2300	.0000	.0000	.0000	.0000	.0000	.0121	.0104	.0087	.0066	.0048
2600	.0000	.0000	.0000	.0000	.0000	.0115	.0098	.0081	.0061	.0044
3000	.0000	.0000	.0000	.0000	.0000	.0108	.0091	.0075	.0056	.0040
3500	.0000	.0000	.0000	.0000	.0000	.0101	.0084	.0068	.0050	.0036
4000	.0000	.0000	.0000	.0000	0.0000	.0094	.0077	.0062	.0045	.0032
4500	.0000	.0000	.0000	.0000	0.0000	.0088	.0072	.0057	.0041	.0030
5000	.0000	.0000	.0000	.0000	0.0000	.0083	.0067	.0052	.0038	.0027
6000	.0000	.0000	0.0000	0.0000	0.0000	.0074	.0059	.0046	.0033	.0024
7000	.0000	0.0000	0.0000	0.0000	0.0000	.0066	.0052	.0040	.0029	.0021
8000	.0000	0.0000	0.0000	0.0000	0.0000	.0059	.0047	.0036	.0026	.0019
9000	0.0000	0.0000	0.0000	0.0000	0.0000	.0054	.0042	.0033	.0024	.0017
10000	0.0000	0.0000	0.0000	0.0000	0.0000	.0049	.0038	.0030	.0022	.0016
12000	0.0000	0.0000	0.0000	0.0000	0.0000	.0042	.0032	.0025	.0018	.0013
14000	0.0000	0.0000	0.0000	0.0000	0.0000	.0037	.0028	.0022	.0016	.0011
17000	0.0000	0.0000	0.0000	0.0000	0.0000	.0030	.0024	.0018	.0013	.0009
20000	0.0000	0.0000	0.0000	0.0000	0.0000	.0026	.0020	.0016	.0011	.0008
23000	0.0000	0.0000	0.0000	0.0000	0.0000	.0022	.0018	.0014	.0010	.0007
26000	0.0000	0.0000	0.0000	0.0000	0.0000	.0019	.0016	.0012	.0009	.0006
30000	0.0000	0.0000	0.0000	0.0000	0.0000	.0017	.0014	.0010	.0007	.0005
35000	0.0000	0.0000	0.0000	0.0000	0.0000	.0014	.0011	.0009	.0006	.0004
40000	0.0000	0.0000	0.0000	0.0000	0.0000	.0012	.0010	.0008	.0005	.0004
45000	0.0000	0.0000	0.0000	0.0000	0.0000	.0011	.0009	.0007	.0005	.0003
50000	0.0000	0.0000	0.0000	0.0000	0.0000	.0010	.0007	.0006	.0004	.0003

TA ON AL, M2/M1=0.149

TABLE 18-10

E, KEV	IONIZATION DEFICIENCY K/KL=					SPUTTERING YIELD ALPHA				
	0.5	0.65	0.8	1.0	1.3	0.5	0.65	0.8	1.0	1.3
5.0	.0017	.0021	.0025	.0030	.0036	.1204	.1167	.1132	.1087	.1024
6.0	.0017	.0022	.0026	.0030	.0037	.1201	.1162	.1126	.1080	.1016
7.0	.0018	.0022	.0026	.0031	.0038	.1197	.1158	.1121	.1074	.1009
8.0	.0018	.0023	.0027	.0032	.0038	.1195	.1155	.1117	.1069	.1003
9.0	.0018	.0023	.0027	.0032	.0039	.1192	.1151	.1113	.1065	.0998
10	.0019	.0023	.0028	.0033	.0039	.1190	.1148	.1109	.1060	.0993
12	.0019	.0024	.0029	.0034	.0040	.1186	.1143	.1103	.1053	.0984
14	.0020	.0025	.0029	.0034	.0041	.1183	.1139	.1098	.1047	.0976
17	.0021	.0026	.0030	.0036	.0042	.1179	.1134	.1092	.1039	.0967
20	.0022	.0027	.0031	.0037	.0043	.1176	.1130	.1087	.1033	.0959
23	.0022	.0027	.0032	.0037	.0044	.1174	.1127	.1082	.1028	.0952
26	.0023	.0028	.0033	.0038	.0045	.1172	.1124	.1079	.1023	.0947
30	.0023	.0029	.0034	.0039	.0046	.1170	.1121	.1075	.1018	.0940
35	.0024	.0030	.0035	.0040	.0047	.1168	.1118	.1070	.1012	.0933
40	.0025	.0031	.0036	.0041	.0048	.1168	.1116	.1067	.1007	.0927
45	.0026	.0031	.0036	.0042	.0049	.1167	.1114	.1064	.1003	.0921
50	.0026	.0032	.0037	.0043	.0050	.1168	.1113	.1062	.1000	.0917
60	.0028	.0034	.0039	.0045	.0051	.1171	.1111	.1059	.0994	.0909
70	.0029	.0035	.0040	.0046	.0053	.1175	.1110	.1056	.0990	.0902
80	.0030	.0036	.0041	.0047	.0054	.1178	.1110	.1054	.0986	.0897
90	.0031	.0037	.0043	.0048	.0055	.1181	.1110	.1053	.0983	.0892
100	.0032	.0038	.0044	.0049	.0056	.1184	.1110	.1051	.0981	.0888
120	.0033	.0040	.0045	.0051	.0057	.1188	.1111	.1050	.0976	.0881
140	.0035	.0042	.0047	.0053	.0059	.1191	.1113	.1049	.0973	.0876
170	.0037	.0043	.0049	.0055	.0060	.1195	.1115	.1048	.0970	.0869
200	.0038	.0045	.0050	.0056	.0062	.1199	.1118	.1049	.0968	.0864
230	.0040	.0046	.0052	.0057	.0063	.1201	.1120	.1050	.0966	.0861
260	.0041	.0048	.0053	.0058	.0064	.1202	.1122	.1051	.0965	.0857
300	.0042	.0049	.0054	.0059	.0065	.1205	.1125	.1053	.0964	.0854
350	.0044	.0050	.0055	.0060	.0066	.1209	.1129	.1055	.0963	.0851
400	.0045	.0051	.0056	.0060	.0067	.1213	.1133	.1058	.0963	.0849
450	.0046	.0052	.0057	.0061	.0067	.1217	.1137	.1061	.0962	.0848
500	.0047	.0053	.0057	.0061	.0067	.1222	.1142	.1063	.0962	.0846
600	.0048	.0053	.0058	.0062	.0066	.1237	.1153	.1069	.0962	.0845
700	.0049	.0054	.0058	.0062	.0065	.1251	.1163	.1074	.0963	.0844
800	.0050	.0054	.0058	.0062	.0064	.1262	.1171	.1078	.0963	.0844
900	.0050	.0054	.0058	.0061	.0063	.1271	.1178	.1081	.0964	.0844
1000	.0050	.0054	.0057	.0061	.0062	.1278	.1183	.1084	.0965	.0846
1200	.0050	.0054	.0057	.0059	.0059	.1283	.1185	.1084	.0966	.0850
1400	.0049	.0053	.0056	.0057	.0056	.1285	.1186	.1085	.0968	.0854
1700	.0047	.0052	.0054	.0054	.0052	.1287	.1187	.1086	.0972	.0861
2000	.0046	.0051	.0053	.0051	.0049	.1288	.1188	.1090	.0977	.0868
2300	.0045	.0049	.0051	.0048	.0045	.1289	.1195	.1101	.0987	.0876
2600	.0044	.0048	.0049	.0045	.0042	.1291	.1202	.1113	.0997	.0884
3000	.0043	.0046	.0046	.0042	.0039	.1294	.1212	.1127	.1010	.0893
3500	.0041	.0043	.0043	.0038	.0035	.1299	.1224	.1144	.1024	.0902
4000	.0040	.0041	.0040	.0035	.0032	.1305	.1235	.1159	.1036	.0910
4500	.0039	.0039	.0038	.0032	.0029	.1313	.1246	.1172	.1047	.0917
5000	.0038	.0037	.0035	.0029	.0027	.1322	.1257	.1182	.1056	.0923
6000	.0036	.0030	.0031	.0023	.0023	.1352	.1279	.1195	.1065	.0929
7000	.0034	.0026	.0027	.0018	.0020	.1378	.1297	.1204	.1072	.0933
8000	.0032	.0023	.0024	.0016	.0018	.1401	.1312	.1211	.1077	.0936
9000	.0030	.0022	.0021	.0015	.0016	.1419	.1324	.1216	.1081	.0939
10000	.0028	.0022	.0019	.0016	.0014	.1434	.1333	.1220	.1084	.0941
12000	.0024	.0035	.0013	.0029	.0010	.1447	.1342	.1227	.1089	.0945
14000	.0020	.0048	.0009	.0043	.0006	.1453	.1346	.1231	.1092	.0948
17000	.0015	.0063	.0004	.0059	.0003	.1457	.1349	.1235	.1095	.0951
20000	.0011	.0074	0.0000	.0070	0.0000	.1456	.1348	.1237	.1096	.0952
23000	.0008	.0080	0.0000	.0077	0.0000	.1454	.1346	.1237	.1095	.0952
26000	.0005	.0081	0.0000	.0079	0.0000	.1451	.1343	.1236	.1094	.0951
30000	.0003	.0077	0.0000	.0076	0.0000	.1446	.1339	.1234	.1090	.0949
35000	.0001	.0066	0.0000	.0065	0.0000	.1440	.1334	.1229	.1086	.0945
40000	0.0000	.0048	0.0000	.0048	0.0000	.1435	.1330	.1225	.1080	.0940
45000	0.0000	.0026	0.0000	.0026	0.0000	.1430	.1326	.1219	.1074	.0934
50000	0.0000	0.0000	0.0000	0.0000	0.0000	.1427	.1322	.1213	.1067	.0928

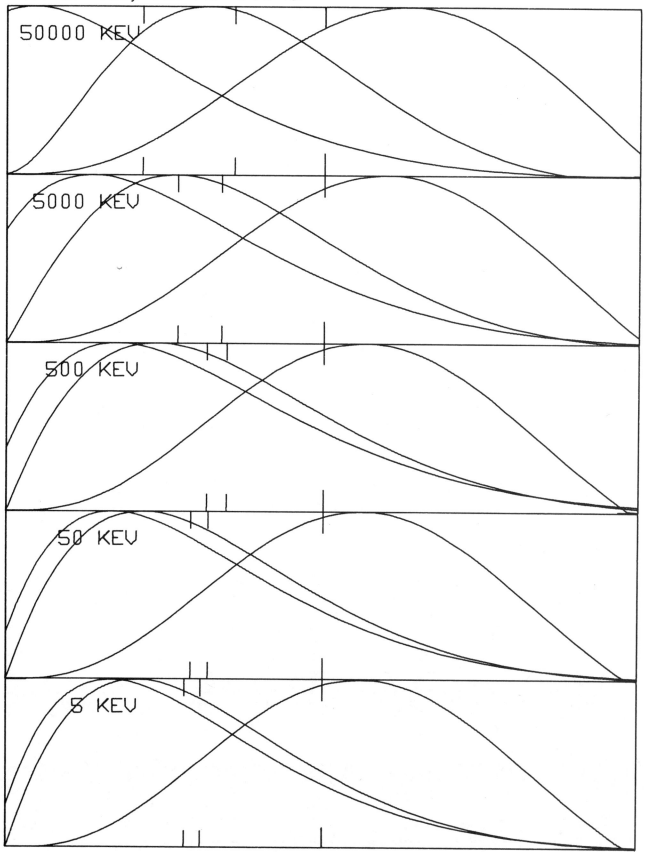

TA ON AL, M2/M1=0.149

50000 KEV

5000 KEV

500 KEV

50 KEV

5 KEV

AG ON C, M2/M1=0.111   TABLE 19- 1

| E,KEV | FRACTIONAL DEPOSITED ENERGY | | | | | MEAN RANGE,MICROGRAM/SQ.CM. | | | | |
K/KL=	0.6	0.80	1.0	1.2	1.4	0.6	0.80	1.0	1.2	1.4
1.0	.8807	.8451	.8114	.7795	.7492	.3684	.3662	.3640	.3618	.3597
1.2	.8772	.8407	.8062	.7735	.7426	.4182	.4156	.4130	.4105	.4079
1.4	.8741	.8368	.8016	.7684	.7369	.4636	.4606	.4576	.4547	.4519
1.7	.8701	.8319	.7958	.7617	.7296	.5265	.5230	.5196	.5162	.5129
2.0	.8667	.8276	.7907	.7560	.7233	.5856	.5816	.5777	.5739	.5701
2.3	.8637	.8238	.7862	.7510	.7177	.6420	.6376	.6332	.6289	.6247
2.6	.8609	.8203	.7822	.7464	.7128	.6964	.6915	.6867	.6819	.6773
3.0	.8577	.8162	.7774	.7410	.7068	.7664	.7609	.7555	.7501	.7448
3.5	.8540	.8117	.7721	.7351	.7003	.8506	.8443	.8381	.8320	.8260
4.0	.8507	.8076	.7674	.7297	.6945	.9316	.9245	.9176	.9107	.9040
4.5	.8477	.8039	.7631	.7249	.6893	1.010	1.002	.9943	.9867	.9792
5.0	.8450	.8005	.7591	.7205	.6845	1.086	1.077	1.069	1.060	1.052
6.0	.8401	.7945	.7521	.7127	.6760	1.227	1.217	1.207	1.197	1.188
7.0	.8358	.7892	.7460	.7059	.6686	1.362	1.351	1.339	1.328	1.318
8.0	.8319	.7844	.7405	.6998	.6620	1.493	1.480	1.468	1.455	1.443
9.0	.8283	.7800	.7354	.6942	.6560	1.620	1.606	1.592	1.578	1.565
10	.8250	.7760	.7308	.6891	.6505	1.744	1.728	1.713	1.697	1.683
12	.8191	.7687	.7225	.6800	.6407	1.974	1.955	1.937	1.919	1.902
14	.8137	.7623	.7152	.6719	.6321	2.196	2.175	2.154	2.134	2.114
17	.8066	.7537	.7055	.6613	.6208	2.520	2.494	2.470	2.445	2.422
20	.8003	.7461	.6969	.6520	.6110	2.833	2.804	2.775	2.746	2.718
23	.7945	.7393	.6892	.6436	.6021	3.117	3.084	3.051	3.020	2.989
26	.7892	.7329	.6821	.6360	.5941	3.397	3.360	3.323	3.288	3.253
30	.7827	.7252	.6735	.6268	.5845	3.765	3.723	3.681	3.641	3.601
35	.7753	.7165	.6639	.6165	.5737	4.221	4.171	4.123	4.076	4.029
40	.7684	.7086	.6551	.6072	.5640	4.672	4.615	4.559	4.504	4.451
45	.7621	.7012	.6471	.5987	.5552	5.118	5.053	4.989	4.927	4.867
50	.7562	.6944	.6397	.5909	.5471	5.559	5.485	5.414	5.344	5.276
60	.7452	.6819	.6261	.5767	.5326	6.364	6.276	6.191	6.109	6.029
70	.7354	.6707	.6141	.5642	.5199	7.177	7.074	6.973	6.876	6.782
80	.7264	.6606	.6033	.5530	.5085	7.998	7.877	7.759	7.646	7.535
90	.7180	.6513	.5935	.5428	.4982	8.824	8.683	8.547	8.415	8.288
100	.7102	.6427	.5844	.5335	.4889	9.651	9.490	9.334	9.184	9.039
120	.6959	.6270	.5680	.5168	.4721	11.16	10.97	10.78	10.60	10.43
140	.6831	.6131	.5536	.5022	.4576	12.72	12.49	12.26	12.05	11.84
170	.6661	.5949	.5348	.4834	.4389	15.17	14.86	14.56	14.27	14.00
200	.6512	.5789	.5185	.4672	.4230	17.70	17.30	16.92	16.55	16.21
230	.6377	.5647	.5042	.4531	.4091	19.91	19.46	19.03	18.61	18.21
260	.6255	.5520	.4914	.4406	.3967	22.22	21.70	21.21	20.73	20.27
300	.6108	.5367	.4762	.4257	.3822	25.47	24.84	24.23	23.64	23.08
350	.5944	.5199	.4595	.4096	.3665	29.78	28.95	28.16	27.41	26.69
400	.5798	.5050	.4448	.3954	.3528	34.31	33.23	32.22	31.28	30.39
450	.5666	.4917	.4317	.3828	.3407	39.00	37.64	36.38	35.22	34.14
500	.5545	.4797	.4198	.3714	.3298	43.80	42.13	40.61	39.20	37.91
600	.5331	.4587	.3990	.3515	.3112	52.71	50.57	48.60	46.79	45.13
700	.5144	.4407	.3812	.3345	.2954	62.25	59.46	56.92	54.61	52.50
800	.4979	.4249	.3656	.3197	.2818	72.27	68.69	65.48	62.58	59.97
900	.4830	.4108	.3519	.3067	.2698	82.67	78.17	74.19	70.66	67.48
1000	.4695	.3981	.3395	.2949	.2591	93.34	87.82	83.01	78.78	75.02
1200	.4455	.3754	.3180	.2744	.2404	114.1	106.7	100.3	94.75	89.75
1400	.4249	.3561	.2999	.2571	.2249	135.7	126.0	117.8	110.7	104.4
1700	.3986	.3316	.2772	.2356	.2056	169.3	155.6	144.2	134.6	126.2
2000	.3764	.3112	.2585	.2181	.1899	203.8	185.4	170.5	158.1	147.6
2300	.3570	.2936	.2426	.2036	.1769	237.9	214.6	196.4	181.0	168.5
2600	.3400	.2784	.2289	.1912	.1659	272.2	243.7	222.0	203.4	188.8
3000	.3203	.2608	.2133	.1773	.1535	318.2	282.3	255.6	232.6	215.3
3500	.2992	.2422	.1969	.1628	.1406	375.6	330.1	296.7	268.1	247.3
4000	.2812	.2265	.1831	.1508	.1300	432.6	377.2	336.7	302.4	278.2
4500	.2655	.2129	.1713	.1405	.1210	488.9	423.4	375.6	335.6	307.9
5000	.2517	.2010	.1611	.1317	.1133	544.3	468.9	413.4	367.8	336.7
6000	.2285	.1812	.1442	.1173	.1006	652.6	557.1	486.2	429.4	391.5
7000	.2095	.1653	.1308	.1060	.0907	757.2	642.0	555.3	487.6	443.1
8000	.1937	.1522	.1199	.0968	.0828	858.6	723.6	621.6	542.9	491.9
9000	.1803	.1413	.1108	.0892	.0763	955.7	802.2	684.1	595.6	538.4
10000	.1688	.1319	.1032	.0828	.0708	1050.	878.1	744.5	645.9	582.7

K/KL=	MEAN DAMAGE DEPTH, MICROGRAM/SQ.CM.					MEAN IONZN.DEPTH, MICROGRAM/SQ.CM.				
E,KEV	0.6	0.80	1.0	1.2	1.4	0.6	0.80	1.0	1.2	1.4
1.0	.2163	.2146	.2128	.2112	.2095	.1993	.1974	.1954	.1935	.1917
1.2	.2464	.2442	.2422	.2401	.2382	.2270	.2246	.2223	.2200	.2178
1.4	.2734	.2710	.2686	.2663	.2641	.2519	.2492	.2465	.2439	.2414
1.7	.3108	.3080	.3052	.3025	.2999	.2863	.2831	.2800	.2770	.2741
2.0	.3460	.3427	.3396	.3365	.3336	.3187	.3150	.3115	.3081	.3047
2.3	.3797	.3761	.3726	.3692	.3658	.3497	.3456	.3417	.3378	.3341
2.6	.4124	.4084	.4045	.4007	.3970	.3798	.3753	.3709	.3666	.3625
3.0	.4548	.4503	.4458	.4415	.4373	.4188	.4137	.4087	.4039	.3992
3.5	.5062	.5009	.4958	.4909	.4860	.4662	.4602	.4544	.4489	.4435
4.0	.5561	.5501	.5442	.5386	.5331	.5121	.5053	.4987	.4924	.4863
4.5	.6046	.5978	.5913	.5849	.5788	.5568	.5491	.5418	.5347	.5278
5.0	.6518	.6443	.6370	.6300	.6232	.6003	.5918	.5836	.5757	.5681
6.0	.7394	.7305	.7219	.7137	.7058	.6810	.6709	.6613	.6520	.6431
7.0	.8248	.8144	.8045	.7949	.7857	.7597	.7480	.7368	.7260	.7157
8.0	.9084	.8964	.8850	.8741	.8636	.8369	.8234	.8105	.7982	.7864
9.0	.9903	.9767	.9638	.9514	.9395	.9126	.8972	.8826	.8687	.8553
10	1.071	1.055	1.041	1.027	1.014	.9870	.9697	.9533	.9376	.9227
12	1.220	1.202	1.184	1.168	1.152	1.125	1.104	1.085	1.066	1.048
14	1.367	1.346	1.325	1.306	1.287	1.261	1.237	1.214	1.192	1.171
17	1.586	1.558	1.533	1.509	1.486	1.464	1.433	1.404	1.377	1.351
20	1.801	1.767	1.736	1.707	1.680	1.665	1.627	1.591	1.558	1.527
23	1.997	1.958	1.922	1.889	1.857	1.848	1.804	1.763	1.724	1.689
26	2.193	2.148	2.107	2.069	2.033	2.031	1.980	1.933	1.889	1.848
30	2.454	2.401	2.352	2.307	2.265	2.276	2.215	2.159	2.107	2.059
35	2.782	2.717	2.658	2.603	2.553	2.585	2.510	2.441	2.378	2.320
40	3.111	3.033	2.962	2.897	2.838	2.896	2.804	2.722	2.647	2.579
45	3.439	3.347	3.264	3.189	3.120	3.208	3.098	3.000	2.913	2.834
50	3.766	3.658	3.563	3.477	3.399	3.519	3.390	3.277	3.176	3.086
60	4.373	4.237	4.118	4.012	3.917	4.099	3.935	3.793	3.668	3.556
70	4.989	4.822	4.677	4.550	4.435	4.689	4.486	4.311	4.159	4.024
80	5.612	5.411	5.239	5.088	4.953	5.286	5.039	4.830	4.649	4.490
90	6.238	6.002	5.801	5.626	5.470	5.887	5.593	5.347	5.137	4.953
100	6.866	6.592	6.362	6.162	5.985	6.489	6.146	5.862	5.621	5.411
120	8.032	7.686	7.401	7.157	6.942	7.625	7.185	6.828	6.528	6.270
140	9.227	8.803	8.460	8.167	7.910	8.780	8.236	7.800	7.438	7.129
170	11.07	10.52	10.08	9.708	9.385	10.54	9.829	9.268	8.809	8.419
200	12.95	12.27	11.73	11.27	10.88	12.32	11.43	10.74	10.18	9.708
230	14.60	13.83	13.20	12.67	12.23	13.95	12.90	12.09	11.44	10.90
260	16.31	15.44	14.71	14.10	13.60	15.61	14.38	13.46	12.71	12.09
300	18.68	17.65	16.79	16.06	15.48	17.85	16.39	15.29	14.42	13.69
350	21.77	20.53	19.47	18.58	17.88	20.71	18.94	17.62	16.57	15.71
400	24.97	23.50	22.23	21.17	20.34	23.60	21.52	19.97	18.74	17.73
450	28.26	26.53	25.04	23.80	22.83	26.52	24.11	22.32	20.91	19.75
500	31.60	29.61	27.88	26.45	25.33	29.44	26.70	24.68	23.08	21.76
600	37.78	35.33	33.22	31.47	30.08	34.90	31.58	29.13	27.20	25.63
700	44.32	41.34	38.77	36.64	34.95	40.49	36.54	33.64	31.35	29.48
800	51.16	47.56	44.47	41.92	39.89	46.17	41.56	38.17	35.49	33.30
900	58.22	53.94	50.29	47.28	44.88	51.91	46.61	42.70	39.61	37.08
1000	65.44	60.43	56.17	52.67	49.89	57.68	51.67	47.22	43.70	40.81
1200	79.29	72.97	67.60	63.15	59.62	68.71	61.45	56.07	51.75	48.03
1400	93.76	85.89	79.24	73.73	69.38	79.90	71.26	64.83	59.62	55.07
1700	116.3	105.8	96.95	89.66	84.01	96.90	85.96	77.73	71.10	65.29
2000	139.6	126.1	114.8	105.6	98.52	114.0	100.5	90.33	82.17	75.13
2300	162.3	145.9	132.2	121.1	112.6	131.2	115.1	102.6	92.84	84.65
2600	185.4	165.8	149.5	136.4	126.5	148.3	129.4	114.6	103.1	93.84
3000	216.5	192.3	172.4	156.5	144.8	170.7	148.0	130.0	116.3	105.6
3500	255.6	225.4	200.7	181.2	167.0	198.2	170.5	148.5	132.0	119.6
4000	294.7	258.2	228.4	205.3	188.7	224.9	192.3	166.3	146.9	132.9
4500	333.5	290.6	255.7	228.8	209.8	250.8	213.3	183.3	161.2	145.5
5000	372.0	322.5	282.4	251.7	230.3	276.1	233.6	199.7	174.8	157.6
6000	447.6	384.8	334.1	295.8	269.7	324.5	272.2	230.6	200.4	180.4
7000	521.1	445.0	383.7	337.9	307.2	370.3	308.6	259.6	224.2	201.5
8000	592.5	503.1	431.4	378.1	342.9	414.0	342.9	286.9	246.4	221.2
9000	661.7	559.3	477.2	416.6	377.1	455.6	375.5	312.7	267.4	239.8
10000	728.9	613.5	521.3	453.6	409.9	495.4	406.6	337.2	287.2	257.4

K/KL= E,KEV	RELATIVE RANGE STRAGGLING					RELATIVE DAMAGE STRAGGLING				
	0.6	0.80	1.0	1.2	1.4	0.6	0.80	1.0	1.2	1.4
1.0	.0542	.0538	.0533	.0529	.0525	.3055	.3045	.3036	.3028	.3020
1.2	.0542	.0537	.0533	.0528	.0524	.3054	.3044	.3035	.3026	.3018
1.4	.0541	.0537	.0532	.0527	.0523	.3054	.3043	.3033	.3024	.3016
1.7	.0541	.0536	.0531	.0527	.0522	.3052	.3041	.3031	.3021	.3012
2.0	.0541	.0536	.0531	.0526	.0521	.3052	.3040	.3029	.3019	.3010
2.3	.0540	.0535	.0530	.0525	.0520	.3051	.3039	.3027	.3017	.3007
2.6	.0540	.0535	.0530	.0525	.0520	.3051	.3038	.3026	.3015	.3005
3.0	.0540	.0534	.0529	.0524	.0519	.3051	.3037	.3025	.3013	.3003
3.5	.0540	.0534	.0529	.0523	.0518	.3052	.3037	.3024	.3011	.3000
4.0	.0539	.0534	.0528	.0523	.0517	.3053	.3037	.3023	.3010	.2998
4.5	.0539	.0533	.0528	.0523	.0517	.3055	.3038	.3023	.3009	.2996
5.0	.0539	.0533	.0527	.0522	.0516	.3057	.3039	.3023	.3008	.2995
6.0	.0539	.0532	.0526	.0521	.0515	.3062	.3042	.3023	.3007	.2993
7.0	.0538	.0532	.0526	.0520	.0514	.3068	.3045	.3025	.3007	.2991
8.0	.0538	.0532	.0525	.0519	.0513	.3074	.3049	.3027	.3008	.2990
9.0	.0538	.0531	.0525	.0518	.0512	.3082	.3054	.3030	.3009	.2990
10	.0538	.0531	.0524	.0518	.0512	.3090	.3060	.3034	.3011	.2991
12	.0537	.0530	.0523	.0517	.0510	.3108	.3073	.3042	.3016	.2993
14	.0537	.0530	.0523	.0516	.0509	.3127	.3086	.3051	.3021	.2995
17	.0537	.0529	.0522	.0515	.0508	.3157	.3107	.3066	.3030	.3000
20	.0536	.0529	.0521	.0513	.0506	.3187	.3129	.3080	.3039	.3005
23	.0536	.0528	.0520	.0512	.0505	.3222	.3154	.3098	.3051	.3012
26	.0536	.0528	.0519	.0512	.0504	.3256	.3178	.3115	.3063	.3019
30	.0536	.0527	.0519	.0510	.0502	.3299	.3208	.3136	.3076	.3027
35	.0536	.0526	.0517	.0509	.0501	.3349	.3243	.3159	.3091	.3036
40	.0535	.0526	.0516	.0507	.0499	.3394	.3273	.3179	.3103	.3042
45	.0535	.0525	.0515	.0506	.0497	.3436	.3301	.3196	.3114	.3048
50	.0534	.0524	.0514	.0505	.0496	.3474	.3325	.3211	.3122	.3051
60	.0534	.0523	.0512	.0502	.0493	.3549	.3371	.3239	.3138	.3057
70	.0533	.0522	.0510	.0500	.0490	.3610	.3406	.3259	.3147	.3058
80	.0532	.0520	.0508	.0497	.0487	.3659	.3433	.3272	.3151	.3056
90	.0531	.0518	.0506	.0495	.0484	.3699	.3453	.3279	.3151	.3051
100	.0530	.0517	.0504	.0492	.0481	.3729	.3468	.3283	.3147	.3045
120	.0528	.0514	.0500	.0487	.0475	.3766	.3481	.3274	.3126	.3022
140	.0526	.0510	.0496	.0482	.0469	.3785	.3482	.3258	.3100	.2997
170	.0522	.0505	.0489	.0475	.0461	.3788	.3469	.3226	.3057	.2958
200	.0518	.0499	.0483	.0467	.0452	.3772	.3443	.3189	.3014	.2919
230	.0513	.0494	.0476	.0459	.0444	.3727	.3397	.3143	.2969	.2878
260	.0508	.0488	.0469	.0451	.0435	.3677	.3349	.3096	.2926	.2840
300	.0502	.0480	.0460	.0441	.0425	.3606	.3284	.3037	.2873	.2792
350	.0493	.0470	.0448	.0429	.0412	.3518	.3205	.2968	.2812	.2738
400	.0485	.0460	.0438	.0418	.0399	.3432	.3131	.2904	.2758	.2690
450	.0477	.0450	.0427	.0406	.0388	.3351	.3062	.2847	.2709	.2646
500	.0469	.0441	.0417	.0396	.0377	.3274	.2999	.2795	.2665	.2607
600	.0452	.0422	.0397	.0375	.0355	.3133	.2888	.2707	.2593	.2542
700	.0436	.0405	.0378	.0355	.0335	.3011	.2794	.2634	.2532	.2487
800	.0421	.0388	.0361	.0338	.0318	.2906	.2713	.2571	.2480	.2440
900	.0406	.0373	.0346	.0322	.0302	.2816	.2644	.2517	.2435	.2399
1000	.0393	.0359	.0331	.0307	.0287	.2738	.2584	.2470	.2396	.2363
1200	.0367	.0332	.0304	.0281	.0261	.2625	.2495	.2398	.2334	.2303
1400	.0344	.0309	.0281	.0258	.0239	.2538	.2425	.2340	.2283	.2254
1700	.0314	.0279	.0251	.0229	.0211	.2438	.2343	.2270	.2220	.2192
2000	.0288	.0253	.0227	.0206	.0189	.2361	.2278	.2214	.2169	.2141
2300	.0266	.0232	.0207	.0187	.0171	.2299	.2224	.2166	.2123	.2096
2600	.0247	.0214	.0190	.0171	.0156	.2247	.2178	.2124	.2083	.2056
3000	.0225	.0194	.0171	.0153	.0139	.2190	.2127	.2075	.2037	.2010
3500	.0202	.0173	.0152	.0135	.0123	.2133	.2073	.2024	.1987	.1961
4000	.0183	.0155	.0136	.0121	.0109	.2085	.2028	.1981	.1944	.1918
4500	.0166	.0141	.0122	.0109	.0098	.2045	.1989	.1943	.1906	.1881
5000	.0153	.0128	.0111	.0099	.0089	.2010	.1955	.1909	.1873	.1847
6000	.0130	.0108	.0094	.0083	.0075	.1952	.1897	.1851	.1816	.1790
7000	.0113	.0094	.0081	.0071	.0064	.1904	.1848	.1803	.1768	.1742
8000	.0100	.0082	.0071	.0063	.0057	.1863	.1807	.1762	.1726	.1701
9000	.0090	.0074	.0063	.0056	.0051	.1827	.1771	.1726	.1690	.1666
10000	.0082	.0067	.0058	.0051	.0047	.1794	.1739	.1693	.1658	.1634

TABLE 19- 4

AG ON C, M2/M1=0.111

E,KEV	RELATIVE IONIZATION STRAGGLING					RELATIVE TRANSVERSE RANGE STRAGGLING				
K/KL=	0.6	0.80	1.0	1.2	1.4	0.6	0.80	1.0	1.2	1.4
1.0	.3493	.3491	.3490	.3489	.3489	.4620	.4630	.4639	.4649	.4658
1.2	.3494	.3492	.3490	.3489	.3489	.4619	.4629	.4638	.4648	.4658
1.4	.3495	.3493	.3491	.3490	.3489	.4617	.4627	.4637	.4648	.4658
1.7	.3496	.3493	.3491	.3490	.3489	.4615	.4625	.4636	.4646	.4657
2.0	.3498	.3494	.3492	.3490	.3489	.4612	.4623	.4634	.4645	.4656
2.3	.3499	.3495	.3492	.3490	.3489	.4610	.4621	.4632	.4643	.4654
2.6	.3501	.3496	.3493	.3491	.3490	.4607	.4619	.4630	.4642	.4653
3.0	.3503	.3498	.3495	.3492	.3490	.4604	.4616	.4628	.4639	.4651
3.5	.3507	.3501	.3497	.3493	.3491	.4600	.4612	.4624	.4636	.4648
4.0	.3512	.3505	.3499	.3495	.3493	.4595	.4608	.4621	.4633	.4645
4.5	.3517	.3508	.3502	.3497	.3494	.4591	.4604	.4617	.4630	.4642
5.0	.3522	.3512	.3505	.3500	.3496	.4587	.4600	.4613	.4626	.4639
6.0	.3532	.3520	.3511	.3504	.3499	.4579	.4593	.4606	.4620	.4633
7.0	.3543	.3529	.3518	.3510	.3503	.4571	.4585	.4599	.4613	.4627
8.0	.3555	.3538	.3525	.3515	.3508	.4563	.4578	.4592	.4607	.4621
9.0	.3568	.3548	.3533	.3521	.3512	.4555	.4570	.4585	.4600	.4615
10	.3581	.3559	.3542	.3528	.3517	.4547	.4563	.4578	.4594	.4609
12	.3609	.3581	.3559	.3542	.3528	.4532	.4549	.4565	.4581	.4597
14	.3638	.3604	.3577	.3556	.3539	.4517	.4535	.4552	.4569	.4586
17	.3683	.3639	.3604	.3577	.3556	.4496	.4515	.4533	.4551	.4569
20	.3727	.3673	.3630	.3597	.3571	.4476	.4495	.4515	.4534	.4552
23	.3776	.3711	.3660	.3620	.3589	.4456	.4476	.4497	.4517	.4536
26	.3824	.3747	.3687	.3641	.3605	.4436	.4458	.4479	.4500	.4521
30	.3884	.3792	.3722	.3667	.3625	.4412	.4435	.4457	.4479	.4501
35	.3954	.3844	.3760	.3696	.3647	.4383	.4408	.4431	.4455	.4478
40	.4017	.3890	.3794	.3721	.3666	.4356	.4382	.4407	.4432	.4456
45	.4076	.3931	.3824	.3743	.3682	.4331	.4358	.4384	.4410	.4436
50	.4129	.3969	.3850	.3762	.3695	.4306	.4335	.4363	.4390	.4417
60	.4233	.4038	.3899	.3796	.3719	.4259	.4290	.4321	.4350	.4380
70	.4318	.4093	.3936	.3822	.3735	.4217	.4250	.4283	.4315	.4347
80	.4388	.4137	.3964	.3840	.3747	.4178	.4214	.4249	.4284	.4317
90	.4446	.4173	.3985	.3852	.3754	.4143	.4181	.4219	.4255	.4291
100	.4492	.4201	.4001	.3860	.3758	.4111	.4151	.4191	.4229	.4267
120	.4556	.4236	.4011	.3855	.3754	.4051	.4095	.4139	.4182	.4223
140	.4595	.4255	.4010	.3843	.3745	.3999	.4048	.4096	.4142	.4187
170	.4623	.4263	.3998	.3819	.3728	.3935	.3990	.4043	.4095	.4145
200	.4625	.4255	.3977	.3792	.3708	.3883	.3944	.4003	.4059	.4114
230	.4594	.4222	.3943	.3762	.3685	.3839	.3905	.3969	.4030	.4090
260	.4555	.4185	.3909	.3733	.3662	.3802	.3873	.3942	.4008	.4072
300	.4496	.4133	.3865	.3697	.3635	.3763	.3841	.3916	.3988	.4057
350	.4419	.4068	.3813	.3658	.3605	.3727	.3813	.3895	.3974	.4050
400	.4343	.4007	.3767	.3624	.3581	.3702	.3795	.3885	.3970	.4052
450	.4271	.3951	.3726	.3596	.3561	.3685	.3786	.3882	.3973	.4060
500	.4202	.3900	.3690	.3572	.3545	.3675	.3782	.3885	.3982	.4075
600	.4068	.3809	.3631	.3536	.3525	.3668	.3789	.3904	.4012	.4115
700	.3956	.3734	.3587	.3513	.3514	.3676	.3810	.3936	.4054	.4166
800	.3863	.3676	.3554	.3498	.3509	.3694	.3840	.3976	.4104	.4225
900	.3786	.3629	.3530	.3490	.3509	.3720	.3877	.4023	.4160	.4288
1000	.3723	.3593	.3514	.3487	.3513	.3751	.3919	.4075	.4219	.4355
1200	.3651	.3561	.3511	.3502	.3535	.3825	.4014	.4187	.4348	.4496
1400	.3605	.3547	.3519	.3524	.3562	.3909	.4117	.4306	.4480	.4641
1700	.3568	.3545	.3543	.3563	.3605	.4045	.4279	.4489	.4681	.4856
2000	.3553	.3556	.3573	.3604	.3648	.4186	.4443	.4672	.4879	.5067
2300	.3553	.3574	.3603	.3641	.3687	.4330	.4609	.4855	.5076	.5275
2600	.3562	.3595	.3633	.3677	.3725	.4475	.4773	.5034	.5266	.5475
3000	.3580	.3626	.3673	.3722	.3772	.4665	.4987	.5265	.5511	.5730
3500	.3608	.3665	.3721	.3774	.3825	.4896	.5244	.5541	.5800	.6030
4000	.3640	.3705	.3766	.3822	.3873	.5121	.5490	.5802	.6073	.6312
4500	.3672	.3742	.3807	.3865	.3917	.5338	.5726	.6051	.6332	.6578
5000	.3705	.3779	.3846	.3905	.3957	.5548	.5952	.6288	.6577	.6829
6000	.3767	.3845	.3916	.3975	.4028	.5946	.6377	.6732	.7032	.7293
7000	.3824	.3905	.3977	.4037	.4088	.6320	.6773	.7140	.7449	.7715
8000	.3877	.3958	.4030	.4090	.4141	.6673	.7142	.7519	.7833	.8104
9000	.3924	.4006	.4078	.4137	.4187	.7006	.7488	.7872	.8190	.8464
10000	.3968	.4049	.4120	.4179	.4227	.7323	.7815	.8204	.8525	.8799

E,KEV	RELATIVE TRANSV. DAMAGE STRAGGLING					RELATIVE TRANSV. IONZN. STRAGGLING				
K/KL=	0.6	0.80	1.0	1.2	1.4	0.6	0.80	1.0	1.2	1.4
1.0	.2901	.2851	.2803	.2757	.2712	.2536	.2483	.2433	.2385	.2338
1.2	.2908	.2856	.2806	.2758	.2711	.2541	.2487	.2435	.2385	.2337
1.4	.2914	.2861	.2809	.2760	.2712	.2548	.2492	.2438	.2386	.2337
1.7	.2925	.2869	.2815	.2764	.2714	.2557	.2499	.2443	.2389	.2338
2.0	.2935	.2877	.2822	.2768	.2717	.2567	.2507	.2449	.2393	.2340
2.3	.2945	.2886	.2828	.2773	.2721	.2577	.2515	.2455	.2398	.2343
2.6	.2956	.2894	.2835	.2779	.2725	.2587	.2522	.2461	.2402	.2346
3.0	.2969	.2905	.2844	.2786	.2730	.2600	.2533	.2470	.2409	.2351
3.5	.2986	.2919	.2856	.2795	.2738	.2615	.2546	.2480	.2417	.2357
4.0	.3001	.2933	.2867	.2805	.2745	.2631	.2559	.2491	.2426	.2364
4.5	.3017	.2946	.2878	.2814	.2752	.2645	.2571	.2501	.2434	.2370
5.0	.3031	.2958	.2889	.2822	.2760	.2660	.2583	.2511	.2442	.2377
6.0	.3060	.2983	.2910	.2841	.2774	.2688	.2608	.2531	.2459	.2390
7.0	.3087	.3006	.2930	.2857	.2788	.2715	.2630	.2551	.2475	.2403
8.0	.3112	.3028	.2948	.2873	.2801	.2739	.2651	.2568	.2490	.2416
9.0	.3135	.3048	.2965	.2887	.2813	.2762	.2671	.2585	.2504	.2427
10	.3157	.3066	.2981	.2901	.2825	.2783	.2689	.2600	.2517	.2438
12	.3197	.3102	.3012	.2927	.2846	.2824	.2724	.2630	.2542	.2458
14	.3233	.3132	.3038	.2949	.2865	.2859	.2755	.2656	.2563	.2476
17	.3277	.3171	.3071	.2977	.2888	.2904	.2794	.2689	.2591	.2498
20	.3314	.3203	.3098	.2999	.2906	.2942	.2826	.2716	.2613	.2516
23	.3345	.3229	.3121	.3018	.2921	.2973	.2853	.2739	.2632	.2531
26	.3370	.3252	.3139	.3033	.2933	.2999	.2876	.2758	.2648	.2544
30	.3398	.3276	.3159	.3049	.2945	.3028	.2900	.2779	.2664	.2556
35	.3425	.3298	.3177	.3062	.2954	.3057	.2925	.2798	.2679	.2566
40	.3445	.3315	.3190	.3071	.2959	.3078	.2943	.2812	.2689	.2572
45	.3460	.3327	.3198	.3076	.2960	.3095	.2956	.2822	.2695	.2575
50	.3471	.3335	.3203	.3077	.2959	.3107	.2966	.2829	.2698	.2574
60	.3481	.3341	.3205	.3073	.2949	.3120	.2975	.2833	.2697	.2568
70	.3484	.3340	.3199	.3063	.2934	.3125	.2977	.2831	.2690	.2556
80	.3482	.3334	.3189	.3048	.2915	.3125	.2974	.2824	.2678	.2540
90	.3476	.3325	.3175	.3030	.2893	.3122	.2968	.2813	.2663	.2521
100	.3467	.3313	.3158	.3009	.2868	.3116	.2959	.2800	.2646	.2500
120	.3448	.3285	.3120	.2962	.2814	.3102	.2937	.2769	.2606	.2451
140	.3424	.3253	.3078	.2912	.2757	.3083	.2911	.2734	.2563	.2401
170	.3382	.3198	.3010	.2834	.2671	.3049	.2865	.2676	.2495	.2323
200	.3335	.3139	.2940	.2755	.2587	.3011	.2815	.2614	.2425	.2246
230	.3296	.3073	.2863	.2671	.2499	.2973	.2759	.2547	.2349	.2165
260	.3237	.3008	.2787	.2589	.2414	.2934	.2704	.2480	.2274	.2087
300	.3169	.2922	.2690	.2485	.2309	.2880	.2630	.2393	.2179	.1989
350	.3085	.2818	.2575	.2365	.2188	.2812	.2539	.2289	.2067	.1875
400	.3002	.2718	.2468	.2253	.2078	.2742	.2451	.2190	.1963	.1772
450	.2920	.2624	.2367	.2151	.1978	.2673	.2366	.2097	.1866	.1677
500	.2839	.2533	.2273	.2056	.1886	.2604	.2284	.2008	.1776	.1589
600	.2671	.2354	.2091	.1879	.1720	.2456	.2119	.1835	.1604	.1429
700	.2516	.2195	.1932	.1727	.1580	.2318	.1969	.1682	.1456	.1293
800	.2374	.2053	.1794	.1596	.1460	.2189	.1835	.1548	.1328	.1176
900	.2245	.1927	.1673	.1483	.1356	.2069	.1713	.1429	.1216	.1075
1000	.2127	.1814	.1567	.1384	.1267	.1957	.1603	.1324	.1119	.0988
1200	.1914	.1622	.1395	.1230	.1127	.1741	.1409	.1150	.0964	.0851
1400	.1733	.1465	.1257	.1108	.1017	.1555	.1247	.1009	.0841	.0744
1700	.1510	.1275	.1094	.0966	.0889	.1322	.1049	.0842	.0698	.0620
2000	.1333	.1126	.0968	.0858	.0792	.1133	.0894	.0713	.0590	.0527
2300	.1205	.1019	.0877	.0778	.0720	.0999	.0784	.0623	.0515	.0460
2600	.1101	.0933	.0805	.0715	.0662	.0889	.0696	.0552	.0456	.0407
3000	.0990	.0841	.0727	.0648	.0601	.0771	.0603	.0478	.0395	.0353
3500	.0882	.0752	.0653	.0583	.0542	.0657	.0514	.0408	.0338	.0302
4000	.0796	.0683	.0596	.0533	.0497	.0568	.0445	.0355	.0295	.0264
4500	.0728	.0627	.0549	.0493	.0460	.0497	.0392	.0314	.0261	.0234
5000	.0672	.0582	.0512	.0461	.0430	.0440	.0348	.0281	.0235	.0210
6000	.0587	.0512	.0453	.0410	.0382	.0354	.0284	.0231	.0195	.0174
7000	.0525	.0461	.0410	.0372	.0347	.0296	.0239	.0197	.0167	.0149
8000	.0480	.0423	.0376	.0342	.0319	.0255	.0207	.0171	.0146	.0130
9000	.0445	.0392	.0349	.0317	.0296	.0227	.0184	.0151	.0129	.0115
10000	.0418	.0368	.0326	.0297	.0276	.0207	.0167	.0136	.0115	.0103

K/KL=	RANGE SKEWNESS					DAMAGE SKEWNESS				
E,KEV	0.6	0.80	1.0	1.2	1.4	0.6	0.80	1.0	1.2	1.4
1.0	.4052	.3996	.3942	.3888	.3834	.4154	.4068	.3988	.3912	.3841
1.2	.4045	.3988	.3932	.3876	.3821	.4169	.4077	.3990	.3910	.3834
1.4	.4040	.3981	.3923	.3866	.3809	.4179	.4082	.3991	.3907	.3828
1.7	.4032	.3971	.3912	.3853	.3794	.4192	.4088	.3992	.3903	.3819
2.0	.4025	.3963	.3902	.3841	.3781	.4206	.4096	.3995	.3900	.3813
2.3	.4019	.3955	.3892	.3830	.3769	.4222	.4106	.3999	.3900	.3808
2.6	.4014	.3948	.3884	.3821	.3758	.4241	.4119	.4006	.3902	.3806
3.0	.4007	.3940	.3874	.3809	.3745	.4270	.4138	.4018	.3907	.3806
3.5	.3999	.3930	.3862	.3795	.3730	.4311	.4168	.4037	.3918	.3809
4.0	.3991	.3921	.3851	.3783	.3716	.4357	.4201	.4060	.3932	.3815
4.5	.3984	.3912	.3841	.3771	.3703	.4406	.4237	.4085	.3948	.3823
5.0	.3978	.3904	.3832	.3760	.3690	.4458	.4276	.4113	.3966	.3833
6.0	.3965	.3889	.3814	.3741	.3669	.4558	.4351	.4168	.4003	.3856
7.0	.3954	.3876	.3798	.3723	.3648	.4668	.4435	.4229	.4046	.3883
8.0	.3943	.3863	.3784	.3706	.3630	.4785	.4524	.4296	.4094	.3915
9.0	.3933	.3850	.3769	.3690	.3612	.4908	.4618	.4366	.4144	.3949
10	.3923	.3839	.3756	.3675	.3595	.5035	.4715	.4439	.4197	.3986
12	.3905	.3817	.3731	.3647	.3564	.5305	.4924	.4597	.4315	.4069
14	.3888	.3797	.3708	.3621	.3536	.5576	.5133	.4755	.4432	.4154
17	.3863	.3768	.3675	.3584	.3495	.5979	.5441	.4988	.4605	.4277
20	.3839	.3740	.3643	.3549	.3457	.6372	.5738	.5212	.4769	.4394
23	.3817	.3715	.3615	.3517	.3422	.6801	.6066	.5459	.4953	.4528
26	.3796	.3690	.3587	.3486	.3388	.7207	.6374	.5690	.5124	.4652
30	.3768	.3658	.3551	.3446	.3345	.7712	.6753	.5972	.5331	.4798
35	.3734	.3619	.3507	.3398	.3292	.8282	.7176	.6283	.5555	.4955
40	.3700	.3580	.3464	.3351	.3241	.8791	.7549	.6553	.5745	.5084
45	.3667	.3543	.3422	.3305	.3192	.9246	.7878	.6787	.5907	.5190
50	.3634	.3506	.3381	.3261	.3143	.9655	.8169	.6990	.6044	.5276
60	.3573	.3436	.3304	.3176	.3053	1.040	.8683	.7345	.6282	.5414
70	.3513	.3368	.3229	.3095	.2965	1.099	.9083	.7608	.6443	.5494
80	.3453	.3301	.3155	.3015	.2880	1.147	.9392	.7795	.6544	.5530
90	.3394	.3236	.3083	.2937	.2796	1.185	.9629	.7923	.6595	.5531
100	.3336	.3171	.3013	.2861	.2715	1.215	.9805	.8003	.6607	.5504
120	.3225	.3047	.2877	.2714	.2559	1.254	1.000	.8012	.6488	.5350
140	.3116	.2926	.2745	.2573	.2409	1.276	1.007	.7929	.6305	.5153
170	.2959	.2751	.2556	.2370	.2194	1.286	1.001	.7701	.5969	.4820
200	.2807	.2584	.2375	.2177	.1991	1.279	.9828	.7400	.5600	.4470
230	.2659	.2421	.2198	.1990	.1793	1.254	.9478	.6993	.5178	.4083
260	.2516	.2264	.2029	.1810	.1605	1.223	.9090	.6574	.4765	.3708
300	.2333	.2064	.1814	.1583	.1367	1.174	.8548	.6022	.4237	.3235
350	.2116	.1826	.1560	.1315	.1088	1.109	.7867	.5361	.3626	.2691
400	.1909	.1601	.1321	.1064	.0826	1.042	.7205	.4743	.3068	.2199
450	.1712	.1388	.1094	.0826	.0580	.9751	.6571	.4169	.2561	.1752
500	.1524	.1184	.0879	.0601	.0348	.9088	.5968	.3639	.2099	.1348
600	.1165	.0796	.0469	.0175	-.0092	.7646	.4789	.2673	.1298	.0658
700	.0834	.0441	.0095	-.0212	-.0489	.6361	.3762	.1851	.0625	.0078
800	.0527	.0113	-.0248	-.0566	-.0851	.5231	.2874	.1148	.0052	-.0418
900	.0240	-.0192	-.0565	-.0892	-.1183	.4243	.2104	.0544	-.0438	-.0847
1000	-.0028	-.0476	-.0860	-.1195	-.1490	.3381	.1438	.0023	-.0861	-.1221
1200	-.0524	-.0998	-.1398	-.1742	-.2044	.2128	.0473	-.0734	-.1491	-.1803
1400	-.0969	-.1464	-.1875	-.2226	-.2531	.1146	-.0279	-.1326	-.1989	-.2273
1700	-.1564	-.2081	-.2504	-.2861	-.3167	.0010	-.1153	-.2016	-.2578	-.2841
2000	-.2090	-.2622	-.3052	-.3410	-.3716	-.0857	-.1825	-.2556	-.3047	-.3299
2300	-.2463	-.3003	-.3434	-.3791	-.4094	-.1523	-.2369	-.3014	-.3457	-.3697
2600	-.2811	-.3356	-.3787	-.4140	-.4439	-.2063	-.2819	-.3402	-.3809	-.4041
3000	-.3258	-.3805	-.4234	-.4583	-.4876	-.2645	-.3316	-.3838	-.4211	-.4436
3500	-.3810	-.4356	-.4779	-.5121	-.5407	-.3220	-.3819	-.4290	-.4635	-.4853
4000	-.4363	-.4907	-.5324	-.5658	-.5937	-.3679	-.4229	-.4667	-.4994	-.5208
4500	-.4923	-.5462	-.5872	-.6199	-.6471	-.4055	-.4575	-.4991	-.5305	-.5516
5000	-.5489	-.6022	-.6425	-.6746	-.7011	-.4373	-.4871	-.5272	-.5579	-.5788
6000	-.6637	-.7157	-.7546	-.7854	-.8107	-.4885	-.5360	-.5746	-.6043	-.6251
7000	-.7798	-.8304	-.8681	-.8976	-.9219	-.5286	-.5753	-.6133	-.6427	-.6634
8000	-.8965	-.9456	-.9820	-1.011	-1.034	-.5615	-.6081	-.6461	-.6754	-.6959
9000	-1.013	-1.061	-1.096	-1.124	-1.146	-.5895	-.6363	-.6745	-.7038	-.7242
10000	-1.129	-1.175	-1.209	-1.236	-1.258	-.6138	-.6611	-.6995	-.7290	-.7492

		IONIZATION SKEWNESS					RANGE KURTOSIS			
K/KL=	0.6	0.80	1.0	1.2	1.4	0.6	0.80	1.0	1.2	1.4
E,KEV										
1.0	.5217	.5148	.5085	.5027	.4974	3.170	3.161	3.152	3.143	3.135
1.2	.5237	.5152	.5094	.5031	.4975	3.169	3.159	3.150	3.142	3.133
1.4	.5252	.5172	.5100	.5034	.4974	3.168	3.158	3.149	3.140	3.131
1.7	.5272	.5186	.5108	.5038	.4975	3.166	3.156	3.147	3.138	3.129
2.0	.5292	.5200	.5118	.5044	.4976	3.165	3.155	3.145	3.136	3.127
2.3	.5314	.5217	.5129	.5051	.4979	3.164	3.154	3.144	3.134	3.125
2.6	.5339	.5235	.5143	.5059	.4984	3.163	3.153	3.142	3.133	3.123
3.0	.5375	.5263	.5163	.5073	.4993	3.162	3.151	3.141	3.131	3.121
3.5	.5426	.5302	.5192	.5094	.5006	3.160	3.149	3.139	3.128	3.119
4.0	.5480	.5344	.5224	.5117	.5022	3.159	3.148	3.137	3.126	3.116
4.5	.5538	.5389	.5258	.5143	.5040	3.158	3.146	3.135	3.125	3.114
5.0	.5598	.5436	.5294	.5169	.5060	3.157	3.145	3.134	3.123	3.112
6.0	.5715	.5528	.5365	.5223	.5099	3.154	3.142	3.131	3.120	3.109
7.0	.5840	.5627	.5442	.5282	.5143	3.152	3.140	3.128	3.117	3.106
8.0	.5972	.5730	.5523	.5344	.5189	3.150	3.138	3.126	3.114	3.103
9.0	.6108	.5837	.5606	.5408	.5237	3.148	3.136	3.123	3.111	3.100
10	.6247	.5946	.5691	.5473	.5287	3.147	3.134	3.121	3.109	3.097
12	.6542	.6179	.5873	.5615	.5395	3.143	3.130	3.117	3.104	3.092
14	.6834	.6408	.6053	.5754	.5501	3.140	3.126	3.113	3.100	3.088
17	.7260	.6741	.6311	.5953	.5653	3.136	3.121	3.107	3.094	3.081
20	.7668	.7057	.6556	.6140	.5795	3.132	3.116	3.102	3.089	3.076
23	.8106	.7397	.6818	.6343	.5949	3.128	3.112	3.097	3.083	3.070
26	.8516	.7714	.7062	.6529	.6090	3.124	3.108	3.093	3.079	3.065
30	.9019	.8099	.7357	.6753	.6258	3.119	3.102	3.087	3.072	3.058
35	.9579	.8525	.7679	.6995	.6437	3.113	3.096	3.080	3.065	3.051
40	1.007	.8896	.7957	.7200	.6586	3.107	3.090	3.073	3.058	3.043
45	1.051	.9220	.8196	.7374	.6710	3.101	3.084	3.067	3.051	3.036
50	1.089	.9503	.8402	.7521	.6812	3.096	3.078	3.061	3.045	3.030
60	1.155	.9983	.8748	.7763	.6974	3.086	3.067	3.049	3.033	3.017
70	1.207	1.035	.9002	.7932	.7077	3.076	3.056	3.038	3.021	3.006
80	1.248	1.063	.9186	.8043	.7136	3.067	3.046	3.028	3.011	2.995
90	1.280	1.084	.9312	.8108	.7158	3.058	3.037	3.018	3.001	2.985
100	1.305	1.100	.9394	.8138	.7153	3.049	3.028	3.009	2.991	2.975
120	1.335	1.115	.9415	.8074	.7035	3.033	3.011	2.991	2.973	2.957
140	1.350	1.119	.9353	.7950	.6878	3.018	2.995	2.975	2.957	2.941
170	1.355	1.112	.9169	.7706	.6611	2.997	2.974	2.954	2.936	2.920
200	1.346	1.095	.8925	.7428	.6332	2.979	2.956	2.936	2.918	2.903
230	1.325	1.067	.8599	.7099	.6036	2.961	2.938	2.918	2.901	2.887
260	1.299	1.037	.8265	.6774	.5752	2.945	2.922	2.902	2.886	2.873
300	1.260	.9950	.7825	.6360	.5399	2.926	2.904	2.885	2.870	2.858
350	1.208	.9421	.7301	.5882	.5000	2.906	2.885	2.867	2.854	2.844
400	1.154	.8906	.6813	.5448	.4646	2.890	2.869	2.854	2.842	2.834
450	1.101	.8412	.6362	.5057	.4332	2.876	2.857	2.843	2.833	2.827
500	1.049	.7941	.5948	.4704	.4053	2.864	2.847	2.835	2.827	2.823
600	.9332	.6969	.5166	.4074	.3576	2.844	2.830	2.823	2.820	2.820
700	.8314	.6137	.4519	.3566	.3198	2.830	2.821	2.818	2.819	2.823
800	.7428	.5432	.3988	.3158	.2898	2.821	2.817	2.818	2.824	2.832
900	.6663	.4839	.3552	.2832	.2660	2.816	2.816	2.822	2.832	2.844
1000	.6006	.4343	.3198	.2573	.2471	2.815	2.820	2.830	2.843	2.859
1200	.5083	.3716	.2795	.2311	.2274	2.819	2.833	2.852	2.873	2.896
1400	.4390	.3276	.2538	.2163	.2164	2.830	2.854	2.881	2.909	2.937
1700	.3635	.2830	.2307	.2057	.2087	2.855	2.893	2.931	2.968	3.004
2000	.3105	.2542	.2182	.2024	.2067	2.890	2.938	2.986	3.031	3.073
2300	.2763	.2347	.2091	.1996	.2058	2.922	2.981	3.037	3.089	3.137
2600	.2519	.2215	.2039	.1991	.2068	2.959	3.028	3.091	3.148	3.202
3000	.2298	.2105	.2009	.2007	.2099	3.014	3.094	3.165	3.229	3.289
3500	.2130	.2036	.2010	.2049	.2155	3.089	3.182	3.262	3.332	3.397
4000	.2038	.2014	.2037	.2106	.2220	3.171	3.273	3.360	3.436	3.505
4500	.1994	.2021	.2079	.2169	.2290	3.256	3.368	3.461	3.541	3.612
5000	.1981	.2046	.2131	.2236	.2362	3.345	3.464	3.562	3.645	3.718
6000	.2014	.2128	.2248	.2373	.2505	3.527	3.658	3.764	3.853	3.925
7000	.2089	.2231	.2371	.2508	.2642	3.713	3.852	3.964	4.057	4.127
8000	.2181	.2341	.2493	.2637	.2772	3.899	4.046	4.161	4.257	4.323
9000	.2282	.2452	.2611	.2758	.2894	4.085	4.236	4.354	4.452	4.512
10000	.2383	.2560	.2724	.2873	.3008	4.269	4.424	4.543	4.643	4.696

TABLE 19- 8

AG ON C, M2/M1=0.111

E,KEV / K/KL=	DAMAGE KURTOSIS					IONIZATION KURTOSIS				
	0.6	0.80	1.0	1.2	1.4	0.6	0.80	1.0	1.2	1.4
1.0	2.866	2.838	2.813	2.789	2.768	2.965	2.936	2.910	2.886	2.864
1.2	2.875	2.844	2.817	2.792	2.768	2.974	2.943	2.915	2.889	2.866
1.4	2.881	2.849	2.819	2.793	2.769	2.981	2.948	2.918	2.891	2.867
1.7	2.889	2.854	2.823	2.795	2.770	2.991	2.955	2.923	2.895	2.869
2.0	2.897	2.860	2.828	2.798	2.771	3.000	2.963	2.929	2.898	2.871
2.3	2.906	2.867	2.832	2.801	2.773	3.011	2.971	2.935	2.902	2.873
2.6	2.915	2.874	2.838	2.805	2.776	3.022	2.979	2.941	2.907	2.877
3.0	2.929	2.885	2.846	2.811	2.779	3.038	2.992	2.951	2.914	2.882
3.5	2.949	2.900	2.857	2.819	2.785	3.060	3.009	2.964	2.924	2.889
4.0	2.970	2.917	2.870	2.828	2.792	3.084	3.028	2.978	2.935	2.897
4.5	2.992	2.934	2.883	2.838	2.799	3.109	3.047	2.993	2.946	2.905
5.0	3.016	2.952	2.896	2.848	2.806	3.135	3.067	3.009	2.958	2.914
6.0	3.059	2.986	2.923	2.869	2.821	3.185	3.107	3.040	2.982	2.932
7.0	3.107	3.023	2.952	2.890	2.838	3.238	3.148	3.072	3.007	2.951
8.0	3.158	3.062	2.982	2.913	2.855	3.294	3.192	3.106	3.033	2.971
9.0	3.211	3.103	3.013	2.937	2.872	3.352	3.237	3.140	3.060	2.991
10	3.265	3.144	3.044	2.961	2.890	3.411	3.282	3.176	3.086	3.011
12	3.380	3.232	3.112	3.012	2.929	3.537	3.379	3.251	3.144	3.056
14	3.495	3.320	3.179	3.063	2.967	3.661	3.475	3.324	3.200	3.098
17	3.668	3.450	3.276	3.136	3.021	3.842	3.612	3.428	3.280	3.158
20	3.836	3.575	3.369	3.205	3.072	4.015	3.742	3.526	3.354	3.214
23	4.022	3.712	3.471	3.280	3.128	4.202	3.882	3.632	3.433	3.273
26	4.199	3.842	3.566	3.351	3.180	4.377	4.012	3.729	3.505	3.327
30	4.420	4.002	3.682	3.435	3.241	4.590	4.170	3.846	3.592	3.391
35	4.669	4.180	3.811	3.527	3.306	4.826	4.343	3.973	3.685	3.458
40	4.892	4.338	3.922	3.606	3.361	5.032	4.492	4.081	3.763	3.514
45	5.093	4.477	4.019	3.673	3.407	5.212	4.621	4.173	3.829	3.560
50	5.273	4.600	4.103	3.730	3.445	5.369	4.733	4.252	3.884	3.598
60	5.603	4.818	4.251	3.831	3.509	5.633	4.916	4.381	3.973	3.656
70	5.869	4.988	4.362	3.901	3.550	5.837	5.054	4.473	4.033	3.693
80	6.082	5.121	4.442	3.947	3.573	5.993	5.157	4.537	4.071	3.712
90	6.252	5.223	4.498	3.973	3.583	6.112	5.232	4.580	4.091	3.719
100	6.388	5.301	4.535	3.983	3.582	6.202	5.286	4.606	4.098	3.715
120	6.557	5.385	4.542	3.939	3.533	6.290	5.328	4.601	4.058	3.665
140	6.651	5.419	4.515	3.873	3.470	6.321	5.329	4.567	3.999	3.602
170	6.790	5.409	4.439	3.758	3.369	6.303	5.283	4.486	3.893	3.500
200	6.677	5.352	4.341	3.638	3.269	6.238	5.204	4.386	3.781	3.398
230	6.578	5.233	4.212	3.517	3.175	6.134	5.086	4.262	3.664	3.301
250	6.452	5.100	4.082	3.403	3.089	6.014	4.960	4.138	3.552	3.211
300	6.262	4.917	3.914	3.262	2.984	5.844	4.790	3.978	3.411	3.100
350	6.011	4.691	3.719	3.107	2.869	5.626	4.582	3.791	3.253	2.976
400	5.756	4.474	3.542	2.971	2.770	5.410	4.386	3.619	3.112	2.868
450	5.505	4.270	3.384	2.854	2.684	5.200	4.201	3.463	2.986	2.773
500	5.261	4.080	3.241	2.752	2.610	4.997	4.028	3.322	2.875	2.689
600	4.730	3.713	2.999	2.593	2.494	4.570	3.695	3.067	2.684	2.548
700	4.273	3.406	2.805	2.472	2.405	4.199	3.414	2.859	2.533	2.436
800	3.885	3.153	2.651	2.379	2.336	3.881	3.179	2.691	2.412	2.347
900	3.561	2.944	2.527	2.307	2.283	3.611	2.984	2.554	2.317	2.275
1000	3.292	2.775	2.429	2.252	2.243	3.382	2.823	2.443	2.241	2.218
1200	2.981	2.588	2.329	2.201	2.205	3.083	2.630	2.325	2.167	2.155
1400	2.772	2.470	2.272	2.177	2.186	2.866	2.497	2.251	2.124	2.116
1700	2.571	2.365	2.230	2.167	2.177	2.632	2.364	2.185	2.090	2.080
2000	2.448	2.308	2.216	2.172	2.179	2.468	2.278	2.146	2.074	2.060
2300	2.362	2.265	2.202	2.174	2.171	2.337	2.206	2.104	2.048	2.036
2600	2.306	2.238	2.196	2.180	2.166	2.236	2.152	2.072	2.027	2.018
3000	2.259	2.220	2.198	2.192	2.166	2.135	2.099	2.039	2.006	2.000
3500	2.230	2.214	2.207	2.211	2.172	2.046	2.053	2.010	1.987	1.984
4000	2.219	2.218	2.222	2.232	2.185	1.985	2.022	1.990	1.974	1.974
4500	2.220	2.228	2.239	2.254	2.201	1.944	2.000	1.976	1.965	1.967
5000	2.227	2.242	2.258	2.276	2.221	1.917	1.985	1.966	1.959	1.963
6000	2.250	2.274	2.297	2.319	2.267	1.892	1.968	1.956	1.954	1.960
7000	2.277	2.307	2.335	2.360	2.321	1.892	1.961	1.954	1.954	1.962
8000	2.303	2.339	2.371	2.399	2.378	1.907	1.959	1.956	1.959	1.968
9000	2.327	2.368	2.404	2.436	2.438	1.932	1.961	1.961	1.966	1.976
10000	2.348	2.395	2.436	2.470	2.500	1.963	1.964	1.968	1.975	1.985

	REFLECTION COEFFICIENT					SPUTTERING EFFICIENCY				
K/KL=	0.6	0.80	1.0	1.2	1.4	0.6	0.80	1.0	1.2	1.4
E,KEV										
1.0	.0000	.0000	.0000	.0000	.0000	.0154	.0147	.0139	.0132	.0126
1.2	.0000	.0000	.0000	.0000	.0000	.0154	.0146	.0138	.0131	.0125
1.4	.0000	.0000	.0000	.0000	.0000	.0154	.0145	.0138	.0130	.0124
1.7	.0000	.0000	.0000	.0000	.0000	.0153	.0145	.0137	.0129	.0122
2.0	.0000	.0000	.0000	.0000	.0000	.0153	.0144	.0136	.0128	.0121
2.3	.0000	.0000	.0000	.0000	.0000	.0153	.0144	.0136	.0128	.0121
2.6	.0000	.0000	.0000	.0000	.0000	.0153	.0144	.0135	.0127	.0120
3.0	.0000	.0000	.0000	.0000	.0000	.0153	.0143	.0135	.0126	.0119
3.5	.0000	.0000	.0000	.0000	.0000	.0154	.0143	.0134	.0126	.0118
4.0	.0000	.0000	.0000	.0000	.0000	.0154	.0143	.0134	.0125	.0117
4.5	.0000	.0000	.0000	.0000	.0000	.0154	.0143	.0134	.0125	.0117
5.0	.0000	.0000	.0000	.0000	.0000	.0155	.0143	.0133	.0124	.0116
6.0	.0000	.0000	.0000	.0000	.0000	.0156	.0144	.0133	.0124	.0116
7.0	.0000	.0000	.0000	.0000	.0000	.0157	.0144	.0133	.0124	.0115
8.0	.0000	.0000	.0000	.0000	.0000	.0158	.0145	.0133	.0123	.0114
9.0	.0000	.0000	.0000	.0000	.0000	.0159	.0146	.0134	.0123	.0114
10	.0000	.0000	.0000	.0000	.0000	.0160	.0146	.0134	.0123	.0114
12	.0000	.0000	.0000	.0000	.0000	.0162	.0147	.0135	.0123	.0113
14	.0000	.0000	.0000	.0000	.0000	.0164	.0149	.0135	.0123	.0113
17	.0000	.0000	.0000	.0000	.0000	.0166	.0150	.0136	.0123	.0113
20	.0000	.0000	.0000	.0000	.0000	.0168	.0151	.0136	.0123	.0112
23	.0000	.0000	.0000	.0000	.0000	.0170	.0152	.0137	.0123	.0112
26	.0000	.0000	.0000	.0000	.0000	.0171	.0152	.0137	.0123	.0111
30	.0000	.0000	.0000	.0000	.0000	.0172	.0153	.0137	.0123	.0111
35	.0000	.0000	.0000	.0000	.0000	.0173	.0153	.0137	.0122	.0110
40	.0000	.0000	.0000	.0000	.0000	.0173	.0153	.0136	.0121	.0109
45	.0000	.0000	.0000	.0000	.0000	.0173	.0153	.0136	.0121	.0108
50	.0000	.0000	.0000	.0000	.0000	.0174	.0152	.0135	.0120	.0107
60	.0000	.0000	.0000	.0000	.0000	.0173	.0151	.0133	.0118	.0105
70	.0000	.0000	.0000	.0000	.0000	.0172	.0149	.0131	.0116	.0103
80	.0000	.0000	.0000	.0000	.0000	.0171	.0148	.0129	.0114	.0101
90	.0000	.0000	.0000	.0000	.0000	.0170	.0146	.0127	.0112	.0099
100	.0000	.0000	.0000	.0000	.0000	.0169	.0144	.0125	.0110	.0097
120	.0000	.0000	.0000	.0000	.0000	.0166	.0141	.0121	.0105	.0092
140	.0000	.0000	.0000	.0000	.0000	.0162	.0138	.0117	.0101	.0088
170	.0000	.0000	.0000	.0000	.0000	.0158	.0133	.0112	.0095	.0083
200	.0000	.0000	.0000	.0000	.0000	.0153	.0129	.0108	.0090	.0079
230	.0000	.0000	.0000	.0000	.0000	.0149	.0125	.0104	.0086	.0075
260	.0000	.0000	.0000	.0000	.0000	.0145	.0121	.0101	.0083	.0072
300	.0000	.0000	.0000	.0000	.0000	.0139	.0117	.0097	.0079	.0068
350	.0000	.0000	.0000	.0000	.0000	.0134	.0112	.0092	.0074	.0064
400	.0000	.0000	.0000	.0000	.0000	.0128	.0108	.0088	.0070	.0060
450	.0000	.0000	.0000	.0000	.0000	.0124	.0104	.0084	.0066	.0057
500	.0000	.0000	.0000	.0000	.0000	.0119	.0100	.0080	.0063	.0054
600	.0000	.0000	.0000	.0000	.0000	.0112	.0093	.0073	.0056	.0049
700	.0000	.0000	.0000	.0000	.0000	.0106	.0086	.0066	.0050	.0044
800	.0000	.0000	.0000	.0000	.0000	.0100	.0081	.0060	.0046	.0040
900	.0000	.0000	.0000	.0000	.0000	.0095	.0076	.0055	.0042	.0037
1000	.0000	.0000	.0000	.0000	.0000	.0091	.0071	.0051	.0038	.0034
1200	.0000	.0000	0.0000	0.0000	0.0000	.0083	.0063	.0045	.0034	.0030
1400	.0000	.0000	0.0000	0.0000	0.0000	.0075	.0057	.0041	.0031	.0027
1700	.0000	0.0000	0.0000	0.0000	0.0000	.0067	.0049	.0035	.0027	.0023
2000	.0000	0.0000	0.0000	0.0000	0.0000	.0059	.0043	.0031	.0024	.0021
2300	0.0000	0.0000	0.0000	0.0000	0.0000	.0053	.0039	.0028	.0022	.0018
2600	0.0000	0.0000	0.0000	0.0000	0.0000	.0048	.0035	.0025	.0020	.0016
3000	0.0000	0.0000	0.0000	0.0000	0.0000	.0043	.0031	.0023	.0018	.0014
3500	0.0000	0.0000	0.0000	0.0000	0.0000	.0037	.0027	.0020	.0015	.0012
4000	0.0000	0.0000	0.0000	0.0000	0.0000	.0032	.0024	.0018	.0014	.0010
4500	0.0000	0.0000	0.0000	0.0000	0.0000	.0028	.0021	.0016	.0012	.0008
5000	0.0000	0.0000	0.0000	0.0000	0.0000	.0025	.0019	.0014	.0011	.0007
6000	0.0000	0.0000	0.0000	0.0000	0.0000	.0021	.0016	.0012	.0009	.0006
7000	0.0000	0.0000	0.0000	0.0000	0.0000	.0018	.0013	.0010	.0008	.0005
8000	0.0000	0.0000	0.0000	0.0000	0.0000	.0015	.0012	.0009	.0007	.0004
9000	0.0000	0.0000	0.0000	0.0000	0.0000	.0014	.0011	.0008	.0006	.0004
10000	0.0000	0.0000	0.0000	0.0000	0.0000	.0013	.0010	.0007	.0005	.0004

K/KL= E,KEV	IONIZATION DEFICIENCY					SPUTTERING YIELD ALPHA				
	0.6	0.80	1.0	1.2	1.4	0.6	0.80	1.0	1.2	1.4
1.0	.0017	.0021	.0025	.0029	.0033	.1195	.1153	.1112	.1074	.1037
1.2	.0017	.0022	.0026	.0030	.0033	.1191	.1147	.1105	.1066	.1029
1.4	.0017	.0022	.0027	.0031	.0034	.1187	.1142	.1100	.1059	.1021
1.7	.0018	.0023	.0027	.0031	.0035	.1181	.1135	.1092	.1051	.1012
2.0	.0019	.0024	.0028	.0032	.0036	.1177	.1130	.1085	.1043	.1004
2.3	.0019	.0024	.0029	.0033	.0036	.1173	.1125	.1080	.1037	.0997
2.6	.0019	.0025	.0029	.0033	.0037	.1169	.1120	.1074	.1031	.0990
3.0	.0020	.0025	.0030	.0034	.0038	.1165	.1115	.1068	.1024	.0983
3.5	.0021	.0026	.0031	.0035	.0039	.1161	.1109	.1061	.1016	.0975
4.0	.0021	.0027	.0031	.0036	.0039	.1157	.1104	.1056	.1010	.0967
4.5	.0022	.0027	.0032	.0036	.0040	.1153	.1100	.1050	.1004	.0961
5.0	.0022	.0028	.0033	.0037	.0041	.1151	.1096	.1046	.0999	.0955
6.0	.0023	.0029	.0034	.0038	.0042	.1145	.1089	.1037	.0989	.0945
7.0	.0024	.0030	.0035	.0039	.0043	.1141	.1084	.1031	.0981	.0936
8.0	.0025	.0031	.0036	.0040	.0044	.1138	.1079	.1025	.0975	.0928
9.0	.0026	.0032	.0037	.0041	.0045	.1135	.1075	.1019	.0969	.0922
10	.0026	.0033	.0038	.0042	.0046	.1132	.1071	.1015	.0963	.0916
12	.0028	.0034	.0040	.0044	.0048	.1128	.1065	.1007	.0954	.0905
14	.0029	.0036	.0041	.0045	.0049	.1125	.1060	.1000	.0946	.0897
17	.0031	.0037	.0043	.0047	.0051	.1122	.1054	.0992	.0936	.0886
20	.0032	.0039	.0045	.0049	.0053	.1120	.1049	.0986	.0929	.0877
23	.0034	.0041	.0046	.0051	.0054	.1119	.1046	.0980	.0922	.0869
26	.0035	.0042	.0048	.0052	.0056	.1118	.1043	.0976	.0916	.0862
30	.0036	.0043	.0049	.0054	.0057	.1117	.1039	.0971	.0910	.0855
35	.0038	.0045	.0051	.0055	.0059	.1116	.1036	.0965	.0903	.0847
40	.0039	.0047	.0052	.0057	.0060	.1116	.1033	.0961	.0897	.0840
45	.0040	.0048	.0054	.0058	.0061	.1116	.1031	.0957	.0892	.0834
50	.0042	.0049	.0055	.0059	.0062	.1117	.1029	.0954	.0888	.0829
60	.0044	.0051	.0057	.0061	.0064	.1118	.1026	.0948	.0881	.0821
70	.0045	.0053	.0058	.0062	.0065	.1119	.1024	.0944	.0875	.0814
80	.0047	.0054	.0059	.0063	.0066	.1121	.1022	.0940	.0870	.0809
90	.0048	.0055	.0060	.0064	.0066	.1122	.1021	.0937	.0866	.0804
100	.0049	.0056	.0061	.0064	.0067	.1123	.1020	.0935	.0862	.0799
120	.0051	.0058	.0062	.0064	.0067	.1126	.1020	.0930	.0853	.0790
140	.0052	.0059	.0062	.0064	.0067	.1128	.1020	.0926	.0845	.0783
170	.0054	.0060	.0063	.0063	.0066	.1130	.1020	.0921	.0837	.0776
200	.0055	.0061	.0063	.0062	.0066	.1132	.1021	.0918	.0831	.0771
230	.0055	.0061	.0063	.0061	.0065	.1133	.1021	.0915	.0828	.0770
260	.0055	.0061	.0063	.0061	.0064	.1134	.1021	.0913	.0827	.0770
300	.0055	.0061	.0062	.0060	.0063	.1136	.1021	.0911	.0827	.0772
350	.0055	.0060	.0061	.0058	.0061	.1137	.1021	.0911	.0828	.0775
400	.0054	.0059	.0060	.0057	.0059	.1138	.1022	.0912	.0831	.0779
450	.0054	.0058	.0059	.0055	.0057	.1138	.1023	.0914	.0835	.0784
500	.0053	.0057	.0057	.0053	.0055	.1139	.1025	.0918	.0839	.0788
600	.0052	.0055	.0054	.0048	.0050	.1135	.1027	.0933	.0852	.0799
700	.0051	.0054	.0051	.0044	.0046	.1133	.1031	.0947	.0863	.0809
800	.0050	.0052	.0048	.0040	.0042	.1134	.1036	.0960	.0874	.0818
900	.0049	.0051	.0045	.0037	.0039	.1137	.1042	.0971	.0883	.0826
1000	.0048	.0049	.0043	.0034	.0036	.1142	.1049	.0982	.0891	.0832
1200	.0047	.0046	.0040	.0031	.0031	.1167	.1072	.0995	.0900	.0843
1400	.0046	.0043	.0037	.0029	.0027	.1191	.1093	.1005	.0907	.0850
1700	.0044	.0039	.0033	.0026	.0024	.1225	.1120	.1016	.0915	.0857
2000	.0041	.0035	.0030	.0024	.0023	.1252	.1142	.1025	.0921	.0859
2300	.0033	.0031	.0026	.0022	.0029	.1262	.1151	.1030	.0925	.0850
2600	.0026	.0028	.0023	.0020	.0035	.1272	.1157	.1034	.0929	.0839
3000	.0017	.0024	.0019	.0017	.0042	.1280	.1161	.1037	.0933	.0827
3500	.0010	.0019	.0015	.0014	.0049	.1286	.1165	.1041	.0936	.0814
4000	.0005	.0015	.0012	.0012	.0053	.1290	.1166	.1043	.0938	.0804
4500	.0002	.0012	.0009	.0010	.0056	.1292	.1167	.1044	.0940	.0798
5000	.0002	.0009	.0007	.0008	.0057	.1294	.1167	.1045	.0941	.0795
6000	.0006	.0005	.0004	.0005	.0054	.1296	.1167	.1047	.0942	.0797
7000	.0016	.0003	.0002	.0003	.0045	.1297	.1167	.1047	.0942	.0808
8000	.0030	.0001	.0001	.0002	.0033	.1299	.1168	.1048	.0941	.0826
9000	.0047	.0000	.0000	.0001	.0018	.1300	.1170	.1048	.0940	.0848
10000	.0066	0.0000	0.0000	0.0000	0.0000	.1302	.1172	.1049	.0938	.0875

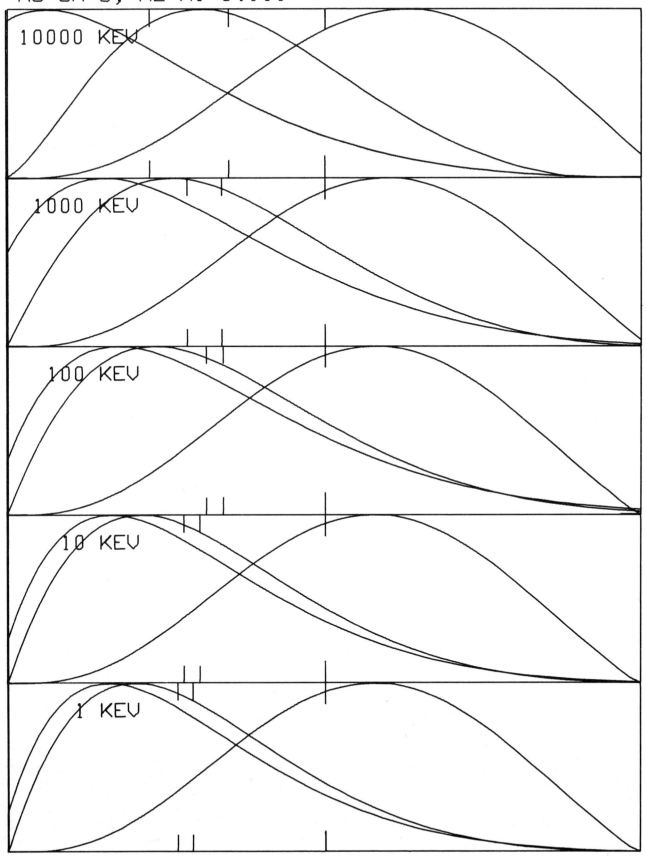

AG ON C, M2/M1=0.111

10000 KEV

1000 KEV

100 KEV

10 KEV

1 KEV

TABLE 20- 1

AG ON RE, M2/M1=0.0833

K/KL =	FRACTIONAL DEPOSITED ENERGY					MEAN RANGE,MICROGRAM/SQ.CM.				
E,KEV	0.65	0.80	1.0	1.2	1.45	0.65	0.80	1.0	1.2	1.45
1.0	.8817	.8465	.8130	.7813	.7440	.3800	.3778	.3756	.3734	.3707
1.2	.8782	.8420	.8077	.7753	.7372	.4316	.4289	.4263	.4237	.4205
1.4	.8751	.8380	.8031	.7700	.7312	.4784	.4754	.4724	.4695	.4659
1.7	.8710	.8329	.7970	.7632	.7235	.5434	.5399	.5364	.5330	.5288
2.0	.8674	.8284	.7918	.7572	.7168	.6045	.6005	.5965	.5926	.5878
2.3	.8642	.8244	.7871	.7519	.7109	.6628	.6583	.6538	.6495	.6441
2.6	.8613	.8208	.7828	.7472	.7056	.7190	.7140	.7091	.7043	.6984
3.0	.8578	.8164	.7778	.7415	.6992	.7914	.7858	.7803	.7749	.7682
3.5	.8539	.8116	.7721	.7351	.6922	.8786	.8722	.8659	.8597	.8520
4.0	.8503	.8072	.7670	.7295	.6859	.9625	.9553	.9482	.9412	.9327
4.5	.8471	.8033	.7624	.7243	.6802	1.044	1.036	1.028	1.020	1.011
5.0	.8442	.7997	.7582	.7197	.6750	1.122	1.113	1.105	1.096	1.086
6.0	.8391	.7933	.7509	.7114	.6660	1.268	1.258	1.248	1.238	1.226
7.0	.8346	.7877	.7444	.7042	.6580	1.409	1.397	1.386	1.375	1.361
8.0	.8304	.7826	.7384	.6976	.6508	1.545	1.532	1.519	1.506	1.491
9.0	.8265	.7778	.7330	.6916	.6443	1.677	1.663	1.648	1.634	1.617
10	.8228	.7734	.7279	.6860	.6382	1.806	1.790	1.774	1.759	1.740
12	.8160	.7651	.7185	.6758	.6271	2.045	2.026	2.008	1.990	1.968
14	.8098	.7576	.7101	.6666	.6173	2.277	2.256	2.235	2.214	2.188
17	.8013	.7475	.6988	.6544	.6043	2.616	2.590	2.564	2.539	2.509
20	.7936	.7385	.6887	.6436	.5928	2.945	2.914	2.884	2.855	2.819
23	.7866	.7302	.6796	.6338	.5825	3.242	3.207	3.174	3.141	3.101
26	.7800	.7226	.6712	.6249	.5731	3.535	3.497	3.459	3.423	3.378
30	.7720	.7133	.6609	.6140	.5619	3.922	3.879	3.836	3.794	3.743
35	.7627	.7027	.6494	.6019	.5493	4.404	4.353	4.302	4.253	4.194
40	.7542	.6930	.6390	.5910	.5381	4.882	4.823	4.764	4.708	4.639
45	.7463	.6841	.6294	.5811	.5279	5.357	5.288	5.222	5.157	5.079
50	.7389	.6758	.6206	.5719	.5186	5.827	5.749	5.674	5.601	5.513
60	.7253	.6607	.6046	.5554	.5019	6.683	6.591	6.502	6.415	6.310
70	.7130	.6472	.5904	.5410	.4874	7.554	7.444	7.338	7.236	7.111
80	.7019	.6350	.5778	.5282	.4746	8.437	8.308	8.183	8.062	7.916
90	.6916	.6240	.5664	.5166	.4631	9.328	9.178	9.032	8.892	8.723
100	.6822	.6138	.5559	.5061	.4527	10.23	10.05	9.884	9.722	9.529
120	.6654	.5955	.5372	.4874	.4344	11.86	11.65	11.45	11.25	11.02
140	.6504	.5794	.5210	.4714	.4187	13.56	13.30	13.06	12.83	12.54
170	.6303	.5587	.5002	.4508	.3987	16.24	15.90	15.57	15.26	14.89
200	.6123	.5409	.4824	.4333	.3819	19.03	18.58	18.16	17.76	17.28
230	.5943	.5254	.4670	.4181	.3676	21.52	20.96	20.48	20.02	19.48
260	.5779	.5118	.4533	.4047	.3551	24.13	23.45	22.89	22.36	21.72
300	.5581	.4956	.4373	.3890	.3406	27.79	26.92	26.24	25.58	24.80
350	.5365	.4780	.4199	.3721	.3250	32.62	31.50	30.60	29.75	28.75
400	.5176	.4627	.4048	.3574	.3115	37.66	36.27	35.12	34.04	32.79
450	.5011	.4490	.3914	.3445	.2997	42.85	41.19	39.75	38.41	36.88
500	.4864	.4367	.3795	.3330	.2892	48.15	46.21	44.45	42.84	41.00
600	.4638	.4152	.3589	.3134	.2709	57.93	55.70	53.40	51.30	48.92
700	.4449	.3970	.3415	.2969	.2557	68.36	65.67	62.70	60.00	56.97
800	.4287	.3811	.3265	.2827	.2426	79.29	76.02	72.24	68.86	65.11
900	.4144	.3671	.3133	.2704	.2313	90.61	86.63	81.95	77.82	73.27
1000	.4016	.3546	.3015	.2594	.2213	102.2	97.42	91.76	86.81	81.43
1200	.3787	.3328	.2812	.2405	.2043	125.0	118.5	111.0	104.5	97.39
1400	.3592	.3144	.2642	.2249	.1904	148.6	140.1	130.5	122.2	113.2
1700	.3347	.2915	.2432	.2056	.1734	185.0	173.0	159.7	148.5	136.6
2000	.3143	.2725	.2259	.1900	.1596	222.0	206.2	188.7	174.3	159.4
2300	.2968	.2563	.2114	.1770	.1483	258.6	239.0	217.2	199.3	181.5
2600	.2815	.2423	.1990	.1659	.1387	295.2	271.6	245.2	223.8	203.0
3000	.2639	.2263	.1848	.1534	.1279	343.8	314.7	281.9	255.6	230.9
3500	.2453	.2094	.1701	.1404	.1168	403.9	367.5	326.5	294.0	264.5
4000	.2295	.1952	.1577	.1297	.1076	463.1	419.2	369.8	331.0	296.7
4500	.2158	.1830	.1473	.1206	.0999	521.2	469.7	411.8	366.7	327.6
5000	.2039	.1724	.1382	.1128	.0933	578.2	518.9	452.6	401.2	357.5
6000	.1840	.1548	.1233	.1001	.0826	688.7	613.9	530.7	466.9	414.2
7000	.1679	.1408	.1116	.0902	.0742	794.6	704.4	604.6	528.7	467.4
8000	.1546	.1293	.1020	.0822	.0675	896.3	790.9	674.9	587.2	517.6
9000	.1435	.1198	.0942	.0756	.0621	994.1	873.8	742.0	642.8	565.3
10000	.1339	.1117	.0876	.0702	.0576	1088.	953.3	806.1	695.9	610.6

E,KEV	MEAN DAMAGE DEPTH,MICROGRAM/SQ.CM.					MEAN IONZN.DEPTH,MICROGRAM/SQ.CM.				
K/KL=	0.65	0.80	1.0	1.2	1.45	0.65	0.80	1.0	1.2	1.45
1.0	.2146	.2129	.2113	.2097	.2078	.1974	.1955	.1937	.1919	.1897
1.2	.2450	.2429	.2409	.2390	.2366	.2255	.2231	.2209	.2186	.2160
1.4	.2722	.2698	.2675	.2653	.2626	.2505	.2478	.2452	.2427	.2396
1.7	.3097	.3070	.3043	.3017	.2986	.2850	.2818	.2788	.2758	.2723
2.0	.3451	.3419	.3389	.3360	.3324	.3174	.3139	.3104	.3071	.3030
2.3	.3791	.3756	.3722	.3689	.3649	.3487	.3447	.3408	.3371	.3325
2.6	.4123	.4084	.4046	.4009	.3964	.3792	.3747	.3704	.3662	.3611
3.0	.4555	.4510	.4466	.4424	.4373	.4189	.4138	.4089	.4041	.3983
3.5	.5081	.5028	.4977	.4928	.4869	.4674	.4614	.4556	.4501	.4433
4.0	.5595	.5534	.5475	.5418	.5351	.5148	.5079	.5012	.4948	.4871
4.5	.6097	.6027	.5960	.5896	.5819	.5612	.5533	.5457	.5384	.5296
5.0	.6588	.6509	.6434	.6362	.6276	.6066	.5977	.5891	.5809	.5712
6.0	.7497	.7404	.7314	.7228	.7126	.6905	.6799	.6697	.6599	.6483
7.0	.8392	.8280	.8175	.8074	.7954	.7733	.7607	.7486	.7371	.7235
8.0	.9273	.9143	.9019	.8902	.8764	.8553	.8403	.8262	.8128	.7970
9.0	1.014	.9991	.9849	.9715	.9557	.9363	.9189	.9026	.8872	.8691
10	1.100	1.083	1.066	1.051	1.033	1.016	.9964	.9778	.9602	.9397
12	1.260	1.239	1.219	1.200	1.179	1.165	1.141	1.118	1.097	1.072
14	1.418	1.393	1.369	1.347	1.321	1.314	1.285	1.257	1.231	1.202
17	1.655	1.622	1.592	1.564	1.531	1.538	1.499	1.463	1.431	1.393
20	1.891	1.849	1.811	1.776	1.737	1.763	1.713	1.667	1.627	1.580
23	2.107	2.057	2.012	1.972	1.926	1.969	1.910	1.857	1.808	1.754
26	2.323	2.265	2.213	2.166	2.113	2.178	2.107	2.045	1.989	1.926
30	2.613	2.542	2.479	2.423	2.360	2.458	2.372	2.296	2.228	2.153
35	2.977	2.889	2.812	2.743	2.667	2.813	2.703	2.608	2.525	2.434
40	3.342	3.235	3.142	3.061	2.972	3.170	3.034	2.919	2.820	2.711
45	3.706	3.579	3.471	3.376	3.273	3.528	3.365	3.228	3.111	2.985
50	4.069	3.921	3.796	3.688	3.570	3.886	3.693	3.534	3.400	3.256
60	4.747	4.558	4.401	4.266	4.116	4.561	4.311	4.108	3.938	3.760
70	5.432	5.199	5.008	4.847	4.666	5.243	4.931	4.681	4.475	4.260
80	6.123	5.844	5.618	5.428	5.219	5.930	5.551	5.254	5.010	4.758
90	6.816	6.489	6.228	6.009	5.773	6.618	6.171	5.824	5.542	5.251
100	7.509	7.134	6.836	6.588	6.327	7.306	6.787	6.389	6.069	5.741
120	8.791	8.324	7.956	7.654	7.369	8.603	7.944	7.448	7.053	6.652
140	10.10	9.538	9.099	8.740	8.426	9.910	9.107	8.510	8.040	7.565
170	12.11	11.40	10.85	10.41	10.04	11.89	10.86	10.11	9.526	8.939
200	14.17	13.30	12.64	12.10	11.67	13.87	12.62	11.72	11.01	10.31
230	15.98	14.96	14.23	13.65	13.14	15.68	14.22	13.18	12.37	11.57
260	17.86	16.68	15.87	15.23	14.64	17.51	15.83	14.65	13.74	12.84
300	20.44	19.05	18.13	17.40	16.68	19.97	18.02	16.65	15.59	14.56
350	23.80	22.16	21.06	20.19	19.30	23.09	20.78	19.17	17.93	16.72
400	27.27	25.39	24.08	23.04	21.96	26.24	23.58	21.72	20.29	18.90
450	30.81	28.70	27.17	25.94	24.65	29.39	26.40	24.29	22.66	21.09
500	34.41	32.08	30.30	28.86	27.36	32.54	29.22	26.85	25.03	23.27
600	40.96	38.43	36.20	34.36	32.48	38.37	34.43	31.73	29.65	27.61
700	47.92	45.11	42.34	40.03	37.72	44.33	39.79	36.67	34.24	31.88
800	55.21	52.04	48.65	45.83	43.04	50.40	45.26	41.63	38.81	36.07
900	62.76	59.14	55.09	51.71	48.42	56.54	50.80	46.61	43.33	40.17
1000	70.49	66.36	61.61	57.64	53.81	62.73	56.40	51.56	47.78	44.19
1200	85.57	80.38	74.31	69.25	64.37	74.83	67.57	61.30	56.37	51.77
1400	101.3	94.78	87.22	80.94	74.93	87.02	78.73	70.92	64.76	59.09
1700	125.6	116.9	106.8	98.50	90.68	105.3	95.32	85.04	76.95	69.69
2000	150.5	139.3	126.4	115.9	106.2	123.5	111.6	98.78	88.72	79.86
2300	174.9	161.2	145.6	132.9	121.2	141.6	127.4	112.2	100.2	89.81
2600	199.5	183.1	164.6	149.7	136.0	159.3	142.9	125.1	111.3	99.39
3000	232.4	212.3	189.7	171.6	155.3	182.5	162.9	141.8	125.5	111.6
3500	273.5	248.4	220.6	198.4	178.7	210.5	187.1	161.8	142.3	126.1
4000	314.3	284.2	250.8	224.4	201.5	237.6	210.3	180.9	158.3	139.8
4500	354.6	319.3	280.3	249.8	223.5	263.7	232.6	199.1	173.5	152.9
5000	394.4	353.9	309.2	274.4	245.0	289.0	254.1	216.6	188.1	165.3
6000	472.1	421.0	365.0	321.8	286.1	337.1	295.1	249.6	215.3	188.5
7000	547.2	485.7	418.3	366.8	325.1	382.3	333.5	280.3	240.6	210.0
8000	619.8	548.0	469.4	409.8	362.2	425.1	369.8	309.1	264.1	230.0
9000	690.0	608.0	518.4	450.8	397.7	465.7	404.2	336.3	286.2	248.7
10000	757.9	665.9	565.5	490.2	431.6	504.5	436.9	362.1	307.1	266.4

TABLE 20- 3

AG ON BE, M2/M1=0.0833

E,KEV	RELATIVE RANGE STRAGGLING					RELATIVE DAMAGE STRAGGLING				
K/KL=	0.65	0.80	1.0	1.2	1.45	0.65	0.80	1.0	1.2	1.45
1.0	.0413	.0410	.0407	.0404	.0400	.2958	.2951	.2944	.2938	.2931
1.2	.0413	.0410	.0406	.0403	.0399	.2957	.2949	.2941	.2935	.2927
1.4	.0413	.0409	.0406	.0403	.0398	.2956	.2947	.2939	.2932	.2923
1.7	.0413	.0409	.0406	.0402	.0398	.2954	.2944	.2935	.2928	.2919
2.0	.0413	.0409	.0405	.0402	.0397	.2952	.2942	.2933	.2924	.2915
2.3	.0412	.0409	.0405	.0401	.0397	.2952	.2941	.2931	.2922	.2912
2.6	.0412	.0408	.0404	.0401	.0396	.2952	.2940	.2929	.2920	.2909
3.0	.0412	.0408	.0404	.0400	.0395	.2953	.2940	.2928	.2918	.2906
3.5	.0412	.0408	.0404	.0400	.0395	.2956	.2941	.2928	.2916	.2903
4.0	.0412	.0407	.0403	.0399	.0394	.2960	.2943	.2928	.2915	.2900
4.5	.0412	.0407	.0403	.0399	.0394	.2966	.2946	.2929	.2915	.2898
5.0	.0411	.0407	.0403	.0398	.0393	.2971	.2950	.2931	.2915	.2897
6.0	.0411	.0407	.0402	.0398	.0393	.2981	.2955	.2933	.2914	.2893
7.0	.0411	.0406	.0402	.0397	.0392	.2994	.2964	.2938	.2915	.2891
8.0	.0411	.0406	.0401	.0397	.0391	.3010	.2974	.2944	.2918	.2891
9.0	.0411	.0406	.0401	.0396	.0391	.3027	.2986	.2952	.2923	.2893
10	.0411	.0406	.0401	.0396	.0390	.3047	.3000	.2961	.2929	.2895
12	.0411	.0405	.0400	.0395	.0389	.3096	.3036	.2988	.2948	.2907
14	.0411	.0405	.0400	.0395	.0388	.3145	.3072	.3014	.2967	.2920
17	.0411	.0405	.0399	.0394	.0387	.3216	.3124	.3053	.2996	.2938
20	.0411	.0405	.0399	.0393	.0386	.3281	.3174	.3089	.3022	.2955
23	.0411	.0404	.0398	.0392	.0385	.3346	.3226	.3127	.3049	.2971
26	.0411	.0404	.0398	.0392	.0385	.3405	.3273	.3161	.3073	.2985
30	.0411	.0404	.0397	.0391	.0384	.3476	.3330	.3201	.3101	.3001
35	.0411	.0404	.0397	.0390	.0382	.3553	.3391	.3242	.3129	.3017
40	.0411	.0403	.0396	.0389	.0381	.3619	.3443	.3276	.3150	.3028
45	.0410	.0403	.0395	.0388	.0380	.3676	.3487	.3303	.3165	.3035
50	.0410	.0402	.0395	.0387	.0379	.3725	.3524	.3324	.3176	.3040
60	.0410	.0402	.0393	.0386	.0376	.3811	.3584	.3353	.3183	.3041
70	.0410	.0401	.0392	.0384	.0374	.3873	.3626	.3368	.3180	.3036
80	.0409	.0400	.0391	.0382	.0372	.3917	.3652	.3374	.3172	.3026
90	.0409	.0399	.0389	.0380	.0369	.3946	.3668	.3372	.3160	.3014
100	.0408	.0397	.0388	.0378	.0367	.3963	.3674	.3365	.3144	.2999
120	.0407	.0395	.0384	.0374	.0362	.3956	.3652	.3329	.3101	.2958
140	.0405	.0393	.0381	.0370	.0358	.3928	.3616	.3286	.3056	.2916
170	.0402	.0388	.0376	.0364	.0350	.3866	.3550	.3218	.2990	.2856
200	.0398	.0384	.0370	.0358	.0343	.3793	.3478	.3150	.2929	.2801
230	.0395	.0379	.0365	.0351	.0336	.3704	.3399	.3083	.2872	.2753
260	.0391	.0374	.0359	.0345	.0329	.3615	.3322	.3021	.2820	.2710
300	.0385	.0367	.0351	.0337	.0320	.3501	.3226	.2944	.2758	.2659
350	.0378	.0359	.0342	.0326	.0309	.3371	.3118	.2859	.2690	.2604
400	.0371	.0351	.0333	.0317	.0299	.3254	.3022	.2785	.2632	.2556
450	.0365	.0343	.0324	.0308	.0289	.3150	.2937	.2720	.2581	.2514
500	.0358	.0335	.0316	.0299	.0280	.3057	.2861	.2664	.2537	.2477
600	.0344	.0320	.0299	.0281	.0262	.2910	.2745	.2577	.2470	.2419
700	.0330	.0305	.0284	.0266	.0246	.2791	.2650	.2508	.2416	.2372
800	.0318	.0292	.0270	.0251	.0232	.2693	.2573	.2451	.2371	.2331
900	.0306	.0279	.0257	.0239	.0219	.2611	.2508	.2402	.2333	.2296
1000	.0295	.0268	.0245	.0227	.0208	.2542	.2453	.2361	.2299	.2265
1200	.0274	.0246	.0224	.0206	.0187	.2447	.2374	.2298	.2245	.2214
1400	.0256	.0228	.0206	.0188	.0170	.2376	.2314	.2248	.2201	.2171
1700	.0232	.0204	.0183	.0166	.0149	.2298	.2245	.2188	.2147	.2116
2000	.0211	.0184	.0164	.0148	.0132	.2238	.2192	.2141	.2102	.2071
2300	.0194	.0168	.0149	.0134	.0119	.2187	.2145	.2097	.2061	.2030
2600	.0179	.0154	.0136	.0122	.0108	.2144	.2105	.2060	.2025	.1994
3000	.0162	.0139	.0122	.0109	.0096	.2097	.2059	.2016	.1982	.1952
3500	.0145	.0123	.0107	.0095	.0084	.2048	.2011	.1970	.1936	.1906
4000	.0130	.0110	.0095	.0085	.0074	.2006	.1970	.1929	.1896	.1866
4500	.0118	.0099	.0085	.0076	.0067	.1971	.1935	.1894	.1861	.1831
5000	.0107	.0089	.0077	.0068	.0060	.1940	.1904	.1863	.1830	.1800
6000	.0091	.0075	.0065	.0057	.0050	.1887	.1850	.1809	.1775	.1745
7000	.0078	.0064	.0055	.0049	.0043	.1842	.1805	.1763	.1730	.1700
8000	.0069	.0056	.0049	.0043	.0038	.1804	.1766	.1724	.1690	.1661
9000	.0061	.0051	.0043	.0038	.0034	.1769	.1731	.1689	.1656	.1626
10000	.0056	.0046	.0040	.0035	.0031	.1737	.1700	.1658	.1625	.1596

	RELATIVE IONIZATION STRAGGLING					RELATIVE TRANSVERSE RANGE STRAGGLING				
K/KL=	0.65	0.80	1.0	1.2	1.45	0.65	0.80	1.0	1.2	1.45
E,KEV										
1.0	.3381	.3381	.3382	.3383	.3385	.4523	.4533	.4543	.4553	.4565
1.2	.3383	.3382	.3382	.3383	.3385	.4521	.4532	.4542	.4552	.4564
1.4	.3384	.3383	.3382	.3383	.3384	.4519	.4530	.4540	.4551	.4564
1.7	.3385	.3383	.3382	.3382	.3383	.4516	.4527	.4538	.4549	.4562
2.0	.3386	.3383	.3382	.3381	.3382	.4514	.4525	.4536	.4547	.4561
2.3	.3388	.3385	.3382	.3381	.3382	.4511	.4522	.4534	.4545	.4559
2.6	.3391	.3386	.3384	.3382	.3382	.4508	.4520	.4531	.4543	.4557
3.0	.3395	.3390	.3386	.3384	.3382	.4504	.4516	.4528	.4540	.4555
3.5	.3403	.3396	.3390	.3386	.3384	.4498	.4511	.4524	.4536	.4552
4.0	.3412	.3403	.3395	.3390	.3386	.4493	.4506	.4519	.4532	.4548
4.5	.3422	.3410	.3401	.3394	.3389	.4488	.4502	.4515	.4528	.4544
5.0	.3434	.3419	.3408	.3399	.3392	.4483	.4497	.4511	.4524	.4541
6.0	.3456	.3437	.3422	.3410	.3400	.4474	.4488	.4502	.4516	.4534
7.0	.3481	.3457	.3437	.3422	.3408	.4464	.4479	.4494	.4508	.4526
8.0	.3509	.3478	.3454	.3435	.3418	.4455	.4470	.4485	.4500	.4519
9.0	.3537	.3500	.3471	.3448	.3427	.4445	.4461	.4477	.4493	.4512
10	.3567	.3523	.3489	.3462	.3437	.4436	.4453	.4469	.4485	.4505
12	.3631	.3573	.3528	.3493	.3460	.4419	.4436	.4453	.4470	.4491
14	.3695	.3622	.3566	.3522	.3481	.4402	.4420	.4438	.4455	.4477
17	.3788	.3693	.3620	.3564	.3511	.4377	.4396	.4415	.4434	.4457
20	.3876	.3759	.3671	.3602	.3539	.4353	.4374	.4394	.4414	.4438
23	.3973	.3831	.3724	.3644	.3568	.4330	.4352	.4373	.4394	.4420
26	.4062	.3897	.3773	.3681	.3594	.4308	.4331	.4353	.4375	.4402
30	.4171	.3975	.3831	.3724	.3625	.4280	.4304	.4328	.4351	.4379
35	.4290	.4059	.3892	.3769	.3656	.4248	.4273	.4298	.4323	.4353
40	.4392	.4130	.3942	.3805	.3680	.4217	.4244	.4271	.4297	.4329
45	.4480	.4190	.3984	.3834	.3699	.4188	.4217	.4245	.4272	.4306
50	.4555	.4240	.4017	.3856	.3714	.4161	.4191	.4220	.4249	.4284
60	.4684	.4313	.4061	.3880	.3726	.4109	.4141	.4173	.4205	.4243
70	.4777	.4364	.4089	.3892	.3730	.4062	.4097	.4132	.4166	.4207
80	.4843	.4398	.4105	.3896	.3729	.4019	.4057	.4095	.4131	.4175
90	.4886	.4419	.4113	.3895	.3724	.3981	.4022	.4061	.4100	.4147
100	.4913	.4432	.4114	.3891	.3717	.3946	.3989	.4031	.4072	.4122
120	.4896	.4425	.4095	.3870	.3695	.3882	.3929	.3976	.4021	.4076
140	.4854	.4403	.4067	.3845	.3670	.3827	.3879	.3930	.3979	.4039
170	.4769	.4355	.4017	.3803	.3634	.3760	.3819	.3876	.3931	.3997
200	.4675	.4298	.3965	.3761	.3600	.3707	.3772	.3835	.3895	.3968
230	.4580	.4226	.3904	.3710	.3567	.3663	.3733	.3802	.3867	.3946
260	.4488	.4155	.3846	.3662	.3537	.3627	.3703	.3777	.3847	.3932
300	.4373	.4066	.3775	.3604	.3504	.3590	.3673	.3753	.3830	.3923
350	.4244	.3966	.3699	.3544	.3470	.3557	.3649	.3737	.3821	.3922
400	.4131	.3878	.3634	.3494	.3443	.3535	.3635	.3731	.3823	.3931
450	.4031	.3801	.3579	.3455	.3424	.3522	.3630	.3733	.3831	.3948
500	.3943	.3734	.3534	.3423	.3409	.3516	.3632	.3742	.3846	.3969
600	.3798	.3632	.3476	.3394	.3399	.3518	.3649	.3772	.3888	.4025
700	.3687	.3556	.3436	.3380	.3399	.3535	.3681	.3814	.3942	.4091
800	.3600	.3499	.3411	.3376	.3406	.3562	.3721	.3866	.4003	.4164
900	.3533	.3458	.3396	.3379	.3417	.3596	.3768	.3923	.4070	.4240
1000	.3482	.3428	.3389	.3386	.3430	.3635	.3819	.3984	.4140	.4320
1200	.3437	.3412	.3401	.3417	.3470	.3725	.3932	.4116	.4288	.4486
1400	.3419	.3414	.3424	.3453	.3511	.3823	.4050	.4253	.4439	.4652
1700	.3419	.3436	.3467	.3508	.3571	.3979	.4233	.4459	.4665	.4897
2000	.3435	.3467	.3513	.3562	.3628	.4138	.4417	.4663	.4885	.5133
2300	.3457	.3498	.3554	.3609	.3677	.4298	.4602	.4866	.5101	.5362
2600	.3482	.3531	.3594	.3653	.3722	.4457	.4783	.5062	.5309	.5582
3000	.3519	.3575	.3644	.3706	.3776	.4664	.5016	.5314	.5574	.5860
3500	.3567	.3627	.3701	.3767	.3837	.4916	.5296	.5613	.5887	.6185
4000	.3613	.3677	.3754	.3821	.3890	.5158	.5563	.5895	.6180	.6488
4500	.3658	.3723	.3801	.3869	.3938	.5392	.5817	.6162	.6457	.6773
5000	.3700	.3766	.3845	.3913	.3982	.5617	.6060	.6415	.6718	.7040
6000	.3776	.3843	.3923	.3990	.4057	.6045	.6515	.6888	.7201	.7533
7000	.3843	.3910	.3989	.4056	.4121	.6445	.6935	.7320	.7641	.7978
8000	.3901	.3968	.4046	.4112	.4176	.6821	.7326	.7720	.8046	.8385
9000	.3953	.4020	.4096	.4162	.4224	.7177	.7692	.8093	.8421	.8761
10000	.3998	.4065	.4141	.4205	.4266	.7514	.8036	.8442	.8771	.9111

AG ON BE, M2/M1=0.0833   TABLE 20- 5
RELATIVE TRANSV. DAMAGE STRAGGLING RELATIVE TRANSV. IONZN. STRAGGLING

K/KL= E,KEV	0.65	0.80	1.0	1.2	1.45	0.65	0.80	1.0	1.2	1.45
1.0	.2569	.2518	.2470	.2423	.2367	.2273	.2221	.2171	.2123	.2066
1.2	.2587	.2534	.2483	.2434	.2376	.2290	.2235	.2183	.2133	.2073
1.4	.2604	.2549	.2496	.2446	.2385	.2307	.2250	.2196	.2144	.2082
1.7	.2631	.2573	.2517	.2464	.2400	.2333	.2273	.2216	.2161	.2095
2.0	.2657	.2596	.2537	.2481	.2415	.2359	.2296	.2235	.2178	.2110
2.3	.2683	.2618	.2557	.2499	.2430	.2384	.2318	.2255	.2195	.2124
2.6	.2707	.2640	.2576	.2516	.2444	.2408	.2339	.2273	.2211	.2138
3.0	.2738	.2668	.2601	.2538	.2463	.2439	.2366	.2297	.2232	.2155
3.5	.2775	.2701	.2631	.2564	.2486	.2475	.2398	.2326	.2257	.2177
4.0	.2810	.2732	.2658	.2589	.2507	.2509	.2429	.2353	.2281	.2197
4.5	.2842	.2761	.2685	.2612	.2527	.2541	.2458	.2379	.2304	.2217
5.0	.2874	.2789	.2710	.2635	.2547	.2572	.2485	.2403	.2326	.2235
6.0	.2936	.2845	.2760	.2680	.2586	.2632	.2538	.2451	.2368	.2271
7.0	.2991	.2895	.2805	.2720	.2621	.2685	.2586	.2493	.2406	.2304
8.0	.3040	.2939	.2845	.2756	.2652	.2733	.2629	.2532	.2440	.2333
9.0	.3084	.2979	.2881	.2788	.2679	.2776	.2668	.2566	.2471	.2359
10	.3123	.3015	.2913	.2816	.2704	.2814	.2703	.2598	.2499	.2383
12	.3191	.3076	.2968	.2866	.2746	.2884	.2766	.2655	.2550	.2427
14	.3247	.3127	.3013	.2906	.2781	.2941	.2819	.2702	.2592	.2463
17	.3314	.3188	.3068	.2955	.2822	.3011	.2882	.2760	.2643	.2507
20	.3367	.3236	.3111	.2992	.2853	.3064	.2932	.2804	.2683	.2540
23	.3406	.3274	.3146	.3023	.2878	.3100	.2967	.2837	.2711	.2563
26	.3438	.3304	.3173	.3047	.2898	.3128	.2994	.2862	.2734	.2581
30	.3471	.3336	.3202	.3072	.2917	.3157	.3022	.2888	.2756	.2599
35	.3501	.3365	.3227	.3093	.2933	.3183	.3048	.2911	.2776	.2613
40	.3523	.3385	.3245	.3107	.2942	.3201	.3066	.2927	.2788	.2621
45	.3538	.3399	.3256	.3115	.2944	.3214	.3079	.2938	.2796	.2624
50	.3548	.3408	.3263	.3118	.2943	.3224	.3089	.2945	.2800	.2625
60	.3560	.3415	.3265	.3113	.2930	.3241	.3103	.2956	.2805	.2623
70	.3562	.3413	.3257	.3100	.2909	.3249	.3109	.2957	.2800	.2612
80	.3557	.3405	.3243	.3080	.2882	.3250	.3108	.2952	.2789	.2596
90	.3548	.3392	.3225	.3056	.2852	.3246	.3102	.2941	.2773	.2574
100	.3535	.3377	.3204	.3029	.2819	.3239	.3093	.2927	.2754	.2549
120	.3504	.3346	.3156	.2966	.2746	.3213	.3068	.2888	.2705	.2484
140	.3468	.3309	.3103	.2900	.2671	.3181	.3035	.2842	.2650	.2416
170	.3406	.3244	.3017	.2799	.2559	.3128	.2979	.2767	.2563	.2312
200	.3340	.3173	.2928	.2698	.2452	.3073	.2918	.2689	.2474	.2211
230	.3271	.3088	.2827	.2588	.2339	.3024	.2851	.2603	.2376	.2108
260	.3201	.3002	.2728	.2483	.2233	.2975	.2784	.2519	.2281	.2009
300	.3109	.2890	.2602	.2351	.2103	.2908	.2694	.2411	.2162	.1888
350	.2995	.2756	.2455	.2201	.1956	.2823	.2584	.2282	.2023	.1751
400	.2885	.2629	.2320	.2065	.1826	.2737	.2478	.2162	.1896	.1628
450	.2777	.2509	.2196	.1943	.1710	.2651	.2375	.2049	.1780	.1518
500	.2673	.2397	.2082	.1831	.1606	.2564	.2277	.1944	.1673	.1419
600	.2452	.2178	.1872	.1634	.1432	.2366	.2072	.1739	.1476	.1246
700	.2255	.1987	.1693	.1470	.1289	.2184	.1890	.1562	.1310	.1103
800	.2080	.1822	.1540	.1331	.1170	.2019	.1729	.1409	.1170	.0983
900	.1924	.1677	.1410	.1214	.1069	.1869	.1586	.1276	.1050	.0882
1000	.1786	.1550	.1297	.1114	.0984	.1733	.1459	.1161	.0947	.0796
1200	.1560	.1351	.1127	.0967	.0858	.1497	.1249	.0983	.0795	.0669
1400	.1378	.1192	.0996	.0856	.0762	.1301	.1080	.0844	.0679	.0573
1700	.1162	.1008	.0846	.0732	.0656	.1065	.0879	.0684	.0550	.0467
2000	.0997	.0870	.0736	.0641	.0578	.0882	.0726	.0566	.0455	.0390
2300	.0884	.0773	.0657	.0574	.0520	.0759	.0624	.0485	.0390	.0336
2600	.0796	.0698	.0596	.0523	.0474	.0663	.0544	.0423	.0340	.0294
3000	.0705	.0621	.0532	.0469	.0427	.0563	.0463	.0360	.0290	.0252
3500	.0620	.0549	.0473	.0419	.0382	.0470	.0388	.0303	.0245	.0213
4000	.0556	.0494	.0429	.0382	.0348	.0401	.0332	.0261	.0213	.0185
4500	.0506	.0452	.0394	.0352	.0322	.0348	.0290	.0229	.0188	.0163
5000	.0466	.0418	.0366	.0328	.0300	.0306	.0257	.0205	.0169	.0147
6000	.0407	.0367	.0324	.0292	.0266	.0246	.0208	.0169	.0140	.0122
7000	.0364	.0330	.0293	.0264	.0241	.0205	.0175	.0143	.0120	.0104
8000	.0332	.0302	.0268	.0242	.0221	.0176	.0151	.0124	.0105	.0091
9000	.0306	.0278	.0247	.0223	.0204	.0155	.0133	.0109	.0092	.0080
10000	.0284	.0258	.0229	.0207	.0189	.0139	.0118	.0096	.0080	.0070

K/KL= E,KEV	RANGE SKEWNESS					DAMAGE SKEWNESS				
	0.65	0.80	1.0	1.2	1.45	0.65	0.80	1.0	1.2	1.45
1.0	.3601	.3554	.3507	.3461	.3403	.3426	.3349	.3279	.3213	.3137
1.2	.3596	.3547	.3498	.3451	.3392	.3442	.3355	.3276	.3203	.3120
1.4	.3591	.3541	.3491	.3442	.3381	.3459	.3364	.3278	.3200	.3111
1.7	.3584	.3532	.3481	.3430	.3368	.3487	.3384	.3289	.3204	.3107
2.0	.3579	.3525	.3472	.3420	.3356	.3522	.3409	.3307	.3214	.3109
2.3	.3573	.3519	.3464	.3411	.3346	.3563	.3440	.3329	.3228	.3115
2.6	.3568	.3512	.3457	.3403	.3336	.3610	.3475	.3354	.3245	.3124
3.0	.3562	.3505	.3448	.3392	.3324	.3678	.3527	.3392	.3271	.3137
3.5	.3555	.3496	.3438	.3381	.3310	.3771	.3597	.3443	.3306	.3156
4.0	.3549	.3488	.3428	.3370	.3297	.3871	.3672	.3497	.3343	.3175
4.5	.3543	.3481	.3420	.3359	.3286	.3975	.3749	.3553	.3381	.3194
5.0	.3537	.3473	.3411	.3350	.3275	.4082	.3829	.3610	.3419	.3213
6.0	.3526	.3460	.3396	.3332	.3255	.4258	.3948	.3683	.3455	.3211
7.0	.3516	.3448	.3382	.3316	.3236	.4460	.4092	.3779	.3511	.3228
8.0	.3507	.3437	.3369	.3301	.3219	.4686	.4256	.3893	.3586	.3262
9.0	.3497	.3426	.3356	.3287	.3203	.4932	.4438	.4025	.3676	.3313
10	.3489	.3415	.3344	.3273	.3187	.5193	.4635	.4171	.3781	.3376
12	.3472	.3396	.3322	.3248	.3159	.5838	.5141	.4567	.4089	.3598
14	.3457	.3378	.3301	.3225	.3132	.6471	.5639	.4959	.4396	.3823
17	.3434	.3352	.3270	.3191	.3094	.7365	.6342	.5511	.4830	.4142
20	.3413	.3326	.3242	.3159	.3058	.8181	.6981	.6011	.5221	.4427
23	.3393	.3303	.3215	.3130	.3026	.8998	.7590	.6487	.5591	.4693
26	.3373	.3280	.3190	.3101	.2994	.9733	.8137	.6910	.5918	.4923
30	.3347	.3259	.3156	.3064	.2952	1.059	.8776	.7399	.6290	.5182
35	.3315	.3214	.3115	.3019	.2902	1.149	.9451	.7907	.6669	.5436
40	.3284	.3178	.3075	.2975	.2854	1.222	1.001	.8318	.6966	.5628
45	.3253	.3142	.3035	.2932	.2806	1.282	1.047	.8650	.7196	.5769
50	.3222	.3107	.2997	.2889	.2759	1.330	1.085	.8916	.7372	.5867
60	.3164	.3041	.2923	.2809	.2671	1.385	1.141	.9266	.7560	.5928
70	.3106	.2976	.2852	.2731	.2586	1.420	1.178	.9471	.7634	.5909
80	.3049	.2912	.2781	.2654	.2503	1.442	1.202	.9572	.7630	.5837
90	.2993	.2849	.2712	.2579	.2421	1.454	1.216	.9596	.7569	.5728
100	.2937	.2787	.2644	.2506	.2342	1.460	1.223	.9562	.7466	.5594
120	.2830	.2667	.2512	.2364	.2188	1.463	1.214	.9309	.7115	.5227
140	.2724	.2550	.2384	.2227	.2040	1.453	1.191	.8971	.6716	.4841
170	.2571	.2380	.2200	.2030	.1829	1.420	1.144	.8391	.6092	.4270
200	.2424	.2217	.2024	.1842	.1630	1.376	1.089	.7782	.5478	.3730
230	.2280	.2059	.1852	.1660	.1435	1.309	1.021	.7109	.4851	.3226
260	.2141	.1906	.1688	.1485	.1250	1.240	.9523	.6462	.4263	.2761
300	.1963	.1711	.1479	.1264	.1017	1.147	.8644	.5653	.3544	.2196
350	.1752	.1481	.1233	.1005	.0744	1.036	.7615	.4735	.2747	.1569
400	.1552	.1263	.1001	.0762	.0490	.9322	.6673	.3917	.2050	.1016
450	.1361	.1057	.0782	.0533	.0252	.8351	.5813	.3190	.1441	.0525
500	.1180	.0860	.0575	.0317	.0027	.7450	.5030	.2543	.0907	.0087
600	.0833	.0487	.0181	-.0091	-.0395	.5758	.3649	.1492	.0089	-.0670
700	.0515	.0146	-.0176	-.0460	-.0774	.4329	.2500	.0636	-.0567	-.1287
800	.0221	-.0167	-.0502	-.0796	-.1117	.3122	.1540	-.0071	-.1106	-.1797
900	-.0053	-.0457	-.0803	-.1104	-.1431	.2104	.0733	-.0662	-.1556	-.2222
1000	-.0309	-.0727	-.1082	-.1388	-.1719	.1245	.0052	-.1159	-.1937	-.2578
1200	-.0785	-.1224	-.1592	-.1905	-.2240	.0105	-.0867	-.1859	-.2501	-.3047
1400	-.1208	-.1665	-.2040	-.2357	-.2692	-.0734	-.1549	-.2385	-.2937	-.3399
1700	-.1767	-.2240	-.2623	-.2942	-.3275	-.1651	-.2303	-.2982	-.3445	-.3803
2000	-.2253	-.2736	-.3122	-.3440	-.3770	-.2319	-.2862	-.3438	-.3847	-.4124
2300	-.2699	-.3187	-.3571	-.3885	-.4208	-.2867	-.3338	-.3843	-.4212	-.4460
2600	-.3096	-.3586	-.3966	-.4276	-.4592	-.3309	-.3730	-.4186	-.4528	-.4758
3000	-.3563	-.4051	-.4426	-.4729	-.5036	-.3785	-.4161	-.4572	-.4890	-.5109
3500	-.4062	-.4546	-.4913	-.5207	-.5504	-.4255	-.4596	-.4974	-.5273	-.5488
4000	-.4488	-.4965	-.5323	-.5609	-.5896	-.4631	-.4952	-.5310	-.5599	-.5816
4500	-.4855	-.5324	-.5673	-.5951	-.6230	-.4942	-.5252	-.5600	-.5883	-.6104
5000	-.5173	-.5633	-.5974	-.6245	-.6517	-.5207	-.5512	-.5853	-.6133	-.6359
6000	-.5696	-.6137	-.6463	-.6720	-.6980	-.5642	-.5943	-.6282	-.6559	-.6794
7000	-.6105	-.6525	-.6838	-.7083	-.7335	-.5993	-.6295	-.6636	-.6912	-.7152
8000	-.6428	-.6829	-.7129	-.7363	-.7611	-.6290	-.6595	-.6938	-.7213	-.7455
9000	-.6687	-.7067	-.7357	-.7582	-.7826	-.6550	-.6856	-.7201	-.7475	-.7714
10000	-.6894	-.7254	-.7535	-.7753	-.7994	-.6783	-.7090	-.7436	-.7706	-.7938

	IONIZATION SKEWNESS					RANGE KURTOSIS				
K/KL=	0.65	0.80	1.0	1.2	1.45	0.65	0.80	1.0	1.2	1.45
E,KEV										
1.0	.4554	.4494	.4440	.4392	.4339	3.137	3.130	3.124	3.117	3.109
1.2	.4592	.4522	.4460	.4405	.4345	3.136	3.129	3.122	3.116	3.107
1.4	.4619	.4541	.4473	.4413	.4347	3.135	3.128	3.121	3.114	3.106
1.7	.4655	.4568	.4492	.4425	.4352	3.134	3.127	3.120	3.113	3.104
2.0	.4694	.4598	.4513	.4439	.4359	3.133	3.126	3.118	3.111	3.102
2.3	.4740	.4633	.4539	.4458	.4369	3.133	3.125	3.117	3.110	3.101
2.6	.4792	.4673	.4570	.4480	.4383	3.132	3.124	3.116	3.109	3.099
3.0	.4870	.4734	.4617	.4515	.4406	3.131	3.123	3.115	3.107	3.098
3.5	.4980	.4821	.4684	.4566	.4440	3.130	3.121	3.113	3.105	3.096
4.0	.5102	.4917	.4759	.4623	.4480	3.129	3.120	3.112	3.104	3.094
4.5	.5232	.5020	.4839	.4685	.4523	3.128	3.119	3.110	3.102	3.092
5.0	.5368	.5127	.4924	.4751	.4570	3.127	3.118	3.109	3.101	3.091
6.0	.5653	.5354	.5103	.4891	.4671	3.125	3.116	3.107	3.098	3.088
7.0	.5949	.5589	.5289	.5037	.4777	3.123	3.114	3.104	3.096	3.085
8.0	.6250	.5828	.5478	.5185	.4885	3.122	3.112	3.102	3.093	3.082
9.0	.6553	.6068	.5667	.5334	.4994	3.120	3.110	3.100	3.091	3.080
10	.6855	.6307	.5856	.5482	.5102	3.119	3.108	3.099	3.089	3.078
12	.7510	.6825	.6265	.5804	.5339	3.116	3.105	3.095	3.085	3.074
14	.8125	.7319	.6647	.6105	.5560	3.113	3.102	3.092	3.082	3.070
17	.8963	.7971	.7167	.6512	.5857	3.109	3.098	3.087	3.077	3.065
20	.9706	.8556	.7626	.6870	.6117	3.106	3.094	3.083	3.072	3.060
23	1.041	.9116	.8068	.7218	.6376	3.102	3.090	3.079	3.068	3.055
26	1.104	.9611	.8459	.7523	.6601	3.099	3.087	3.075	3.064	3.050
30	1.176	1.018	.8906	.7870	.6853	3.095	3.082	3.070	3.058	3.045
35	1.251	1.077	.9364	.8221	.7101	3.090	3.076	3.064	3.052	3.038
40	1.313	1.125	.9729	.8496	.7287	3.085	3.071	3.058	3.046	3.032
45	1.363	1.163	1.002	.8708	.7422	3.080	3.066	3.053	3.040	3.026
50	1.403	1.194	1.024	.8869	.7515	3.075	3.061	3.047	3.035	3.020
60	1.454	1.230	1.048	.9015	.7582	3.066	3.051	3.037	3.024	3.009
70	1.487	1.252	1.061	.9068	.7488	3.058	3.042	3.027	3.014	2.999
80	1.508	1.264	1.066	.9062	.7410	3.050	3.033	3.018	3.005	2.989
90	1.521	1.271	1.066	.9015	.7313	3.042	3.025	3.010	2.996	2.980
100	1.528	1.272	1.062	.8939	.7204	3.034	3.017	3.002	2.988	2.972
120	1.532	1.268	1.049	.8727	.6986	3.020	3.002	2.986	2.972	2.956
140	1.524	1.255	1.028	.8473	.6756	3.007	2.988	2.972	2.958	2.943
170	1.497	1.226	.9908	.8056	.6404	2.989	2.970	2.954	2.940	2.925
200	1.459	1.188	.9491	.7626	.6056	2.973	2.955	2.939	2.925	2.911
230	1.401	1.134	.8945	.7112	.5644	2.958	2.939	2.924	2.911	2.897
260	1.341	1.080	.8410	.6621	.5257	2.944	2.926	2.911	2.899	2.886
300	1.261	1.010	.7735	.6016	.4785	2.928	2.911	2.896	2.885	2.875
350	1.166	.9269	.6967	.5345	.4269	2.912	2.895	2.882	2.873	2.864
400	1.077	.8507	.6284	.4762	.3828	2.898	2.883	2.872	2.864	2.858
450	.9945	.7814	.5680	.4259	.3452	2.887	2.873	2.864	2.858	2.854
500	.9188	.7183	.5147	.3825	.3134	2.877	2.865	2.858	2.854	2.852
600	.7787	.6047	.4287	.3184	.2696	2.861	2.853	2.850	2.850	2.854
700	.6624	.5119	.3613	.2701	.2376	2.851	2.847	2.848	2.853	2.861
800	.5662	.4363	.3082	.2335	.2139	2.846	2.846	2.851	2.860	2.872
900	.4868	.3749	.2665	.2058	.1963	2.844	2.849	2.858	2.870	2.887
1000	.4215	.3251	.2338	.1849	.1835	2.846	2.855	2.868	2.883	2.903
1200	.3426	.2693	.2013	.1671	.1724	2.858	2.876	2.897	2.919	2.946
1400	.2897	.2342	.1837	.1601	.1687	2.876	2.902	2.930	2.958	2.991
1700	.2390	.2034	.1723	.1596	.1701	2.908	2.946	2.983	3.018	3.059
2000	.2086	.1874	.1699	.1646	.1754	2.944	2.991	3.036	3.078	3.125
2300	.1880	.1757	.1672	.1673	.1798	2.977	3.033	3.084	3.131	3.183
2600	.1750	.1692	.1671	.1712	.1851	3.011	3.075	3.132	3.182	3.238
3000	.1654	.1657	.1697	.1775	.1928	3.058	3.131	3.194	3.249	3.309
3500	.1609	.1663	.1754	.1864	.2029	3.119	3.201	3.271	3.330	3.395
4000	.1614	.1701	.1825	.1956	.2129	3.181	3.271	3.346	3.408	3.476
4500	.1648	.1757	.1903	.2048	.2227	3.244	3.340	3.419	3.484	3.555
5000	.1699	.1823	.1984	.2139	.2322	3.306	3.408	3.491	3.558	3.631
6000	.1826	.1968	.2146	.2312	.2499	3.431	3.542	3.629	3.700	3.776
7000	.1963	.2115	.2302	.2473	.2661	3.554	3.671	3.762	3.835	3.913
8000	.2098	.2256	.2449	.2621	.2808	3.675	3.796	3.889	3.964	4.043
9000	.2226	.2388	.2585	.2758	.2943	3.793	3.917	4.011	4.087	4.166
10000	.2343	.2510	.2712	.2885	.3067	3.908	4.034	4.128	4.205	4.285

TABLE 20- 8

	DAMAGE KURTOSIS					IONIZATION KURTOSIS				
K/KL=	0.65	0.80	1.0	1.2	1.45	0.65	0.80	1.0	1.2	1.45
E,KEV										
1.0	2.701	2.676	2.654	2.634	2.612	2.799	2.773	2.750	2.729	2.706
1.2	2.713	2.685	2.659	2.636	2.611	2.817	2.787	2.761	2.737	2.711
1.4	2.724	2.693	2.665	2.640	2.613	2.831	2.798	2.769	2.743	2.715
1.7	2.741	2.706	2.676	2.649	2.619	2.849	2.812	2.780	2.752	2.721
2.0	2.759	2.722	2.688	2.659	2.627	2.868	2.827	2.792	2.761	2.727
2.3	2.780	2.738	2.702	2.670	2.636	2.889	2.844	2.805	2.771	2.734
2.6	2.802	2.756	2.717	2.682	2.645	2.912	2.863	2.820	2.782	2.742
3.0	2.833	2.782	2.737	2.698	2.657	2.947	2.890	2.841	2.799	2.755
3.5	2.876	2.815	2.764	2.719	2.672	2.995	2.928	2.871	2.823	2.772
4.0	2.921	2.850	2.790	2.739	2.686	3.047	2.969	2.904	2.848	2.790
4.5	2.968	2.886	2.817	2.760	2.699	3.103	3.013	2.938	2.875	2.809
5.0	3.016	2.922	2.845	2.780	2.712	3.161	3.059	2.974	2.903	2.829
6.0	3.099	2.982	2.887	2.807	2.726	3.282	3.154	3.048	2.961	2.872
7.0	3.191	3.049	2.934	2.839	2.744	3.408	3.252	3.125	3.021	2.916
8.0	3.291	3.122	2.986	2.875	2.765	3.535	3.350	3.202	3.081	2.959
9.0	3.397	3.199	3.042	2.915	2.789	3.663	3.449	3.278	3.140	3.002
10	3.508	3.280	3.101	2.957	2.815	3.790	3.548	3.354	3.199	3.045
12	3.776	3.472	3.244	3.063	2.888	4.070	3.761	3.519	3.326	3.137
14	4.035	3.661	3.385	3.169	2.961	4.331	3.960	3.671	3.443	3.221
17	4.393	3.930	3.586	3.320	3.067	4.683	4.229	3.876	3.600	3.333
20	4.715	4.178	3.772	3.459	3.164	4.991	4.463	4.055	3.736	3.430
23	5.003	4.438	3.966	3.607	3.271	5.278	4.682	4.223	3.864	3.523
26	5.260	4.673	4.142	3.741	3.366	5.530	4.874	4.369	3.976	3.602
30	5.561	4.951	4.349	3.896	3.477	5.816	5.092	4.535	4.100	3.690
35	5.882	5.245	4.565	4.058	3.590	6.109	5.313	4.701	4.224	3.775
40	6.153	5.489	4.743	4.187	3.679	6.341	5.488	4.831	4.319	3.837
45	6.382	5.690	4.887	4.291	3.747	6.526	5.626	4.931	4.390	3.881
50	6.579	5.854	5.002	4.371	3.798	6.671	5.732	5.006	4.442	3.909
60	6.903	6.060	5.145	4.461	3.836	6.823	5.828	5.064	4.471	3.898
70	7.141	6.196	5.231	4.504	3.843	6.906	5.876	5.085	4.470	3.869
80	7.314	6.284	5.275	4.516	3.830	6.945	5.895	5.083	4.453	3.832
90	7.437	6.337	5.290	4.506	3.804	6.954	5.895	5.067	4.424	3.791
100	7.519	6.366	5.283	4.479	3.770	6.942	5.883	5.042	4.390	3.750
120	7.571	6.375	5.197	4.359	3.665	6.878	5.863	4.992	4.320	3.684
140	7.546	6.335	5.081	4.223	3.555	6.781	5.816	4.925	4.240	3.617
170	7.424	6.216	4.884	4.015	3.396	6.603	5.712	4.804	4.109	3.516
200	7.242	6.055	4.680	3.816	3.249	6.407	5.579	4.666	3.973	3.414
230	6.966	5.799	4.456	3.624	3.120	6.184	5.374	4.469	3.796	3.284
260	6.681	5.540	4.244	3.450	3.004	5.965	5.167	4.277	3.627	3.160
300	6.307	5.207	3.985	3.244	2.871	5.684	4.900	4.034	3.419	3.009
350	5.865	4.823	3.699	3.026	2.731	5.355	4.590	3.760	3.189	2.843
400	5.460	4.478	3.454	2.845	2.616	5.054	4.309	3.519	2.992	2.702
450	5.093	4.173	3.245	2.696	2.521	4.778	4.056	3.309	2.823	2.582
500	4.761	3.905	3.066	2.573	2.443	4.525	3.830	3.126	2.680	2.481
600	4.183	3.487	2.816	2.427	2.347	4.043	3.449	2.856	2.491	2.353
700	3.717	3.161	2.631	2.328	2.281	3.646	3.145	2.649	2.352	2.261
800	3.342	2.906	2.491	2.259	2.235	3.321	2.900	2.488	2.249	2.191
900	3.042	2.705	2.386	2.211	2.203	3.055	2.704	2.363	2.170	2.139
1000	2.804	2.548	2.306	2.178	2.180	2.840	2.547	2.265	2.111	2.099
1200	2.569	2.394	2.231	2.148	2.160	2.606	2.382	2.169	2.055	2.053
1400	2.436	2.311	2.197	2.141	2.155	2.457	2.282	2.116	2.027	2.026
1700	2.336	2.256	2.183	2.150	2.162	2.321	2.197	2.077	2.011	2.005
2000	2.297	2.242	2.191	2.168	2.177	2.242	2.151	2.061	2.009	1.995
2300	2.253	2.219	2.189	2.179	2.190	2.166	2.098	2.031	1.991	1.980
2600	2.227	2.208	2.193	2.191	2.204	2.110	2.058	2.007	1.976	1.968
3000	2.210	2.205	2.203	2.209	2.224	2.056	2.019	1.983	1.961	1.957
3500	2.207	2.211	2.221	2.233	2.251	2.010	1.985	1.961	1.948	1.947
4000	2.214	2.224	2.241	2.258	2.278	1.980	1.963	1.946	1.939	1.942
4500	2.226	2.241	2.262	2.282	2.305	1.959	1.947	1.937	1.933	1.939
5000	2.242	2.260	2.284	2.306	2.331	1.946	1.937	1.930	1.930	1.938
6000	2.277	2.299	2.327	2.352	2.380	1.932	1.928	1.926	1.929	1.941
7000	2.311	2.336	2.368	2.396	2.427	1.928	1.926	1.927	1.933	1.948
8000	2.343	2.372	2.406	2.437	2.471	1.928	1.929	1.933	1.941	1.957
9000	2.371	2.404	2.442	2.476	2.512	1.932	1.935	1.942	1.952	1.968
10000	2.396	2.433	2.476	2.513	2.550	1.936	1.943	1.953	1.964	1.980

	REFLECTION COEFFICIENT					SPUTTERING EFFICIENCY				
K/KL =	0.65	0.80	1.0	1.2	1.45	0.65	0.80	1.0	1.2	1.45
E,KEV										
1.0	.0000	.0000	.0000	.0000	.0000	.0145	.0137	.0131	.0124	.0117
1.2	.0000	.0000	.0000	.0000	.0000	.0145	.0137	.0130	.0123	.0115
1.4	.0000	.0000	.0000	.0000	.0000	.0145	.0137	.0129	.0122	.0115
1.7	.0000	.0000	.0000	.0000	.0000	.0145	.0137	.0129	.0122	.0114
2.0	.0000	.0000	.0000	.0000	.0000	.0146	.0137	.0129	.0121	.0113
2.3	.0000	.0000	.0000	.0000	.0000	.0147	.0137	.0129	.0121	.0112
2.6	.0000	.0000	.0000	.0000	.0000	.0148	.0138	.0129	.0121	.0112
3.0	.0000	.0000	.0000	.0000	.0000	.0149	.0139	.0129	.0121	.0112
3.5	.0000	.0000	.0000	.0000	.0000	.0150	.0139	.0130	.0121	.0111
4.0	.0000	.0000	.0000	.0000	.0000	.0152	.0140	.0130	.0121	.0111
4.5	.0000	.0000	.0000	.0000	.0000	.0153	.0141	.0131	.0121	.0111
5.0	.0000	.0000	.0000	.0000	.0000	.0155	.0142	.0131	.0121	.0111
6.0	.0000	.0000	.0000	.0000	.0000	.0158	.0145	.0133	.0122	.0110
7.0	.0000	.0000	.0000	.0000	.0000	.0161	.0147	.0134	.0123	.0110
8.0	.0000	.0000	.0000	.0000	.0000	.0164	.0148	.0135	.0123	.0110
9.0	.0000	.0000	.0000	.0000	.0000	.0166	.0150	.0136	.0124	.0110
10	.0000	.0000	.0000	.0000	.0000	.0167	.0151	.0137	.0124	.0110
12	.0000	.0000	.0000	.0000	.0000	.0170	.0153	.0138	.0124	.0110
14	.0000	.0000	.0000	.0000	.0000	.0172	.0154	.0138	.0125	.0109
17	.0000	.0000	.0000	.0000	.0000	.0173	.0155	.0139	.0125	.0109
20	.0000	.0000	.0000	.0000	.0000	.0174	.0155	.0139	.0125	.0109
23	.0000	.0000	.0000	.0000	.0000	.0170	.0155	.0139	.0124	.0109
26	.0000	.0000	.0000	.0000	.0000	.0167	.0155	.0139	.0124	.0108
30	.0000	.0000	.0000	.0000	.0000	.0163	.0155	.0138	.0124	.0108
35	.0000	.0000	.0000	.0000	.0000	.0158	.0155	.0137	.0123	.0107
40	.0000	.0000	.0000	.0000	.0000	.0154	.0154	.0136	.0122	.0106
45	.0000	.0000	.0000	.0000	.0000	.0152	.0153	.0135	.0121	.0105
50	.0000	.0000	.0000	.0000	.0000	.0150	.0152	.0134	.0119	.0104
60	.0000	.0000	.0000	.0000	.0000	.0153	.0149	.0131	.0116	.0101
70	.0000	.0000	.0000	.0000	.0000	.0156	.0146	.0128	.0113	.0098
80	.0000	.0000	.0000	.0000	.0000	.0159	.0143	.0125	.0110	.0096
90	.0000	.0000	.0000	.0000	.0000	.0161	.0141	.0122	.0107	.0093
100	.0000	.0000	.0000	.0000	.0000	.0162	.0138	.0120	.0104	.0091
120	.0000	.0000	.0000	.0000	.0000	.0159	.0134	.0115	.0100	.0086
140	.0000	.0000	.0000	.0000	.0000	.0154	.0131	.0111	.0095	.0082
170	.0000	.0000	.0000	.0000	.0000	.0148	.0127	.0105	.0090	.0077
200	.0000	.0000	.0000	.0000	.0000	.0141	.0123	.0100	.0085	.0072
230	.0000	.0000	.0000	.0000	.0000	.0135	.0118	.0096	.0080	.0068
260	.0000	.0000	.0000	.0000	.0000	.0130	.0114	.0093	.0076	.0064
300	.0000	.0000	.0000	.0000	.0000	.0124	.0109	.0088	.0071	.0060
350	.0000	.0000	.0000	.0000	.0000	.0117	.0103	.0084	.0065	.0055
400	.0000	.0000	.0000	.0000	.0000	.0111	.0098	.0079	.0061	.0051
450	.0000	.0000	.0000	.0000	.0000	.0105	.0093	.0075	.0057	.0047
500	.0000	.0000	.0000	.0000	.0000	.0101	.0089	.0071	.0053	.0044
600	.0000	.0000	.0000	.0000	.0000	.0093	.0081	.0064	.0047	.0040
700	.0000	.0000	.0000	.0000	0.0000	.0087	.0074	.0057	.0041	.0036
800	.0000	.0000	.0000	.0000	0.0000	.0081	.0068	.0051	.0037	.0033
900	.0000	.0000	.0000	.0000	0.0000	.0077	.0063	.0047	.0034	.0030
1000	.0000	.0000	.0000	.0000	0.0000	.0072	.0059	.0043	.0031	.0028
1200	.0000	0.0000	0.0000	0.0000	0.0000	.0065	.0052	.0037	.0027	.0025
1400	0.0000	0.0000	0.0000	0.0000	0.0000	.0058	.0046	.0033	.0025	.0022
1700	0.0000	0.0000	0.0000	0.0000	0.0000	.0050	.0040	.0029	.0022	.0019
2000	0.0000	0.0000	0.0000	0.0000	0.0000	.0044	.0035	.0026	.0020	.0016
2300	0.0000	0.0000	0.0000	0.0000	0.0000	.0039	.0031	.0023	.0018	.0015
2600	0.0000	0.0000	0.0000	0.0000	0.0000	.0035	.0028	.0021	.0016	.0013
3000	0.0000	0.0000	0.0000	0.0000	0.0000	.0031	.0025	.0018	.0014	.0012
3500	0.0000	0.0000	0.0000	0.0000	0.0000	.0027	.0021	.0016	.0012	.0010
4000	0.0000	0.0000	0.0000	0.0000	0.0000	.0023	.0019	.0014	.0011	.0009
4500	0.0000	0.0000	0.0000	0.0000	0.0000	.0021	.0017	.0013	.0010	.0008
5000	0.0000	0.0000	0.0000	0.0000	0.0000	.0019	.0015	.0011	.0009	.0007
6000	0.0000	0.0000	0.0000	0.0000	0.0000	.0015	.0013	.0010	.0007	.0006
7000	0.0000	0.0000	0.0000	0.0000	0.0000	.0013	.0011	.0008	.0006	.0005
8000	0.0000	0.0000	0.0000	0.0000	0.0000	.0011	.0010	.0007	.0005	.0004
9000	0.0000	0.0000	0.0000	0.0000	0.0000	.0010	.0008	.0006	.0005	.0004
10000	0.0000	0.0000	0.0000	0.0000	0.0000	.0010	.0008	.0006	.0004	.0003

TABLE 20-10

AG ON BE, M2/M1=0.0833

K/KL=	IONIZATION DEFICIENCY					SPUTTERING YIELD ALPHA				
E,KEV	0.65	0.80	1.0	1.2	1.45	0.65	0.80	1.0	1.2	1.45
1.0	.0015	.0020	.0024	.0027	.0031	.1198	.1155	.1115	.1077	.1032
1.2	.0015	.0020	.0024	.0028	.0032	.1192	.1149	.1108	.1069	.1023
1.4	.0016	.0021	.0025	.0029	.0033	.1188	.1143	.1101	.1061	.1014
1.7	.0016	.0022	.0026	.0030	.0034	.1181	.1135	.1092	.1051	.1003
2.0	.0017	.0023	.0027	.0031	.0035	.1176	.1129	.1084	.1043	.0994
2.3	.0018	.0023	.0027	.0031	.0036	.1171	.1123	.1077	.1035	.0986
2.6	.0018	.0024	.0028	.0032	.0036	.1166	.1117	.1071	.1028	.0978
3.0	.0019	.0025	.0029	.0033	.0037	.1161	.1111	.1064	.1020	.0969
3.5	.0020	.0026	.0030	.0034	.0038	.1156	.1105	.1057	.1012	.0960
4.0	.0020	.0027	.0031	.0035	.0039	.1152	.1099	.1050	.1004	.0951
4.5	.0021	.0027	.0032	.0036	.0040	.1148	.1094	.1044	.0998	.0944
5.0	.0022	.0028	.0033	.0037	.0041	.1144	.1089	.1039	.0992	.0937
6.0	.0023	.0030	.0034	.0039	.0043	.1140	.1083	.1030	.0982	.0926
7.0	.0024	.0031	.0036	.0040	.0044	.1136	.1077	.1023	.0974	.0917
8.0	.0025	.0032	.0037	.0041	.0046	.1133	.1072	.1017	.0966	.0908
9.0	.0026	.0033	.0038	.0043	.0047	.1130	.1067	.1011	.0959	.0900
10	.0027	.0034	.0040	.0044	.0048	.1127	.1063	.1005	.0953	.0893
12	.0028	.0036	.0042	.0046	.0050	.1122	.1053	.0994	.0939	.0878
14	.0030	.0038	.0044	.0048	.0052	.1116	.1045	.0983	.0927	.0865
17	.0031	.0040	.0046	.0050	.0055	.1108	.1034	.0970	.0912	.0848
20	.0033	.0042	.0048	.0053	.0057	.1099	.1025	.0959	.0900	.0835
23	.0035	.0044	.0050	.0054	.0059	.1081	.1019	.0952	.0892	.0825
26	.0036	.0045	.0051	.0056	.0060	.1064	.1014	.0945	.0885	.0818
30	.0038	.0047	.0053	.0057	.0062	.1045	.1009	.0939	.0877	.0809
35	.0039	.0049	.0055	.0059	.0063	.1025	.1004	.0932	.0869	.0800
40	.0041	.0051	.0056	.0061	.0064	.1010	.1000	.0926	.0863	.0793
45	.0042	.0052	.0057	.0062	.0065	.1000	.0996	.0921	.0857	.0786
50	.0044	.0053	.0058	.0063	.0066	.0994	.0993	.0916	.0851	.0781
60	.0047	.0055	.0060	.0064	.0067	.1009	.0985	.0907	.0840	.0772
70	.0049	.0057	.0061	.0064	.0068	.1025	.0979	.0900	.0831	.0764
80	.0050	.0058	.0062	.0065	.0069	.1039	.0975	.0894	.0823	.0757
90	.0051	.0059	.0063	.0065	.0069	.1052	.0971	.0889	.0816	.0751
100	.0052	.0060	.0064	.0066	.0069	.1061	.0968	.0884	.0810	.0746
120	.0050	.0062	.0065	.0067	.0070	.1064	.0968	.0877	.0800	.0734
140	.0048	.0063	.0066	.0068	.0070	.1063	.0968	.0871	.0792	.0724
170	.0046	.0064	.0066	.0069	.0070	.1058	.0969	.0865	.0783	.0713
200	.0044	.0064	.0066	.0069	.0069	.1051	.0970	.0861	.0778	.0707
230	.0045	.0064	.0066	.0068	.0068	.1043	.0966	.0856	.0775	.0707
260	.0047	.0064	.0066	.0066	.0066	.1035	.0962	.0852	.0773	.0709
300	.0048	.0064	.0065	.0064	.0064	.1026	.0957	.0849	.0773	.0713
350	.0050	.0063	.0064	.0062	.0061	.1017	.0954	.0847	.0774	.0720
400	.0052	.0062	.0062	.0059	.0058	.1011	.0952	.0847	.0777	.0727
450	.0053	.0061	.0060	.0056	.0055	.1006	.0951	.0850	.0780	.0734
500	.0054	.0060	.0059	.0053	.0052	.1004	.0952	.0853	.0784	.0742
600	.0053	.0057	.0054	.0047	.0046	.1007	.0963	.0869	.0795	.0758
700	.0052	.0054	.0049	.0041	.0041	.1014	.0975	.0885	.0806	.0771
800	.0050	.0052	.0045	.0036	.0036	.1022	.0987	.0899	.0815	.0782
900	.0049	.0050	.0042	.0033	.0033	.1032	.0997	.0912	.0824	.0791
1000	.0048	.0048	.0039	.0029	.0030	.1042	.1008	.0922	.0831	.0798
1200	.0045	.0044	.0036	.0027	.0025	.1070	.1026	.0936	.0843	.0800
1400	.0043	.0041	.0033	.0026	.0023	.1096	.1042	.0946	.0852	.0799
1700	.0040	.0037	.0030	.0026	.0021	.1129	.1061	.0957	.0862	.0796
2000	.0037	.0033	.0027	.0025	.0022	.1154	.1074	.0965	.0869	.0793
2300	.0032	.0029	.0024	.0023	.0030	.1165	.1081	.0969	.0873	.0795
2600	.0027	.0025	.0022	.0020	.0038	.1171	.1085	.0972	.0877	.0796
3000	.0021	.0021	.0018	.0018	.0047	.1176	.1088	.0975	.0880	.0798
3500	.0016	.0016	.0015	.0014	.0057	.1178	.1091	.0977	.0882	.0800
4000	.0013	.0013	.0012	.0012	.0063	.1179	.1092	.0979	.0883	.0802
4500	.0012	.0010	.0010	.0009	.0067	.1179	.1092	.0979	.0884	.0804
5000	.0013	.0007	.0008	.0007	.0068	.1178	.1092	.0980	.0884	.0805
6000	.0017	.0003	.0005	.0004	.0064	.1177	.1092	.0980	.0884	.0806
7000	.0025	.0001	.0003	.0002	.0054	.1176	.1092	.0980	.0883	.0806
8000	.0035	.0000	.0001	.0001	.0040	.1175	.1092	.0981	.0881	.0805
9000	.0048	0.0000	.0000	.0000	.0021	.1176	.1093	.0981	.0880	.0804
10000	.0061	0.0000	0.0000	0.0000	0.0000	.1178	.1094	.0981	.0878	.0802

AG ON BE, M2/M1=0.0833

10000 KEV

1000 KEV

100 KEV

10 KEV

1 KEV

E,KEV	FRACTIONAL DEPOSITED ENERGY					MEAN RANGE,MICROGRAM/SQ.CM.				
K/KL=	0.6	0.7	0.85	1.0	1.25	0.6	0.7	0.85	1.0	1.25
2.0	.8788	.8605	.8340	.8086	.7684	.5931	.5914	.5887	.5861	.5818
2.3	.8761	.8575	.8305	.8046	.7638	.6549	.6529	.6499	.6469	.6419
2.6	.8737	.8547	.8273	.8010	.7595	.7119	.7096	.7062	.7029	.6974
3.0	.8708	.8514	.8234	.7966	.7545	.7827	.7802	.7764	.7726	.7664
3.5	.8675	.8477	.8191	.7918	.7489	.8657	.8628	.8585	.8543	.8473
4.0	.8646	.8444	.8153	.7876	.7440	.9443	.9411	.9364	.9317	.9239
4.5	.8619	.8414	.8119	.7837	.7395	1.020	1.016	1.011	1.006	.9974
5.0	.8595	.8387	.8087	.7802	.7355	1.093	1.089	1.083	1.078	1.068
6.0	.8552	.8339	.8031	.7739	.7282	1.233	1.228	1.222	1.215	1.205
7.0	.8514	.8296	.7982	.7684	.7220	1.367	1.362	1.354	1.347	1.334
8.0	.8481	.8258	.7939	.7636	.7164	1.495	1.489	1.481	1.473	1.459
9.0	.8450	.8224	.7900	.7592	.7114	1.619	1.613	1.603	1.594	1.579
10	.8423	.8193	.7864	.7552	.7069	1.739	1.732	1.722	1.712	1.695
12	.8375	.8139	.7802	.7483	.6990	1.963	1.955	1.943	1.931	1.912
14	.8333	.8092	.7748	.7423	.6921	2.177	2.168	2.154	2.141	2.119
17	.8278	.8030	.7676	.7343	.6831	2.484	2.473	2.457	2.442	2.416
20	.8228	.7974	.7613	.7273	.6752	2.777	2.765	2.746	2.728	2.699
23	.8182	.7923	.7554	.7209	.6679	3.047	3.033	3.013	2.992	2.959
26	.8139	.7875	.7501	.7150	.6613	3.308	3.293	3.270	3.248	3.211
30	.8087	.7817	.7435	.7078	.6534	3.647	3.630	3.604	3.578	3.537
35	.8027	.7750	.7360	.6997	.6444	4.057	4.037	4.007	3.978	3.931
40	.7971	.7690	.7292	.6923	.6363	4.454	4.432	4.398	4.366	4.312
45	.7920	.7633	.7230	.6856	.6289	4.841	4.815	4.778	4.742	4.682
50	.7872	.7581	.7172	.6793	.6221	5.217	5.189	5.148	5.107	5.042
60	.7785	.7485	.7066	.6679	.6098	5.913	5.880	5.832	5.785	5.708
70	.7706	.7400	.6972	.6578	.5989	6.591	6.553	6.498	6.444	6.355
80	.7634	.7321	.6886	.6487	.5892	7.255	7.212	7.149	7.088	6.987
90	.7566	.7248	.6806	.6403	.5803	7.906	7.858	7.788	7.718	7.606
100	.7503	.7180	.6733	.6325	.5721	8.546	8.492	8.414	8.337	8.211
120	.7387	.7055	.6598	.6184	.5573	9.724	9.662	9.569	9.479	9.332
140	.7282	.6943	.6478	.6058	.5443	10.89	10.82	10.71	10.61	10.44
170	.7141	.6793	.6319	.5893	.5273	12.63	12.54	12.41	12.28	12.07
200	.7015	.6660	.6179	.5749	.5125	14.36	14.25	14.09	13.94	13.69
230	.6898	.6535	.6052	.5619	.4994	15.89	15.77	15.59	15.42	15.13
260	.6792	.6421	.5937	.5502	.4877	17.43	17.30	17.10	16.90	16.59
300	.6663	.6285	.5799	.5362	.4737	19.53	19.37	19.13	18.91	18.54
350	.6519	.6135	.5646	.5208	.4584	22.20	22.01	21.72	21.44	21.00
400	.6389	.6003	.5510	.5071	.4449	24.92	24.68	24.34	24.01	23.48
450	.6271	.5886	.5387	.4949	.4329	27.67	27.39	26.99	26.60	25.97
500	.6163	.5782	.5276	.4838	.4221	30.44	30.11	29.64	29.19	28.47
600	.5970	.5620	.5079	.4644	.4031	35.45	35.06	34.49	33.94	33.06
700	.5802	.5479	.4910	.4478	.3870	40.72	40.24	39.54	38.86	37.79
800	.5654	.5352	.4762	.4333	.3730	46.20	45.61	44.75	43.93	42.62
900	.5521	.5235	.4631	.4204	.3608	51.85	51.13	50.09	49.10	47.54
1000	.5401	.5125	.4513	.4089	.3499	57.62	56.76	55.52	54.34	52.50
1200	.5190	.4899	.4310	.3891	.3316	68.26	67.19	65.65	64.18	61.88
1400	.5009	.4698	.4138	.3724	.3163	79.58	78.23	76.28	74.43	71.54
1700	.4778	.4440	.3922	.3515	.2973	97.68	95.75	93.00	90.42	86.46
2000	.4584	.4222	.3741	.3340	.2815	116.8	114.1	110.4	106.9	101.7
2300	.4416	.4053	.3587	.3191	.2679	134.4	131.3	126.8	122.7	116.4
2600	.4270	.3908	.3451	.3061	.2561	152.9	149.1	143.8	138.8	131.2
3000	.4098	.3742	.3294	.2911	.2425	178.7	173.9	167.0	160.7	151.2
3500	.3912	.3565	.3126	.2751	.2280	212.5	206.0	196.9	188.7	176.5
4000	.3752	.3413	.2981	.2613	.2156	247.6	239.1	227.4	217.0	201.8
4500	.3610	.3281	.2854	.2494	.2049	283.5	272.8	258.3	245.5	227.1
5000	.3484	.3163	.2741	.2388	.1955	319.9	306.8	289.3	274.0	252.2
6000	.3264	.2953	.2546	.2207	.1797	391.1	373.1	350.1	329.8	301.2
7000	.3079	.2777	.2384	.2058	.1667	463.9	440.4	410.9	385.2	349.4
8000	.2920	.2627	.2246	.1931	.1557	537.5	508.2	471.5	440.0	396.7
9000	.2781	.2496	.2126	.1822	.1464	611.4	576.0	531.6	494.0	443.0
10000	.2658	.2380	.2021	.1727	.1383	685.3	643.7	591.0	547.2	488.4
12000	.2447	.2183	.1843	.1568	.1248	831.9	777.5	707.4	650.9	576.1
14000	.2273	.2021	.1699	.1438	.1140	976.1	908.7	820.4	750.9	660.1
17000	.2057	.1824	.1524	.1284	.1012	1186.	1100.	983.3	894.4	779.7
20000	.1882	.1665	.1386	.1162	.0912	1389.	1284.	1139.	1031.	892.5

	MEAN DAMAGE DEPTH,MICROGRAM/SQ.CM.					MEAN IONZN.DEPTH,MICROGRAM/SQ.CM.				
K/KL=	0.6	0.7	0.85	1.0	1.25	0.6	0.7	0.85	1.0	1.25
E,KEV										
2.0	.3119	.3108	.3092	.3076	.3050	.2861	.2849	.2830	.2812	.2782
2.3	.3456	.3443	.3424	.3405	.3375	.3170	.3155	.3133	.3112	.3077
2.6	.3762	.3748	.3727	.3706	.3672	.3451	.3434	.3410	.3385	.3346
3.0	.4141	.4125	.4101	.4077	.4039	.3798	.3779	.3751	.3724	.3680
3.5	.4583	.4565	.4538	.4511	.4468	.4202	.4181	.4150	.4119	.4069
4.0	.5002	.4982	.4952	.4923	.4875	.4586	.4562	.4528	.4494	.4439
4.5	.5406	.5384	.5351	.5319	.5267	.4955	.4929	.4891	.4854	.4794
5.0	.5798	.5774	.5738	.5703	.5646	.5314	.5286	.5244	.5204	.5138
6.0	.6556	.6528	.6486	.6445	.6378	.6007	.5974	.5926	.5878	.5801
7.0	.7286	.7253	.7205	.7158	.7081	.6674	.6636	.6581	.6526	.6438
8.0	.7993	.7955	.7900	.7847	.7761	.7321	.7278	.7215	.7153	.7053
9.0	.8679	.8637	.8576	.8516	.8419	.7949	.7901	.7830	.7761	.7649
10	.9348	.9301	.9233	.9166	.9059	.8561	.8507	.8428	.8351	.8228
12	1.059	1.054	1.046	1.038	1.025	.9699	.9635	.9543	.9452	.9307
14	1.180	1.174	1.164	1.155	1.141	1.080	1.073	1.062	1.052	1.035
17	1.356	1.348	1.337	1.325	1.308	1.241	1.232	1.219	1.206	1.186
20	1.527	1.517	1.503	1.490	1.469	1.398	1.387	1.371	1.356	1.331
23	1.683	1.672	1.656	1.641	1.617	1.540	1.528	1.510	1.493	1.465
26	1.836	1.824	1.806	1.789	1.762	1.681	1.667	1.647	1.627	1.596
30	2.038	2.024	2.003	1.983	1.952	1.866	1.850	1.826	1.803	1.767
35	2.287	2.270	2.245	2.222	2.184	2.096	2.076	2.047	2.020	1.976
40	2.533	2.513	2.483	2.455	2.412	2.323	2.299	2.265	2.232	2.182
45	2.775	2.751	2.717	2.685	2.635	2.547	2.519	2.479	2.441	2.383
50	3.014	2.987	2.947	2.910	2.853	2.769	2.737	2.690	2.647	2.580
60	3.458	3.424	3.376	3.331	3.261	3.181	3.141	3.083	3.030	2.948
70	3.900	3.859	3.800	3.745	3.662	3.594	3.544	3.474	3.409	3.310
80	4.340	4.290	4.220	4.156	4.057	4.006	3.945	3.861	3.784	3.667
90	4.778	4.719	4.636	4.561	4.447	4.418	4.345	4.245	4.155	4.019
100	5.212	5.143	5.048	4.961	4.832	4.828	4.743	4.626	4.521	4.366
120	6.025	5.938	5.818	5.710	5.550	5.600	5.491	5.343	5.211	5.017
140	6.840	6.732	6.586	6.455	6.263	6.377	6.241	6.057	5.895	5.661
170	8.066	7.924	7.735	7.568	7.324	7.550	7.367	7.126	6.916	6.616
200	9.292	9.114	8.879	8.673	8.376	8.725	8.491	8.188	7.927	7.559
230	10.40	10.19	9.910	9.670	9.324	9.810	9.524	9.163	8.853	8.420
250	11.53	11.27	10.95	10.67	10.27	10.90	10.56	10.14	9.777	9.277
300	13.04	12.73	12.35	12.02	11.55	12.37	11.95	11.45	11.01	10.42
350	14.96	14.58	14.12	13.72	13.16	14.22	13.71	13.08	12.56	11.85
400	16.90	16.45	15.90	15.44	14.79	16.08	15.47	14.73	14.11	13.28
450	18.85	18.34	17.70	17.17	16.42	17.93	17.23	16.36	15.65	14.70
500	20.80	20.23	19.50	18.89	18.05	19.78	18.98	17.99	17.19	16.11
600	24.35	23.68	22.80	22.05	21.05	23.21	22.23	20.99	20.03	18.73
700	28.03	27.26	26.20	25.30	24.13	26.69	25.54	24.04	22.91	21.38
800	31.82	30.93	29.69	28.64	27.28	30.22	28.89	27.13	25.81	24.05
900	35.69	34.68	33.25	32.03	30.47	33.78	32.26	30.25	28.74	26.73
1000	39.62	38.48	36.85	35.46	33.69	37.35	35.65	33.38	31.66	29.41
1200	46.83	45.46	43.53	41.87	39.71	43.95	41.94	39.27	37.16	34.51
1400	54.44	52.80	50.51	48.53	45.93	50.75	48.40	45.28	42.76	39.65
1700	66.51	64.41	61.48	58.94	55.57	61.26	58.36	54.52	51.31	47.45
2000	79.19	76.55	72.89	69.71	65.46	72.05	68.55	63.92	59.98	55.28
2300	90.76	87.73	83.53	79.87	74.88	82.15	78.17	72.86	68.33	62.81
2600	102.9	99.37	94.53	90.29	84.47	92.48	87.96	81.90	76.73	70.34
3000	119.8	115.6	109.7	104.6	97.49	106.6	101.3	94.10	88.00	80.34
3500	142.0	136.7	129.3	122.9	114.0	124.6	118.2	109.5	102.1	92.76
4000	165.1	158.5	149.4	141.5	130.7	143.0	135.3	125.0	116.2	105.0
4500	188.8	180.8	169.9	160.4	147.5	161.5	152.5	140.4	130.2	117.1
5000	212.9	203.4	190.5	179.3	164.2	180.1	169.7	155.8	144.0	129.1
6000	259.6	247.4	230.6	216.2	196.8	216.6	203.4	185.9	171.1	152.4
7000	307.8	292.4	271.3	253.1	229.1	253.0	236.9	215.6	197.5	174.9
8000	356.8	337.9	312.1	290.0	261.1	289.2	270.0	244.6	223.2	196.6
9000	406.3	383.7	352.8	326.6	292.7	324.9	302.5	273.0	248.2	217.5
10000	456.1	429.4	393.3	362.9	323.7	360.2	334.5	300.8	272.4	237.8
12000	555.4	520.3	473.4	434.0	384.3	429.3	396.8	354.4	319.1	276.4
14000	653.7	609.7	551.6	503.2	442.8	496.1	456.9	405.7	363.3	312.8
17000	798.0	740.3	665.3	603.1	526.8	592.2	542.9	478.6	425.8	363.8
20000	938.0	866.5	774.4	698.6	606.6	683.6	624.4	547.2	484.3	411.2

TABLE 21- 3

AU ON C, M2/M1=0.061

	RELATIVE RANGE STRAGGLING					RELATIVE DAMAGE STRAGGLING				
K/KL =	0.6	0.7	0.85	1.0	1.25	0.6	0.7	0.85	1.0	1.25
E,KEV										
2.0	.0306	.0305	.0303	.0301	.0298	.2943	.2942	.2941	.2940	.2938
2.3	.0306	.0305	.0303	.0301	.0298	.2940	.2939	.2938	.2936	.2935
2.6	.0306	.0304	.0302	.0301	.0297	.2937	.2936	.2934	.2933	.2931
3.0	.0306	.0304	.0302	.0300	.0297	.2933	.2932	.2930	.2929	.2927
3.5	.0305	.0304	.0302	.0300	.0297	.2928	.2927	.2925	.2924	.2922
4.0	.0305	.0304	.0302	.0300	.0296	.2923	.2922	.2920	.2919	.2917
4.5	.0305	.0304	.0302	.0299	.0296	.2919	.2917	.2916	.2914	.2912
5.0	.0305	.0303	.0301	.0299	.0296	.2914	.2913	.2911	.2910	.2907
6.0	.0305	.0303	.0301	.0299	.0295	.2907	.2905	.2903	.2901	.2899
7.0	.0305	.0303	.0301	.0299	.0295	.2900	.2898	.2896	.2894	.2891
8.0	.0304	.0303	.0301	.0298	.0295	.2893	.2892	.2889	.2887	.2884
9.0	.0304	.0303	.0300	.0298	.0294	.2888	.2886	.2883	.2880	.2877
10	.0304	.0303	.0300	.0298	.0294	.2883	.2880	.2877	.2874	.2871
12	.0304	.0302	.0300	.0297	.0293	.2873	.2870	.2866	.2863	.2858
14	.0304	.0302	.0300	.0297	.0293	.2864	.2861	.2856	.2853	.2848
17	.0304	.0302	.0299	.0297	.0292	.2855	.2851	.2845	.2840	.2834
20	.0304	.0302	.0299	.0296	.0292	.2850	.2844	.2837	.2831	.2823
23	.0303	.0301	.0299	.0296	.0291	.2846	.2839	.2830	.2823	.2813
26	.0303	.0301	.0298	.0296	.0291	.2844	.2836	.2826	.2817	.2806
30	.0303	.0301	.0298	.0295	.0291	.2844	.2834	.2822	.2811	.2797
35	.0303	.0301	.0298	.0295	.0290	.2848	.2835	.2819	.2806	.2789
40	.0303	.0301	.0298	.0295	.0290	.2854	.2839	.2819	.2803	.2783
45	.0303	.0301	.0298	.0294	.0289	.2862	.2844	.2821	.2802	.2779
50	.0303	.0301	.0297	.0294	.0289	.2872	.2851	.2824	.2802	.2775
60	.0303	.0301	.0297	.0294	.0288	.2899	.2871	.2836	.2808	.2774
70	.0303	.0300	.0297	.0293	.0288	.2927	.2892	.2849	.2815	.2773
80	.0303	.0300	.0297	.0293	.0287	.2955	.2913	.2862	.2822	.2774
90	.0303	.0300	.0296	.0293	.0287	.2981	.2933	.2875	.2829	.2774
100	.0303	.0300	.0296	.0292	.0286	.3007	.2953	.2887	.2836	.2775
120	.0303	.0300	.0296	.0292	.0285	.3062	.2994	.2912	.2850	.2777
140	.0303	.0300	.0296	.0291	.0285	.3111	.3029	.2934	.2862	.2778
170	.0303	.0300	.0295	.0291	.0283	.3170	.3072	.2959	.2874	.2777
200	.0303	.0299	.0294	.0290	.0282	.3217	.3104	.2976	.2881	.2773
230	.0303	.0299	.0294	.0289	.0281	.3251	.3125	.2983	.2881	.2766
260	.0303	.0299	.0294	.0288	.0280	.3277	.3139	.2986	.2878	.2757
300	.0302	.0298	.0293	.0287	.0279	.3300	.3150	.2985	.2871	.2744
350	.0302	.0298	.0292	.0286	.0277	.3316	.3154	.2979	.2859	.2727
400	.0301	.0297	.0290	.0284	.0275	.3322	.3151	.2968	.2844	.2709
450	.0301	.0296	.0289	.0283	.0273	.3319	.3142	.2955	.2828	.2692
500	.0300	.0295	.0288	.0281	.0271	.3309	.3129	.2939	.2811	.2674
600	.0298	.0293	.0285	.0278	.0267	.3264	.3087	.2900	.2769	.2637
700	.0296	.0290	.0282	.0274	.0263	.3212	.3041	.2860	.2730	.2603
800	.0294	.0288	.0279	.0271	.0258	.3159	.2995	.2822	.2693	.2572
900	.0292	.0285	.0276	.0267	.0254	.3106	.2950	.2784	.2659	.2544
1000	.0289	.0282	.0273	.0264	.0250	.3055	.2907	.2749	.2627	.2519
1200	.0284	.0277	.0266	.0256	.0242	.2953	.2823	.2680	.2571	.2477
1400	.0279	.0271	.0259	.0249	.0234	.2862	.2748	.2620	.2523	.2441
1700	.0271	.0262	.0250	.0239	.0223	.2746	.2651	.2543	.2462	.2396
2000	.0263	.0253	.0241	.0229	.0213	.2649	.2572	.2479	.2412	.2359
2300	.0254	.0245	.0231	.0219	.0202	.2581	.2515	.2434	.2375	.2329
2600	.0246	.0236	.0222	.0210	.0193	.2525	.2467	.2395	.2344	.2302
3000	.0236	.0226	.0212	.0199	.0182	.2464	.2414	.2353	.2309	.2271
3500	.0225	.0214	.0199	.0187	.0169	.2402	.2361	.2310	.2272	.2238
4000	.0214	.0203	.0188	.0175	.0158	.2353	.2318	.2274	.2241	.2209
4500	.0204	.0193	.0178	.0165	.0148	.2312	.2282	.2243	.2214	.2183
5000	.0195	.0183	.0169	.0156	.0139	.2278	.2251	.2216	.2189	.2160
6000	.0178	.0167	.0152	.0140	.0124	.2226	.2202	.2171	.2146	.2116
7000	.0164	.0152	.0138	.0127	.0112	.2187	.2163	.2134	.2109	.2079
8000	.0151	.0140	.0126	.0116	.0101	.2154	.2131	.2102	.2077	.2046
9000	.0140	.0129	.0116	.0106	.0092	.2126	.2103	.2074	.2049	.2017
10000	.0130	.0120	.0107	.0097	.0085	.2102	.2079	.2049	.2023	.1990
12000	.0114	.0104	.0093	.0084	.0072	.2061	.2036	.2006	.1978	.1943
14000	.0101	.0092	.0081	.0073	.0063	.2026	.2000	.1968	.1939	.1903
17000	.0086	.0078	.0069	.0062	.0052	.1981	.1953	.1919	.1890	.1852
20000	.0075	.0068	.0059	.0053	.0044	.1940	.1912	.1877	.1847	.1809

K/KL=	RELATIVE IONIZATION STRAGGLING					RELATIVE TRANSVERSE RANGE STRAGGLING				
E,KEV	0.6	0.7	0.85	1.0	1.25	0.6	0.7	0.85	1.0	1.25
2.0	.3330	.3333	.3338	.3343	.3351	.4455	.4460	.4468	.4475	.4488
2.3	.3329	.3332	.3336	.3341	.3349	.4454	.4460	.4467	.4475	.4488
2.6	.3327	.3330	.3334	.3339	.3348	.4454	.4459	.4467	.4475	.4489
3.0	.3324	.3327	.3331	.3336	.3345	.4453	.4459	.4467	.4475	.4489
3.5	.3320	.3323	.3328	.3333	.3342	.4452	.4458	.4466	.4475	.4489
4.0	.3316	.3319	.3324	.3329	.3338	.4451	.4457	.4466	.4474	.4489
4.5	.3313	.3316	.3321	.3326	.3335	.4450	.4456	.4465	.4474	.4488
5.0	.3309	.3312	.3317	.3323	.3332	.4449	.4455	.4464	.4473	.4488
6.0	.3303	.3306	.3311	.3317	.3326	.4447	.4453	.4463	.4472	.4487
7.0	.3298	.3301	.3306	.3311	.3321	.4444	.4451	.4461	.4470	.4486
8.0	.3294	.3297	.3302	.3307	.3316	.4442	.4449	.4459	.4468	.4485
9.0	.3290	.3293	.3298	.3303	.3312	.4440	.4446	.4457	.4467	.4483
10	.3287	.3290	.3294	.3299	.3308	.4437	.4444	.4454	.4465	.4482
12	.3281	.3283	.3287	.3291	.3300	.4432	.4440	.4450	.4461	.4479
14	.3277	.3278	.3282	.3286	.3294	.4427	.4435	.4446	.4457	.4475
17	.3274	.3274	.3276	.3279	.3286	.4420	.4428	.4439	.4451	.4470
20	.3274	.3273	.3274	.3275	.3281	.4413	.4421	.4433	.4445	.4464
23	.3276	.3274	.3272	.3273	.3277	.4406	.4414	.4426	.4439	.4459
26	.3279	.3276	.3273	.3272	.3274	.4399	.4407	.4420	.4433	.4453
30	.3286	.3281	.3275	.3272	.3271	.4389	.4398	.4412	.4425	.4446
35	.3299	.3290	.3280	.3274	.3270	.4378	.4387	.4401	.4415	.4437
40	.3313	.3301	.3287	.3278	.3270	.4367	.4376	.4391	.4405	.4428
45	.3330	.3314	.3295	.3283	.3271	.4356	.4366	.4380	.4395	.4419
50	.3348	.3328	.3305	.3289	.3273	.4345	.4355	.4370	.4385	.4410
60	.3391	.3363	.3330	.3306	.3281	.4324	.4335	.4351	.4367	.4393
70	.3434	.3397	.3354	.3323	.3290	.4304	.4315	.4332	.4349	.4376
80	.3476	.3431	.3378	.3340	.3298	.4285	.4297	.4314	.4332	.4360
90	.3516	.3463	.3401	.3356	.3307	.4266	.4278	.4297	.4315	.4345
100	.3555	.3493	.3422	.3371	.3314	.4248	.4261	.4280	.4299	.4330
120	.3635	.3556	.3466	.3400	.3330	.4213	.4227	.4247	.4267	.4301
140	.3704	.3610	.3503	.3426	.3342	.4180	.4195	.4217	.4238	.4273
170	.3791	.3676	.3548	.3456	.3357	.4135	.4151	.4175	.4198	.4236
200	.3860	.3728	.3581	.3478	.3367	.4094	.4111	.4137	.4162	.4203
230	.3911	.3764	.3602	.3490	.3372	.4054	.4072	.4099	.4126	.4170
260	.3951	.3790	.3617	.3498	.3373	.4017	.4036	.4065	.4094	.4140
300	.3990	.3815	.3628	.3503	.3373	.3972	.3993	.4024	.4055	.4104
350	.4021	.3833	.3635	.3503	.3369	.3922	.3945	.3979	.4012	.4065
400	.4038	.3840	.3634	.3499	.3364	.3878	.3903	.3939	.3974	.4032
450	.4043	.3840	.3629	.3492	.3357	.3838	.3865	.3903	.3941	.4002
500	.4040	.3834	.3620	.3482	.3349	.3803	.3831	.3872	.3912	.3977
600	.3996	.3798	.3587	.3448	.3326	.3738	.3769	.3815	.3859	.3931
700	.3944	.3758	.3552	.3415	.3305	.3685	.3719	.3769	.3818	.3896
800	.3891	.3716	.3518	.3385	.3286	.3642	.3678	.3732	.3785	.3869
900	.3837	.3674	.3485	.3357	.3271	.3605	.3645	.3703	.3759	.3850
1000	.3785	.3633	.3454	.3333	.3258	.3575	.3618	.3679	.3739	.3835
1200	.3682	.3551	.3395	.3292	.3239	.3528	.3575	.3644	.3711	.3818
1400	.3593	.3479	.3346	.3261	.3228	.3496	.3548	.3624	.3697	.3814
1700	.3482	.3391	.3289	.3228	.3221	.3468	.3527	.3613	.3695	.3826
2000	.3395	.3324	.3249	.3209	.3222	.3458	.3524	.3618	.3709	.3852
2300	.3342	.3288	.3233	.3208	.3235	.3460	.3532	.3635	.3734	.3890
2600	.3303	.3263	.3225	.3213	.3251	.3471	.3549	.3661	.3767	.3934
3000	.3266	.3242	.3224	.3226	.3275	.3496	.3582	.3704	.3820	.4001
3500	.3239	.3231	.3232	.3247	.3307	.3539	.3633	.3768	.3895	.4091
4000	.3224	.3229	.3246	.3272	.3340	.3591	.3694	.3839	.3976	.4187
4500	.3220	.3235	.3264	.3299	.3372	.3649	.3760	.3916	.4061	.4285
5000	.3222	.3245	.3284	.3325	.3404	.3711	.3829	.3995	.4149	.4385
6000	.3249	.3282	.3331	.3381	.3463	.3845	.3976	.4161	.4331	.4589
7000	.3284	.3322	.3379	.3433	.3519	.3983	.4127	.4328	.4512	.4789
8000	.3321	.3363	.3425	.3482	.3569	.4122	.4279	.4494	.4690	.4983
9000	.3359	.3404	.3469	.3528	.3616	.4262	.4428	.4657	.4864	.5171
10000	.3396	.3443	.3510	.3571	.3659	.4400	.4576	.4816	.5032	.5351
12000	.3467	.3516	.3586	.3648	.3735	.4671	.4864	.5123	.5355	.5694
14000	.3531	.3582	.3652	.3714	.3801	.4933	.5139	.5416	.5661	.6015
17000	.3616	.3668	.3738	.3800	.3886	.5307	.5531	.5828	.6088	.6459
20000	.3688	.3740	.3810	.3872	.3957	.5661	.5899	.6212	.6484	.6866

TABLE 21- 5

AU ON C, M2/M1=0.061

	RELATIVE TRANSV. DAMAGE STRAGGLING					RELATIVE TRANSV. IONZN. STRAGGLING				
K/KL=	0.6	0.7	0.85	1.0	1.25	0.6	0.7	0.85	1.0	1.25
E,KEV										
2.0	.1778	.1757	.1728	.1699	.1654	.1596	.1575	.1545	.1516	.1469
2.3	.1785	.1764	.1733	.1704	.1657	.1603	.1581	.1550	.1520	.1472
2.6	.1792	.1770	.1739	.1709	.1661	.1610	.1588	.1555	.1524	.1475
3.0	.1801	.1779	.1746	.1715	.1665	.1619	.1596	.1562	.1530	.1479
3.5	.1813	.1789	.1755	.1723	.1671	.1630	.1606	.1571	.1538	.1485
4.0	.1824	.1800	.1765	.1731	.1678	.1642	.1617	.1580	.1546	.1491
4.5	.1835	.1810	.1774	.1739	.1684	.1653	.1627	.1590	.1554	.1497
5.0	.1847	.1821	.1783	.1748	.1691	.1664	.1637	.1599	.1562	.1504
6.0	.1869	.1842	.1802	.1764	.1705	.1686	.1658	.1617	.1578	.1517
7.0	.1891	.1862	.1821	.1781	.1718	.1708	.1679	.1636	.1595	.1530
8.0	.1913	.1883	.1839	.1797	.1732	.1730	.1699	.1654	.1611	.1544
9.0	.1935	.1903	.1857	.1813	.1746	.1751	.1719	.1672	.1627	.1557
10	.1956	.1922	.1875	.1829	.1759	.1772	.1738	.1689	.1642	.1570
12	.1997	.1961	.1910	.1861	.1786	.1814	.1777	.1724	.1674	.1596
14	.2037	.1998	.1943	.1891	.1811	.1853	.1814	.1757	.1703	.1621
17	.2092	.2050	.1990	.1934	.1847	.1909	.1866	.1804	.1746	.1656
20	.2144	.2098	.2034	.1973	.1880	.1962	.1914	.1848	.1785	.1688
23	.2195	.2146	.2076	.2011	.1912	.2014	.1963	.1891	.1824	.1721
26	.2243	.2190	.2116	.2047	.1941	.2063	.2009	.1932	.1860	.1751
30	.2301	.2244	.2165	.2091	.1978	.2123	.2065	.1982	.1905	.1789
35	.2367	.2306	.2220	.2140	.2018	.2191	.2128	.2039	.1956	.1830
40	.2426	.2361	.2269	.2183	.2054	.2252	.2184	.2089	.2001	.1867
45	.2479	.2410	.2313	.2223	.2087	.2307	.2235	.2135	.2041	.1901
50	.2528	.2455	.2353	.2259	.2116	.2356	.2282	.2176	.2078	.1931
60	.2617	.2538	.2427	.2324	.2169	.2447	.2366	.2251	.2145	.1985
70	.2692	.2608	.2489	.2379	.2213	.2522	.2436	.2314	.2201	.2030
80	.2755	.2666	.2540	.2424	.2249	.2585	.2495	.2367	.2247	.2067
90	.2808	.2715	.2584	.2462	.2279	.2639	.2545	.2411	.2286	.2098
100	.2854	.2757	.2621	.2494	.2303	.2685	.2587	.2449	.2319	.2123
120	.2923	.2821	.2677	.2542	.2339	.2754	.2653	.2508	.2371	.2163
140	.2975	.2869	.2718	.2576	.2363	.2807	.2702	.2551	.2408	.2190
170	.3030	.2919	.2759	.2609	.2382	.2862	.2754	.2596	.2445	.2214
200	.3066	.2950	.2783	.2625	.2388	.2900	.2789	.2624	.2466	.2224
230	.3089	.2972	.2795	.2629	.2383	.2924	.2809	.2638	.2474	.2227
260	.3103	.2984	.2798	.2626	.2371	.2939	.2821	.2644	.2475	.2222
300	.3111	.2989	.2793	.2613	.2348	.2949	.2827	.2643	.2467	.2207
350	.3108	.2983	.2776	.2587	.2312	.2951	.2824	.2632	.2447	.2179
400	.3097	.2968	.2751	.2555	.2272	.2945	.2813	.2613	.2421	.2143
450	.3078	.2945	.2721	.2518	.2228	.2933	.2796	.2589	.2390	.2102
500	.3056	.2918	.2686	.2478	.2183	.2917	.2776	.2562	.2356	.2057
600	.3001	.2846	.2603	.2386	.2078	.2883	.2732	.2504	.2279	.1937
700	.2939	.2769	.2517	.2295	.1977	.2840	.2681	.2441	.2201	.1823
800	.2873	.2691	.2432	.2206	.1884	.2792	.2624	.2375	.2123	.1719
900	.2805	.2613	.2350	.2121	.1797	.2740	.2565	.2307	.2046	.1626
1000	.2736	.2536	.2269	.2039	.1716	.2685	.2504	.2238	.1972	.1541
1200	.2587	.2381	.2110	.1880	.1572	.2563	.2365	.2086	.1820	.1414
1400	.2446	.2237	.1965	.1738	.1448	.2442	.2232	.1944	.1682	.1308
1700	.2251	.2044	.1774	.1554	.1291	.2270	.2047	.1752	.1500	.1177
2000	.2077	.1874	.1610	.1400	.1162	.2110	.1880	.1584	.1344	.1068
2300	.1914	.1722	.1473	.1275	.1058	.1948	.1725	.1441	.1214	.0966
2600	.1769	.1588	.1353	.1169	.0971	.1800	.1586	.1315	.1102	.0876
3000	.1599	.1432	.1217	.1050	.0873	.1623	.1423	.1170	.0975	.0775
3500	.1418	.1268	.1077	.0928	.0775	.1431	.1249	.1020	.0844	.0669
4000	.1266	.1132	.0961	.0829	.0695	.1268	.1102	.0895	.0737	.0583
4500	.1138	.1017	.0865	.0748	.0630	.1128	.0978	.0791	.0649	.0512
5000	.1029	.0921	.0785	.0680	.0575	.1009	.0873	.0703	.0575	.0453
6000	.0877	.0787	.0673	.0586	.0499	.0837	.0723	.0581	.0475	.0375
7000	.0762	.0686	.0591	.0517	.0443	.0705	.0610	.0490	.0401	.0318
8000	.0673	.0608	.0527	.0464	.0401	.0602	.0522	.0421	.0346	.0275
9000	.0603	.0547	.0477	.0422	.0367	.0520	.0452	.0366	.0302	.0242
10000	.0545	.0497	.0436	.0389	.0340	.0453	.0396	.0322	.0268	.0216
12000	.0460	.0423	.0376	.0338	.0298	.0355	.0313	.0258	.0217	.0176
14000	.0401	.0371	.0333	.0302	.0267	.0289	.0257	.0214	.0182	.0149
17000	.0344	.0319	.0288	.0263	.0234	.0229	.0204	.0172	.0147	.0122
20000	.0310	.0287	.0259	.0236	.0210	.0200	.0177	.0148	.0125	.0103

E, KEV	RANGE SKEWNESS K/KL= 0.6	0.7	0.85	1.0	1.25	DAMAGE SKEWNESS 0.6	0.7	0.85	1.0	1.25
2.0	.3143	.3123	.3092	.3061	.3011	.2875	.2859	.2837	.2816	.2782
2.3	.3140	.3119	.3087	.3056	.3005	.2865	.2848	.2824	.2801	.2765
2.6	.3137	.3115	.3083	.3051	.2999	.2854	.2836	.2810	.2786	.2749
3.0	.3133	.3111	.3078	.3046	.2992	.2839	.2820	.2793	.2767	.2728
3.5	.3130	.3107	.3073	.3040	.2985	.2819	.2799	.2771	.2745	.2705
4.0	.3126	.3103	.3068	.3034	.2978	.2801	.2780	.2751	.2724	.2683
4.5	.3123	.3099	.3064	.3029	.2972	.2784	.2763	.2732	.2704	.2662
5.0	.3120	.3095	.3060	.3024	.2966	.2768	.2746	.2715	.2686	.2642
6.0	.3114	.3089	.3052	.3016	.2956	.2742	.2717	.2684	.2652	.2605
7.0	.3109	.3084	.3046	.3008	.2946	.2720	.2693	.2656	.2622	.2571
8.0	.3105	.3079	.3040	.3001	.2938	.2702	.2672	.2632	.2595	.2540
9.0	.3101	.3074	.3034	.2995	.2931	.2688	.2655	.2610	.2570	.2510
10	.3097	.3070	.3029	.2989	.2924	.2676	.2640	.2591	.2547	.2482
12	.3090	.3062	.3020	.2979	.2911	.2643	.2601	.2544	.2494	.2421
14	.3083	.3054	.3011	.2969	.2899	.2626	.2576	.2510	.2452	.2370
17	.3075	.3045	.3000	.2956	.2884	.2626	.2563	.2480	.2408	.2308
20	.3067	.3036	.2990	.2944	.2870	.2653	.2574	.2471	.2382	.2262
23	.3059	.3027	.2980	.2934	.2858	.2697	.2602	.2478	.2373	.2232
26	.3053	.3020	.2972	.2924	.2846	.2757	.2644	.2499	.2376	.2212
30	.3044	.3010	.2961	.2911	.2831	.2859	.2720	.2542	.2394	.2197
35	.3034	.2999	.2948	.2897	.2814	.3011	.2837	.2615	.2433	.2194
40	.3024	.2988	.2935	.2883	.2798	.3184	.2972	.2704	.2486	.2203
45	.3014	.2978	.2923	.2870	.2783	.3370	.3119	.2804	.2549	.2221
50	.3005	.2968	.2912	.2857	.2768	.3567	.3276	.2912	.2619	.2246
60	.2989	.2949	.2891	.2834	.2741	.4042	.3661	.3189	.2812	.2339
70	.2973	.2932	.2872	.2812	.2716	.4469	.4031	.3455	.2998	.2429
80	.2957	.2915	.2852	.2791	.2691	.4930	.4379	.3704	.3171	.2513
90	.2942	.2898	.2834	.2770	.2668	.5333	.4704	.3935	.3331	.2589
100	.2927	.2882	.2815	.2750	.2645	.5711	.5007	.4149	.3478	.2658
120	.2899	.2851	.2781	.2713	.2602	.6442	.5593	.4558	.3758	.2785
140	.2871	.2821	.2748	.2677	.2561	.7066	.6088	.4898	.3985	.2880
170	.2830	.2777	.2699	.2623	.2500	.7826	.6682	.5298	.4240	.2972
200	.2790	.2734	.2651	.2571	.2442	.8412	.7130	.5590	.4411	.3014
230	.2752	.2693	.2606	.2522	.2387	.8834	.7422	.5780	.4489	.2991
260	.2714	.2652	.2561	.2473	.2332	.9151	.7629	.5907	.4521	.2943
300	.2664	.2598	.2503	.2410	.2261	.9449	.7808	.6001	.4512	.2853
350	.2601	.2532	.2430	.2332	.2175	.9671	.7917	.6029	.4441	.2715
400	.2540	.2466	.2359	.2255	.2090	.9770	.7935	.5986	.4326	.2558
450	.2479	.2402	.2289	.2180	.2008	.9778	.7888	.5891	.4180	.2392
500	.2419	.2338	.2220	.2107	.1927	.9716	.7793	.5757	.4013	.2220
600	.2303	.2215	.2087	.1964	.1771	.9369	.7467	.5327	.3578	.1836
700	.2190	.2095	.1958	.1827	.1622	.8925	.7071	.4868	.3142	.1471
800	.2081	.1979	.1833	.1695	.1478	.8433	.6638	.4407	.2721	.1129
900	.1974	.1866	.1712	.1567	.1341	.7917	.6188	.3956	.2320	.0812
1000	.1870	.1756	.1595	.1443	.1208	.7392	.5729	.3520	.1942	.0517
1200	.1667	.1543	.1368	.1203	.0951	.6216	.4686	.2656	.1226	-.0016
1400	.1476	.1342	.1154	.0979	.0713	.5127	.3723	.1887	.0600	-.0476
1700	.1206	.1060	.0856	.0668	.0384	.3694	.2460	.0903	-.0191	-.1057
2000	.0956	.0799	.0582	.0382	.0084	.2490	.1404	.0093	-.0835	-.1532
2300	.0717	.0550	.0320	.0111	-.0198	.1617	.0662	-.0478	-.1287	-.1886
2600	.0492	.0316	.0076	-.0140	-.0458	.0890	.0049	-.0949	-.1660	-.2184
3000	.0213	.0028	-.0224	-.0449	-.0776	.0092	-.0620	-.1464	-.2068	-.2518
3500	-.0106	-.0302	-.0565	-.0798	-.1134	-.0704	-.1286	-.1978	-.2478	-.2864
4000	-.0399	-.0603	-.0875	-.1114	-.1456	-.1336	-.1815	-.2391	-.2812	-.3154
4500	-.0669	-.0879	-.1158	-.1402	-.1747	-.1846	-.2245	-.2731	-.3091	-.3404
5000	-.0920	-.1135	-.1419	-.1666	-.2014	-.2263	-.2599	-.3017	-.3331	-.3623
6000	-.1377	-.1600	-.1890	-.2139	-.2486	-.2810	-.3094	-.3451	-.3726	-.4002
7000	-.1781	-.2008	-.2301	-.2550	-.2895	-.3207	-.3463	-.3788	-.4044	-.4317
8000	-.2143	-.2373	-.2666	-.2914	-.3255	-.3512	-.3753	-.4063	-.4311	-.4586
9000	-.2470	-.2701	-.2994	-.3240	-.3576	-.3759	-.3993	-.4296	-.4541	-.4821
10000	-.2768	-.2999	-.3291	-.3535	-.3865	-.3966	-.4198	-.4499	-.4745	-.5031
12000	-.3296	-.3525	-.3813	-.4050	-.4369	-.4308	-.4540	-.4844	-.5094	-.5392
14000	-.3752	-.3978	-.4259	-.4489	-.4798	-.4593	-.4828	-.5136	-.5390	-.5697
17000	-.4336	-.4555	-.4825	-.5044	-.5337	-.4969	-.5205	-.5514	-.5771	-.6084
20000	-.4832	-.5041	-.5300	-.5507	-.5786	-.5313	-.5544	-.5847	-.6100	-.6411

TABLE 21- 7

AU ON C, M2/M1=0.061

K/KL=	IONIZATION SKEWNESS					RANGE KURTOSIS				
E,KEV	0.6	0.7	0.85	1.0	1.25	0.6	0.7	0.85	1.0	1.25
2.0	.3973	.3968	.3961	.3955	.3948	3.107	3.104	3.100	3.096	3.090
2.3	.3972	.3965	.3956	.3949	.3940	3.106	3.103	3.099	3.096	3.089
2.6	.3967	.3959	.3949	.3941	.3931	3.106	3.103	3.099	3.095	3.089
3.0	.3957	.3949	.3938	.3929	.3918	3.105	3.102	3.098	3.094	3.088
3.5	.3943	.3934	.3922	.3913	.3901	3.105	3.102	3.098	3.093	3.087
4.0	.3929	.3920	.3907	.3897	.3885	3.104	3.101	3.097	3.093	3.086
4.5	.3916	.3906	.3893	.3883	.3870	3.104	3.101	3.096	3.092	3.085
5.0	.3905	.3894	.3880	.3869	.3855	3.103	3.100	3.096	3.091	3.084
6.0	.3887	.3874	.3858	.3845	.3829	3.103	3.099	3.095	3.090	3.083
7.0	.3875	.3860	.3841	.3826	.3806	3.102	3.099	3.094	3.089	3.082
8.0	.3869	.3851	.3828	.3810	.3787	3.101	3.098	3.093	3.088	3.081
9.0	.3867	.3846	.3819	.3797	.3770	3.101	3.097	3.092	3.088	3.080
10	.3868	.3843	.3812	.3786	.3755	3.100	3.097	3.092	3.087	3.079
12	.3868	.3837	.3797	.3765	.3725	3.099	3.096	3.090	3.085	3.077
14	.3882	.3843	.3794	.3754	.3704	3.098	3.095	3.089	3.084	3.076
17	.3927	.3874	.3807	.3752	.3684	3.097	3.093	3.088	3.082	3.074
20	.3996	.3925	.3837	.3765	.3676	3.096	3.092	3.086	3.081	3.072
23	.4080	.3991	.3880	.3790	.3678	3.095	3.091	3.085	3.079	3.070
26	.4178	.4069	.3932	.3823	.3686	3.094	3.090	3.084	3.078	3.069
30	.4323	.4185	.4014	.3876	.3706	3.092	3.088	3.082	3.076	3.067
35	.4523	.4347	.4128	.3954	.3739	3.091	3.087	3.080	3.074	3.065
40	.4735	.4520	.4253	.4041	.3780	3.089	3.085	3.079	3.073	3.063
45	.4956	.4700	.4384	.4134	.3827	3.088	3.084	3.077	3.071	3.061
50	.5181	.4885	.4520	.4231	.3877	3.087	3.082	3.076	3.069	3.059
60	.5708	.5319	.4842	.4467	.4009	3.084	3.079	3.073	3.066	3.055
70	.6201	.5726	.5144	.4688	.4133	3.082	3.077	3.070	3.063	3.052
80	.6657	.6101	.5423	.4891	.4247	3.079	3.074	3.067	3.060	3.049
90	.7076	.6447	.5678	.5078	.4351	3.077	3.072	3.065	3.057	3.046
100	.7461	.6763	.5912	.5247	.4445	3.075	3.070	3.062	3.055	3.043
120	.8172	.7350	.6346	.5562	.4618	3.071	3.065	3.057	3.050	3.038
140	.8767	.7839	.6704	.5819	.4755	3.067	3.061	3.053	3.045	3.033
170	.9481	.8420	.7124	.6113	.4904	3.061	3.055	3.047	3.038	3.026
200	1.002	.8857	.7430	.6321	.4999	3.056	3.049	3.041	3.032	3.019
230	1.040	.9148	.7623	.6443	.5039	3.050	3.044	3.035	3.026	3.013
260	1.067	.9357	.7752	.6516	.5050	3.045	3.038	3.029	3.020	3.006
300	1.093	.9541	.7851	.6558	.5029	3.039	3.032	3.022	3.013	2.999
350	1.112	.9660	.7890	.6544	.4963	3.031	3.024	3.014	3.004	2.990
400	1.120	.9692	.7862	.6478	.4867	3.023	3.016	3.006	2.996	2.982
450	1.121	.9660	.7786	.6376	.4749	3.017	3.009	2.998	2.989	2.974
500	1.116	.9582	.7675	.6248	.4617	3.010	3.002	2.992	2.982	2.968
600	1.091	.9275	.7313	.5857	.4238	2.997	2.989	2.978	2.968	2.954
700	1.057	.8910	.6926	.5462	.3880	2.985	2.977	2.966	2.956	2.943
800	1.018	.8519	.6536	.5082	.3555	2.975	2.967	2.956	2.946	2.933
900	.9765	.8120	.6155	.4725	.3264	2.965	2.957	2.947	2.937	2.924
1000	.9339	.7720	.5788	.4391	.3005	2.957	2.949	2.938	2.929	2.917
1200	.8348	.6844	.5050	.3773	.2607	2.940	2.933	2.923	2.915	2.905
1400	.7428	.6049	.4402	.3249	.2293	2.927	2.920	2.911	2.904	2.896
1700	.6222	.5024	.3591	.2616	.1935	2.911	2.905	2.898	2.893	2.887
2000	.5217	.4185	.2948	.2131	.1675	2.900	2.895	2.890	2.886	2.884
2300	.4507	.3611	.2543	.1849	.1520	2.891	2.887	2.884	2.883	2.884
2600	.3926	.3151	.2231	.1645	.1411	2.885	2.882	2.881	2.882	2.887
3000	.3305	.2668	.1919	.1453	.1315	2.880	2.879	2.881	2.885	2.894
3500	.2709	.2214	.1642	.1299	.1246	2.878	2.880	2.885	2.892	2.906
4000	.2259	.1880	.1452	.1207	.1215	2.879	2.884	2.893	2.903	2.921
4500	.1917	.1633	.1321	.1156	.1207	2.884	2.891	2.903	2.916	2.938
5000	.1658	.1451	.1234	.1131	.1215	2.891	2.900	2.915	2.931	2.956
6000	.1404	.1284	.1170	.1138	.1262	2.909	2.923	2.944	2.964	2.996
7000	.1274	.1213	.1169	.1182	.1329	2.932	2.950	2.975	3.000	3.038
8000	.1215	.1195	.1199	.1243	.1405	2.958	2.979	3.009	3.037	3.080
9000	.1200	.1209	.1247	.1313	.1483	2.986	3.010	3.044	3.075	3.121
10000	.1210	.1242	.1305	.1386	.1562	3.015	3.042	3.079	3.113	3.162
12000	.1270	.1332	.1428	.1531	.1714	3.077	3.108	3.151	3.188	3.243
14000	.1353	.1433	.1550	.1666	.1855	3.140	3.175	3.221	3.262	3.320
17000	.1481	.1577	.1715	.1845	.2045	3.235	3.274	3.325	3.369	3.431
20000	.1593	.1700	.1852	.1995	.2209	3.329	3.371	3.425	3.471	3.535

TABLE 21- 8

AU ON C, M2/M1=0.061

E,KEV	DAMAGE KURTOSIS					IONIZATION KURTOSIS				
K/KL=	0.6	0.7	0.85	1.0	1.25	0.6	0.7	0.85	1.0	1.25
2.0	2.482	2.477	2.471	2.464	2.455	2.574	2.570	2.564	2.558	2.550
2.3	2.484	2.479	2.471	2.464	2.454	2.576	2.571	2.565	2.559	2.550
2.6	2.485	2.479	2.471	2.464	2.453	2.577	2.572	2.565	2.558	2.549
3.0	2.485	2.479	2.470	2.463	2.451	2.577	2.572	2.564	2.557	2.547
3.5	2.484	2.478	2.469	2.461	2.449	2.577	2.571	2.563	2.556	2.545
4.0	2.483	2.477	2.468	2.460	2.448	2.576	2.570	2.562	2.554	2.543
4.5	2.483	2.476	2.467	2.459	2.446	2.575	2.569	2.560	2.553	2.542
5.0	2.482	2.476	2.466	2.458	2.445	2.575	2.569	2.560	2.552	2.540
6.0	2.483	2.476	2.465	2.456	2.442	2.576	2.569	2.559	2.550	2.538
7.0	2.485	2.477	2.466	2.456	2.441	2.579	2.571	2.560	2.550	2.536
8.0	2.488	2.479	2.467	2.456	2.440	2.583	2.574	2.561	2.551	2.535
9.0	2.492	2.482	2.468	2.456	2.439	2.589	2.578	2.564	2.552	2.535
10	2.497	2.486	2.471	2.458	2.439	2.595	2.583	2.567	2.553	2.535
12	2.505	2.492	2.475	2.460	2.439	2.605	2.591	2.572	2.556	2.535
14	2.517	2.502	2.481	2.464	2.440	2.620	2.603	2.580	2.561	2.536
17	2.541	2.520	2.494	2.472	2.443	2.649	2.626	2.597	2.573	2.541
20	2.569	2.543	2.510	2.483	2.447	2.685	2.656	2.618	2.587	2.548
23	2.595	2.563	2.524	2.491	2.449	2.726	2.689	2.643	2.605	2.557
26	2.625	2.587	2.539	2.501	2.451	2.771	2.726	2.670	2.625	2.568
30	2.671	2.623	2.564	2.517	2.457	2.836	2.779	2.709	2.653	2.584
35	2.736	2.675	2.600	2.541	2.468	2.921	2.849	2.760	2.691	2.605
40	2.809	2.733	2.641	2.569	2.481	3.010	2.922	2.814	2.730	2.627
45	2.886	2.795	2.684	2.599	2.496	3.102	2.996	2.869	2.770	2.650
50	2.968	2.860	2.731	2.632	2.514	3.194	3.071	2.924	2.810	2.673
60	3.168	3.022	2.850	2.720	2.567	3.408	3.245	3.051	2.903	2.728
70	3.359	3.176	2.962	2.802	2.617	3.605	3.406	3.168	2.989	2.778
80	3.538	3.320	3.066	2.878	2.662	3.786	3.552	3.275	3.067	2.822
90	3.705	3.453	3.162	2.946	2.701	3.950	3.685	3.372	3.136	2.862
100	3.861	3.576	3.249	3.008	2.736	4.100	3.806	3.460	3.199	2.898
120	4.154	3.804	3.405	3.116	2.791	4.372	4.023	3.618	3.311	2.959
140	4.404	3.997	3.535	3.203	2.833	4.595	4.201	3.747	3.401	3.007
170	4.709	4.229	3.689	3.301	2.877	4.857	4.410	3.897	3.503	3.059
200	4.947	4.407	3.804	3.371	2.905	5.051	4.564	4.007	3.576	3.094
230	5.129	4.541	3.889	3.413	2.916	5.171	4.667	4.080	3.623	3.114
260	5.269	4.643	3.952	3.439	2.920	5.255	4.738	4.129	3.653	3.124
300	5.405	4.738	4.007	3.456	2.915	5.324	4.799	4.168	3.673	3.126
350	5.512	4.810	4.043	3.456	2.901	5.362	4.833	4.184	3.674	3.115
400	5.568	4.843	4.051	3.440	2.879	5.364	4.834	4.175	3.656	3.093
450	5.584	4.846	4.038	3.412	2.852	5.339	4.812	4.147	3.624	3.064
500	5.571	4.827	4.010	3.376	2.823	5.296	4.772	4.105	3.582	3.028
600	5.424	4.697	3.881	3.262	2.745	5.136	4.613	3.951	3.439	2.920
700	5.244	4.545	3.743	3.147	2.670	4.963	4.444	3.794	3.297	2.815
800	5.051	4.384	3.604	3.038	2.602	4.790	4.279	3.642	3.163	2.719
900	4.855	4.223	3.471	2.936	2.539	4.623	4.120	3.501	3.040	2.632
1000	4.662	4.065	3.345	2.842	2.483	4.464	3.972	3.369	2.928	2.554
1200	4.255	3.735	3.107	2.676	2.390	4.173	3.707	3.144	2.741	2.430
1400	3.892	3.443	2.903	2.538	2.314	3.917	3.479	2.954	2.588	2.331
1700	3.435	3.078	2.654	2.376	2.228	3.593	3.195	2.724	2.407	2.216
2000	3.071	2.790	2.464	2.255	2.166	3.328	2.968	2.547	2.271	2.131
2300	2.856	2.626	2.362	2.196	2.135	3.128	2.809	2.436	2.195	2.085
2600	2.694	2.505	2.289	2.156	2.116	2.964	2.681	2.352	2.142	2.052
3000	2.533	2.388	2.223	2.124	2.102	2.785	2.545	2.267	2.091	2.022
3500	2.393	2.289	2.172	2.104	2.095	2.610	2.416	2.192	2.051	1.998
4000	2.299	2.225	2.143	2.096	2.096	2.474	2.317	2.137	2.024	1.983
4500	2.234	2.183	2.127	2.097	2.100	2.367	2.242	2.098	2.007	1.973
5000	2.191	2.157	2.120	2.101	2.107	2.282	2.182	2.067	1.995	1.967
6000	2.160	2.140	2.121	2.112	2.121	2.182	2.109	2.025	1.972	1.951
7000	2.155	2.143	2.131	2.128	2.138	2.118	2.062	1.998	1.958	1.942
8000	2.163	2.154	2.147	2.145	2.156	2.073	2.031	1.981	1.949	1.935
9000	2.176	2.169	2.163	2.163	2.174	2.042	2.009	1.969	1.943	1.931
10000	2.191	2.185	2.180	2.180	2.191	2.020	1.992	1.960	1.939	1.928
12000	2.218	2.213	2.210	2.211	2.223	1.989	1.971	1.948	1.934	1.925
14000	2.237	2.235	2.235	2.238	2.252	1.968	1.956	1.940	1.930	1.924
17000	2.248	2.252	2.260	2.270	2.290	1.945	1.938	1.929	1.924	1.923
20000	2.238	2.253	2.273	2.293	2.322	1.924	1.920	1.917	1.917	1.923

AU ON C, M2/M1=0.061

TABLE 21- 9

E,KEV	REFLECTION COEFFICIENT K/KL= 0.6	0.7	0.85	1.0	1.25	SPUTTERING EFFICIENCY 0.6	0.7	0.85	1.0	1.25
2.0	.0000	.0000	.0000	.0000	.0000	.0129	.0126	.0121	.0117	.0111
2.3	.0000	.0000	.0000	.0000	.0000	.0129	.0125	.0121	.0116	.0110
2.6	.0000	.0000	.0000	.0000	.0000	.0128	.0124	.0120	.0115	.0109
3.0	.0000	.0000	.0000	.0000	.0000	.0127	.0124	.0120	.0115	.0108
3.5	.0000	.0000	.0000	.0000	.0000	.0127	.0123	.0119	.0114	.0107
4.0	.0000	.0000	.0000	.0000	.0000	.0126	.0122	.0118	.0113	.0107
4.5	.0000	.0000	.0000	.0000	.0000	.0126	.0122	.0117	.0112	.0106
5.0	.0000	.0000	.0000	.0000	.0000	.0125	.0121	.0117	.0112	.0105
6.0	.0000	.0000	.0000	.0000	.0000	.0124	.0121	.0116	.0111	.0104
7.0	.0000	.0000	.0000	.0000	.0000	.0124	.0120	.0115	.0110	.0103
8.0	.0000	.0000	.0000	.0000	.0000	.0124	.0120	.0114	.0109	.0102
9.0	.0000	.0000	.0000	.0000	.0000	.0124	.0120	.0114	.0109	.0101
10	.0000	.0000	.0000	.0000	.0000	.0124	.0120	.0113	.0108	.0100
12	.0000	.0000	.0000	.0000	.0000	.0124	.0120	.0113	.0108	.0099
14	.0000	.0000	.0000	.0000	.0000	.0125	.0120	.0113	.0107	.0099
17	.0000	.0000	.0000	.0000	.0000	.0126	.0121	.0113	.0107	.0098
20	.0000	.0000	.0000	.0000	.0000	.0127	.0121	.0114	.0107	.0097
23	.0000	.0000	.0000	.0000	.0000	.0129	.0122	.0114	.0106	.0096
26	.0000	.0000	.0000	.0000	.0000	.0131	.0124	.0115	.0106	.0095
30	.0000	.0000	.0000	.0000	.0000	.0133	.0125	.0115	.0106	.0094
35	.0000	.0000	.0000	.0000	.0000	.0135	.0127	.0116	.0106	.0093
40	.0000	.0000	.0000	.0000	.0000	.0138	.0129	.0117	.0107	.0092
45	.0000	.0000	.0000	.0000	.0000	.0140	.0130	.0118	.0107	.0092
50	.0000	.0000	.0000	.0000	.0000	.0142	.0132	.0119	.0107	.0092
60	.0000	.0000	.0000	.0000	.0000	.0146	.0135	.0122	.0110	.0093
70	.0000	.0000	.0000	.0000	.0000	.0149	.0138	.0124	.0111	.0094
80	.0000	.0000	.0000	.0000	.0000	.0151	.0140	.0126	.0113	.0095
90	.0000	.0000	.0000	.0000	.0000	.0152	.0142	.0127	.0114	.0095
100	.0000	.0000	.0000	.0000	.0000	.0153	.0143	.0128	.0115	.0096
120	.0000	.0000	.0000	.0000	.0000	.0153	.0143	.0128	.0114	.0095
140	.0000	.0000	.0000	.0000	.0000	.0153	.0142	.0127	.0114	.0094
170	.0000	.0000	.0000	.0000	.0000	.0151	.0140	.0125	.0112	.0092
200	.0000	.0000	.0000	.0000	.0000	.0149	.0138	.0123	.0110	.0090
230	.0000	.0000	.0000	.0000	.0000	.0147	.0136	.0121	.0108	.0088
260	.0000	.0000	.0000	.0000	.0000	.0146	.0135	.0119	.0106	.0086
300	.0000	.0000	.0000	.0000	.0000	.0143	.0132	.0117	.0104	.0084
350	.0000	.0000	.0000	.0000	.0000	.0141	.0130	.0114	.0101	.0081
400	.0000	.0000	.0000	.0000	.0000	.0138	.0127	.0111	.0098	.0079
450	.0000	.0000	.0000	.0000	.0000	.0135	.0125	.0108	.0096	.0076
500	.0000	.0000	.0000	.0000	.0000	.0133	.0122	.0106	.0093	.0074
600	.0000	.0000	.0000	.0000	.0000	.0127	.0118	.0102	.0088	.0069
700	.0000	.0000	.0000	.0000	.0000	.0122	.0113	.0098	.0084	.0065
800	.0000	.0000	.0000	.0000	.0000	.0118	.0109	.0095	.0080	.0061
900	.0000	.0000	.0000	.0000	.0000	.0114	.0105	.0091	.0076	.0057
1000	.0000	.0000	.0000	.0000	.0000	.0110	.0102	.0088	.0073	.0054
1200	.0000	.0000	.0000	.0000	.0000	.0104	.0095	.0081	.0065	.0049
1400	.0000	.0000	.0000	.0000	.0000	.0099	.0090	.0074	.0059	.0044
1700	.0000	.0000	.0000	.0000	.0000	.0092	.0082	.0066	.0051	.0039
2000	.0000	.0000	.0000	.0000	.0000	.0086	.0076	.0059	.0045	.0034
2300	.0000	.0000	.0000	.0000	0.0000	.0079	.0069	.0053	.0040	.0031
2600	.0000	.0000	.0000	0.0000	0.0000	.0074	.0064	.0049	.0037	.0029
3000	.0000	.0000	.0000	0.0000	0.0000	.0067	.0058	.0044	.0034	.0027
3500	.0000	.0000	0.0000	0.0000	0.0000	.0059	.0051	.0039	.0031	.0024
4000	.0000	.0000	0.0000	0.0000	0.0000	.0053	.0046	.0035	.0028	.0022
4500	.0000	.0000	0.0000	0.0000	0.0000	.0048	.0041	.0032	.0026	.0021
5000	.0000	.0000	0.0000	0.0000	0.0000	.0044	.0038	.0030	.0025	.0020
6000	0.0000	0.0000	0.0000	0.0000	0.0000	.0039	.0034	.0027	.0022	.0018
7000	0.0000	0.0000	0.0000	0.0000	0.0000	.0036	.0031	.0025	.0020	.0016
8000	0.0000	0.0000	0.0000	0.0000	0.0000	.0034	.0029	.0023	.0019	.0015
9000	0.0000	0.0000	0.0000	0.0000	0.0000	.0032	.0027	.0022	.0018	.0014
10000	0.0000	0.0000	0.0000	0.0000	0.0000	.0030	.0026	.0021	.0017	.0013
12000	0.0000	0.0000	0.0000	0.0000	0.0000	.0027	.0023	.0019	.0015	.0011
14000	0.0000	0.0000	0.0000	0.0000	0.0000	.0024	.0021	.0017	.0013	.0010
17000	0.0000	0.0000	0.0000	0.0000	0.0000	.0021	.0018	.0014	.0011	.0008
20000	0.0000	0.0000	0.0000	0.0000	0.0000	.0017	.0015	.0012	.0009	.0007

| K/KL= | IONIZATION DEFICIENCY | | | | | SPUTTERING YIELD ALPHA | | | | |
E,KEV	0.6	0.7	0.85	1.0	1.25	0.6	0.7	0.85	1.0	1.25
2.0	.0014	.0017	.0020	.0022	.0027	.1239	.1217	.1184	.1153	.1105
2.3	.0015	.0017	.0020	.0023	.0027	.1235	.1212	.1179	.1147	.1099
2.6	.0015	.0017	.0020	.0023	.0028	.1231	.1207	.1174	.1142	.1092
3.0	.0015	.0018	.0021	.0024	.0028	.1225	.1201	.1168	.1135	.1085
3.5	.0016	.0018	.0021	.0024	.0029	.1219	.1194	.1161	.1127	.1076
4.0	.0016	.0018	.0021	.0025	.0029	.1213	.1188	.1154	.1119	.1068
4.5	.0016	.0019	.0022	.0025	.0030	.1207	.1182	.1148	.1113	.1060
5.0	.0017	.0019	.0022	.0025	.0030	.1202	.1177	.1142	.1107	.1053
6.0	.0017	.0019	.0023	.0026	.0031	.1194	.1168	.1131	.1096	.1042
7.0	.0018	.0020	.0023	.0027	.0031	.1186	.1160	.1122	.1086	.1031
8.0	.0018	.0020	.0024	.0027	.0032	.1180	.1153	.1114	.1078	.1022
9.0	.0018	.0021	.0024	.0028	.0032	.1174	.1147	.1108	.1071	.1014
10	.0019	.0021	.0025	.0028	.0033	.1169	.1142	.1101	.1065	.1007
12	.0019	.0022	.0026	.0029	.0034	.1162	.1134	.1092	.1054	.0995
14	.0020	.0023	.0026	.0029	.0034	.1156	.1127	.1085	.1045	.0985
17	.0021	.0024	.0027	.0030	.0035	.1148	.1118	.1075	.1034	.0972
20	.0022	.0025	.0028	.0031	.0036	.1142	.1110	.1066	.1024	.0961
23	.0023	.0026	.0029	.0033	.0037	.1134	.1102	.1056	.1014	.0949
26	.0024	.0027	.0030	.0034	.0038	.1127	.1094	.1048	.1004	.0938
30	.0025	.0028	.0032	.0035	.0039	.1119	.1085	.1037	.0992	.0925
35	.0027	.0030	.0033	.0036	.0041	.1109	.1075	.1026	.0980	.0911
40	.0028	.0031	.0035	.0038	.0042	.1101	.1066	.1016	.0969	.0898
45	.0030	.0032	.0036	.0039	.0043	.1094	.1058	.1007	.0960	.0888
50	.0031	.0034	.0037	.0041	.0044	.1088	.1051	.0999	.0951	.0878
60	.0033	.0036	.0040	.0043	.0047	.1079	.1041	.0988	.0939	.0866
70	.0035	.0038	.0042	.0045	.0049	.1072	.1032	.0978	.0929	.0855
80	.0036	.0040	.0044	.0047	.0050	.1065	.1025	.0969	.0920	.0846
90	.0038	.0041	.0046	.0049	.0052	.1059	.1018	.0962	.0912	.0838
100	.0039	.0043	.0047	.0050	.0053	.1054	.1012	.0955	.0904	.0831
120	.0041	.0045	.0049	.0052	.0056	.1044	.1000	.0942	.0890	.0815
140	.0043	.0047	.0051	.0054	.0057	.1036	.0991	.0931	.0878	.0802
170	.0045	.0049	.0053	.0056	.0059	.1025	.0978	.0917	.0862	.0786
200	.0047	.0051	.0055	.0058	.0061	.1017	.0969	.0906	.0850	.0772
230	.0049	.0052	.0057	.0060	.0063	.1012	.0963	.0897	.0840	.0763
260	.0050	.0054	.0058	.0061	.0064	.1007	.0958	.0889	.0833	.0755
300	.0051	.0055	.0060	.0063	.0065	.1003	.0952	.0881	.0824	.0746
350	.0052	.0057	.0061	.0064	.0066	.0998	.0947	.0873	.0816	.0738
400	.0053	.0058	.0062	.0065	.0067	.0994	.0943	.0867	.0809	.0732
450	.0054	.0058	.0063	.0065	.0068	.0990	.0940	.0862	.0804	.0727
500	.0054	.0059	.0063	.0066	.0068	.0987	.0937	.0858	.0800	.0724
600	.0053	.0058	.0064	.0065	.0068	.0980	.0933	.0850	.0793	.0720
700	.0052	.0056	.0063	.0065	.0068	.0973	.0929	.0846	.0789	.0718
800	.0051	.0055	.0063	.0064	.0067	.0969	.0927	.0844	.0786	.0717
900	.0051	.0055	.0062	.0063	.0065	.0965	.0925	.0843	.0785	.0717
1000	.0051	.0054	.0062	.0061	.0063	.0962	.0923	.0844	.0785	.0717
1200	.0053	.0056	.0060	.0057	.0057	.0958	.0921	.0852	.0791	.0719
1400	.0055	.0057	.0059	.0053	.0051	.0956	.0920	.0861	.0798	.0722
1700	.0057	.0058	.0057	.0048	.0043	.0957	.0921	.0874	.0809	.0726
2000	.0059	.0059	.0054	.0044	.0036	.0961	.0925	.0886	.0817	.0730
2300	.0059	.0058	.0052	.0041	.0034	.0973	.0934	.0891	.0822	.0733
2600	.0058	.0056	.0049	.0038	.0032	.0985	.0943	.0896	.0825	.0736
3000	.0056	.0054	.0046	.0035	.0030	.1001	.0955	.0901	.0829	.0740
3500	.0054	.0050	.0042	.0032	.0029	.1019	.0970	.0906	.0833	.0744
4000	.0052	.0047	.0039	.0030	.0027	.1036	.0983	.0911	.0836	.0748
4500	.0049	.0044	.0036	.0028	.0027	.1050	.0994	.0915	.0840	.0751
5000	.0047	.0042	.0033	.0027	.0026	.1063	.1004	.0920	.0844	.0755
6000	.0043	.0037	.0029	.0025	.0020	.1080	.1019	.0932	.0855	.0762
7000	.0039	.0033	.0026	.0024	.0016	.1093	.1031	.0943	.0864	.0768
8000	.0035	.0029	.0023	.0023	.0014	.1104	.1040	.0952	.0872	.0774
9000	.0032	.0026	.0021	.0021	.0014	.1111	.1047	.0959	.0878	.0778
10000	.0029	.0023	.0019	.0020	.0015	.1118	.1053	.0965	.0883	.0782
12000	.0024	.0018	.0016	.0018	.0021	.1126	.1062	.0974	.0891	.0787
14000	.0020	.0014	.0013	.0016	.0030	.1131	.1067	.0979	.0895	.0791
17000	.0015	.0010	.0010	.0012	.0048	.1133	.1070	.0982	.0897	.0793
20000	.0010	.0006	.0008	.0008	.0070	.1132	.1070	.0980	.0895	.0793

AU ON C, M2/M1=0.061

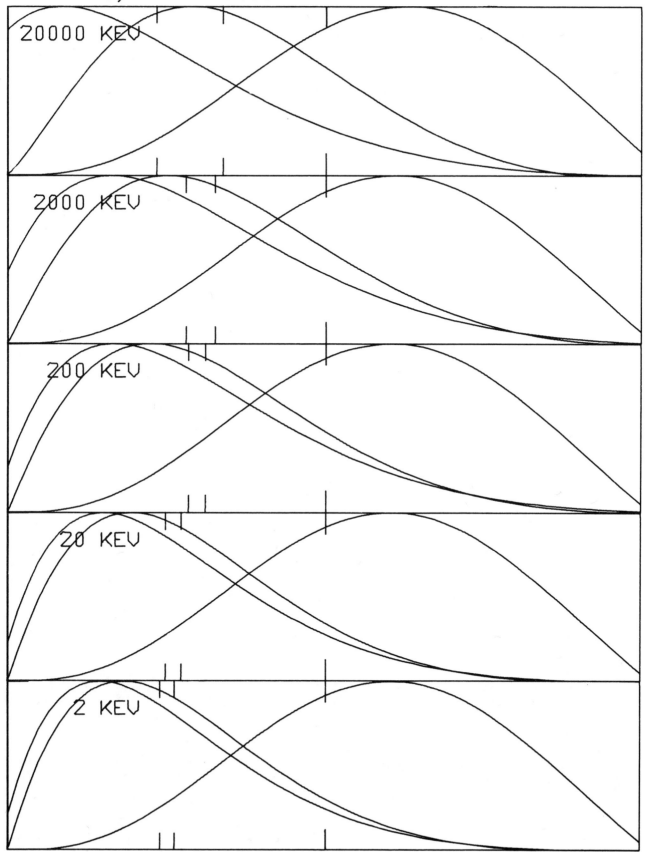

E,KEV	FRACTIONAL DEPOSITED ENERGY					MEAN RANGE,MICROGRAM/SQ.CM.				
K/KL=	0.7	0.85	1.0	1.2	1.4	0.7	0.85	1.0	1.2	1.4
5.0	.8392	.8093	.7809	.7450	.7113	1.141	1.135	1.129	1.122	1.114
6.0	.8341	.8035	.7743	.7376	.7032	1.295	1.288	1.281	1.272	1.263
7.0	.8296	.7983	.7685	.7312	.6962	1.435	1.427	1.420	1.409	1.399
8.0	.8255	.7936	.7633	.7253	.6898	1.567	1.559	1.550	1.539	1.527
9.0	.8218	.7893	.7586	.7200	.6841	1.694	1.684	1.675	1.662	1.650
10	.8183	.7853	.7542	.7151	.6788	1.816	1.805	1.795	1.781	1.768
12	.8120	.7782	.7462	.7063	.6693	2.049	2.037	2.025	2.010	1.994
14	.8064	.7718	.7392	.6986	.6609	2.273	2.260	2.246	2.228	2.210
17	.7989	.7633	.7299	.6883	.6500	2.595	2.578	2.562	2.541	2.520
20	.7922	.7558	.7216	.6793	.6404	2.902	2.883	2.864	2.839	2.815
23	.7864	.7492	.7144	.6714	.6319	3.184	3.163	3.142	3.114	3.087
26	.7810	.7431	.7077	.6642	.6242	3.458	3.434	3.411	3.380	3.351
30	.7743	.7356	.6996	.6554	.6150	3.812	3.786	3.760	3.725	3.691
35	.7667	.7271	.6904	.6455	.6046	4.243	4.212	4.182	4.142	4.104
40	.7597	.7193	.6820	.6365	.5952	4.660	4.626	4.591	4.547	4.503
45	.7531	.7121	.6743	.6282	.5866	5.067	5.028	4.990	4.940	4.891
50	.7469	.7053	.6670	.6205	.5786	5.463	5.420	5.378	5.323	5.269
60	.7352	.6925	.6535	.6063	.5639	6.195	6.144	6.095	6.031	5.968
70	.7245	.6810	.6413	.5936	.5508	6.910	6.852	6.795	6.721	6.649
80	.7147	.6705	.6303	.5821	.5392	7.611	7.546	7.481	7.397	7.315
90	.7057	.6608	.6202	.5717	.5286	8.301	8.227	8.154	8.059	7.967
100	.6973	.6519	.6109	.5622	.5190	8.979	8.896	8.815	8.710	8.607
120	.6818	.6356	.5941	.5450	.5017	10.23	10.13	10.03	9.911	9.790
140	.6680	.6211	.5793	.5300	.4868	11.47	11.35	11.24	11.10	10.96
170	.6497	.6022	.5600	.5105	.4675	13.32	13.18	13.05	12.87	12.70
200	.6337	.5857	.5433	.4939	.4511	15.17	15.01	14.84	14.62	14.42
230	.6194	.5710	.5285	.4793	.4367	16.81	16.62	16.43	16.19	15.96
260	.6064	.5578	.5154	.4664	.4240	18.47	18.25	18.04	17.77	17.51
300	.5909	.5421	.4998	.4511	.4091	20.72	20.47	20.22	19.90	19.59
350	.5739	.5250	.4828	.4345	.3930	23.60	23.30	23.00	22.61	22.24
400	.5588	.5100	.4680	.4201	.3791	26.55	26.18	25.82	25.35	24.90
450	.5454	.4967	.4549	.4074	.3670	29.53	29.09	28.66	28.11	27.59
500	.5332	.4847	.4431	.3960	.3561	32.54	32.02	31.52	30.88	30.27
600	.5118	.4640	.4228	.3764	.3380	38.01	37.38	36.77	35.99	35.24
700	.4937	.4465	.4058	.3599	.3228	43.76	42.98	42.23	41.27	40.35
800	.4779	.4313	.3910	.3457	.3098	49.76	48.79	47.87	46.69	45.57
900	.4640	.4181	.3781	.3334	.2985	55.93	54.76	53.63	52.21	50.88
1000	.4516	.4062	.3667	.3225	.2885	62.24	60.83	59.49	57.81	56.23
1200	.4306	.3862	.3473	.3041	.2713	73.99	72.22	70.54	68.42	66.42
1400	.4129	.3695	.3311	.2889	.2571	86.43	84.19	82.05	79.37	76.87
1700	.3909	.3486	.3110	.2703	.2396	106.2	103.0	100.0	96.32	92.91
2000	.3726	.3313	.2945	.2550	.2252	127.0	122.6	118.6	113.7	109.2
2300	.3571	.3166	.2806	.2420	.2132	146.6	141.3	136.4	130.5	125.0
2600	.3436	.3037	.2685	.2308	.2029	166.9	160.5	154.6	147.5	140.9
3000	.3280	.2889	.2546	.2180	.1911	195.0	186.8	179.3	170.4	162.1
3500	.3113	.2731	.2399	.2045	.1788	231.4	220.6	210.8	199.2	188.8
4000	.2970	.2596	.2274	.1931	.1684	268.8	254.9	242.6	228.1	215.3
4500	.2845	.2478	.2165	.1832	.1594	306.8	289.6	274.5	256.9	241.6
5000	.2734	.2375	.2070	.1746	.1516	345.2	324.4	306.3	285.5	267.4
6000	.2542	.2198	.1907	.1600	.1384	420.5	392.4	368.6	341.4	318.3
7000	.2383	.2052	.1773	.1480	.1278	496.3	460.3	430.3	396.1	367.6
8000	.2247	.1929	.1660	.1380	.1189	572.2	527.9	491.1	449.6	415.7
9000	.2129	.1822	.1563	.1295	.1113	647.7	594.7	550.9	501.8	462.5
10000	.2026	.1729	.1479	.1221	.1048	722.6	660.7	609.6	552.7	508.1
12000	.1850	.1571	.1338	.1099	.0941	869.6	790.2	723.5	650.6	595.7
14000	.1707	.1444	.1225	.1001	.0855	1013.	915.4	833.0	744.1	679.0
17000	.1533	.1290	.1089	.0886	.0754	1220.	1095.	989.6	876.8	797.0
20000	.1395	.1169	.0983	.0796	.0676	1419.	1266.	1138.	1002.	907.8
23000	.1281	.1070	.0897	.0723	.0614	1610.	1429.	1279.	1120.	1012.
26000	.1186	.0988	.0826	.0664	.0563	1793.	1585.	1413.	1232.	1111.
30000	.1081	.0898	.0749	.0600	.0507	2026.	1782.	1583.	1374.	1236.
35000	.0976	.0808	.0672	.0536	.0453	2302.	2015.	1783.	1540.	1382.
40000	.0891	.0737	.0611	.0486	.0411	2564.	2233.	1971.	1696.	1519.
45000	.0822	.0679	.0562	.0446	.0377	2812.	2440.	2149.	1843.	1647.
50000	.0765	.0631	.0522	.0414	.0350	3049.	2637.	2318.	1983.	1769.

	MEAN DAMAGE DEPTH,MICROGRAM/SQ.CM.					MEAN IONZN.DEPTH,MICROGRAM/SQ.CM.				
K/KL=	0.7	0.85	1.0	1.2	1.4	0.7	0.85	1.0	1.2	1.4
E,KEV										
5.0	.5892	.5857	.5822	.5777	.5734	.5381	.5340	.5300	.5247	.5196
6.0	.6739	.6694	.6650	.6594	.6540	.6158	.6104	.6052	.5986	.5922
7.0	.7502	.7449	.7398	.7333	.7271	.6852	.6790	.6730	.6654	.6580
8.0	.8216	.8158	.8101	.8028	.7958	.7503	.7433	.7367	.7281	.7199
9.0	.8903	.8838	.8775	.8695	.8617	.8127	.8051	.7977	.7883	.7792
10	.9570	.9499	.9431	.9342	.9257	.8734	.8651	.8570	.8467	.8367
12	1.087	1.079	1.070	1.060	1.050	.9917	.9819	.9723	.9600	.9482
14	1.214	1.204	1.194	1.182	1.170	1.107	1.096	1.084	1.070	1.056
17	1.399	1.387	1.375	1.359	1.345	1.277	1.263	1.248	1.230	1.213
20	1.580	1.565	1.550	1.532	1.514	1.444	1.425	1.408	1.386	1.365
23	1.747	1.729	1.712	1.691	1.671	1.596	1.575	1.555	1.529	1.505
26	1.912	1.891	1.872	1.847	1.824	1.746	1.722	1.699	1.670	1.643
30	2.129	2.104	2.081	2.052	2.025	1.946	1.917	1.889	1.855	1.823
35	2.397	2.367	2.339	2.304	2.271	2.194	2.158	2.124	2.083	2.044
40	2.663	2.626	2.593	2.551	2.513	2.440	2.397	2.356	2.306	2.260
45	2.924	2.882	2.842	2.794	2.749	2.685	2.633	2.584	2.526	2.472
50	3.183	3.133	3.087	3.032	2.981	2.927	2.866	2.809	2.742	2.681
60	3.666	3.603	3.546	3.477	3.414	3.382	3.303	3.232	3.147	3.071
70	4.146	4.068	3.999	3.915	3.840	3.836	3.737	3.649	3.546	3.454
80	4.622	4.529	4.446	4.347	4.259	4.289	4.169	4.063	3.939	3.831
90	5.095	4.985	4.888	4.773	4.672	4.740	4.597	4.472	4.327	4.201
100	5.562	5.435	5.324	5.193	5.078	5.188	5.020	4.876	4.710	4.566
120	6.438	6.277	6.137	5.975	5.834	6.037	5.821	5.637	5.428	5.250
140	7.312	7.115	6.946	6.751	6.583	6.884	6.616	6.391	6.138	5.924
170	8.621	8.368	8.153	7.909	7.699	8.149	7.800	7.511	7.189	6.921
200	9.924	9.614	9.353	9.058	8.806	9.405	8.972	8.617	8.226	7.902
230	11.10	10.73	10.43	10.09	9.804	10.56	10.05	9.627	9.171	8.795
260	12.27	11.86	11.51	11.13	10.81	11.72	11.12	10.63	10.11	9.683
300	13.87	13.37	12.98	12.53	12.15	13.26	12.54	11.98	11.37	10.87
350	15.88	15.30	14.83	14.30	13.86	15.18	14.33	13.65	12.93	12.35
400	17.92	17.24	16.69	16.09	15.57	17.11	16.11	15.33	14.50	13.82
450	19.98	19.20	18.57	17.88	17.30	19.02	17.89	17.00	16.06	15.29
500	22.03	21.17	20.46	19.68	19.02	20.92	19.65	18.66	17.60	16.75
600	25.76	24.74	23.89	22.95	22.18	24.35	22.84	21.68	20.45	19.44
700	29.64	28.45	27.45	26.33	25.44	27.87	26.11	24.76	23.34	22.17
800	33.64	32.28	31.11	29.80	28.78	31.45	29.44	27.90	26.26	24.93
900	37.75	36.20	34.85	33.34	32.17	35.09	32.82	31.07	29.21	27.71
1000	41.92	40.18	38.65	36.93	35.60	38.75	36.23	34.27	32.17	30.49
1200	49.63	47.57	45.76	43.71	42.10	45.64	42.70	40.34	37.81	35.83
1400	57.77	55.34	53.18	50.73	48.79	52.75	49.36	46.56	43.54	41.23
1700	70.70	67.58	64.82	61.67	59.13	63.78	59.61	56.09	52.27	49.39
2000	84.27	80.36	76.89	72.91	69.71	75.09	70.08	65.76	61.06	57.57
2300	96.89	92.36	88.34	83.69	79.85	85.83	80.09	75.12	69.65	65.54
2600	110.0	104.8	100.1	94.67	90.14	96.77	90.23	84.53	78.23	73.47
3000	128.3	121.9	116.2	109.6	104.0	111.6	103.9	97.14	89.64	83.95
3500	152.1	144.0	136.8	128.5	121.5	130.5	121.1	112.9	103.8	96.87
4000	176.7	166.6	157.7	147.5	139.1	149.6	138.4	128.6	117.8	109.6
4500	201.7	189.6	178.8	166.6	156.6	168.8	155.6	144.2	131.5	122.0
5000	227.1	212.7	200.0	185.7	174.1	188.0	172.8	159.6	145.0	134.2
6000	276.7	257.9	241.3	222.7	207.9	225.8	206.4	189.7	171.4	157.8
7000	327.1	303.4	282.6	259.4	241.3	263.1	239.4	219.0	196.8	180.5
8000	378.0	349.0	323.7	295.7	274.1	299.8	271.6	247.4	221.3	202.2
9000	429.0	394.5	364.4	331.4	306.3	335.8	303.0	275.0	244.8	223.1
10000	479.8	439.6	404.7	366.5	337.9	371.0	333.6	301.7	267.6	243.1
12000	579.4	527.5	482.8	434.3	398.7	439.4	392.7	352.9	310.8	281.2
14000	677.7	613.6	558.8	499.8	457.3	504.8	448.9	401.4	351.4	316.6
17000	821.8	739.3	669.1	594.3	541.5	597.8	528.2	469.3	408.0	366.3
20000	961.8	860.6	775.0	684.5	621.6	685.1	602.3	532.5	460.4	412.1
23000	1098.	977.8	876.8	770.9	698.2	767.5	672.0	591.6	509.2	454.7
26000	1229.	1091.	974.9	853.8	771.6	845.7	737.8	647.3	554.9	494.6
30000	1399.	1236.	1100.	959.4	864.9	944.1	820.3	716.9	612.0	544.4
35000	1601.	1409.	1249.	1084.	975.3	1059.	916.5	797.8	678.0	602.0
40000	1794.	1574.	1391.	1203.	1080.	1167.	1006.	872.8	739.1	655.3
45000	1979.	1732.	1526.	1315.	1179.	1268.	1090.	943.1	796.2	705.0
50000	2157.	1882.	1655.	1423.	1273.	1363.	1169.	1009.	849.7	751.7

AU ON BE, M2/M1=0.0457 TABLE 22- 3

	RELATIVE RANGE STRAGGLING					RELATIVE DAMAGE STRAGGLING				
K/KL=	0.7	0.85	1.0	1.2	1.4	0.7	0.85	1.0	1.2	1.4
E,KEV										
5.0	.0230	.0228	.0227	.0225	.0223	.2869	.2869	.2869	.2869	.2869
6.0	.0230	.0228	.0226	.0224	.0222	.2856	.2856	.2855	.2855	.2855
7.0	.0230	.0228	.0226	.0224	.0222	.2845	.2843	.2843	.2842	.2842
8.0	.0229	.0228	.0226	.0224	.0222	.2834	.2832	.2831	.2830	.2830
9.0	.0229	.0228	.0226	.0224	.0221	.2824	.2822	.2821	.2820	.2819
10	.0229	.0227	.0226	.0223	.0221	.2816	.2813	.2812	.2810	.2809
12	.0229	.0227	.0225	.0223	.0221	.2802	.2798	.2796	.2793	.2792
14	.0229	.0227	.0225	.0223	.0220	.2792	.2787	.2783	.2780	.2777
17	.0229	.0227	.0225	.0222	.0220	.2783	.2775	.2769	.2763	.2759
20	.0229	.0227	.0225	.0222	.0220	.2778	.2768	.2759	.2751	.2745
23	.0229	.0227	.0224	.0222	.0219	.2779	.2764	.2753	.2742	.2734
26	.0229	.0226	.0224	.0222	.0219	.2782	.2764	.2749	.2735	.2725
30	.0228	.0226	.0224	.0221	.0219	.2788	.2765	.2747	.2729	.2716
35	.0228	.0225	.0224	.0221	.0218	.2801	.2771	.2748	.2724	.2707
40	.0228	.0226	.0224	.0221	.0218	.2816	.2779	.2750	.2722	.2701
45	.0228	.0226	.0224	.0220	.0217	.2832	.2788	.2755	.2721	.2697
50	.0228	.0226	.0223	.0220	.0217	.2850	.2799	.2760	.2722	.2694
60	.0228	.0226	.0223	.0220	.0217	.2897	.2832	.2781	.2732	.2697
70	.0228	.0226	.0223	.0219	.0216	.2941	.2862	.2801	.2742	.2700
80	.0228	.0225	.0223	.0219	.0216	.2981	.2888	.2818	.2750	.2702
90	.0228	.0225	.0222	.0219	.0215	.3017	.2911	.2832	.2757	.2704
100	.0228	.0225	.0222	.0219	.0215	.3048	.2931	.2844	.2762	.2704
120	.0228	.0225	.0222	.0218	.0214	.3103	.2960	.2859	.2765	.2700
140	.0228	.0225	.0222	.0218	.0214	.3145	.2981	.2868	.2764	.2693
170	.0228	.0225	.0221	.0217	.0213	.3190	.3000	.2874	.2758	.2681
200	.0228	.0224	.0221	.0216	.0212	.3217	.3008	.2872	.2748	.2667
230	.0228	.0224	.0220	.0215	.0211	.3221	.3005	.2861	.2730	.2650
260	.0228	.0224	.0220	.0214	.0210	.3216	.2997	.2847	.2712	.2633
300	.0227	.0223	.0219	.0213	.0208	.3203	.2981	.2826	.2688	.2611
350	.0227	.0222	.0218	.0212	.0207	.3177	.2958	.2799	.2658	.2585
400	.0226	.0221	.0216	.0210	.0205	.3147	.2931	.2771	.2630	.2561
450	.0225	.0220	.0215	.0209	.0203	.3114	.2904	.2744	.2604	.2540
500	.0225	.0219	.0214	.0207	.0201	.3080	.2875	.2718	.2580	.2520
600	.0223	.0217	.0211	.0204	.0197	.3001	.2811	.2664	.2537	.2486
700	.0221	.0214	.0208	.0201	.0194	.2928	.2752	.2616	.2501	.2457
800	.0219	.0212	.0205	.0197	.0190	.2862	.2699	.2574	.2469	.2432
900	.0217	.0209	.0203	.0194	.0186	.2802	.2652	.2537	.2442	.2411
1000	.0214	.0207	.0200	.0191	.0183	.2748	.2610	.2505	.2418	.2391
1200	.0209	.0201	.0193	.0184	.0176	.2663	.2545	.2456	.2383	.2361
1400	.0204	.0195	.0187	.0178	.0169	.2593	.2493	.2417	.2354	.2336
1700	.0197	.0187	.0179	.0169	.0159	.2509	.2430	.2370	.2320	.2304
2000	.0190	.0180	.0171	.0160	.0151	.2444	.2381	.2333	.2292	.2278
2300	.0183	.0172	.0163	.0152	.0143	.2397	.2344	.2304	.2268	.2254
2600	.0176	.0165	.0156	.0145	.0135	.2360	.2314	.2279	.2247	.2233
3000	.0167	.0156	.0147	.0136	.0126	.2319	.2281	.2250	.2222	.2208
3500	.0158	.0146	.0137	.0126	.0117	.2279	.2247	.2220	.2194	.2179
4000	.0149	.0137	.0128	.0117	.0108	.2247	.2218	.2194	.2169	.2154
4500	.0141	.0129	.0120	.0109	.0100	.2219	.2193	.2171	.2147	.2130
5000	.0133	.0122	.0113	.0102	.0094	.2196	.2171	.2150	.2126	.2109
6000	.0120	.0109	.0100	.0091	.0083	.2156	.2132	.2111	.2087	.2068
7000	.0109	.0099	.0090	.0081	.0074	.2122	.2098	.2077	.2052	.2033
8000	.0100	.0090	.0082	.0073	.0066	.2093	.2069	.2046	.2021	.2001
9000	.0092	.0082	.0075	.0066	.0060	.2068	.2042	.2019	.1993	.1972
10000	.0085	.0076	.0068	.0061	.0055	.2045	.2018	.1994	.1967	.1946
12000	.0074	.0065	.0059	.0052	.0047	.2002	.1974	.1949	.1921	.1899
14000	.0065	.0057	.0051	.0045	.0041	.1964	.1935	.1910	.1881	.1858
17000	.0054	.0048	.0043	.0038	.0034	.1915	.1885	.1858	.1829	.1806
20000	.0046	.0041	.0036	.0032	.0029	.1873	.1842	.1815	.1785	.1762
23000	.0040	.0035	.0031	.0028	.0025	.1835	.1804	.1776	.1746	.1723
26000	.0035	.0031	.0028	.0024	.0022	.1802	.1770	.1742	.1711	.1688
30000	.0030	.0027	.0024	.0021	.0019	.1761	.1729	.1701	.1671	.1648
35000	.0026	.0022	.0020	.0018	.0016	.1717	.1684	.1657	.1626	.1604
40000	.0022	.0020	.0017	.0015	.0014	.1677	.1645	.1617	.1588	.1566
45000	.0020	.0017	.0016	.0014	.0012	.1641	.1610	.1583	.1553	.1532
50000	.0018	.0016	.0014	.0013	.0011	.1608	.1578	.1551	.1523	.1502

AU ON BE, M2/M1=0.0457 — TABLE 22-4

E,KEV	RELATIVE IONIZATION STRAGGLING K/KL= 0.7	0.85	1.0	1.2	1.4	RELATIVE TRANSVERSE RANGE STRAGGLING 0.7	0.85	1.0	1.2	1.4
5.0	.3255	.3262	.3269	.3279	.3288	.4399	.4408	.4418	.4430	.4441
6.0	.3246	.3252	.3259	.3268	.3278	.4397	.4406	.4415	.4428	.4440
7.0	.3236	.3242	.3249	.3258	.3268	.4394	.4404	.4413	.4426	.4439
8.0	.3228	.3233	.3240	.3249	.3259	.4391	.4401	.4411	.4424	.4437
9.0	.3221	.3226	.3232	.3241	.3250	.4389	.4399	.4409	.4422	.4436
10	.3214	.3219	.3225	.3233	.3243	.4386	.4396	.4407	.4421	.4434
12	.3206	.3209	.3214	.3221	.3230	.4381	.4391	.4402	.4416	.4430
14	.3200	.3202	.3206	.3212	.3220	.4375	.4386	.4397	.4412	.4427
17	.3199	.3197	.3198	.3202	.3209	.4367	.4379	.4390	.4405	.4421
20	.3202	.3197	.3195	.3196	.3201	.4359	.4371	.4383	.4399	.4414
23	.3211	.3202	.3196	.3194	.3196	.4351	.4363	.4376	.4392	.4408
26	.3223	.3209	.3200	.3194	.3193	.4343	.4356	.4369	.4385	.4402
30	.3242	.3221	.3207	.3196	.3191	.4333	.4346	.4359	.4377	.4394
35	.3267	.3238	.3217	.3200	.3191	.4320	.4334	.4348	.4366	.4384
40	.3294	.3256	.3229	.3205	.3192	.4307	.4322	.4336	.4355	.4374
45	.3322	.3275	.3241	.3212	.3194	.4295	.4310	.4325	.4344	.4364
50	.3351	.3295	.3255	.3219	.3197	.4283	.4299	.4314	.4334	.4354
60	.3417	.3341	.3287	.3238	.3207	.4260	.4276	.4292	.4314	.4335
70	.3477	.3383	.3316	.3255	.3217	.4238	.4255	.4272	.4294	.4316
80	.3531	.3421	.3342	.3270	.3225	.4216	.4234	.4252	.4276	.4299
90	.3579	.3454	.3364	.3283	.3232	.4196	.4215	.4233	.4258	.4282
100	.3622	.3483	.3384	.3294	.3238	.4176	.4196	.4215	.4241	.4266
120	.3696	.3532	.3416	.3312	.3246	.4138	.4159	.4180	.4207	.4234
140	.3754	.3569	.3439	.3324	.3252	.4103	.4125	.4147	.4176	.4205
170	.3816	.3606	.3461	.3334	.3255	.4054	.4079	.4103	.4134	.4165
200	.3856	.3628	.3472	.3336	.3254	.4010	.4037	.4063	.4097	.4130
230	.3866	.3626	.3465	.3325	.3243	.3968	.3996	.4024	.4060	.4096
260	.3865	.3618	.3453	.3312	.3231	.3929	.3959	.3989	.4027	.4065
300	.3854	.3601	.3434	.3293	.3216	.3883	.3915	.3947	.3988	.4029
350	.3831	.3575	.3408	.3269	.3198	.3831	.3867	.3901	.3946	.3990
400	.3803	.3547	.3382	.3247	.3183	.3787	.3824	.3862	.3910	.3957
450	.3771	.3519	.3357	.3227	.3169	.3747	.3788	.3827	.3879	.3929
500	.3738	.3492	.3334	.3209	.3158	.3712	.3755	.3797	.3852	.3905
600	.3669	.3442	.3295	.3181	.3146	.3649	.3697	.3744	.3804	.3863
700	.3603	.3398	.3261	.3159	.3139	.3599	.3652	.3703	.3769	.3833
800	.3543	.3358	.3232	.3142	.3134	.3559	.3616	.3672	.3743	.3812
900	.3487	.3322	.3208	.3129	.3132	.3527	.3588	.3648	.3724	.3798
1000	.3436	.3289	.3187	.3120	.3132	.3502	.3567	.3630	.3712	.3790
1200	.3341	.3228	.3152	.3109	.3134	.3464	.3537	.3608	.3699	.3786
1400	.3263	.3180	.3127	.3105	.3141	.3442	.3523	.3600	.3700	.3795
1700	.3174	.3128	.3104	.3109	.3156	.3430	.3521	.3608	.3719	.3825
2000	.3112	.3095	.3094	.3120	.3175	.3435	.3535	.3632	.3754	.3870
2300	.3088	.3090	.3104	.3142	.3202	.3452	.3562	.3667	.3800	.3925
2600	.3078	.3094	.3120	.3167	.3230	.3478	.3597	.3710	.3852	.3987
3000	.3076	.3107	.3145	.3202	.3266	.3522	.3652	.3775	.3929	.4075
3500	.3087	.3132	.3180	.3245	.3311	.3587	.3730	.3864	.4032	.4190
4000	.3106	.3161	.3216	.3287	.3354	.3659	.3814	.3960	.4139	.4308
4500	.3129	.3192	.3252	.3327	.3394	.3737	.3903	.4058	.4249	.4426
5000	.3155	.3224	.3288	.3365	.3432	.3817	.3994	.4158	.4359	.4545
6000	.3213	.3287	.3355	.3435	.3502	.3984	.4182	.4362	.4587	.4783
7000	.3270	.3347	.3416	.3497	.3564	.4152	.4369	.4564	.4807	.5013
8000	.3325	.3403	.3472	.3553	.3620	.4318	.4552	.4759	.5018	.5233
9000	.3376	.3454	.3523	.3604	.3671	.4482	.4730	.4948	.5221	.5442
10000	.3424	.3502	.3570	.3651	.3717	.4643	.4903	.5131	.5415	.5643
12000	.3505	.3583	.3652	.3732	.3797	.4963	.5235	.5480	.5777	.6017
14000	.3576	.3654	.3723	.3802	.3866	.5264	.5547	.5805	.6112	.6363
17000	.3668	.3744	.3812	.3890	.3953	.5680	.5983	.6256	.6572	.6837
20000	.3745	.3820	.3888	.3964	.4025	.6059	.6386	.6670	.6991	.7267
23000	.3812	.3886	.3952	.4027	.4086	.6405	.6760	.7052	.7377	.7662
26000	.3870	.3943	.4008	.4081	.4139	.6723	.7111	.7408	.7735	.8028
30000	.3937	.4008	.4072	.4143	.4200	.7112	.7546	.7848	.8176	.8479
35000	.4008	.4077	.4140	.4209	.4263	.7549	.8047	.8352	.8680	.8993
40000	.4068	.4136	.4197	.4264	.4315	.7943	.8508	.8814	.9142	.9462
45000	.4120	.4186	.4245	.4310	.4360	.8301	.8937	.9241	.9568	.9895
50000	.4164	.4230	.4287	.4351	.4399	.8629	.9338	.9640	.9965	1.030

TABLE 22- 5

AU ON BE, M2/M1=0.0457

K/KL=	RELATIVE TRANSV. DAMAGE STRAGGLING					RELATIVE TRANSV. IONZN. STRAGGLING				
E,KEV	0.7	0.85	1.0	1.2	1.4	0.7	0.85	1.0	1.2	1.4
5.0	.1565	.1527	.1491	.1446	.1404	.1436	.1397	.1360	.1313	.1270
6.0	.1601	.1560	.1521	.1473	.1428	.1472	.1430	.1390	.1340	.1293
7.0	.1638	.1594	.1552	.1500	.1452	.1510	.1464	.1421	.1368	.1318
8.0	.1674	.1627	.1583	.1527	.1476	.1548	.1499	.1453	.1396	.1342
9.0	.1710	.1660	.1613	.1554	.1499	.1585	.1533	.1484	.1423	.1366
10	.1745	.1692	.1642	.1580	.1522	.1622	.1566	.1514	.1449	.1390
12	.1812	.1752	.1697	.1628	.1565	.1691	.1629	.1571	.1499	.1434
14	.1874	.1809	.1748	.1673	.1605	.1756	.1688	.1624	.1546	.1475
17	.1959	.1886	.1818	.1735	.1659	.1845	.1768	.1697	.1610	.1531
20	.2037	.1956	.1882	.1791	.1708	.1925	.1841	.1763	.1667	.1581
23	.2114	.2026	.1945	.1846	.1757	.2003	.1911	.1826	.1723	.1629
26	.2185	.2090	.2003	.1897	.1801	.2074	.1975	.1884	.1773	.1673
30	.2269	.2166	.2071	.1957	.1853	.2159	.2052	.1953	.1833	.1725
35	.2361	.2249	.2146	.2022	.1910	.2252	.2136	.2029	.1899	.1782
40	.2440	.2321	.2211	.2077	.1958	.2332	.2208	.2094	.1956	.1831
45	.2509	.2383	.2266	.2125	.1999	.2401	.2272	.2151	.2005	.1873
50	.2569	.2437	.2314	.2166	.2034	.2462	.2327	.2201	.2048	.1910
60	.2665	.2523	.2391	.2231	.2088	.2563	.2420	.2286	.2121	.1973
70	.2740	.2591	.2451	.2281	.2129	.2642	.2493	.2352	.2178	.2021
80	.2801	.2645	.2499	.2320	.2160	.2706	.2552	.2405	.2223	.2058
90	.2851	.2689	.2537	.2351	.2184	.2757	.2599	.2447	.2259	.2087
100	.2891	.2724	.2567	.2374	.2202	.2798	.2637	.2481	.2286	.2109
120	.2952	.2778	.2612	.2408	.2225	.2851	.2686	.2523	.2319	.2131
140	.2994	.2814	.2641	.2428	.2235	.2886	.2717	.2549	.2336	.2141
170	.3033	.2844	.2663	.2438	.2234	.2918	.2743	.2568	.2345	.2139
200	.3053	.2856	.2667	.2432	.2220	.2935	.2755	.2573	.2340	.2127
230	.3061	.2855	.2656	.2412	.2193	.2958	.2766	.2576	.2333	.2110
260	.3060	.2844	.2637	.2385	.2161	.2973	.2770	.2571	.2319	.2089
300	.3049	.2821	.2604	.2344	.2113	.2983	.2765	.2556	.2294	.2056
350	.3023	.2782	.2556	.2286	.2052	.2983	.2748	.2528	.2255	.2009
400	.2988	.2736	.2502	.2226	.1989	.2970	.2722	.2492	.2210	.1958
450	.2948	.2686	.2445	.2165	.1927	.2949	.2688	.2450	.2161	.1906
500	.2904	.2634	.2387	.2103	.1866	.2921	.2649	.2404	.2109	.1853
600	.2802	.2516	.2260	.1972	.1742	.2835	.2545	.2287	.1985	.1730
700	.2697	.2401	.2139	.1851	.1629	.2742	.2439	.2173	.1866	.1616
800	.2594	.2290	.2026	.1739	.1526	.2648	.2336	.2063	.1755	.1511
900	.2493	.2184	.1920	.1637	.1433	.2555	.2236	.1959	.1652	.1415
1000	.2394	.2085	.1821	.1543	.1349	.2464	.2141	.1861	.1556	.1328
1200	.2190	.1890	.1638	.1377	.1204	.2287	.1957	.1679	.1384	.1177
1400	.2006	.1718	.1479	.1236	.1082	.2124	.1792	.1519	.1237	.1050
1700	.1766	.1499	.1280	.1063	.0933	.1901	.1578	.1316	.1054	.0894
2000	.1564	.1319	.1119	.0925	.0816	.1702	.1397	.1148	.0907	.0770
2300	.1405	.1181	.1000	.0826	.0730	.1496	.1248	.1018	.0798	.0677
2600	.1270	.1066	.0902	.0745	.0661	.1312	.1120	.0908	.0709	.0602
3000	.1119	.0939	.0795	.0659	.0586	.1100	.0975	.0788	.0612	.0522
3500	.0966	.0813	.0690	.0575	.0514	.0880	.0828	.0667	.0518	.0443
4000	.0843	.0712	.0608	.0509	.0458	.0704	.0710	.0571	.0444	.0381
4500	.0743	.0631	.0542	.0457	.0413	.0564	.0613	.0494	.0385	.0332
5000	.0662	.0565	.0488	.0415	.0377	.0454	.0533	.0431	.0338	.0293
6000	.0553	.0476	.0414	.0356	.0325	.0373	.0430	.0348	.0274	.0239
7000	.0476	.0413	.0362	.0313	.0287	.0329	.0356	.0290	.0229	.0201
8000	.0419	.0366	.0323	.0282	.0259	.0304	.0301	.0247	.0197	.0173
9000	.0376	.0330	.0293	.0258	.0238	.0288	.0260	.0215	.0173	.0152
10000	.0341	.0302	.0270	.0238	.0220	.0277	.0228	.0189	.0154	.0136
12000	.0290	.0259	.0233	.0207	.0192	.0231	.0181	.0152	.0124	.0110
14000	.0255	.0229	.0207	.0185	.0172	.0194	.0150	.0127	.0105	.0093
17000	.0219	.0198	.0180	.0162	.0151	.0152	.0120	.0102	.0085	.0075
20000	.0196	.0177	.0162	.0146	.0135	.0120	.0101	.0086	.0072	.0064
23000	.0179	.0162	.0148	.0134	.0124	.0096	.0088	.0075	.0063	.0056
26000	.0166	.0150	.0137	.0125	.0115	.0078	.0079	.0067	.0056	.0050
30000	.0153	.0138	.0126	.0115	.0105	.0061	.0070	.0059	.0049	.0044
35000	.0139	.0126	.0115	.0106	.0096	.0048	.0061	.0052	.0043	.0038
40000	.0128	.0116	.0105	.0099	.0088	.0042	.0053	.0045	.0037	.0033
45000	.0117	.0106	.0097	.0093	.0081	.0041	.0045	.0038	.0032	.0028
50000	.0106	.0097	.0089	.0088	.0074	.0043	.0037	.0032	.0027	.0024

TABLE 22- 6

AU ON BE, M2/M1=0.0457

K/KL= E,KEV	RANGE SKEWNESS					DAMAGE SKEWNESS				
	0.7	0.85	1.0	1.2	1.4	0.7	0.85	1.0	1.2	1.4
5.0	.2722	.2691	.2661	.2621	.2582	.2306	.2283	.2262	.2236	.2214
6.0	.2716	.2685	.2654	.2612	.2572	.2271	.2239	.2211	.2180	.2153
7.0	.2712	.2679	.2647	.2605	.2563	.2228	.2191	.2159	.2123	.2093
8.0	.2707	.2674	.2641	.2598	.2555	.2188	.2146	.2110	.2070	.2037
9.0	.2703	.2669	.2636	.2592	.2548	.2154	.2107	.2067	.2022	.1985
10	.2700	.2665	.2631	.2586	.2542	.2129	.2075	.2030	.1980	.1939
12	.2693	.2657	.2622	.2575	.2530	.2100	.2031	.1973	.1910	.1859
14	.2686	.2650	.2613	.2566	.2519	.2101	.2011	.1937	.1858	.1795
17	.2678	.2640	.2602	.2553	.2504	.2145	.2019	.1916	.1807	.1723
20	.2670	.2630	.2592	.2541	.2490	.2230	.2060	.1924	.1782	.1673
23	.2663	.2622	.2582	.2530	.2478	.2358	.2137	.1960	.1777	.1639
26	.2656	.2615	.2574	.2520	.2467	.2507	.2230	.2011	.1786	.1618
30	.2648	.2605	.2563	.2507	.2453	.2725	.2374	.2097	.1815	.1606
35	.2637	.2593	.2550	.2493	.2437	.3018	.2573	.2225	.1871	.1611
40	.2628	.2582	.2537	.2479	.2421	.3323	.2785	.2366	.1942	.1632
45	.2619	.2572	.2526	.2465	.2406	.3633	.3005	.2516	.2024	.1665
50	.2609	.2561	.2514	.2452	.2392	.3946	.3228	.2671	.2113	.1708
60	.2593	.2543	.2493	.2429	.2366	.4701	.3794	.3092	.2392	.1887
70	.2577	.2525	.2473	.2406	.2341	.5375	.4296	.3464	.2637	.2043
80	.2561	.2507	.2454	.2384	.2317	.5963	.4732	.3784	.2843	.2171
90	.2546	.2490	.2435	.2363	.2293	.6472	.5104	.4052	.3012	.2271
100	.2531	.2473	.2416	.2342	.2270	.6911	.5421	.4276	.3147	.2345
120	.2503	.2442	.2382	.2304	.2228	.7565	.5866	.4564	.3285	.2381
140	.2475	.2411	.2348	.2266	.2187	.8047	.6177	.4747	.3349	.2365
170	.2434	.2365	.2298	.2211	.2127	.8537	.6469	.4889	.3354	.2285
200	.2393	.2320	.2249	.2157	.2068	.8830	.6616	.4923	.3288	.2164
230	.2354	.2277	.2202	.2106	.2012	.8937	.6621	.4842	.3134	.1993
250	.2315	.2234	.2156	.2055	.1958	.8951	.6562	.4718	.2958	.1813
300	.2264	.2178	.2096	.1989	.1887	.8873	.6418	.4512	.2707	.1573
350	.2201	.2109	.2021	.1908	.1800	.8674	.6173	.4221	.2386	.1282
400	.2138	.2041	.1948	.1829	.1715	.8404	.5886	.3912	.2072	.1007
450	.2076	.1974	.1876	.1751	.1632	.8091	.5575	.3598	.1769	.0750
500	.2015	.1908	.1806	.1676	.1552	.7751	.5252	.3286	.1480	.0511
600	.1897	.1780	.1669	.1528	.1395	.6930	.4508	.2616	.0916	.0080
700	.1782	.1656	.1537	.1387	.1246	.6136	.3813	.2011	.0422	-.0292
800	.1670	.1537	.1411	.1252	.1104	.5392	.3177	.1471	-.0007	-.0616
900	.1562	.1421	.1288	.1122	.0968	.4702	.2600	.0992	-.0380	-.0899
1000	.1457	.1309	.1171	.0998	.0837	.4065	.2079	.0567	-.0704	-.1147
1200	.1253	.1092	.0942	.0756	.0585	.2942	.1218	-.0088	-.1177	-.1531
1400	.1062	.0889	.0728	.0532	.0352	.1989	.0505	-.0618	-.1550	-.1843
1700	.0794	.0606	.0434	.0224	.0035	.0816	-.0360	-.1251	-.1991	-.2223
2000	.0547	.0347	.0165	-.0055	-.0252	-.0116	-.1044	-.1747	-.2339	-.2533
2300	.0312	.0102	-.0089	-.0316	-.0518	-.0763	-.1537	-.2126	-.2628	-.2801
2600	.0093	-.0126	-.0323	-.0556	-.0763	-.1283	-.1938	-.2440	-.2873	-.3034
3000	-.0177	-.0406	-.0610	-.0849	-.1059	-.1838	-.2371	-.2785	-.3151	-.3304
3500	-.0484	-.0722	-.0932	-.1177	-.1389	-.2375	-.2800	-.3135	-.3444	-.3593
4000	-.0762	-.1008	-.1223	-.1471	-.1684	-.2795	-.3143	-.3423	-.3693	-.3843
4500	-.1017	-.1269	-.1486	-.1736	-.1950	-.3132	-.3425	-.3667	-.3911	-.4063
5000	-.1253	-.1508	-.1727	-.1978	-.2192	-.3409	-.3664	-.3879	-.4105	-.4261
6000	-.1676	-.1934	-.2153	-.2402	-.2616	-.3818	-.4044	-.4238	-.4449	-.4607
7000	-.2047	-.2305	-.2523	-.2768	-.2980	-.4128	-.4345	-.4533	-.4741	-.4900
8000	-.2376	-.2633	-.2848	-.3088	-.3297	-.4377	-.4595	-.4785	-.4993	-.5156
9000	-.2672	-.2926	-.3138	-.3374	-.3577	-.4587	-.4810	-.5005	-.5217	-.5382
10000	-.2940	-.3191	-.3400	-.3630	-.3827	-.4771	-.5001	-.5201	-.5419	-.5585
12000	-.3392	-.3635	-.3835	-.4053	-.4229	-.5117	-.5350	-.5552	-.5771	-.5939
14000	-.3786	-.4020	-.4213	-.4418	-.4574	-.5410	-.5645	-.5850	-.6072	-.6241
17000	-.4302	-.4523	-.4706	-.4896	-.5027	-.5781	-.6022	-.6230	-.6455	-.6624
20000	-.4757	-.4967	-.5140	-.5317	-.5428	-.6097	-.6341	-.6552	-.6780	-.6948
23000	-.5169	-.5368	-.5532	-.5699	-.5797	-.6373	-.6619	-.6832	-.7063	-.7230
26000	-.5547	-.5737	-.5894	-.6053	-.6141	-.6620	-.6868	-.7082	-.7313	-.7479
30000	-.6012	-.6192	-.6340	-.6490	-.6572	-.6914	-.7163	-.7377	-.7608	-.7774
35000	-.6544	-.6713	-.6852	-.6995	-.7077	-.7240	-.7488	-.7701	-.7930	-.8094
40000	-.7033	-.7193	-.7325	-.7464	-.7553	-.7532	-.7776	-.7986	-.8213	-.8375
45000	-.7489	-.7642	-.7768	-.7905	-.8007	-.7797	-.8036	-.8242	-.8465	-.8626
50000	-.7918	-.8065	-.8185	-.8324	-.8441	-.8041	-.8274	-.8475	-.8693	-.8853

TABLE 22- 7

AU ON BE, M2/M1=0.0457

E,KEV	IONIZATION SKEWNESS					RANGE KURTOSIS				
K/KL=	0.7	0.85	1.0	1.2	1.4	0.7	0.85	1.0	1.2	1.4
5.0	.3508	.3502	.3499	.3499	.3503	3.079	3.075	3.072	3.068	3.064
6.0	.3488	.3475	.3466	.3461	.3460	3.078	3.075	3.071	3.067	3.062
7.0	.3464	.3445	.3432	.3422	.3419	3.077	3.074	3.070	3.066	3.062
8.0	.3444	.3419	.3402	.3388	.3381	3.077	3.073	3.070	3.065	3.061
9.0	.3430	.3399	.3377	.3358	.3348	3.076	3.073	3.069	3.064	3.060
10	.3425	.3387	.3358	.3333	.3318	3.076	3.072	3.069	3.064	3.059
12	.3437	.3380	.3337	.3297	.3272	3.075	3.071	3.067	3.063	3.058
14	.3477	.3397	.3336	.3278	.3240	3.074	3.070	3.066	3.061	3.057
17	.3578	.3457	.3364	.3274	.3212	3.073	3.069	3.065	3.060	3.055
20	.3716	.3548	.3419	.3292	.3204	3.072	3.068	3.064	3.058	3.053
23	.3919	.3691	.3516	.3343	.3222	3.071	3.067	3.063	3.057	3.052
26	.4135	.3846	.3623	.3403	.3249	3.070	3.066	3.062	3.056	3.051
30	.4429	.4059	.3773	.3491	.3293	3.069	3.065	3.060	3.055	3.049
35	.4796	.4327	.3965	.3607	.3355	3.068	3.063	3.059	3.053	3.047
40	.5155	.4591	.4156	.3725	.3420	3.067	3.062	3.057	3.051	3.045
45	.5501	.4848	.4342	.3841	.3487	3.065	3.061	3.056	3.050	3.044
50	.5835	.5095	.4523	.3955	.3553	3.064	3.059	3.054	3.048	3.042
60	.6544	.5629	.4919	.4214	.3713	3.062	3.057	3.052	3.045	3.039
70	.7160	.6094	.5264	.4439	.3851	3.060	3.055	3.049	3.043	3.036
80	.7692	.6494	.5560	.4630	.3968	3.058	3.052	3.047	3.040	3.034
90	.8148	.6836	.5812	.4791	.4064	3.056	3.050	3.045	3.038	3.031
100	.8538	.7128	.6026	.4926	.4143	3.054	3.048	3.043	3.035	3.029
120	.9117	.7561	.6339	.5116	.4242	3.050	3.044	3.038	3.031	3.024
140	.9543	.7874	.6560	.5242	.4300	3.047	3.041	3.034	3.027	3.020
170	.9981	.8187	.6770	.5347	.4336	3.042	3.035	3.029	3.021	3.013
200	1.025	.8369	.6878	.5381	.4331	3.037	3.030	3.024	3.015	3.008
230	1.036	.8420	.6875	.5330	.4280	3.032	3.025	3.018	3.010	3.002
260	1.040	.8411	.6827	.5252	.4213	3.028	3.020	3.013	3.005	2.997
300	1.036	.8340	.6721	.5121	.4110	3.022	3.014	3.007	2.998	2.990
350	1.024	.8192	.6548	.4937	.3971	3.015	3.007	3.000	2.991	2.983
400	1.007	.8005	.6350	.4745	.3828	3.009	3.001	2.993	2.984	2.976
450	.9861	.7796	.6141	.4551	.3686	3.003	2.995	2.987	2.978	2.970
500	.9633	.7575	.5927	.4360	.3547	2.997	2.989	2.981	2.972	2.964
600	.9192	.7129	.5497	.3986	.3273	2.986	2.978	2.970	2.961	2.953
700	.8714	.6676	.5081	.3641	.3021	2.976	2.968	2.960	2.952	2.944
800	.8216	.6226	.4684	.3324	.2790	2.968	2.959	2.952	2.944	2.937
900	.7707	.5784	.4308	.3034	.2578	2.960	2.952	2.945	2.937	2.931
1000	.7193	.5352	.3952	.2768	.2384	2.953	2.945	2.938	2.931	2.926
1200	.5934	.4375	.3205	.2253	.2007	2.940	2.933	2.927	2.921	2.917
1400	.4802	.3517	.2568	.1828	.1695	2.930	2.924	2.919	2.915	2.912
1700	.3381	.2460	.1800	.1333	.1332	2.919	2.914	2.911	2.909	2.908
2000	.2269	.1649	.1226	.0975	.1072	2.912	2.909	2.907	2.907	2.909
2300	.1710	.1264	.0973	.0835	.0971	2.907	2.906	2.907	2.909	2.913
2600	.1319	.1006	.0816	.0760	.0918	2.904	2.906	2.908	2.913	2.919
3000	.0970	.0790	.0699	.0721	.0895	2.904	2.908	2.913	2.920	2.929
3500	.0708	.0648	.0644	.0729	.0909	2.907	2.914	2.921	2.932	2.944
4000	.0568	.0591	.0646	.0772	.0950	2.912	2.922	2.932	2.946	2.960
4500	.0501	.0583	.0679	.0830	.1003	2.920	2.933	2.945	2.962	2.978
5000	.0482	.0604	.0728	.0896	.1063	2.930	2.944	2.959	2.978	2.996
6000	.0521	.0671	.0814	.0997	.1166	2.954	2.972	2.991	3.014	3.036
7000	.0613	.0770	.0919	.1104	.1273	2.980	3.001	3.023	3.050	3.074
8000	.0725	.0880	.1027	.1210	.1379	3.007	3.032	3.056	3.085	3.112
9000	.0842	.0992	.1134	.1314	.1481	3.035	3.062	3.088	3.120	3.148
10000	.0955	.1099	.1236	.1412	.1579	3.063	3.092	3.120	3.154	3.183
12000	.1100	.1256	.1403	.1587	.1756	3.115	3.148	3.179	3.215	3.245
14000	.1231	.1398	.1554	.1745	.1916	3.167	3.203	3.236	3.275	3.305
17000	.1407	.1589	.1755	.1956	.2129	3.245	3.285	3.320	3.361	3.391
20000	.1565	.1758	.1933	.2140	.2315	3.323	3.366	3.403	3.445	3.475
23000	.1710	.1911	.2092	.2304	.2479	3.400	3.445	3.483	3.527	3.556
26000	.1843	.2050	.2235	.2450	.2625	3.476	3.523	3.562	3.606	3.635
30000	.2008	.2219	.2408	.2624	.2798	3.576	3.624	3.664	3.708	3.738
35000	.2196	.2409	.2598	.2814	.2985	3.697	3.747	3.787	3.831	3.862
40000	.2368	.2580	.2767	.2980	.3147	3.816	3.865	3.905	3.949	3.982
45000	.2527	.2736	.2919	.3126	.3288	3.930	3.979	4.020	4.063	4.097
50000	.2676	.2879	.3057	.3257	.3414	4.041	4.089	4.131	4.173	4.209

K/KL=	DAMAGE KURTOSIS					IONIZATION KURTOSIS				
E,KEV	0.7	0.85	1.0	1.2	1.4	0.7	0.85	1.0	1.2	1.4
5.0	2.377	2.369	2.362	2.354	2.346	2.473	2.465	2.459	2.452	2.446
6.0	2.385	2.373	2.364	2.353	2.345	2.480	2.469	2.461	2.451	2.444
7.0	2.388	2.375	2.364	2.352	2.342	2.484	2.471	2.461	2.450	2.441
8.0	2.390	2.376	2.364	2.351	2.340	2.488	2.473	2.461	2.449	2.439
9.0	2.393	2.378	2.365	2.350	2.338	2.494	2.477	2.463	2.448	2.437
10	2.398	2.381	2.366	2.350	2.337	2.501	2.481	2.465	2.449	2.436
12	2.413	2.390	2.372	2.353	2.337	2.522	2.496	2.475	2.453	2.436
14	2.434	2.405	2.382	2.358	2.339	2.551	2.516	2.488	2.460	2.439
17	2.477	2.436	2.403	2.369	2.344	2.606	2.555	2.516	2.476	2.447
20	2.530	2.473	2.429	2.385	2.352	2.671	2.602	2.549	2.496	2.458
23	2.594	2.518	2.460	2.403	2.361	2.759	2.666	2.595	2.526	2.476
26	2.663	2.566	2.493	2.423	2.371	2.850	2.732	2.643	2.556	2.495
30	2.759	2.634	2.541	2.453	2.387	2.970	2.820	2.707	2.597	2.520
35	2.884	2.723	2.604	2.492	2.410	3.117	2.927	2.785	2.648	2.552
40	3.010	2.814	2.669	2.533	2.436	3.258	3.030	2.860	2.696	2.582
45	3.138	2.905	2.735	2.575	2.463	3.392	3.128	2.931	2.742	2.611
50	3.264	2.996	2.801	2.617	2.492	3.519	3.222	2.999	2.786	2.639
60	3.562	3.216	2.969	2.733	2.579	3.785	3.418	3.143	2.879	2.701
70	3.828	3.411	3.116	2.833	2.654	4.012	3.586	3.265	2.958	2.752
80	4.060	3.581	3.241	2.915	2.717	4.204	3.728	3.369	3.025	2.793
90	4.262	3.727	3.347	2.981	2.767	4.365	3.847	3.456	3.080	2.826
100	4.437	3.852	3.435	3.033	2.807	4.501	3.947	3.528	3.125	2.851
120	4.710	4.041	3.547	3.077	2.842	4.690	4.085	3.631	3.188	2.872
140	4.914	4.178	3.621	3.095	2.856	4.820	4.179	3.700	3.228	2.879
170	5.125	4.314	3.682	3.095	2.855	4.941	4.264	3.760	3.258	2.874
200	5.253	4.389	3.705	3.075	2.839	5.000	4.302	3.784	3.265	2.858
230	5.281	4.390	3.684	3.042	2.807	4.983	4.286	3.765	3.243	2.836
260	5.272	4.365	3.648	3.003	2.771	4.943	4.252	3.732	3.211	2.810
300	5.226	4.309	3.588	2.949	2.722	4.871	4.191	3.674	3.161	2.772
350	5.136	4.219	3.504	2.880	2.662	4.767	4.103	3.591	3.092	2.723
400	5.025	4.118	3.417	2.813	2.605	4.659	4.009	3.503	3.020	2.673
450	4.906	4.014	3.331	2.750	2.554	4.552	3.914	3.415	2.949	2.623
500	4.782	3.910	3.247	2.690	2.507	4.451	3.822	3.327	2.880	2.574
600	4.522	3.698	3.079	2.571	2.428	4.318	3.659	3.146	2.730	2.458
700	4.277	3.506	2.932	2.471	2.363	4.182	3.505	2.983	2.600	2.358
800	4.051	3.334	2.805	2.389	2.310	4.041	3.360	2.840	2.488	2.275
900	3.844	3.182	2.697	2.322	2.267	3.896	3.224	2.714	2.392	2.206
1000	3.656	3.047	2.604	2.267	2.233	3.748	3.095	2.603	2.311	2.149
1200	3.323	2.833	2.478	2.212	2.194	3.346	2.822	2.430	2.207	2.095
1400	3.048	2.663	2.385	2.179	2.169	2.984	2.589	2.295	2.131	2.062
1700	2.725	2.470	2.287	2.152	2.147	2.534	2.308	2.144	2.053	2.033
2000	2.484	2.331	2.220	2.139	2.135	2.189	2.099	2.038	2.001	2.016
2300	2.356	2.253	2.179	2.125	2.125	2.060	2.015	1.986	1.969	1.992
2600	2.268	2.199	2.151	2.117	2.118	1.984	1.964	1.954	1.947	1.973
3000	2.191	2.153	2.127	2.110	2.113	1.931	1.927	1.929	1.929	1.953
3500	2.135	2.121	2.112	2.107	2.112	1.908	1.910	1.914	1.916	1.934
4000	2.106	2.105	2.105	2.108	2.114	1.908	1.908	1.910	1.910	1.921
4500	2.094	2.099	2.105	2.112	2.118	1.919	1.914	1.911	1.907	1.911
5000	2.091	2.100	2.108	2.117	2.124	1.932	1.921	1.914	1.907	1.905
6000	2.095	2.108	2.119	2.131	2.139	1.906	1.909	1.910	1.909	1.906
7000	2.112	2.124	2.134	2.146	2.157	1.890	1.901	1.909	1.913	1.909
8000	2.133	2.142	2.151	2.163	2.175	1.880	1.898	1.910	1.917	1.914
9000	2.154	2.161	2.168	2.179	2.193	1.874	1.896	1.911	1.920	1.918
10000	2.175	2.180	2.186	2.196	2.211	1.870	1.895	1.912	1.923	1.921
12000	2.199	2.206	2.214	2.228	2.244	1.871	1.891	1.906	1.917	1.918
14000	2.220	2.231	2.242	2.258	2.276	1.874	1.890	1.901	1.911	1.915
17000	2.251	2.266	2.281	2.301	2.321	1.882	1.890	1.897	1.905	1.913
20000	2.280	2.299	2.318	2.342	2.363	1.891	1.892	1.895	1.902	1.913
23000	2.308	2.331	2.354	2.381	2.403	1.899	1.896	1.896	1.902	1.916
26000	2.335	2.362	2.387	2.417	2.440	1.906	1.900	1.899	1.905	1.920
30000	2.371	2.401	2.430	2.462	2.487	1.914	1.907	1.905	1.912	1.929
35000	2.414	2.449	2.480	2.515	2.542	1.922	1.916	1.917	1.926	1.944
40000	2.457	2.493	2.526	2.564	2.593	1.926	1.926	1.931	1.943	1.962
45000	2.498	2.536	2.571	2.610	2.640	1.929	1.937	1.947	1.962	1.982
50000	2.538	2.578	2.613	2.653	2.685	1.930	1.947	1.964	1.984	2.003

	REFLECTION COEFFICIENT					SPUTTERING EFFICIENCY				
K/KL=	0.7	0.85	1.0	1.2	1.4	0.7	0.85	1.0	1.2	1.4
E,KEV										
5.0	.0000	.0000	.0000	.0000	.0000	.0113	.0109	.0105	.0099	.0095
6.0	.0000	.0000	.0000	.0000	.0000	.0112	.0108	.0103	.0098	.0093
7.0	.0000	.0000	.0000	.0000	.0000	.0112	.0107	.0102	.0097	.0092
8.0	.0000	.0000	.0000	.0000	.0000	.0112	.0107	.0101	.0096	.0091
9.0	.0000	.0000	.0000	.0000	.0000	.0112	.0107	.0101	.0095	.0090
10	.0000	.0000	.0000	.0000	.0000	.0113	.0107	.0101	.0094	.0089
12	.0000	.0000	.0000	.0000	.0000	.0114	.0107	.0100	.0094	.0088
14	.0000	.0000	.0000	.0000	.0000	.0115	.0108	.0100	.0093	.0087
17	.0000	.0000	.0000	.0000	.0000	.0118	.0109	.0101	.0093	.0086
20	.0000	.0000	.0000	.0000	.0000	.0120	.0110	.0102	.0093	.0085
23	.0000	.0000	.0000	.0000	.0000	.0124	.0113	.0104	.0093	.0085
26	.0000	.0000	.0000	.0000	.0000	.0127	.0116	.0105	.0094	.0085
30	.0000	.0000	.0000	.0000	.0000	.0131	.0119	.0108	.0095	.0085
35	.0000	.0000	.0000	.0000	.0000	.0135	.0122	.0110	.0096	.0085
40	.0000	.0000	.0000	.0000	.0000	.0139	.0125	.0112	.0097	.0085
45	.0000	.0000	.0000	.0000	.0000	.0142	.0127	.0114	.0098	.0086
50	.0000	.0000	.0000	.0000	.0000	.0144	.0129	.0115	.0099	.0086
60	.0000	.0000	.0000	.0000	.0000	.0145	.0130	.0117	.0101	.0088
70	.0000	.0000	.0000	.0000	.0000	.0145	.0131	.0118	.0102	.0089
80	.0000	.0000	.0000	.0000	.0000	.0145	.0131	.0119	.0103	.0090
90	.0000	.0000	.0000	.0000	.0000	.0144	.0131	.0119	.0103	.0091
100	.0000	.0000	.0000	.0000	.0000	.0144	.0130	.0119	.0103	.0091
120	.0000	.0000	.0000	.0000	.0000	.0142	.0129	.0116	.0100	.0089
140	.0000	.0000	.0000	.0000	.0000	.0141	.0127	.0114	.0097	.0087
170	.0000	.0000	.0000	.0000	.0000	.0138	.0124	.0110	.0093	.0084
200	.0000	.0000	.0000	.0000	.0000	.0135	.0121	.0107	.0089	.0080
230	.0000	.0000	.0000	.0000	.0000	.0132	.0117	.0104	.0085	.0077
260	.0000	.0000	.0000	.0000	.0000	.0129	.0115	.0101	.0083	.0074
300	.0000	.0000	.0000	.0000	.0000	.0124	.0111	.0098	.0079	.0071
350	.0000	.0000	.0000	.0000	.0000	.0119	.0107	.0094	.0075	.0066
400	.0000	.0000	.0000	.0000	.0000	.0115	.0103	.0090	.0072	.0063
450	.0000	.0000	.0000	.0000	.0000	.0111	.0100	.0087	.0069	.0059
500	.0000	.0000	.0000	.0000	.0000	.0108	.0097	.0084	.0066	.0057
600	.0000	.0000	.0000	.0000	.0000	.0103	.0092	.0078	.0059	.0051
700	.0000	.0000	.0000	.0000	.0000	.0100	.0088	.0073	.0054	.0047
800	.0000	.0000	.0000	.000C	.0000	.0096	.0084	.0069	.0050	.0044
900	.0000	.0000	.0000	.0000	.0000	.0093	.0081	.0065	.0046	.0041
1000	.0000	.0000	.0000	.0000	.0000	.0089	.0078	.0062	.0043	.0038
1200	.0000	.0000	.0000	.0000	.0000	.0082	.0070	.0056	.0039	.0035
1400	.0000	.0000	.0000	.0000	.0000	.0075	.0064	.0051	.0036	.0032
1700	.0000	.0000	.0000	.0000	.0000	.0067	.0055	.0044	.0033	.0029
2000	.0000	.0000	.0000	.0000	.0000	.0059	.0049	.0039	.0030	.0027
2300	.0000	0.0000	0.0000	0.0000	0.0000	.0054	.0044	.0036	.0028	.0025
2600	.0000	0.0000	0.0000	0.0000	0.0000	.0049	.0040	.0033	.0026	.0023
3000	0.0000	0.0000	0.0000	0.0000	0.0000	.0044	.0036	.0030	.0024	.0021
3500	0.0000	0.0000	0.0000	0.0000	0.0000	.0038	.0032	.0027	.0022	.0019
4000	0.0000	0.0000	0.0000	0.0000	0.0000	.0034	.0029	.0025	.0020	.0018
4500	0.0000	0.0000	0.0000	0.0000	0.0000	.0031	.0026	.0023	.0019	.0016
5000	0.0000	0.0000	0.0000	0.0000	0.0000	.0028	.0024	.0021	.0017	.0015
6000	0.0000	0.0000	0.0000	0.0000	0.0000	.0025	.0022	.0019	.0015	.0013
7000	0.0000	0.0000	0.0000	0.0000	0.0000	.0023	.0020	.0017	.0014	.0012
8000	0.0000	0.0000	0.0000	0.0000	0.0000	.0022	.0019	.0016	.0013	.0011
9000	0.0000	0.0000	0.0000	0.0000	0.0000	.0021	.0018	.0014	.0012	.0010
10000	0.0000	0.0000	0.0000	0.0000	0.0000	.0020	.0017	.0013	.0011	.0009
12000	0.0000	0.0000	0.0000	0.0000	0.0000	.0018	.0015	.0012	.0009	.0008
14000	0.0000	0.0000	0.0000	0.0000	0.0000	.0016	.0013	.0010	.0008	.0007
17000	0.0000	0.0000	0.0000	0.0000	0.0000	.0014	.0011	.0009	.0007	.0006
20000	0.0000	0.0000	0.0000	0.0000	0.0000	.0012	.0009	.0007	.0006	.0005
23000	0.0000	0.0000	0.0000	0.0000	0.0000	.0010	.0008	.0007	.0005	.0004
26000	0.0000	0.0000	0.0000	0.0000	0.0000	.0009	.0007	.0006	.0004	.0004
30000	0.0000	0.0000	0.0000	0.0000	0.0000	.0008	.0006	.0005	.0004	.0003
35000	0.0000	0.0000	0.0000	0.0000	0.0000	.0007	.0005	.0004	.0003	.0003
40000	0.0000	0.0000	0.0000	0.0000	0.0000	.0006	.0004	.0004	.0003	.0002
45000	0.0000	0.0000	0.0000	0.0000	0.0000	.0005	.0004	.0003	.0002	.0002
50000	0.0000	0.0000	0.0000	0.0000	0.0000	.0004	.0003	.0003	.0002	.0002

K/KL=	IONIZATION DEFICIENCY					SPUTTERING YIELD ALPHA				
E,KEV	0.7	0.85	1.0	1.2	1.4	0.7	0.85	1.0	1.2	1.4
5.0	.0018	.0021	.0024	.0028	.0031	.1180	.1146	.1111	.1067	.1028
6.0	.0018	.0021	.0025	.0028	.0032	.1168	.1133	.1096	.1052	.1013
7.0	.0019	.0022	.0025	.0029	.0032	.1158	.1121	.1084	.1040	.1000
8.0	.0019	.0023	.0026	.0029	.0033	.1149	.1112	.1074	.1029	.0988
9.0	.0020	.0023	.0026	.0030	.0033	.1142	.1104	.1065	.1019	.0978
10	.0020	.0024	.0027	.0030	.0034	.1136	.1097	.1057	.1011	.0968
12	.0022	.0025	.0028	.0031	.0035	.1126	.1084	.1044	.0996	.0952
14	.0023	.0026	.0029	.0032	.0035	.1117	.1074	.1032	.0983	.0938
17	.0024	.0027	.0030	.0034	.0037	.1106	.1061	.1018	.0966	.0920
20	.0026	.0029	.0032	.0035	.0038	.1096	.1049	.1006	.0953	.0906
23	.0027	.0031	.0033	.0036	.0039	.1086	.1039	.0995	.0942	.0893
26	.0029	.0032	.0035	.0038	.0040	.1078	.1030	.0986	.0932	.0883
30	.0031	.0034	.0037	.0040	.0042	.1067	.1019	.0975	.0920	.0870
35	.0033	.0037	.0039	.0042	.0044	.1056	.1007	.0962	.0907	.0856
40	.0035	.0039	.0041	.0044	.0046	.1045	.0996	.0951	.0896	.0845
45	.0036	.0040	.0043	.0046	.0047	.1036	.0986	.0941	.0885	.0834
50	.0038	.0042	.0045	.0047	.0049	.1027	.0977	.0931	.0876	.0825
60	.0040	.0044	.0047	.0050	.0051	.1011	.0959	.0914	.0858	.0808
70	.0042	.0047	.0050	.0052	.0054	.0997	.0945	.0898	.0842	.0793
80	.0044	.0048	.0052	.0054	.0056	.0985	.0932	.0885	.0827	.0781
90	.0045	.0050	.0053	.0056	.0057	.0975	.0921	.0873	.0815	.0770
100	.0047	.0051	.0055	.0057	.0058	.0967	.0911	.0862	.0803	.0760
120	.0049	.0054	.0057	.0060	.0060	.0955	.0896	.0844	.0783	.0742
140	.0051	.0055	.0059	.0061	.0060	.0945	.0884	.0830	.0766	.0727
170	.0053	.0057	.0061	.0063	.0061	.0933	.0869	.0812	.0745	.0710
200	.0054	.0058	.0062	.0064	.0060	.0923	.0857	.0798	.0729	.0696
230	.0054	.0058	.0062	.0064	.0060	.0912	.0846	.0785	.0717	.0686
260	.0054	.0058	.0062	.0064	.0059	.0901	.0836	.0775	.0708	.0678
300	.0053	.0057	.0061	.0064	.0058	.0889	.0826	.0763	.0699	.0670
350	.0052	.0056	.0060	.0063	.0056	.0876	.0815	.0752	.0690	.0663
400	.0052	.0055	.0059	.0061	.0054	.0866	.0806	.0745	.0684	.0658
450	.0051	.0055	.0058	.0059	.0051	.0857	.0799	.0739	.0680	.0654
500	.0051	.0054	.0056	.0057	.0049	.0851	.0794	.0735	.0677	.0651
600	.0052	.0054	.0052	.0051	.0041	.0841	.0785	.0734	.0676	.0647
700	.0054	.0054	.0049	.0045	.0035	.0836	.0779	.0735	.0676	.0645
800	.0054	.0053	.0045	.0041	.0029	.0833	.0777	.0738	.0678	.0644
900	.0054	.0052	.0042	.0037	.0025	.0832	.0776	.0741	.0680	.0644
1000	.0053	.0050	.0040	.0033	.0022	.0832	.0777	.0744	.0682	.0644
1200	.0047	.0045	.0036	.0031	.0023	.0841	.0787	.0750	.0686	.0644
1400	.0041	.0039	.0032	.0029	.0026	.0851	.0799	.0755	.0689	.0645
1700	.0033	.0032	.0029	.0028	.0029	.0865	.0815	.0763	.0693	.0647
2000	.0026	.0026	.0026	.0027	.0032	.0877	.0828	.0769	.0698	.0649
2300	.0025	.0024	.0025	.0026	.0031	.0883	.0833	.0772	.0701	.0651
2600	.0024	.0023	.0024	.0025	.0030	.0887	.0836	.0775	.0704	.0653
3000	.0023	.0022	.0023	.0024	.0028	.0891	.0839	.0779	.0707	.0655
3500	.0023	.0022	.0022	.0023	.0025	.0895	.0842	.0782	.0711	.0658
4000	.0023	.0022	.0022	.0022	.0023	.0901	.0845	.0786	.0714	.0661
4500	.0023	.0022	.0021	.0021	.0021	.0906	.0848	.0789	.0717	.0663
5000	.0022	.0021	.0021	.0021	.0019	.0911	.0851	.0792	.0720	.0666
6000	.0018	.0020	.0020	.0018	.0016	.0928	.0863	.0799	.0725	.0670
7000	.0015	.0018	.0019	.0016	.0014	.0943	.0873	.0806	.0729	.0674
8000	.0012	.0017	.0017	.0016	.0013	.0955	.0881	.0811	.0732	.0676
9000	.0010	.0015	.0016	.0017	.0014	.0965	.0888	.0816	.0734	.0679
10000	.0008	.0014	.0015	.0019	.0016	.0973	.0894	.0819	.0736	.0681
12000	.0007	.0011	.0013	.0032	.0029	.0980	.0899	.0823	.0739	.0683
14000	.0007	.0009	.0010	.0044	.0042	.0984	.0901	.0825	.0741	.0684
17000	.0008	.0007	.0007	.0059	.0057	.0985	.0902	.0827	.0742	.0684
20000	.0008	.0005	.0005	.0068	.0067	.0985	.0902	.0827	.0742	.0684
23000	.0008	.0003	.0003	.0073	.0073	.0983	.0901	.0827	.0742	.0684
26000	.0008	.0002	.0002	.0074	.0074	.0981	.0899	.0826	.0741	.0683
30000	.0007	.0001	.0000	.0070	.0071	.0977	.0897	.0824	.0740	.0682
35000	.0006	0.0000	0.0000	.0060	.0061	.0973	.0894	.0821	.0737	.0680
40000	.0004	0.0000	0.0000	.0044	.0044	.0968	.0890	.0818	.0734	.0678
45000	.0002	0.0000	0.0000	.0023	.0024	.0964	.0888	.0815	.0731	.0677
50000	0.0000	0.0000	0.0000	0.0000	0.0000	.0960	.0885	.0812	.0728	.0675

AU ON BE, M2/M1=0.0457

50000 KEV

5000 KEV

500 KEV

50 KEV

5 KEV

TABLE 23- 1

U ON BE, M2/M1=0.0378

| E,KEV | FRACTIONAL DEPOSITED ENERGY | | | | | MEAN RANGE, MICROGRAM/SQ.CM. | | | | |

K/KL=	0.65	0.80	1.0	1.2	1.4	0.65	0.80	1.0	1.2	1.4
2.0	.8747	.8488	.8158	.7845	.7547	.6304	.6277	.6242	.6207	.6173
2.3	.8718	.8454	.8117	.7799	.7496	.6959	.6928	.6888	.6848	.6809
2.6	.8692	.8423	.8081	.7758	.7451	.7563	.7529	.7484	.7440	.7396
3.0	.8661	.8386	.8038	.7709	.7397	.8315	.8277	.8226	.8177	.8127
3.5	.8627	.8346	.7990	.7655	.7337	.9196	.9153	.9096	.9039	.8984
4.0	.8596	.8310	.7948	.7607	.7285	1.003	.9983	.9920	.9857	.9796
4.5	.8568	.8277	.7910	.7563	.7237	1.083	1.078	1.071	1.064	1.057
5.0	.8543	.8247	.7874	.7524	.7193	1.161	1.155	1.147	1.140	1.133
6.0	.8497	.8194	.7811	.7453	.7116	1.309	1.303	1.294	1.285	1.277
7.0	.8456	.8146	.7756	.7391	.7048	1.451	1.443	1.433	1.424	1.414
8.0	.8419	.8103	.7706	.7335	.6987	1.587	1.579	1.567	1.556	1.546
9.0	.8386	.8064	.7661	.7284	.6932	1.718	1.709	1.696	1.684	1.672
10	.8355	.8028	.7619	.7238	.6882	1.845	1.835	1.821	1.808	1.795
12	.8299	.7964	.7544	.7154	.6791	2.082	2.070	2.054	2.039	2.024
14	.8249	.7906	.7478	.7081	.6712	2.308	2.295	2.277	2.259	2.242
17	.8182	.7829	.7390	.6983	.6607	2.632	2.616	2.595	2.575	2.554
20	.8123	.7760	.7311	.6897	.6515	2.941	2.923	2.899	2.875	2.851
23	.8068	.7698	.7241	.6820	.6432	3.226	3.205	3.178	3.152	3.125
26	.8019	.7642	.7176	.6750	.6357	3.501	3.479	3.449	3.419	3.390
30	.7957	.7572	.7098	.6664	.6267	3.857	3.832	3.798	3.765	3.732
35	.7887	.7492	.7008	.6568	.6165	4.287	4.258	4.219	4.181	4.144
40	.7822	.7419	.6927	.6480	.6073	4.703	4.670	4.626	4.583	4.541
45	.7761	.7352	.6852	.6400	.5989	5.107	5.070	5.021	4.973	4.927
50	.7705	.7288	.6783	.6326	.5912	5.499	5.458	5.404	5.352	5.300
60	.7600	.7172	.6655	.6190	.5771	6.225	6.178	6.115	6.054	5.994
70	.7505	.7067	.6541	.6070	.5647	6.930	6.875	6.804	6.734	6.665
80	.7417	.6972	.6437	.5962	.5536	7.618	7.556	7.474	7.395	7.317
90	.7337	.6883	.6343	.5864	.5436	8.290	8.220	8.129	8.040	7.954
100	.7261	.6802	.6256	.5773	.5344	8.947	8.870	8.769	8.671	8.575
120	.7122	.6653	.6098	.5611	.5180	10.16	10.07	9.954	9.839	9.727
140	.6998	.6520	.5958	.5468	.5037	11.36	11.25	11.12	10.98	10.85
170	.6832	.6344	.5775	.5283	.4852	13.13	13.00	12.83	12.67	12.51
200	.6684	.6189	.5616	.5123	.4693	14.87	14.72	14.52	14.33	14.14
230	.6548	.6048	.5473	.4980	.4552	16.42	16.25	16.03	15.81	15.60
260	.6424	.5920	.5344	.4852	.4427	17.98	17.79	17.54	17.29	17.06
300	.6275	.5767	.5190	.4701	.4279	20.07	19.85	19.56	19.28	19.00
350	.6109	.5599	.5022	.4535	.4119	22.72	22.45	22.10	21.77	21.44
400	.5961	.5450	.4874	.4391	.3979	25.39	25.07	24.66	24.27	23.88
450	.5829	.5316	.4743	.4263	.3855	28.08	27.70	27.23	26.76	26.32
500	.5708	.5196	.4625	.4149	.3745	30.77	30.34	29.79	29.26	28.75
600	.5497	.4987	.4421	.3952	.3553	35.65	35.13	34.47	33.83	33.22
700	.5316	.4809	.4248	.3786	.3393	40.73	40.10	39.30	38.52	37.78
800	.5159	.4654	.4099	.3644	.3256	45.98	45.22	44.24	43.31	42.42
900	.5020	.4519	.3969	.3519	.3138	51.35	50.44	49.27	48.16	47.11
1000	.4896	.4398	.3853	.3408	.3035	56.81	55.73	54.36	53.06	51.83
1200	.4686	.4192	.3656	.3221	.2869	66.85	65.53	63.84	62.24	60.72
1400	.4510	.4020	.3493	.3065	.2734	77.47	75.82	73.72	71.73	69.84
1700	.4289	.3806	.3290	.2873	.2568	94.39	92.07	89.16	86.44	83.90
2000	.4104	.3630	.3124	.2717	.2432	112.2	109.0	105.2	101.6	98.25
2300	.3940	.3481	.2984	.2585	.2311	128.4	124.8	120.2	115.9	111.9
2600	.3796	.3353	.2864	.2471	.2205	145.5	141.1	135.6	130.5	125.8
3000	.3630	.3204	.2725	.2342	.2084	169.3	163.7	156.8	150.4	144.5
3500	.3453	.3047	.2579	.2207	.1956	200.6	193.1	184.0	175.7	168.2
4000	.3301	.2912	.2455	.2093	.1847	233.1	223.4	211.8	201.4	192.0
4500	.3171	.2795	.2347	.1996	.1753	266.6	254.4	239.9	227.2	215.9
5000	.3056	.2691	.2252	.1911	.1672	300.7	285.5	268.2	253.0	239.7
6000	.2871	.2513	.2091	.1776	.1538	366.8	346.7	323.4	303.3	285.9
7000	.2719	.2366	.1958	.1665	.1429	434.9	408.7	378.8	353.5	331.7
8000	.2589	.2240	.1845	.1571	.1338	504.2	471.3	434.2	403.2	376.7
9000	.2476	.2132	.1748	.1489	.1260	574.3	534.0	489.3	452.3	421.0
10000	.2376	.2036	.1663	.1417	.1193	644.6	596.6	543.8	500.6	464.5
12000	.2203	.1873	.1520	.1292	.1081	785.1	720.5	651.0	595.1	549.0
14000	.2057	.1740	.1404	.1186	.0991	924.0	842.1	755.3	686.5	630.3
17000	.1873	.1576	.1263	.1051	.0884	1128.	1019.	906.1	817.7	746.5
20000	.1716	.1444	.1150	.0936	.0800	1326.	1190.	1050.	942.6	856.5

	MEAN DAMAGE DEPTH,MICROGRAM/SQ.CM.					MEAN IONZN.DEPTH,MICROGRAM/SQ.CM.				
K/KL=	0.65	0.80	1.0	1.2	1.4	0.65	0.80	1.0	1.2	1.4
E,KEV										
2.0	.3116	.3103	.3085	.3067	.3050	.2848	.2831	.2810	.2789	.2769
2.3	.3455	.3438	.3417	.3396	.3376	.3157	.3137	.3112	.3087	.3063
2.6	.3762	.3744	.3720	.3696	.3673	.3437	.3415	.3387	.3359	.3332
3.0	.4142	.4121	.4094	.4067	.4042	.3783	.3758	.3726	.3695	.3664
3.5	.4585	.4562	.4531	.4501	.4472	.4187	.4159	.4123	.4087	.4053
4.0	.5006	.4980	.4946	.4913	.4881	.4570	.4539	.4498	.4459	.4421
4.5	.5411	.5383	.5346	.5309	.5274	.4938	.4904	.4860	.4817	.4776
5.0	.5805	.5774	.5734	.5694	.5655	.5296	.5259	.5211	.5165	.5119
6.0	.6567	.6531	.6483	.6437	.6391	.5989	.5946	.5889	.5835	.5781
7.0	.7302	.7260	.7205	.7151	.7099	.6657	.6607	.6542	.6479	.6417
8.0	.8014	.7966	.7903	.7842	.7783	.7304	.7247	.7173	.7101	.7032
9.0	.8707	.8653	.8582	.8513	.8447	.7934	.7870	.7786	.7705	.7627
10	.9382	.9321	.9242	.9166	.9092	.8548	.8476	.8383	.8293	.8205
12	1.064	1.057	1.047	1.038	1.030	.9689	.9604	.9494	.9388	.9285
14	1.186	1.178	1.167	1.156	1.146	1.080	1.070	1.057	1.045	1.033
17	1.364	1.354	1.340	1.327	1.315	1.242	1.230	1.214	1.198	1.184
20	1.537	1.525	1.508	1.492	1.477	1.400	1.385	1.365	1.347	1.329
23	1.696	1.682	1.663	1.645	1.628	1.544	1.527	1.504	1.483	1.463
26	1.853	1.836	1.814	1.794	1.775	1.686	1.666	1.641	1.617	1.594
30	2.058	2.038	2.013	1.989	1.966	1.874	1.850	1.820	1.792	1.765
35	2.311	2.287	2.256	2.228	2.200	2.106	2.077	2.041	2.006	1.974
40	2.561	2.532	2.495	2.461	2.429	2.337	2.301	2.257	2.216	2.179
45	2.807	2.772	2.730	2.690	2.653	2.564	2.522	2.470	2.422	2.378
50	3.050	3.009	2.960	2.914	2.873	2.789	2.739	2.679	2.624	2.574
60	3.502	3.451	3.390	3.334	3.282	3.210	3.146	3.070	3.002	2.940
70	3.951	3.888	3.813	3.745	3.684	3.629	3.550	3.456	3.373	3.298
80	4.396	4.320	4.230	4.150	4.078	4.047	3.950	3.837	3.738	3.650
90	4.836	4.747	4.641	4.549	4.465	4.463	4.347	4.213	4.097	3.995
100	5.272	5.168	5.046	4.941	4.846	4.876	4.739	4.584	4.451	4.334
120	6.092	5.959	5.806	5.674	5.558	5.659	5.482	5.283	5.115	4.970
140	6.907	6.742	6.556	6.398	6.259	6.439	6.218	5.973	5.769	5.594
170	8.122	7.908	7.671	7.470	7.297	7.606	7.313	6.995	6.735	6.514
200	9.326	9.061	8.770	8.527	8.318	8.763	8.395	8.001	7.682	7.415
230	10.42	10.11	9.764	9.484	9.239	9.834	9.390	8.921	8.547	8.236
260	11.52	11.15	10.76	10.44	10.16	10.90	10.38	9.836	9.406	9.049
300	12.98	12.55	12.08	11.71	11.39	12.32	11.70	11.05	10.55	10.13
350	14.83	14.31	13.75	13.31	12.93	14.10	13.34	12.57	11.97	11.47
400	16.68	16.07	15.43	14.92	14.47	15.87	14.98	14.08	13.38	12.81
450	18.53	17.83	17.10	16.52	16.02	17.63	16.60	15.57	14.78	14.14
500	20.37	19.59	18.77	18.11	17.56	19.37	18.22	17.06	16.17	15.45
600	23.71	22.77	21.81	21.02	20.40	22.57	21.16	19.79	18.75	17.89
700	27.15	26.06	24.94	24.00	23.30	25.81	24.15	22.56	21.34	20.34
800	30.68	29.42	28.14	27.04	26.24	29.08	27.18	25.35	23.95	22.81
900	34.26	32.84	31.38	30.13	29.23	32.36	30.23	28.15	26.57	25.29
1000	37.89	36.30	34.66	33.24	32.23	35.65	33.28	30.96	29.18	27.76
1200	44.51	42.68	40.72	39.05	37.82	41.69	38.97	36.19	34.05	32.41
1400	51.49	49.37	47.04	45.07	43.58	47.92	44.81	41.54	39.01	37.13
1700	62.54	59.92	56.94	54.45	52.52	57.58	53.85	49.77	46.61	44.31
2000	74.13	70.93	67.21	64.13	61.69	67.52	63.11	58.18	54.33	51.56
2300	84.58	80.98	76.76	73.23	70.36	76.70	71.76	66.17	61.77	58.56
2600	95.54	91.45	86.63	82.56	79.22	86.14	80.60	74.29	69.29	65.60
3000	110.9	106.0	100.2	95.32	91.27	99.13	92.67	85.28	79.39	74.99
3500	131.2	125.1	117.8	111.7	106.6	115.9	108.1	99.22	92.09	86.73
4000	152.3	144.8	135.9	128.3	122.2	133.1	123.8	113.3	104.8	98.40
4500	174.1	165.0	154.2	145.2	137.8	150.5	139.7	127.3	117.4	110.0
5000	196.3	185.5	172.8	162.1	153.4	168.1	155.6	141.4	130.0	121.4
6000	239.2	225.2	208.7	194.9	183.8	202.5	186.8	168.9	154.5	143.7
7000	283.7	265.9	245.2	227.9	214.1	237.3	218.0	196.1	178.6	165.4
8000	329.3	307.4	281.9	260.9	244.2	272.1	249.0	222.8	202.0	186.4
9000	375.6	349.2	318.6	293.6	274.0	306.9	279.7	249.1	224.8	206.8
10000	422.3	391.1	355.2	326.0	303.4	341.4	310.0	274.7	246.9	226.6
12000	516.1	474.6	427.5	389.6	360.9	409.3	369.3	324.5	289.6	264.6
14000	609.3	557.2	498.3	451.4	416.6	475.6	426.7	372.3	330.2	300.5
17000	746.9	678.3	601.3	540.9	496.9	571.6	509.3	440.3	387.6	351.2
20000	880.9	795.6	700.4	626.4	573.5	663.7	588.0	504.6	441.4	398.4

U ON BE, M2/M1=0.0378

TABLE 23- 3

	RELATIVE RANGE STRAGGLING					RELATIVE DAMAGE STRAGGLING				
K/KL=	0.65	0.80	1.0	1.2	1.4	0.65	0.80	1.0	1.2	1.4
E,KEV										
2.0	.0192	.0191	.0190	.0188	.0187	.2971	.2972	.2974	.2975	.2977
2.3	.0192	.0191	.0190	.0188	.0187	.2966	.2967	.2968	.2970	.2971
2.6	.0192	.0191	.0190	.0188	.0187	.2960	.2961	.2962	.2964	.2965
3.0	.0192	.0191	.0189	.0188	.0186	.2952	.2953	.2954	.2956	.2957
3.5	.0192	.0191	.0189	.0188	.0186	.2941	.2943	.2944	.2946	.2948
4.0	.0192	.0191	.0189	.0188	.0186	.2932	.2933	.2935	.2937	.2939
4.5	.0192	.0191	.0189	.0187	.0186	.2923	.2924	.2926	.2928	.2930
5.0	.0192	.0190	.0189	.0187	.0186	.2914	.2915	.2917	.2919	.2922
6.0	.0192	.0190	.0189	.0187	.0185	.2898	.2899	.2901	.2904	.2906
7.0	.0192	.0190	.0188	.0187	.0185	.2883	.2885	.2887	.2889	.2892
8.0	.0191	.0190	.0188	.0186	.0185	.2870	.2872	.2874	.2876	.2879
9.0	.0191	.0190	.0188	.0186	.0185	.2858	.2860	.2862	.2864	.2867
10	.0191	.0190	.0188	.0186	.0184	.2848	.2849	.2851	.2853	.2855
12	.0191	.0190	.0188	.0186	.0184	.2827	.2828	.2830	.2832	.2834
14	.0191	.0190	.0188	.0186	.0184	.2810	.2811	.2812	.2814	.2816
17	.0191	.0189	.0187	.0185	.0183	.2790	.2789	.2789	.2790	.2792
20	.0191	.0189	.0187	.0185	.0183	.2775	.2772	.2771	.2771	.2772
23	.0191	.0189	.0187	.0185	.0183	.2763	.2759	.2756	.2755	.2755
26	.0191	.0189	.0187	.0185	.0182	.2754	.2748	.2743	.2741	.2740
30	.0191	.0189	.0187	.0184	.0182	.2746	.2738	.2730	.2725	.2723
35	.0191	.0189	.0186	.0184	.0182	.2742	.2729	.2717	.2710	.2706
40	.0190	.0189	.0186	.0184	.0182	.2741	.2723	.2707	.2697	.2692
45	.0190	.0189	.0186	.0184	.0181	.2743	.2721	.2701	.2688	.2680
50	.0190	.0188	.0186	.0184	.0181	.2748	.2721	.2696	.2680	.2670
60	.0190	.0188	.0186	.0183	.0181	.2767	.2729	.2693	.2670	.2655
70	.0190	.0188	.0186	.0183	.0180	.2788	.2739	.2694	.2664	.2645
80	.0190	.0188	.0185	.0183	.0180	.2811	.2751	.2696	.2660	.2636
90	.0190	.0188	.0185	.0182	.0180	.2833	.2762	.2698	.2657	.2630
100	.0190	.0188	.0185	.0182	.0179	.2854	.2774	.2702	.2655	.2624
120	.0190	.0188	.0185	.0182	.0179	.2901	.2800	.2711	.2654	.2616
140	.0190	.0188	.0185	.0182	.0179	.2943	.2823	.2719	.2654	.2610
170	.0191	.0188	.0184	.0181	.0178	.2993	.2851	.2728	.2652	.2601
200	.0191	.0188	.0184	.0181	.0177	.3032	.2871	.2733	.2649	.2593
230	.0191	.0188	.0184	.0180	.0177	.3056	.2883	.2731	.2640	.2582
260	.0191	.0188	.0184	.0180	.0176	.3072	.2890	.2726	.2631	.2572
300	.0191	.0187	.0183	.0179	.0176	.3085	.2894	.2717	.2617	.2558
350	.0191	.0187	.0183	.0179	.0175	.3090	.2892	.2704	.2600	.2541
400	.0190	.0187	.0182	.0178	.0174	.3088	.2885	.2690	.2583	.2526
450	.0190	.0186	.0182	.0177	.0173	.3079	.2875	.2675	.2567	.2512
500	.0190	.0186	.0181	.0176	.0172	.3065	.2861	.2659	.2551	.2498
600	.0189	.0185	.0180	.0174	.0170	.3019	.2821	.2624	.2518	.2474
700	.0189	.0184	.0178	.0173	.0168	.2969	.2780	.2592	.2489	.2452
800	.0188	.0183	.0177	.0171	.0166	.2920	.2740	.2562	.2463	.2432
900	.0187	.0182	.0175	.0169	.0163	.2873	.2702	.2534	.2440	.2415
1000	.0186	.0180	.0174	.0167	.0161	.2828	.2667	.2509	.2420	.2399
1200	.0184	.0178	.0170	.0163	.0157	.2744	.2602	.2465	.2391	.2372
1400	.0181	.0175	.0167	.0159	.0153	.2670	.2547	.2428	.2367	.2350
1700	.0178	.0170	.0162	.0154	.0147	.2578	.2477	.2382	.2337	.2321
2000	.0174	.0166	.0157	.0149	.0141	.2503	.2422	.2344	.2313	.2297
2300	.0169	.0161	.0151	.0143	.0135	.2452	.2383	.2318	.2292	.2278
2600	.0165	.0157	.0146	.0138	.0130	.2411	.2352	.2296	.2273	.2260
3000	.0160	.0151	.0140	.0131	.0123	.2367	.2319	.2272	.2252	.2240
3500	.0153	.0144	.0133	.0124	.0116	.2325	.2285	.2247	.2228	.2217
4000	.0147	.0137	.0126	.0117	.0109	.2291	.2258	.2225	.2207	.2197
4500	.0142	.0132	.0120	.0111	.0103	.2263	.2235	.2207	.2189	.2179
5000	.0136	.0126	.0115	.0105	.0098	.2240	.2216	.2190	.2172	.2161
6000	.0126	.0116	.0104	.0095	.0088	.2206	.2183	.2159	.2143	.2128
7000	.0117	.0107	.0096	.0087	.0080	.2179	.2157	.2132	.2117	.2099
8000	.0109	.0099	.0088	.0080	.0073	.2156	.2134	.2109	.2094	.2072
9000	.0102	.0092	.0081	.0073	.0067	.2137	.2114	.2087	.2073	.2048
10000	.0095	.0086	.0076	.0068	.0062	.2120	.2096	.2068	.2053	.2026
12000	.0085	.0075	.0066	.0059	.0053	.2091	.2064	.2033	.2016	.1987
14000	.0076	.0067	.0058	.0052	.0047	.2064	.2035	.2002	.1982	.1953
17000	.0065	.0057	.0049	.0044	.0039	.2028	.1997	.1961	.1936	.1909
20000	.0057	.0050	.0043	.0038	.0034	.1995	.1962	.1924	.1893	.1871

E,KEV	RELATIVE IONIZATION STRAGGLING					RELATIVE TRANSVERSE RANGE STRAGGLING				
K/KL=	0.65	0.80	1.0	1.2	1.4	0.65	0.80	1.0	1.2	1.4
2.0	.3336	.3343	.3352	.3362	.3371	.4378	.4385	.4395	.4405	.4415
2.3	.3331	.3338	.3348	.3358	.3367	.4377	.4385	.4395	.4406	.4416
2.6	.3326	.3333	.3343	.3353	.3363	.4377	.4385	.4395	.4406	.4416
3.0	.3319	.3326	.3336	.3347	.3357	.4377	.4385	.4396	.4406	.4417
3.5	.3310	.3318	.3328	.3339	.3349	.4376	.4384	.4396	.4406	.4417
4.0	.3301	.3309	.3320	.3331	.3342	.4376	.4384	.4395	.4407	.4418
4.5	.3293	.3301	.3312	.3323	.3335	.4375	.4384	.4395	.4407	.4418
5.0	.3285	.3293	.3305	.3316	.3328	.4374	.4383	.4395	.4406	.4418
6.0	.3270	.3279	.3291	.3303	.3315	.4373	.4382	.4394	.4406	.4418
7.0	.3257	.3266	.3278	.3291	.3303	.4371	.4380	.4393	.4405	.4417
8.0	.3246	.3255	.3267	.3280	.3292	.4369	.4379	.4392	.4404	.4417
9.0	.3235	.3244	.3257	.3269	.3282	.4367	.4377	.4390	.4403	.4416
10	.3226	.3235	.3247	.3260	.3273	.4365	.4375	.4389	.4402	.4415
12	.3208	.3217	.3229	.3242	.3256	.4362	.4372	.4386	.4400	.4413
14	.3194	.3202	.3214	.3227	.3240	.4358	.4369	.4383	.4397	.4411
17	.3178	.3185	.3196	.3208	.3221	.4352	.4363	.4378	.4393	.4407
20	.3167	.3172	.3182	.3193	.3206	.4346	.4358	.4373	.4388	.4403
23	.3160	.3163	.3171	.3181	.3193	.4341	.4353	.4369	.4384	.4399
26	.3156	.3157	.3162	.3171	.3182	.4335	.4348	.4364	.4380	.4395
30	.3156	.3153	.3154	.3160	.3170	.4328	.4340	.4357	.4374	.4390
35	.3159	.3151	.3148	.3151	.3158	.4319	.4332	.4349	.4366	.4383
40	.3167	.3153	.3145	.3144	.3149	.4309	.4323	.4341	.4359	.4376
45	.3178	.3158	.3144	.3140	.3142	.4300	.4314	.4333	.4351	.4369
50	.3191	.3165	.3145	.3137	.3137	.4291	.4306	.4325	.4344	.4362
60	.3230	.3188	.3155	.3138	.3133	.4274	.4289	.4309	.4329	.4348
70	.3268	.3213	.3166	.3141	.3131	.4258	.4273	.4294	.4315	.4335
80	.3306	.3237	.3178	.3145	.3130	.4241	.4258	.4279	.4301	.4322
90	.3342	.3260	.3190	.3150	.3130	.4226	.4243	.4265	.4287	.4309
100	.3377	.3282	.3201	.3154	.3130	.4210	.4228	.4251	.4274	.4297
120	.3446	.3326	.3224	.3164	.3132	.4180	.4199	.4224	.4248	.4272
140	.3505	.3363	.3243	.3173	.3134	.4152	.4172	.4198	.4224	.4249
170	.3577	.3407	.3265	.3182	.3137	.4113	.4134	.4163	.4190	.4218
200	.3632	.3440	.3280	.3188	.3138	.4076	.4099	.4130	.4159	.4188
230	.3667	.3459	.3287	.3189	.3137	.4040	.4065	.4097	.4128	.4159
260	.3693	.3471	.3290	.3188	.3136	.4007	.4033	.4067	.4100	.4132
300	.3715	.3479	.3289	.3185	.3133	.3966	.3994	.4030	.4065	.4100
350	.3729	.3480	.3285	.3179	.3128	.3920	.3949	.3988	.4026	.4064
400	.3732	.3475	.3277	.3171	.3123	.3878	.3910	.3951	.3992	.4032
450	.3727	.3466	.3266	.3163	.3116	.3840	.3874	.3918	.3961	.4003
500	.3717	.3452	.3254	.3154	.3110	.3805	.3841	.3887	.3933	.3977
600	.3672	.3408	.3220	.3131	.3088	.3740	.3780	.3831	.3881	.3930
700	.3622	.3363	.3188	.3110	.3070	.3685	.3728	.3785	.3839	.3892
800	.3571	.3322	.3159	.3092	.3055	.3638	.3685	.3746	.3805	.3862
900	.3521	.3283	.3133	.3076	.3045	.3599	.3649	.3714	.3777	.3838
1000	.3473	.3248	.3110	.3063	.3037	.3564	.3618	.3687	.3754	.3818
1200	.3373	.3189	.3075	.3043	.3037	.3507	.3566	.3643	.3717	.3789
1400	.3287	.3141	.3049	.3029	.3042	.3463	.3529	.3613	.3695	.3773
1700	.3185	.3087	.3023	.3019	.3057	.3419	.3493	.3588	.3678	.3766
2000	.3110	.3049	.3010	.3019	.3076	.3391	.3473	.3578	.3678	.3774
2300	.3088	.3032	.3013	.3033	.3097	.3376	.3466	.3580	.3689	.3793
2600	.3078	.3024	.3021	.3051	.3120	.3371	.3468	.3591	.3708	.3820
3000	.3077	.3021	.3037	.3078	.3150	.3375	.3482	.3616	.3743	.3864
3500	.3087	.3027	.3062	.3114	.3187	.3394	.3511	.3659	.3797	.3928
4000	.3101	.3040	.3090	.3150	.3223	.3422	.3550	.3710	.3859	.4000
4500	.3118	.3057	.3118	.3185	.3257	.3458	.3596	.3767	.3927	.4076
5000	.3135	.3076	.3147	.3218	.3290	.3500	.3647	.3829	.3997	.4155
6000	.3145	.3123	.3204	.3281	.3352	.3592	.3757	.3959	.4146	.4318
7000	.3160	.3171	.3258	.3337	.3408	.3695	.3876	.4096	.4297	.4483
8000	.3179	.3218	.3308	.3389	.3459	.3803	.3998	.4235	.4450	.4646
9000	.3201	.3263	.3355	.3437	.3506	.3914	.4123	.4374	.4601	.4807
10000	.3226	.3305	.3399	.3481	.3550	.4028	.4249	.4513	.4750	.4965
12000	.3280	.3382	.3478	.3560	.3627	.4256	.4498	.4786	.5041	.5270
14000	.3340	.3451	.3547	.3629	.3694	.4484	.4743	.5052	.5321	.5561
17000	.3435	.3541	.3636	.3717	.3781	.4818	.5099	.5434	.5721	.5974
20000	.3534	.3617	.3711	.3792	.3855	.5141	.5440	.5796	.6097	.6360

TABLE 23- 5

U ON BE, M2/M1=0.0378

K/KL= E,KEV	RELATIVE TRANSV. DAMAGE STRAGGLING					RELATIVE TRANSV. IONZN. STRAGGLING				
	0.65	0.80	1.0	1.2	1.4	0.65	0.80	1.0	1.2	1.4
2.0	.1151	.1129	.1102	.1076	.1051	.1055	.1033	.1005	.0978	.0952
2.3	.1161	.1138	.1109	.1082	.1057	.1066	.1042	.1013	.0984	.0958
2.6	.1170	.1147	.1117	.1089	.1062	.1075	.1051	.1020	.0991	.0963
3.0	.1182	.1158	.1127	.1097	.1069	.1088	.1062	.1030	.0999	.0971
3.5	.1198	.1172	.1139	.1108	.1079	.1103	.1076	.1042	.1010	.0980
4.0	.1212	.1185	.1151	.1118	.1088	.1119	.1090	.1055	.1021	.0989
4.5	.1227	.1199	.1163	.1129	.1097	.1134	.1104	.1067	.1032	.0999
5.0	.1242	.1212	.1175	.1140	.1107	.1149	.1118	.1079	.1042	.1008
6.0	.1271	.1239	.1198	.1161	.1125	.1179	.1145	.1103	.1064	.1027
7.0	.1300	.1265	.1222	.1181	.1144	.1209	.1173	.1127	.1085	.1046
8.0	.1328	.1291	.1245	.1201	.1161	.1238	.1199	.1151	.1106	.1064
9.0	.1356	.1316	.1267	.1221	.1179	.1268	.1226	.1174	.1126	.1082
10	.1383	.1341	.1289	.1241	.1196	.1296	.1252	.1197	.1146	.1100
12	.1437	.1390	.1332	.1279	.1230	.1353	.1303	.1242	.1186	.1135
14	.1489	.1437	.1373	.1315	.1261	.1407	.1352	.1285	.1224	.1168
17	.1562	.1502	.1430	.1365	.1306	.1484	.1422	.1346	.1277	.1214
20	.1630	.1564	.1484	.1412	.1346	.1557	.1487	.1402	.1326	.1257
23	.1698	.1625	.1537	.1458	.1386	.1630	.1552	.1458	.1375	.1299
26	.1762	.1682	.1586	.1500	.1423	.1698	.1613	.1511	.1420	.1339
30	.1841	.1752	.1646	.1552	.1468	.1783	.1689	.1576	.1476	.1387
35	.1930	.1831	.1714	.1611	.1519	.1878	.1773	.1648	.1538	.1440
40	.2009	.1902	.1775	.1662	.1563	.1963	.1849	.1713	.1594	.1488
45	.2081	.1966	.1829	.1709	.1603	.2039	.1917	.1772	.1644	.1531
50	.2146	.2024	.1878	.1751	.1638	.2108	.1979	.1825	.1689	.1569
60	.2265	.2129	.1967	.1826	.1701	.2232	.2090	.1920	.1770	.1638
70	.2363	.2216	.2041	.1887	.1753	.2335	.2183	.1999	.1837	.1695
80	.2446	.2289	.2102	.1939	.1795	.2421	.2260	.2066	.1893	.1741
90	.2517	.2352	.2154	.1981	.1830	.2494	.2326	.2122	.1940	.1780
100	.2578	.2405	.2198	.2017	.1859	.2555	.2382	.2169	.1980	.1812
120	.2672	.2489	.2267	.2073	.1903	.2648	.2467	.2242	.2040	.1861
140	.2743	.2551	.2318	.2113	.1933	.2717	.2530	.2296	.2083	.1894
170	.2819	.2617	.2369	.2150	.1959	.2792	.2598	.2352	.2126	.1926
200	.2872	.2660	.2400	.2170	.1970	.2843	.2644	.2387	.2151	.1941
230	.2904	.2685	.2413	.2173	.1964	.2881	.2673	.2407	.2166	.1944
260	.2925	.2698	.2417	.2169	.1952	.2907	.2691	.2417	.2172	.1940
300	.2940	.2704	.2411	.2153	.1929	.2930	.2704	.2419	.2168	.1926
350	.2946	.2698	.2392	.2124	.1893	.2945	.2707	.2409	.2151	.1900
400	.2942	.2683	.2364	.2087	.1854	.2949	.2701	.2390	.2125	.1869
450	.2932	.2661	.2331	.2046	.1812	.2945	.2689	.2364	.2091	.1833
500	.2917	.2635	.2294	.2003	.1770	.2936	.2672	.2334	.2051	.1795
600	.2896	.2577	.2206	.1897	.1680	.2914	.2640	.2262	.1939	.1706
700	.2860	.2511	.2117	.1796	.1595	.2879	.2597	.2186	.1831	.1619
800	.2811	.2440	.2029	.1702	.1515	.2834	.2544	.2108	.1730	.1537
900	.2754	.2367	.1946	.1615	.1441	.2782	.2486	.2032	.1638	.1460
1000	.2689	.2293	.1865	.1535	.1371	.2726	.2423	.1957	.1553	.1388
1200	.2508	.2122	.1709	.1397	.1244	.2583	.2261	.1802	.1419	.1256
1400	.2331	.1962	.1569	.1279	.1133	.2441	.2104	.1661	.1305	.1142
1700	.2086	.1747	.1389	.1131	.0994	.2239	.1887	.1474	.1162	.0997
2000	.1870	.1563	.1238	.1010	.0881	.2055	.1697	.1315	.1043	.0877
2300	.1698	.1414	.1117	.0912	.0795	.1881	.1538	.1182	.0936	.0782
2600	.1550	.1287	.1015	.0828	.0723	.1723	.1399	.1068	.0842	.0701
3000	.1380	.1143	.0900	.0736	.0644	.1538	.1240	.0938	.0737	.0613
3500	.1204	.0995	.0784	.0642	.0565	.1339	.1073	.0805	.0628	.0523
4000	.1060	.0874	.0691	.0567	.0502	.1172	.0936	.0697	.0540	.0452
4500	.0940	.0775	.0614	.0506	.0452	.1031	.0822	.0608	.0468	.0395
5000	.0840	.0693	.0551	.0455	.0410	.0912	.0726	.0534	.0408	.0347
6000	.0700	.0583	.0465	.0388	.0351	.0745	.0592	.0435	.0332	.0285
7000	.0596	.0502	.0403	.0339	.0309	.0621	.0493	.0363	.0278	.0241
8000	.0518	.0440	.0357	.0303	.0277	.0524	.0418	.0309	.0239	.0208
9000	.0456	.0391	.0321	.0275	.0252	.0448	.0359	.0267	.0208	.0183
10000	.0407	.0351	.0292	.0252	.0232	.0388	.0312	.0235	.0184	.0163
12000	.0335	.0288	.0249	.0219	.0202	.0299	.0243	.0187	.0150	.0134
14000	.0287	.0241	.0220	.0195	.0181	.0240	.0198	.0155	.0126	.0114
17000	.0242	.0187	.0189	.0170	.0157	.0187	.0156	.0124	.0102	.0092
20000	.0216	.0144	.0169	.0151	.0140	.0161	.0133	.0105	.0086	.0076

| E,KEV | RANGE SKEWNESS | | | | | DAMAGE SKEWNESS | | | | |
K/KL=	0.65	0.80	1.0	1.2	1.4	0.65	0.80	1.0	1.2	1.4
2.0	.2528	.2505	.2475	.2445	.2415	.2575	.2566	.2553	.2541	.2529
2.3	.2526	.2502	.2471	.2440	.2410	.2557	.2546	.2532	.2518	.2505
2.6	.2523	.2499	.2468	.2436	.2405	.2535	.2523	.2508	.2494	.2480
3.0	.2521	.2496	.2463	.2431	.2399	.2503	.2491	.2475	.2461	.2448
3.5	.2518	.2492	.2459	.2426	.2393	.2461	.2449	.2434	.2420	.2407
4.0	.2515	.2489	.2455	.2421	.2388	.2420	.2409	.2394	.2381	.2368
4.5	.2512	.2486	.2451	.2417	.2383	.2380	.2369	.2355	.2342	.2330
5.0	.2510	.2483	.2448	.2413	.2378	.2343	.2332	.2318	.2306	.2293
6.0	.2506	.2478	.2442	.2406	.2370	.2273	.2262	.2249	.2236	.2225
7.0	.2502	.2474	.2436	.2399	.2363	.2212	.2200	.2186	.2173	.2161
8.0	.2499	.2470	.2431	.2394	.2356	.2157	.2143	.2128	.2114	.2101
9.0	.2496	.2466	.2427	.2388	.2350	.2109	.2092	.2074	.2059	.2046
10	.2493	.2463	.2423	.2383	.2345	.2066	.2046	.2025	.2008	.1993
12	.2488	.2456	.2415	.2375	.2335	.1975	.1952	.1927	.1907	.1891
14	.2483	.2451	.2409	.2367	.2326	.1906	.1876	.1844	.1820	.1802
17	.2477	.2443	.2400	.2357	.2314	.1840	.1793	.1746	.1712	.1686
20	.2471	.2437	.2392	.2347	.2303	.1810	.1741	.1673	.1624	.1589
23	.2466	.2431	.2384	.2339	.2294	.1800	.1707	.1616	.1551	.1505
26	.2461	.2425	.2378	.2331	.2285	.1814	.1693	.1574	.1492	.1433
30	.2455	.2418	.2369	.2321	.2274	.1861	.1698	.1540	.1430	.1354
35	.2448	.2410	.2360	.2310	.2262	.1957	.1735	.1522	.1376	.1274
40	.2441	.2402	.2350	.2300	.2250	.2083	.1798	.1526	.1340	.1212
45	.2435	.2395	.2342	.2290	.2239	.2232	.1881	.1547	.1319	.1163
50	.2429	.2388	.2333	.2280	.2228	.2397	.1978	.1580	.1310	.1126
60	.2418	.2375	.2318	.2263	.2209	.2839	.2259	.1713	.1345	.1094
70	.2407	.2363	.2304	.2247	.2191	.3271	.2538	.1851	.1390	.1077
80	.2397	.2351	.2291	.2232	.2174	.3681	.2805	.1986	.1439	.1068
90	.2387	.2340	.2277	.2217	.2157	.4066	.3057	.2115	.1487	.1063
100	.2377	.2328	.2265	.2202	.2141	.4426	.3292	.2236	.1533	.1061
120	.2359	.2308	.2241	.2176	.2112	.5130	.3757	.2485	.1637	.1073
140	.2342	.2288	.2218	.2150	.2084	.5723	.4146	.2689	.1719	.1078
170	.2316	.2259	.2185	.2112	.2042	.6428	.4605	.2919	.1800	.1070
200	.2290	.2230	.2152	.2076	.2002	.6950	.4938	.3071	.1836	.1041
230	.2266	.2203	.2121	.2042	.1965	.7241	.5121	.3117	.1805	.0971
260	.2242	.2176	.2091	.2009	.1929	.7431	.5234	.3120	.1749	.0891
300	.2210	.2141	.2051	.1965	.1881	.7576	.5306	.3080	.1652	.0777
350	.2171	.2097	.2002	.1911	.1822	.7635	.5307	.2982	.1508	.0629
400	.2131	.2053	.1954	.1858	.1765	.7602	.5239	.2849	.1348	.0479
450	.2092	.2010	.1906	.1805	.1709	.7508	.5125	.2694	.1179	.0331
500	.2053	.1968	.1858	.1754	.1654	.7370	.4978	.2525	.1007	.0186
600	.1978	.1885	.1767	.1655	.1547	.6998	.4566	.2121	.0617	-.0109
700	.1904	.1804	.1678	.1558	.1444	.6565	.4127	.1724	.0253	-.0380
800	.1831	.1725	.1591	.1465	.1345	.6100	.3684	.1344	-.0081	-.0629
900	.1759	.1648	.1507	.1374	.1248	.5622	.3248	.0985	-.0386	-.0857
1000	.1689	.1572	.1424	.1285	.1155	.5139	.2826	.0647	-.0664	-.1066
1200	.1552	.1423	.1262	.1113	.0973	.4041	.1957	.0006	-.1143	-.1442
1400	.1420	.1281	.1109	.0949	.0802	.3037	.1184	-.0549	-.1545	-.1764
1700	.1232	.1080	.0892	.0721	.0563	.1733	.0199	-.1241	-.2037	-.2168
2000	.1055	.0891	.0691	.0509	.0343	.0658	-.0603	-.1797	-.2428	-.2499
2300	.0883	.0708	.0496	.0306	.0133	-.0063	-.1141	-.2164	-.2697	-.2745
2600	.0720	.0535	.0314	.0116	-.0062	-.0646	-.1575	-.2458	-.2915	-.2953
3000	.0516	.0320	.0088	-.0118	-.0302	-.1270	-.2038	-.2771	-.3153	-.3187
3500	.0280	.0072	-.0172	-.0385	-.0575	-.1873	-.2488	-.3078	-.3392	-.3432
4000	.0061	-.0156	-.0408	-.0628	-.0821	-.2339	-.2838	-.3322	-.3589	-.3639
4500	-.0141	-.0366	-.0626	-.0850	-.1046	-.2706	-.3118	-.3524	-.3757	-.3820
5000	-.0331	-.0562	-.0827	-.1055	-.1252	-.3000	-.3348	-.3694	-.3905	-.3980
6000	-.0682	-.0923	-.1195	-.1425	-.1623	-.3379	-.3675	-.3978	-.4176	-.4266
7000	-.0994	-.1242	-.1518	-.1748	-.1945	-.3646	-.3919	-.4206	-.4402	-.4508
8000	-.1275	-.1526	-.1804	-.2033	-.2228	-.3848	-.4114	-.4397	-.4599	-.4717
9000	-.1529	-.1783	-.2061	-.2289	-.2481	-.4011	-.4277	-.4564	-.4772	-.4902
10000	-.1761	-.2016	-.2293	-.2519	-.2709	-.4150	-.4419	-.4712	-.4929	-.5068
12000	-.2171	-.2427	-.2701	-.2921	-.3104	-.4385	-.4665	-.4972	-.5203	-.5358
14000	-.2526	-.2780	-.3047	-.3262	-.3438	-.4593	-.4881	-.5199	-.5440	-.5606
17000	-.2980	-.3227	-.3484	-.3689	-.3856	-.4886	-.5177	-.5500	-.5749	-.5923
20000	-.3364	-.3603	-.3848	-.4043	-.4200	-.5176	-.5457	-.5771	-.6016	-.6193

TABLE 23- 7

U ON BE, M2/M1=0.0378

K/KL=	IONIZATION SKEWNESS					RANGE KURTOSIS				
	0.65	0.80	1.0	1.2	1.4	0.65	0.80	1.0	1.2	1.4
E,KEV										
2.0	.3708	.3715	.3727	.3738	.3749	3.070	3.068	3.065	3.062	3.059
2.3	.3698	.3704	.3712	.3722	.3733	3.070	3.068	3.064	3.061	3.058
2.6	.3681	.3687	.3695	.3705	.3716	3.070	3.067	3.064	3.061	3.058
3.0	.3656	.3661	.3670	.3680	.3692	3.069	3.067	3.064	3.060	3.057
3.5	.3621	.3628	.3638	.3649	.3661	3.069	3.067	3.063	3.060	3.057
4.0	.3587	.3594	.3606	.3618	.3631	3.069	3.066	3.063	3.059	3.056
4.5	.3554	.3562	.3575	.3588	.3602	3.069	3.066	3.062	3.059	3.056
5.0	.3523	.3532	.3545	.3559	.3575	3.068	3.066	3.062	3.059	3.055
6.0	.3467	.3476	.3490	.3506	.3522	3.068	3.065	3.061	3.058	3.054
7.0	.3418	.3427	.3441	.3457	.3474	3.067	3.065	3.061	3.057	3.054
8.0	.3378	.3385	.3397	.3412	.3430	3.067	3.064	3.060	3.057	3.053
9.0	.3343	.3347	.3357	.3372	.3388	3.067	3.064	3.060	3.056	3.052
10	.3314	.3315	.3322	.3334	.3350	3.066	3.063	3.059	3.055	3.052
12	.3249	.3246	.3249	.3259	.3275	3.066	3.063	3.059	3.055	3.051
14	.3208	.3196	.3192	.3197	.3210	3.065	3.062	3.058	3.054	3.050
17	.3184	.3153	.3131	.3125	.3130	3.064	3.061	3.057	3.053	3.049
20	.3196	.3140	.3094	.3072	.3068	3.064	3.060	3.056	3.052	3.047
23	.3238	.3151	.3076	.3035	.3018	3.063	3.060	3.055	3.051	3.046
26	.3302	.3180	.3073	.3010	.2979	3.063	3.059	3.054	3.050	3.046
30	.3413	.3240	.3086	.2993	.2940	3.062	3.058	3.053	3.049	3.044
35	.3581	.3342	.3125	.2989	.2908	3.061	3.057	3.052	3.048	3.043
40	.3772	.3463	.3180	.3001	.2891	3.060	3.056	3.051	3.047	3.042
45	.3979	.3599	.3249	.3024	.2883	3.059	3.055	3.050	3.045	3.041
50	.4197	.3745	.3326	.3055	.2884	3.059	3.055	3.049	3.044	3.040
60	.4752	.4132	.3555	.3178	.2936	3.057	3.053	3.048	3.043	3.038
70	.5269	.4495	.3772	.3296	.2988	3.056	3.052	3.046	3.041	3.036
80	.5740	.4826	.3969	.3404	.3035	3.055	3.050	3.045	3.039	3.034
90	.6165	.5124	.4147	.3500	.3077	3.054	3.049	3.043	3.038	3.032
100	.6547	.5393	.4306	.3584	.3111	3.052	3.048	3.042	3.036	3.031
120	.7190	.5841	.4563	.3708	.3145	3.050	3.045	3.039	3.033	3.028
140	.7711	.6202	.4765	.3801	.3164	3.048	3.043	3.036	3.030	3.025
170	.8315	.6617	.4992	.3897	.3175	3.045	3.040	3.033	3.027	3.021
200	.8759	.6919	.5148	.3955	.3173	3.042	3.036	3.029	3.023	3.017
230	.9074	.7139	.5264	.4002	.3187	3.039	3.033	3.026	3.019	3.013
260	.9299	.7292	.5338	.4026	.3190	3.036	3.030	3.023	3.016	3.010
300	.9494	.7419	.5388	.4028	.3181	3.032	3.026	3.019	3.012	3.005
350	.9614	.7485	.5389	.3994	.3148	3.028	3.022	3.014	3.007	3.000
400	.9635	.7475	.5340	.3928	.3097	3.024	3.017	3.009	3.002	2.995
450	.9584	.7409	.5254	.3839	.3031	3.020	3.013	3.005	2.998	2.991
500	.9478	.7301	.5140	.3731	.2953	3.016	3.009	3.001	2.993	2.986
600	.9038	.6886	.4750	.3387	.2697	3.009	3.001	2.993	2.985	2.978
700	.8557	.6448	.4359	.3055	.2454	3.002	2.994	2.986	2.978	2.971
800	.8072	.6016	.3987	.2746	.2231	2.995	2.988	2.979	2.971	2.965
900	.7599	.5602	.3639	.2465	.2030	2.989	2.982	2.973	2.966	2.959
1000	.7145	.5211	.3319	.2211	.1851	2.984	2.976	2.968	2.960	2.954
1200	.6282	.4483	.2750	.1784	.1565	2.973	2.966	2.957	2.951	2.945
1400	.5522	.3854	.2275	.1439	.1338	2.964	2.957	2.949	2.943	2.938
1700	.4558	.3076	.1710	.1044	.1082	2.953	2.946	2.939	2.934	2.930
2000	.3774	.2462	.1286	.0761	.0901	2.944	2.938	2.932	2.928	2.926
2300	.3194	.2055	.1052	.0633	.0813	2.937	2.932	2.927	2.924	2.923
2600	.2722	.1740	.0889	.0556	.0760	2.930	2.926	2.923	2.922	2.923
3000	.2218	.1419	.0743	.0506	.0724	2.925	2.922	2.921	2.922	2.924
3500	.1738	.1132	.0636	.0492	.0714	2.920	2.919	2.920	2.924	2.928
4000	.1378	.0930	.0581	.0510	.0728	2.918	2.919	2.922	2.928	2.934
4500	.1108	.0790	.0558	.0545	.0755	2.917	2.920	2.926	2.933	2.941
5000	.0906	.0693	.0557	.0589	.0789	2.918	2.923	2.931	2.940	2.950
6000	.0725	.0622	.0587	.0670	.0862	2.924	2.933	2.945	2.958	2.970
7000	.0644	.0614	.0645	.0760	.0943	2.934	2.945	2.962	2.977	2.992
8000	.0619	.0640	.0716	.0852	.1027	2.944	2.959	2.979	2.997	3.014
9000	.0626	.0684	.0792	.0943	.1111	2.957	2.974	2.997	3.017	3.035
10000	.0653	.0737	.0869	.1030	.1193	2.970	2.990	3.014	3.037	3.057
12000	.0734	.0856	.1018	.1193	.1350	2.997	3.021	3.050	3.075	3.098
14000	.0828	.0973	.1156	.1339	.1495	3.026	3.101	3.053	3.085	3.113	3.137
17000	.0961	.1132	.1337	.1530	.1691	3.069	3.101	3.136	3.165	3.192
20000	.1074	.1265	.1489	.1692	.1866	3.112	3.147	3.184	3.215	3.243

E,KEV \ K/KL=	DAMAGE KURTOSIS					IONIZATION KURTOSIS				
	0.65	0.80	1.0	1.2	1.4	0.65	0.80	1.0	1.2	1.4
2.0	2.339	2.335	2.331	2.326	2.322	2.445	2.443	2.440	2.438	2.436
2.3	2.340	2.336	2.330	2.325	2.321	2.446	2.443	2.440	2.437	2.434
2.6	2.340	2.335	2.329	2.324	2.319	2.445	2.442	2.439	2.436	2.433
3.0	2.338	2.333	2.328	2.322	2.317	2.444	2.440	2.437	2.433	2.431
3.5	2.336	2.331	2.325	2.320	2.315	2.440	2.437	2.434	2.431	2.428
4.0	2.333	2.329	2.323	2.318	2.313	2.437	2.434	2.431	2.428	2.425
4.5	2.331	2.326	2.321	2.316	2.311	2.434	2.431	2.428	2.425	2.422
5.0	2.329	2.324	2.319	2.313	2.308	2.431	2.428	2.425	2.422	2.420
6.0	2.326	2.321	2.315	2.310	2.305	2.426	2.423	2.420	2.417	2.415
7.0	2.323	2.318	2.312	2.306	2.301	2.423	2.420	2.416	2.413	2.410
8.0	2.322	2.317	2.310	2.304	2.298	2.422	2.418	2.413	2.409	2.406
9.0	2.322	2.315	2.308	2.301	2.295	2.422	2.416	2.410	2.406	2.403
10	2.323	2.315	2.306	2.299	2.293	2.422	2.416	2.409	2.403	2.400
12	2.320	2.312	2.302	2.295	2.288	2.421	2.412	2.404	2.397	2.393
14	2.322	2.312	2.300	2.291	2.284	2.424	2.413	2.401	2.393	2.387
17	2.333	2.317	2.301	2.289	2.280	2.439	2.421	2.403	2.391	2.382
20	2.352	2.328	2.306	2.289	2.277	2.463	2.435	2.409	2.391	2.379
23	2.372	2.342	2.312	2.291	2.276	2.495	2.456	2.419	2.395	2.379
26	2.398	2.358	2.321	2.295	2.276	2.533	2.481	2.432	2.401	2.379
30	2.439	2.385	2.335	2.301	2.278	2.589	2.518	2.453	2.410	2.382
35	2.499	2.425	2.357	2.312	2.282	2.666	2.570	2.483	2.426	2.388
40	2.565	2.469	2.382	2.325	2.287	2.749	2.626	2.515	2.443	2.396
45	2.636	2.516	2.408	2.339	2.294	2.835	2.685	2.549	2.462	2.405
50	2.711	2.566	2.437	2.354	2.301	2.922	2.745	2.585	2.482	2.415
60	2.896	2.689	2.509	2.395	2.322	3.137	2.893	2.674	2.534	2.444
70	3.071	2.806	2.576	2.434	2.343	3.334	3.029	2.757	2.583	2.471
80	3.235	2.915	2.638	2.468	2.361	3.510	3.152	2.831	2.626	2.495
90	3.386	3.015	2.695	2.500	2.378	3.668	3.262	2.897	2.665	2.517
100	3.526	3.106	2.747	2.528	2.393	3.808	3.359	2.956	2.700	2.536
120	3.789	3.277	2.841	2.578	2.420	4.043	3.525	3.056	2.757	2.566
140	4.009	3.418	2.917	2.617	2.441	4.229	3.656	3.134	2.801	2.589
170	4.271	3.584	3.005	2.660	2.463	4.437	3.803	3.223	2.851	2.615
200	4.467	3.705	3.066	2.688	2.477	4.581	3.906	3.284	2.886	2.633
230	4.590	3.777	3.099	2.699	2.477	4.651	3.961	3.324	2.912	2.655
260	4.676	3.825	3.118	2.703	2.472	4.690	3.994	3.349	2.929	2.672
300	4.748	3.861	3.127	2.700	2.463	4.709	4.012	3.365	2.940	2.685
350	4.792	3.875	3.122	2.686	2.448	4.699	4.008	3.364	2.941	2.691
400	4.798	3.865	3.105	2.665	2.431	4.665	3.983	3.346	2.930	2.686
450	4.778	3.839	3.078	2.639	2.412	4.616	3.945	3.316	2.910	2.673
500	4.740	3.801	3.046	2.610	2.394	4.558	3.898	3.278	2.883	2.653
600	4.593	3.674	2.951	2.528	2.357	4.402	3.757	3.144	2.788	2.564
700	4.430	3.542	2.857	2.452	2.321	4.249	3.619	3.014	2.694	2.478
800	4.262	3.413	2.767	2.383	2.289	4.103	3.488	2.893	2.607	2.399
900	4.096	3.289	2.684	2.321	2.260	3.969	3.368	2.784	2.527	2.329
1000	3.934	3.173	2.606	2.267	2.233	3.845	3.258	2.686	2.456	2.268
1200	3.601	2.955	2.466	2.185	2.182	3.645	3.078	2.533	2.341	2.184
1400	3.309	2.769	2.350	2.123	2.140	3.476	2.929	2.411	2.249	2.119
1700	2.948	2.545	2.214	2.055	2.093	3.265	2.748	2.270	2.142	2.049
2000	2.668	2.373	2.115	2.010	2.059	3.093	2.607	2.167	2.064	2.000
2300	2.518	2.279	2.074	1.996	2.046	2.950	2.508	2.111	2.025	1.974
2600	2.411	2.212	2.049	1.991	2.041	2.829	2.430	2.073	1.999	1.957
3000	2.312	2.150	2.034	1.993	2.039	2.694	2.348	2.039	1.977	1.944
3500	2.233	2.103	2.030	2.004	2.043	2.558	2.270	2.013	1.960	1.934
4000	2.185	2.077	2.035	2.018	2.051	2.450	2.210	1.997	1.951	1.930
4500	2.156	2.064	2.044	2.033	2.060	2.361	2.164	1.986	1.946	1.927
5000	2.140	2.060	2.056	2.048	2.070	2.289	2.126	1.979	1.942	1.926
6000	2.132	2.074	2.075	2.071	2.088	2.193	2.071	1.960	1.928	1.917
7000	2.137	2.095	2.094	2.092	2.106	2.126	2.032	1.946	1.918	1.911
8000	2.149	2.118	2.113	2.111	2.123	2.076	2.003	1.936	1.911	1.907
9000	2.163	2.140	2.131	2.129	2.140	2.038	1.981	1.929	1.906	1.904
10000	2.176	2.159	2.147	2.146	2.155	2.009	1.964	1.923	1.902	1.901
12000	2.197	2.190	2.174	2.174	2.183	1.967	1.939	1.913	1.896	1.898
14000	2.209	2.209	2.195	2.197	2.207	1.939	1.921	1.905	1.892	1.896
17000	2.210	2.218	2.216	2.225	2.238	1.911	1.902	1.895	1.889	1.896
20000	2.194	2.210	2.229	2.247	2.264	1.893	1.888	1.885	1.887	1.893

U ON BE, M2/M1=0.0378

TABLE 23- 9

K/KL=	REFLECTION COEFFICIENT					SPUTTERING EFFICIENCY				
E,KEV	0.65	0.80	1.0	1.2	1.4	0.65	0.80	1.0	1.2	1.4
2.0	.0000	.0000	.0000	.0000	.0000	.0117	.0113	.0108	.0104	.0099
2.3	.0000	.0000	.0000	.0000	.0000	.0117	.0113	.0108	.0103	.0099
2.6	.0000	.0000	.0000	.0000	.0000	.0116	.0112	.0107	.0102	.0098
3.0	.0000	.0000	.0000	.0000	.0000	.0115	.0111	.0106	.0101	.0097
3.5	.0000	.0000	.0000	.0000	.0000	.0115	.0110	.0105	.0100	.0096
4.0	.0000	.0000	.0000	.0000	.0000	.0114	.0109	.0104	.0099	.0096
4.5	.0000	.0000	.0000	.0000	.0000	.0113	.0109	.0103	.0098	.0095
5.0	.0000	.0000	.0000	.0000	.0000	.0112	.0108	.0102	.0097	.0094
6.0	.0000	.0000	.0000	.0000	.0000	.0111	.0107	.0101	.0096	.0092
7.0	.0000	.0000	.0000	.0000	.0000	.0110	.0106	.0100	.0095	.0091
8.0	.0000	.0000	.0000	.0000	.0000	.0110	.0105	.0099	.0094	.0089
9.0	.0000	.0000	.0000	.0000	.0000	.0109	.0104	.0098	.0093	.0088
10	.0000	.0000	.0000	.0000	.0000	.0109	.0103	.0097	.0092	.0087
12	.0000	.0000	.0000	.0000	.0000	.0108	.0102	.0095	.0091	.0085
14	.0000	.0000	.0000	.0000	.0000	.0107	.0101	.0094	.0089	.0084
17	.0000	.0000	.0000	.0000	.0000	.0107	.0101	.0093	.0088	.0082
20	.0000	.0000	.0000	.0000	.0000	.0108	.0101	.0093	.0087	.0080
23	.0000	.0000	.0000	.0000	.0000	.0110	.0102	.0093	.0086	.0079
26	.0000	.0000	.0000	.0000	.0000	.0113	.0103	.0093	.0085	.0078
30	.0000	.0000	.0000	.0000	.0000	.0116	.0105	.0094	.0085	.0077
35	.0000	.0000	.0000	.0000	.0000	.0120	.0107	.0095	.0084	.0076
40	.0000	.0000	.0000	.0000	.0000	.0124	.0110	.0095	.0084	.0076
45	.0000	.0000	.0000	.0000	.0000	.0127	.0112	.0096	.0084	.0075
50	.0000	.0000	.0000	.0000	.0000	.0130	.0114	.0097	.0084	.0075
60	.0000	.0000	.0000	.0000	.0000	.0134	.0118	.0100	.0086	.0075
70	.0000	.0000	.0000	.0000	.0000	.0138	.0122	.0102	.0087	.0075
80	.0000	.0000	.0000	.0000	.0000	.0140	.0124	.0104	.0087	.0075
90	.0000	.0000	.0000	.0000	.0000	.0142	.0126	.0106	.0088	.0075
100	.0000	.0000	.0000	.0000	.0000	.0143	.0127	.0107	.0089	.0075
120	.0000	.0000	.0000	.0000	.0000	.0143	.0127	.0107	.0088	.0074
140	.0000	.0000	.0000	.0000	.0000	.0142	.0126	.0106	.0088	.0073
170	.0000	.0000	.0000	.0000	.0000	.0139	.0124	.0105	.0087	.0072
200	.0000	.0000	.0000	.0000	.0000	.0136	.0121	.0103	.0085	.0070
230	.0000	.0000	.0000	.0000	.0000	.0134	.0119	.0101	.0084	.0068
260	.0000	.0000	.0000	.0000	.0000	.0132	.0117	.0099	.0082	.0067
300	.0000	.0000	.0000	.0000	.0000	.0130	.0115	.0097	.0080	.0064
350	.0000	.0000	.0000	.0000	.0000	.0128	.0112	.0094	.0077	.0062
400	.0000	.0000	.0000	.0000	.0000	.0125	.0109	.0091	.0074	.0059
450	.0000	.0000	.0000	.0000	.0000	.0122	.0106	.0089	.0071	.0057
500	.0000	.0000	.0000	.0000	.0000	.0120	.0103	.0086	.0068	.0054
600	.0000	.0000	.0000	.0000	.0000	.0114	.0099	.0081	.0062	.0051
700	.0000	.0000	.0000	.0000	.0000	.0109	.0094	.0076	.0057	.0048
800	.0000	.0000	.0000	.0000	.0000	.0104	.0090	.0072	.0053	.0045
900	.0000	.0000	.0000	.0000	.0000	.0100	.0087	.0068	.0049	.0042
1000	.0000	.0000	.0000	.0000	.0000	.0096	.0083	.0064	.0045	.0040
1200	.0000	.0000	.0000	.0000	.0000	.0089	.0076	.0057	.0039	.0036
1400	.0000	.0000	.0000	.0000	.0000	.0084	.0070	.0050	.0035	.0033
1700	.0000	.0000	.0000	.0000	.0000	.0077	.0062	.0042	.0029	.0029
2000	.0000	.0000	.0000	.0000	.0000	.0071	.0055	.0036	.0026	.0026
2300	.0000	.0000	.0000	0.0000	0.0000	.0066	.0050	.0032	.0023	.0024
2600	.0000	.0000	.0000	0.0000	0.0000	.0060	.0045	.0030	.0022	.0022
3000	.0000	.0000	0.0000	0.0000	0.0000	.0055	.0040	.0027	.0020	.0020
3500	.0000	.0000	0.0000	0.0000	0.0000	.0048	.0035	.0025	.0019	.0019
4000	.0000	.0000	0.0000	0.0000	0.0000	.0044	.0032	.0023	.0018	.0018
4500	.0000	.0000	0.0000	0.0000	0.0000	.0040	.0029	.0022	.0018	.0017
5000	.0000	.0000	0.0000	0.0000	0.0000	.0036	.0026	.0021	.0017	.0016
6000	0.0000	0.0000	0.0000	0.0000	0.0000	.0033	.0025	.0020	.0016	.0015
7000	0.0000	0.0000	0.0000	0.0000	0.0000	.0031	.0024	.0019	.0016	.0014
8000	0.0000	0.0000	0.0000	0.0000	0.0000	.0030	.0023	.0018	.0015	.0013
9000	0.0000	0.0000	0.0000	0.0000	0.0000	.0029	.0022	.0017	.0014	.0012
10000	0.0000	0.0000	0.0000	0.0000	0.0000	.0028	.0022	.0017	.0014	.0011
12000	0.0000	0.0000	0.0000	0.0000	0.0000	.0025	.0021	.0015	.0013	.0010
14000	0.0000	0.0000	0.0000	0.0000	0.0000	.0023	.0019	.0014	.0011	.0009
17000	0.0000	0.0000	0.0000	0.0000	0.0000	.0020	.0016	.0012	.0009	.0008
20000	0.0000	0.0000	0.0000	0.0000	0.0000	.0016	.0013	.0010	.0008	.0006

U ON BE, M2/M1=0.0378

TABLE 23-10

	IONIZATION DEFICIENCY					SPUTTERING YIELD ALPHA				
K/KL=	0.65	0.80	1.0	1.2	1.4	0.65	0.80	1.0	1.2	1.4
E,KEV										
2.0	.0014	.0017	.0021	.0024	.0027	.1285	.1252	.1211	.1171	.1133
2.3	.0014	.0017	.0021	.0024	.0028	.1279	.1246	.1203	.1162	.1125
2.6	.0015	.0018	.0021	.0025	.0028	.1273	.1239	.1196	.1153	.1118
3.0	.0015	.0018	.0022	.0025	.0028	.1265	.1231	.1187	.1144	.1108
3.5	.0015	.0018	.0022	.0026	.0029	.1256	.1221	.1176	.1133	.1097
4.0	.0016	.0019	.0022	.0026	.0029	.1248	.1212	.1167	.1123	.1087
4.5	.0016	.0019	.0023	.0026	.0030	.1240	.1204	.1158	.1114	.1078
5.0	.0016	.0019	.0023	.0027	.0030	.1233	.1196	.1150	.1106	.1069
6.0	.0016	.0020	.0024	.0027	.0031	.1220	.1183	.1135	.1092	.1053
7.0	.0017	.0020	.0024	.0028	.0031	.1209	.1171	.1123	.1080	.1039
8.0	.0017	.0020	.0024	.0028	.0032	.1199	.1160	.1111	.1068	.1026
9.0	.0017	.0021	.0025	.0028	.0032	.1191	.1151	.1101	.1058	.1015
10	.0018	.0021	.0025	.0029	.0032	.1183	.1142	.1092	.1049	.1005
12	.0018	.0021	.0026	.0029	.0033	.1169	.1127	.1075	.1031	.0986
14	.0019	.0022	.0026	.0030	.0033	.1157	.1114	.1061	.1016	.0970
17	.0020	.0023	.0027	.0031	.0034	.1142	.1098	.1044	.0997	.0950
20	.0021	.0024	.0028	.0031	.0035	.1130	.1084	.1029	.0980	.0933
23	.0022	.0025	.0029	.0032	.0036	.1121	.1075	.1018	.0967	.0919
26	.0023	.0026	.0030	.0033	.0036	.1114	.1066	.1008	.0956	.0907
30	.0025	.0028	.0031	.0034	.0037	.1105	.1057	.0996	.0942	.0892
35	.0027	.0030	.0033	.0036	.0038	.1096	.1046	.0984	.0928	.0877
40	.0028	.0031	.0034	.0037	.0039	.1087	.1036	.0973	.0916	.0864
45	.0030	.0033	.0036	.0038	.0040	.1079	.1027	.0963	.0905	.0853
50	.0031	.0035	.0037	.0039	.0041	.1071	.1019	.0954	.0895	.0843
60	.0034	.0037	.0040	.0042	.0043	.1053	.1000	.0936	.0878	.0825
70	.0036	.0040	.0043	.0044	.0045	.1037	.0983	.0920	.0863	.0810
80	.0038	.0042	.0045	.0047	.0047	.1023	.0969	.0907	.0850	.0798
90	.0040	.0044	.0047	.0049	.0049	.1010	.0955	.0894	.0838	.0786
100	.0042	.0046	.0049	.0050	.0050	.0999	.0943	.0882	.0828	.0776
120	.0044	.0049	.0052	.0054	.0054	.0980	.0923	.0861	.0808	.0758
140	.0047	.0051	.0055	.0057	.0056	.0964	.0906	.0843	.0791	.0742
170	.0050	.0054	.0058	.0060	.0060	.0945	.0885	.0820	.0770	.0723
200	.0052	.0057	.0061	.0063	.0062	.0930	.0869	.0802	.0752	.0708
230	.0053	.0058	.0063	.0065	.0065	.0920	.0857	.0788	.0739	.0696
260	.0053	.0059	.0064	.0067	.0067	.0913	.0847	.0776	.0728	.0687
300	.0054	.0059	.0065	.0068	.0069	.0904	.0836	.0763	.0716	.0676
350	.0055	.0060	.0066	.0070	.0071	.0895	.0825	.0750	.0705	.0666
400	.0055	.0060	.0067	.0071	.0072	.0887	.0816	.0741	.0695	.0657
450	.0056	.0061	.0067	.0071	.0072	.0880	.0808	.0733	.0688	.0650
500	.0056	.0061	.0067	.0072	.0072	.0874	.0801	.0727	.0682	.0644
600	.0059	.0062	.0067	.0071	.0067	.0857	.0786	.0722	.0675	.0635
700	.0060	.0063	.0066	.0071	.0063	.0843	.0776	.0718	.0669	.0628
800	.0062	.0063	.0065	.0070	.0058	.0832	.0768	.0716	.0665	.0623
900	.0064	.0064	.0064	.0068	.0054	.0825	.0763	.0714	.0661	.0618
1000	.0065	.0064	.0062	.0066	.0050	.0819	.0761	.0713	.0658	.0615
1200	.0066	.0065	.0057	.0061	.0043	.0820	.0767	.0709	.0647	.0609
1400	.0068	.0066	.0053	.0055	.0038	.0824	.0776	.0706	.0638	.0604
1700	.0069	.0066	.0048	.0049	.0032	.0833	.0788	.0704	.0629	.0600
2000	.0070	.0066	.0043	.0043	.0028	.0843	.0799	.0702	.0622	.0598
2300	.0073	.0065	.0041	.0041	.0026	.0850	.0800	.0703	.0621	.0598
2600	.0076	.0063	.0040	.0039	.0026	.0857	.0799	.0705	.0622	.0599
3000	.0079	.0061	.0038	.0037	.0026	.0866	.0798	.0708	.0625	.0600
3500	.0081	.0058	.0037	.0036	.0026	.0876	.0798	.0712	.0631	.0603
4000	.0081	.0056	.0036	.0035	.0026	.0885	.0798	.0716	.0637	.0606
4500	.0081	.0053	.0035	.0034	.0026	.0893	.0800	.0721	.0644	.0609
5000	.0080	.0051	.0035	.0033	.0026	.0901	.0802	.0726	.0651	.0611
6000	.0070	.0046	.0033	.0031	.0024	.0916	.0821	.0738	.0670	.0618
7000	.0061	.0042	.0032	.0029	.0022	.0929	.0837	.0749	.0686	.0623
8000	.0052	.0039	.0030	.0027	.0020	.0939	.0851	.0758	.0699	.0628
9000	.0045	.0036	.0029	.0025	.0019	.0947	.0863	.0765	.0709	.0631
10000	.0039	.0033	.0028	.0023	.0017	.0953	.0872	.0771	.0717	.0635
12000	.0029	.0028	.0025	.0021	.0014	.0961	.0885	.0780	.0725	.0640
14000	.0023	.0023	.0021	.0018	.0012	.0965	.0891	.0784	.0725	.0643
17000	.0018	.0018	.0017	.0015	.0009	.0965	.0891	.0786	.0717	.0646
20000	.0018	.0014	.0012	.0012	.0006	.0960	.0883	.0783	.0700	.0646

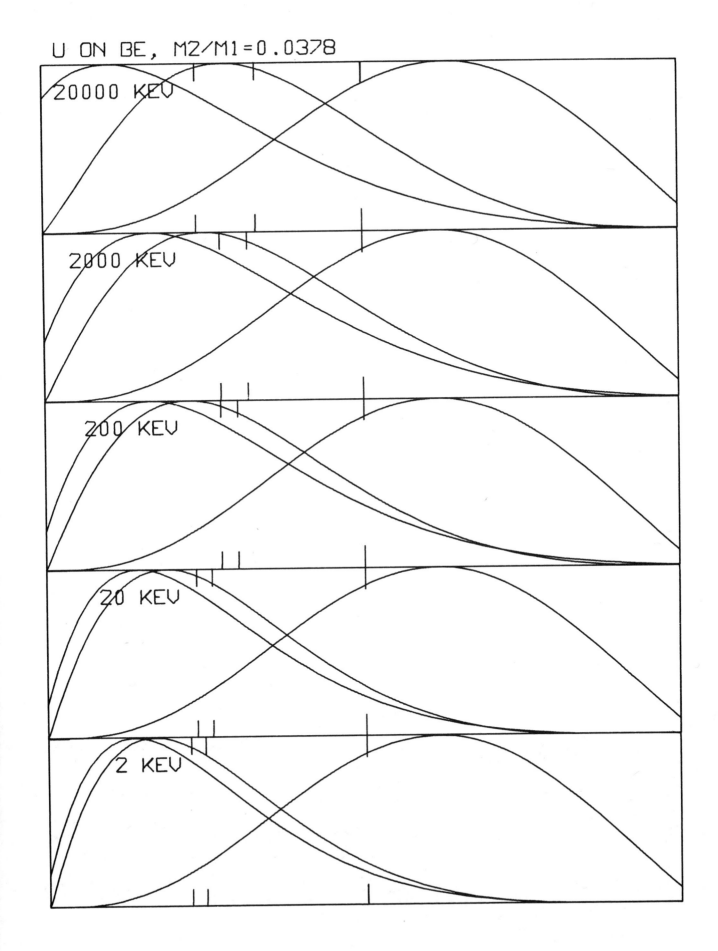

U ON BE, M2/M1=0.0378

20000 KEV

2000 KEV

200 KEV

20 KEV

2 KEV